ALLE · ZEIT · WACH · 1842

J. Reitner H. Keupp (Eds.)

# Fossil and Recent Sponges

With 233 Figures

Springer-Verlag
Berlin Heidelberg New York
London Paris Tokyo
Hong Kong Barcelona
Budapest

Dr. JOACHIM REITNER
Prof. Dr. HELMUT KEUPP

Institut für Paläontologie
Freie Universität Berlin
Schwendener Straße 8
1000 Berlin 33, FRG

*Cover illustration:* The cover shows a multiple recolonialization of the calcareous basal skeleton of *Spirastrella (Acanthochaetetes) wellsi* (Hartman and Goreau 1975). Collected from the Phillipines by Dr. R. van Soest. Length 5 cm. Collection J. Reitner, Institute of Palaeontology, Free University, Berlin

ISBN-13: 978-3-642-75658-0 e-ISBN-13:978-3-642-75656-6
DOI:10.1007/978-3-642-75656-6

Library of Congress Cataloging-in-Publication Data. Fossil and recent sponges / J. Reitner, H. Keupp (eds.). p. cm. Proceedings of a symposium held Sept. 26–28, 1988, at the Institut für Paläontologie, Freie Universität Berlin. 1. Sponges – Congresses. 2. Sponges, Fossil – Congresses. I. Reitner, J. (Joachim), II. Keupp, Helmut, 1949–. III. Freie Universität Berlin. Institut für Paläontologie. QL370.7.F67 1990 593.4 – dc20 90-41848


Softcover reprint of the hardcover 1st edition 1991



Typesetting: International Typesetters Inc., Makati, Philippines
32/3145-543210 – Printed on acid-free paper

# Preface

Sponges, filter-feeding organisms in all main types of aquatic environments, have been known since the latest Precambrian. Through the whole Phanerozoic they play an important role as constructors in buildups, and live on soft bottoms as well as on hard substrates in different water depths and environments.

During Paleozoic and Mesozoic times highly diverse sponge communities with rigid skeletons are primary frame builders, which were replaced subsequently in the late Mesozoic by rapidly growing, rigid skeletal-forming cnidarians. Sponges with rigid skeletons are *lithistids* – a desma-bearing polyphyletic group of demosponges, dictyid hexactinellids, and *sclerosponges* (stromatoporoids, chaetetids, sphinctozoans) – a polyphyletic group of demosponges which secrete different types of secondary calcareous skeletons. Today these types of sponges are restricted to cryptic and deeper water environments. *Soft sponges* are mostly reported in fossil record by isolated spicules. Complete specimens are only known from fossillagerstätten which are characterized by rapid sedimentation rates as seen in the Middle Cambrian Burgess and Wheeler Shales (North America) as well as in the early Devonian Hunsrück slate of West Germany.

In connection with the Phanerozoic $CO_2$-cycles, the main faunal changes of the sponges are observed during the Carboniferous, the Triassic/Jurassic boundary, and the Late Cretaceous. The most important change in sponge taxa occurred at the end of the Devonian. Most of the modern sponge lineages probably derive from stem groups of the early Carboniferous. Only few earlier Paleozoic sponge taxa exhibit clear relationships to modern ones. Most of the older groups became extinct in the late Devonian (Frasne/Famenene) worldwide extinction events. But, despite the enormous post-Devonian fossil material, the phylogenetic, paleobiological, and paleoecological relationships between fossil and Recent taxa are poorly known. The reasons for this are often poorly preserved fossil sponge remains and the highly convergent nature of many well-preserved characters such as megascleres and secondary basal skeletons.

According to these facts, for a long period, paleontologists and neontologists worked in different ways because both groups had different scientific goals.

As a result of the rediscovery of sponges with rigid calcareous skeleton by Willard Hartman and Thomas Goreau and Jean Vacelet in the 1960's, the communication between paleospongiologists and neospongiologists has been increasing within the last decade, both scientific groups developing more and more common goals.

Our aim is to confirm this development. We believe that taxonomic, phylogenetic, and (paleo-)ecological questions could be more easily solved if both scientific groups collaborated closely. We therefore organized a conference on Fossil and Recent Sponges in Berlin in September 26–28, 1988.

This conference was a forum where more than 70 biologists and paleontologists discussed present scientific questions dealing with fossil and modern sponges. The positive feedback of this symposium is reflected by the 43 accepted contributions for this Proceedings Volume.

The organizers thank all participants of the symposium for presenting their papers and posters and for constructive discussions. We are indebted to all referees who have done good work in reviewing all the papers presented here.

For financial support, the following institutions are gratefully acknowledged: Senator für Wissenschaft und Forschung Berlin, Deutsche Forschungsgemeinschaft (DFG), and Deutscher Akademischer Austauschdienst (DAAD). Last but not least our special thanks to all at the Institut für Paläontologie, Freie Universität Berlin, for help and the perfect organization which was a garantee for a successful conference.

Berlin, Spring 1991

JOACHIM REITNER
HELMUT KEUPP

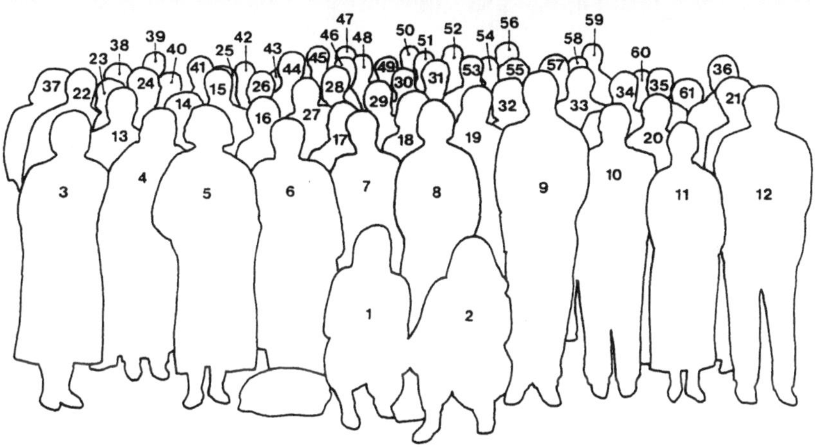


Most of the participants of the International Conference on Fossil and Recent Sponges, Berlin 1988, September 26–28 are shown in front of the Institut für Paläontologie der Freien Universität Berlin, in the order indicated in the drawing.

*1* M.C. Diaz *2* E. Villamizar *3* S. Pomponi *4* M.-J. Uriz *5* R. Desqueyroux-Faundez *6* E. Gaino *7* N. Boury-Esnault *8* R. Manconi *9* T. Koch *10* R. Kott *11* R. Wood *12* R. Pronzato *13* M. Sarà *14* W.C. Jones *15* K. Rigby *16* D. Mehl (pro: Salomon) *17* C. Volkmer-Ribeiro *18* B. Spiegler *19* W.H. de Weerdt *20* J. Kaźmierczak *21* H. Keupp *22* J. Reitner *23* G. Gruber *24* N. Weissenfels *25* D. Barthel *26* A. Bodzioch *27* B. Senowbari-Daryan *28* P. Willenz *29* T. Steiger *30* A. Pisera *31* B. Lang *32* D. Rosell *33* T. Engeser *34* P. Riedel *35* U. Scheer *36* C. Errenst *37* P. Wolz *38* U. Schweitzer *39* M. Rauer *40* J. Vacelet *41* P.-F. Langenbruch *42* S. Stone *43* P. Schäfer *44* U. Saller *45* A. Sole-Cava *46* C. Levi *47* M. Schudack *48* W. Zimmerle *49* C. Gaillard *50* F. Grothe *51* R. van Soest *52* W. Gasse *53* J. Pickett *54* G. Corriero *55* T. Brachert *56* R. West *57* F.-H. Brugger *58* R. Kohring *59* J. Hooper *60* H.M. Reiswig *61* M. Pansini

# Contents

## Taxonomic and Phylogenetic Aspects

*Calcarea*



## Aspects of Sponge Biology

## (Paleo-)Ecology and Diagenesis of Sponges

# List of Contributors


BARTHEL, DAGMAR: Institut für Meereskunde, Abt. Meeresbotanik, Düsternbrooker Weg 20, 2300 Kiel 1, FRG

BODZIOCH, ADAM: Uniwersytet im. A. Mickiewicza, Katedra Geologii, Zaklad Geologii Dynamicznej i Regionalnej, ul. R. Strzalkowskiego 5, 60-854 Poznan, Poland

BRACHERT, THOMAS C.: Institut für Geowissenschaften, Johannes-Gutenberg-Universität, Saarstraße 21, 6500 Mainz, FRG

CAPON, ROBERT J.: University of Melbourne, Dept. of Organic Chemistry, Parkville, Victoria 3052, Australia

CICOGNA, F.: Centro Lubrense Esplorazioni Marine, Via Castelle 11, 80056 Massalubrense, Italy

CORRIERO, GUISEPPE: Istituto di Zoologia, Università di Genova, Via Balbi 5, 16126 Genova, Italy

CUIF, JEAN-PIERRE: Lab. Pétrologie sédimentaire et Paléontologie, Bat. 504, Université, 91405 Orsay, France

DEUSCH, M.: Institut für Geologie und Paläontologie der Universität, Böblinger Straße 72, 7000 Stuttgart 10, FRG

DE WEERDT, WALENTINA H.: Institute of Taxonomic Zoology (Zoölogisch Museum), University of Amsterdam, P.O. Box 4766, 1009 AT Amsterdam, The Netherlands

DIAZ, MARIA CRISTINA: University of California, Marine Sciences Office, 273 Applied Sciences, UCSC/Santa Cruz, CA 95064, USA

FRIEBE, AXEL: Institut für Geologie und Paläontologie der Universität, Böblinger Straße 72, 7000 Stuttgart 10, FRG

GAINO, ELDA: Istituto di Zoologia, Università di Genova, Via Balbi 5, 16126 Genova, Italy

GASSE, WOLFGANG: c/o Firma FEMA-Salzgitter, Steinfurter Straße 145, 4400 Münster, FRG

GAUTRET, PASCALE: Laboratoire Pétrologie sédimentaire et Paléontologie, Bat. 504, Université, 91405 Orsay, France

GÖCKE, RUDOLF: Institut für Medizinische Physik der Universität Münster, Hüfferstraße 68, 4400 Münster, FRG

HILPERT, KARL-HEINZ: Wilhelmstraße 24, 4354 Datteln, FRG

HOOPER, JOHN N. A.: Northern Territory Museum of Arts and Sciences, P.O. Box 4646, Darwin NT 0801, Australia

JONES, W. CLIFFORD: School of Animal Biology, University College of North Wales, Bangor, Gwynedd, LL57 2UW, UK

KAŹMIERCZAK, JOSEF: Institute of Paleobiology, Polish Academy of Sciences, Al. Zwirki i Wigury 93, 02-089 Warszawa, Poland

KEENAN, CLIVE P.: Southern Fisheries Research Centre, P.O. Box 76, Deception Bay, Queensland 4508, Australia

KERSHAW, S.: West London Institute, Borough Road, Isleworth, Middlesex TWF 5DU, UK

KEUPP, HELMUT: Institut für Paläontologie, Freie Universität Berlin, Schwendener Str. 8, 1000 Berlin 33, FRG

KRAUTTER, MANFRED: Institut für Geologie und Paläontologie der Universität, Böblinger Straße 72, 7000 Stuttgart 10, FRG

LANG, BERNHARD: Institut für Paläontologie, Universität, Loewenichstraße 28, 8520 Erlangen, FRG

LANGENBRUCH, PAUL-FRIEDRICH: Zoologisches Institut (entwicklungsgeschichtliche Abteilung) der Universität Bonn, Poppelsdorfer Schloß, 5300 Bonn 1, FRG

LAUGHLIN, ROGER A.: Universidad Central de Venezuela, Aparado 47058, Caracas 1041-A, Venezuela

LEDGER, P. W.: Alzna Corporation, 950 Page Mill Road, P.O. Box 10950, Palo Alto, California 94303-0802, USA

LEVI, CLAUDE: Lab. de Biologie des Invertébrés marins et Malacologie Museum National d'Histoire naturelle, 57 Rue Cuvier, Paris Cedex 05, France

MANARA, ELENA: Istituto di Zoologia, Università di Genova, Via Balbi 5, 16126 Genova, Italy

MANCONI, RENATA: Istituto di Zoologia dell'Università, Via Muroni 25, 07100 Sassari, Italy

MEHL, DORTE (pro: SALOMON) Institut für Paläontologie, Freie Universität Berlin, Schwendener Straße 8, 1000 Berlin 33, FRG

MÜLLER, WALTER: Waldackerweg 68, 7300 Esslingen, FRG

PANSINI, MAURIZIO: Istituto di Zoologia, Università di Genova, Via Balbi 5, 16126 Genova, Italy

PARRY, DAVID L.: University College of the Northern Territory, P.O. Box 1341, Darwin, NT 0801, Australia

PICKETT, JOHN: Geological Survey of New South Wales, Building B 11 A, University of NSW, P.O. Box 1, Kensington NSW. 2033, Australia

PISERA, ANDRZEJ: Zaklad Paleobiologii, Polish Academy of Sciences, Zwirki i Wiguri 93, 02-089 Warszawa, Poland

POMPONI, SHIRLEY A.: Harbor Branch Oceanographic Institution, Inc., 5600 Old Dixie Highway, Fort Pierce, Florida 34946, USA

PRONZATO, ROBERTO: Istituto di Zoologia, Università di Genova, Via Balbi 5, 16126 Genova, Italy

REINHARDT, JOACHIM W.: Institut für Paläontologie, Universität, Loewenichstraße 28, 8520 Erlangen, FRG

REISWIG, HENRY M.: Redpath Museum and Biological Department, McGill University, 859 Sherbrooke St. W., Montreal, Quebec, Canada H3A 2K6

REITNER, JOACHIM: Institut für Paläontologie, Freie Universität Berlin, Schwendener Straße 8, 1000 Berlin 33, FRG

RIEDEL, PETER: Institut für Paläontologie, Universität, Loewenichstraße 28, 8520 Erlangen, FRG

RIGBY, J. KEITH: Brigham Young University, Dept. of Geology, Provo, Utah 84602, USA

SALLER, UWE: Zoologisches Institut (entwicklungsgeschichtliche Abteilung), Universität Bonn, Poppelsdorfer Schloß, 5300 Bonn 1, FRG

SARÀ, MICHELE: Istituto di Zoologia, Università di Genova, Via Balbi 5, 16126 Genova, Italy

SENOWBARI-DARYAN, BABA: Institut für Paläontologie, Universität, Loewenichstraße 28, 8520 Erlangen, FRG

SOLÉ-CAVA, ANTONIO M.: Port Erin Marine Laboratory, University of Liverpool, Port Erin, Isle of Man, UK

TABACHNICK, KONSTANTIN R.: Institute of Oceanology, Academy of Science USSR, Dept. of Bottom Fauna, Krasikova 23, 117218 Moscow, USSR

THORPE, J. P.: Port Erin Marine Laboratory, University of Liverpool, Port Erin, Isle of Man, UK

TRAMMER, JERZY: Institute of Geology, Warsaw University, Al. Zwirki i Wigury 93, 02-089 Warszawa, Poland

URIZ, MARIA J.: Centro de Estudios Avanzados de Blanes, Cami de Santa Bàrbara, 17300 Blanes (Girona), Spain

VACELET, JEAN: Centre d'Océanologie de Marseille, Station Marine d'Endoume, 13007 Marseille, France

VAN SOEST, ROB W. M.: Institute of Taxonomic Zoology, University of Amsterdam, P.O. Box 4766, 1009 AT Amsterdam, The Netherlands

VILLAMIZAR, ESTRELLA: Fundación Cientifica Los Roques, Aparado 1139, Carmelitas, Caracas 1010-A, Venezuela

VOLKMER-RIBEIRO, CECILIA: Museu de Ciencias Naturais, Fundacao Zoobotanica do Rio Grande do Sul, Caixa Postal 1188, 90000 Porto Alegre, Brazil

WEST, RON R.: Dept. of Geology, Thompson Hall, Kansas State University, Manhattan, Kansas 66506, USA

WOOD, RACHEL: Dept. Earth Sciences, University of Cambridge, Downing Street, Cambridge CB2 3EQ, UK

WRIGHT, AMY E.: Harbor Branch Oceanographic Institution, Inc., 5600 Old Dixie Highway, Fort Pierce, Florida 34946, USA

ZIMMERLE, WINFRIED: Prinzengarten 6, 3100 Celle, FRG

# Taxonomic and Phylogenetic Aspects

# Introduction

Taxonomic and phylogenetic aspects of sponges and related groups represent a wide field of research. Therefore, many different concepts have been discussed over the last decade. A key group to these topics concerning the demosponges are the "sclerosponges" and the "lithistids"; both sponge groups are characterized by only single (calcareous skeletons, desma spicules, respectively), but convergent characters (Vacelet 1985; Burton 1929). This knowledge leads to a new phylogenetic and taxonomic framework. At the conference in Berlin, 20 talks (7 on fossil sponges) dealt with this topic and demonstrated the heterogeny and weakly established taxonomic and phylogenetic models.

Today, the most widely accepted taxonomic framework of demosponges is based on larval behavior (Levi 1973; Bergquist 1978; Hartman 1982). The demosponges are divided into the subclass Tetractinomorpha with oviparous parenchymella larvae and into the subclass Ceractinomorpha with incubated vivaparous parenchymella larvae. The Tetractinomorpha includes more or less all types of tetraxon spicule-bearing sponges, and the Ceractinomopha includes mostly monaxonic sponges. Nevertheless, this classification is artificial and does not respect monophyletic groups. The reason for this is the emphasis on single characters, e.g. vivipary or ovipary, while ignoring conflicting features, e.g. common microscleres, axial condensation, etc. Soest (this Vol.) established ideas of a new taxonomic framework based on monophyletic groups (phylogenetic systematics). Using this method rigorously the single character ovipary or vivipary of larvae is reduced to synapomorphies of minor clades of demosponges. This hypothesis leads to a complete new grouping of major sponge taxa. At the conference different contributions were made on this topic using the rules of cladism strictly. Grothe pointed out the synapomorphies of Calcarea and Homoscleromorpha and Reitner demonstrated the convergent character of Recent and fossil calcareous basal skeletons of the monophyletic Hadromerida. De Weerdt used cladistic methods to reconstruct former endemic areas with area cladograms (areal cladistics). Besides these new taxonomic concepts of the major sponge groups, the always important systematic revision of minor sponge groups is represented by a number of papers, e.g. Diaz et al., who revised the Halichondrida of the Central-Atlantic area. Many of the talks concentrated on fossil sponge taxa and their modern relatives. An important overview on modern "lithistids" from the Central Pacific was given by Levi. "Lithistids" are a polyphyletic group of demosponges which are characterized by desmas and may represent a relict stem group of certain demosponges. Modern lithistids exhibit a close relationship to Mesozoic taxa, as pointed out by Levi. The Palaeozoic history of demosponges and heteractinid sponges was reviewed by Rigby.

Many discussions concentrated on sponges with calcareous, re-inforced skeletons, such as "coralline sponges", "sclerosponges", "chaetetids", "stromatoporoids", "pharetronids", and "sphinctozoans", in older terms. These morphological terms represent a characteristic skeletal type. Within the fossil record these organisms are important framebuilders in different organism accumulations. They first occur in the Early Cambrian (Archaecyathida) and play an important role during the whole Palaeozoic and Early Mesozoic. Most of these organisms exhibit no spicule remains and, therefore, the systematic relationships are problematic, and the sponge affinities are based mostly on the preserved excurrent canal systems (astrorhize system). Using new technical methods and staining techniques as well as studying numerous thin sections spicule remains of coralline sponges were detected (Kazmierczak 1974; Gray 1980; Wood 1987; Reitner 1987). But in many cases the taxonomic affinities are still open. By strictly comparing the fossil spicule skeletons with modern ones and using phylogenetic systematics in some cases, the relationships could be solved (Reitner, this Vol.). The preserved spicule remains are not widely accepted in all cases. Particularly, the curved, randomly oriented spicule remains could be interpreted as traces of microborings. Kazmierczak's chapter (this Vol.) on the sponge affinities of *Favosites* demonstrates this problem as well as Senowbari Daryan's "Overview on sphinctozoans". Cuif and Gautred, using the different characteristic micro- and ultrastructures of the calcareous skeletons for phylogenetic reconstructions, found another way to recognize phylogenetic relationships. The importance of spicules in the reconstruction of monophyletic lineages was widely accepted by the participants of the conference in Berlin.

Modern sponges with re-inforced calcareous skeletons are rare and relict groups. Vacelet has given an overview on modern "pharetronids", calcareous sponges including both subclasses, the Calcaronea and Calcinea, with some new important biogeographical data. He first reported deep-water (ca. 2000 m) pharetronids from the Atlantic Ocean. He made further important contributions and discoveries on demosponges with a secondary calcareous skeleton. Important in this respect are both the discovery of colonial deeper water sphinctozoans of the *Vaceletia* type near New Caledonia and the description of a new *Merlia* species, together with Uriz, from the Mediterranean Balearic Islands. Fossil sponges without a re-inforced skeleton have only a slight chance of preservation. Exceptions are e.g. the famous locality of the Middle Cambrian Burgess Shale in which turbiditic mudflows very rapidly cover the organisms. A diageneticaly, very early silification also presents a chance to preserve entire sponges. This rapid silification preserved the freshwater sponges from the Aptian of Argentina, first described by Ott and Volkheimer (1972). The material was restudied by Volkmer Ribero and Reitner. Important new data are the discovery of preserved oocytes in *Palaeospongilla chubutensis* Ott & Volkheimer 1972 and the recognition of a gemmule-bearing new species, *Spongilla patagonia*, which exhibits a strong relationship to the extant species *S. alba*. Freshwater sponges, particularly complete ones, are extremely rare within the fossil report. Early diagenesis and rapid cementation by decaying cyanobacteria play an important role in the preservation of the entire spicule arrangement in the excavations of the fossil haplosclerid boring sponge *Aka* (Reitner and Keupp, this Vol.). Recent and fossil *Aka* are important eroders of calcareous material. But *Aka* is also found in skeletons of hexactinellids and lithistids in Jurassic sponge mounds. Therefore, it is a good indicator of a very early alteration of siliceous skeletal material into calcite.

Last, but not least, the story of hexactinellid sponges is presented. This sponge group differs in many characters from the demosponges and the phylogenetic affinities are unknown. Most spongiologists accept the presence of spicules and choanoflagellate cells as synapomorphies of both major sponge taxa. Only few people deal with this group, most of them are palaeontologists. At the conference in Berlin, four presentations dealt with hexactinellids, only one with a modern taxon. Reiswig gives an emendation of the modern taxon *Dactylocalyx* and presents new TEM micrographs of the soft tissue. An important contribution to the phylogeny of hexactinellids is given by Mehl. She has created a new taxonomic and phylogenetic framework of hexactinellids, including fossil taxa, demonstrating that many modern hexactinellids are characterized only by symplesiomorphies, and she reconstructed a hypothetical hexactinellid sponge. Gasse and Müller are revising important Mesozoic hexactinellid taxa.

The whole session on phylogeny and taxonomy demonstrated the controversy of many these ideas. Nevertheless, the methods of phylogenetic systematics become more and more accepted. During the "Third Conference on the Biology of Sponges" in Woods Hole 1985, only one paper dealt with cladistics. Working with monophyletic groups is the only possibility to understand phylogenetic trends within the sponges. Important in this respect is to incorporate the time factor and extinct taxa in modern phylogenetic trees. The conference in Berlin, where neontologists and palaeontologists worked on the same topics and questions, has demonstrated this new way.

## References


Bergquist PR (1978) Sponges. Hutchinson, Pennsylvania, 268 pp

Burton M (1929) Lithistidae. Fish Mar Biol Survey, Spec Rep 7:3–12

Gray D (1980) Spicule pseudomorphs in a new Paleozoic chaetetid, and its sclerosponge affinities. Palaeontology 23:803–820

Hartman WD (1982) Porifera. In: Parker SP (ed) Synopsis and classification of living organisms. McGraw Hill, New York, pp 640–666

Kazmierczak J (1974) Lower Cretaceous sclerosponge from the Slovakian Tatra Mountains. Palaeontology 17:341–347

Levi C (1973) Systématique de la classe Démospongiaria (Démosponges). In: Grasse PP (ed) Traité de Zoologie III (1). Masson, Paris, pp 577–631

Ott E, Volkheimer W (1972) *Palaeospongilla chubutensis* n.g. and n.sp.- ein Süßwasserschwamm aus der Kreide Patagoniens. Neues Jahrb Geol Paläontol Abh 140:49–63

Reitner J (1987) *Euzkadiella erenoensis* n. gen. n.sp. ein Stromatopore mit spiculärem Skelett aus dem

Oberapt von Ereño (Prov. Guipuzcoa, Nordspanien) und die systematische Stellung der Stromatoporen. Paläontol Z 61:203–222

Vacelet J (1985) Coralline sponge and the evolution of Porifera. In: Conway Morris S, George JD, Gibson R, Platt HM (eds) The origins and relationships of lower invertebrates. Syst Asoc Spec 28:1–13

Wood R (1987) Biology and revised systematics of some late Mesozoic stromatoporoids. Spec Pap Palaeontol 37:89 pp

# New Perspectives on the Hexactinellid Genus *Dactylocalyx* Stutchbury

H.M. Reiswig

## Introduction

In spite of its important position in the early taxonomic history of the Hexactinellida, *Dactylocalyx* remains a poorly known genus. As Schulze (1887) pointed out, *D. pumiceus* was most likely the first hexactinellid referred to in 1780; when described by Stutchbury in 1841, it was the first dictyonine and the fourth recognized species of the group that eventually formed the class Hexactinellida. It served repeatedly as the basis for the Dactylocalycidae in the taxonomic schemes of Gray (1867) and Ijima (1903), although the type specimen was never depicted and its characters barely appreciated. *Dactylocalyx pumiceus* remains important as the largest and most common hexactinellid throughout the western tropical Atlantic. Interest in the genus was renewed with Reid's (1957, 1962) resurrection of the Dactylocalycidae and the surprising transfer of the group to the Lychniscosa. I provide here a short review of the history of the genus, and present result of skeletal and soft-tissue investigations on recently collected specimens.

## Historical Account

As Schulze (1887) and Reid (1957) have given extensive accounts of the history of *Dactylocalyx*, I will present here only significant points deserving attention. The genus has at one time or another encompassed 23 species names, most of these being lithistids erected by Bowerbank and eventually transferred; inappropriate hexactinellids have been moved to other genera or synonymized with *D. pumiceus*. Reid (1957) accepts only two species names as probably valid: *D. pumiceus* Stutchbury, 1841 and *D. subglobosus* Gray, 1867, and even these may be merely varieties. It should be noted that many of the transfers suggested by Ijima (1927) should be viewed with caution. The proposals to transfer *D. subglobosa* Gray to *Myliusia* Gray, and to recognize *D. ingalli* Bowerbank separately, together with *D. subglobosa* of Schmidt, 1880, and Schulze, 1887, appear to be totally unfounded.

J. Reitner and H. Keupp (Eds.)
Fossil and Recent Sponges

## *Types*

In his original description of *Dactylocalyx pumiceus* from Barbados in the Bristol Institution Museum, Stutchbury failed to provide any indication of the size, shape, or number of specimens examined. Both Bowerbank (1858) and Gray (1859) imply that there was a single type specimen of which Gray, prior to 1858, obtained one-half plus a fragment of the remaining half for the British Museum. The type specimen was known only as "magnificent" until Bowerbank (1869, p. 78) gave dimensions which allowed certain determination of the half specimen and the fragment during my recent visit to the British Museum. The half in the British Museum as of June 1988 is depicted here for the first time (Fig. 1) [BM (NH) Porifera Reg. No. 1867.2.19.1]. The status of the specimen as the Holotype remains questionable, since Sollas (1879) and later Schulze (1887) state that Stutchbury saw two specimens. The second specimen was intact in the Bristol Museum when Sollas (1879) described it as *D. pumiceus* var. *stutchburyi*. The basis for Sollas' claim that Stutchbury was aware of the second specimen remains unsupported. Labels on the specimen should be examined.

• The type specimen of *Dactylocalyx subglobosa* Gray, 1867 also resided in the British Museum [BM (NH) Reg. No. 1840.5.20.13] as of June 1988. Its original location recorded as "Malacca (?)" is almost certainly erroneous, as indicated by an anonymous addition to the label noting that the attached coral is common in the West Indies.

## *Major Features*

The genus has been diagnosed by many workers, including Bowerbank (1862, 1869), Gray (1867), Schulze (1887), and Reid (1957). The general features most consistently noted are the conspicuous, deep depressions in the form of circular pits or grooves on both inner and outer surfaces, the irregular network of fine (ca. 1 mm diameter) anastomosing tubules which compose the walls and the general irregularity of the dictyonal frame of the tubule walls. The most extensive illustrated description of skeletal organization of *D. pumiceus* remains that of Sollas (1879), where he noted the abundance of synapticular strands, common addition of secondary "darning" fibers and the occasional presence of apparent octahedral nodes. One of the latter is shown in a very small line drawing. In his outstanding but text-only review, Reid (1957) stressed the importance of these rare but apparently "genuine" lychniscs by transfer of the genus to the re-established family Dactylocalycidae Gray within the order Lychniscosa Schrammen. No illustrations of purported lychniscs were provided. In spite of the

→

**Fig. 1.** Type specimen (1/2) of *Dactylocalyx pumiceus* Stutchbury in the British Museum (Natural History) in upper surface (**a**) and side view (**b**). The upper surface (**c**) and lower surface (**d**) are enlarged to the same scale to show depressions and openings of the tubule systems

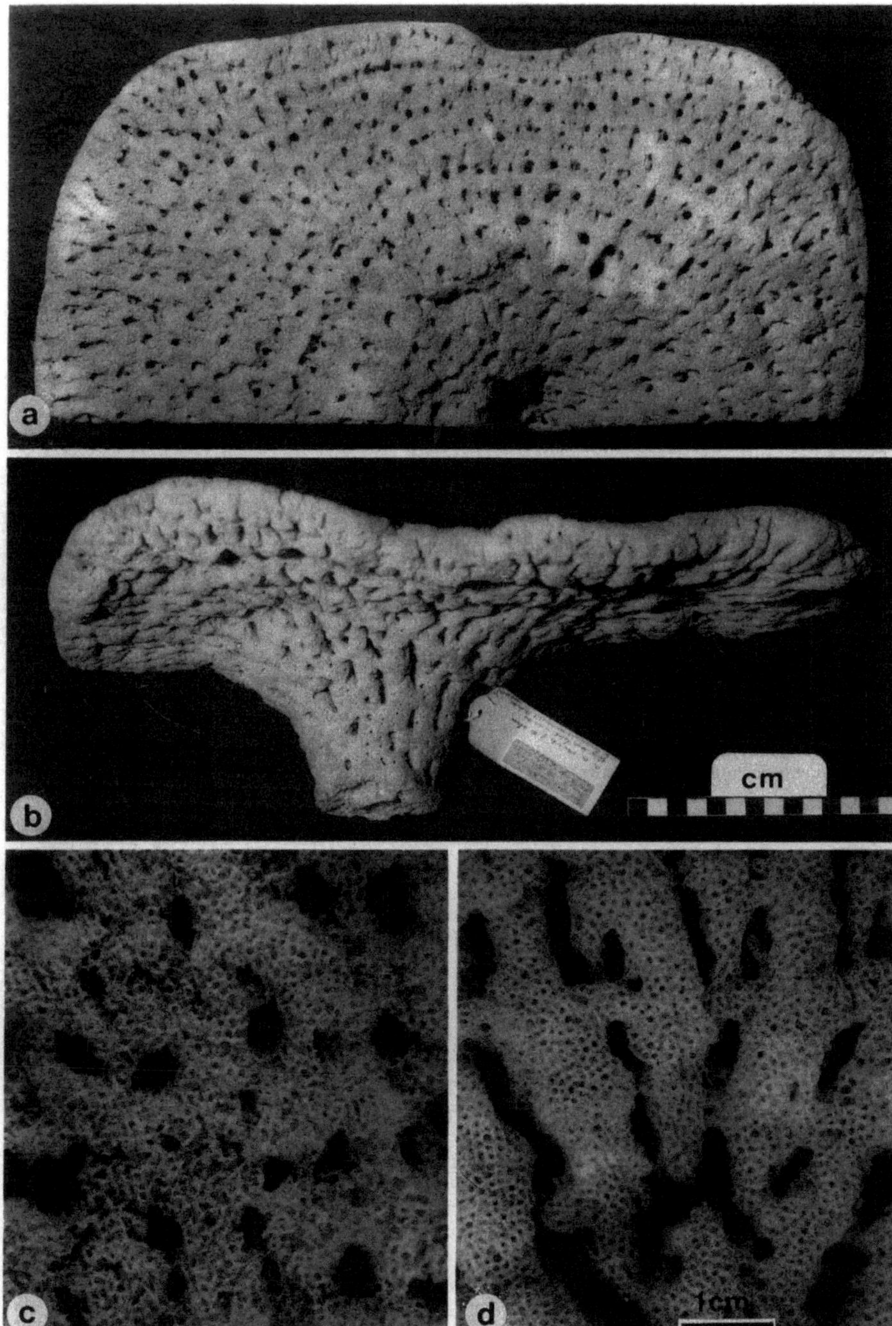
a
b
cm
c
d
1cm

prominence of the tubule system as a generic character, the sizes of these elements are only casually alluded to in most papers; no satisfactory measurements are provided. A major uncertainty also remains with regard to the existence of an inhalant system of intercanal spaces, a major feature of the hexactinosan family Meandrospongidae Zittel to which *Dactylocalyx* has belonged through much of its history. Schulze (1887) considered the intercanal system to be present in *Dactylocalyx*, while Reid (1957) doubted its existence here.

## *Spicules*

The free spicules of *Dactylocalyx* species have been rather poorly studied, as indicated in the summary of Table 1. Large rhabdodiactins without a central cross and without any evidence of barbation have been reported from both species. Ijima (1927) took these as uncinates and assigned the genus accordingly. Megascleres including four-, five- and six-rayed megascleres have been described from *D. pumiceus* but tetractins (stauracts) have not been seen in *D. subglobosus*. The megascleres have been described as highly variable in form, with paratangential rays curved and occasionally coiled (Sollas 1879), although regular straight-ray forms have also been noted as common in dermal and intermedial locations. The tips of rays have generally been described as inflated, rounded or clavate, but acute roughened tips are not rare. Accounts of the hexasters of the two

**Table 1.** Presence and measurements of *Dactylocalyx* spicules

| Ref.[a] | Megasclere length (mm) | | | | Hexaster diam. (μm) | |
|---|---|---|---|---|---|---|
| | Monaxon | Tetractin | Pentactin | Hexactin | Discohex. | Oxyhex. |
| *D. pumiceus*[b] | | | | | | |
| 1-T | – | – | – | + | + | + |
| 2-T | + | – | – | – | + | – |
| 3-T | to 12.5 | + | + | + | – | – |
| 4 | 2–12 | – | 0.8–2.4 | 0.8–2.4 | 30–100 | 0 |
| 5 | – | – | 0 | 0.26–1.25 | 41–78 P | 35–61 |
| 7 CNMs | 0.3–8+ | 0.7[c] | 0.31 | 0.41 | 59 ± 11 | 0 |
| | | (0.6–0.8) | (0.1–0.8) | (0.1–1.0) | (25–84) | |
| 7 CNM1 | to 11.2 | 0.6 | 0.41 | 0.52 | 72 ± 18 | 56 ± 11* |
| | | (0.2–0.7) | (0.1–0.6) | (0.2–1.2) | (38–111) | (46–80) |
| 7 RM | to 12 | + | 0.25 ± 0.14 | 0.47 ± 0.26 | 46 ± 8 | 63 ± 6* |
| | | | (0.05–0.72) | (0.13–1.1) | (25–77) | (57–75) |
| *D. subglobosus* | | | | | | |
| 2-T | + | – | – | – | + P | – |
| 6 | – | – | + | + | 50–100 | – |

[a] References: 1. Bowerbank 1858; 2. Carter 1873; 3. Sollas 1879; 4. Reid 1957; 5. Mothes de Moraes 1977; 6. Schulze 1887; 7. this study.

[b] Symbols: + = noted as present; – = not mentioned; 0 = noted as absent; T = type specimen; p = pappiform, with unequal ray length; * = onychohexasters.

[c] Mean ± SD with ranges in parenthesis below.

species are often contradictory: oxyhexasters have been specifically reported from *D. pumiceus* by two workers, and particularly noted as being absent by two others, or not remarked upon by two more. Oxyhexasters have not been seen in *D. subglobosa*, but the data are few and the quality of those observations is questionable. Schulze's specimen of *D. subglobosus* obtained through O. Schmidt was from St. Lucia, West Indies, far from the supposed type locality of this species. Clearly there is great variability in the small amount of spicule data available. The question arises: do these reports indicate true differences between specimens or do they reflect different techniques used by the investigators?

### *Soft Tissues*

Nothing whatsoever has been reported about the living tissues of any presently recognized member of the genus.

### New Specimens

Three specimens collected relatively recently from the *D. pumiceus* type locality, Barbados, served as new study material. Two well-preserved, wet specimens of *Dactylocalyx* were discovered in the unidentified and uncatalogued collections of the National Museum of Canada. The specimens are both small, measuring 24 by 35 mm and 40 by 41 mm in diameter and height respectively, and could be assigned to either of the two accepted species on the basis of form and limited external sculpture. They were collected 10 July 1965 by dredge from 187 to 250 m off the west coast at Holetown by D. Pocock, and are now catalogued as NMCI 1987–0227, and denoted here as NMCs and NMCl.

A third fresh specimen of *Dactylocalyx* was obtained personally by dredging in 170 to 200 m off Holetown (13°11.4′N, 59°39.2′W) on 13 July 1987. Immediately upon retrieval of the dredge, fragments were cut from the less soiled, internal sites of the slightly mud-fouled, broken specimen and fixed in a cold, freshly prepared combination of 1% osmium tetroxide, 6.25% glutaraldehyde, 1% NaCl in 0.13 M s-collidine buffer for 15 min. The fragments were rinsed, postfixed in the same cocktail, re-rinsed six times in cold buffer and dehydrated to 95% ethanol. Tissues reported here were transported in ethanol to Montreal for desilicification in 4% HF for 45 min, then rinsed, dehydrated, and embedded in Spurr epoxy. Semi-thin (0.7 $\mu$m) and ultrathin sections were cut on a Sorvall microtome and viewed by phase contrast light microscopy and, after staining with uranyl acetate and lead citrate, by transmission electron microscopy with a Phillips 410 EM. The remainder of the specimen was partly Bouin's fixed and partly dried for skeletal studies. The large fragments represent 3/4 of the original conical specimen ca. 9 cm high by 13 cm diameter. It conforms to the shape and depression pattern of the *D. pumiceus* described by Sollas (1879). The specimen is deposited in the Redpath Museum, McGill University as Cat. No. 87-7-13.4 and referred to here as specimen RM.

***Skeletal Nodes***

The dictyonal frames of all three specimens are similar, consisting of the meandering tubule system characteristic of *D. pumiceus*. The organization of the tubule system and the possible existence of an intercanal space network has not yet been investigated. The irregular frameworks all contain the perforated nodes stressed by Reid (1957). These are common in some surface areas, but they rarely approach the regularity of classic lychniscs of *Ventriculites* Mantell or *Neoaulocystis* (Schulze) Zhuravleva – a hexactine center surrounded by a set of 12 symmetrical synapticular struts. Here the nodes are extremely irregular in form, often consisting of a lacy set of synapticulae joining adjacent rays of recently connected pentactine and hexactine megascleres to main framework (Fig. 2A,B). They are more often planar rather than three-dimensional (Fig. 2C,D). In some areas of the smaller specimens, the irregular foramina between synapticulae

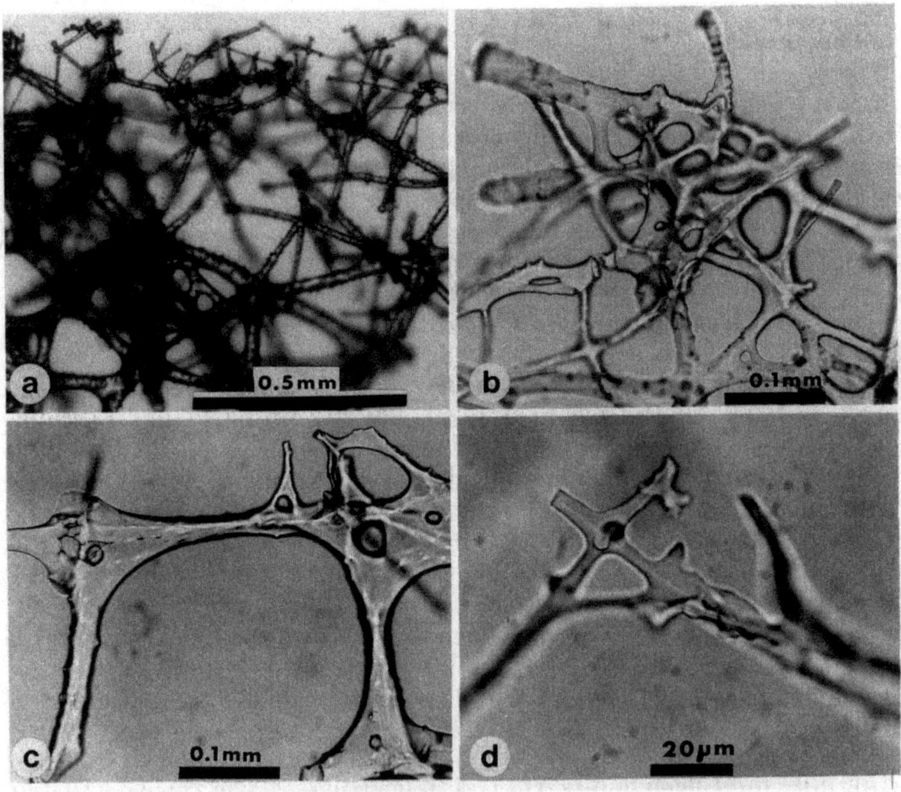


**Fig. 2. a** Surface-attached spicule network in the RM specimen of *D. pumiceus*. **b** Irregular synapticular network of heavily ensheathed spicules and strands in the surface skeleton of the RM specimen. **c** Perforate hexactine nodes in the surface of the RM specimen. **d** A broken hexactine node in early stages of synapticular strut development while attached to the main framework of the RM specimen of *D. pumiceus*

and main rays persist several mm into the framework where beam and nodal thickening are at intermediate levels; in deeper, older areas the foramina are occluded but their "ghosts" remain visible within the solid nodal masses. They are not characteristic of most of the nodes surveyed.

### *Spicules*

Spicule preparations were made following the procedure outlined in Reiswig and Browman (1987) to avoid loss of rare microscleres. Where standard deviations are given in Table 1, at least 50, and usually 100, measurements were made. The spicule complements of the three specimens from the same locality showed surprising variation (Table 1). All possess robust rhabdodiactins which show no evidence of either a central cross or barbation, and similar sets of megascleres. They all contain at least two types of discohexasters (Fig. 3A), but those of the CNM1 specimen are clearly larger. Both the CNM1 and the RM specimen have a spicule type, an onychohexaster (Fig. 3B), not previously reported for this genus; it is included in Table 1 in the oxyhexaster column, although it would technically be classed as a variety of discohexaster. Proper oxyhexasters were not seen, but rare developmental stages of the discohexasters may easily be confused with them. On the basis of spicules, framework, and general morphology, all three specimens are assigned to *Dactylocalyx pumiceus*.

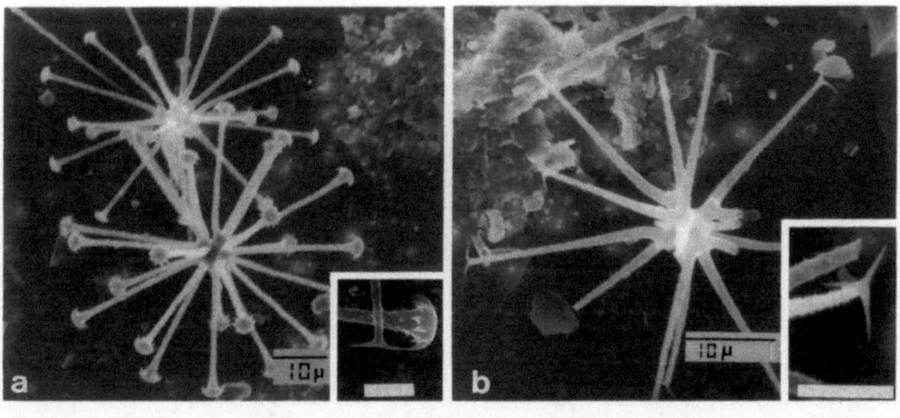


**Fig. 3. a** Discohexasters from specimen NMCI (SEM). The insert shows terminal tips of a discohexaster and onychohexaster in side view (scale = 5 µm). **b** Partly damaged onychohexaster from specimen NMCI with secondary rays ending in two claws (SEM). The filtration procedure causes common breakage of the delicate secondary rays of this spicule type. The *insert* shows a terminal tip with three claws (scale = 5 µm)

### *Soft Tissues*

In general, tissues are well preserved for study at the light microscope level, while ultrastructurally some distortion and considerable extraction are evident in places. The preserved soft tissues of *Dactylocalyx pumiceus* consist mainly of flagellated chambers; sparse mesohyle containing cellular elements is concentrated mainly in cushions between adjacent chambers (Fig. 4A). The most striking

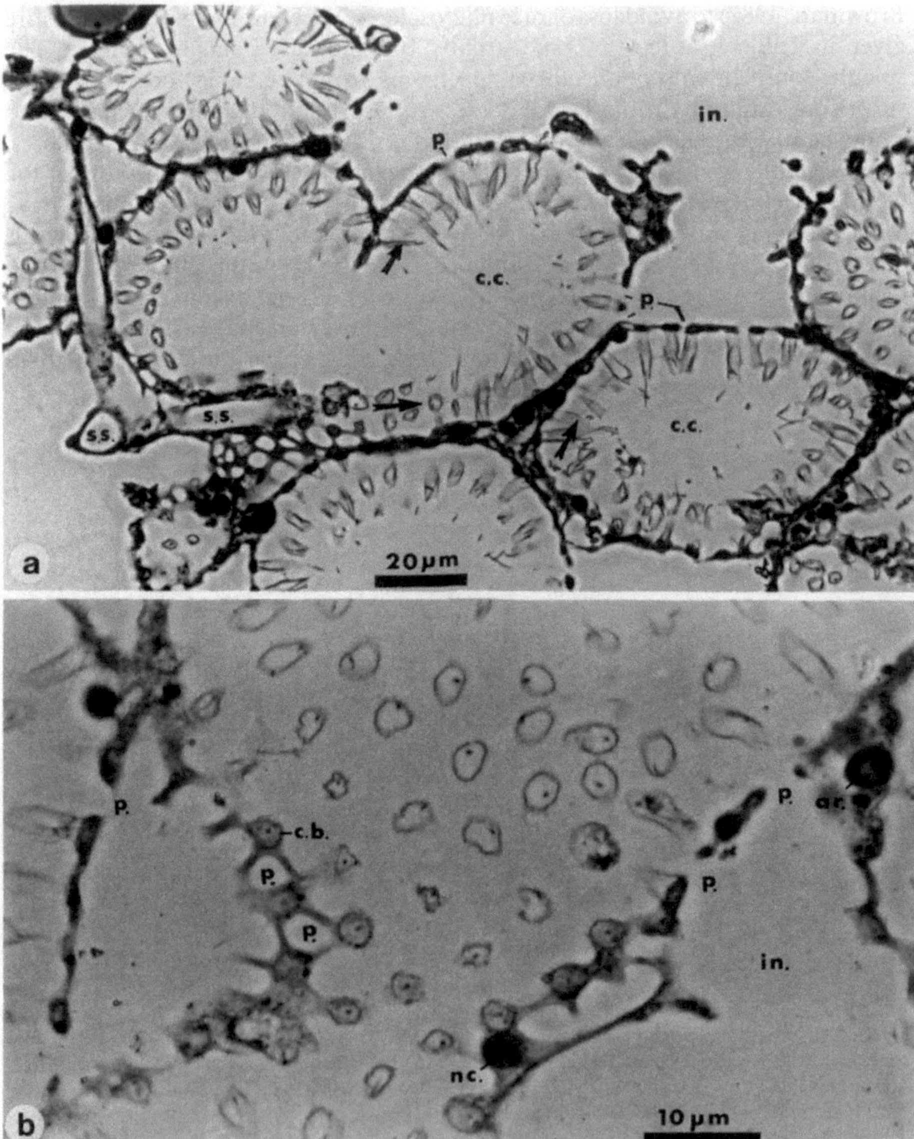

feature of the tissue relative to other hexatinellid tissues examined to date is the absence of a secondary reticulum at the mid-level of the collars. Most chambers are rather standard for the hexactinellid-type, being ovoid and measuring 50×75 μm, with some irregular in shape and extending to 160 μm. Apopyles are wide gaps in chamber walls, 30 to 50 μm in diameter. The circular to oval prosopyles of the membrana reticularis (reticulate portion of chamber wall) vary from 1 to 5 μm diameter (mean: 3.6 μm) (Fig. 4B). Collars of typical poriferan form are 10 μm in length and project vertically from the collar bodies of the chamber wall toward the chamber center; the flagella extend a further 6 to 7 μm. The collars are circular in section and consist of 37 to 43 microvilli (mean = 40). Even with the limitations of phase microscopy, it is fairly obvious that the collar bodies are anucleate nodes of the very thin membrana reticularis; they are spaced at mean nearest-neighbor distance of 6.4 μm (5.2–7.5 μm).

The arrangement of the chamber walls is clearer in low-power electron micrographs (Fig. 5). The absence of a secondary reticulum is confirmed, but a fibrous network is easily noted at the expected location; this is also detectable with the light microscope (Fig. 6A, arrows). Both of the chamber wall surfaces, whether shared by adjacent chambers or forming the boundary between an inhalant space and a chamber (membrana reticularis as those figured) are bounded by a continuous syncytium with gaps only at the surfaces occupied by the separate collar bodies. The syncytium contains occasional nucleolate nuclei 2.5–3.0 μm in diameter (Figs. 4, 5), numerous small dense mitochondria ca. 0.25 μm in diameter (Fig. 6B), and numerous empty and filled vesicles of wide variation in size. This syncytium appears to be continuous between inner and outer wall surfaces (note continuity around prosopyles in Fig. 5) and between walls of adjacent chambers; it may be continuous throughout the specimen. The syncytium completely surrounds a thin reticulum of "internal" space – a mesohyle – containing a continuous fibrillar mesolammella which is probably collagenous.

In addition to the mesolamella network, the mesohyle contains many living tissue elements. The stolon systems connecting collar bodies extend through the mesohyle space; neither cytoplasmic connections nor perforate plugs have been detected between this choanosyncytial network and other elements. The collar bodies appear wedged into tight sockets of the main chamber wall syncytium which stains rather heavily at the region of closest contact. The choanosyncytial system of nodes and stolons thus appears to be totally without connection to nucleated structures, and further appears to lack mitochondria. The glycocalyx fibril pad on the surface of the collar bodies and around the base of the flagellum appears similar to those of choanocytes of cellular sponges and is probably of the same material composing the fibrous network located outside the collars.

**Fig. 4. a** Semi-thin section of the choanosome of the RM specimen of *D. pumiceus* in phase contrast. *in.* inhalent space; *s.s.* spicule space; *P.* prosopyle; *c.c.* chamber center; *t.* thesocyte (?); *arrow* fibrous network. **b** Semi-thin section nearly tangential to a flagellated chamber wall showing an inhalant space (*in.*) leading to prosopyles (*P.*) of the membrana reticularis, occasional nuclei (*nc.*) of the main syncytium, collar bodies (*c.b.*) and an archaeocyte (*ar.*) in a chamber wall cushion

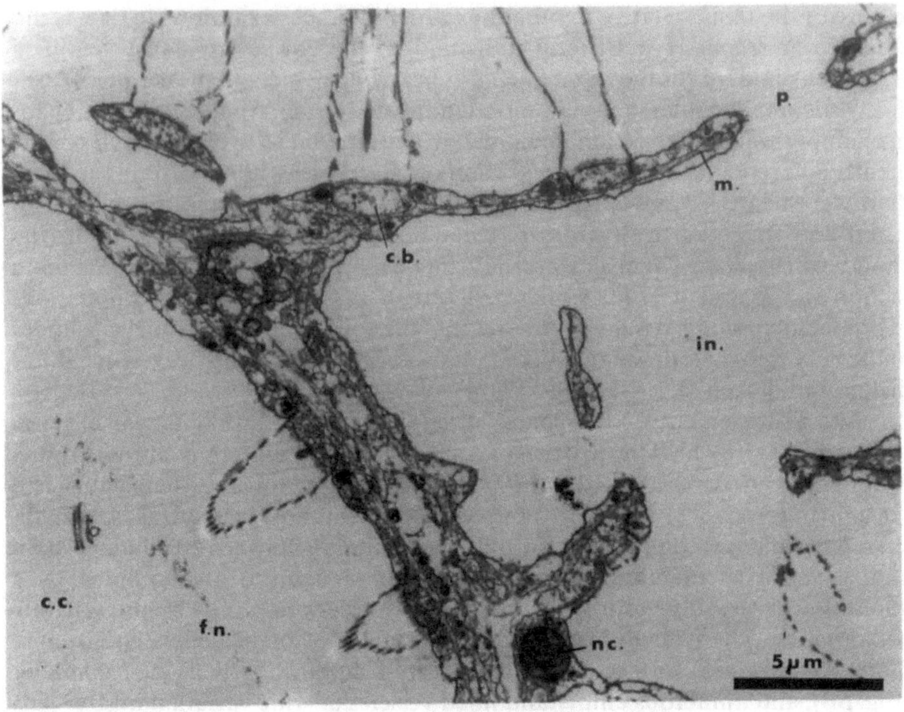


**Fig. 5.** Thin section through flagellated chamber walls showing inhalant space (*in.*), porocyte (*P.*), collar bodies (*c.b.*), thin mesohyle space (*m.*), nucleus of the main syncytium (*nc.*), fibrous network (*f.n.*), and chamber centre (*c.c.*)

The mesohyle also contains a sparse population of archaeocytes often located in small groups in the flat or triangular cusions between chamber walls. These cells may be nucleolate or not, but are easily distinguished from the wall syncytium by their larger and less densely staining mitochondria, ca. 0.45 μm diameter, and the presence of endoplasmic reticular membranes. Although close contacts are commonly encountered between adjacent archaeocytes as well as between archaeocytes and the general wall syncytium, the distinctive perforate plugs expected to be detected have not been seen at all. In the absence of connections between the choanosyncytium and these nucleated elements, "choanoblasts" in the sense of Mackie and Singla (1983) cannot be designated. Rare symbiotic bacteria (Fig. 6B) and large thesocytes (?) (Fig. 4A top left edge) have been detected as additional mesohyle elements, but have not yet been given thorough examination. Special chord syncytia and spherulous cells have not been detected here.

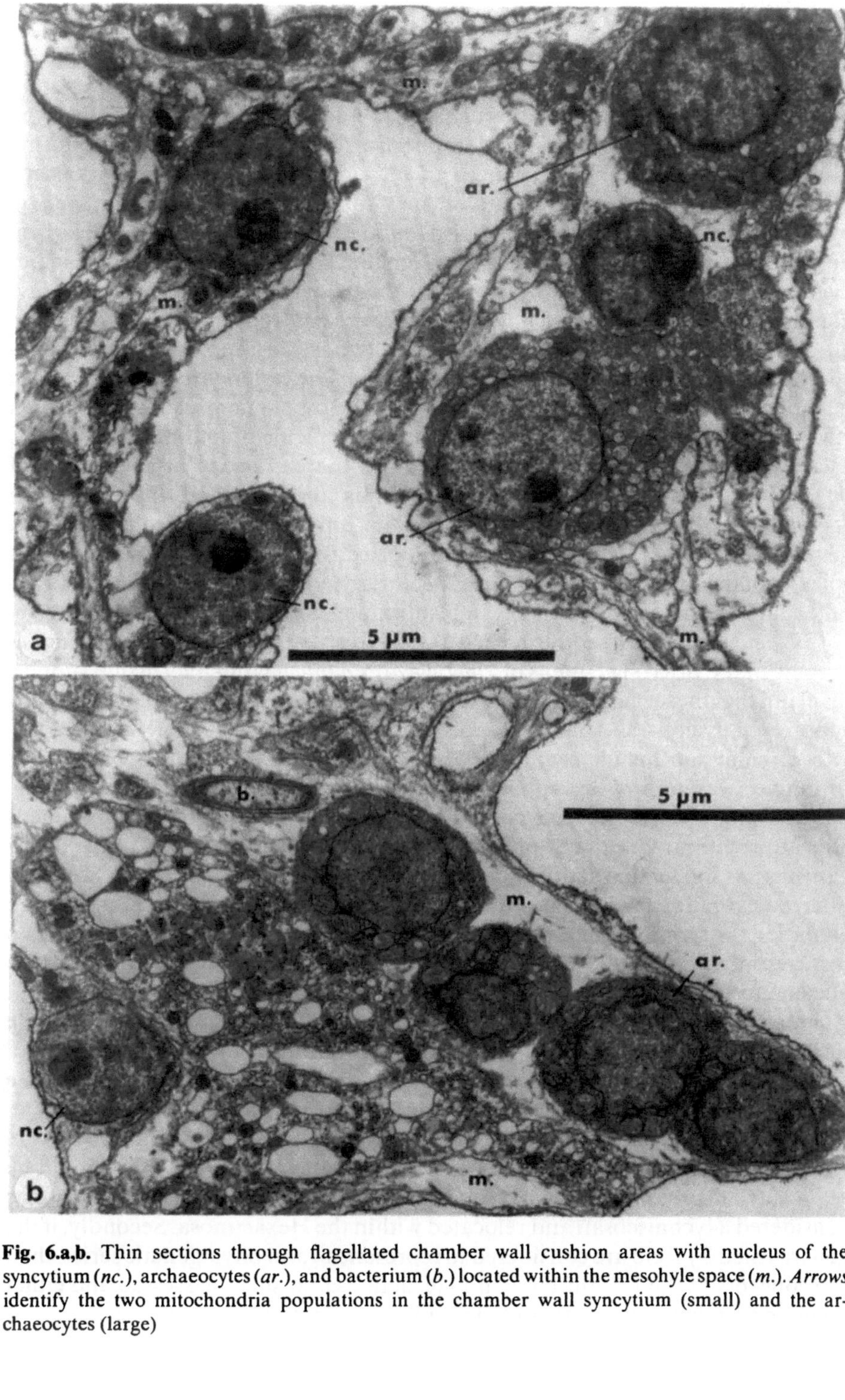


**Fig. 6.a,b.** Thin sections through flagellated chamber wall cushion areas with nucleus of the syncytium (*nc.*), archaeocytes (*ar.*), and bacterium (*b.*) located within the mesohyle space (*m.*). *Arrows* identify the two mitochondria populations in the chamber wall syncytium (small) and the archaeocytes (large)

## Discussion

The assignment of the new specimens to *D. pumiceus* is based on negative (in relation to *D. subglobosus*) as well as positive characters. Carter (1873) reported that *D. pumiceus* has regular discohexasters and oxyhexasters while the type specimen of *D. subglobosa* has regular spherical discohexasters and "pappiform" discohexasters (with unequal length of terminal rays), but no oxyhexasters. Schulze (1887), reviewing fragments of "new" material assigned to *D. subglobosa* by O. Schmidt (West Indian origin), also found no oxyhexasters, but he makes no mention of discohexasters of unequal ray length. According to Schulze, the major distinction between the two species is the presence of oxyhexasters in *D. pumiceus* and their absence in *D. subglobosus*. The three recent specimens from Barbados all lack "pappiform" discohexasters; the two larger specimens have onychohexasters that are indistinguishable from oxyhexasters except at very high magnification. Onychohexasters comprise about 30% of microscleres in specimen RM, only about 1% in CNMl, and are absent in spicule preparations of CNMs. The overlap in other spicule size ranges and the variability in abundance of onychohexasters in the specimens possessing them suggest that the three specimens belong to a single population with quite variable spiculation. Their general pattern is consistent with that reported for *D. pumiceus*. A resurvey of spicules of the type specimens of both species is needed, as well as a careful survey of the large number of specimens in institutional collections.

The new specimens confirm Reid's (1957) discovery of rare perforated nodes in some older, dried specimens of *Dactylocalyx pumiceus* but they do not support his claim that these include genuine octahedra or that the nodal system represents suppressed lychniscs. This does not negate Reid's resurrection of the Dactylocalycidae nor his inclusion of *Dactylocalyx* within the Lychniscosa. At our primitive awareness of hexactinellid history, the definition of major taxa on one or a few major characters has proven advantageous for categorizing specimens and information. Greater sophistication of knowledge in other groups of organisms has shown that simple definitions of major taxa are inadequate to reflect inferred evolutionary lineages. In his attempt to reveal his interpretation of an evolutionary series of genera associated with *Dactylocalyx*, Reid was forced to disregard the importance accorded the single-character distinction between the Hexactinosa and Lychniscosa – the form of the dictyonal nodes. His claim that lychnisc suppression has led to the development of forms appearing as simple dictyonines essentially negates the utility of Schrammen's Lychniscosa as a concept. He offered no formal emendation nor replacement of the order. There appear to be two possible resolutions to the problem created by Reid's action. Firstly, if genera included within the Dactylocalycidae by Reid (1957) are reexamined and considered not to form an evolutionary series, they should be reallocated to groups with most similar genera. *Dactylocalyx* will probably not be considered a lychniscosan and relocated within the Hexactinosa. Secondly, if the genera listed by Reid are considered to represent a valid phylogenetic series, then the concepts of the Lychniscosa and Hexactinosa must be reassessed and either emended or new concepts and definitions proposed. Since the structure of

*Dactylocalyx* remains so poorly known, I do not support a change in the taxonomic status of any of those genera at this time. Actions based on partial information are unlikely to lead to a stable solution of the status of this important genus.

In the area of ultrastructural analysis of soft tissue, *Dactylocalyx pumiceus* maintains its prominent position. The ultrastructure of only one species, *Rhabdocalyptus dawsoni*, has been published to date (Mackie and Singla 1983). While tissue of *D. pumiceus* conforms to the essential syncytial nature of hexactinellids, it contravenes the pattern known in three hexasterophora: the lyssacinosan above (*R. d.*) and the hexactinosans *Aphrocallistes vastus* and *Farrea occa* (unpublished observations). All three have a well-developed secondary reticulum extending from the main syncytium into the flagellated chambers and all three have common conspicuous perforated plugs between cellular and syncytial elements. The absence of these features may be characteristic of only *Dactylocalyx* or it may represent a major organizational diversion of a major portion of the Hexasterophora. Before proposals are made to modify the criteria defining the Symplasma (Reiswig and Mackie 1983), a more thorough search should be made for the plugs and secondary reticulum in other specimens of *D. pumiceus* and other lychniscosans.

*Acknowledgments.* I wish to thank S. Stone and P. Frank for providing access to collections in their charge. I also wish to acknowledge the aid of D. Lachance, Dr. W. Hunte and the entire staff of the Bellairs Research Institute in enabling the dredging operations at Barbados. This work was supported by an operating grant from the Natural Sciences and Engineering Research Council of Canada.

## References

Bowerbank JS (1858) On the anatomy and physiology of the Spongiadae. Part 1: On the spicules. Philos Trans R Soc Lond 148:279–332

Bowerbank JS (1862) On the anatomy and physiology of the Spongiadae. Part III. Philos Trans R Soc Lond 152:1087–1135

Bowerbank JS (1869) A monograph of the siliceo-fibrous sponges. (Part I). Proc Zool Soc Lond 1869:66–100

Carter HJ (1873) On the Hexactinellidae and Lithistidae generally, and particularly on the Aphrocallistidae, *Aulodictyon*, and Farreae, together with facts elicited from their deciduous structures and descriptions respectively of three new species. Ann Mag Nat Hist (4) 12:349–373

Gray JE (1859) Description on *MacAndrewia* and *Myliusia*, two new forms of sponges. Proc Zool Soc Lond XXVII:437–440

Gray JE (1867) Notes on the arrangement of sponges, with the description of some new genera. Proc Zool Soc Lond 1867:492–558

Ijima I (1903) Studies on the Hexactinellida. Contribution III. (*Placosoma*, a new Euplectellida; Leucopsacidae and Caulophacidae). J Coll Sci Imp Univ Tokyo 18:1–124

Ijima I (1927) The Hexactinellida of the Siboga Expedition. Siboga Exped Reports vol 6. p 383

Mackie GO, Singla CL (1983) Studies on hexactinellid sponges. I. Histology of *Rhabdocalyptus dawsoni* (Lambe 1873). Philos Trans R Soc Lond B 301:365–400

Mothes de Moraes B (1977) Ocorrencia de *Dactylocalyx pumiceus* Stutchbury, 1841, no litoral do Rio Grande do Sul (Porifera, Hexactinellida). Inheringia (Zool) 50:41–50

Reid REH (1957) Notes on hexactinellid sponges. II. *Dactylocalyx* Stutchbury and the family Dactylocalycidae Gray. Ann Mag Nat Hist (12) 10:821–826

Reid REH (1962) Notes on hexactinellid sponges. IV. Nine cretaceous Lychniscosa. Ann Mag Nat Hist (13) 5:33–45

Reiswig HM, Browman HI (1987) Use of membrane filters for microscopic preparations of sponge spicules. Trans Am Microsc Soc 106:10–20

Reiswig HM, Mackie GO (1983) Studies on hexactinellid sponges. III. The taxonomic status of Hexactinellida within the Porifera. Philos Trans R Soc Lond B 301:419–428

Schulze FE (1887) Report on the Hexactinellida collected by H.M.S. Challenger during the years 1873–1876. Rep Sci Res Challenger Zool vol 21. p 513

Sollas WJ (1879) Observations on *Dactylocalyx pumiceus*, with a description of a new variety, *D. stutchburyi*. J R Microsc Soc 2:122–133

Stutchbury S (1841) Description of a new sponge from Barbados. Proc Zool Soc Lond 9:86–87

# The Hexactinellid Sponge Genus *Becksia* Schlüter, 1868 from the Campanian of the NW Münsterland (Upper Cretaceous, NW Germany)

W. GASSE, R. GÖCKE, and K.-H. HILPERT

## Introduction

In the district of Coesfeld near the villages of Holtwick, Osterwick, Darup and Lette, northwest area of the Münster Cretaceous Basin (Fig. 1), extraordinary associations of hexactinellid and lithistid sponges are exposed. These occur in Late Senonian strata [late Lower Campanian (Campanian 2–3)]. The regional so-called Osterwicker Schichten include alteration layers of glauconitic-sandy, argillaceous and calcareous marls, and marly limestones (Table 1).

A first systematic review indicates a dominance of hexactinellid sponges as opposed to lithistid ones. The Lithistida are represented by species of the genera *Callopegma*, *Seliscothon*, *Verruculina*, *Jerea* and *Siphonia*. The Hexactinellida comprise species of the genera *Aphrocallistes*, *Becksia*, *Camerospongia*, *Coeloptychium*, *Coscinopora*, *Lepidospongia*, *Leptophragma*, *Plocoscyphia*, *Sporadoscinia*, *Tremabolites* and *Ventriculites* (Gasse et al. 1988).

Well known from this region is the occurrence of specimens of the genus *Becksia*. Also from this area the type species *Becksia soekelandi* was erected by Schlüter (1868). Recent collected and well-prepared individuals of this species have stimulated the following remarks.

## Stratigraphy

In the NW Münsterland the stratigraphical biozonation (see Table 1) of the Late Senonian was originated by Schlüter in the years 1868, 1870, 1872, 1876 and 1877. He first named it Coeloptychien-Kreide with the *Becksia soekelandi*-zone [= Obere Quadraten-Kreide with the belemnite index species *Gonioteuthis quadrata* (Blainville, 1827)] in the lower part. The upper part was characterized by the *Lepidospongia rugosa*-zone [= Untere Mucronaten-Schichten with the belemnite index species *Belemnitella mucronata* (Schlotheim, 1813)].

Since Wehrli's (1949) investigations, these strata are known as Osterwicker Mergel/Schichten (= Osterwick marls/formation). In 1964, Arnold modified this biozonation. He defined the substages Campanian 2 with *Becksia soekelandi* and Campanian 3 with *Lepidospongia rugosa* lacking the ammonite species *Hoplitoplacenticeras vari* (Schlüter, 1876) within the Lower Campanian. However, recent collections imply that the stratigraphical range of both sponge species

J. Reitner and H. Keupp (Eds.)
Fossil and Recent Sponges

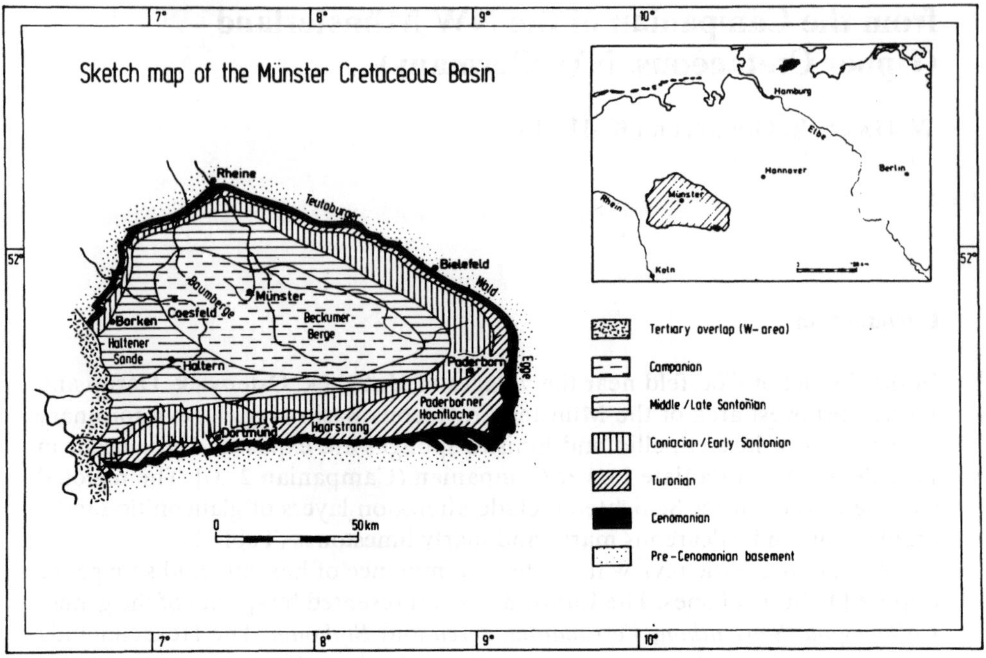


**Fig. 1.** Sketch map of the Münster Cretaceous Basin (Gasse, after Kaever 1980)

range through the respective overlying strata. Therefore, the biozonation by *Becksia soekelandi* and *Lepidospongia rugosa* has to be amended into acme zones.

## Type Species: *Becksia soekelandi* Schlüter, 1868

The type species *Becksia soekelandi* was erected by Schlüter, who named it in honor of Dr. F.C. Becks (1805–1847), Professor of Mineralogy and Natural History at the Academy in Münster, and B. Soekeland (1797–1845), Director of the Gymnasium in Coesfeld.

Commonly, its body is cup-shaped – like a Roman glass – with distinct roots at the base and a wall comprising laterally joined vertical tubes (Figs. 2i-k,4). The dictyonal skeleton consists of a rigid framework of lychniscid scleres with an external cover layer and roots (Fig. 5).

Basically, there are two diverse morphological types of *Becksia soekelandi* with intermediate stages in the Coesfeld district, especially in the areas of Holtwick and south of Coesfeld. The first is a flat, cup-shaped one having a small indistinct central cavity, a folded rim and irregular tubes (south Coesfeld, Fig. 4g,h). The second is higher, also cup-shaped, with a large and deep central cavity, an almost smooth rim and radial-symmetrical tubes (Holtwick, Fig. 4a-f).

**Table 1.** Bio- and lithostratigraphy of the Campanian in the Coesfeld district, NW Münsterland. (After Arnold 1964)

| Stage | | Biozones<br>Belemnites | <br>Ammonites/ sponges | Lithological units | Lithology |
|---|---|---|---|---|---|
| | 5/6 | *Belemnitella mucronata* | *Bostrychoceras polyplocum* | Baumberge formation | Marls<br>Calcareous, fine-grained sandstone (Baumberge-sandstone)<br>Marls<br>Glauconitic, sandy limestone (Billerbeck-greensandstone) |
| | 4 | | *Hoplitoplacenticeras vari* | Coesfeld formation | Marly limestones |
| Campanian | 3 | | *Lepidospongia rugosa* | Osterwick formation | Nodular, fine-grained glauconitic-sandy/ argillaceous/calcareous marls<br>Fine-grained glauconitic-sandy/argillaceous/ siliceous/calcareous marls |
| | 2 | *Gonioteuthis quadrata* | *Becksia soekelandi* | | Fine-grained glauconitic-sandy marls |
| | 1 | | *Scaphites binodosus* | Dülmen formation | Sandy marls<br>Calcareous marls<br>Calcareous sandstone of Dülmen |

In order to explain the morphological variability, a few parameters were established by measurement (see Table 2). An interesting index is the relation between diameter and height of one specimen (Diameter/Height-Index = DHI). The average DHI of the twelve measured specimen amounts 1,5. Significant deviation values are calculated for the specimens documented in Fig. 4i (DHI = 2.1), Fig. 4k (DHI = 0.7) and Fig. 4l (DHI = 0.9).

Consequently, the following initial inferences are appropriate. The appreciable morphological variability of *Becksia soekelandi* – flat to high cup-shaped forms – is the result of different ecological habitats (see Fig. 4a-h,j?). The position of the other specimens (Fig. 4i,k,l) is problematical. At present, the authors are unable to decide definitely whether these individuals represent further ecological varieties or different ontogenetical respectively phylogenetic stages of *Becksia soekelandi*. It is possible these forms have to be separated systematically as new species.

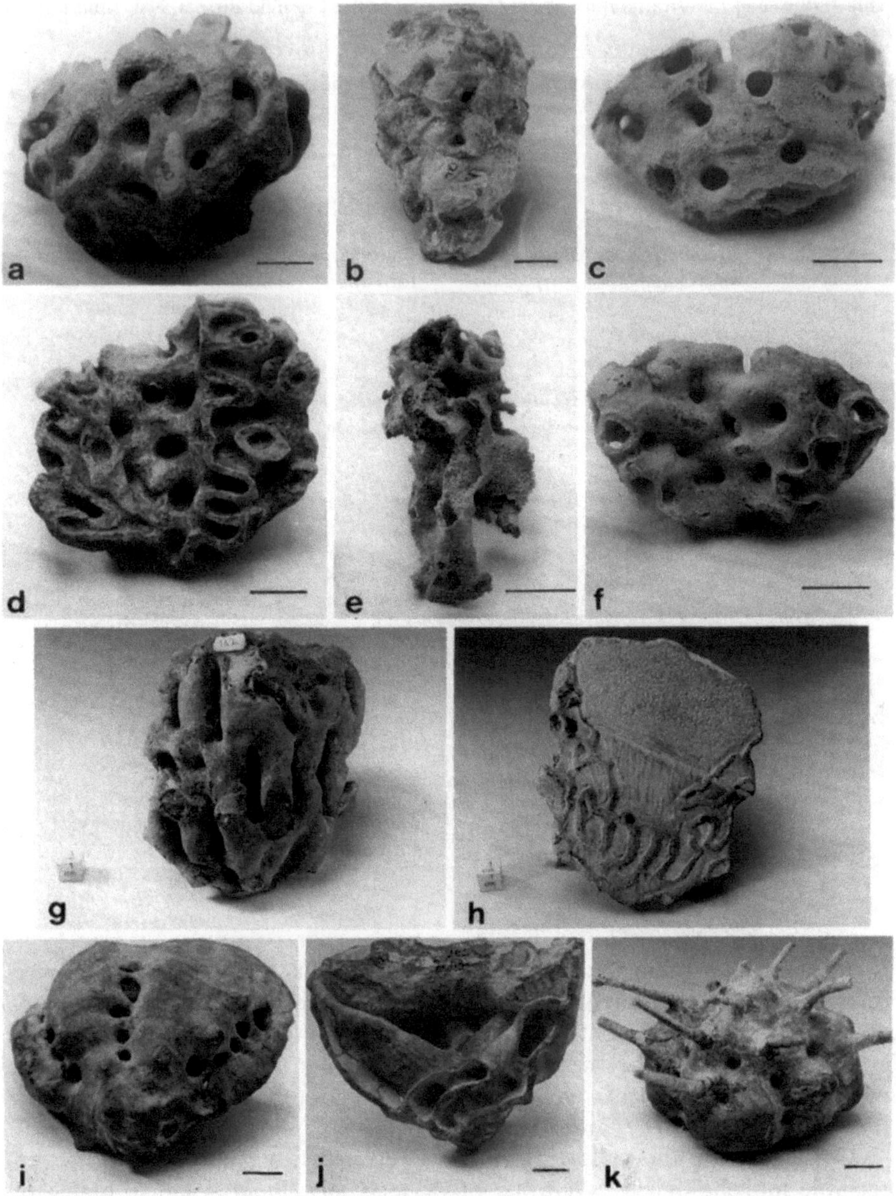

**Fig. 2.** Macrographs. **a,d** *Becksia nidiformis* (Leonhard, 1897), **a** external side; **d** internal side. **b** *Becksia crispata* (Quenstedt, 1878), external side. **c,f** *Becksia feuerwehri* Schrammen, 1912, etched specimen, **c** external side; **f** internal side. **e** *Becksia arborea* Schrammen, 1912, fragmentary etched specimen. **g,h** *Becksia augustae* Schrammen, 1902 **g** external side; **h** internal view, cross-section. **i-k** *Becksia soekelandi* Schlüter, 1868 **i** external side; **j** internal side, cross-section; **k** external side. Bar = 1 cm

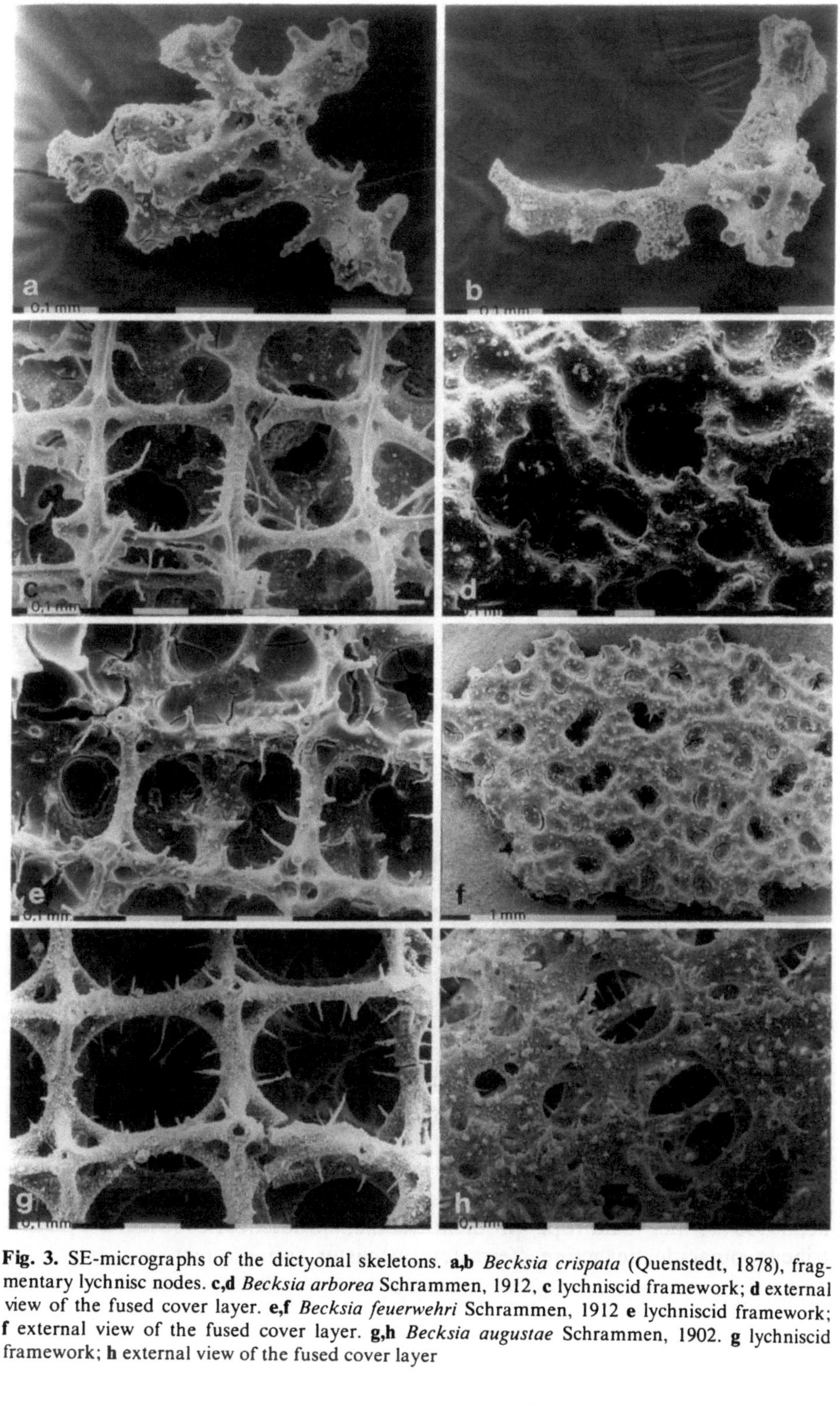

**Fig. 3.** SE-micrographs of the dictyonal skeletons. **a,b** *Becksia crispata* (Quenstedt, 1878), fragmentary lychnisc nodes. **c,d** *Becksia arborea* Schrammen, 1912, **c** lychniscid framework; **d** external view of the fused cover layer. **e,f** *Becksia feuerwehri* Schrammen, 1912 **e** lychniscid framework; **f** external view of the fused cover layer. **g,h** *Becksia augustae* Schrammen, 1902. **g** lychniscid framework; **h** external view of the fused cover layer

**Table 2.** Dimensions of *Becksia soekelandi* Schlüter, 1868 from the Coesfeld district, mainly from Holtwick

| | 2i,j | 2k | 4a | 4b,e | 4c,f | 4d | 4g | 4h | 4i | 4j | 4k | 4l |
|---|---|---|---|---|---|---|---|---|---|---|---|---|
| Dimensions [in mm] Fig. no. | | 11 | 20 | 21,24 | 22,25 | 23 | 26 | 27 | 28 | 29 | 30 | 31 |
| Height of specimen | 65 | 44 | 45 | 60 | 60 | 37 | 45 | 29 | 40 | 40 | 87 | 44 |
| Diameter of specimen | 80 | 59 | 74 | 100 | 115 | 48 | 65 | 55 | 85 | 75 | 59 | 41 |
| Diameter/height-index | 1,2 | 1,3 | 1,6 | 1,7 | 1,9 | 1,3 | 1,4 | 1,9 | 2,1 | 1,9 | 0,7 | 0,9 |
| Depth of central cavity | 45 | – | 37 | 55 | 55 | 29 | 25 | 24 | – | 30 | 69 | 22 |
| Thickness of cup walls | 2 | 2 | 2–4 | 2–3 | 2–3 | 1–2 | 2–3 | 2–3 | 2–3 | 2–3 | 2–3 | 2–3 |
| Thickness of tube walls | 2 | 1–2 | – | – | – | – | 2–3 | 2–3 | 2–3 | 2–3 | – | – |
| Diameter of tube apertures | – | – | – | – | – | – | 10–15 | 10–15 | 6–8 | – | – | – |
| Number of tubes at the case of central cavity | – | – | 6 | 8 | 10 | 8 | – | – | – | – | – | – |
| External diameter of tubes at the base of central cavity | – | – | 11–13 | 12–14 | 10–15 | 5–8 | – | – | – | – | – | – |
| Diameter of tube apertures | – | – | – | – | – | – | 10–15 | 10–15 | 6–8 | – | – | – |
| Number of vertical ostia rows outside | 5 | 3–5 | – | 5 | 5 | 3 | – | – | – | – | 5 | 3–4 |
| Number of vertical ostia | 2–5 | 2–4 | – | 5 | 5 | 2–3 | – | – | – | – | 2–4 | 2–4 |
| Number of root niveaus | 3 | 4 | 2 | 3 | 4 | 2 | 3 | 2 | 2 | – | 2 | 2 |
| Number of roots | – | 7 | 7 | 8 | 12 | 7 | 8 | 7 | 8 | – | 8 | 7 |
| Length of roots | – | 31 | 46 | 40 | 48 | 12 | 30 | 35 | – | – | 25 | 8 |
| Diameter of roots | 4 | 2–4 | 3–4 | 3–5 | 3–5 | 3–4 | 3–5 | 2–3 | 4 | 4–5 | 3–5 | 2 |

## Additional *Becksia* Species

In addition to the type species *Becksia soekelandi* Schlüter, 1868 to date eight species of the genus *Becksia* have been described from the European Upper Cretaceous (Table 3). Thanks to the curators of the Geological and Paleontological Collections of the Universities of Bonn, Göttingen, Tübingen and the Roemer-Museum, Hildesheim, it was possible to study the holotypes of *Becksia soekelandi* (Fig. 2i-k), *B. crispata* (Fig. 2b), *B. augustae* (Fig. 2g,h), *B. feuerwehri* (Fig. 2c,f), *B. arborea* (Fig. 2e) and a paratype of *B. nidiformis* (Fig. 2a,d). As may be observed, the external disposition of these species are distinctly different, but the parenchymal skeletons are approximately identical. In general it is possible to differentiate a rigid framework of lychnisc scleres (dictyonal skeleton) from a porous cover layer.

The study of isolated scleres gave an ambivalent result, and therefore their taxonomic usability is problematical. For example, the examination of the residue of one etched specimen of *B. soekelandi* showed many different types of scleres (Gasse et al. 1988). The residue from etching comprises an allochthonous mixture of lithistid and hexactinellid scleres (mega-/microsclere types: megaclone, dicranoclone, tetraclone, style, acanthoxe, ophirhabd, amphistrongyle, acanthotriactine, calthrop, prodichotriaene, prodichocriccotriaene, phyllotriaene, pinacide, hexactine, sterraster, pycnaster, sphaeraster).

**Table 3.** List of the described Upper Cretaceous nine *Becksia* species

| Species | Type stratum | Locality |
|---|---|---|
| ?*Becksia plicosa* Schlüter, 1895 | Lower Campanian | Coesfeld/NW Germany |
| *Becksia soekelandi* Schlüter, 1868 | Lower Campanian | Holtwick/NW Germany |
| *Becksia arborea* Schrammen, 1912 | Lower Campanian | Oberg/NW Germany |
| *Becksia feuerwehri* Schrammen, 1912 | Lower Campanian | Oberg / NW Germany |
| *Becksia augustae* Schrammen, 1902 | Lower Campanian | Misburg/NW Germany |
| *Becksia ojcoviensis* Malecki, 1980 | Santonian | Ojców/Poland |
| *Becksia crispata* Quenstedt, 1877 | Upper Turonian | Dörnten/NW Germany |
| *Becksia nidiformis* Leonhard, 1897 | Middle Turonian | Opole/Poland |
| *Becksia haugi* Moret, 1925 | Cenomanian | Le Havre/France |

## Bathymetry

The bathymetric distribution of modern Hexactinellida is widespread. The greatest abundance and diversity of this group is developed in water depths between 300 and 2000 m. But in the Antarctic Sea, some Hexactinellids, which in other parts of the world live in deep-water environments, have colonized coastal waters to depths of about 10 m. On the other hand, some species ranges to depths of 6000 m or more (Marshall 1979).

The bathymetric interpretation of fossil sponge communities is difficult (cf. Reid 1968). Empirically, fossil hexactinellids colonized environments of lower water depths. The Lychniscida are referred to by Defretin-Lefranc (1958) as bathyal dwellers. Becksiidae in particular seem to prefer shallower marine biotopes. According to Schrammen (1924), Wagner (1963) and Ulbrich (1974) *Becksia soekelandi* predominates in the littoral and neritic zones. However, on the basis of functional morphological investigations in Lychniscida, Reif and Robinson (1976) stated that the special architecture of the lychniscid sponges (» lanternal nodes») is not a major advantage for living in more energetic environments.

As regards the external disposition of *Becksia soekelandi* from the Holtwick area, the high-cup-shaped specimens with radial symmetrical tubes are indicative of a deep water milieu characterized by quite constant ecological conditions [analogous to recent observations (cf. Arndt 1943; Marshall 1979); Fig. 4a-f]. These facilitate the fuller growth of the sponge skeleton. In contrast, the specimens from the locality south of Coesfeld reflect a rather turbulent environment influenced by changing ecological factors (Fig. 4g,h). For these bathymetric interpretations, the strong roots developed in both morphological types constitute complicated components. In deeper water environments such distinct roots are not necessary for settling. Another point of view is represented by epizoism, e.g. at the outer surface of the cup-shaped body by settlement of an oyster (Fig. 4c). The occurrence of the oyster indicates a quite turbulent environment. It is very difficult for a definite explanation of these complex ecological conditions to be given now, but investigations are still in progress.

**Fig. 4.** Macrographs, diverse external views of *Becksia soekelandi* Schlüter, 1868. Bar = 1 cm

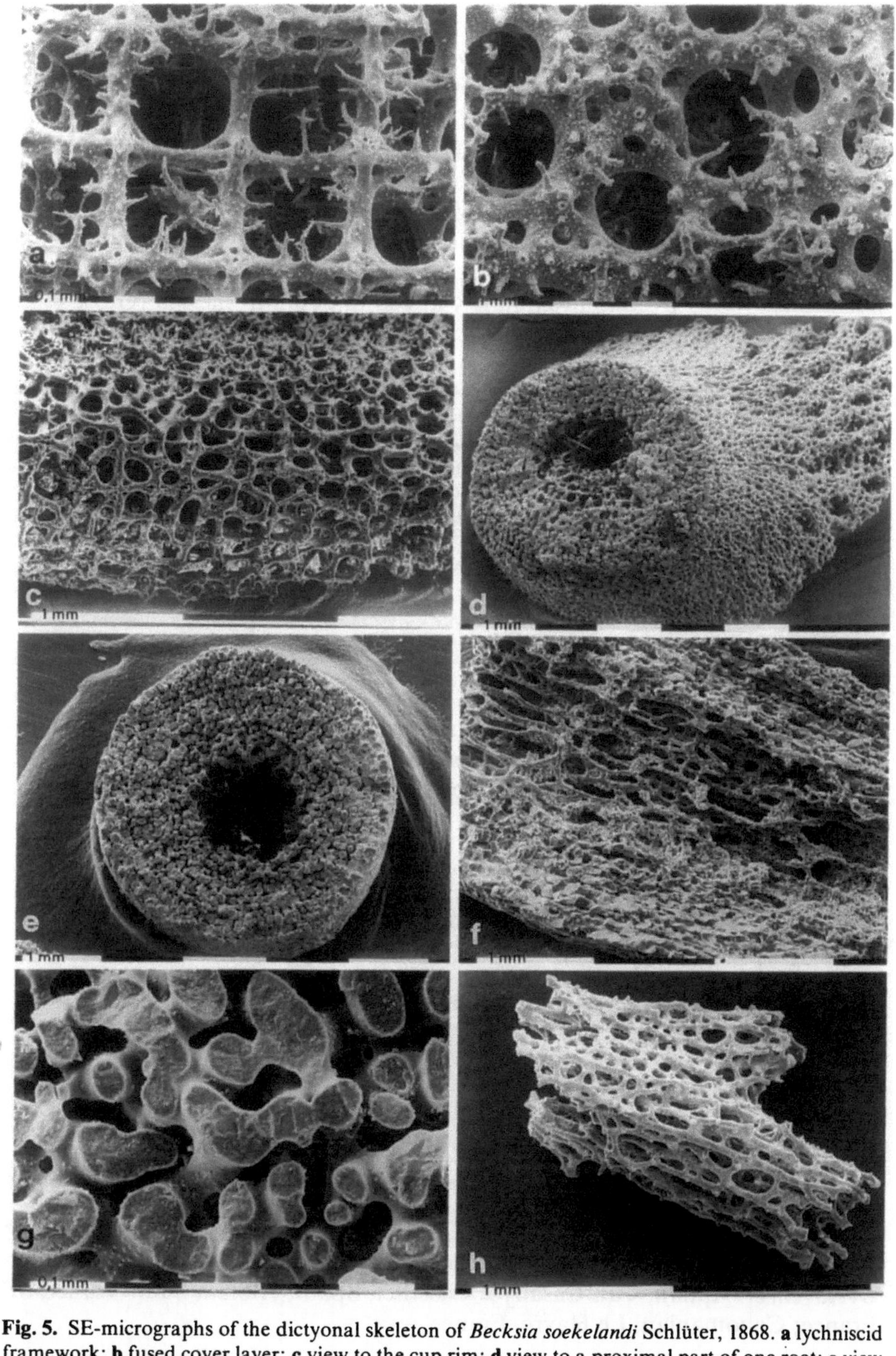


**Fig. 5.** SE-micrographs of the dictyonal skeleton of *Becksia soekelandi* Schlüter, 1868. **a** lychniscid framework; **b** fused cover layer; **c** view to the cup rim; **d** view to a proximal part of one root; **e** view to a cross-section of one root; **f** view to a longitudinal section of one root (**g**, cf. **e**) magnified detail of the root skeleton; **h** branched distal part of one root

Apropos the Lower Campanian sponge communities in the Coesfeld district in general, it is interesting that there is a predominance of caulophytical specimens such as *Coeloptychium*, i.e., individuals characterized by long stems and distinct rooted basal faces (see Rauff 1893, 1894; Müller 1978; Ulbrich 1974). Commonly the sponge bodies are funnel-shaped or fungoid. Empirically, sponges of this form preferred deeper-water environments with slow currents and muddy substrata ideal for colonization (see Nestler 1961). Hence, the Lower Campanian sponge communities of the northwestern Münsterland typify lower neritic ones which lived approximately at 200 m depth in the sea (Gasse et al. 1988).

*Acknowledgements.* We gratefully acknowledge the Deutsche Forschungsgemeinschaft for financial sponsoring of our research. Special thanks are due to Professor Dr. M.J. Kaever, who encouraged the current investigations with scientific and organizational advice. We also thank the following persons for technical help and kindness, making it possible to study original sponge material from diverse collections: Priv.-Doz. Dr. K. Oekentorp, Professor Dr. F. Hillenkamp, Dipl.-Geol. J. Heinisch, Münster; Professor Dr. H. Remy, Bonn; M. Boetzkes, M.A., Hildesheim; Dr. H. Jahnke, Göttingen; Dr. A. Liebau, Tübingen.

## Systematics

The general classification is based on the taxonomic recommendations of de Laubenfels (1955). The species are arranged according to their first stratigraphical appearance. The listed dimensions concern the specimens which are shown in Figs. 2 and 4. Further detailed descriptions and differential diagnosis will be published elsewhere.

Class Hexactinellida Schmidt, 1870
Order Lychniscida Schrammen, 1902
Family Becksiidae Schrammen, 1912
Genus *Becksia* Schlüter, 1868
[Type species *Becksia soekelandi* Schlüter, 1868]

### *Becksia haugi* Moret, 1925

1925 *Becksia haugi* nov. sp. – Moret: 228, Pl. 23, Fig. 2–3
1942 *Becksia haugi* Moret – Herenger: 170, Text-Fig. 7b-b$_1$, Pl. 2, Fig. 5
1944 *Becksia haugi* Moret – Herenger: 99, Fig. 7c

*Remark:* To date only fragments are known.
*Occurrence:* Aptian, Can Casanyes Castellet, Spain; Albian, Saint-Vallier, SE France; Cenomanian, Le Havre, France.

***Becksia nidiformis* (Leonhard, 1897) (Fig. 2a,d)**

1897 *Plocoscyphia nidiformis* n.sp. – Leonhard: 35, Text-Fig. 5a-b
1897 *Plocoscyphia crassilobata* – Leonhard: 35–36, Pl. 3, Fig. 6
1912 *Becksia nidiformis* (Leonhard, 1897) – Schrammen: 296
1919 *Becksia nidiformis* (Leonhard, 1897) – O'Connel: 117–118, 131

*Dimensions* (in mm) of the specimen Fig. 2a,d:
Height of specimen: 45
Diameter of specimen: 51
Depth of central cavity: 40
Thickness of cup/tubes wall: 1–3
Diameter of tube ostia: 4–7

*Occurrence*: Turonian (*Scaphites*-Schichten), Opole, Poland.
*Collection*: Fig. 2a,d, paratype, Geologisch-Paläontologisches Institut und Museum der Universität Göttingen (FRG).

***Becksia crispata* (Quenstedt, 1878) (Fig. 2b, 3a-b)**

1878 *Becksia crispata* – Quenstedt: 482, Pl. 138, Fig. 7
1912 *Becksia crispata* (Quenstedt, 1878) – Schrammen: 296
1958 *Becksia crispata* – Defretin-Lefranc: 86, Fig. 26; Pl. 12, Fig. 1–2

*Dimensions* (in mm) of specimen Fig. 2b:
Height of specimen: 65
Diameter of specimen: 40
Diameter of ostia: 2–5

*Occurrence:* Turonian (Oberer Pläner), Dörnten, NW Germany; Early Senonian (»Emscherian»), N France.
*Collection*: Fig. 2b,3a-b, holotype, Institut und Museum für Geologie und Paläontologie der Universität Tübingen (FRG).

***Becksia ojcoviensis* Malecki, 1980**

1980 *Becksia ojcoviensis* n.sp. – Malecki: 419–421, Text-Fig. 7a-c, Pl. 9, Fig. 1a-b

*Average dimensions* (in mm) of cup- and pear-shaped specimens:
Height of specimen: 69
Diameter of specimen: 95
Diameter of sponge tubes: 10
Thickness of tube walls: 2
Depth of central cavity: 33

*Occurrence*: Santonian, Korzkiew, Poland.

***Becksia soekelandi* Schlüter, 1868 (Figs. 2i-k, 4a-l, 5a-h, Table 2)**

1868 *Becksia soekelandi* n.sp. – Schlüter: 93
1872 *Becksia soekelandi* Schlüter, 1868 – Schlüter: 20, Pl. 1, Fig. 5–7
1878 *Becksia soekelandi* – Quenstedt: 489, Pl. 138, Fig. 14
1877 *Becksia soekelandi* – Zittel: 58
1883 *Becksia soekelandi* – Hinde: 144
1889 *Becksia soekelandi* – Griepenkerl: 23
1900 *Becksia soekelandi* – Wollemann: 8
1912 *Becksia soekelandi* – Schrammen: 297
1919 *Becksia soekelandi* – O'Connel: 181–182
1924 *Becksia soekelandi* – Schrammen: 140
1974 *Becksia soekelandi* – Ulbrich: 66, Pl. 12, Fig. 2–3
1986 *Becksia soekelandi* – Kaever: 75, Pl. 4, Fig. 4

*Remark*: The great morphological variability is discussed in the section on Type Species

*Occurrence*: Lower Campanian, Holtwick, Coesfeld district, Westphalia; Lower Campanian, Boimstorf, Saxony; Lower Campanian, Biewende, Saxony; Lower Campanian, Ilsenburg, subhercynian area.
*Collection*: Fig. 2i-k, originals from Holtwick collected by C. Schlüter, Institut für Paläontologie der Universität Bonn; Fig. 4a-f, 4i-l, Holtwick; Fig. 4g,h, south of Coesfeld, K.-H. Hilpert, Datteln [FRG].

***?Becksia plicosa* Schlüter, 1895**

1895 *Becksia plicosa* n.sp. – Schlüter: 204–205

*Remark:* The systematic validity of this species is doubtful. Unfortunately C. Schlüter (1895) gave no figures of the described specimen. The existence of the holotype is unknown.

*Dimensions* (in mm) after Schlüter (1895):
Height of specimen: 45
Diameter of specimen: 115
Thickness of wall: 1,5
Length of roots (fragments): 5–10
Diameter of roots: 2–6

*Occurrence:* Lower/?Upper Campanian, Coesfeld, Westphalia.

***Becksia augustae* (Schrammen, 1902 (Fig. 2g,h, 3g,h)**

1902 *Becksia augustae* n.sp. – Schrammen: 1, Pl. 2, Fig. 2
1912 *Becksia augustae* – Schrammen: 298, textpl. 14, Fig. 5; Pl. 40, Fig. 1
1919 *Becksia augustae* – O'Connel: 182, 192

*Dimensions* (in mm) of the specimen Fig. 2g,h
Height of specimen: 120
Diameter of specimen: 92
Depth of central cavity: 75
Thickness of wall: 2–4
Diameter of tubes: 10–17

*Occurrence:* Lower Campanian, Misburg, NW Germany.
*Collection:* Fig. 2g,h, holotype; Fig. 3g,h, paratype, Dr. A. Schrammen, ROEMER-Museum, Hildesheim (FRG).

### ***Becksia feuerwehri*** **Schrammen, 1912 (Fig. 2c,f, 3e-f)**

1912 *Becksia feuerwehri* n.sp. – Schrammen: 298, Pl. 40, Fig. 4
1919 *Becksia feuerwehri* – O'Connel: 182

*Dimensions* (in mm) of the specimen Fig. 2c,f:
Height of specimen: 29
Diameter of specimen: 46
Thickness of cup wall: 10
Thickness of walls: 1
Diameter of ostia: 2–4

*Occurrence:* Lower Campanian, Oberg, NW-Germany.
*Collection:* Fig. 2c,f, holotype, Fig. 3e-f, paratype, Dr. A. Schrammen, ROEMER-Museum, Hildesheim.

### ***Becksia arborea*** **Schrammen, 1912 (Fig. 2e, 3c-d)**

1912 *Becksia arborea* n.sp. – Schrammen: 299, Pl. 40, Fig. 2

*Dimensions* (in mm) of the specimen Fig. 2e:
Height of specimen: 43
Diameter of specimen: 20
Thickness of wall: 1
Diameter of ostia: 2–4

*Occurrence:* Lower Campanian, Oberg, NW-Germany.
*Collection:* Fig. 2e, holotype, Fig. 3c-d, paratype, Dr. A. Schrammen, ROEMER-Museum, Hildesheim.

## References

Arndt W (1943) Die tiergeographische Gliederung der Schwammfauna der Nord- und Ostsee. Arch Naturgesch NF 12, 2:303-349 (Leipzig)

Arnold H (1964) Die höhere Oberkreide im nordwestlichen Münsterland. Fortschr Geol Rhein Westfalen 7:649-678

Defretin-Lefranc S (1958) Contribution à l'étude des spongiaires siliceux du Crétacé supérieur du Nord de la France. Thèses faculté des sciences de Lille, pp 173

de Laubenfels MW (1955) Porifera. In: Moore RC (ed) Treatise on Invertebrate Paleontology. Part E Geol Soc Am Lawrence, Kansas, pp 21-112

Gasse W, Göcke R, Hilpert KH (1988) Oberkretazische Spongien des NW-Münsterlandes – ein Überblick. Beitr Elektronenmikrosk Direktabb Oberfl 21:385-396 (Münster)

Griepenkerl O (1889) Die Versteinerungen der Senonen Kreide von Koenigslutter im Herzogthum Braunschweig. Palaeontol Abh (4) 5:305-149 (Berlin)

Herénger L (1942) Contribution a l'étude des spongiaires siliceux du Jurassique et Cretacé de Catalogne. Trav Lab Geol Fac Sci Univ Grenoble 23:143-190

Herénger L (1944) Spongiaires siliceux du Cretacé du Sud-Est de la France. Trav Lab Geol Fac Sci Univ Grenoble 24:79-111 (Grenoble)

Hinde GJ (1884) Catalogue of the fossil sponges of the British Museum. London, pp 248

Kaever M (1986) Porifera. In: Kaever M, Oekentorp K, Siegfried P (eds) Fossilien Westfalens. Teil I. Invertebraten der Kreide. Münst Forsch Geol Paläont 33/34, pp 59-96

Kaever M (1980) Geologie des westlichen Münsterlandes. Führer vor- und frühgeschichtlicher Denkmäler 45:20-39 (Mainz) Römisch-Germanisches Zentralmuseum

Leonhard R (1897) Die Fauna der Kreideformation in Oberschlesien. Palaeontographica 44, 1-2:11-70 (Stuttgart)

Malecki J (1980) Santonian siliceous sponges from Korzkiew near Kraków (Poland). Ann Soc Géol Pologne 50:409-431 (Kraków)

Marshall NB (1979) Developments in Deep-Sea Biology. Blandford Press, Poole, Dorset, pp 566

Moret L (1925) Contribution à l'étude des spongiarires siliceux du Crétace supérieur francais. Mém Soc Geol Fr NS 5:1-322 (Paris)

Müller W (1978) Beobachtungen zur Ökologie von Kieselspongien aus dem Weißen Jura der Schwäbischen Alb. Stuttg Beitr Naturkd Ser B 16:1-143 (Stuttgart)

Nestler H (1961) Spongien aus der Weißen Schreibkreide (Unt. Maastricht) der Insel Rügen (Ostsee). Paläontol Abh I, 1:1-70 (Berlin)

O'Connel M (1919) The Schrammen Collection of Cretaceous Silicispongiae in the American Museum of Natural History. Bull Am Mus Nat Hist 41:1-261 (New York)

Quenstedt FA (1878) Petrefactenkunde Deutschlands. Bd 5: Schwämme. Fues's Verlag, Leipzig, pp 612

Rauff H (1893/1894) Palaeospongiologie, Erster oder allgemeiner Theil und zweiter Theil, erste Hälfte. Palaeontographica 40:1-346 (Stuttgart)

Reid REH (1968) Bathymetric distribution of Calcarea and Hexactinellida in the present and the past. Geol Mag 105:546-559 (Caxton Hill)

Reif WE, Robinson JA (1976) On functional morphology of the skeleton in lychnisc sponges (Porifera, Hexactinellida). Paläontol Z 50, 1/2:57-69 (Stuttgart)

Schlüter C (1868) Über die jüngsten Schichten der unteren Senon-Bildungen und deren Verbreitung. Verh nat-hist Ver Preuß. Rhld u Westph, Sitz-Ber 25 Jg, 3 Folge, 5 Jg:92-93; Bonn

Schlüter C (1870) Über die Spongitarienbänke der unteren Mucronaten- und oberen Quadratenschichten und über Lepidospongia rugosa insbesondere. Verh Nat Hist Ver Preuß Rheinl Westf Sitzungsber 3, 7:139-141 (Bonn)

Schlüter C (1872) Über die Spongitarien-Bänke der oberen Quadratenschichten des Münsterlandes. Festschr 20 Hauptvers dt geol Geol Ges, Bonn, pp 38

Schlüter C (1876) Verbreitung der Cephalopoden der oberen Kreide Norddeutschlands. Z Dt Geol Ges 28:457-518 (Berlin)

Schlüter C (1877) Verbreitung der Inoceramen in den Zonen der norddeutschen Kreide. Z Dt Geol Ges 29:735-742 (Berlin)

Schlüter C (1895) Über einige Spongien aus der Kreide Westphalens. Z Dt Geol Ges 47:194-210 (Berlin)

Schrammen A (1902) Neue Hexactinelliden aus der oberen Kreide von Nordwestdeutschland. Mitt Mus Hildesheim 15:1-26

Schrammen A (1910-1912) Die Kieselspongien der oberen Kreide von Nordwestdeutschland. I-II. Palaeontogr Suppl 5:1-385 (Stuttgart)

Schrammen A (1924) Die Kieselspongien der oberen Kreide von Nordwestdeutschland. III. Monogr Geol Palaeontol 1, 2:1-159 (Berlin)

Ulbrich H (1974) Die Spongien der Ilsenburg-Entwicklung (oberes Unter-Campan) der Subherzynen Kreidemulde. Freib Forsch-H C 291:1-121 (Leipzig)

Wagner W (1963) Die Schwammfauna der Oberkreide von Neuburg (Donau). Palaeontographica A 122:166-250 (Stuttgart)

Wehrli H (1949) Erläuterungen zur geologischen Karte 1:25.000, Blatt Nottuln, Nr 4010. Arch Geol L-Amt NW: Krefeld (unpubl.)

Wollemann A (1901) Die Fauna des Senons von Biewende bei Wolfenbüttel. Jahrb Preuß Geol L Anst 21 (1900):1-30 (Berlin)

Zittel KA (1877) Studien über fossile Spongien. I. Hexactinellida. Abh Kgl Bayer Akad Wiss Math Phys Cl 13, 1:1-63 (München)

# Observations on the Jurassic Hexactinellid Sponge *Tremadictyon radicatum* (Quenstedt)

W. MÜLLER

## Introduction

From an analysis of the fauna from an Oxfordian reef near Auerbach in Oberpfalz R. Meyer and L. Übelacker found a great number of irregularly shaped clumps (Meyer et al. 1987). These clumps were determined to be basal fragments of *Tremadictyon radicatum* (Fig. 1). This large siliceous sponge is the only described form from the Upper Jurassic which built root knots to anchor itself in soft sediment. This species may be one of the hypothetical "pioneer sponges" which contributed to the establishment of new reefs. The following description supports this supposition.

## Systematics

Order: Dictyida (Zittel 1877)
Family: Staurodermatidae (Zittel 1877)
Genus: *Tremadictyon* (Zittel 1877)
Species typica: *T. reticulatum* (Goldfuss 1826), established by de Laubenfels (1955, p. 79).

Diagnosis of genus: (after Schrammen 1936, p. 22, varied). Tubular to funnel-shaped, large, relatively thin-walled sponge in which the oscula and postika are arranged in quincunx. The canals are corresponding to those of modification 4 after Rauff (1894, p. 129). The sponge body comprises a regular skeleton composed of hexactins with 0.3-mm-long rays. An overlying layer of stauractins exists which extends over the whole ostia with no discernible openings.

*Tremadictyon radicatum* (Quenstedt).
1858 *Spongites reticulatus* – Quenstedt, p. 694; Pl. 84, Fig. 2–3.
1877 *Scyphia reticulata* – Quenstedt, p. 29; Pl. 115, Fig. 3.
– *Scyphia reticulata radicata* – Quenstedt, p. 31; Pl. 115, Fig. 12.
– *Retispongia radicata* – Quenstedt, p. 34; Pl. 115, Fig. 20.
1880 *Tremadictyon reticulatum* Goldf. – Zittel, p. 175; Text-Fig. 89.
1910 *Tremadictyon radicatum* Quenst. sp. – Kolb, p. 155.
1937 *Tremadictyon radicatum* Quenst. sp. – Schrammen, p. 24, Pl. 3, Fig. 2 (Dictyonalia); Pl. 4, Fig. 1 (Inner surface of the skeleton).

J. Reitner and H. Keupp (Eds.)
Fossil and Recent Sponges

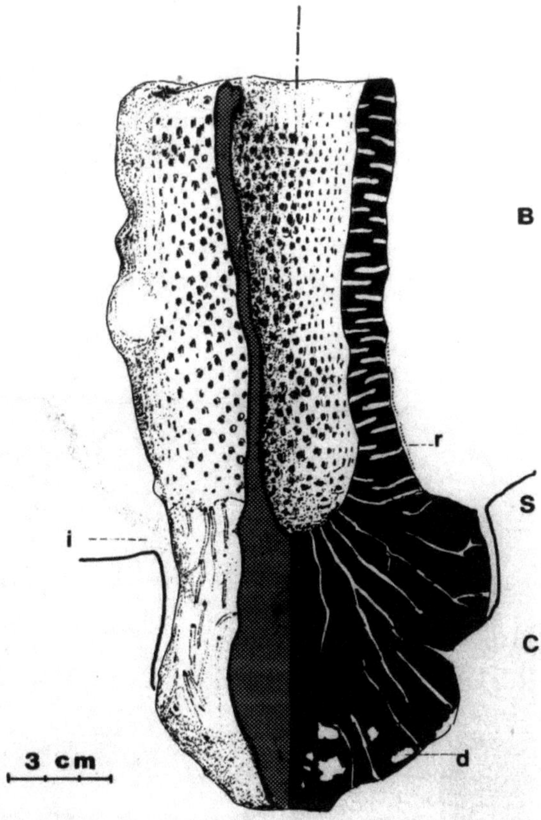


**Fig. 1.** *Tremadictyon radicatum* (Quenstedt). Vertical section, schematic. *B* Main sponge body; *S* surface of the sediment; *r* rests of the layer of stauractins; *d* debris collected from the sponge and built in into the root clod; *c* outer surface of the root clod showing the channels of the root system; *i* interstice between the sponge and the sediment. (Observed by Wiedenmayer 1978)

Lectotypus: Original from Quenstedt (1877, Pl. 115, Fig. 20; established here, deposited at Geolog. Paläontolog. Institute of University Tübingen.

Locus typicus: Oberdigisheim near Balingen, Baden-Württemberg, West Germany.

Stratum typicum: Upper Jurassic, Kimmeridgian, (Weißjura Beta/ Gamma).

Diagnosis: Large, tubular to funnel-shaped representatives of the genus, able to produce a shapeless root clod for to anchoring itself in soft mud. The construction of the skeleton is different in both main body and root clod. The boundary between is uneven but clear.

Description of the figured specimen (Fig. 2 Staatl. Museum f. Naturkde. Stuttgart Inv. Nr. 21877): The silicified sponge was dissolved out of the Malm limestone with HCl so that, with the exception of the paragaster all details of the skeleton and canal systems were exposed.

**Fig. 2.** *Tremadictyon radicatum* (Quenstedt), habitus. Upper Kimmeridgian (Weissjura Epsilon/Zeta). Sontheim, West Germany. Staatl. Museum f. Naturkde. Stuttgart, Inv. Nr. 21877. For details see text

The sponge is altogether 190 mm high with 70 mm of it composed of the root system. The cross-section is circular with an average diameter of 70 mm for the main sponge body, 60 mm at the boundary between the body and the root system, and 90 mm at the root system. The wall is 8–10 mm thick.

The root clod is lumpy, shapeless with short rounded protrusions. In certain places foreign material appears to be included in the clod. The boundary between the main sponge body and the root system is uneven, but sharp. The paragaster is as in other examples tubular and ends up in the form of a hollow half-ball.

The canal system of the sponge body and the root system differ from each other (Fig. 3). The outer surface of the sponge body contains ostia which occur at distances of ca 2 mm in oblique rows (9 ostia $cm^{-2}$). The ostia near the root system are round, but become elongated toward the top of the sponge body. The canal system is composed of these 2-mm-wide pores belonging to short epirhysis which enter horizontally into the wall and finally end. Epirhysis and ostia are lined by a smooth siliceous "skin" which is penetrated by diminutive pores and is built of crooked flattened hexactins.

In contrast, the root system is composed of many fine 0.5 mm wide channels, which penetrate the clod upwards in unorganized manner ending with round postika in the bottom of the paragaster (Fig. 4). These channels appear as fine troughs on the outer surface and may show starlike figures. Evidently the channels of the root had function other than that in the main sponge body.

The skeletal components of the root system are also different than that in the main sponge body. Hexactins are found within both skeletons, which combine to form a fixed structure. However, in the root system the structure is entirely irregular and narrow meshed. In the main sponge body the dictyonal skeleton is regularly right-angled coarsely meshed. An overlying layer of stauractins is here and there preserved. The stauractins covered the entire sponge body and seemingly covered all ostia-like openings at a distance of 1 mm from the body. Thus, between the sponge and the outer layer a narrow empty space existed which can compared to the "vestibule" in Recent sponges.

The flanks of the figured specimen were colonized after death by a few placunopsids, as well as a small serpulid. The root system is, however, like all other examined root systems, free from colonizers.

Occurrence: *Tremadictyon radicatum* is ascribed from the Oxfordian/Kimmeridgian boundary to the beginning Tithonian by Schrammen (1937, p. 25). With his report from the lower Oxfordian, the form occurs within the entire Upper Jurassic of southern Germany.

Samples: 16 root clods: Private collection Übelacker, Bachetsfeld: two whole specimens and six fragments: Staatl. Museum f. Naturkunde in Stuttgart.

Taxonomic note: *Tremadictyon radicatum* and *Tremadictyon reticulatum* differ only by possession of a root clod. Perhaps both species should be united. In this case *Tremadictyon radicatum* would receive the status of a younger synonym.

Discussion: Most of the siliceous sponges of the Upper Jurassic required a hard surface on which to attach as shown by Müller (1978). The sponges colonized already calcified fossil sponges or algal crusts. They would have had difficulty in anchoring themselves in soft muddy sediment. This is documented by the almost complete absence of sponges in the muddy Dogger, except in association with hard surfaces such as oyster banks. How, then, did sponge reefs establish themselves in the marls of the Lower Oxfordian?

Gaillard (1983, p. 139) hypothesized the existence of a "pioneer sponge" but had no organic evidence.

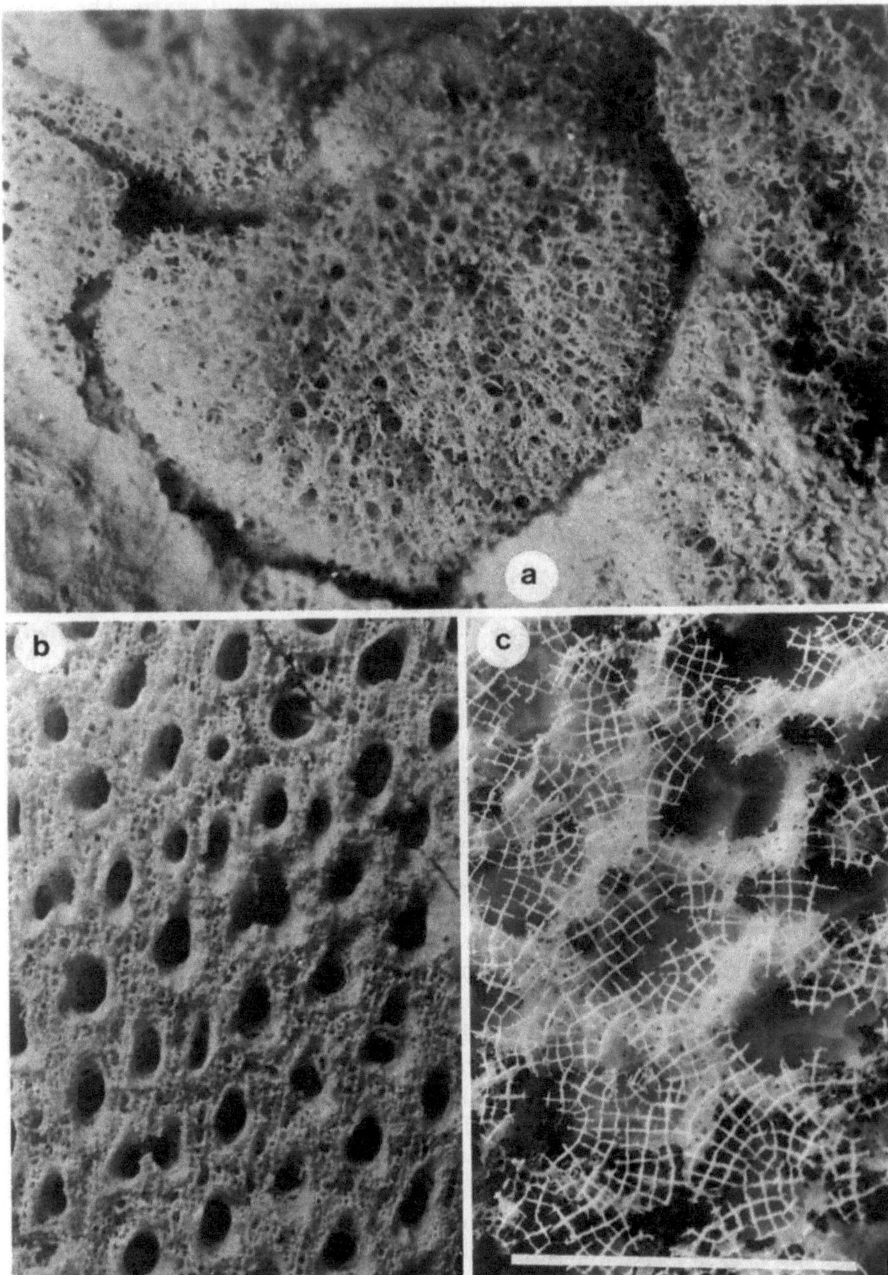


**Fig. 3.** **a** *Tremadictyon radicatum*. The same specimen as Fig. 2. View of the top of a root system; the outside a fine-meshed skeleton with numerous fine channels. **b** *Tremadictyon radicatum*. The same specimen as Fig. 2. Ostia occurring in oblique rows on the outer surface of the main sponge body. **c** *Tremadictyon* sp. Kimmeridgian (Weissjura Gamma). Erkenbrechtsweiler Steige. A delicate layer of stauractins covering the entire sponge body. Scale = 1 cm

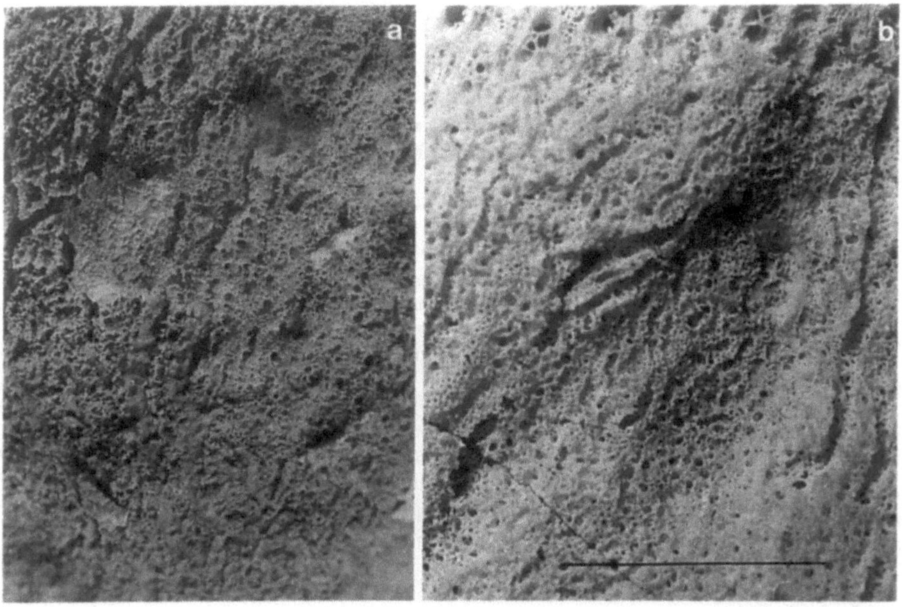

**Fig. 4.** *Tremadictyon radicatum* (Quenstedt), the same specimen as Fig. 2. Two different views (**a,b**) on the outer side of the root clod showing the complex intake system with irregular throughs and fine pores. Scale = 1 cm

Förster and Schairer (1987) dug out an entire Oxfordian reef near Biburg (Frankenalb) and found that, directly below the bioherme, isolated rests of great plate-shaped sponges occur. Hence, they are of the opinion that the growth of the bioherme, or its lateral expansion, began with the colonization by a few plate-shaped sponges, for example *Stauroderma* or *Tyroidium*. However, neither genus possesses any organs qualified to settle in a soft sediment, added to which, *Tremadictyon* and *Stauroderma*, when badly preserved, cannot be differentiated.

Very important are the reports from Wiedenmayer (1978) concerning the examination of living sponge reefs of the Great Bahama Bank: "Some species of sponges, chiefly Hauromenids, are believed to be responsible for early lithifications" (Müller: penetrating the sediment with its base) "*Spheciospongia vesparium* (Lamarck) is by far the most important of these species."

In a comparison with the exact information of Wiedenmayer (1978), the sponge *Tremadictyon* described here shows some correspondence:

*Spheciospongia* as well as *Tremadictyon* are able to anchor themselves in soft sediment. Both sponges built debris into its base. Both sponges have developed two different mechanisms for water intake into the base or into the main sponge body. Both sponges were able to found new reefs. Therefore *Tremadictyon radicatum* may be a "pioneer sponge". But further investigations are necessary.

## References

De Laubenfels MW (1955) Porifera. In: Moore RC (ed) Treatise on invertebrate paleontology, Part E. Geol Soc Am and Univ, Kansas Press, pp E21–E112

Förster R, Schairer G (1987) Faunen- und Faziesanalyse des oberjurassischen Algen-Schwamm-Bioherms von Biburg, Frankenalb. Jahresber 1986 Mitt Freunde Bayer Staatsslg Paläont Hist Geol, pp 14–32

Gaillard C (1983) Les Bioherms a Spongiaires et leur Environnement dans l'Oxfordien du Jura meridional. Doc Lab Géol Lyon 90:429

Goldfuss A (1826–1833) Petrefacta Germaniae. List & Francke, Düsseldorf, pp 252

Kolb R (1910) Die Kieselspongien des Schwäbischen Weißen Jura. Palaeontographica 57:141–256

Meyer RKF, Müller W, Übelacker L (1987) Die Schwamm und Begleitfauna eines Schwammriffes des unteren Malm in der Oberpfalz. Geol Bl Nordost-Bayern 37:63–88

Müller W (1978) Beobachtungen zur Okologie von Kieselspongien aus dem Weißen Jura der Schwäbischen Alb. Stuttg Beitr Naturkd B 37:1–10

Quenstedt FA (1856–1858) Der Jura. (1856: 1–368, pl 1–48; 1857: 369–842, pl 49–100; 1858 Titelblatt). Laupp und Siebeck Tübingen, pp 842

Quenstedt FA (1877–1878) Petrefactenkunde Deutschlands, 5. Schwämme, 1–612, 28 pl (1877: 1–448, pl 115–139; 1878: 449–612, pl 137–142). Laupp und Siebeck Leipzig

Rauff H (1893–1894) Palaeospongiologie Palaeontographica 40:346

Schrammen A (1936–1937) Die Kieselspongien des oberen Jura von Süddeutschland. Palaeontographica 84/85:149–194; 14–23

Wiedenmayer F (1978) Modern sponge bioherms of the Great Bahama Bank. Ecol Geol Helv 71/3:699–744

Zittel KA (1876–1880) Handbuch der Palaeontologie, I. Band. Palaeozoologie, 1, Abtheilung Protozoa, Echinodermata und Molluscoidea. R Oldenbourg München, pp 1765:558

Zittel KA (1877) Studien über fossile Spongien. Abh Königl Bayr Akad Wiss (Math-Phys VI) 13(1):1–63

# Are Protospongiidae the Stem Group of Modern Hexactinellida?

D. Mehl*

## Introduction

The family Protospongiidae Hinde, comprising the genera *Protospongia* Salter and *Diagonella* Rauff – both first found in rocks of Middle Cambrian age – has long been considered to be the stem group of all later Hexactinellida. They are small, conical to subcylindrical sponges, whose skeletal wall apparently consisted of a single layer of stauractins only (Fig. 1). These sponges have been considered as primitive among Hexactinellida (Rigby 1986) and their skeletons of nonfused stauractins were supposed to have supported a simple unfolded choanosome layer of ascon type (Finks 1960, 1970, 1983). On account of studies done by Okada (1928) on spicule-bearing parenchymella larvae of the recent hexactinellid *Farrea sollasii* Schulze, stauractins have been assumed to be the original hexactinellid spicule from which hexactins developed by addition of radial rays (this hypothesis was originally introduced by Minchin 1905). Mainly isolated stauractins have been found in the Lower Cambrian (the first whole hexactinellids are from the Middle Cambrian; only a questionable fragment of ?*Protospongia hicksi* Hinde is described by Rigby (1987) from the Early Cambrian Lower Parker Slate of North America), but some isolated hexactins have also been reported (Finks 1983; Rigby 1986; and others). Recently, rocks of the lowermost Cambrian of China (Sinian-Cambrian boundary strata) have been studied for fossil content (Wei-Ming and Qian 1988). Isolated triaxial spicules (hexactins and pentactins) were found in these strata, but no stauractins. A purpose of this chapter is to discuss the status of the Protospongiidae and the probability that they represent the stem group of modern hexactinellids. Also, Okada's ontogenetic studies done on *Farrea sollasii* and their phylogenetic significance will be discussed.

The first definitely dictyonal hexactinellids are described from the Upper Devonian of the Holy Cross Mts. of Poland by Rigby et al. (1981). The figures of Rigby et al. show a "modern" type dictyonal Hexactinellida (parenchymalia consisting of fused hexactins, not diactins or other derivatives) with a channel system similar to that of *Laocaetis* Pomel. From the lowermost Upper Devonian of Belgium Termier et al. (1981) reported a dictyonal sponge, *Pseudopemmatites feurmarieri* Fraiport. They referred to it as a lyssacine sponge, but, as also stated by Rigby (pers. comm.), the pictures of Termier et al. show an obvious dictyonine reticulation (e.g. Table 2, Figs. 2, 3). An Upper Ordovician hexactinellid with

*Name changed through marriage; previously Salomon

J. Reitner and H. Keupp (Eds.)
Fossil and Recent Sponges

**Fig. 1.** *Diagonella ovata* Rigby. Scale 0.5 cm. (Original at the Institut für Paläontologie, F U Berlin)

partly fused spicules is described by Rigby and Webby (1988); *Wareembaia concentrica* Rigby and Webby has dermal spicules which are dictyonally fused in regular vertical rays and horizontally connected by synapticulae. (However, this sponge does not represent a modern dictyonal reticulation; the fused spicules are not parenchymal hexactins but dermal rhabdodiactins).

Hexasterophora Schulze and Amphidiscophora Schulze are very old in origin. Ampidiscs have been reported from the late Carboniferous *Uralonema* Librovitcs (1929), (the microscleres discussed below are illustrated in Fig. 2). Whole amphi- and hemidiscophoran sponges, *Itararella gracilis* and *Micro-hemidiscia ortmanni*, from Late Carboniferous strata of Uruguay are documented by Kling and Reif (1969). Isolated amphidiscs were found in Upper Silurian rocks of Scandinavia by Mostler (1986). Hemidiscs have been reported even from Upper Cambrian limestones of Iran (Mostler and Mosleh-Yazdi 1976). The oldest hexasters known are from the Lower Ordovician of Sweden.

In addition, oxy- and hemioxyhexasters are reported from the Mississippian *Griphodictya epiphanes*, in which they occur together with paraclavules (Hall and Clarke 1899).

Because of this documentation of Hexasterophora and Amphidiscophora from the early Palaeozoic, Mostler (1986) suggested that the development of these taxa is to be regarded as independent of that of the Protospongiidae. His new

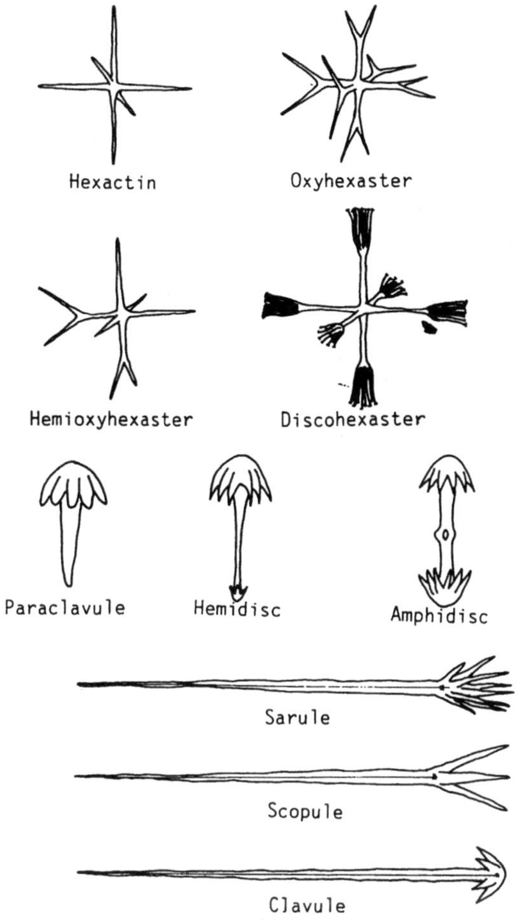


**Fig. 2.** Hexactinellid spicules. Discohexaster after Ijima (1927). Other spicules modified after Reid (1958)

paleontological findings lead him to conclude that the Hexactinellida are polyphyletic in origin. This statement and also the hypothesis of Finks (1970) will be critically re-examined in this chapter.

## Discussion

What would an Early Cambrian hexactinellid have been like? To answer this question it is necessary to examine the main autapomorphies of Hexactinellida. They are as follows:

Soft body: Flagellated chambers contain a choano-syncytium instead of choanocytes as in other sponges. At a mature state this syncytium consists of enucleate, flagellated collar bodies connected by plugged junctions to a basic main reticulum in which nuclei are unevenly scattered (Reiswig 1979; Mackie and Singla 1983). The choano-syncytium is separated from dermal and gastral membranes by a loose network of trabecular tissues supported by collagenous mesolamella. However, the presence of a secondary reticulum, as described from the species *Aphrocallistes vastus* Schulze, *Chonelasma calyx* Schulze by Reiswig (1979), and *Rhabdocalyptus dawsoni* (Lambe) by Mackie and Singla (1983), seems not to be a general feature of all hexactinellids. According to new histological investigations done by Reiswig (this Vol.), a secondary reticulum seems not to be present in *Dactylocalyx pumiceus* Stutchbury. Also, perforated plugs between cellular and enucleate elements seem to be absent in *Dactylocalyx*.

The wall is without a pinacoderm and pores are noncontractile (Tuzet 1973). Hexactinellids possess an ability for impulse conduction and interruption of flagellar activity, and thus can arrest the flow of the feeding current as a response to certain stimuli [shown for *Rhabdocalyptus dawsoni* by Mackie et al. (1983)].

Skeleton: Three-axial spicules (originally hexactins). The spicules could have supported the choanosome (as the basic functional feature of a sponge) or substained the open nature of the trabecular network, as suggested by Kirkpatrick (1909). A differentiation into mega- and microscleres might have been present at a very early stage. This would have resulted in formation of mega- and microhexactins (as reported by Okada 1928, hexasters ontogenetically originate from microhexactins). These features, the syncytial organization of tissues, the capability of impulse conduction, and the triaxial symmetry of the spicules, are autapomorphies constituting the Hexactinellida as a monophylum. Accordingly, these characters belong to the basic pattern of the taxon Hexactinellida, which is probably the monophylum best-established within the Porifera!

The spicule association found in sponges from Lower Cambrian rocks may give rise to two different hypotheses:

1. The spicules originated from different types of sponges: stauractins from a protosponge and hexactins from another hexactinellid sponge, which possessed a hexactin-bearing parenchymal layer. In this case, the latter may have belonged to the same clade as the Middle Cambrian *Stephenospongia* Rigby. This sponge was found in rocks actually older that the ones containing Protospongiidae (Rigby 1986).
2. Both sorths of spicules originated from the same type of sponges whose skeletons were differentiated into a parenchymal layer (hexactins and maybe derivatives like stauractins or rhabdodiactins) and a dermal layer (stauractins). In this case, the Lower Cambrian hexactinellid would have possessed an at least double-layered wall supporting probably a folded, leucon-type choanosome. Such an hypothetical Early Cambrian hexactinellid is shown in Figs. 3 and 4. Evolutionary reconstructions by Hadzi (1963) suggest that the ascon-type choanosome is not the plesiomorphic organisation of sponges but rather a result of a secondary union of flagellated chambers.

The fact that early Paleozoic hexactinellids are preserved as entire body fossils at all, in spite of their lyssacine character (no visible fusion of spicules), is remarkable. This gives rise to the assumption that their skeletons may have contained more collagen than do recent hexactinellids: after death something must have

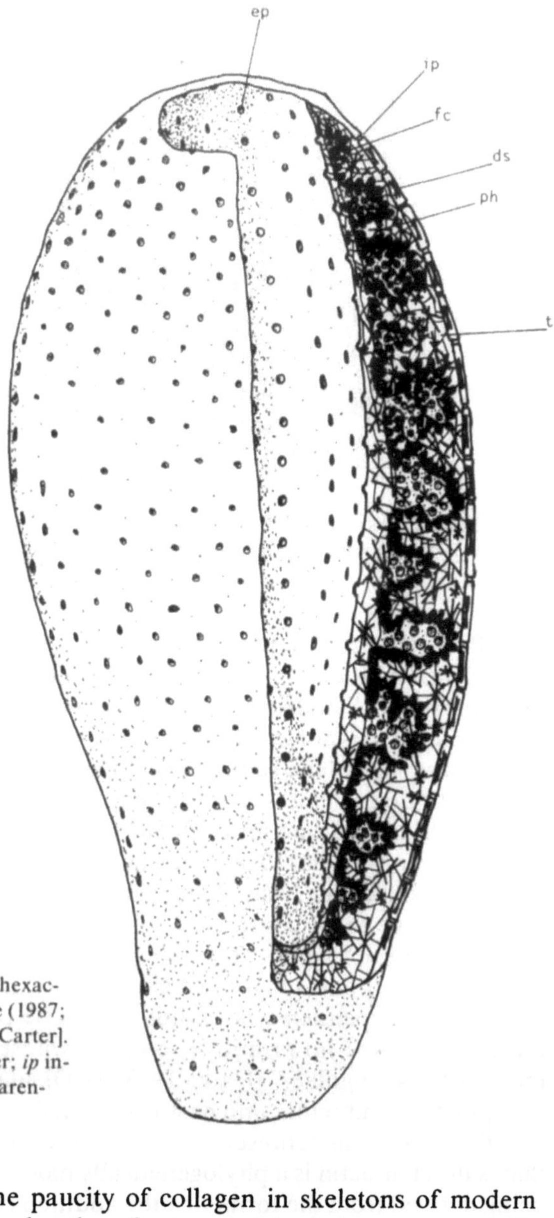


**Fig. 3.** Hypothetical Lower Cambrian hexactinellid [based on a drawing by Schulze (1987; Table LV, Fig. 1) of *Rosella antarctica* Carter]. *ep* exhalant pore; *fc* flagellated chamber; *ip* inhalant pore; *ds* dermal stauractin; *ph* parenchymal hexactin; *tt* trabecular tissue

kept their spicules in place. The paucity of collagen in skeletons of modern hexactinellids might be a derived rather than an original character. A well-developed collagen skeleton, however, might have obviated the necessity of silicious spicules and given rise to their reduction in some genera (analogous to the reduction which can be observed in keratose demosponges). Thus, the Protospongiidae may have been sponges whose parenchymalia (though present in the first place) had been reduced due to collagen development and with only the dermal stauractins left over.

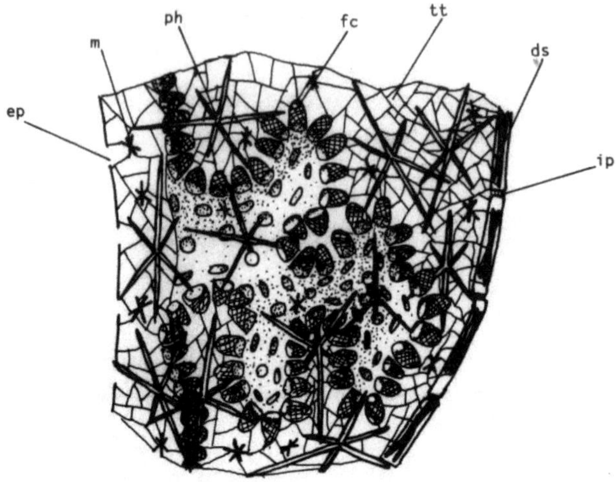


**Fig. 4.** Magnified section of Fig. 3. *ep* exhalant pore; *ph* parenchymal hexactin; *fc* flagellated chamber; *m* microhexactin; *tt* trabecular tissue; *ds* dermal stauractin; *ip* inhalant pore

Many Mesozoic hexactinellids possessed dermal stauractins (e.g., the Jurassic *Cypellia* Zittel, Fig. 5), and they are known also from recent Hexactinellida, e.g., *Bathydorus* Schulze. Stauractine autodermalia are present in many hexactinellids. Thus, it is not too surprising to find them in hexactinellid larvae (though, as in *Farrea sollasii* the stauractins are not present during the later ontogenesis. The adult *Farrea* possesses dermal pentactins, no stauractins). The scleroblasts, from which the stauractins of the larvae originate, can be traced back to their dermal layer (Okada 1928). However, phylogenetic interpretations, such as recapitulation, on the basis of these first-formed larval spicules (e.g., Tabachnick this Vol.) have to be considered with caution. According to Hadzi such interpretations are not valid: "The freely living larva, the spongula, which is characteristic for spongiae and which actually has a very manifold structure, developed independently and thus cannot be considered to be a recapitulated form. There were no primitive spongiae whose form and structure resembled those of the spongulae." (Hadzi 1963: 7). Okada himself was not certain how to interpret the stauractins, and he warned: "It does *not* necessarily follow from this that the stauractins represent the most primitive form of hexactinellid spicules that is the stauractin is a phylogenetically older type than the hexactin, the latter being derived from the former by the addition of the radial rays." (Okada 1828: 12, my italics). The dermal nature of stauractins in *Protospongia* was suggested already by Sollas (1880). This possibility was taken into consideration also by Reid (1958).

When listing the skeletal features characteristic of Hexactinellida, it can be seen that the Protospongiidae lack most of them (see Table 1). For this reason, *Protospongia* appears to be unsuitable for consideration as an ancestral form, but can more reasonably be viewed as a peripheral clade of the Hexactinellida.

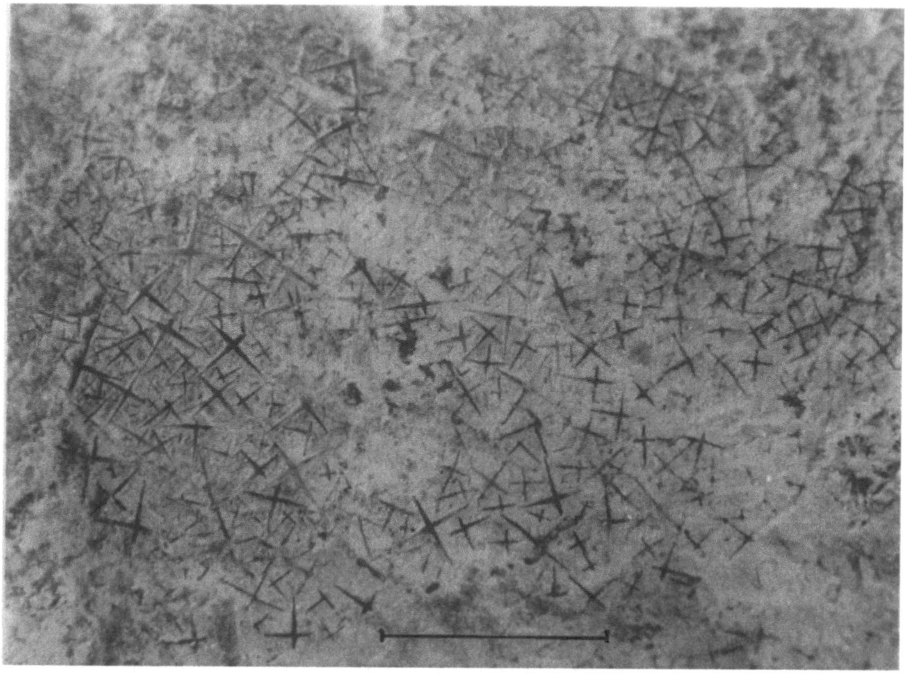

**Fig. 5.** Dermal layer of stauractins in *Cypellia rugosa* Quenstedt. Scale 0.5 cm. (Original at Staatliches Museum für Naturkunde, Stuttgart, BRD, Inv no 61953)

According to Finks (1960: 116) "It is likely that many if not all dictyonines arose from that line of the protospongioids in which the primitive parallel orientation of spicules was maintained, namely, the Protospongiidae and their descendants, the Dictyospongiidae." If Protospongiidae are considered to be the stem group of modern Hexactinellida, this would infer that the hexactine spicule was developed at least twice during the history of Hexactinellida: hexactins were first present in the Lower Cambrian. *Protospongia* and *Diagonella* seem to contain no hexactins, but they are present in other Middle Cambrian sponges [e.g., *Testiispongia* (Rigby) and *Hintzespongia* Ribgy & Gutschick] and in some Dictyospongiidae as well. Since all modern hexactinellids possess hexactins – which were never developed by any other sponge – it is justified to consider hexactins as an autapomorphic character of the Hexactinellida.

Hexactins have existed at least since the lowermost Cambrian. The Middle Cambrian Protospongiidae probably were a highly specialized clade of the Hexactinellida, rather than the stem group. Some Late Devonian Hexactinellida actually correspond to the requirements for a stem group of the modern dictyonal hexactinellids (Hexactinosa and Lychniscosa if we consider these taxa to be monophyletic ): *Gabelia* Rigby & Murphy is a Late Devonian Hexactinallid, whose skeleton consists of regularly oriented hexactins arranged in quadrangles

**Table 1.** Distribution of characters within different hexactinellid groups. *Hintze-spongia* and *Stephenospongia* are Middle Cambrian hexactinellids

| Skeletal characters: | Protospongiidae | Dictyospongiidae | *Hintzespongia* | Amphidiscophorida | Lyssacinosa | Hexactinosa | Scopularia | Clavularia | Lychniscosa | *Stephenospongia* |
|---|---|---|---|---|---|---|---|---|---|---|
| Hexactins | – | + | + | + | + | + | + | + | + | + |
| Stauractins | + | + | + | ? | + | + | + | + | + | + |
| Hexasters | – | + | – | – | + | + | + | + | + | – |
| Amphidiscs | – | ? | – | + | – | – | – | – | –-.. | – |
| Paraclavules | – | + | – | ? | + | – | – | – | – | – |
| Lychniscs | – | – | – | – | – | – | – | – | + | – |
| Dictyonal skeleton | – | – | – | – | – | + | + | + | + | – |
| Reduction of parenchymalia (Rhabdodiactins) | ? | + | ? | + | + | – | – | – | – | ? |
| Regular parallel rows (spiculation) | + | + | +/– | + | +/– | + | + | + | + | – |
| Sarules | – | – | – | – | – | + | – | + | – | – |
| Scopules | – | – | – | – | – | + | + | – | – | – |
| Clavules | – | – | – | – | – | + | – | + | – | – |

(Rigby and Murphy 1983). Another Upper Devonian hexactinellid, *Matta-spongia* Rigby, has a double-layered wall, the inner layer consisting of hexactins organized regularly in vertical and horizontal rows (Rigby 1970). However, no microscleres are known from these sponges, so their taxonomic status remains uncertain.

Another possibility is that hexactins might be found in Middle Cambrian Protospongiida by critical re-examination of the fossils. They might not all be single-walled. The *Protospongia hicksi* Hinde, as figured by Rigby (1986, Pl. 18, Fig. 1), seems to contain several layers. At least four ranks of stauractins can be seen partly overlapping each other. According to Sollas (1980), *Protospongia fenestrata* Salter appears to have consisted of more that one layer of spicules, though he was not certain whether this impression was due to the original morphology of the sponge or to compaction during fossilisation.

*Testiispongia* Rigby is a Middle Cambrian ?protosponge, which sometimes contains hexactins (Rigby 1986). The Middle Cambrian *Hintzespongia* Rigby and Gutschick "has a body wall composed of an inner irregularly arranged net and an outer layer of regularly arranged protospongelike spicules" (Rigby 1976, p. 57). The inner layer contains hexactins. Also, some later Paleozoic protosponge-like Hexactinellida possessed hexactine parenchymal spicules.

Hexasters are reported from the Mississippian dictyosponge *Griphodictya epiphanes* Hall and Clarke, which also has paraclavules (Hall and Clarke 1899). The Dictyospongiidae Hall and Clarke are Paleozoic Hexactinellida whose main diagnostic character seems to be the paraclavule. Whether they can be considered as a monophylum will have to be critically re-examined. Paraclavules have been reported from seven genera of the Dictyospongiidae only. Their only other diagnostic character might appear to be their parenchymal layers of bundled rhapdodiactins, which is not a well-established synapomorphy, since the reduction of hexactins to diactins occurred convergently in several clades of the Hexactinellida, e.g., in the Amphidiscophora and in the Lyssacinosa (lyssacine Hexasterophora). The fact that paraclavules and hexasters occur together in *Griphodictya* led Finks (1983, p. 109) to conclude that Hexasterophora and Amphidiscophora diverged during the Paleozoic from a taxon in which amphidiscs and hexasters occurred together. Based on the proposal of Finks that the Dictyospongiidae contain representatives of both Hexasterophora and Amphidiscophora, Mostler (1986) suggested to eliminate the taxon Dictyospongiidae in its present sense and to assign its representatives to the Hexasterophora and Amphidiscophora, respectively. However, *Griphodictya epihanes* Hall and Clarke contains both hexasters and Paraclavules, the latter supposedly homologous with amphidiscs according to Finks (1970, 1983). It is, however, very unlikely that amphidiscs developed through hemidiscs from paraclavules by addition of the "missing" end. The three types of spicules probably developed independently of each other. Another example of such a convergence are the sceptrules (sarules, scopules, and clavules), which are an autapomorphy of some Hexactinosa. They are very similar to the cuspidates formed by some Amphidiscophorida, e.g., *Pheronema* Leidy and *Sericolophus* (Ijima), (Ijima and Okada 1935).

The hexasters reported from the paratype of *Itararella gracilis* (Kling and Reif 1969) probably are foreign in origin (as suggested by the authors). No reason exists that the stem group of the sister groups Hexasterophora adn Amphidiscophora might have possessed hexasters and amphidiscs as well. Hexasters or amphidiscs are well established autapomorphies of the adelphotaxa.

## Conclusions

1. Hexactines are a constitutive character of the monophylum Hexactinellida.

2. The Protospongiidae were not the "primitive" ancestors of all later Hexactinellida, but more reasonably a highly specialized clade of the taxon. The characters required to establish the stem group of at least some modern Hexactinellida dictyonal sponges are present in various Later Paleozoic hexactinellids such as the Late Devonian *Gabelia* or *Mattaspongia*.

3. The stauractins occurring first in the larvae of, e.g., *Farrea sollasii*, are probably of dermal origin. Therefore, their value for the interpretation suggested by most authors as a recapitulation of parenchymal spicules is questionable. Present knowledge of the ontogenetic development of sponges (especially that of hexactinellid sponges) is far too scarce to allow any such phylogenetic interpretations.

4. From rocks of the lowermost Cambrian (Vendian) of China, triaxial spicules have been reported, but no stauractins. This seems to support the hypothesis that triaxons are the original hexactinellid spicules and all other spicules were developed from the simple hexactin. The presence of different stages of reduction indicates that stauractins arose convergently from hexactins within all main clades of the Hexactinellida.

5. So far, we know very little about the phylogenetic transition between the Paleozoic Hexactinellida and the modern (Mesozoic and Recent) ones. For this reason alone, it seems more than speculative to conclude anything about the origin of the Hexactinellida on the basis of scarce larval studies done on a few recent hexactinellid species. The origin of the different hexactinellid clades needs to be studied, e.g., to trace back the development of dictyonal skeletons. Also it is absolutely necessary to study the ontogenetic development on hexactinellid sponges carefully, e.g., to learn more about the functions of their larval characters (such as spiculation).

*Acknowledgments.* I am indebted to Dr. J.K. Rigby, Provo, Utah, USA, for providing pertinent literature, for providing a specimen of *Diagonella*, and for valuable discussions. Thanks are due also to Dr. J. Reiner (Berlin, Germany) and to Dr. H.M. Reiswig (Montreal, Canada) for discussions. Dr. H.M. Reiswig gave useful comments on the manuscript. I also thank Deutsche Forschungsgemeinschaft for financially supporting my research on the Hexactinellida.

## References

Finks RM (1960) Late Paleozoic sponge faunas of the Texas region. The siliceous sponges. Bull Am Mus Nat Hist 120:1–160

Finks RM (1970) The evolution and ecologic history of sponges during Palaeozoic times. Symp Zool Soc Lond 25 (1970):3–22

Finks RM (1983) Fossil Hexactinellida. In: Rigby JK, Stearn CW (eds) Sponges and Spongiomorphs. Notes for a short course. Broadhead, Indiana, pp 270

Hadzi J (1963) The evolution of the Metazoa. Pergamon, Oxford, pp 499

Hall J, Clarke JM (1899) A memoir on the Palaeozoic reticulate sponges constituting the family Dictyospongiidae. Mem N Y State Mus 2:343–448

Hinde GJ (1888) A monograph of the British fossil sponges. Palaeontographical Society London 1877 (Sponges of Palaeozoic group) pp 93–188

Ijima I (1927) The Hexactinellida of the Siboga Expedition, Vitkomsten op Zoologisch, Botanisch, Oceanographisch en Geologisch Gebeid verzameld in Nederlandsch Oost-Indie 1899–1900 4, Leiden, pp 1-382 E J Brill

Ijima I, Okada Y (1935) Studies on Hexactinellida. Contribution V. Pheronematidae and Hyalonematidae. J Fac Sci Univ Tokyo, Sect 4 (Zool), 4 (1):41–469

Kirkpatrick R (1909) On the regular hexactine spicule of Hexactinellida. Ann Mag Nat Hist Lond 8 (4):505–509

Kling SA, Reif W-E (1969) The paleozoic history of amphidisc and hemidisc sponges: new evidence from the Carboniferous of Uruguay. J Paleontol 43 (6):1429–1434

Librovitch LS (1929) *Uralonema karpinskii* nov gen, sp i drugie kremnevie gubki iz kamennougolnikh otlozhenii vostochnogo sklona Urala. Tr Geol Kom-ta nov ser 179:11–57

Mackie GO, Singla CL (1983) Studies on hexactinellid sponges I. Histology of *Rhabdocalyptus dawsoni* (Lambe 1873). Philos Trans R Soc Lond B 301:365–400

Mackie GO, Lawn ID, Pavans de Cecatty M (1983) Studies on hexactinellid sponges II. Excitability, conduction and coordination of responses in *Rhabdocalyptus dawsoni* (Lambe 1873). Philos Trans R Soc Lond B 301:401–418

Minchin EA (1905) A speculation on the phylogeny of the hexactinellid sponges. Zool Anz 28:439–448

Mostler H (1986) Beitrag zur stratigraphischen Verbreitung und phylogenetischen Stellung der Amphidiscophora und Hexasterophora (Hexactinellida, Porifera). Mitt Österr. Geol Ges 78 (1985):319–359

Mostler H, Mosleh-Yazdi A (1976) Neue Poriferen aus oberkambrischen Gesteinen der Milaformation im Elburzgebirge (Iran). Geol Paläont Mitt Innsbruck 5 (1):1–36

Okada Y (1928) On the development of a hexactinellid sponge, *Farrea sollasii*. Tokyo Univ J Fac Sci Sect 4 Zool 2:1–27

Reid REH (1958) Upper Cretaceous Hexactinellida of Great Britain and Northern Ireland. Palaeontographical Society 1957 I-XLVIII:1–48

Reiswig HM (1979) Histology of Hexactinellida (Porifera) Coll Int Cent Nat Res Sci 291:73–180

Rigby JK (1970) Two Upper Devonian hexactinellid sponges from Alberta. J Paleontol 44:7–16

Rigby JK (1976) Some observations on occurrences of Cambrian porifera in western North America and their evolution. Brigham Young Univ Geol Stud 23 (2):51–60

Rigby JK (1986) Sponges of the Burgess Shale (Middle Cambrian), British Columbia, Palaeontogr Canad 1986(2):1–105

Rigby JK (1987) Early Cambrian sponges from Vermont and Pennsylvania the only ones described from North America. J Paleontol 61(3):451–461

Rigby JK, Murphy MA (1983) *Gabelia*, a new lyssacid protosponge from the Roberts Mountains, Nevada. J Paleontol 57 (2):797–803

Rigby JK, Webby BD (1988) Late Ordovician sponges from the Malongulli Formation of central New South Wales, Australia. Palaeontogr Am 56:1–147

Rigby JK, Racki G, Wrzolek T (1981) Occurrence of dictyid hexactinellid sponges in the Upper Devonian Holy Cross Mts. Acta Geol Pol 31 (3–4):363–369

Schulze FE (1987) Report on the Hexactinellida. Challenger Rep 21:1–513

Sollas WJ (1980) On the structure and affinities of the genus *Protospongia* (Salter). Q J Geol Soc Lond 36:362–367

Termier H, Termier G, Tsien HH (1981) Spongaires des calcaires des recifaux du Frasnien de L'Ardenne. Soc Belg Geol Bull 90:287–298

Tuzet O (1973) Hexactinellides ou hyalosponges (Triaxonia F E Schulze). Traite de Zoologie; anatomie, systematique, biologie 3(1). Masson, Paris, pp 633–690

Wei-Ming D, Qian Y (1988) Late Sinian to Early Cambrian small shelly fossils from Yangjiaping, Shimen, Hunan, Acta Micropaleontol Sin 5(1):50–67

# Demosponge Higher Taxa Classification Re-Examined

R.W.M. van Soest

## Introduction

Thirty-five years ago the foundation was laid for a classification of the Demonspongiae that in essence is still in use among virtually all neontologists. I am referring to Lévi's (1953) system based on reproductive characters. Unfortunately, when examined closely, this scheme of classification cannot endure a test of logical consistence. Especially the proposed subclass Tetractinomorpha is entirely based on a single primitive feature, viz. the absence of the viviparous reproduction. Furthermore, the synapomorphy for the major other subclass, Ceractinomorpha, the viviparous behavior, was insufficiently established, as would appear later when several ceractinomorph groups were found to be oviparous. The application of these characters induced Lévi to search for and emphasize supporting characters, such as the axial condensation.

Although previously formulated classifications were certainly not in agreement with each other, more or less clearly defined natural groups were at the time recognized by classifiers such as Topsent, Burton, and De Laubenfels. Some of these natural groups were completely disrupted by Lévi's proposals, notably the Halichondrida and the Poecilosclerida.

How was it possible that such a fundamentally unsound classification (even Lévi himself expressed his doubts over the validity of the tetractinomorphs) could persist largely unchallenged until the present day? The reason for it must be sought in the support Lévi's system received from Bergquist and Hartman, who went to great lengths to find additional characters for its firm establishment notwithstanding their initial grave criticisms of the system. Characters in seeming support (mostly biochemical, reproductive, ultrahistological, etc.: all studied in only a few examples and subsequently extrapolated to whole orders) were broadly advertised; conflicting characters were largely ignored, or at the most used to point out that the classification was not yet quite consistent.

Since there are apparently few generalists among modern sponge taxonomists, i.e., those studying all sponges worldwide, instead of particular groups or regions, the Lévi-Bergquist-Hartman system (which will be hereafter abbreviated as LBH) has not been seriously challenged and has grown into a genuine paradigm.

It is the purpose of the present paper to investigate to what extent already existing classifications were unjustifiably emended by the Lévi system and later additions by Bergquist (1978, 1980a) and Hartman (1982).

J. Reitner and H. Keupp (Eds.)
Fossil and Recent Sponges

## Methods

I will not present a whole new set of characters to take priority over characters already in use. A new analysis will be made building on previous exercises (see Van Soest 1987) with the same traditional characters (including even reproductive behavior); the only difference between my approach and the LBH approach is that I make use of cladistic principles (the best general treatment of phylogenetic classification using these principles is probably Wiley's 1981; sponge examples are outlined in a previous symposium volume, see Van Soest 1991; a good example of application of these principles to the classification of Haplosclerida is found in De Weerdt 1989).

Briefly my methods are:

- outgroup comparison, character compatibility and parsimony are used to determine primitive vs. advanced nature of a character (polarity).
- no a priori weighting of a character is practiced.
- choice of characters is determined by their occurrence in two or more of the studied groups (a group might have any number of important or less important characters, but only those that occur also outside the group are used).

Regarding character/state scoring: many characters/states may not be found in all members of the studied groups; in obvious cases of secondary loss (e.g., chelae in many species and genera of Poecilosclerida) the groups to which these sponges belong are still scored as possessing the characters/states (these are then so-called underlying synapomorphies).

Regarding outgroups: most of these used below are nonextant fictional groups composed from a combination of different extant outgroups (no group existing in the present can serve as an ancestral model); outgroup comparison is not done formally according to methods proposed by Maddison et al. (1984), but the validity of using more than one extant outgroup is recognized and applied as far as possible.

Regarding parsimony: there are three different approaches to parsimony: *Wagner* or simple parsimony which allows for parallel development of a character (-state) and reversal or loss to be equally likely; *Camin-Sokal* or irreversible parsimony which predicts the unlikelihood of reversals or losses (e.g., if a character is not present in a taxon it is much more likely that it has never been there in the first place than that it has been lost secondarily); *Dollo* parsimony which takes the reverse attitude (losses are much more easily acquired than new parallel developments).

Parsimony plays an important role in polarity decisions of sponge characters, because the majority of the potential synapomorphies are underlying.

As in my previous attempts (Van Soest 1987), I constructed matrices of taxa and characters, which are then analyzed without and with computer assistance. The latter frequently consists of constructing so-called Wagner Trees, networks determining the distance and graphically depicting the relationships between taxa and hypothesized ancestral taxa; alternative methods of analysis, e.g., the group compatibility method (Zandee 1987) yield closely comparable results. The programs, of which several kinds are generally available (e.g., PAUP, PHYLIPP, TREE TOOLS, MACCLADE, CAFCA), compute tree length (number of character transformation steps) and consistency index (quotient of the theoretical minimum length of a tree and the actual length) for different possible trees. Some programs actively seek to find the best tree, which is assumed to be the most likely representation of the phylogenetic relationships; others allow you to construct alternative trees which are then compared and evaluated. Several of the mentioned programs have the three parsimony options, and most allow weighting of characters (if you decide, however, to use that option then the primary advantage of using the programs diminishes substantially). The programs, of course, do not necessarily produce better classifications than the ones made by hand; one major advantage, however, is that they show you how many different equally parsimonious solutions exist for any given character distribution.

Of paramount importance for any kind of analysis to work properly is that homologies of characters/states are established. In my previous attempt (Van Soest 1987) this was not always firmly established due to the fact that a large number of often widely diverging sponge groups were treated in the same analysis with unavoidable problems in scoring homologous characters/states. In the present analysis, demosponge subgroups (related orders and/or families) were studied separately allowing more accurate scoring of character distribution.

## Demosponge Higher Taxa Used in This Study

An immediate problem I encountered when setting out on this project is the delimitation of the Demospongiae. To avoid duplicating discussions generated by others in this volume, I prefer to confine myself to the non-Homosclerophorid demosponges, most of which share the presence of monaxone siliceous spicules and spongin or related collagenous substances in some form produced by spongocytes. Whether or not Homosclerophorid spicules are derived from monaxones or independently developed from polyaxone spicules will not be discussed here. Whether or not organic fibrillar material found in Homosclerophorida, Calcarea, and Hexactinellida is homologous with spongin remains to be demonstrated. Even if the virtual absence of organic cementing material of the Hexactinellids, Calcarea, and Homosclerophorida is eventually established, the problem remains whether this absence indicates a synapomorphy or three homoplasic developments. In the following I will occasionally refer to Homosclerophorids in the process of outgroup analysis.

Likewise I will not enter into a discussion of sclerosponges; I explained in a previous paper (Van Soest 1984) that I can only see them as polyphyletic relicts, each easily assigned to recent nonsclerosponge orders.

Terminal taxa used in the various analyses are generally accepted as monophyletic. For an evaluation of this, one is referred to various definitions given in reviews of classification of sponges such as those of Lévi (1973), Bergquist (1978), and Hartman (1982); this will not be further discussed.

I will concentrate on analyzing the key problem presented by the Axinellida, and will only briefly discuss the results obtained with the remaining monophyletic groups.

## The Axinellid Problem

By putting so much emphasis on reproductive behavior and axial condensation of the skeleton, Lévi (1953, 1973) created a substantial problem in the group of the Halichondrida in the sense used by Topsent, De Laubenfels, and Burton (which included the family Axinellidae s.l.). Viviparous genera such as *Halichondria* and *Hymeniacidon* were transferred to the Ceractinomorpha, oviparous genera were transferred to the Tetractinomorpha. The same was done with Raspailiids (which were formerly considered close to Microcionids). Aster-bearing axially condensed genera were removed from the Hadromerids and incorporated in the Axinellida. *Sigmaxinella* was also assigned to the Axinellida, following Hallmann (1917), but close relatives such as *Desmacella* and *Biemna* were left in the Poecilosclerida.

In the following I will compare the LBH classification with an alternative classification which is more or less a combination of that of Topsent/Burton/De Laubenfels with some modern amendments, derived from my previous exercises (Van Soest 1987) and from Hooper's (1991) studies on Raspailiid-Microcionid relationships; this classification will be further referred to as TBL.

The following groups of ordinal/familial level are included in the analysis:

Tetractinellida-Hadromerida-Hemiasterellidae-Axinellidae-Halichondrida-Raspailiidae-Microcionidae-Myxillidae s.l. (including Hymedesmiidae, Anchinoidae, Crellidae, Coelosphaeridae)-Sigmaxinellidae-Desmacellidae.

A further undefined outgroup showing many characters in common with Homosclerophorida is added.

The following characters were used:

1. Presence of spongin: recognizable in most of the above-mentioned groups, perhaps with the exception of Tetractinellida, although fibrous conditions in certain genera such as *Cinachyra* are probably caused by the presence of spongin.
2. Vivipary is considered an advanced character state (ovipary being the alternative) in spite of the fact that outgroups such as Calcarea and Homoscleromorpha are viviparous, because there are several oviparous groups among the alleged viviparous ones, pointing to the probable homoplasic nature of vivipary (see further below).
3. Axial condensation of the choanosomal skeleton: in branching forms there is a central column of closely packed spicules.
4. Anisotropic choanosomal reticulation: the spicules or fibers are arranged in ladder-like structures with primary lines and interconnecting ones.
5. Radiate architecture: arrangement of the spicules strictly perpendicular to

the surface without reticulation; this may be the alternative condition of an anisotropic reticulation (and as such would then be a state of the same character); on the other hand, in the outgroup neither condition is found (Calcarea and Homosclerophorida have largely unorganized masses of spicules; the "reticulate" condition of the Hexactinellida skeleton seems nonhomologous to the anisotropic arrangement).
6. Presence of acanthostyles (see further below).
7. Euasters (see further below).
8. Sigmatose microscleres (either sigmata, chelae, toxa, or any combination of them) (see further below).
9. Isonitriles: secondary metabolites found in some sponge genera belonging to the Axinellidae and Halichondrida [information of J.-C. Braekman, Univ. of Brussels, based on reports published in the (biochemical) literature].

The two trees and their character support are presented in Figs. 1 and 2.

When the taxon-character matrix (Table 1) is examined, it is immediately apparent that conflicting character distributions occur, e.g., the distribution of asters, sigmatose microscleres, and acanthostyles conflict with the axial condensation of the skeleton. To find which of the two trees represents the "best", the following analytical options were used: first simply counting support and conflict of character distribution in the two alternative trees (character compatibility), then Wagner Tree analysis under various parsimony assumptions: unweighted reversible, irreversible, and dollo.

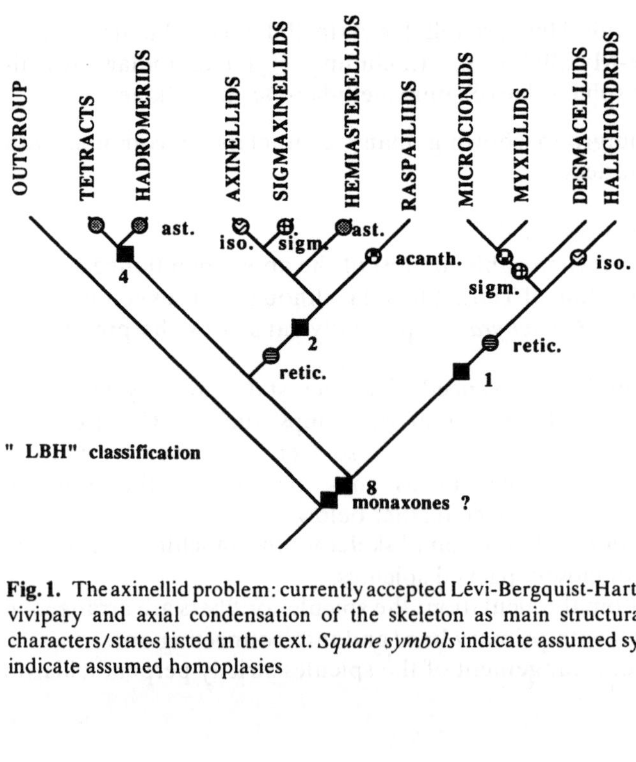


**Fig. 1.** The axinellid problem: currently accepted Lévi-Bergquist-Hartman classification (LBH), with vivipary and axial condensation of the skeleton as main structural features. *Numbers* refer to characters/states listed in the text. *Square symbols* indicate assumed synapomorphies, *round symbols* indicate assumed homoplasies

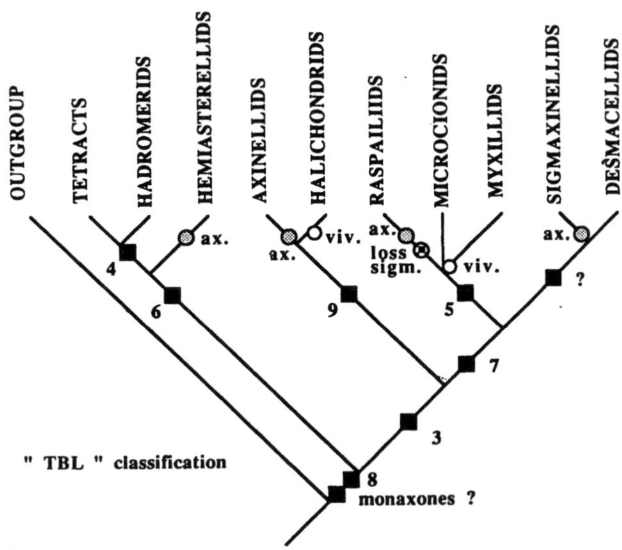


**Fig. 2.** The axinellid problem: classification combined from points of view of older authors (Topsent-Burton-De Laubenfels) (TBL) and some new insights. *Round symbols* indicate assumed homoplasies

**Table 1.** Taxon-character matrix for the evaluation of the Axinellid problem

| | Outgroup | Tetractinellida | Hadromerida | Hemiasterellidae | Axinellidae | Sigmaxinellidae | Raspailiidae | Microcionidae | Myxillidae | Desmacellidae | Halichondriidae |
|---|---|---|---|---|---|---|---|---|---|---|---|
| 1. Spongin | 0 | ? | 1 | 1 | 1 | 1 | 1 | 1 | 1 | 1 | 1 |
| 2. Vivipary | 0 | 0 | 0 | 0 | 0 | 0 | 0 | 1 | 1 | 0 | 1 |
| 3. Axial condensation | 0 | 0 | 0 | 1 | 1 | 1 | 1 | 0 | 0 | 0 | 0 |
| 4. Anisotropic reticulation | 0 | 0 | 0 | 0 | 1 | 1 | 1 | 1 | 1 | 1 | ? |
| 5. Radiate architecture | 0 | 1 | 1 | 0 | 0 | 0 | 0 | 0 | 0 | 0 | 0 |
| 6. Acanthostyles | 0 | 0 | 0 | 0 | 0 | 0 | 1 | 1 | 1 | 0 | 0 |
| 7. Euasters | 0 | 1 | 1 | 1 | 0 | 0 | 0 | 0 | 0 | 0 | 0 |
| 8. Sigmatose microscleres | 0 | 0 | 0 | 0 | 0 | 1 | 0 | 1 | 1 | 1 | 0 |
| 9. Isonitriles | 0 | 0 | 0 | 0 | 1 | 0 | 0 | 0 | 0 | 0 | 1 |

Character compatibility analysis yields a six-character support and three-character conflict for the TBL classification, against a four-character support and five-character conflict for the LBH classification.

The results of the Wagner Tree analysis show (Table 2) that no matter which parsimony option is used, the TBL classification is invariably more consistent and shorter than the LBH classification. Moreover, together with a few slightly different equally parsimonious variations, the former is also always the most consistent and shortest possible classification. The only way to have the LBH classification come out on top is by either leaving out (ignoring) characters such as the asters, acanthostyles, and sigmatose microscleres, or by assigning heavy weights to characters such as vivipary and axial condensation, and low weights to above mentioned characters.

**Table 2.** Statistics (tree length and consistency index) under various parsimony assumptions of cladograms presented in Figs. 1, 2, 4, 6

| Cladogram | Tree length | | | Consistency Index | | |
|---|---|---|---|---|---|---|
| | Wagner | Camin-Sokal | Dollo | Wagner | Camin-Sokal | Dollo |
| Axinellid problem | | | | | | |
| LBH tree (Fig. 1) | 16 | 16 | 28 | 0.56 | 0.56 | 0.32 |
| TBL tree (Fig. 2) | 14 | 14 | 17 | 0.64 | 0.64 | 0.53 |
| Tetractinellida/ Hadromerida tree (Fig. 4) | 26 | 29 | 32 | 0.54 | 0.48 | 0.44 |
| Halichondrida-Ceractinomorpha tree (Fig. 5) | 33 | 40 | 39 | 0.42 | 0.35 | 0.36 |
| Keratosa tree (Fig. 6) | 14 | 14 | 19 | 0.71 | 0.71 | 0.53 |
| (Dollo-alternative | – | – | 17 | – | – | 0.59) |

## *Why Should Vivipary Be of Such Significance?*

At the time Lévi, after long years of study, constructed his new classification, all available information (Lévi 1956) indeed pointed towards a neat distribution of incubated solid parenchymella larvae over the Poecilosclerida, Haplosclerida, and Keratosa. (Homosclerophorida and Calcarea, of course, also incubate larvae, but these are coeloblastulae.) In the course of time, so many exceptions to this distribution were found (e.g., Verongids, Agelasids, Petrosids, Desmacellids) that the strength of this character as a synapomorphy is greatly diminished. Moreover, larval morphology, where known in the different groups, is different (e.g., Wapstra and Van Soest 1987). It seems now uncertain that vivipary was developed only once; it is even likely, in view of its significance for life strategy, that it is easily independently developed (vivipary-ovipary issues are also found in, e.g., coelenterates, molluscs, and ascidians, and there is no higher taxon phylogenetic relationships attached to it there).

### *Why Should the Axial Condensation Be of Such Significance?*

It is indeed striking to see the similarities of *Axinella* with aster-bearing *Stelligera*, sigma-bearing *Sigmaxinella*, or acanthostyle-bearing *Raspailia*. However, many forms exist which share undubitable synapomorphies with these genera (e.g., *Pseudaxinella*, *Timea*, *Desmacella*, and *Clathria*), which do not possess the axial condensation. Three alternative explanations to explain this distribution of characters are apparent:

A) Asters + acanthostyles + sigmatose microscleres are each homoplasic developments (one in Axinellids, one respectively in Hadromerida, Poecilosclerid Desmacellids and Poecilosclerid Microcionids + Myxillids).
B) All these characters are primitive demosponge features lost in most of the close relatives of the mentioned genera (all Axinellids excepting the Hemiasterellids have lost their asters, all Tetractinomorpha excepting the Raspailiids have lost their acanthostyles, all Tetractinomorpha excepting the Sigmaxinellids have lost their sigmatose microscleres).
C) The axial condensation is homoplasic in four different lines. To make this idea more palatable to those who have problems accepting this, consider the exact architecture of, e.g., *Raspailia ramosa* (Fig. 3A,B): basically the axial skeleton is an anisotropic reticulation of primary spicule tracts interconnected by single spicules; were it not so laterally compressed, then it could not be distinguished from the average Poecilosclerid or even Haplosclerid skeleton. It is quite unlike radiating skeletons of, e.g., *Polymastia* or *Tethya*. The Hemiasterellid *Hemiasterella stylifera* (Fig. 3C,D), on the other hand, has an axial skeleton with far less obvious anisotropic arrangements. Similarity between the two genera *Raspailia* and *Hemiasterella* is confined to the overall compressed, packed arrangement of the spicules in the central column.

A fourth possibility would be the primitive nature of the axial condensation, but that does not help to solve the dilemma. The third explanation is not only the most consistent when the distribution of other characters is concerned, but is also easily explained as a highly functional response of branching sponges for the problem of obtaining rigidity.

### *Euasters, Acanthostyles, and Sigmatose Microscleres as Synapomorphies*

It is a very striking and significant observation that no species are known which combine the possession of asters with that of acanthostyles. Asters are confined to the group of Astrophorids-Hadromerids-Hemiasterellids, and there is no reason

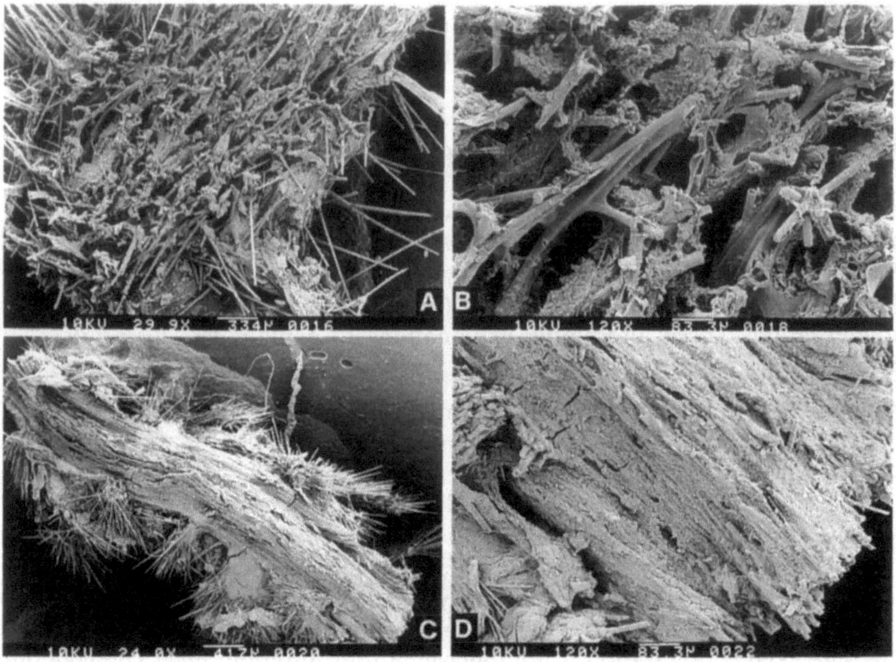


**Fig. 3.** SEM photographs in two magnifications of longitudinal sections through the axis of *Raspailia ramosa* (**A,B**) and *Hemiasterella stylifera* (**C,D**) to show differences (and thus presumed absence of homology) in architecture of "axially condensed" skeletons

not to consider them a synapomorphy uniting these three groups. Acanthostyles present a few problems: they are seemingly confined to Raspailiidae-Microcionidae-Myxillidae s.l.-Agelasidae. However, a synapomorphy found in the group comprised by Microcionids-Myxillids-Desmacellids-Mycalids-Petrosiids-Haplosclerids are the sigmatose microscleres, but these are not found in Raspailiidae and Agelasidae. In the case of Raspailiidae a parsimonious solution is the subsequent loss of them, in the case of Agelasidae there is some doubt over the homologous status of the acanthostyles. Moreover, biochemical evidence points to a close relationship between Agelasidae and Axinellidae (see also below).

## *Axinellid-Halichondrid Synapomorphies*

Elsewhere in this volume attention will be drawn towards the Halichondrids, so it will be sufficient to point out that the underlying synapomorphy used here (possession of isonitriles) can be corroborated by at least one other, namely the indiscriminate occurrence of stylote and oxeote spicules. Apart from the vivipary of two of about a dozen genera of Halichondrids (*Hymeniacidon* and *Hali-*

*chondria*) there is very little in common between them and the Raspailiid-Poecilosclerid-Haplosclerid assemblage.

To summarize: the currently accepted subclass classification of Ceractinomorpha vs. Tetractinomorpha including a tetractinomorph order Axinellida is less consistent with the distribution of available characters than a classification based on a combination of those of previous authors, who recognized other assemblages within the Demospongiae. In fact this alternative classification resembles to a large extent that of Reid (1968). Next to the Homoscleromorpha, which may or may not be related to Calcarea, these assemblages seem to be: Tetractinellida + Hadromerida + Hemiasterellidae/Axinellidae + Poecilosclerida + Haplosclerida/Keratosa. The synapomorphies for these assemblages are mostly underlying, presumably due to widespread loss of ancestral characters and the search for additional synapomorphies should go on. Until these are found I think we should refrain from using subclass or other higher taxa names for ordinal groups (contra, e.g., Möhn 1984). I propose to formally suppress the Axinellida as an undesirable polyphyletic ordinal taxon.

In the following the phylogenetic relationships within the above mentioned Demosponge assemblages will be further examined.

## Tetractinellid-Hadromerid-Hemiasterellid Relationships

The possession of euasters and the basic radiate architecture of the skeleton are considered synapomorphies; the former is absent from Spirophoridae and Polymastiidae/Suberitidae, as well as from individual species and genera, the latter is lacking in Chondrosida and dubiously developed in Lithistids and Hemiasterellids.

Taxa used are: Astrophorida-Desmophorida-remaining Lithistids (excepting Desmanthids and sublithids)-Spirophorida-asteriferous Hadromerida-asterless Hadromerida (family Polymastiidae)-Tethyidae-Hemiasterellidae-Chondrosida. The outgroup is undefined but shows many similarities with Homosclerophorida.

Characters used are (Table 3):

1. Radiate architecture: supposedly a synapomorphy of the group, but not expressed clearly in Lithistidae, Hemiasterellidae, and Chondrosida.
2. Radiate oxeotes: not found in Polymastiidae, Hemiasterellidae, and most asteriferous Hadromerida.
3. Stylotes: found in those groups not having oxeotes and as such may be an alternative character state; however, both are found in the Tethyidae.
4. Triaenes (long-shafted): a synapomorphy of the Tetractinellida (sometimes secondary loss has occurred); calthrops-type triaenes found in, e.g., *Pachastrella* are considered shortened triaenes not necessarily homologous to Homosclerophorid calthrops; ancestral state has been put as "?" to avoid heavy impact of this decision.
5. Tylostyles: a synapomorphy of Hadromerida.

**Table 3.** Tetractinellida-Hadromerida taxon-character matrix

| | Outgroup | Astrophorida | Desmophorida | Spirophorida | Remaining Lithistids | Tethya | Chondrosida | Astrose Hadromerida | Polymastiidae | Hemiasterellidae |
|---|---|---|---|---|---|---|---|---|---|---|
| 1. Radiate archit. | 0 | 1 | ? | 1 | ? | 1 | 0 | 1 | 1 | 0 |
| 2. Radiate oxeotes | 0 | 1 | 1 | 1 | 1 | 1 | 0 | 0 | 0 | 0 |
| 3. Stylotes | 0 | 0 | 0 | 0 | 0 | 1 | 0 | 1 | 1 | 1 |
| 4. Triaenes | ? | 1 | 1 | 1 | 0 | 0 | 0 | 0 | 0 | 0 |
| 5. Tylostyles | 0 | 0 | 0 | 0 | 0 | 0 | 0 | 1 | 1 | 0 |
| 6. Desmata | ? | 0 | 1 | 0 | 1 | 0 | 0 | 0 | 0 | 0 |
| 7. Organic cortex | 0 | 1 | ? | 1 | ? | 1 | 1 | 0 | 0 | 0 |
| 8. Ectosomal crust of microscleres | 0 | 1 | 1 | 1 | 1 | 1 | 1 | 1 | 0 | 1 |
| 9. Euasters (1 = only sphaerasters, 2 = diverse) | 0 | 2 | 0 | 0 | 0 | 2 | 1 | 2 | 0 | 2 |
| 10. Spirasters | 0 | 0 | 1 | 0 | 0 | 0 | 0 | 1 | 0 | 0 |
| 11. Amphiasters | 0 | 1 | 1 | 0 | 0 | 0 | 0 | 0 | 0 | 0 |
| 12. Microrhabds | 0 | 1 | 1 | 0 | 1 | 0 | 0 | 1 | 0 | 0 |
| 13. Sigmaspires | 0 | 0 | 0 | 1 | 1 | 0 | 0 | 0 | 0 | 0 |

6. Desmata: this character may be a synapomorphy for all Lithistids, but it is not compatible with other character distributions (e.g., astrose microscleres and sigmaspires). A distinct possibility is that desmata are an ancestral character lost in many modern forms but present in relict forms such as the Desmophorida and remaining Lithistids and sub-Lithistids (see also Van Soest and Zea 1986; Van Soest 1987).
7. Organic cortex: possibly a synapomorphy for a large part of the studied group; since it is normally only detected when special histological sections are made, or when there is a zonation of spicules, this character may be more widespread.
8. Ectosomal crust of microscleres: except for Polymastiidae this is a synapomorphy for the whole group.
9. Euasters: several states of this character are recognized: 1 = only sphaerasters present, 2 = diversity of euastrose microscleres.
10. Spirasters: found in Desmophorida and asteriferous Hadromerida, which may or may not be a homoplasious development.
11. Amphiasters and metasters: a synapomorphy of Astrophorida and Desmophorida.
12. Microrhabds: these may not be all homologous in the different groups.
13. Sigmaspires: a strong synapomorphy for Spirophorida and (some) remaining Lithistids (e.g., *Scleritoderma*).

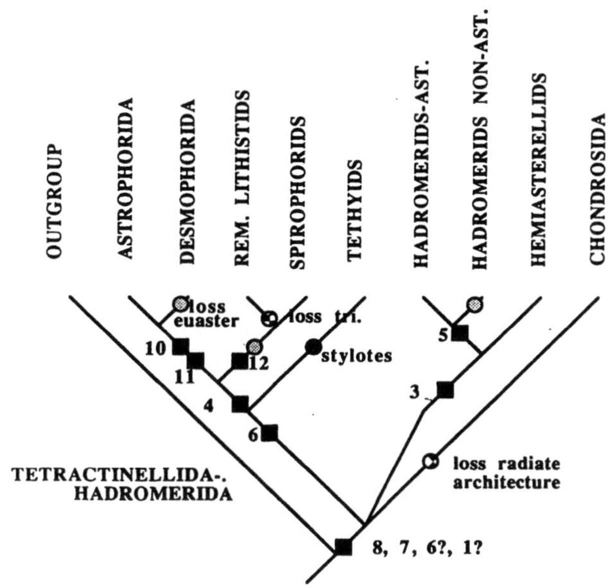


**Fig. 4.** Best cladogram as determined by MACCLADE, using various parsimony options, assumed to represent the most likely phylogenetic relationship of the tetractinellid-hadromerid groups of sponges. Note that trichotomous parts of the tree are in fact unsolved relationships between groups. *Round symbols* indicate assumed homoplasies

The result of the taxon-character analysis (Fig. 4) is more or less as expected: Tetractinellida and Hadromerida are clearly distinct groups. The position of the Tethyids and the Chondrosiids remains unsolved with the present character matrix. The noncompatibility of desmata with the distribution of sigmaspires and euasters prevents solution of the relationships between Lithistids and tetractinellids.

## Halichondrid-Poecilosclerid-Haplosclerid-Keratosa Relationships

As discussed already, the (underlying) synapomorphy is the anisotropic skeleton (not in Halichondriidae, and in numerous individual species and genera). There is a possibility that the anisotropic skeleton represents a plesiomorphous character uniting all non-Homosclerophorid Demosponges which was secondarily modified into a radial skeleton in Tetractinellids. So far character analysis favors the first option. The reason for inclusion of the Keratosa in the present assemblage is that no synapomorphies are recognized to group the Halichondrid-Poecilosclerid-Haplosclerid assemblage, mainly due to the ambiguous position of the Agelasida.

Taxa included in the analysis are: Axinellidae-Halichondriidae-Agelasidae-Microcionidae-Myxillidae s.l.-Raspailiidae-Sigmaxinellidae-Desmacellidae - Mycalidae s.l. - Desmacididae - Spongillidae - Petrosiidae - Haplosclerida-Keratosa.

Characters used are (Table 4):

1. Anisotropic reticulation (see above).
2. Absence of spicules of any kind: an (aut-)apomorphy of the Keratosa.
3. Subtylostyles: a synapomorphy of a large group of Poecilosclerida including the Raspailiidae (although individual genera and species may have other megascleres).
4. Oxeotes only: a synapomorphy of Haplosclerida (including Petrosida) and some Poecilosclerid families as Spongillidae and Desmacididae. The present content of the latter family accepts also style-bearing genera (e.g., *Esperiopsis*), but these may be better placed in other families. In fact, the family Desmacididae may be entirely artificial and better be abandoned.
5. Presence of special ectosomal megascleres (subtylostyles, (aniso-)tornotes, tylotes or strongyles), differing from those of the choanosomal skeleton: a synapomorphy of Microcionidae, Raspailiidae, and Myxillidae.
6. Acanthostyles: a possible synapomorphy of the same group of families; a probable homoplasy is found in Agelasidae (and sclerosponge Ceratoporellidae), since the acanthostyles are provided with verticils of spines.
7. Sigmata: a plausible synapomorphy for most of the studied taxa with the exception of Axinellidae, Halichondriidae, Agelasidae, and Keratosa. If so,

**Table 4.** Halichondrida-"Ceractinomorpha" taxon-character matrix

| | Outgroup | Axinellidae | Halichondriidae | Raspailiidae | Microcionidae | Myxillidae | Sigmaxinellidae | Desmacellidae | Mycalidae | Spongillidae | Desmacididae | Petrosida | Haplosclerida | Keratosa | Agelasida |
|---|---|---|---|---|---|---|---|---|---|---|---|---|---|---|---|
| 1. Anisodictyal reticulation | 0 | 1 | 1 | 1 | 1 | 1 | 1 | 1 | 1 | 1 | 1 | 0 | 1 | 1 | 1 |
| 2. Spicules absent | 0 | 0 | 0 | 0 | 0 | 0 | 0 | 0 | 0 | 0 | 0 | 0 | 0 | 1 | 0 |
| 3. Subtylostyles | 0 | 0 | 0 | 1 | 1 | 1 | 1 | 1 | 1 | 0 | 0 | 0 | 0 | 0 | 0 |
| 4. Oxeotes only | 0 | 0 | 0 | 0 | 0 | 0 | 0 | 0 | 0 | 1 | 0 | 1 | 1 | 0 | 0 |
| 5. Special ectosomal megascleres | 0 | 0 | 0 | 1 | 1 | 1 | 0 | 0 | 0 | 0 | 0 | 0 | 0 | 0 | 0 |
| 6. Acanthostyles | 0 | 0 | 0 | 1 | 1 | 1 | 0 | 0 | 0 | 0 | 0 | 0 | 0 | 0 | 1 |
| 7. Sigmata | 0 | 0 | 0 | 0 | 0 | 1 | 1 | 1 | 1 | 0 | 1 | 1 | 1 | 0 | 0 |
| 8. Chelae | 0 | 0 | 0 | 0 | 1 | 1 | 0 | 0 | 1 | 0 | 1 | 0 | 0 | 0 | 0 |
| 9. Toxa | 0 | 0 | 0 | 0 | 1 | 1 | 0 | 1 | 1 | 0 | 1 | 1 | 1 | 0 | 0 |
| 10. Vivipary | 0 | 0 | 1 | 0 | 1 | 1 | 0 | 0 | 1 | 1 | 1 | 0 | 1 | 1 | 0 |
| 11. Larva with bare posterior pole | 0 | ? | 0 | ? | 1 | 1 | ? | ? | 1 | 1 | 1 | ? | 1 | 0 | ? |
| 12. Proline + ornithine | 0 | 1 | 0 | 0 | 0 | 0 | 0 | 0 | 0 | 0 | 0 | 0 | 0 | 0 | 1 |
| 13. Isonitriles | 0 | 1 | 1 | 0 | 0 | 0 | 0 | 0 | 0 | 0 | 0 | 0 | 0 | 0 | 0 |
| 14. Taurin synthesis via cysteic acid | 0 | 0 | 0 | 0 | 1 | 0 | 0 | 0 | 1 | 1 | 1 | 0 | 0 | 1 | 0 |

it is most parsimonious to assume the secondary loss in Raspailiidae, Microcionidae, Spongillidae, and many individual genera and species.

8. Chelae: again a plausible synapomorphy for most of the studied taxa excepting the ones mentioned above; again it must then be assumed to have been secondarily lost in Raspailiidae, Sigmaxinellidae, Desmacellidae, Spongillidae and Haplosclerida.
9. Toxa: the same pattern again; presumably secondarily lost in Raspailiidae, Sigmaxinellidae, and Spongillidae.
10. Vivipary.
11. Larva with bare posterior pole: a synapomorphy (for as far as larvae are known) for Poecilosclerida and Haplosclerida.
12. Proline + ornithine derivatives: biochemical compounds found in some representatives of Axinellidae and Agelasidae (pers. commun. J.-C. Braekman, Université Libre de Bruxelles)
13. Isonitriles.
14. Taurin synthesis through cysteic acid: confined to some Poecilosclerida and some Keratosa.

The result of the taxon-character analysis (Fig. 5) shows a higher consistency count and lower tree length count for the confinement of acanthostyles to the group of Microcionids-Raspailiids-Myxillids with a probable homoplasic

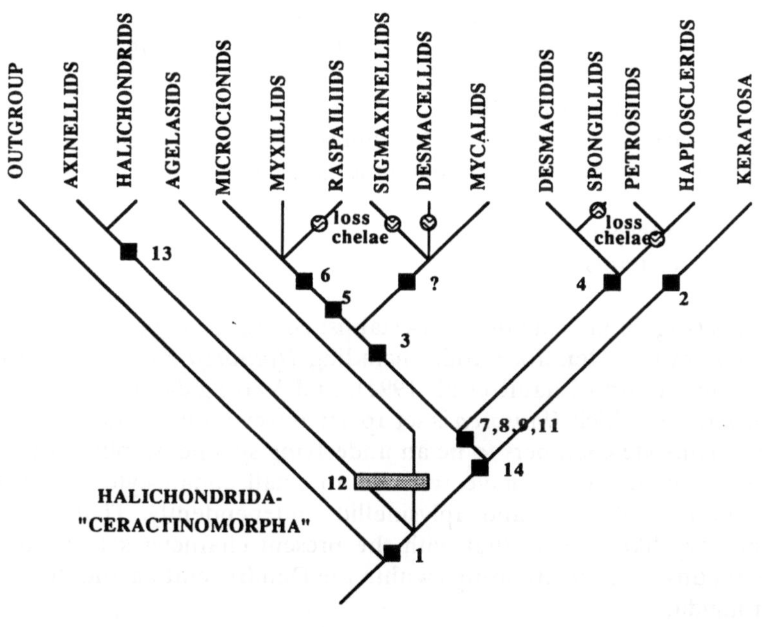


**Fig. 5.** Best cladogram as determined by MACCLADE, using various parsimony options, assumed to represent the most likely phylogenetic relationship of the Halichondrida-"Ceractinomorpha" groups of sponges. *Round symbols* indicate assumed homoplasies

development in Agelasids, as opposed to the primitive nature of this character and subsequent loss in many Poecilosclerids and all Haplosclerids (even the dollo option recognizes this as better).

There are five groups recognized in this assemblage, viz. Axinellids-Halichondrids/Microcionids-Myxillids-Raspailiids/Sigmaxinellids-Desmacellids-Mycalids s.l./Desmacidids-Spongillids-Petrosiids-Haplosclerids/Keratosa. Note that in the present classification there is no synapomorphy indicated for the group Desmacididae-Spongillidae, and that furthermore there are two unsolved trichotomies: Microcionidae-Raspailiidae-Myxillidae(s.l.) and Mycalidae-Desmacellidae-Sigmaxinellidae.

Only the Keratosa has an unequivocal synapomorphy, viz. the complete loss of spicules, for which reason we can safely examine the character distribution within the group.

## Phylogenetic Relationships Within the Keratosa

The group has been recently revised by Bergquist (1980b) and character distributions are derived from the study. Taxa included are: Dictyoceratida-Dysideidae - Dendroceratida - Halisarcidae - Ianthellidae - Aplysinellidae-Aplysinidae.

Characters used are (Table 5):

1. Larval ciliation (0 = small, overall ciliated, 1 = large, tufted)
2. Choanocyte chamber size (0 = $< 30\ \mu m$, 1 = $> 30\ \mu m$).
3. Reticulation of spongin fibres (0 = anisodictyal, 1 = dendritic, 2 = isodictyal, 3 = absent).
4. Presence of organic material within the fibers (cell masses).
5. Incorporation of sand and other inorganic foreign material within the fibers.
6. Aerophobic color change.
7. Rich in terpenes.
8. Rich in sterols.

Results (Fig. 6) indicate the well-established groups to be more or less correct: Dictyoceratida, Dendroceratida including *Halisarca* and *Dysidea* (see for the latter also Boury-Esnault et al. 1991), and Verongida. There is a Dollo best alternative in which *Halisarca* is set apart; however this assumes the large size of the choanocyte chambers to be an underlying synapomorphy for all Keratosa, which situation would have reverted to small choanocyte chambers in dictyoceratids, aplysinids and aplysinellids independently. That does not seem to be very likely. Note that with the present character set two trichotomies remain unsolved: relationships within the Dendroceratida and those within the Verongida.

There is no objective argument for the independent ordinal status of Dictyoceratida, Dendroceratida, and Verongida. In spite of their various apomorphous characters they are firmly united by the loss of spicules and the develop-

**Table 5.** Keratosa taxon-character matrix

| | Outgroup | Dictyoceratida | Dysideidae | Dendroceratida | Halisarcidae | Aplysinidae | Ianthellidae | Aplysinellidae |
|---|---|---|---|---|---|---|---|---|
| 1. Larval ciliation (0 = uniform, 1 = tufted) | 0 | 1 | 0 | 1 | 0 | ? | ? | ? |
| 2. Choanocyte chamb. > 30 $\mu$m | 0 | 0 | 1 | 1 | 1 | 0 | 1 | 0 |
| 3. Retic. (1 = dendr., 2 = isodict. 3 = absent) | 0 | 0 | 0 | 1 | 3 | 2 | 0 | 1 |
| 4. Organic pith | 0 | 0 | 0 | 0 | 0 | 1 | 1 | 1 |
| 5. Sand in fibers | 0 | 1 | 1 | 1 | 0 | 0 | 0 | 0 |
| 6. Aerophobic colour change | 0 | 0 | 0 | 1 | 0 | 1 | 1 | 1 |
| 7. Rich in terpenes | 0 | 1 | 1 | 1 | ? | 0 | 0 | 0 |
| 8. Rich in sterols | 0 | 0 | 0 | 0 | 0 | 1 | ? | 1 |

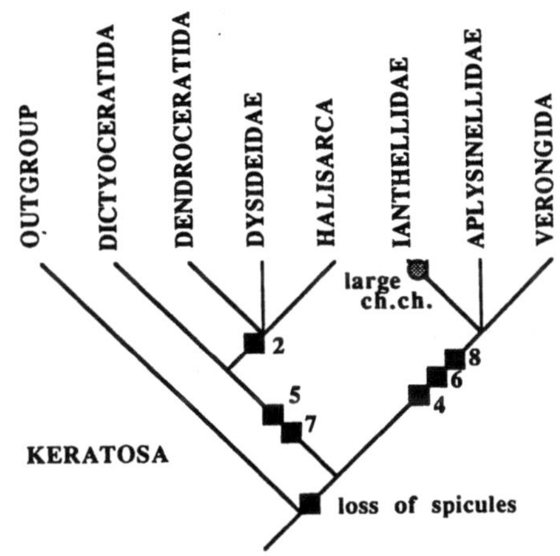


**Fig. 6.** Best cladogram as determined by MACCLADE, using various parsimony options, assumed to represent the most likely phylogenetic relationship of the Keratosa. *Round symbol* indicates an assumed homoplasy. There is a "Dollo" best alternative discussed in the text

ment of the pithed spongin fiber skeleton. Suborder status would seem to be the most proper rank.

## Conclusions

Interrelationships of the above-treated major groups are still not completely solved, and also the relationships with Homosclerophorida, Calcarea, and Hexactinellida. A consensus classification that emerges is (Fig. 7):

Hexactinellids are assumed to be the sister group of the remaining sponges (this cannot be discussed here at length, but see Reiswig and Mackie, 1983, and

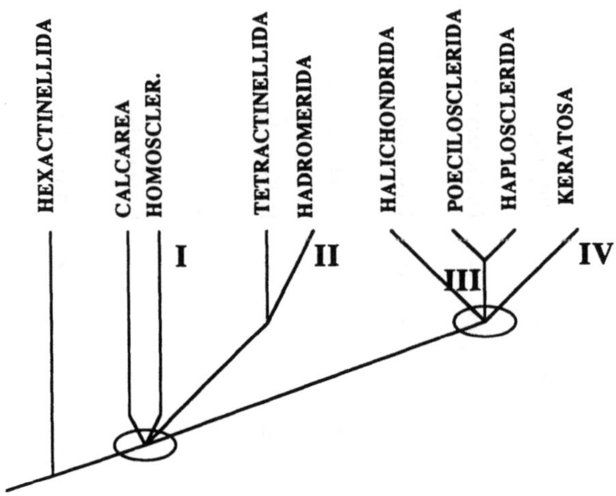


**Fig. 7.** Presumed consensus of phylogenetic relationships between major groups (*I-IV*) of Demospongia-Calcarea orders

Bergquist 1985); based on many mostly organic and cellular synapomorphies Calcarea, Homosclerophorida, and the remaining Demosponges form a monophyletic group, but internally they form a trichotomy, which in effect means that their relationship is unsolved: in a preliminary character analysis it was found that Calcarea separate from Homosclerophorida + Demospongiae is equally parsimonious as Calcarea + Homosclerophorida in one group; from the above-presented analysis it appeared that Tetractinellida and Hadromerida form a monophyletic group in the remaining Demosponges, as do the Halichondrida, "Ceractinomorpha", and Keratosa; the latter three form a polytomy largely due to the ambiguous position of the Agelasidae. I will refrain from speculating further about possible relationships until more detailed taxon-character analyses have been carried out; at that time, too, a formal classification can be proposed.

*Acknowledgements.* In developing my ideas on sponge classification I received immaterial support from Dr. W.H. De Weerdt (Amsterdam) during many hours of discussion. Dr. J.N.A. Hooper (Darwin) allowed me a preview of his voluminous thesis on axinellids. Dr. J.-C. Braekman (Brussels) contributed in discussions on secondary metabolite distribution in sponges. Dr. S.A. Pomponi and Dr. M.C. Diaz (Fort Pierce) joined me in a study of Halichondrids including some former Axinellida. Mrs. W. van Ginkel made the SEM photos. All these contributions are gratefully acknowledged.

## References

Bergquist PR (1978) Sponges. Hutchinson University Library, London

Bergquist PR (1980a) The ordinal and subclass classification of the Demospongiae (Porifera): appraisal of the present arrangement, and proposal of a new order. NZ J Zool 7:1–16

Bergquist PR (1980b) A revision of the supraspecific classification of the orders Dictyoceratida, Dendroceratida, and Verongida (class Demospongiae). NZ J Zool 7:443–503

Bergquist PR (1985) Poriferan relationships. In: Morris SC, George JD, Gibson R, Platt HM (eds) The systematics association special volume 28

Boury-Esnault N, De Vos L, Donadey C, Vacelet J (1991) Ultrastructure of choanosome and sponge classification. In: Rützler K, Hartman WD (eds) Proc 3rd Int Conf Biol Sponges Woods Hole

Burton M (1956) The sponges of West Africa. Atl Rep 4:111–147

De Laubenfels MW (1936) A discussion of the sponge fauna of the Dry Tortugas in particular and the West Indies in general, with material for a revision of the families and orders of the Porifera. Publ. Carnegie Inst Wash Pap Tortugas Lab 30:1–225

De Weerdt WH (1989) Phylogeny and vicariance biogeography of North Atlantic Chalinidae (Haposclerida, Demospongiae). Beaufortia, 39(3):55–95

Hallmann EF (1917) A revision of the genera with microscleres included, or provisionally included in the family Axinellidae; with descriptions of some Australian species. Part II. Proc Linn Soc NSW 41(3):495–552

Hartman WD (1982) Porifera. In: Parker SP (ed) Synopsis and classification of living organisms (1). McGraw-Hill, New York

Hooper JNA (1991) Character stability and affinities between Microcionidae (Poecloscierida) and Axinellida. In: Rützler K, Hartman WD (eds) Proc 3rd Int Conf Biol Sponges, Woods Hole (in press)

Lévi C (1953) Sur une nouvelle classification des Démosponges. CR Acad Sci Paris 236:853–855

Lévi C (1956) Etude des Halisarca de Roscoff. Embryologie et systématique des Démosponges. Arch Zool Exp Gén 93(1):1–181

Lévi C (1973) Systématique de la classe de Demospongiae (Démosponges). In: Grassé PP (ed) Traité Zool 3(1):577–631

Maddison WP, Donoghue MJ, Maddison DR (1984) Outgroup analysis and parsimony. Syst Zool 33:83–103

Möhn E (1984) System und Phylogenie der Lebewesen. Band 1. Schweizerbart, Stuttgart

Reid REH (1968) Microscleres in Demosponge classification. Univ Kansas Paleontol Contrib 35:1–37

Reiswig HM, Mackie GO (1983) Studies on hexactinellid sponges III The taxonomic status of Hexactinellida within the Porifera. Philos Trans R Soc Lond Biol Sci 301:419–428

Topsent E (1928) Spongiaires de l'Atlantique et de la Méditerranée provenant des croisiéres de Prince Albert ler de Monaco. Rés Camp Sci Prince Albert ler Monaco 74:1–376

Van Soest RWM (1984) Deficient *Merlia normani* Kirkpatrick, 1908, from the Curaçao reefs, with a discussion on the phylogenetic interpretation of sclerosponges. Bijdr Dierk 54:211–219

Van Soest RWM (1987) Phylogenetic exercises with monophyletic groups of sponges. In: Vacelet J, Boury-Esnault N (eds) Taxonomy of the Porifera from the N.E. Atlantic and Mediterranean Sea. NATO ASI Ser G13. Springer, Berlin Heidelberg New York Tokyo

Van Soest RWM (1991) Towards a phylogenetic classification of sponges. In: Rützler K, Hartman WD (eds) Proc 3rd Int Conf Biol Sponges, Woods Hole (in press)

Van Soest RWM, Zea S (1986) A new sublithistid sponge *Monanthus ciocalyptoides* n.sp. (Porifera, Halichondrida), from the West Indian region. Bull Zool Mus Univ Amst 10(4):201–205

Wapstra M, Van Soest RWM (1987) Sexual reproduction, larval morphology and behaviour of Demosponges from the south west coast of the Netherlands. In: Vacelet J, Boury-Esnault N (eds) Taxonomy of Porifera from the N.E. Atlantic and Mediterranean Sea. NATO ASI Ser G13 Springer, Berlin Heidelberg New York Tokyo

Wiley EO (1981) Phylogenetics: the theory and practice of phylogenetic systematics. Wiley, New York

Zandee M (1987) A computing environment for cladistic analysis. Preliminaries of a user manual for CAFCA/PC. In: Hovenkamp P, Gittenberger E, Hennipman E, de Jong R, Roos MC, Sluys R, Zandee M (eds) Systematics and evolution: a matter of diversity. Utrecht University

# Lithistid Sponges from the Norfolk Rise. Recent and Mesozoic Genera

C. Levi

## Introduction

Numerous genera have been erected for Recent sponges which bear desmas, and various attempts have been made to somehow relate these taxa to the paleontological classification; this classification is based predominantly on the study of Mesozoic forms. It is based exclusively on a number of skeletal characters, including the monaxial, tetraxial, or anaxial origin of desmas, the growth and ornamentation of desmas, the organization of desmas within the skeleton, and occasionally on the type of ectosomal megascleres (dermalia). The order Lithistida Schmidt is usually subdivided into nine suborders, which contain two groups of Paleozoic species (Orchocladina Rauff 1893 and Tricranocladina Reid 1967). The others are partly or exclusively Mesozoic or Cenozoic, and these include Sphaerocladina Schrammen 1910, Rhizomorina Zittel 1878, Dicranocladina Schrammen 1924, Megamorina Zittel 1878, Tetracladina Zittel 1878, Didymorina Rauff 1893 and Helomorina Schrammen 1924. The progressive discovery of additional living species during the 19th century allowed us to associate extinct and extant forms within the suborders Tetracladina, Dicranocladina, and Megamorina in more detail. In contrast, there is still some doubt as to whether or not there are any living representatives of Didymorina and Sphaerocladina, although the genus *Vetulina* Schmidt has been attributed to the latter suborder.

The diversity of living Lithistida is relatively poor in comparison to the Upper Cretaceous fauna in particular. Nevertheless, two suborders which contain living species appear to have retained a relatively high generic diversity. These are the Tetracladina consisting of the genera *Theonella* Gray, *Discodermia* Bocage, *Racodiscula* Zittel, *Kaliapsis* Bowerbank, *Neosiphonia* Sollas, *Siliquariospongia* Hoshino, *Brachiaster* Wilson, and the Rhizomorina. The majority of the Recent species may be found on the plateaus and shelves which surround continents and islands, and species may have bathymetric distributions extending from the littoral to approximately 200 m depth. Epibathyal species may be found living at much greater depths, up to approximately 600 m. The order is known from tropical and subtropical regions. However, with the exception of the genera *Theonella* and *Discodermia*, which contain a rich diversity of species, most other genera have been recorded only rarely, and most are relatively poorly known. Similarly, the Dicranocladina have been infrequently recorded also, although the most typical

J. Reitner and H. Keupp (Eds.)
Fossil and Recent Sponges

genus *Corallistes* Schmidt is known to occur in the Atlantic, Banda Sea, and Indian Ocean. This genus is epibathyal and lives between 200 and 600 m depth. Other genera with widespread distributions, such as *Macandrewia* Gray, *Callipelta* Sollas, and *Homophymia* Vacelet and Vasseur, are also assigned to this suborder but they are placed there with much less confidence. The Megamorina are known only from two fragments, which were attributed by Schmidt and Sollas to two different genera, *Lyidium* and *Pleroma*, both epibathyal, from Cuba and Fiji respectively. In contrast, living species of Rhizomorina appear to be fairly numerous, but the homogeneity of that group is highly disputable. Two typical genera of that suborder are relatively common: *Leiodermatium* Schmidt and *Microscleroderma* Kirkpatrick. That group also includes numerous genera of Schmidt, such as *Aciculites*, *Scleritoderma*, *Gastrophanella*, *Siphonidium*, etc. Other genera with monocrepidial desmas are more uncertainly placed in the suborder. These anomalies include *Neopelta* Sollas, *Desmanthus* Topsent, *Lophacanthus* Hentschel, *Monanthus* Kirkpatrick, and *Petromica* Topsent. Finally, there are several sponge genera which possess relatively aberrant desmas, but generally these taxa are easier to place on the basis of the remainder of their spiculation. Examples include *Desmatiderma* Topsent, *Helophloeina* Topsent, *Lithochela* Burton, and *Crambe* Gray. In short, many uncertainties still exist in the classification of Lithistida, and the systematics group is still relatively poorly known since many genera have been described from only single specimens or even fragments of specimens.

## Discussion

Few principal collections of Lithistida have been studied. West Indies (Schmidt), N.E. Atlantic (Schmidt, Topsent, Carter), southern Africa (Kirkpatrick, Burton), Indian Ocean (Dendy, Burton, von Lendenfeld, Vacelet et al.), southwest Pacific (Döderlein, Sollas, Wilson). However, Reid (1967) was able to provide a comparative study on the distribution of living Lithistida. He has shown that this distribution could be interpreted correctly in relation to the geological history of the mesogean sea (Tethys or Tethys II).

From 1976 to 1985 I had the opportunity to study large and diverse populations of lithistids in the vicinity of southern New Caledonia, on the Norfolk Rise. That fauna was remarkably rich, so that only the Mesozoic strata contained comparable lithistid faunae, especially that of the Cretaceous of Europe. Most of the Recent bathyal sponges were found living in the upper layers of hard-bottom subreefs, on the margins of islands and on the high grounds of the rise. The first work on this remarkable fauna was published in 1979 (Levi 1979), although only a small proportion of its diversity was known at the time, and complete descriptions of other Lithistida species were subsequently given by Levi and Levi (1983, 1988a). Other species collected from the same region during more recent cruises (MUSORSTOM 4 and 5) are currently being studied. This material consists of the following species:

Tetracladina

Discodermiidae

*Discodermia proliferans* Levi and Levi 1983

Phymatellidae

*Aulaxinia clavata* Levi and Levi 1988a (Fig. 1D)

*Reidispongia coerulea* Levi and Levi 1988a (Fig. 1E)

*Neosiphonia superstes* Sollas 1888 (Fig. 1A)

Dicranocladina

Corallistidae

*Corallistes fulvodesmus* Levi and Levi 1983

*Corallistes multituberculatus* Levi and Levi 1983

*Corallistes undulatus* Levi and Levi 1983

*Corallistes microstylifer* Levi and Levi 1983

*Iouea moreti* Levi and Levi 1988a (Fig. 1B)

*Herengeria auriculata* Levi and Levi 1988a (Fig. 1F)

*Macandrewia spinifoliata* Levi and Levi 1983

*Callipelta punctata* Levi and Levi 1983

Neopeltidae

*Neopelta plinthosellina* Levi and Levi 1988a

Megamorina

Pleromidae

*Pleroma menoui* Levi and Levi 1983

*Pleroma turbinatum* Sollas 1888

Dorydermidae

*Anaderma rancureli* Levi and Levi 1983

Rhizomorina

*Aciculites oxytylota* Levi and Levi 1983

*Aciculites papillata* Levi and Levi 1983

*Scleritoderma camusi* Levi and Levi 1983

*Scleritoderma* sp.

*Microscleroderma herdmani* Dendy 1905

*Microscleroderma stoneae* Levi and Levi 1983

*Jereopsis graphidiophora* Levi and Levi 1983

In addition to these Pacific species, we have been able to examine the sponges collected from the southwest islands of the Philippines during the cruises of MUSORSTOM 1 and 2, found between 80–200 m depth. A number of Lithistida were discovered, consisting mostly of *Theonella*. Moreover, we also collected many interesting species known only from Wilson's (1925) descriptions of the ALBATROSS collections from the north Celebes Sea. Those species from the Philippine Islands comprise the subject of a forthcoming publication (Levi and Levi 1989). This present chapter provides an overview and comment on the taxonomy and classification of a group of Porifera which has been traditionally combined within the single order of Lithistida. That taxon is now generally regarded as consisting of a polyphyletic assemblage of species, united by a single common feature of a skeleton composed of "articulated" spicules or desmas,

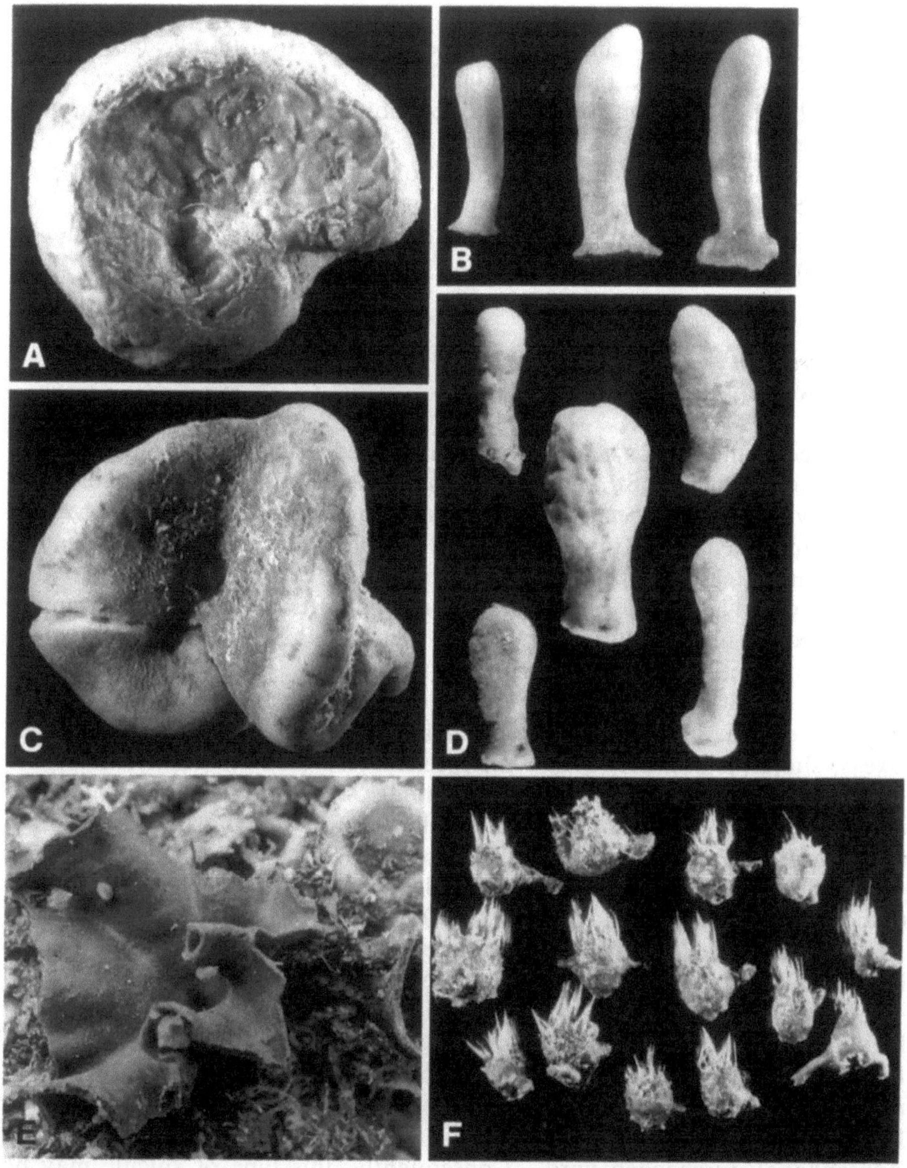

**Fig. 1.** All specimens were collected on the Norfolk Rise, Southern New Caledonia. **A** *Neosiphonia superstes* Sollas 1888. **B** *Iouea moreti* Levi and Levi 1988. **C** *Brachiaster simplex* Wilson 1925. **D** *Aulaxinia clavata* Levi and Levi 1988. **E** *Reidispongia coerulea* Levi and Levi 1988. **F** *Herengeria auriculata* Levi and Levi 1988

which produce a solid skeletal network. That structure also provides a high fossilization potential for most lithistid species, unlike most other demosponge groups. The phylogeny of the Lithistida has been frequently discussed (cf. Reid 1970; Rigby 1983), and we are unable to contribute anything new on the subject, but the pressing need to create a stable classification for these sponges still persists. I will leave for the moment any speculation on the Rhizomorina and living incertae sedis, and I will restrict myself to remarks pertaining to sponges assigned rightly or wrongly to the suborders Tetracladina, Megamorina, Dicranocladina, and Helomorina.

## *The Genera Ophiraphidites Carter 1876 and Costifer Wilson 1925: with oxeas and heloclone desmas*

Carter erected the genus *Ophiraphidites* for isolated vermiform spicules found in the sediment off Cape St. Vincent, but he also noted that those spicules undoubtedly belong to a *Pachastrella*-like demosponge from the abyssal depths. In 1903, Schrammen created the family Ophiraphididae, which grouped together taxa bearing oxeas, calthrops, and vermiform spicules called ophirhabds which are tangled and lack zygoses. On the basis of this feature, he placed the group in the Tetraxonia incertae suborder. In 1925 Wilson described the genus *Costifer*, erected for a specimen dredged from the northwest of the Celebes Sea, which possessed a vasiform growth form and a principal skeleton composed of monocrepidial desmas 2–2.5 mm long, occurring singly or in articulated bundles. Oxeas, short plagiotriaenes, and microscleres are also characteristic of the species, which Wilson placed in the Pleromidae. An examination of the holotype of *Costifer vasiformis* Wilson showed that the spicules were thick, hypersilicified, vermiform, with protuberances and depressions. The axial filament (the "crepis" or crepidial body) is always visible along the length of the spicules, surrounding the coaxial grooves. Homologous spicules found in fossil forms (viz. Helomorina: *Isoraphinia* Zittel, *Heloraphinia* Schrammen, *Pachycoton* Schrammen) are called heloclones and the affinities between the two are obvious.

*Costifer vasiformis* appears to be a living representative of the fossil Helomorina, which was extant during the Jurassic and the Cretaceous in Europe. In addition to their heloclones, the Helomorina have dichotriaenes as superficial megascleres. In his diagnosis of Helomorinidae, Schrammen commented that the heloclone "in seiner ganzen Länge von einem in der Regel gut sichtbaren Achsenkanal durchzogen wird". Consequently, this type of spicule may be regarded as being closer to an oxea than to a desma, since it has a short crepis in which the growth at the extremities is not immediately defined but depends on the prevailing environmental conditions which control the zygoses. The deformations in the cavities or bumps on the heloclones probably develop during the second phase of growth in thickness of the spicule, and this may be related to the effects produced by contiguity of neighboring spicules. If *Costifer* is closely related to Helomorina, as I propose here, following Reid, and is not a Megamorina Pleromidae, then its microxea microscleres and streptosclere asters indicate that the affinities of all these sponges lie with the Pachastrellidae demosponges in part, and with the whole of the Lithistida regrouped by Lagneau-Hérenger (1962) under the name Tetralithistida.

Among the sponges dredged from the southeast of New Caledonia at 310 m depth, we discovered a curved lamellate specimen, in which the skeleton was intermediate between both *Costifer* and *Ophiraphidites*. That skeleton was composed of interlaced vermiform oxeas bearing continuous axial canals. Many had irregular diameter and deformations, although not as well marked as heloclones. In fact those spicules are independent; their deformations occurred more frequently at the extremities, and one could also see some depressions in cupules. The remainder of the spiculation was the same as a Pachastrellidae, containing microxeas in bundles and streptoscleres. It seems probable that numerous intermediate conditions must exist between oxeas and heloclone desmas.

### *The Genera Brachiaster Wilson, Pachastrella Schmidt and Propachastrella Schrammen: with Triaenes and Tetraxial Desmas*

In the collection of sponges obtained by the ALBATROSS expedition from the northwest of the Celebes Sea, Wilson (1925) found a macerated fragment which lacked microscleres, in which the skeleton was composed of small dichotriaenes, dichomesotriaenes, and other spicules which were more complex than either dicho- or dichomesotriaene desmas. He named the species *Brachiaster simplex* and placed it in Lithistida incertae sedis. We recollected *B. simplex* from a station in the southwest of the Philippine Islands (Fig. 1C). We initially considered that the species was something like a *Pachastrella* with very large triaenes, but the principal skeleton was identical to that of *Brachiaster*, of which we have reexamined the holotype. Therefore this specimen is remarkable because it contains all the intermediate stages between small mesotriaenes, which are very abundant in the sponge, near the surface, and large mesotriaenes becoming articulated desmas. From a certain length, approximately 250–300 μm, the clades of the mesotriaenes become divided, whereas the deuteroclades remain very short and sometimes the division is immediate. At that stage the growth of axial filaments ends, and any increment in length depends on a corresponding increase in spicule thickness around the aborted or completed axis, similar to growth around the typical crepis of a desma. The remainder of the spiculation is similar to that of *Pachastrella* Schmidt, and *Brachiaster* could perhaps be considered to be a *Pachastrella* with desmas and mesotriaenes. These latter spicules have many characteristics found in the enigmatic genus *Triptolemus* Sollas, which Topsent (1902) compared to *Pachastrella* on the basis of examination of some mesodichotriaenes in *P. abyssi*. These have since been found in *Pachataxa* also (Levi and Levi 1983).

Schrammen (1910) established the genus *Propachastrella* for *Pachastrella primaeva* Zittel 1878. The principal spiculation in that species consists of calthrops, which are also capable of having fairly large intraspecific variability. The spicules figured by Hurcewicz from a specimen in the lower Campanian of Pologne, and by Moret from a specimen collected from Santonian in Provence show a tendency to bifurcate. The short deuteroclades seem to undergo an onset of desmogenesis and these vasiform, foliaceous sponges also resemble large specimen of *Pachastrella* from the Philippine Islands.

### *The Genera Neopelta Schmidt and Plinthosella Zittel*

The comparison between these two genera is important in understanding the systematic position of Lithistida bearing superficial siliceous discs. Sollas (1888) created the family Neopeltidae for Lithistida with monocrepidial desmas, monocrepidial ectosomal discs, microxeas, and streptoscleres. Sollas suggested that *N. imperfecta* from Barbados had a skeletal structure which was very similar to that of the Tetracladina, although the desmas of that species were strictly monocrepidial. Schmidt (1879) also noted that the thick ectosomal discs appeared to resemble sculptures. We collected a sponge from the Norfolk rise which was identical to *Neopelta*: *N. plinthosellina*, but which had monocrepidial desmas of complex morphology (flattened close to the surface of the sponge and more branching in deeper layers). The genus *Plinthosellina* was initially created for rare fossil lithistids containing siliceous plaques in the superficial region, and Zittel observed tetracrepidial desmas in material of *P. squamosa* Zittel. However, since Zittel's period, many authors have described other species and from those results it is possible that the tetracrepidial structure of desmas is erroneous. This is the case for *P. massiliensis* Moret, *P. cenomanensis* Moret, *P. acanthodes* Defretin-Lefranc, *P. punctata* Lagneau-Hérenger, and *P. elegans* Hurcewicz; their desmas are often described as resembling tetracrepidial forms, although that is not always easy to demonstrate, and their morphology is equally as variable as those desmas found in *N. plinthosellina*. A comparison between the desmas of Mesozoic species and those of *Neopelta* needs to be redone, and the existence of the family Plinthosellidae Schrammen amongst the Tetracladina requires re-evaluation.

### *The Genera Corallistes Schmidt and Iouea de Laubenfels (= Phalangium Schrammen): Corallistidae or Pachinionidae*

The family Corallistidae Sollas contains Lithistida with ectosomal triaenes and tuberculate monocrepidial desmas. Lagneau-Hérenger (1962) considered that the suborder Dicranocladina was a synonym of Corallistidae, and proposed that the junior name should therefore be suppressed. At that time the suborder contained only the family Pachinionidae Schrammen 1924, to which Lagneau-Hérenger added two families of uncertain affinity: Acrochordoniidae Schrammen and Phrissospongiidae n.fam. Under this scheme, the genus *Corallistes* would fit with the Pachinionidae. The discovery of *Iouea moreti* Levi and Levi (Fig. 1B) provides additional evidence. This species is very similar to *Phalangium* and *Pachinion*, being a typical Pachinionidae with dichotriaenes at the surface and streptoscleres, similar to the Corallistidae. If one considers that this combination of spicules is a good character to discriminate the family, then the family Corallistidae (= Pachinionidae Schrammen) should be retained. The genus *Corallistes* Schmidt contains species with dichotriaenes bearing spined or indented clades. It includes *C. nolitangere* Schmidt, *C. thomasi* Sollas and *C. microstylifer* Levi and Levi. Similar dichotriaenes with spined clades have been described in some fossil forms also. The genera *Gignouxia* Moret (leaf-like or conical) and the genus *Phrissospongia* Moret (considered by Moret to be very close to the genus *Phalangium*) both have dichotriaenes bearing small or large spines. Living species of that group should be included in the genus, or subgenus, *Gignouxia*. Those two genera are Corallistidae, whereas the family Phrissospongiidae proposed by Lagneau-Hérenger is still not clearly defined.

## *Remarks on the Classification*

The majority of paleontologists accept, to a large extent, the supergroup Lithistida, which is related to the Choristida as defined by Sollas. Lagneau-Hérenger concurred with that opinion and further proposed that the suborders Tetracladina, Megamorina, and Dicranocladina (eventually Helomorina) should be reunited in the order Tetralithistida. The zoological classification of Sollas (1888) places corresponding living or Recent species in three groups: Hoplophora, Triaenosa, and Rhabdosa. It is perfectly clear that with the exception of desmas, all these living and fossil sponges have shared characters, of which the major one is the possession of streptosclere microscleres. This is why Reid (1963) proposed removing the traditional barrier between Choristida and Lithistida in creating the two new orders, Streptosclerophorida and Euasterophorida.

The Streptosclerophorida are divided into four suborders with or without desmas: Tetracladina, Megamorina, Dicranocladina, Streptosclerina (= Asterostreptidae Topsent 1902), and the remaining Helomorina incertae sedis. In our opinion this classification should be adopted. It is the first move towards a new classification of Porifera with desmas. At first view the Tetracladina and Streptosclerina are the closest suborders, since they have tetraxonic principal spicules. As we have already noted from the study of the Philippines specimen of *Brachiaster*, this confirms the very close relationship between this sponge with desmas and *Pachastrella*. But in having calthrops and lacking cortical triaenes, the Pachastrellidae appears to be sufficiently different from the Tetracladina, in which the tetraxial crepis is much shorter. It is more difficult to say whether that difference is simply one of minor biometric variation, or if it represents two different modes of development of desmas.

The suborder Tetracladina contains three families of living species.

1. Phymatellidae Schrammen (= Jereidae and Aulaxiniidae), with tetraclone desmas, with smooth clones and with ectosomal dichotriaenes. Examination of microscleres of *Neosiphonia* (Fig. 1A) and *Aulaxinia* (Fig. 1D) show that there exists in this family two categories of streptoscleres, metasters, spirasters, or amphiasters.
2. Theonellidae Lendenfeld (= Discodermiidae Schrammen), with large te-

traclones, either corrugated wart-like or sometimes smooth, and ectosomal phyllotriaenes (or discotriaenes). Microscleres are either streptosclere metasters and microstrongyles, or microstrongyles and spined microxeas.
3. Phymaraphinidae have tetraclone desmas in which the apex is ornamented with a crown of large tubercles, and ectosomal phyllotriaenes (or discotriaenes). Living *Kaliapsis* have streptosclere amphiasters and microstrongyles.

The suborder Dicranocladina is represented by one typical family, the Corallistidae Sollas (= Coscinospongiidae Lendenfeld, = Pachinionidae Schrammen) (cf. Brimaud and Vachard 1986). The desmas are monocrepidial and tubercular, often tridentate. The ectosomal triaenes are dichotriaenes with smooth or spined clades. The microscleres are streptoscleres of two categories and sometimes spined microxeas.

The suborder Megamorina is represented by two families:
1. Pleromidae Sollas 1888, with megaclone desmas bearing terminal cupules, ectosomal dichotriaenes, and streptosclere and spined microxea microscleres.
2. Dorydermidae Moret 1926 has irregular megaclone desmas sometimes associated in compact tissues or fibers, ectosomal dichotriaenes, and streptosclere and spined microxea microscleres.

The suborder incertae sedis of Helomorina is perhaps represented by the family Isoraphiniidae Schrammen, to which is assigned the genus *Costifer* with heloclone desmas, ectosomal plagiotriaenes, and microscleres consisting of streptoscleres and spined microxeas.

The genera *Callipelta* Sollas, *Macandrewia* Gray, *Daedalopelta* Sollas and *Homophymia* Vacelet and Vasseur have in common an ectosomal layer of phyllotriaenes (or discotriaenes) and streptosclere microscleres or microrhabds. Although it is true that the desmas are always monocrepidial, there is no evidence that they are the same as true dicranoclones.

### *What Is a Desma?*

Sollas (1888) defined a desma as being an ordinary spicule in which successive layers of silica are deposited and in which growth of the axial rod abruptly terminates. Layers of silica deposited after the cessation in growth of the crepis and tubercular expansions become asymmetrical and completely independent of the crepis. The peripheral silica forms the epirhabd of monaxial desmas and the epactine of tetraxial desmas. Adjoining desmas produce a complex skeletal network via entanglement of their extremities, which is aided by their lateral expansions or tubercles.

Observations on numerous Lithistida show that during the course of development the free desmas are particularly numerous in the subectosomal region, lying near the surface of the sponge. These desmas are progressively incorporated into the existing rigid skeleton, and each becomes adapted in shape according to the local spatial structure. Thus the process of organization of desmas in a rigid skeleton poses questions different than those concerning skeletons which contain "normal" spicules. In such skeletons longitudinal growth of "normal spicules is achieved by elongation of the axial filaments, which is usually very rapid and produces a spicule of fixed geometry, whereas true "articulation" between neighboring spicules does not exist. Any apparent "articulation" (or more precisely, rigidity) is solely due to the development of spongin binding spicules together at their nodes. By comparison, in desmas there is a simultaneous growth in length and thickness of the spicule, which follows after the cessation in development of the axial filament, and deposit of this "supplementary" silica is no longer guided by the form of the filament but by local spatial restrictions. One can easily observe all these stages in the terminal growth around the tubercule of a desma clone, or on the epirhabd of neighboring desmas.

Sollas (1888) has presented excellent descriptions and figures on the structure of the axis of these clones, which resemble the stratified pith in spongin fibers of Dendroceratida. However, two problems remain to be resolved: the process of consecutive silicification which follows the arrest in growth of the crepis, and the initial microstructure of desmas.

Sollas considered that secondary (and perhaps also primary) silicification of desmas was due to the action of numerous sclerocytes, functioning like spongocytes of the horny sponges. In "normal" spicules growth in thickness is produced by concentric deposits of silica deposited around the axial filament, and in this respect the difference between the two modes of development is not so fundamental, because in desmas the crepis is also surrounded by numerous siliceous layers, which together comprise nearly the entire structure of the spicule. The true difference, however, lies in the mechanism of clonal elongation, beyond the extremities of the crepis. Examples of these can be seen around the organic grains, guided by the discontinuous movement and abnormalities in sclerocytes (such grains are often visible in the axial pith of clones). In megaclones and dicranoclones of *Pleroma* and Corallistidae one often sees heterogeneous and heavily pigmented clonal pith, of which the nature and origin are still unknown.

In relation to microstructure, Sollas also noted that young desmas (in *Theonella* for example) have one surface which is typically eroded, suggestive of the presence of a siliceous network, which ultimately becomes solid and smooth. In *Pleroma* these young desmas are composed of a layer alternating between smooth and granular silica, in which the smooth silica occurs externally. In fact in all Lithistida that we have observed which contain tetraclones, megaclones, or dicranoclones, the young desmas always have a microgranular eroded structure on the surface around the crepis and on the rough clones. The structural transformation is progressive and the appearance of alternating opaline layers corresponds to the progressive rigidity of spicules (which also conveniently form centrifugal layers). The dynamics of this microstructural evolution requires investigation.

*Acknowledgments.* I thank Dr. J.N.A. Hooper for translating this paper.

## References


Brimaud C, Vachard D (1986) Les Spongiaires siliceux du Tortonien des Bétiques (Miocène de l'Espagne du sud), espèces nouvelles ou peu connues. I. Choristides et Lithistides. Bull Mus Natl Hist Nat Paris (4) 8 (C,3):293–341

Burton M (1929) Descriptions of South African Sponges collected in the South African marine Survey. II. The Lithistidae. S Afr Fish Mar Biol Surv Rep 7:1–12

Carter HJ (1876) Descriptions and figures of deep sea Sponges and their spicules from the Atlantic Ocean dredged up on board HMS Porcupine chiefly in 1869. Ann Mag Nat Hist (4) 18:458–479

Defretin-Lefranc S (1958) Contribution à l'étude des Spongiaires siliceux du Crétacé supérieur du Nord de la France. Thèse Fac Sci Lille, pp 173

de Laubenfels MW (1955) Porifera. In: Moore RC (ed) Treatise of Invertebrate Palaeontology, part E. Geol Soc Am and Univ Kansas, pp 21–112

Hurcewicz H (1966) Siliceous Sponges from the upper Cretaceous of Poland I Tetraxonia. Acta Palaeontol Pol 11:15–129

Kirkpatrick R (1902) Descriptions of South African Sponges. Part 2. Mar Invest S Afr Dept Agric 1:219–232

Lagneau-Hérenger L (1962) Contribution à l'étude des Spongiaires siliceux du Crétacé inférieur. Mém Soc Géol Fr NS 41:1–262

Levi C (1979) The Demosponge Fauna from the new Caledonian area. Proc Int Symp mar biogeogr Evol in the Southern Hemisphere. NZ Oceanogr Inst: 307–315

Levi C, Levi P (1983) Eponges Tetractinellides et Lithistides bathyales de Nouvelle Calédonie. Bull Mus Natl Hist Nat Paris (4) 5 (A,1):101–168

Levi C, Levi P (1988) Nouveaux Spongiaires Lithistides bathyaux à affinités crétacées de la Nouvelle Calédonie. Bull Mus Natl Hist Nat Paris (4) 10 (A,2):241–263

Levi C, Levi P (1989) Spongiares (Musorstom 1 & 2). In: Forest J (ed) Resultats des Campagnes Musorstom, vol 4. Mém Mus Natn Mist Nat, (A) 143:25–103

Moret L (1925) Contribution à l'étude des Spongiaires du Crétacé supérieur français. Mém Soc Géol Fr NS: 1–332

Mostler H (1976) Poriferen Spiculae der alpinen Trias. Geol Paläontol Mitt Innsbruck 6(5):1–42

Rauff H (1893–1894) Palaeospongiologie. Palaeontographica 40:1–346

Reid REH (1963) Preliminary notice of a classification of the Demospongia. Ir Nat J 14(5):90–94

Reid REH (1967) Tethys and the zoogeography of some modern and Mesozoic Porifera. Publ Syst Ass 7:171–181

Reid REH (1970) Tetraxons and demosponge phylogeny. In: Fry WG (ed) The biology of Porifera. London, Academic Press, pp 63–89

Rigby JK (1983) Fossil Demospongia. In: Rigby JK, Stearn W (eds) Sponges and spongiomorphs. Univ Tennessee Dept Geol Sc Stud Geol 7:12–37

Schmidt O (1870) Grundzüge einer Spongien Fauna des atlantischen Gebietes. Engelmann, Leipzig, pp 1–88

Schmidt O (1879–1880) Die Spongien des Meeresbusen von Mexico. Fischer, Jena

Schrammen A (1910) Die Kieselspongien der oberen Kreide von Nordwestdeutschland 1 Tetraxonia, Monaxonia and Silicea incertae sedis. Palaeontogr Suppl 5:1–175

Schrammen A (1924) Kieselspongien der oberen Kreide von Nordwestdeutschland 3. Monograph Geol Paläontol Tübingen 1(2):1–151

Sollas WJ (1888) Tetractinellida. Rep on the Sci Res HMS Challenger Exped Zool 25:1–458

Topsent E (1902) Les Asterostreptidae. Bull Soc Sci Med Ouest Fr 11(2):1–18

Vacelet J, Vasseur P (1971) Eponges des récifs coralliens de Tulear (Madagascar). Tethys Suppl 1:51–126

Wilson HV (1925) Silicious and horny sponges collected by the US Fisheries Steamer Albatross during the Philippine expedition 1907–10. Bull US Nat Mus 100(2):273–532

Zittel KA v (1878) Studien über fossile Spongien. 2 Lithistidae. Abh Bayer Akad Wiss 13:65–154

# Evolution of Paleozoic Heteractinid Calcareous Sponges and Demosponges – Patterns and Records

J.K. Rigby

## Introduction

Paleozoic sponges show many of the classic evolutionary patterns documented in other fossil groups, even though the sponge record is limited (Fig. 1). For example, shifts from thin-walled to thick-walled skeletons show well in both calcareous Heteractinida and in siliceous Demospongea. Increasing skeletal complexity also shows well in both groups. Similar patterns are also evident in the Hexactinellida, but these sponges will be treated elsewhere.

## Heteractinida

The Heteractinida are a relatively minor, exclusively Paleozoic, order of calcareous sponges (Rigby 1983c). These sponges have their principal record in North America and Europe but have been reported also from South America, Australia, Africa, and Asia, largely as isolated spicules. Except for the recently described Cambrian *Jawonya* from Australia (Kruse 1987), the order is basically known from the North American and European cratons (Rigby 1983c). Most genera and species are known only from their type areas, except for *Astraeospongium*, *Regispongia*, and *Eiffelia*, which have somewhat broader geographic ranges.

For many years the oldest heteractinid calcareous genus known was *Eiffelia* (Walcott 1920). It was initially described from the famed Burgess Shale of British Columbia in western Canada, but has also been reported from other localities in North America, Africa, and Europe, and less certainly from South America and Australia (Rigby 1983c). *Eiffelia* is a distinctly regularly spiculed, thin-walled, subspherical genus with a skeleton made of at least three ranks of uniformly placed and oriented sexiradiates (Fig. 2) (Rigby 1986a). On the basis of the early occurrence of the genus in the stratigraphic record and the apparant radiation from thin- to thick-walled sponges within the Heteractinida, two lineages were differentiated (Rigby 1983c). One leads from *Eiffelia* toward *Astraeospongium* and is the major lineage in the order. The other leads toward the Carboniferous *Zangerlispongia* (Fig. 3) (Rigby and Nitecki 1975) and the Ordovician *Toquimiella* (Rigby 1967). The latter genera, like *Eiffelia*, have skeletons that are thin-walled, well-organized, and composed of geometrically predictable and ranked sexiradiates or modified sexiradiates with central nodes of the ray axis but

J. Reitner and H. Keupp (Eds.)
Fossil and Recent Sponges

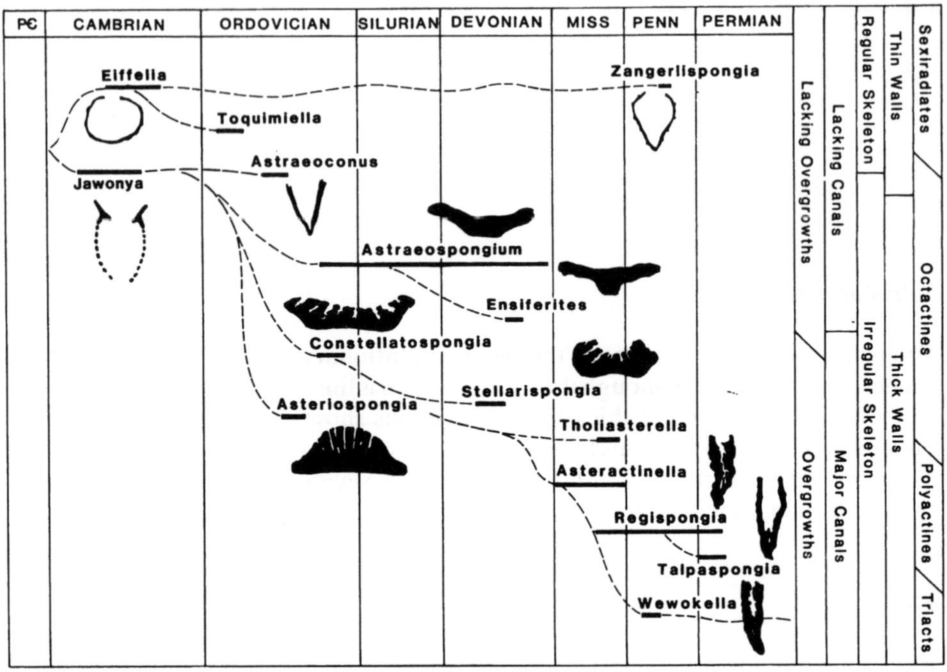


**Fig. 1.** Range chart, growth form, and possible evolutionary development of genera in the Heteractinida. Ranges are shown as *solid black lines*. *Dashed lines* indicate possible relationships. *Outline drawings* are generalized and not to scale. (After Rigby 1983c)

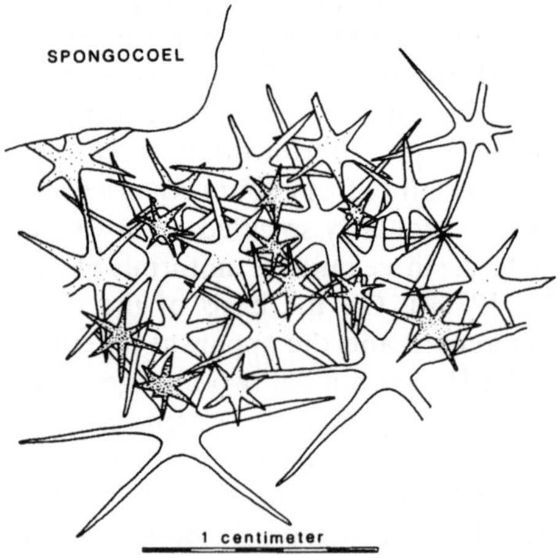


**Fig. 2.** Camera lucida drawing of the skeleton of *Eiffelia globosa* Walcott 1920, from the Middle Cambrian Burgess shale of Canada. Three ranks of spicules are differentiated. First-order spicules are *unpatterned* and third-order spicules are most *closely stippled*. Regular orientation and geometrical placement characterize the genus. (After Rigby 1986a)

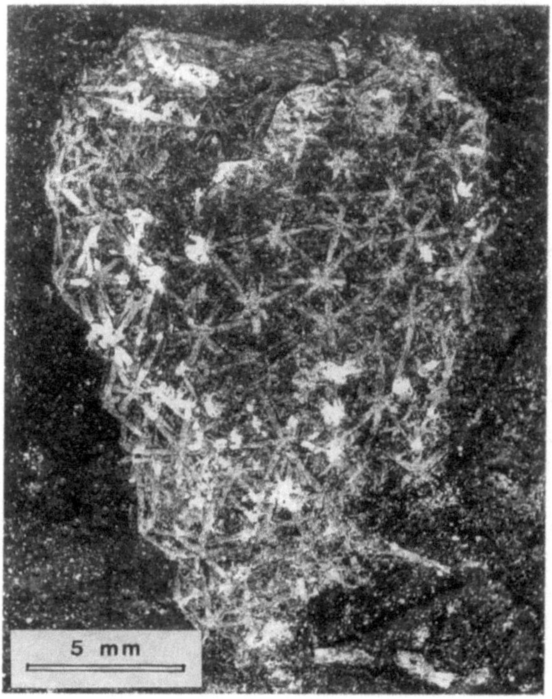


**Fig. 3.** Photomicrograph of *Zangerlispongia richardsoni* Rigby and Nitecki 1975, showing the general well-organized, predictable, thin skeleton preserved as bright pyrite in siliceous shale of the Carbon Formation. Upper Carboniferous, Illinois

no pronounced proximal or distal rays. No intermediates are known to me between *Eiffelia* and *Zangerlispongia*, for example, and *Zangerlispongia* appears to represent the end of that limited slow evolutionary trend.

A by far more diverse and locally successful lineage includes most genera within the order and contains forms with a thickened skeletal wall and diminished regularity of the skeleton from that seen in the Eiffeliidae. Sexiradiates of the thin-walled skeleton are replaced by octactines at first, and then by polyactines as different genera developed thicker walls. Discovery of new Ordovician and Devonian sponges with skeletal overgrowths and partially fused skeletons prompted modification of the evolutionary scheme proposed by Rigby and Nitecki (1975). More recent discovery of spicules in *Jawonya* (Kruse 1987) prompts additional modification of the scheme suggested by Rigby (1983c). The main evolutionary history of the Heteractinida thus principally involves genera included in the Astraeospongiidae and Wewokellidae.

In the summer of 1988, Rachel Wood, on a specimen of *Jawonya* discovered by Kruse, recognized that the sponge, initially considered to be a Middle Cambrian sphinctozoan, has a skeleton made of numerous, small, and irregularly oriented sexiradiates and modified octactines; it is a heteractinid. Their discovery raises the question whether the earliest sponges of the order really were the regular thin-walled *Eiffelia* or whether they were sponges with thin porous walls made of irregularly oriented and spaced small spicules, as in *Jawonya*.

*Eiffelia* is visualized as the stem for the slowly evolving *Zangerlispongia* line, but *Jawonya* is now considered as the likely predecessor of *Astraeoconus* Rietschel (1968) and, ultimately, of the remainder of the Astraeospongiidae and Wewokellidae (Fig. 1). *Jawonya* and *Astraeoconus* both have thin walls built of many layers or of felted, densely packed, small octactines or octactine-derived spicules that lie close together in an unfused, but irregularly oriented, structure. *Jawonya* and *Astraeoconus* could well be placed in a separate family from *Astraeospongium* and genera considered by Rigby (1983c) as members of the Astraeospongiidae.

*Astraeospongium* is one of the most widely recorded sponges within the heteractinids (Richardson 1950). It is a saucer to low bowl-shaped or low obconical sponge with thick walls and a skeleton composed of a felted mass of coarse irregularly oriented octactines (Roemer 1860). The thick walls generally lack canals. The octactines are unmodified, other than having aborted proximal and distal rays in dermal areas.

*Astraeospongium* (Fig. 4) and its close relative, *Ensiferites* (Reimann 1945), generally lack canals and contrast with other Ordovician and younger genera within the family in having smooth-rayed octactines that lack overgrowths. *Ensiferites* is the more specialized of the two (Rigby et al. 1979) and has a

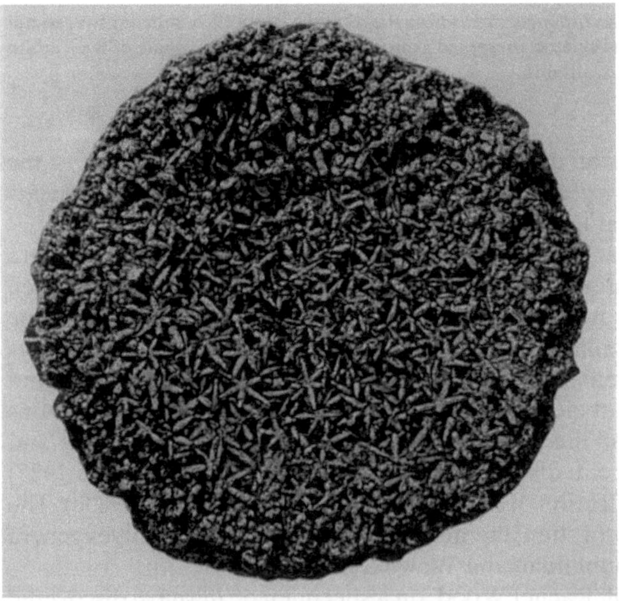

**Fig. 4.** *Astraeospongium meniscus* Roemer 1848, from the Silurian Brownsport Formation of Tennessee. Shallow saucer-shape is characteristic, as is the mat of octactine and octactine-derived spicules, here moderately well oriented on the upper surface. The sponge is approximately 6 cm in diameter

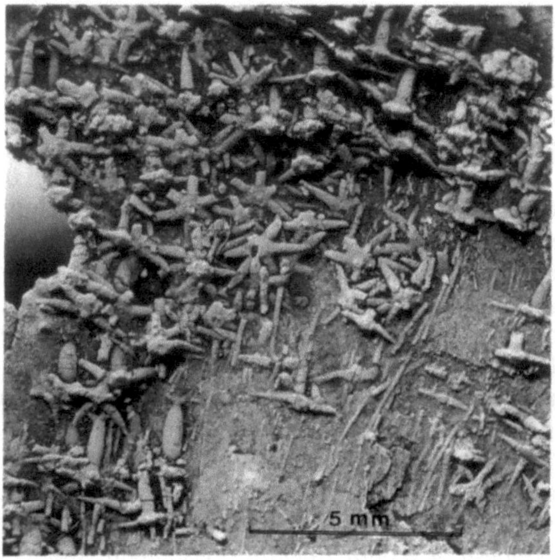


**Fig. 5.** *Ensiferites wattlesi* Reimann 1945, showing the relatively delicate proximal rays and robust distal rays that terminate in numerous supernumerary rays at the tips of the spindles, perhaps best seen at the *lower left*. Approximately ×5

thickened armor-like dermal layer of coarse modified octactines with thin long proximal rays (Fig. 5). *Astraeospongium* and *Ensiferites* appear to represent another dead end within the Heteractinida.

The other three genera, *Asteriospongia*, from the Ordovician of eastern Canada (Rigby 1977a), *Constellatospongia*, from the Ordovician of south-central Canada, near Hudson Bay (Rigby 1977b), and *Stellarispongia* (Fig. 6), from the Devonian of Arctic Canada (Rigby 1976), might well be separated into a distinct family because they contrast with the simpler skeletons of *Astraeospongium* and *Ensiferites* in having prominent canals and various degrees of calcareous overgrowth on the basic octactine spicules in their irregular skeletons. These sponges appear to be transitional into the Wewokellidae of the later Paleozoic, probably through a form like *Constellatospongia* (Fig. 1). The thick-walled *Asteriospongia*

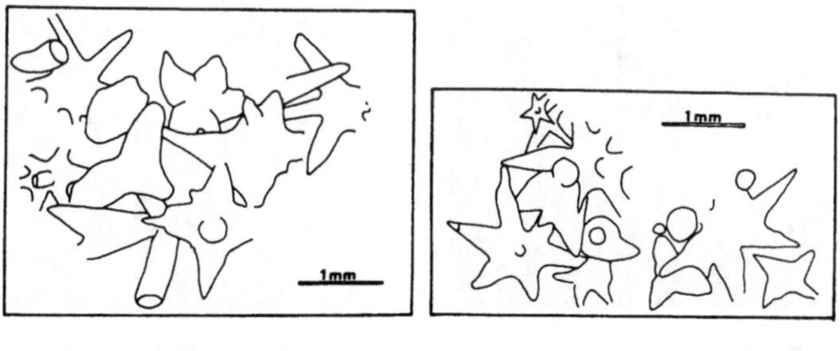


**Fig. 6.** Camera lucida drawings of spicules of *Stellarispongia asteria* Rigby 1976, from the Lower Devonian, Ellesmere Island, Canada. Overgrowths have produced large spicules and mask their initially sharply octactine base. (After Rigby 1976)

and *Stellarispongia* have major spicule overgrowths or have growth forms that would suggest they were not in the lineage leading to the Wewokellidae. I have shown *Tholiasterella*, or a form similar to it, to be the stem for the wewokellids and as having been derived from a form like *Constellatospongia* in the diagram of phylogenies.

Spicule overgrowths are marked in the Devonian *Stellarispongia* (Fig. 6), but they probably reached their maximum development in the Pennsylvanian-Permian *Regispongia*, *Talpaspongia*, and *Wewokella* (Fig. 7) where the basic nature of the choanosomal spicules is obscured by massive grotesque overgrowths. *Tholiasterella* and *Asteractinella* (Fig. 1) were maintained as separate genera, although that separation has been questioned by some workers. Both genera were based on isolated spicules (Hinde 1888), but spicules considered to be characteristic of both have now been found in single specimens (Vandercammen 1950; Reid 1963). The two genera are maintained as separate, for now, because *Tholiasterella* represents structural stages in what is considered to be the gradation from typical astraeospongid octactine-based skeletons with minimal overgrowth into those with maximum overgrowths as in the polyactine-based skeleton of *Asteractinella*, *Regispongia*, and *Talpaspongia*. Spicules in *Tholias-*

**Fig. 7.** *Photomicrographs of Regispongia contorta* (King 1943) (**A**) and *Wewokella solida* Girty 1911 (**B**). **A** from Carboniferous Chainman Shale, western Utah, ×3. **B** from the Pennsylvanian Deese Formation, Oklahoma ×4. Both figures show the gross overgrowths that mask the polyactine or triactine base of the skeletons, respectively

*terella*, figured by Hinde (1888) show low knobby or nodular overgrowths and the beginning of lateral fusion, but they remain relatively simple compared to spicules of *Asteractinella* (Hinde 1888). Polyactine spicules like those in *Asteractinella* (Fig. 8) are obscured beneath massive grotesque overgrowths in *Regispongia* and *Talpaspongia*. The latter genus is typical of late Paleozoic development where the spicular nature of the skeleton is obscured. Relatively new spicules in the dermal or outer choanosomal areas usually clearly demonstrate the basic character of the body spicules.

In addition to shifts in the nature of the basic skeleton structure, the wewokellids also changed their shapes from the dominantly saucer or low, open obconical form of other heteractinids to highly obconical or tubular. *Regispongia* (Fig. 7A) is perhaps characteristic of the Wewokellidae and generally has a twisted subcylindrical form with a skeleton made dominantly of polyactines, but with some sexiradiates and other spicules that suggest a relationship to *Tholiasterella* and *Asteractinella*. *Talpaspongia* may be merely a more tubular growth form of *Regispongia* with a broad central spongocoel. Until these genera can be separately evaluated they are maintained as distinct taxa.

*Wewokella* (Girty 1911) was named from the upper Carboniferous of North America. It has a principal skeleton composed mainly of Y-shaped triactine-based calcareous spicules (Rigby 1978a) but also includes some polyactines, all commonly obscured beneath thick calcareous overgrowths (Fig. 7B). Because of the massive overgrowths, *Wewokella* appears similar to the longer-ranging *Regispongia*, but may be differentiated by differences in the principal spicules. Even in *Wewokella*, however, there are some sexiradiates, polyactines, and other spicules that tie it to the family and order. The coated Y-shaped spicules suggest that it may be a transitional form leading to the inozoans, where such triactines function as coring spicules in tracts covered with calcareous cement.

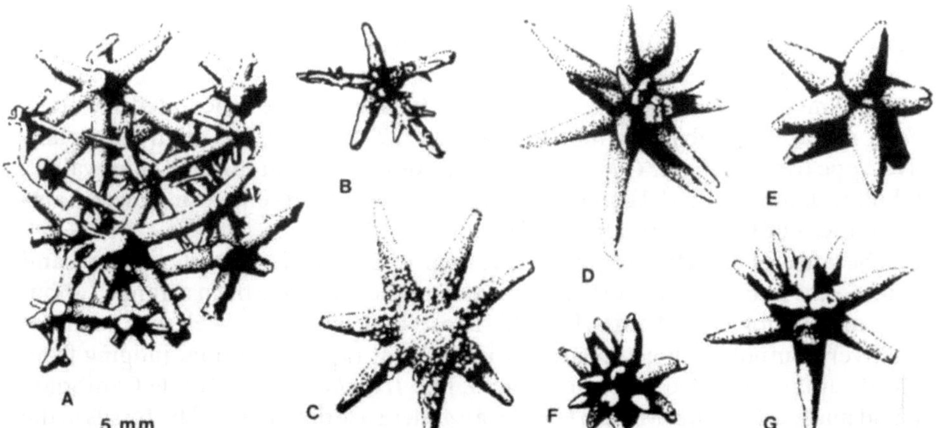


**Fig. 8.** Drawings of spicules of *Tholiasterella* (**A-C**) and *Asteractinella* (**D-G**) from the Carboniferous of Great Britain. (After Hinde 1888; Rigby 1983b). All figures are ×7

*Wewokella* should probably not be shown on the line of evolution of *Tholiasterella* and *Asteractinella* to *Regispongia* because in that line there is a marked complication of many-rayed spicules, whereas if *Wewokella* were derived from any of the other forms there appears to be prominent simplification in the spicules, although *Wewokella* also contains polyactines as minor spicules.

The record of the Heteractinida is still very spotty and the few known genera and species could be arranged in numerous possible phyletic patterns but the general evolution of the group, as outlined by Finks (1970), is that shown in Fig. 1, with the addition of those taxa described after his paper was published. Within the Heteractinida there is a clear separation into at least two broad lineages, both of which have early thin-walled stem forms. One lineage has a regular geometrical skeleton and remained as a relatively primitive-appearing thin-walled group. The other lineage showed marked thickening of the skeleton, presumably to increase the efficiency of circulation and increase the number of choanosomal chambers per unit of volume. In addition there is development of marked overgrowths in some taxa which produced grotesque modifications of the skeleton in late phases of evolution. There is also the suggestion that spicule types that first appear in the dermal or outer part of the skeleton subsequently became principal or choanosomal spicules. For example, triactines developed as dermal elements in early wewokellids and possibly became the dominant interior spicules in later forms, and perhaps developed as the principal spicule in later inozoans.

## Demospongea

The Demospongea have a long, moderately well-documented, Paleozoic history (Finks 1967b) that begins with scattered fossils in the Lower Cambrian and extends with differing degrees of completeness to the Permian and on to the Recent (Rigby 19883b). The origin of the class Demospongea is obscured in the poor record of the Lower Cambrian (Fig. 9), where only a few forms have been described (Rigby 1987b). Isolated spicules of what were probably demosponges are widespread in Lower Cambrian rocks on all continents, but only in eastern North America, Greenland, and the USSR have verifiable articulated skeletons been reported (Rigby 1987a,b; Gorjansky 1977). These are relatively primitive-appearing sponges. *Choia*, *Leptomitus* and *Hazelia* are simple thin-walled, tubular, demosponges. *Leptomitus* (Fig. 9) is well represented in Lower Cambrian rocks of the eastern United States.

Because of skeletal simplicity, *Leptomitus* is considered a stem genus and other forms known from North America, in particular, have been arranged about it in morphologic gradients (Fig. 9). Those that have thin skeletons made of relatively simple spicules apparently underwent rapid evolution, judging from the abruptly expanded record of genera, like *Hazelia*, in the Middle Cambrian. It and many additional Middle Cambrian genera are documented by fossils in the Burgess Shale in western Canada and in the somewhat later Middle Cambrian rocks of Utah (Walcott 1920; Rigby 1978b, 1983a, 1986a).

The probably keratose sponge *Vauxia* has been described from Lower Cambrian specimens from Greenland (Rigby 1987a) and the Middle Cambrian

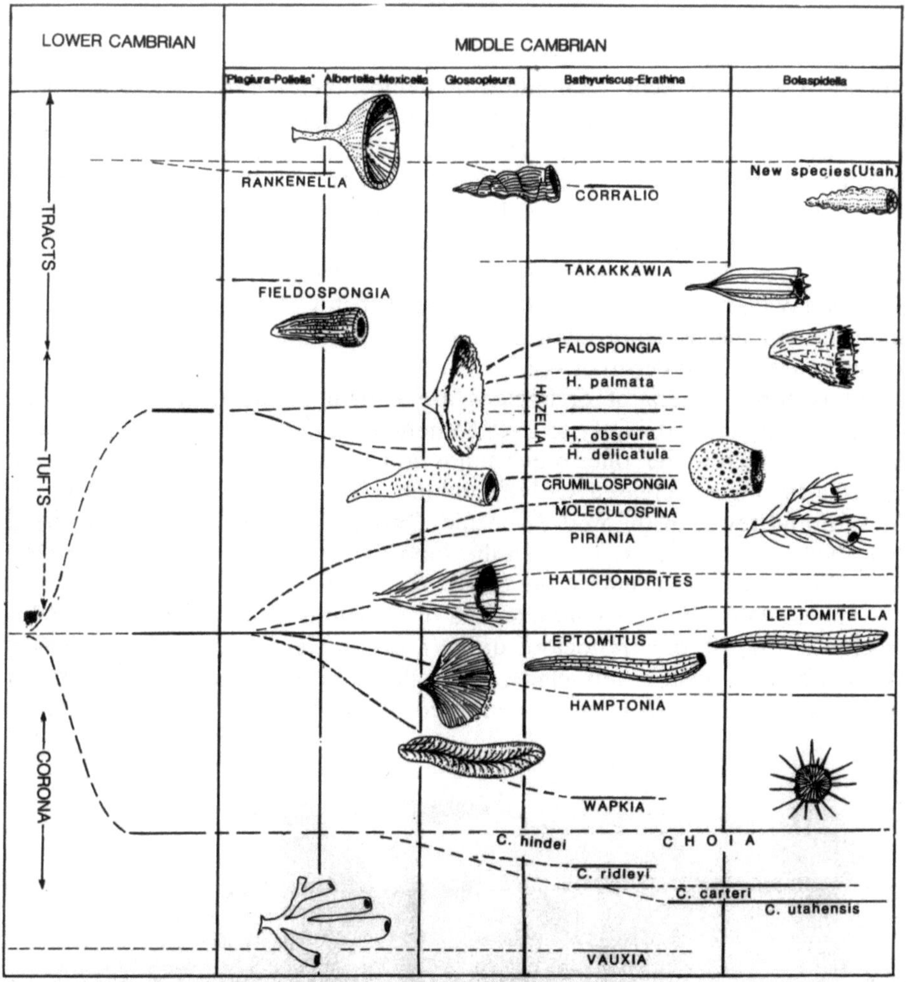


**Fig. 9.** Range chart, growth form, and possible evolutionary development of Cambrian genera in the Demospongia. Known occurrences are shown as *Solid black lines. Dashed lines* indicate possible relationships. Drawings not to scale. (After Rigby 1986a)

of western North America (Rigby 1986a). These sponges occur with *Choia hindei* in the Lower Cambrian. In a diagram constructed several years ago (Rigby 1986a) *Choia hindei* was shown as derived, largely because it is the largest *Choia* that we know. I initially concluded that there was a general size increase through the Middle Cambrian in species of *Choia*. However, newly discovered occurrences of *Choia hindei* in the Lower Cambrian of Greenland indicate that was a wrong interpretation. Figure 9 has been modified to show the new interpretation.

Similarly the Lower Cambrian occurrence of *Hazelia* necessitated some shift in interpretation of relationships of that complex.

## *Orchocladina*

The post-Cambrian record of nonlithistid demosponges of the Paleozoic is poor and most of the fossils included within the demosponges of later periods are lithistids, whose fused skeletons have been selectively preserved. Among the lithistids, several suborders are essentially limited to the Paleozoic and most of these suborders have long ranges. However, most have shorter periods where each group reached its maximum. For example, the Orchocladina have been first recorded in the Middle Cambrian by *Rankenella* from Australia (Kruse 1983), *Corralio* (Walcott 1920; Rigby 1986a) from British Columbia and by an as yet underscribed species from Utah (Fig. 9). The youngest representatives of the suborder are of Late Permian age.

Only three genera of orchoclads are known in the Cambrian but the suborder expanded abruptly to over 35 genera in the Ordovician (Rigby and Webby 1988). However, once other major lithistid sponge groups appeared, the orchoclad sponges decreased. Only approximately ten genera are known from the Silurian and fewer in the mid-Paleozoic. Only six or seven genera have been described from the Permian, when the suborder expanded slightly just before becoming extinct at the end of the Paleozoic.

The Orchocladina are characterized by skeletons made principally of dendroclones (Fig. 10) or chiastoclone desmas, with mixes of other minor spicules. Many spicules appear as tetraclads, although they are acrepid. Evolution in the Orchocladina seems to have concentrated on modifications of overall shape

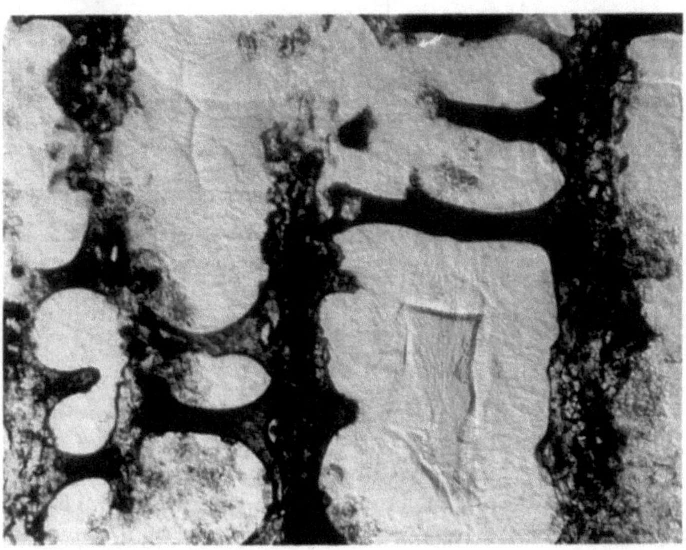

**Fig. 10.** Dendroclones of *Aulocopium* are typical of orchoclad lithistid sponges. They combine at branched ray tips to produce the vertical trabs that result in the ladder-like skeleton, from the Silurian Brownsport Formation, Tennessee, approximately ×75

and shift in canal systems, with some minor modifications of the skeletal structure.

Two groups of Orchoclads can be differentiated within the suborder. Those generally considered as anthaspidellids have relatively regular, simple skeletal structure, typified by such genera as *Calycocoelia* and *Nevadocoelia* (Fig. 11), both described by Bassler (1941) from the Early Ordovician of Nevada in the western United States. In these forms the spicules are relatively large and the canals are simply straight and radial, but in vertically stacked series. This is the major group or lineage within the Orchocladina. Individual spicules appear as "rungs" in the ladder-like structure and merge to produce complex rods that are usually the most well-preserved elements of the skeleton. These rods are commonly cored by monaxons. There seems to be, at present, no easily recognized major evolutionary trends with long-term significance. There is, however, an obvious shift in overall form from the steeply obconical to goblet shape to almost every other shape known, ranging from a simple palmate to broad open fan-shaped, encrusting, massive hemispherical or branching arborescent forms. Such variation suggests adaptations to a variety of environments. Diversity of form is shown well in the late Ordovician assemblage recently described from New South Wales in southeastern Australia by Rigby and Webby (1988).

**Fig. 11.** Characteristic Ordovician orchoclad genera from western United States. **A** *Calycocoelia typicalis* Bassler 1941, from the Ordovician Fort Pena Formation, Texas. *Side view* shows the ladder-like regular skeleton, pierced by pores of the radiating canal system. **B** *Nevadocoelia wistae* Bassler 1941. *Side view* shows the wrinkled exterior and the regular vertical skeleton. Lower Ordovician Pogonip Group, Nevada

The other less extensive lineage has a distinctly less well-organized canal system and skeletal structure. This group might be characterized by *Lissocoelia* (Bassler 1941) from the Ordovician of Nevada, or *Streptosolen* (Bassler 1941), that is also known from western North America and is also a common sponge in eastern North America.

Diversity of structure within the Ordovician genera is particularly well displayed by the diverse assemblage from New South Wales. Simple regular genera, such as *Archaeoscyphia* and *Aulocopium*, are well represented, as are the open-conical forms *Patellispongia* and *Psarodictyum*. Delicate spiculation is developed in *Malongullospongia*, where cavernous canals make the skeleton appear particularly delicate. Modification of characteristic dendroclone spicules have produced upward- and outward-radiating webbed beams in *Gleesonia*. The very regular, but cavernous *Vandonia* is characterized by an almost laminate to stromatoporoid-appearing regular skeleton. In the strongly derived *Vandonia*, horizontal elements of unusually large dendroclones are spaced uniformally with very long ray tips, which produce regular coarse trabs in the lattice-like skeleton when seen in vertical cross-section. These well-preserved silicified sponges have been described by Rigby and Webby (1988) as part of one of the most spectacular fossil sponge faunas known.

Principal development of the suborder was an abrupt, explosive, diversification in the Ordovician, with less diverse continuation into later periods of the Paleozoic of the simple, early, generalized but well-organized vase-shaped to massive sponges. The suborder became extinct near the end of the Paleozoic.

Reasons why one structural element was so successful for a long time and another less successful, or why elements that were so successful for extended periods became unsuccessful, are not known. This is particularly true when we see rediscovery of a skeletal grade or a closely approximated spicule type, particularly across a gap in the record. For example, the evident similarity of *Lithochela* (Burton 1929), a living sublithistid, to the orchocladine dendroclone skeleton is striking. However, the Orchocladina are an entirely Paleozoic group.

### *Sphaerocladina*

The suborder Sphaerocladina appeared in the Ordovician as subspherical to spherical or cup-shaped forms (Fig. 12A-B) whose skeleton is made of desmas that are acrepid sphaeroclones (Fig. 13). These spicules have a globular centrum, from which extend four to six arm-like rays, much like legs of a chair, to articulate and fuse to adjacent centra, in a generally uniformly or radially expanding skeleton. *Astylospongia* (Fig. 12A) and *Caryospongia* (Fig. 12B) are characteristic genera and were among the first of the suborder to be described. *Astylospongia* and its relatives are moderately rare forms in the Ordovician, but they are present in great numbers in Silurian rocks of North America. Literally tens of thousands of individuals have been collected from selected outcrops. The suborder had a major expansion in the Silurian of North America but is unknown from Devonian and younger rocks of that continent. Devonian genera are largely known from Australia (Pickett 1969; Rigby 1986b).

**Fig. 12.** Characteristic sphaeroclone sponges of the lower Paleozoic of North America. **A** *Astylospongia praemorsa* (Goldfuss 1826), Silurian Brownsport Formation, Tennessee. **B** *Caryospongia inciso-lobata* (Roemer 1848); Silurian Brownsport Formation, Tennessee. **C** *Camellaspongia tumula* Rigby and Bayer 1971, Ordovician Maquoketa Formation of Minnesota. **D** *Astylospongia (Palaeomanon) cratera* Roemer 1848, Silurian Brownsport Formation of Tennessee. All figures approximately ×1

Twelve genera have now been included in the suborder, which is not a major group. Their evolution appears to have been moderately conservative. Early forms are subsperical and some later genera developed goblet (Fig. 12C), subcylindrical or palmate shapes like the early orchoclads. After an initial burst of evolution in the Silurian, they then became relatively minor. The suborder probably became extinct toward the end of the Devonian. Permian sponges included in the Astylospongiidae by De Laubenfels (1955) in the Treatise of Invertebrate Paleontology are orchocladine genera. No sphaerocladines are known from the Carboniferous and only the very questionable *Ellesmerespongia* (Rigby 1970) is known from the Permian and it may well be a didymmorine rather than an astylosponge.

Like many groups of organisms, genera in the Astylospongiidae, the family of the Sphaerocladina, show phyletic size increase. Early astylosponges from the Ordovician are generally small, usually less than 1 cm in diameter, whereas those from the Late Ordovician, Silurian, and Devonian are considerably larger, particularly the subspherical species. Later members of the group also become

**Fig. 13.** Sphaeroclone spicules of *Astylospongia praemorsa* (Goldfuss 1826), with a spherical centrum and radiating rays that fuse to adjacent centra to produce the solid skeleton. Silurian Brownsport Formation, Tennessee, ×75

cup-shaped (Fig. 12C,D), or distinctly palmate to irregular. None of these genera was particularly successful, however.

## *Tricranocladina*

The suborder Tricranocladina, or Eutaxicladina, are characterized by tricranocladine spicules that articulate in what has been generally termed the "shoulder" pattern. These commonly four-rayed spicules attached to convex shoulders or around neck-like brachyomes of adjacent spicules and firmly united at ray junctions (Fig. 14). Genera included within this suborder are basically relatives of *Hindia*. This suborder is also a relatively conservative group that began in the Ordovician and lasted until near the end of the Permian. The early described Ordovician *Hindia* and Permian *Scheiia* are both spherical forms with almost ideal radial symmetry. They converge morphologically with the spherical *Carpospongia* of the Sphaerocladina. *Carpospongia* and early *Hindia* commonly occur together. *Hindia* is one of the most widely distributed Paleozoic sponges, which suggests that the spherical forms were considerably more mobile than the associated asymmetrical genera. Perhaps *Hindia* and similar-appearing sponges were floaters.

*Scheiia* appears to have evolved from the spherical *Hindia* and is differentiated largely by loss of the brachyome in the skeletal structure (Fig. 14). Finks (1971) documented differences in symmetry and spicule complements of

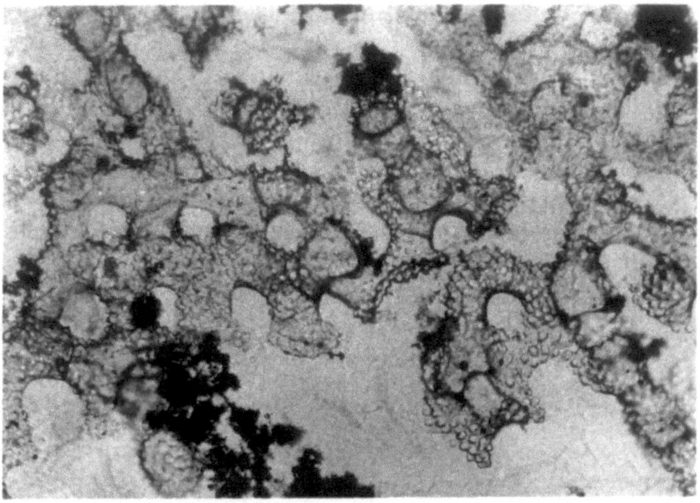

**Fig. 14.** Spicules of *Scheiia tuberosa* Tschernyschew and Stepanov, 1916, show robust tricranoclones with distinctly knobby distal surfaces and ray tips that unite with those knobby surfaces to produce a solid net. Permian Assistance Formation of Arctic Canada, ×75

*Hindia* and the Late Permian *Scheiella* and related that shift to the subsequent origin of the Mesozoic Dicranocladina. He showed one of the few known instances among the sponges where one major taxonomic group convincingly evolved into another, although, even here, there is a Triassic and Lower Jurassic gap. Finks (1971) discussed mosaic evolution of *Hindia* to *Scheiia* and to *Scheiella* and hypothesized that *Scheiella* developed from the Permian *Scheiia*, based on the then known stratigraphic distribution. It was a logical conclusion to derive the more specialized *Scheiella* from *Scheiia*. However, structure of the recently discovered Devonian *Scheielloides* and *Sadleria* in Western Australia (Rigby 1986b) shows that skeletal features which typify *Scheiella* and separate it from the earlier-known genera of the family are now known to extend back to at least the Devonian (Rigby 1986b).

Two lineages of tricranocladine sponges probably developed, one conservative line between *Hindia* and *Scheiia* and the other less conservative line, possibly from *Scheielloides* or *Sadleria* to *Scheiella* and ultimately to the dicranocladines of the Mesozoic. Discovery of the Devonian genera had little impact on the hypothesis that Finks proposed that dicranocladines developed from the tricranocladines, but merely places the lineage away from that of *Hindia* to *Scheiia*.

In general, outside of Australia, the tricranocladines were an exceedingly conservative group, but in the Late Ordovician of Australia, they demonstrated remarkable diversity (Rigby and Webby 1988). Within that Australian Ordovician assemblage an additional five genera of tricranocladine sponges range from arborescent-twiggy to tubular-cylindrical, palmate, bladed and open funnel-shaped to an unusual fenestrate genus with a form not widely developed among Paleozoic sponges, even in other lithistids. The evolutionary patterns within the suborder again show generalized sponges appearing early and persisting for a long time after specialized forms had became extinct. The Tricranocladina also show bursts of almost explosive evolution that are local and commonly of short duration.

The Australian Ordovician genera add many new growth forms previously not known within the suborder and, when combined with the sponges previously described, produce a record of remarkable parallelism between this suborder and others. For example, the genus *Palmatohindia* of the tricranocladines is remarkably similar to the associated *Pseudopalmatohindia* of the orchocladines in terms of occupying morphologic space. Both are Ordovician genera described from Australia (Rigby and Webby 1988).

## *Rhizomorina*

The suborder *Rhizomorina* of the Lithistida is characterized by rhizoclone desmas, which are like irregular monaxial desmas that have laterally directed spinose zygomes which articulate with adjacent spicules. Until a recent discovery of major sponge faunas in the Ordovician of Australia (Rigby and Webby 1988) and the Silurian of Arctic Canada, as yet unpublished, the oldest complete rhizomorine sponges known were Carboniferous *Haplistion* (Finks 1960). Record of complete sponges of the suborder, however, now extends back to the Ordovician. Only locally do sponges of the suborder play more than a moderate role in Paleozoic faunas, although they are major elements of Mesozoic and younger assemblages. Their greatest expansion, apparently, was during the Cretaceous in northern Europe. During the Paleozoic, the rhizomorines were only locally common, for example, *Haplistion* (Fig. 15) occurs only in Carboniferous and Permian rocks of North America and northern Europe. Related genera occur in the Permian reef complexes of West Texas. Finks (1960) reported that isolated rhizoclone spicules are known back to the Cambrian, but the oldest complete rhizomorine sponges known thus far are from the Late Ordovician of Australia (Rigby and Webby 1988).

Like many other groups, the Rhizomorina had an early burst of development in the Ordovician, but then played a relatively minor role through the rest of the Paleozoic. The generalized, three-dimensional gridwork skeleton of *Haplistion* and its relatives was developed early and remained as a moderately conservative structure through the entire Paleozoic history of the suborder. *Taplowia* and *Lewinia*, both described from the Ordovician of Australia, appear strikingly similar to some of the associated obconical orchocladines, even to development of a moderately ladder-like appearing skeletal net. One might argue for at least rudimentary mimicry in the Ordovician assemblages.

**Fig. 15.** *Haplistion arcticum* (Dunikowski 1884), a common Paleozoic rhizomorine sponge showing the regular subcylindrical tracts of rhizoclones making up the gridlike, common skeleton. Permian Assistance Formation, Ellesmere Island, Arctic Canada

### *Megamorina*

The suborder Megamorina appears in the Ordovician, with *Saccospongia* from North America (Finks 1967a) and *Cleifdenospongia* from Australia (Rigby and Webby 1988). Rigby and Dixon (1979) included heloclone-bearing sponges of the early Paleozoic in the Saccospongiidae and described the new Silurian genera *Eochaunactis* and *Haplistionella*. The suborder, however, remained a minor one throughout the Paleozoic and a major break separates these sponges from the late Mesozoic families. One wonders, again, if the Mesozoic spicule shapes were rediscovered from a rhizomorine or nonlithistid group of sponges.

## Conclusions

Paleozoic sponges provide good examples of different evolutionary patterns, such as, for example, the persistence of generalized forms far beyond the extinction of specialized forms; the apparent replacement of one major taxon by another with gradual shift in character; convergence and mimicry across boundaries of major groups of sponges and of sponges with bryozoans. Evidence of rediscovery of structural grade or structural type is apparent in many lineages and rediscovery of spicule patterns is possibly shown in unrelated taxa.

Reasons for spicule diversification are not understood and in most lines, origins of particular groups are obscure and equivocal. Organisms as supposedly simple as the Porifera, however, gradually evolved functioning, long-persisting, solutions to problems of water circulation, skeletal support, and attachment, and underwent spicule, canal, and shape modifications. They demonstrate repeated replacement of momentarily successful faunas by others throughout the Paleozoic record. Sponges of the Paleozoic are an interesting and instructive group of fossils.

## References

Bassler RS (1941) The Nevada Early Ordovician (Pogonip) sponge fauna. Proc US Natl Mus 91(3126):91-102

Burton M (1929) Descriptions of South African sponges collected in the South African Marine Survey, Part II – "The Lithistidae" – with a critical survey of the desma-forming sponges. Union South Africa Fisheries Marine Biol Surv Rep 7, Spec Rep II:1-12

De Laubenfels MW (1955) Porifera. In: Moore RC (ed) Treatise on Invertebrate Paleontology, Part E. Geol Soc Am Univ Kansas Press, Lawrence, Kansas, pp E21-E112

Finks RM (1960) Late Paleozoic sponge faunas of the Texas Region. The siliceous sponges. Bull Am Mus Nat Hist 120, art 1, 160 pp

Finks RM (1967a) The structure of *Saccospongia laxata* Bassler (Ordovician) and the phylogeny of the Demospongea. J Paleontol 41:1137-1149

Finks RM (1967b) Phylum Porifera Grant 1836. In: Harland WB (ed) The fossil record, a symposium with documentation. Geological Society of London, pp 333-341

Finks RM (1970) The evolution and ecologic history of sponges during Palaeozoic times. In: Fry WG (ed) The biology of the Porifera, Symp Zool Soc London 25, Academic Press, London, pp 3-22

Finks RM (1971) A new Permian eutaxicladine demosponge, mosaic evolution, and the origin of the Dicranocladina. J Paleontol 45:977-997

Girty GH (1911) On some new genera and species of Pennsylvanian fossils from the Wewoka Formation of Oklahoma. Annu Rep NY Acad Sci 21:212

Gorjansky V Yu (1977) (First remains of a sponge in the Lower Cambrian of East Siberia) (Minor Reports). Ezheg Sesoyuzni Paleontol Ova 20:274-276 (in Russian)

Hinde GJ (1887-1912) A monograph of the British fossil sponges. Palaeontographical Society, London, part 2, pp 93-188

Kruse PD (1983) Middle Cambrian "Archaeocyathus" from the Georgina Basin is an anthaspidellid sponge. Alcheringa 7:49-58

Kruse PD (1987) Further Australian Cambrian sphinctozoans. Geol Mag 124:543-553

Pickett J (1969) Middle and Upper Palaeozoic sponges from New South Wales. Mem Geol Surv New South Wales, Palaeont 16:24

Reid REH (1963) Preliminary notice of a classification of the Demospongia. Ir Nat J 14:90-94

Reid REH (1968) The Carboniferous sponge "*Doryderma*" *dalryense* Hinde (Class Demospongia) and the origin of heloclones and megaclones. J Paleontol 42:1249-1252

Reimann IG (1945) New Middle Devonian octactinellids. Paleont Contrib Buffalo Soc Nat Hist Bull 19(2):16-21

Richardson ES, Jr (1950) A Middle Devonian octactinellid sponge from New York. Fieldiana: Geology 10:79-88

Rietschel S (1968) Die Octactinellida und ihnen verwandte paläozoische Kalkschwämme (Porifera, Calcarea). Paläontol Z 42:13-32

Rigby JK (1967) A new polyactinal sponge from the Antelope Valley Formation (Ordovician) in the Toquima Range, Nevada. J Paleontol 41:511-515

Rigby JK (1970) *Ellesmerespongia feildeni*, a new Permian sponge from the Canadian Arctic. J Paleontol 44:1143-1145

Rigby JK (1976) A new Devonian heteractinid sponge from southwestern Ellesmere Island, Arctic Canada. Can J Earth Sci 13:120-124

Rigby JK (1977a) Two new Middle Ordovician sponges from Foxe Plain, southeastern District of Franklin. Geol Surv Canada Bull 269:121–129
Rigby JK (1977b) *Constellatospongia*, a new heteractinid astraeosponge from the Upper Ordovician Churchill River Group, Manitoba. Geol Surv Can Bull 269:131–137
Rigby JK (1978a) Two wewokellid calcareous sponges in North America. J Paleontol 52:705–716
Rigby JK (1978b) Porifera of the Middle Cambrian Wheeler Shale, from the Wheeler Amphitheater, House Range, in western Utah. J Paleontol 52:1325–1345
Rigby JK (1983a) Sponges of the Middle Cambrian Marjum Limestone from the House Range and Drum Mountains of western Millard County, Utah. J Paleontol 57:240–270
Rigby JK (1983b) Fossil Demospongia. In: Broadhead TW (ed) Sponges and Spongiomorphs, Notes for a Short Course, Univ Tenn Dept Geol Sci, Stud Geol 7:12–39
Rigby JK (1983c) Heteractinida. In: Broadhead TW (ed) Sponges and Spongiomorphs, Notes for a Short Course, Univ Tenn Dept Geol Sci, Stud Geol 7:70–89
Rigby JK (1986a) Sponges of the Burgess Shale (Middle Cambrian) British Columbia. Palaeontographica Canad Mon 2:105
Rigby JK (1986b) Late Devonian sponges of Western Australia. Geol Surv West Austr Rep 18:59
Rigby JK (1987a) Cambrian and Silurian sponges from northern Greenland. In: Peel JS (ed) North Greenland Lower Paleozoic paleontology and stratigraphy: Shorter Contributions. Grnl Geol Undersogelse, Rapp 132:51–63
Rigby JK (1987b) Early Cambrian sponges from Vermont and Pennsylvania, the only ones described from North America. J Paleontol 61:451–461
Rigby JK, Dixon OA (1979) Sponge fauna of the Upper Silurian Read Bay Formation, Somerset Island, District of Franklin, Arctic Canada. J Paleontol 53:587–627
Rigby JK, Nitecki MW (1975) An unusually well preserved heteractinid sponge from the Pennsylvanian of Illinois and a possible classification and evolutionary scheme for the Heteractinida. J Paleontol 49:329–339
Rigby JK, Webby BD (1988) Late Ordovician sponges from the Malongulli Formation of central New South Wales, Australia. Paleontogr Am 56:1–47
Rigby JK, Schumacher D, Meader SJ (1979) The genus *Ensiferites*, a Devonian astraeosponge of North America. J Paleontol 53:475–493
Roemer CF (1860) Die silurische Fauna des westlichen Tennessee: Eine paleontologische Monographie, Breslau, p 1–100
Vandercammen A (1950) Contribution a l'étude des spongiaires heteractinellides. Inst R Sci Nat Belg Bull 26(9):1–23
Walcott CD (1920) Middle Cambrian Spongiae, Cambrian geology and paleontology. Smithson Misc Collect 67(6):261–364

# The Fossil Record of the Haplosclerid Excavating Sponge *Aka* de Laubenfels

J. REITNER and H. KEUPP

## Introduction

Only few publications deal with Recent species of the genus *Aka* or *Siphonodictyon*, a younger synonym of *Aka* (Johnson 1899; de Laubenfels 1936; Bergquist 1965; Rützler 1971; Thomas 1972; Pang 1973). In the literature, only two publications figure and describe fossil *Aka* spicule arrangements – Müller (1978) and Reitner (1987a). *Aka* spicules are commonly found within burrows in Jurassic and Cretaceous shallow marine carbonates.

The present chapter will document the stratigraphic occurrences, systematic and phylogenetic implications, as well as ecological interpretations, of fossil *Aka* species (Fig. 1).

## Systematic Position of the Genus *Aka*

The excavating sponge *Aka* is classified within the order Haplosclerida, family Adociidae, based on the oxea spicule type and its arrangement in bundles or brushes. The spicules of the inner parts of the sponge exhibit a typical adociid arrangement. Spongin is reduced and cortical spicule tracts are arranged perpendicular to the central spicule bundles. The whole sponge lives wholly inside calcareous material. Only prominent chimneys with apertures protrude from surfaces.

## Systematics and Stratigraphic Distribution of Fossil *Aka*

### *Systematic Description of Early Carnian Aka*

*Aka cassianensis* n.sp. (Fig. 2a-d)

Derivatio nominis: After the Lower Carnian Cassian Formation, named after the type locality, a small village near Cortina d'Ampezzo (Northern Italy).

Holotype: Deposited in the Institut für Paläontologie der Freien Universität Berlin (IPFUB, JR5/89)
Locus typicus: Seelandalpe near Schluderbach

J. Reitner and H. Keupp (Eds.)
Fossil and Recent Sponges

1

**Fig. 1.** Sketch of living *Aka* within Late Jurassic siliceous sponge habitats. Most sponges are dead and covered by bacterial crusts. The entire spicular skeleton is altered to calcareous material. *H* Hexactinellid; *L* lithistid; *A Aka*; *C* micritic crust

Stratum typicum: Triassic, Early Carnian, Cassian Formation
Material: 4 specimens
Diagnosis: Encrusting and excavating *Aka* with relative short, but thick oxeas, arranged in brushes. Spicules connected with their end points as seen in *Adocia*

Description of the holotype (Fig. 2a-d): *Aka* bores in a small chaetetid sponge. Six round-shaped to elongated boring cavities are visible. In two of these cavities cross-cut monaxonic spicules are visible. On the outer surface of the chaetetid sponge, *Aka* is cut obliquely. The sponge protrudes from the surface of the chaetetid sponge. The thick short oxeas are similar to those seen in the cavities. Therefore a close relationship between the excavations and the protruding sponge is probable. The entire sponge is encrusted by dark, micritic, bacterial crusts.

Spicule measurements: length: 620–900 μm, mean 740 μm; width: 80–65 μm, mean: 70 μm.
Diameter of the burrows: 1600–400 μm
Diameter of the entire sponge: ca. 1.5 cm

Description of the paratype (Fig. 3a-c): The described sponge is encrusting. Spicule type and arrangement are typical for *Aka*. The sponge is cut perpendicular to the axis of the spicule bundles and exhibits mostly cross-cut spicules. Only a few spicules retain the entire shape. Within the center of the sponge, a hole is visible. The sponge remains are covered by dark, cloudy, stromatolitic structures. The spicules are smaller than those measured in the holotype.

Spicule measurements: length: 700–850 μm, mean 775 μm; width: 75 μm

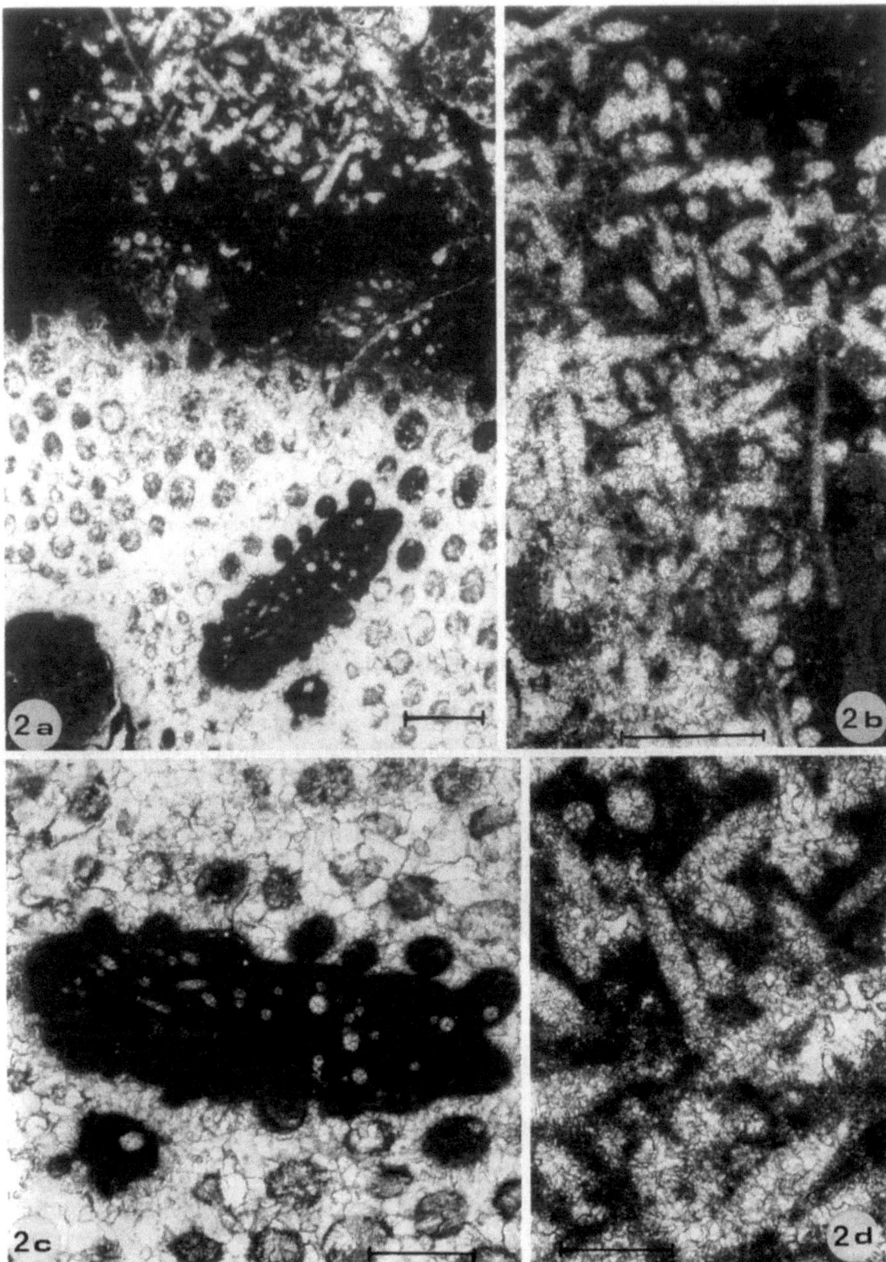


**Fig. 2.** *Aka cassianensis* n.sp. holotype; Cipit Boulders from the Cassian Beds near Seeland Alpe (Schluderbach, Northern Italy). **a** Entire sponge with boring cavities inside a small chaetetid sponge. Scale: 1 mm. **b** Spicular arrangement of the encrusting outer part of the sponge. Scale = 1 mm. **c** Elongated burrowing cavity with cross-sected spicule tracts. Scale = 500 μm. **d** Detail of **b** showing the short thick oxeas. Scale = 500 μm

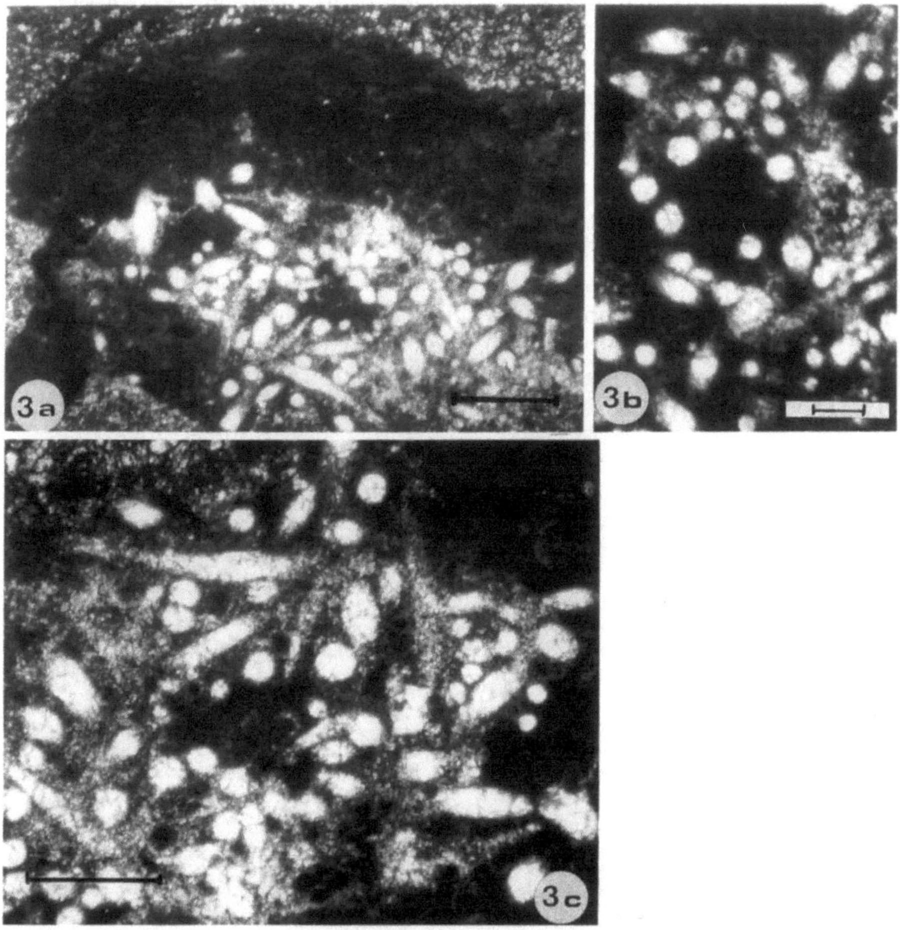


**Fig. 3.** *Aka cassianensis* n.sp. paratype; Cipit Boulders from the Cassian Beds near Seeland Alpe (Schluderbach, northern Italy). Specimen is covered by a dark micritic layer with stromatolitic affinities. Scale = **3a** 400 μm, **3b** 100 μm, **3c** 200 μm

Diameter of the entire sponge: ca. 10 mm
Diameter of the central hole: 5 mm

Differential diagnosis: The new species differs from all modern species by exhibiting significantly larger scleres. In modern *Aka* the spicules size varies from 100–200 μm. Within fossil *Aka* this particular type of short and thick oxea seen in *Aka cassianensis* is never observed.

*Remarks.* From the Lower Carnian Cassian Formation of the Dolomites near Cortina d'Ampezzo (northern Italy), fore-reef debris flows are common containing excellently preserved coralline sponges. Most of these sponges are preserved with their original mineralogy (aragonite, high Mg calcite) (Fürsich and Wendt 1977; Wendt 1979; Reitner 1987b; and others). The associated siliceous sponges are badly preserved because the siliceous spicules are dissolved very early. Only in few thin sections can the characteristic *Aka* spicules arranged in brushes and bundles be seen (Figs. 2–5).

## *Systematic Descriptions of Jurassic and Cretaceous Aka*

*Remarks.* Most post Triassic *Aka* species possess a burrowing ability. Some encrusting specimens are observed within Bathonian siliceous sponge build-ups from St. Aubin-sur-Mer (Normandy). The spicule types and spicule arrangement of the encrusting and boring type are nearly similar. Only two new main types of *Aka* are described here. Based on the lack of characters found in fossil specimens,

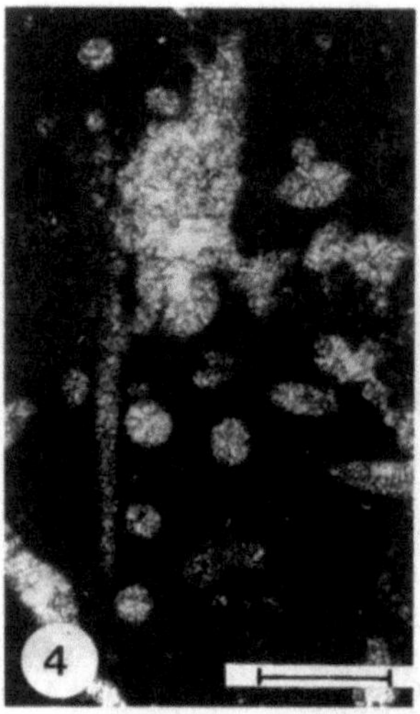


**Fig. 4.** *Aka cassianensis* n.sp.; same locality as Fig. 2. Scale = 200 μm

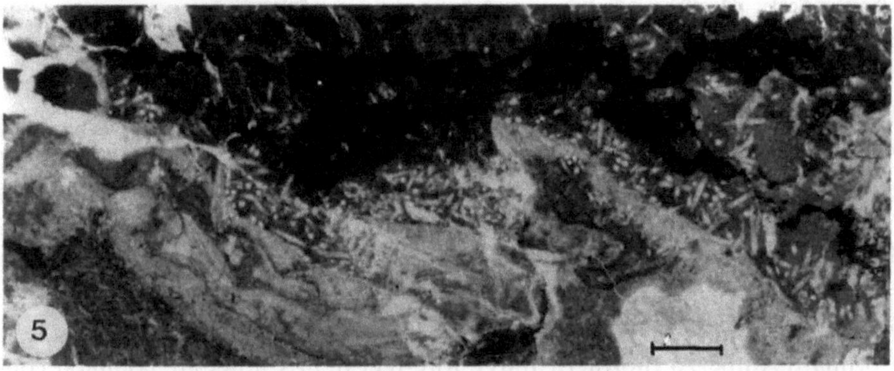


**Fig. 5.** *Aka cassianensis* n.sp.; encrusting specimen. Same locality as Fig. 2. Scale = 2 mm

it is always very difficult to erect new species, but the new observed forms significantly differ from modern ones and new species are justified.

*Aka muelleri* n.sp. (Figs. 6–8)

Derivatio nominis: After Walter Müller, a paleospongiologist who found and described the first, fossil *Aka*.
Holotype: Deposited in the IPFUB, JR6/89
Locus typicus: Roßbach, north part of the Frankenalb in Bavaria
Stratum typicum: Lower Kimmeridgian sponge mound facies
Material: Numerous specimens from Jurassic and Cretaceous carbonate platforms and from calcareous organisms.
Stratigraphic distribution: Bajocian-Late Cretaceous
Geographical distribution: Middle and Western Europe, Arizona/Northern Mexico (Aptian/Albian Mural Limestone).
Diagnosis: Irregularly shaped sponges of various sizes which live exclusively in excavated burrows. The burrows are large and possess generally only one or two openings. The spicules (oxeas) are big and arranged in bundles and brushes. Smaller ectosomal oxeas are observed in few specimens. Within the probable choanosomal area, the spicules exhibit typical haplosclerid arrangements. The sponge bores only in calcareous material, such as corals, diagenetically altered siliceous sponges, coralline red algae, bivalves, and belemnites.

Description of the holotype (Fig. 6): The sponge is located in an oyster shell which is settled by thecidean brachiopods. The microfacies of the host rock is mudstone/wackestone. The micrite is cloudy and exhibits irregularly shaped clasts and tuberoids. Seven spicule tracts are visible which are cut vertically and obliquely. A single spicule tract possesses more than 50 spicules. Two types of oxeas are observed: a rare smaller type in the outer part of the sponge remains, probably representing ectosomal scleres, and a bigger more common one in the center of the sponge. Small spicule types are observed also within the larger spicule tracts (bundles).

Measurements:
Big oxeas (3x): length: 1000–1600 $\mu$m; width: 60 $\mu$m
Small oxeas (2x): length: 800 - 1200 $\mu$m; width: 20–25 $\mu$m

Differential diagnosis: *Aka muelleri* n.sp. differs from *Aka cassianensis* n.sp. in the absence of an encrusting growth form and the possession of much larger scleres. The Recent *Aka* species have much smaller spicules (10 × smaller).

*Aka minima* n.sp. (Figs. 9,10)

Derivatio nominis: After the Latin word "minimus" small
Holotype: IPFUB, JR7/89
Paratype: IPFUB, JR8/89
Locus typicus: Kasendorf, SW of Kulmbach, northern Bavaria (MFS 188); leg. T. Steiger, Munich

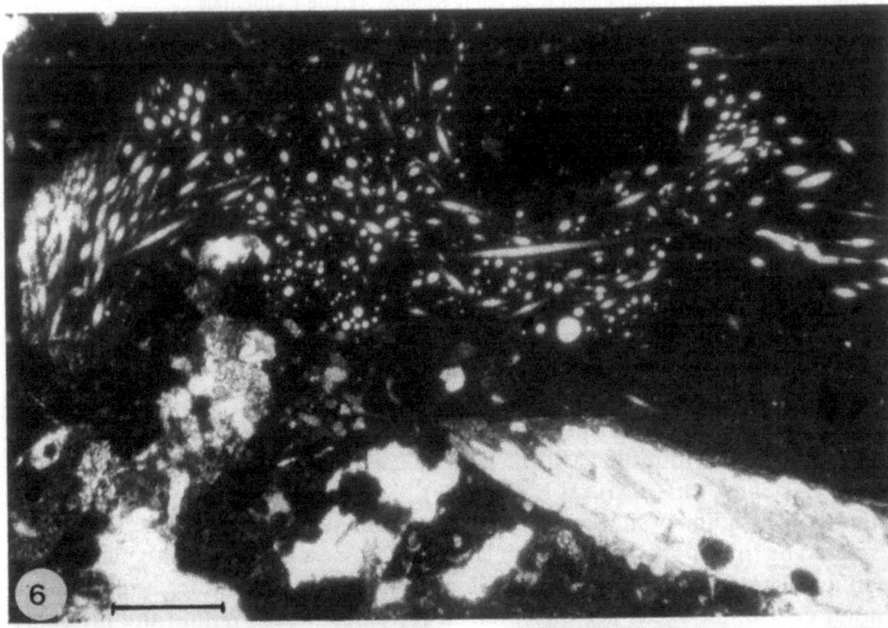

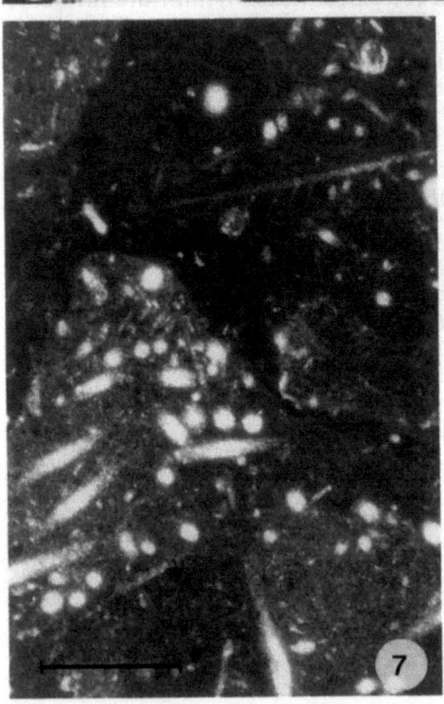

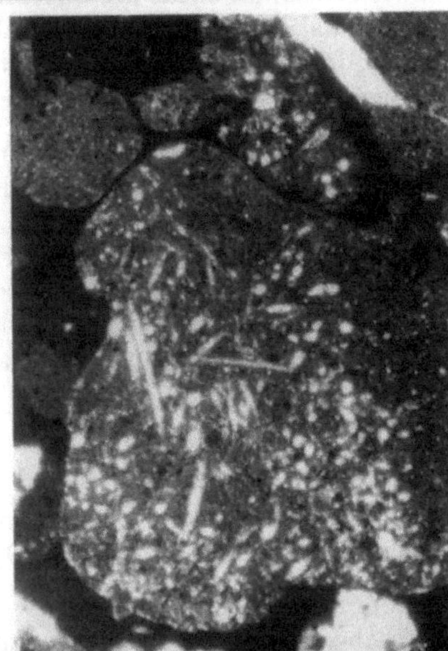

**Fig. 6.** *Aka muelleri* n.sp. holotype; Kimmeridgian sponge mounds from Roßbach northern Frankenalb (southern Germany, Bavaria). Scale = 1 mm

**Fig. 7.** *Aka muelleri* n.sp. from Bajocian/Bathonian sponge biostromes of Vilviestre (Sierra de la Demanda, northern Spain). Scale = 500 μm

**Fig. 8.** *Aka muelleri* n.sp. from a Late Albian reef (Albeniz/Eguino reef) near Araya (Prov. Alava, northern Spain). *Aka* bored within a specimen of *Acanthochaetetes* (deep fore-facies). Scale = 1 mm

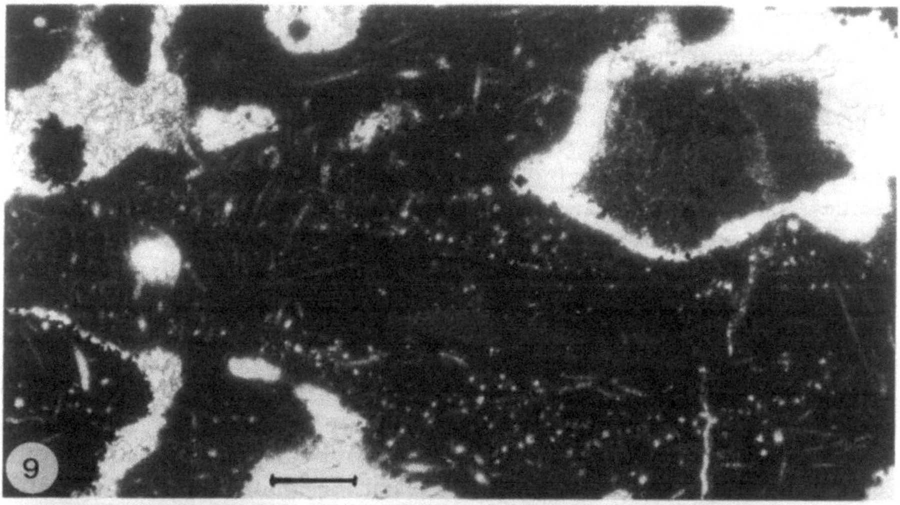


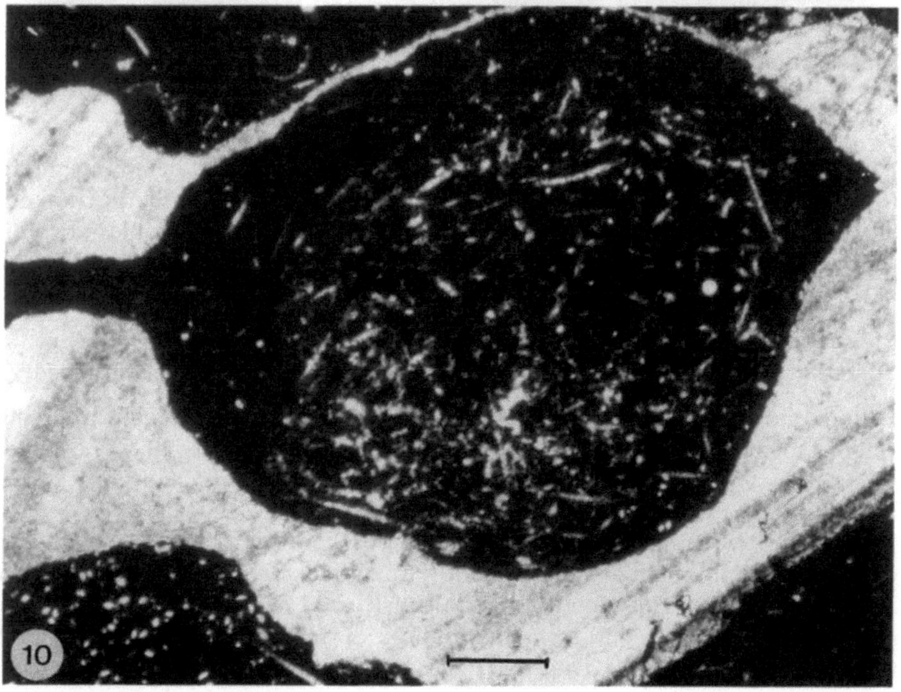


**Fig. 9.** *Aka minima* n.sp. holotype; Kimmeridgian sponge mounds near Kasendorf (Southern Germany, Bavaria). Scale = 200 μm

**Fig. 10.** *Aka minima* n.sp. paratype; Oxfordian/Kimmeridgian from Laibarös (northern Frankenalb, South Germany, Bavaria). *Aka* bored in a belemnite guard. Scale = 500 μm

Stratum typicum: Sponge mounds of the Kimmeridgian
Diagnosis: *Aka* with narrow long oxeas forming spicule bundles. Burrowing cavities elongated and in some cases branched.

Description of the holotype (Fig. 9): *Aka minima* n.sp. is located within a thick brachiopod shell which is totally bored with only thin walls between single sponge cavities. The new species *Aka muelleri* n.sp. is also common. Other fossils include brachiopods and lithistid sponges. The biogenetic pore space of the sponges filled up with pellets. The host microfacies is a tuberolithic filament-bearing wackestone/mudstone. Most bioclasts are surrounded by dark micritic crusts.

The new *Aka* species is located in a large bore cavity of a brachiopod and all cross-sections are visible. The spicules are arranged in loose bundles. The cavity is elongated in contrast to the *Aka muelleri* cavities, which are sack-shaped. The oxeas are thin and relatively long. Other spicule types are not observed. The inner zone of the sponge exhibits a dark micritic halo with rare spicules. The innermost zone of the halo is spicule free. The center of the sponge shows an elongated cavity, probably the main excurrent canal (spongocoel?).

Measurements:
Spicules: length: 300–600 μm, mean 400 μm; width: 15–25 μm, mean: 20 μm
Cavity: length: ca. 10 mm; diameter: 1–5 mm

Paratype (Fig. 10): The paratype is located in a totally bored belemnite guard from the Oxfordian/Kimmeridgian boundary of Laibarös (northern Frankenalb, Bavaria). The sponge is relatively badly preserved. The spicules are concentrated in thick bundles. Only *Aka minima* n.sp. is observed within the belemnite guard.

Measurements:
Spicules: length: 400–640 μm, mean 500 μm; width: 15–25 μm, mean: 20 μm

Differential diagnosis: *Aka minima* n.sp. differs from *Aka muelleri* in having smaller spicules and more elongate burrowing cavities.

*Middle Jurassic Occurrences of Aka*
In the Celtiberic Ranges and the northern part of Spain, vast carbonate platforms of Bajocian/Bathonian age are present (Hinkelbein 1975; Wilde 1988; Scheer 1988; and others). These platforms are constructed of rigid siliceous sponges, lithistid demosponges, and a smaller amount of hexactinellid sponges. Within these sponge build-ups *Aka muelleri* n.sp. (Fig. 7) is a common boring organism.

Measurements:
Bigger oxeas: length: 1000–1500 μm; width: 50–55 μm
Smaller oxeas: length: 850–1000 μm; width: 50 μm

### *Late Jurassic*

Build-ups of siliceous rigid sponges are very common at this time in all of Europe (especially in southern Germany) (Fritz 1958; Wagenplast 1972; Trammer 1982; Gaillard 1983; Flügel and Steiger 1981; Schorr and Koch 1985; and others). The main facies belts are more or less the same as those observed in the Dogger of Spain. The Late Jurassic occurrences are mostly in mud mounds and in some cases in mud banks ("Treuchtlinger Marmor"). The Oxfordian mounds are constructed mainly of hexactinellid sponges. The Kimmeridgian and Tithonian mounds and banks are composed mostly of lithistid demosponges. In both cases *Aka* is a common excavator within sponge skeletons. *Aka muelleri* n.sp. is the most common boring sponge. The smaller *Aka minima* n.sp. is rare and mostly associated together with *Aka muelleri* n.sp.

Müller (1978) has reported *Aka minima* n.sp. within the lithistid sponge *Pyrgochonina acetabulum* (Goldf.) from the Lower Kimmeridgian of the Schwäbische Alb (southern Germany). He has found this new sponge also in hexactinellids, and therefore, he interpreted the new sponge as a parasitic organism. His observation was the first report of fossil *Aka* species.

*Aka* bores exclusively in calcareous material. Therefore *Aka* is a postmortem inhabitant of siliceous sponges after a very early diagenetic replacement of the opal or cristobalite of the siliceous spicules into calcite. *Aka* bores in diagenetically altered sponge skeletons and in belemite guards.

Measurements of *Aka muelleri* n.sp. (excl. holotype):
Bigger oxeas:
length: 1300–1550 μm, mean: 1450 μm; width: 55–60 μm, mean: 50 μm
Smaller oxeas:
length: 700–1000 μm, mean: 900 μm; width: 15–20 μm, mean: 17 μm

Measurements of *Aka minima* n.sp. (excl. holotype):
Oxeas of Müller specimen: length: 350–470 μm, mean 450 μm; width: 15 μm
Oxeas of further specimens: length: 320–620 μm, mean 500 μm; width: 20 μm

### *Early Cretaceous*

In the Lower Cretaceous of the Tethyan realm of Europe, as well as in the north part of the Cretaceous Gulf of Mexico, shallow marine carbonate platforms and reefs are widespread and called "Urgonian". The Urgonian facies of northern Spain is well studied, especially the facies distributions and the sponge communities (Pascal 1985; Reitner 1982, 1986, 1987a, 1987c; Reitner and Engeser 1983, 1985, 1987). The boring sponge *Aka* is common in the deeper part of the coralgal facies belt, within the *Microsolena* deeper water coral facies and within the *Spirastrella* (*Acanthochaetetes*) biofacies (Reitner 1987a, Pl. 15, Figs. 5,6) (Fig. 8). Further occurrences of *Aka* are within micritic mud mounds which were formed by siliceous sponges, comparable with the Jurassic ones, and rarely in rudist mud mounds. Within the beach facies and other permanently wave-agitated facies zones, typical pearl band borings of the hadromerid burrowing sponge *Cliona* are observed. *Cliona* borings are never seen within quiet water facies in which *Aka* is dominant. Spicule type and spicule arrangements are characteristic for the species *Aka muelleri*. The spicules are slightly smaller but they have the same shape and thickness.

Measurements:
Oxeas: length: 800–1000 μm, mean 900 μm; width: 50–55 μm, mean 52 μm

### *Middle Eocene*

Early Tertiary coral reefs are rarely exposed and only few localities are known. Generally the deeper parts of the fore-reefs of these distinct reefs are not preserved or exposed. From the Boltaña anticline in the southern Pyrenees Middle Eocene debris flows are exposed, which bear components of deeper fore-reef mud mound facies. A strong freshwater diagenesis has dissolved out most of the aragonitic organisms, as evidenced by preserved ghost structures. The facies is characterized by flat corals, different encrusting coralline red algae, encrusting rotaliid foraminifera, bryozoa, and sponges. A *Spirastrella* (*Acanthochaetetes*) community composed of small *S.*(*A.*) *eocena* (Rios & Almela), different unidentified lithistid demosponges and *Aka* is observed.

Nearly all larger calcareous organisms are intensively bored by *Aka* and often only a vague relict is preserved. The *Aka* cavities are often elongated or rarely sack-shaped. The living cavities of *Aka* exhibit a length of > 10 mm. Each possesses one narrow opening. The spicules are enormous (1.5–2 mm) and are arranged in bundles. The number of spicule-bundles is highly variable and has therefore no taxonomic significance. Two oxea size categories are observed. A differentiation of spicules to dermal or choanosomal scleres are not seen. The spicule-bundles are covered by dark micritic halos.

The sponge differs from all other fossil *Aka* in cavity size and shape, spicule size, and habitat. Therefore a new species is created.

### *Systematic Description*

*Aka boltanaensis* n.sp. (Figs. 11–13)

Derivatio nominis: After the locus typicus the village Boltaña in the southern Pyrenees.

Holotype: Deposited within IPFUB, JR9/89
Locus typicus: Base of the Boltaña-anticline near a parking lot on the south side of the river.
Stratum typicum: Middle Eocene debris flow
Material: 5 specimens
Diagnosis: Very big *Aka* excavating single elongated or sack-shaped borings, opened at one end. The spicules exhibit two sizes, a large one with a mean size of

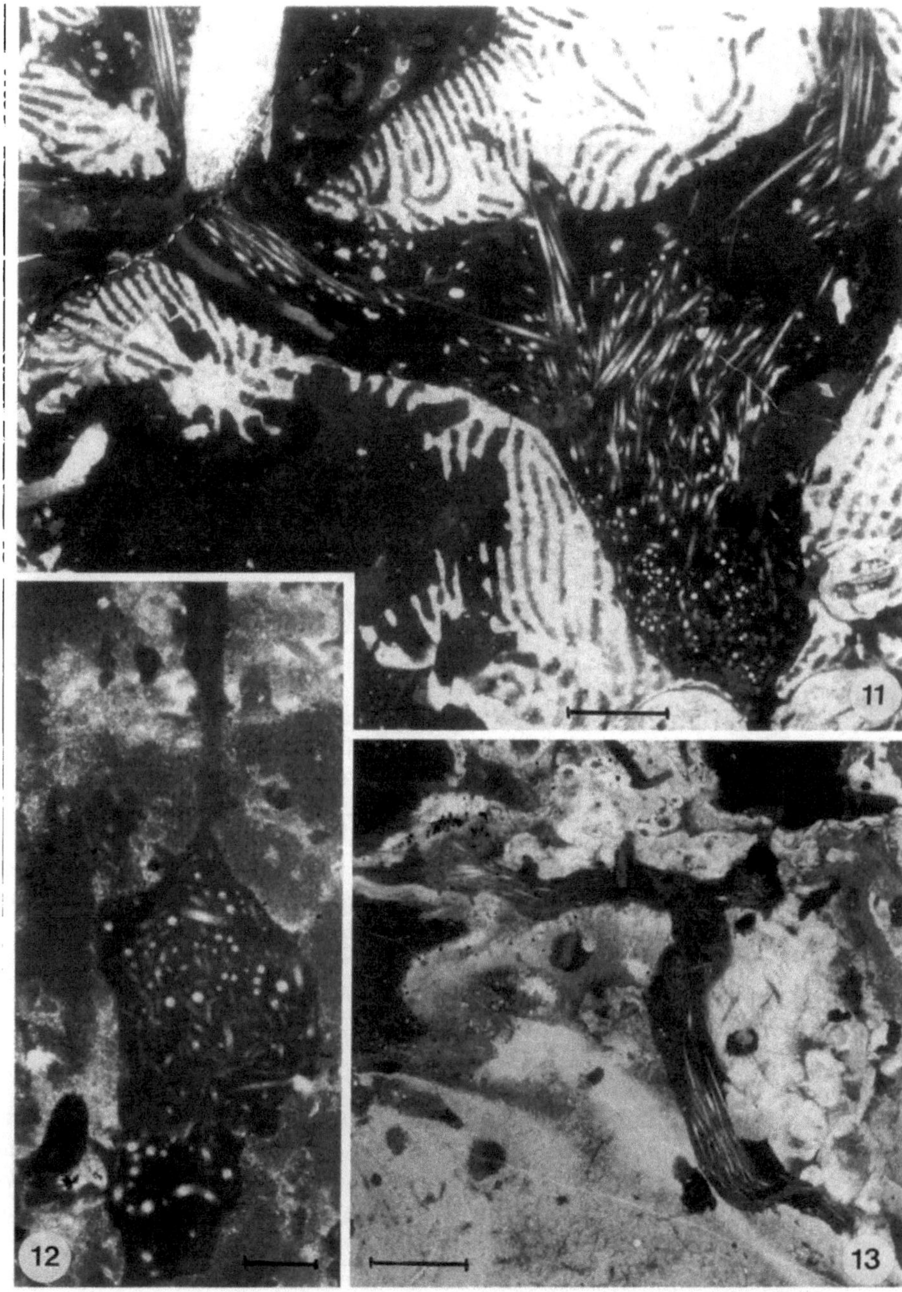


**Fig. 11.** *Aka boltanaensis* n.sp. holotype; Middle Eocene clast within a debris flow of the same age from the Boltaña Anticline near Boltaña (southern Pyrenees). *Aka* boring in a large coral. Scale = 1 mm

**Fig. 12.** *Aka boltanaensis* n.sp. Living cavity with a prominent bottle neck canal which serves as the connection between the outer and inner sponge tissue. Same locality as Fig. 10. Scale = 500 μm

**Fig. 13.** *Aka boltanaensis* n.sp. Characteristic elongated canals in the living cavity of this particular species. Scale = 2 mm

1600 μm and a smaller one with a mean size of 500 μm. The thickness of the spicules varies from 80–20 μm. The ratio of length to thickness shows a long narrow spicule character.

Description of the holotype (Fig. 11): The holotype of *Aka boltanaensis* is in a thin section from the *Spirastrella* (*Acanthochaetetes*) community mud mound facies. Flat coral remains are common, as well as larger rotaliid foraminifera (*Nummulites* sp.), and coralline red algae. The holotype is found in the remains of a thamnasteroid scleractinian coral. The coral is totally bored. The coral skeleton is preserved in granular neomorphic calcite. The big *Aka*-boring (living cavity) has a measured length of 2 cm and a maximum diameter of 9 mm. Several smaller blind cavities are observed extending from the main cavity are observed. These blind cavities are filled with spicules and probably served as mining areas for the sponge immediately before its death. The living cavity of the sponge is a complicated three dimensional network, as evidenced by vertical and oblique cuts of the spicule-bundles and Y-shaped areas connecting the cavities.

Measurements:
Bigger oxeas: length: 1400–2200 μm, mean 1700 μm; width: 60–80 μm, mean 75 μm
Smaller oxeas: length: 300–1000 μm, mean: 560 μm; width: 20–40 μm, mean 35 μm

Differential diagnosis: *Aka boltanaensis* n.sp. differs in first order from all known *Aka* species in having very large, elongated two size categories oxeas. *Aka muelleri* n.sp. has similar-sized spicules and also possessing two sizes of oxeas. The measured dimensions are slightly smaller and the length/thickness ratio is smaller, than in *Aka boltanaensis*. Nevertheless, both species are closely related. The living cavity of *Aka boltanaensis* is much more complicated than seen in other fossil *Aka*. All Recent *Aka* have spicules with length of 100–300 μm.

## *Recent Aka*

*Aka in Petrobiona massiliana* Vacelet and Levi, *1958*
One species is a common boring sponge inside the dead body of a pharetronid sponge *Petrobiona massiliana* Vacelet and Levi (Figs. 16,17) from submarine caves near Marseille (Mediterranean Sea). The spicule shape is typical for *Aka minuta* Thomas, a Pacific species, but the spicule size is bigger (mean 205 μm), as

→

**Fig. 14.** *Aka coralliphaga* (Rützler). Spicule bundles of the central part of an outdoor chimney. Curacao, Playa Kalki (10–20 m) Collected by van Soest. (Scanning electron microscope (SEM) micrograph)

**Fig. 15.** *Aka* aff. *coralliphaga* inside the coralline sponge *Astrosclera willeyana* Lister exhibiting a pitted surface caused by boring activity. The dissolution marks are similar to those observed in *Cliona*. (SEM) micrograph)

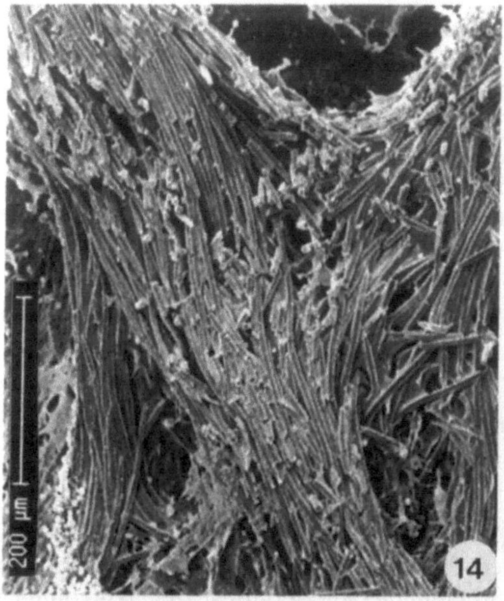


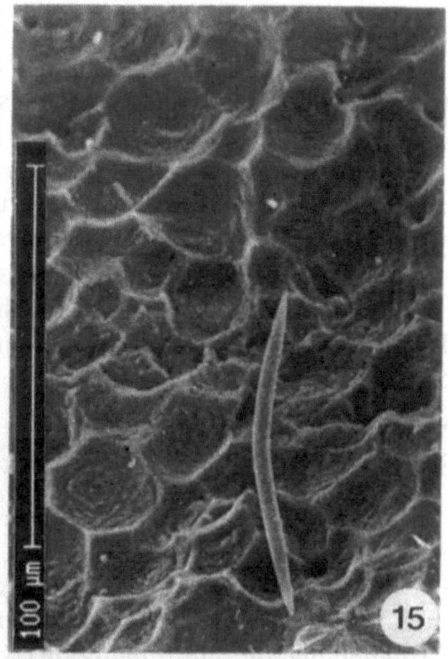

**Fig. 16.** *Aka* cf. *minuta* Thomas inside the dead portion of *Petrobiona massiliana* Vacelet & Levi. Submarine cave near Marseille. (SEM micrograph)

**Fig. 17.** *Aka* cf. *minuta* Thomas inside the basal skeleton of *Petrobiona* with living sponge in an endolithic chamber and narrow incurrent and outcurrent canals. Submarine cave near Marseille. Scale = 200 $\mu m$

observed within the type material of *Aka minuta* (mean size of 120 μm) (Thomas 1972). Therefore *Aka* in *Petrobiona* are classified as *Aka* cf. *minuta*. The spicules are arranged in bundles and brushes and the entire shape is comparable with the fossils species.

Within the dead skeleton of *Petrobiona*, many endolithic sponges are present. Beside *Aka, Cliona* is a rare excavating sponge. In cavities caused by the boring activity of excavating sponges, parasitic nonburrowing sponges settled. In the studied material *Agelas* sp., *Geodia* sp., and *Clathrina* sp. are found. The spicules of *Aka* cf. *minuta* are much smaller than those observed in all described fossil *Aka*. The shape of the living cavities are comparable. In most cases sack-shaped cavities are observed.

Measurements:
Spicules: length: 170–215 μm, mean 200 μm; width: 7.5–10 μm, mean 8 μm

*Aka coralliphaga (Rützler)*
In Caribbean reefs *Aka coralliphaga* (Rützler) dominantly bores in flat corals (Fig. 14). The species was erected by Rützler (1971) and described as "*Siphonodictyon*" *coralliphagum* (p. *Siphonodictyon* Bergquist 1965 is a younger synonym of *Aka* de Laubenfels 1936). *Aka coralliphaga* is also very common inside the dead parts of Indopacific coralline sponges, e.g., *Astrosclera willeyana* (Fig. 15). Rützler (1971) has described different morphotypes of this species which he called "*S.*" *coralliphagum* forma *typica*, forma *obruta*, forma *tubulosa*, and forma *incrustans*. These morphotypes indicate that, as he considered, the various forms are different growth stages outside the living cavities. Beside ontogenetic growth stages an adaptation to water currents is probable. The morphotype forma *tubulosa* is here selected. The forma *tubulosa* is characterized by having a chimney-like main excurrent opening (spongocoel). The spicules in this area are arranged in thick bundles from which perpendicular smaller spicule tracts derive, forming the dermal layer. The center of the sponge exhibits typical adociid spicules connections. The spicules are arranged in bundles. For further descriptions see Rützler (1971).

Measurements:
Spicules: length: 150–200 μm, mean 170 μm; width: 7–9 μm, mean 7.5 μm

## Paleoecology and Biostratinomic Implications

All observed fossil and modern occurrences of *Aka* are more or less found in protected habitats far from very turbulent water. Rützler (1971) noticed that the main depth distribution of *Aka* from the Caribbean is between 10–70 m, below the direct influence of normal wave action. Steady water currents were measured and are probably important for feeding. *Aka* is sometimes found in protected shallower water environments (1–2 m water depth), e.g., *Aka coralliphaga* is often found in boulder-like corals which are bored by *Cliona* spp.

Within the turbulent zones, the most species of the genus *Cliona* occupies the ecological niche of *Aka*. *Cliona* is rarely observed in deeper water (*Cliona viridis*, *Leptoseris* reefs, 90 m), and within West Indies reefs *Cliona delitrix*, *C. agrica*, *C. langue*, and *C. schmidti* is common at 20 m water depth. The main depth distribution of *Cliona* is between 0–10 m (Rützler 1973, 1974; Schwarz 1981). *Cliona* is also observed in cooler, temperate water.

Temperature seems also to be an important limiting factor of the Caribbean *Aka* communities. Within the northern Bermuda reefs *Aka* is not observed. From the Mediterranean, *Aka* is observed only in underwater caves, which are characterized by communities comparable with some Tethyan Cretaceous and modern Indopacific ones. These communities are primary Tethyan relicts (Reitner 1989). Perhaps *Aka* in *Petrobiona* is part of the primary Tethyan sponge community. This idea is supported by the existence of *Aka* as part of the Eocene *Spirastrella*/*Acanthochaetetes* community from the southern Pyrenees. *Aka* in *Petrobiona* is a probable Tethyan survivor and specially adapted to moderate temperate water.

*Aka* bores calcareous material, like *Cliona*. *Aka* produces the same type of boring chip as seen with *Cliona* and its boring method is comparable to *Cliona* (Pomponi 1980). The diameter of the etching chips of *Aka* varies from 14–40 $\mu$m, mean 30 $\mu$m and therefore comparable with the sponge chips of *Cliona* (Pomponi 1980). Similar ecological paleoenvironments of *Aka* are observed in all fossil *Aka* examples. *Aka* is also missing from high turbulent water facies zones. In this particular zone *Cliona* borings rarely exist (Reitner 1987a). *Aka* is common within the reef core, fore-reef, and mud mound facies which are always below normal wave base. In two reefs from the Late Albian Albeniz/Eguino platform in northern Spain, water depth of the coralgal reef core to fore-reef slopes was calculated as 30–100 m (Reitner 1987a) – the main distribution of *Aka*.

The Late Carnian *Aka* are part of the highly diverse coralline sponge community of the Cassian reefs. These sponge communities are normally located in mound structures of the deeper fore-reef. The platforms are constructed of algae and oolite shoals.

The bathymetric position of the Jurassic siliceous sponge mounds is still under discussion. The first studies on these build-ups mentioned a deeper water position based on the lack of true algae, and the presence of micritic sediments. Also the sedimentary megacycle of the Malm of southern Germany indicates a regressive sequence. On the top of sponge mounds, coral reefs were established during the Tithonian and demonstrate shallow marine water conditions. Based on "J. Walther's law" the sponge mounds must be situated in deeper waters (Fritz 1958; Gwinner 1962; Nitzopoulos 1974; Ziegler 1977). New studies on the diagenetic sequences of these mounds and bank facies indicate a contradictory bathymetric position within extremely shallow marine waters (intertidal). These conclusions are mainly based on marine vadose cements (Koch and Schorr 1986). Lang (1988) and Kott (1989) postulated a paleobathymetric position for the Oxfordian and Kimmeridgian sponge biostromes and mounds, based on sedimentological criteria, between normal wave base and storm wave base. All these interpretations, however, indicate a restricted water movement: the optimal environment for *Aka*.

Of special importance is the boring activity of *Aka* in dead siliceous sponges which are often still in original life position. *Aka* and *Cliona* bore only in calcareous material; therefore, the siliceous skeletons of sponges must be altered into calcite before *Aka* starts excavating. Most of the siliceous sponges are surrounded by dark micritic, partially stromatolitic halos. This feature indicates a reduced sedimentation rate; the siliceous sponges are not totally covered by sediment during this stage. The micritic crusts are called "Kalkmumien" (Fritz 1958). These micritic covers are cemented very early before the siliceous spicules are dissolved (cf. Brachert et al. 1987; Lang and Steiger 1984). The micritic layers are probably bacterial crusts decomposing the sponge soft tissue. One characteristic of these bacteria crusts is the synvivo lithification outside their cell walls. The dissolution of the siliceous scleres occurred after this early diagenetic event. Then the organic axial filaments were also decomposed by bacteria. These processes produced high pH values which probably were responsible for dissolving the spicule cristobalite (Land 1976; Reitner 1986). The sclere molds were cemented rapidly by granular or single calcite crystals. After this diagenetic stage, *Aka* was able to bore inside the "siliceous" sponges. *Aka* is an important post mortem organism of siliceous sponge mounds. This diagenetic process is found in Jurassic as well as Cretaceous deeper water mud mound. *Aka* is therefore an excellent indicator for a very early submarine cementation and diagenesis.

The common *Aka* within the Middle Eocene reef rocks also indicate quiet water conditions of this particular biofacies. This biofacies is characterized by *Spirastrella* (*Acanthochaetetes*), a sponge which prefers dark and/or deeper water conditions.

## Conclusions

1. The new Lower Carnian species *Aka cassianensis* is a dominantly encrusting sponge. Borings are only known from the holotype. Spicule shape and arrangement are similar to those in the modern genus *Aka* but the size is different.
2. Jurassic to Recent *Aka* are dominantly excavating sponges.
3. Two new species are described from the Jurassic and Lower Cretaceous, *Aka muelleri*, and *Aka minima*. *Aka muelleri* n.sp. is the most common species.
4. From Middle Eocene fore reef sediments of the southern Pyrenees a further new species is described (*Aka boltanaensis*).
5. Recent species of *Aka* differ in having much smaller oxeas (100–200 μm) in relation to fossil ones (300–2200 μm).
6. *Aka* prefers deeper, moderately agitated water conditions (20–100 m), as evidenced by modern occurrences and by facies analyses of fossil occurrences. The niches for boring sponges in shallower more agitated environ-

ments are preferred occupied by certain species of the hadromerid genus *Cliona*.

7. *Aka* is an excellent indicator for very early carbonate cementation and replacement of the siliceous spicules in calcite in fossil siliceous sponge mud mounds. *Aka* is an important post-mortem faunal element in these particular build-ups (Fig. 1).

*Acknowledgments.* The authors are indebted to Prof. Dr. R. van Soest (Amsterdam) for allowing the authors to study Recent *Aka* species and some donations. We thank Dr. E. Gierlowski-Kordesch (Freie Universität Berlin) for translation assistance and Dr. S. Pomponi (Harbor Branch Inst., Florida) for reviewing the ms. The Deutsche Forschungsgemeinschaft is acknowledged for financing this investigation (Re 6651-1, Ke 322/4).

## References

Berquist PR (1965) The sponges of Micronesia, Part I: the Palao Archipelo. Pac Sci 9:123-204

Brachert T, Dullo WC, Stoffers P (1987) Diagenesis of siliceous sponge limestones from the Pleistocene of the Tyrrhenian Sea (Mediterranean Sea). Facies 17:41-50

De Laubenfels MW (1936) A comparison of the shallow water sponges near the Pacific end of the Panama Canal with those at the Caribbean end. Proc US Natl Mus 83:441-466

Flügel E, Steiger T (1981) An Upper Jurassic sponge-algal buildup from the Northern Frankenalb, West Germany. SEPM Spec Publ 30:371-397

Fritz GK (1958) Schwammstotzen, Tuberolithe und Schuttbreccien in Weißen Jura der Schwäbischen Alb. Arb Geol Paläont Inst TH Stuttgart NF 13:1-118

Fürsich FT, Wendt J (1977) Biostratinomy and palaeoecology of the Cassian Formation (Triassic) of the southern Alps. Palaeogeogr Palaeoclimat Palaeoecol 22:257-323

Gaillard Ch (1983) Les Biohermes à Spongiaires et leur Environment dans l'Oxfordien du Jura Meridional. Doc Lab Geol Lyon 90:1-515

Gwinner MP (1962) Geologie des Weißen Jura der Albhochfläche (Württemberg). N Jb Geol Paläont Abh 115:137-221

Hinkelbein K (1975) Beiträge zur Stratigraphie und Paläontologie des Jura von Ostspanien, VIII: Stratigraphie und Fazies im Mitteljura der zentralen Iberischen Ketten. N Jb Geol Paläont Abh 148:139-184

Johnson JY (1899) Notes on some sponges belonging to the Clionidae obtained in Madeira. J R Microsc Soc 1899:461-463

Koch R, Schorr M (1986) Diagenesis of Upper Jurassic sponge-algal reefs in SW Germany. In: Schroeder JH, Purser BH (eds) Reef Diagenesis. Springer, Berlin Heidelberg New York Tokyo, pp 224-244

Kott R (1989) Fazies und Geochemie des Treuchtlinger Marmors (Unter-und Mittel-Kimmeridge, Südliche Frankenalb). Berliner Geowiss Abh 111:115 pp

Land LS (1976) Early dissolution of sponge spicules from reef sediments, North Jamaica. J Sed Petrol 48:337-344

Lang B (1988) Baffling, binding or debris accumulation? Ecology of Jurassic sponge-bacterial build-ups (Oxfordian, Franconian Alb). Berl Geowiss Abh 100:22 p

Lang B, Steiger T (1984) Paleontology and diagenesis of Upper Jurassic siliceous sponges from the Mazagan Escarpement. Oceanol Acta 1984:93-100

Müller W (1978) Beobachtungen zur Ökologie von Kieselspongien aus dem Weißen Jura der Schwäbischen Alb. Stuttg Beitr Naturkd Ser B 37:15

Nitzopoulos G (1974) Faunistisch-ökologische, stratigraphische und sedimentologische Untersuchungen an Schwammstotzen-Komplexen bei Spielberg am Hahnenkamm. Stuttg Beitr Naturkd B 16:1-143

Pang RK (1973) The systematics of some Jamaican excavating sponges. Postilla Peabody Mus Yale Univ 161:1–75

Pascal A (1985) L'Urgonieu, systèmes biosédimentaires et tectogenèse. Mem Géol Ann Dijon 9:45–72

Pomponi SA (1980) Cytological mechanisms of calcium carbonate excavation by boring sponges. Int Rev Cytol 65:301–319

Reitner J (1982) Die Entwicklung von Inselplattformen und Diapir-Atollen im Alb des Basko-Kantabrikums (Nordspanien). N Jb Geol Paläont Abh 165:87–101

Reitner J (1986) A comparative study of the diagenesis in diapir-influenced-reef atolls and a fault block reef platform in the Late Albian of the Vasco-Cantabrian Basin (Northern Spain). In: Schroeder JH, Purser BH (eds) Reef Diagenesis. Springer, Berlin Heidelberg New York Tokyo, pp 186–209

Reitner J (1987a) Mikrofazielle, palökologische und paläogeographische Analyse ausgewählter Vorkommen flachmariner Karbonate im Basko-Kantabrischen Strike Slip Fault-Becken-System (Nordspanien) an der Wende von der Unterkreide zur Oberkreide. Doc Nat 40:248

Reitner J (1987b) A new calcitic sphinctozoan sponge belonging to the Demospongiae from the Cassian Formation (Lower Carnian; Dolomites, Northern Italy) and its phylogenetic relationship. Geobios 20:571–589

Reitner J (1987c) Phylogenie und Konvergenzen bei rezenten und fossilen Calcarea (Porifera) mit einem kalkigen Basalskelett ("Inozoa, Pharetronida"). Berl Geowiss Abh 86:87–125

Reitner J (1989) Lower and Mid-Cretaceous Coralline Sponge Communities of the Boreal and Tethyan Realms in Comparison with the Modern Ones – Palaeoecological and Palaeogeographic Implications. In: Wiedmann J (ed) Cretaceous of the Western Tethys. Proceedings 3rd International Cretaceous Symposium, Tübingen 1987 pp 851–878. E Schweizerbart'sche Verlagsbuchhandlung Stuttgart

Reitner J, Engeser T (1983) Contributions to the systematics and the palaeoecology of the family Acanthochaetetidae (Fisher 1970) Order Tabulospongida, Class Schlerospongiae. Geobios 16:773–779

Reitner J, Engeser T (1985) Revison der Demospongier mit einem thalamiden aragonitischen Basalskelett und trabekulärer Internstruktur ("Sphinctozoa" pars) Berl Geowiss Abh 60:151–193

Reitner J, Engeser T (1987) Skeletal structures and habitates of Recent and fossil *Acanthochaetetes* (subclass Tetractinomorpha, Demospongiae, Porifera). Coral Reefs 6:151–157

Rützler K (1971) Bredin-Archold Smithsonian Biological Survey of Dominica: Burrowing Sponges, Genus *Siphonodictyon* Bergquist, from the Caribbean. Smith Contrib Zool 77:17

Rützler K (1973) Clionid sponges from the coast of Tunisia. Bull Inst Natl Sci Tech Oceanogr Peche Salammbo 2(4):623–636

Rützler K (1974) The burrowing sponges from Bermuda. Smith Contrib Zool 165:203–216

Scheer U (1988) Influences of the paleogeographic position and sea-level changes on spongiolithic limestones in the lower Bajocian in northern Spain. Berl Geowiss Abh 100:36–37

Schorr M, Koch R (1985) Fazieszonierung eines oberjurassischen Algen-Schwamm-Bioherms (Herrlingen, Schwäbische Alb). Facies 13:227–270

Schwarz A (1981) Zur Taxonomie und Ökologie der Bohrschwämme (Clionidae) im Felslitoral am Capo Caccia (NW-Sardinien) und deren geographische Verbreitung. Hausarbeit der Ersten Staatsprüfung für das Lehramt am Gymnasium Wiss Prüfungsamt Bochum, pp 133

Thomas PA (1972) Boring sponges of the reefs of the Gulf of Mannar and Palk Bay. Proc 1. Symp Corals and Coral Reefs 1969 Mar Biol Assoc India, pp 333–362

Trammer J (1982) Lower to Middle Oxfordian sponges of the Polish Jura. Acta Geol Pol 32:1–39

Valelet J, Levi C (1958) Un cas de survivance en méditerranée du groupe d'eponges fossiles des Pharétronides. CR Acad Sci 246:318–320

Wagenplast P (1972) Paläoökologische Untersuchungen der Fauna aus Bank- und Schwammfazies des Weißen Jura der Schwäbischen Alb. Arb Inst Geol Paläont Univ Stuttg NF 67:1–99

Wendt J (1979) Development of skeletal formation, microstructure, and mineralogy of rigid calcareous sponges from the Late Palaeozoic to Recent. Coll Int CNRS 291:449–457

Wilde S (1988) The Upper Bajocian-Lower Bathonian (Middle Jurassic) of the Northwest Iberian Range (Spain) – a spongiotuberolithic platform environment. Berl Geowiss Abh 100:52

Ziegler B (1977) The "White" (upper) Jurassic in southern Germany. Stuttg Beitr Naturk Ser B 26:79

# Renewed Study of the Type Material of *Palaeospongilla chubutensis* Ott and Volkheimer (1972)

C. Volkmer-Ribeiro and J. Reitner

## Introduction

The first fossilized gemmule-bearing freshwater sponge was described by Ott and Volkheimer (1972) from the Patagonian Lower Cretaceous (Aptian) at the Chubut River valley, in Argentina. However, an accurate reading of the original description shows that the photograph of the gemmules presented in Fig. 3 does not show the thick gemmular wall and the particular arrangement of gemmoscleres shown in Fig. 5. Also the drawings of the gemmoscleres shown in Fig. 6 do not resemble the gemmoscleres photographed in Fig. 5. In a second paper (Volkheimer and Ott 1973), the authors presented a study of the fauna and flora found associated with *P. chubutensis* and concluded that the paleohabitat was a lacustrine one subjected to strong seasonal droughts. This second paper proposed a phyletic relationship of *P. chubutensis* with the extant genus *Spongilla* and, more particularly, with *Spongilla alba* (Carter 1849). Soon after, Racek and Harrison (1975) proposed a strong relationship of *Palaeospongilla* with the extant genus *Radiospongilla*. Only drawings were presented and reference was made to a deposition of the preparations in the Australian Museum (Sydney) and the Smithsonian Institute (Washington). Based on these discrepancies between the illustrations and the phylogenetic ideas given by the original authors and by Racek and Harrison, the present authors decided to undertake a renewed study of these very important fossil materials, which represent the only entirely preserved freshwater sponges.

## Material and Methods

The holotype material of *P. chubutensis* and two paratypes are deposited at the Museo Argentino de Ciencias Naturales Bernardino Rivadavia (Buenos Aires, Argentina). Three pieces of a further paratype are housed with the collection of the Bayerische Staatssammlung für historische Geologie und Paläontologie (Munich, West Germany). One thin section of a paratype (no. Z 4748) is deposited at the Australian Museum (Sydney, Australia). The thin sections, which were made from a topotype (Racek and Harrison 1975) could not be found at the Smithsonian Institute or at the Australian Museum. New thin sections and SEM preparations were made from the paratype deposited in Munich. The

J. Reitner and H. Keupp (Eds.)
Fossil and Recent Sponges

thin section loaned from the Australian Museum was throughly restudied and photographed.

The SEM studies were carried out by the junior author with a Cambridge 360. New preparation methods were used to figure out the spicular skeleton and preserved soft parts as well as preserved gemmules. The probes were etched for 3 h with 5% Titriplex-III-solution ($C_{10}H_{14}N_2O_{18}$*$2H_2O$) (Reitner and Engeser 1987). This method permits etching very fine and sensitive structures, e.g., small gemmoscleres and acanthose microscleres, which are preserved in mono- or paucicrystalline calcite. Under the SEM, backscatter technique (secondary electrons) was used. This method makes it possible to distinguish irregularities on the surface structures caused by special contrast and therefore gives a much better view of spiny structures.

## *Description of the Species*

*1. Palaespongilla chubutensis* Ott and Volkheimer

The renewed study of these fossil specimens revealed that the original description of Ott and Volkheimer and also the redescription presented by Racek and Harrison (1975) apply to an association of two genera. The occurrence of such associations is a common fact among extant freshwater sponges (Volkmer-Ribeiro and de Rosa-Barbosa 1972; Racek 1969).

Part of the material described as *Palaeospongilla chubutensis* is now synonymized under *Spongilla patagonica* n.sp. The main part of the sponge material is *Palaeospongilla chubutensis* sensu strictu.

Palaeospongillidae new family

Type genus: *Palaeospongilla* Ott and Volkheimer 1972, as presently redefined.

Diagnosis: Fossil freshwater sponges with gemmules devoid of gemmoscleres but with a thin inner coat and a thin pneumatic coat.

Gemmules in groups or isolated and protected by an outer armature of tangentially arranged megascleres.

*Palaeospongilla* Ott and Volkheimer 1972, redefined

*Palaeospongilla* Ott and Volkheimer 1972, p. 53 (partim)

Type species: *Palaeospongilla chubutensis* Ott and Volkheimer 1972, as presently redescribed.

Diagnosis: Sponge forming thick crusts. Gemmules without gemmoscleres, with a conspicuous inner coat and thin pneumatic coat. Foramen devoid of a porous tube. Gemmules single but more usually packed in groups of two or several, each package or each single gemmule protected by an outer envelope of tangentially arranged megascleres. Megascleres sparsely spined and curved oxea. Microscleres not detected.

*Palaeospongilla chubutensis* Ott and Volkheimer 1972 (Figs. 1; 2a,b; 3a,b)

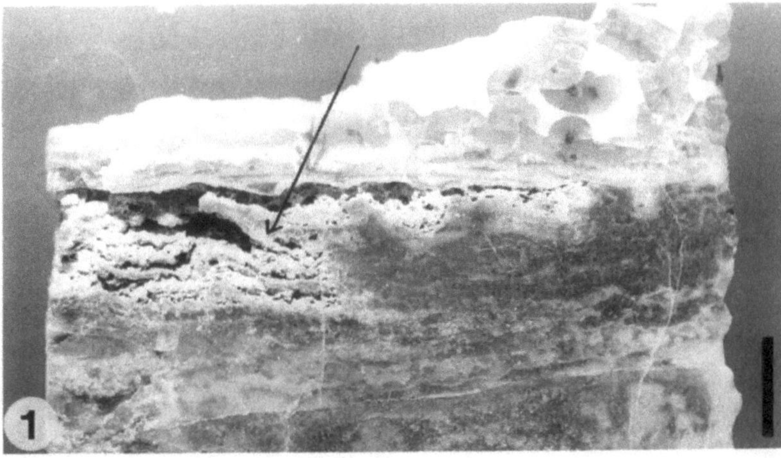


**Fig. 1.** Paratype of *Palaeospongilla chubutensis* deposited at Museo Bernardino Rivadavia, Buenos Aires. *Arrow* points to the sponge fossilized crust. Scale = 1 cm

*Palaeospongilla chubutensis* (Ott and Volkheimer 1972; p. 53 partim and Figs. 2,3,4)
Holotype: As originally designated (Ott and Volkheimer 1972, Fig. 2)
Paratypes: As originally designated (Ott and Volkheimer 1972, p. 53)

### *Redescription*

Sponge forming overlaying crusts around the stems of aquatic plants and plant remains (Fig. 1) (Ott and Volkheimer 1972, Fig. 2). Skeleton consists of main ascending skeletal fibers and of secondary thick fibers placed at right angles to main fibers thus composing a reticulum of square meshes (Ott and Volkheimer 1972, Fig. 3).
Megascleres: Stout, straight to feebly curved, sparsely, abruptly pointed oxea (length, 230–260 μm; width, 15–18 μm).
Microscleres: not detected; Gemmoscleres: absent
Gemmules: As described in the generic definition. Diameters: 483–644 μm
Distribution: Hitherto known only from type locality (valley of the Chubut River, 16 km from locality of Cerro Condor, Chubut Province, Argentina)
Age: Patagonian lower Cretaceous (Aptian), Chubut Group (Ott and Volkheimer 1972, p. 53)

### *Preserved Reproductive Characters*

This exceptionally well-preserved fossil allows study first of the reproductive process of a Cretaceous freshwater sponge. The preserved oocytes, oocyte for-

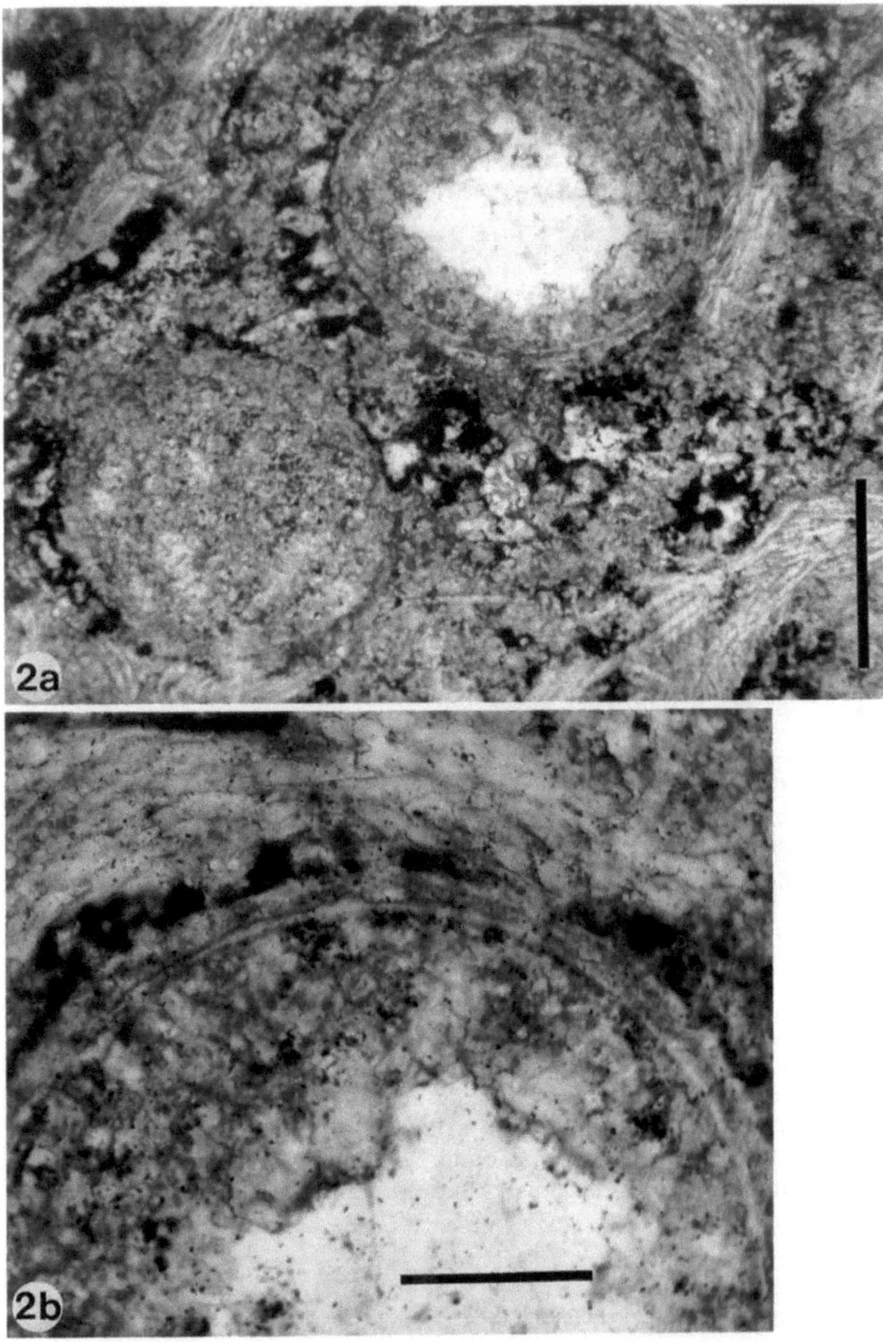


**Fig. 2.** **a** Gemmule formation in *P. chubutensis* photographed from the thin section deposited at the Australian Museum. To the right coats already formed, to the left inner coat in the forming, area of the foraminal tube last to be formed. Notice spongocytes groupped around the coats. Scale = 250 $\mu$m. **b** Detail of right gemmule of **a**. Notice inner gemular coat, pneumatic coat and the tangentially arranged megascleres of the common envelope. Scale = 125 $\mu$m

mation, and gemmule formation are comparable with extant freshwater sponges. The study of the thin section of *Palaeospongilla chubutensis* deposited in the Australian Museum (Sydney) revealed that the sponge was forming gemmules after a period of oocyte production (Fig. 2a,b). One mature oocyte (Fig. 3a,b) can be seen side by side with several empty follicles. The cytological characteristics of this oocyte and of the gemmules of *Palaeospongilla chubutensis* are listed in the legends for the figures (Fig. 2a,b; 3a,b) and generally conform with the literature on this subject (Simpson 1984).

The oocytes of *Palaeospongilla chubutensis* share a large number of identities with the oocytes of Tetractinomorpha according to the descriptions and illustrations of Liaci et al. (1971).

The oocyte was fossilized at the moment when phagocytosis was finished and a follicle epithelium was already isolated from the mesohyle cells (synvivodiagenetic). At this moment nuclear extrusion were taking place and a second nutritional phase to the oocyte was starting with forming of a crown of nurse cells. The nuclei of such nurse cell are conspicuously placed close to the follicle wall and the cell processes are already orientated towards the oocyte surface (Fig.3b).

Diameter of the oocyte plus follicle: 330 $\mu$m; diameter of the nucleus: 133 $\mu$m.

The great number of empty follicles and young gemmules, suggest an alternation of reproductive processes with drastic reduction of the time interval which elapses between sexual reproduction and gemmulation in extant freshwater sponges of temperate regions (Frost et al. 1982). Three hypotheses can tentatively explain this fact:

1. The sponge was living in a temperate region but in a habitat seasonally submerged for short periods.
2. A sudden drought occurred which induced precocious gemmulation.
3. The sponge was living in a tropical habitat subjected to seasonally long periods of drought.

## *Discussion*

*Palaeospongilla chubutensis* displays characteristics not seen in any other presently known extant freshwater sponges. For this reason, a new family is proposed to contain this monotypic fossil genus. Among the extant freshwater sponges only the *Spongilla aspinosa/Spongilla inarmata-Spongilla spoliata* group of species (Volkmer-Ribeiro and Maciel 1983) have the same common character as seen in *Palaeospongilla chubutensis*, that gemmules have no gemmoscleres and are protected by a capsule of normal megascleres. However, in the three extant species each gemmule has its own capsule, which is in extreme close contact with the inner gemmule wall, so that some megascleres could be embedded in the gemmule wall. These capsules also exhibit no distinctive particular arrangement of megascleres which are woven in disorderly fashion around the gemmule (cf. Fig. 2 in Volkmer-Ribeiro and Traveset 1987). The inner gemmular coat of these extant species is quite thin and distinguishing three layers, including a pneumatic

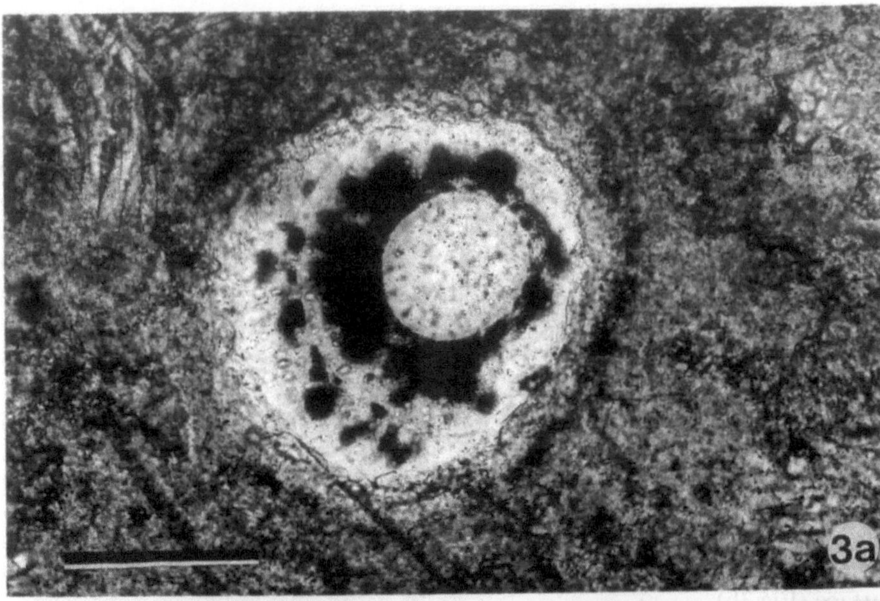
3a

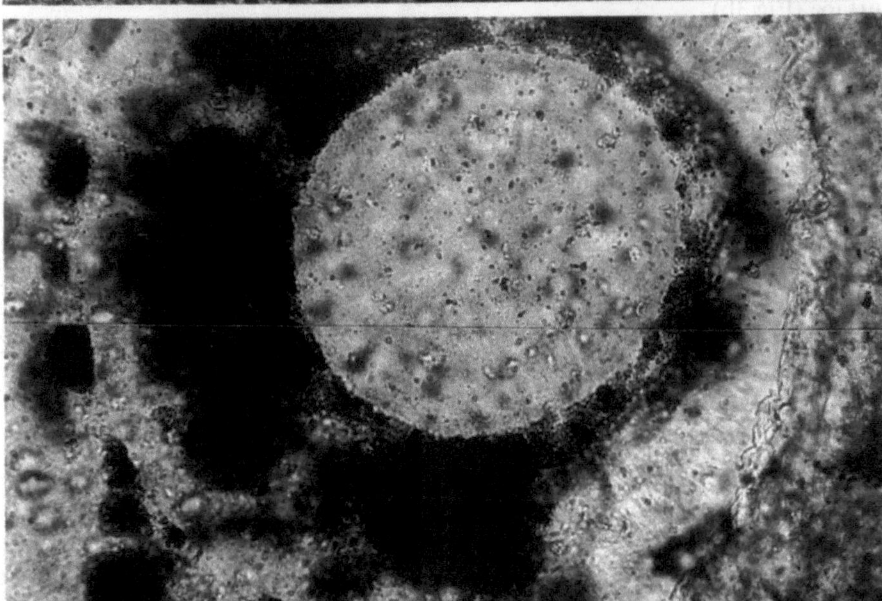
3b

coat is difficult. A very conspicuous foraminal collar is also present in the gemmules of the three extant species. The fossil freshwater sponge has no microscleres, which are present in *Spongilla*. The only microscleres within the described sponges are related to a new species, *Spongilla patagonica* n.sp. (Fig. 5a,b). Also remarkable is the localization of the gemmules of *Palaeospongilla chubutensis* from base to the top layer of the sponge, whilst in the extant *Spongilla* species the gemmules are usually located close to the sponge surface. The three *Spongilla* species form thin skeletal fibers which support waving branches and pierce the sponge surface as hispid projections. *Palaeospongilla chubutensis*, on the contrary, seems to have flat, thick, foliated crusts with lobose contour and a smooth surface (Fig. 2 in Ott and Volkheimer 1972)

2. *Spongilla patagonica n.sp.*
Family Spongillidae Gray (1867), sensu Volkmer-Ribeiro (in press)

*Spongilla* Lamarck (1816) sensu Penney and Racek (1968)

*Palaeospongilla* Ott and Volkheimer 1972, p. 53 (partim)

*Spongilla patagonica* n.sp. (Figs. 4a,b; 5a,b)

*Palaeospongilla chubutensis* Ott and Volkheimer 1972; p. 53 (partim, Fig. 5)
*Palaeospongilla chubutensis* Racek and Harrison 1975; p. 158 (partim)

Derivatio nominis: After Patagonia, a landscape in Argentina
Holotype: The holotype is located on the paratype specimen of *Palaeospongilla chubutensis* which is deposited within the Bayerische Staatssammlung für Historische Geologie und Paläontologie, Munich.
SEM preparation deposited in the Institut für Paläontologie der Freien Universität Berlin IPFUB/JR/89; (Figs. 4a,b; 5a,b).

### *Description*

Megascleres slim, gradually sharp pointed, slightly curved, smooth oxea (length 210–260 μm, width 7–10 μm).
Microscleres: usually straight, spined oxea; some slightly curved oxea occur also (length 100–110 μm, width 9–10 μm).
Gemmoscleres: short, slightly curved to strongly bent, rarely straight, strongly

**Fig. 3.** **a** Maturing oocyte of *P. chubutensis* (thin section deposited at the Australian Museum). Notice surrounding folicle ephitelium and the collagen matrix starting to form outside the epithelium. Nucleolus displaced from the sectioning plane. Scale = 250 μm. **b** Detailed magnification of **a**. Notice dense material being extruded from the nucleus and the lastly phagocytized microgranular cells aligned midway between the follicle epithelium and the nuclear extrusions. The collar of nurse (spherulous) cells starting to differentiate, their nuclei are seen bounding the epithelium cells and their processes are already oriented towards the egg surface. Scale = 125 μm

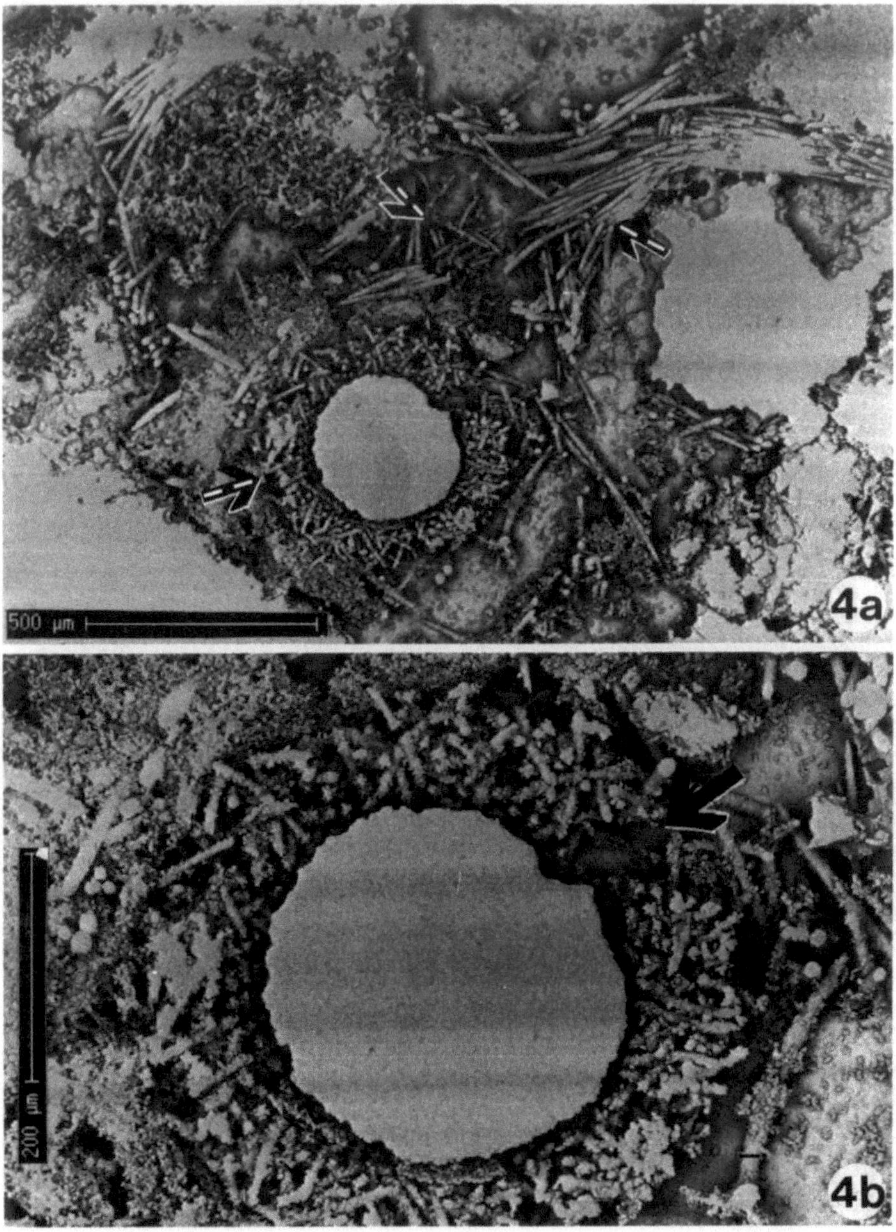


**Fig. 4.** SEM micrographs of *Spongilla patagonica* n.sp. **a** *Bottom arrow* points to one gemmule with gemmoscleres in situ: *left arrow* points to microscleres: *right arrow* indicates the thinner magascleres of *S. patagonica* placed in close contact with the stout megascleres of *P. chubutensis*. **b** Same gemmule as **a** at larger magnification: the *arrow* indicates the place occupied by the long foraminal nipple; the gemmule is seen to have had a conspicuous inner coat and a large pneumatic coat where gemmoscleres were randomly embedded

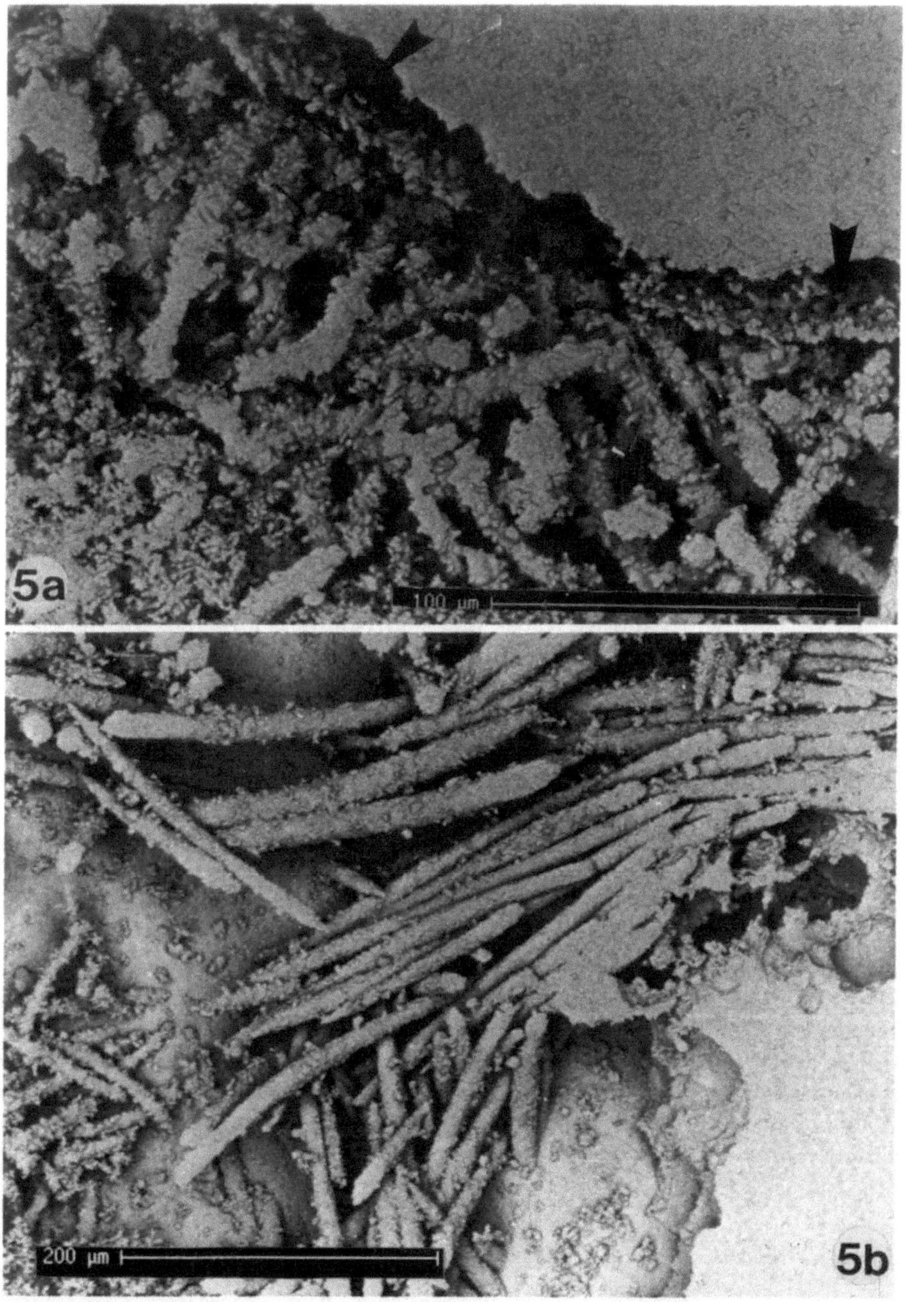


**Fig. 5. a** Same gemmule as Fig. 4a and b, showing details of the gemmular coats and of the gemmoscleres' shape; the *two arrows* indicate the place occupied by the inner gemmular coat. **b** *Spongilla patagonica* n.sp.; Detailed magnification of the area indicated by the left arrow in Fig. 4a shows the megascleres of *P. chubutensis* in the *upper part* and those of *S. patagonica* n.sp. in the *middle* and *bottom parts* of the photograph; *at left* grouping of microscleres of *S. patagonica* n.sp.

spined, cylindrical amphistrongyla, some spines larger and recurved, extremities of the scleres with a starlike profile (length 70–80 μm, width 8 μm).
Gemmules: spherical, with a thick inner- and a large pneumatic coat. The gemmoscleres are within the pneumatic layer randomly orientated and loosely packed, the tips of the spicules sometimes protruding beyond the outer surface of the gemmule coat. Foramen is single, nipple-like, extending to the surface of the pneumatic coat. The gemmule of Fig. 4b has a diameter of 469 μm. The inner coat is 20 μm and the pneumatic coat is 100 μm thick.

Distribution and geological age: Same as for *Palaeospongilla chubutensis*.

### *Discussion*

On exhaustive microscopic scrutiny and SEM preparation of material isolated from one paratype (Munich specimen) of *Palaeospongilla chubutensis*, two gemmules could be detected which are completely different from those of *Palaeospongilla chubutensis*. The gemmule coatings and gemmoscleres (Figs. 4a,b; 5a,b) are almost identical to those of the extant *Spongilla alba* Carter 1849. The gemmoscleres are identical with the ones illustrated by Ott and Volkheimer (1972, p. 58, Fig. 5), and by Racek and Harrison (1975, p. 160, Fig. 6). SEM preparations were made of gemmules of the extant species *Spongilla alba* (Fig. 6a-c) from material collected by E.H. Cordero from Jacarepagua, coastal lagoonal area of Rio de Janeiro, Brazil, in June 31, 1940. [The specimen is deposited in the Porifera collection of Museu de Ciencias Naturais of Fundacao Zoobotanica do Rio Grande do Sul (MCN), no. 1271]. The gemmules of the fossil and the extant species exhibit close affinities of the inner coat, the pneumatic coat, the foraminal area, the gemmoscleres, and the gemmosclere embedding within the pneumatic coat. A further characteristic feature of the two species is the possession of microscleres which are longer and thinner than the gemmoscleres (cf. *Spongilla alba* in Volkmer-Ribeiro and Traveset 1987, p. 227, Fig. 3). The description of a new species is justified by the fact that the fossil material does not allow for a precise perception of the spines covering the microscleres. The fossil species shows dimensions which are smaller than those shown by Penney and Racek (1968) for *Spongilla alba*. The observed dimensions agree with those presented by the same authors for *Spongilla wagneri*, a species placed by Poirrier 1976 into synonymy with *Spongilla alba*. *Spongilla alba* is thus phylogenetically closely related to *Spongilla patagonica* n.sp.

**Fig. 6.** SEM micrograph of a gemmule of the extant *Spongilla alba* Carter 1849 (from specimen MCN no. 1271). **a** Cross-section of gemmule; notice inner coat and thick pneumatic coat; gemmoscleres randomly embedded in the pneumatic coat. **b** Magnification of the area indicated by the *arrow* in **a** shows scanty spongin gluing the gemmoscleres together and the thin outer coat. **c** Detail of the gemmular walls of *S. alba*; notice gemmoscleres protruding from the pneumatic coat and the extremely variable shape of the gemmoscleres' extremities

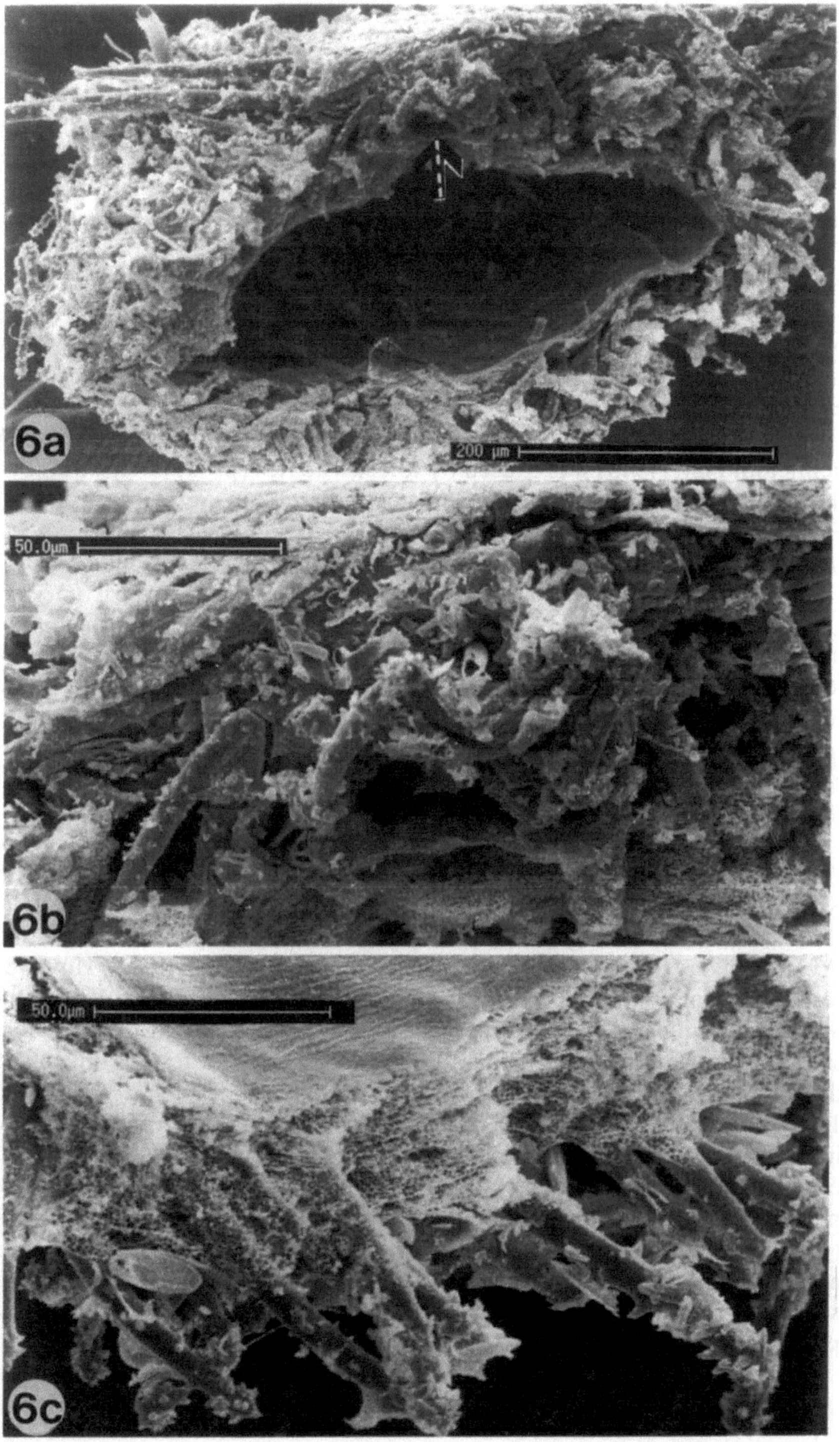
6a
200 µm
50.0µm
6b
50.0µm
6c

## General Remarks

Racek and Harrison (1975) rejected the possibility of a spicular mixture of two distinct species. Both fossil species were silicified synvivodiagenetically and exhibit no overlapping growth. Only one kind of gemmule type was recognized. Epizoism is a common feature among extant marine (Ruetzler 1970) and freshwater sponges Racek 1969; Volkmer-Ribeiro and de Rosa-Barbosa 1972). The degree of skeletal development in the associated species depends strongly on which species is the dominant one. In the present case *Palaeospongilla chubutensis* was the dominant species and the skeletal fibers described by Ott and Volkheimer belong to this species. *Spongilla patagonica* must have formed tiny patches of isolated gemmules covered by skeletal fibers soon overgrown by *Palaeospongilla chubutensis*.

The taxonomic review of *Palaeospongilla chubutensis* proposed by Volkheimer and Ott (1973) was caused by Racek's advice (p. 459) to the authors as to the close relationship between their species and *Spongilla alba* Carter 1849. However, Racek and Harrison (1975) proposed a close relationship of *Palaeospongilla chubutensis* with the extant genus *Radiospongilla* Penny and Racek (1968). They considered *Palaeospongilla chubutensis* a stem group from which the genera *Radiospongilla* and *Spongilla* evolved, each of them inheriting part of the double spicular set of *Palaeospongilla*. The fossil sponge supported Penny and Racek's (1968 p. 63) proposal of a monophyletic origin of all gemmule-producing freshwater sponges, grouped by the authors (op.cit) in the Spongillidae. *Radiospongilla* would lead the rodlike structures up to the birotulated gemmoscleres on account of the incipient rotules found in the gemmoscleres of some species of *Radiospongilla*.

After Marshall's (1883) and Brien's (1967) ideas of a polyphyletic origin for freshwater sponges, Volkmer-Ribeiro (1986, in press) restricted the family Spongillidae sensu Penney and Racek (1968). A large number of genera considered as spongillids by these authors have been transferred into the families Potamolepidae Brien and Metaniidae Volkmer-Ribeiro (1986). However, the restricted family Spongillidae is still considered by Volkmer-Ribeiro as an artificial grouping of genera deserving further studies. The close morphological similarity betwen *Spongilla alba* and *Spongilla patagonica* n.sp. is linked with the distribution and ecology of the taxon *Spongilla alba*, and *Spongilla cenota* (Harrison 1974). According to Harrison (1974), *Spongilla alba* has a circumlitoral distribution in tropical and subtropical areas with a strong tolerance to brackish waters. This ecological preference if extended to *Spongilla patagonica* fits well into the paleohabitats inferred by Ott and Volkheimer (1972) and Volkheimer and Ott (1973) for the fossil fauna and flora associated with *Palaeospongilla chubutensis*, i.e., a lacustrine habitat in a Mediterranean climate. Also the fact that both sponge species were forming gemmules at the moment of fossilization agrees with the proposition of a habitat subjected to periodical droughts.

*Acknowledgments.* The authors are indebted to Dr. Juan Boto, Invertebrate Department, Museo Bernadino Rivadavia, Buenos Aires, for loan of the holotype and the two paratypes of *Palaeospongilla chubutensis*. They acknowledge the kind assistance of Dr. F.W.E. Rowe, Porifera and Echinodermata Section, Australian Museum, Sydney, with the loan of the thin section of *Palaeospongilla chubutensis*. We thank Dr. F. J. Collier, Dept. of Paleobiology, Smithsonian Institute (NMNH), for his interest in looking for the material of *Palaeospongilla chubutensis* within his museum. They acknowledge the encouragement received from Dr. Nicole Boury-Esnault and Dr. Jean Vacelet for presenting the descriptions of the oocyte. The senior author is grateful for funds from the National Research Council of Brazil to support the project (Fellowship no. 306134/76 and Grant no. 409734-88.8). She is also indebted to Dr. H. Keupp and Dr. J. Reitner for funding, which allowed her to participate of the Conference on Fossil and Recent Sponges.

## References

Brien P (1967) Eponges du Luapula et du lac Moero. Result Scient Explor hydrobiol Bangwelo-Luapula Miss JJ Symoens 11, 58:1–52

Frost TM, De Nagy GS, Gilbert JJ (1982) Population dynamics and standing biomass of the freshwater sponge *Spongilla lacustris*. Ecology 63, 5:1203–1210

Harrison FW (1974) Sponges (Porifera: Spongillidae) In: Hart CW, Fuller S (eds) Pollution ecology of freshwater invertebrates. Academic Press, New York, pp 29–66

Liaci L, Scisioli M, Matarrese A (1971) La riproduzine sessuale di alcuni tetractinomorpha (Porifera). Atti Soc Peloritana Sci Fis Mat Nat 17:235–245

Marshall W (1883) On some new siliceous sponges collected by Mr. Pechuel-Loesche in the Congo. Ann Mag Nat Hist 12:301–412

Ott E, Volkheimer W (1972) Palaeospongilla chubutensis n.g. et n.sp. Ein Süsswasserschwamm aus der Kreide Patagoniens. N Jb Geol Paläont Abh 140:49–63

Penny JT, Racek AA (1968) Comprehensive revision of a worldwide collection of freshwater sponges (Porifera: Spongillidae). Bull Smith Inst US Nat Mus 272:1–184

Poirrier MA (1976) A taxonomic study of the *Spongilla alba*, *S. cenota*, *S. wagneri* species group (Porifera: Spongillidae) with ecological observations of *S. alba*. In: Harrison FW, Cowden RR (eds) Aspects of Sponge Biology. Academic Press, New York, pp 203–213

Racek AA (1969) The freshwater sponges of Australia (Porifera: Spongillidae). Aust J Mar Freshwat Res 20:267–310

Racek AA, Harrison FW (1975) The systematic and phylogenetic position of *Palaeospongilla chubutensis* (Porifera: Spongillidae). Proc Linn Soc NS Wales 99, 3:157–165

Reitner J, Engeser T (1987) Skeletal structures and habitats of Recent and fossil Acanthochaetetes (subclass Tetractinomorpha, Demospongiae, Porifera). Coral Reefs 6:13–18

Ruetzler K (1970) Spatial competition among Porifera: solution by epizoism. Oecologia 5:85–95

Simpson TL (1984) The cell biology of sponges. Springer, Berlin Heidelberg New York Tokyo, pp 662

Volkheimer W, Ott E (1973) Esponjas de agua dulce del cretacico de la Patagonia. Nuevos datos acerca de su posicion sistmatica y su importancia paleobiografica y paleoclimatica. Actas Quinto Congr Geol Argent 3:455–461

Volkmer-Ribeiro C (1986) Evolutionary Study of the freshwater sponge Genus Metania Gray, 1867: III Metaniidae, New family. Amazoniana IX (Kiel), 493–509

Volkmer-Ribeiro C (in press) A new insight into the systematics, evolution and taxonomy of freshwater sponges. In: Rützler K, Hartmann WD (eds) New Perspectives in Sponge Biology. Smithsonian Institution Press, Washington

Volkmer-Ribeiro C, De Rosa-Barbosa R (1972) On *Acalle recurvata* and an associated fauna of other freshwater sponges. Rev Bras Biol 32, 3:303–317

Volkmer-Ribeiro C, Maciel SB (1983) New freshwater sponges from Amazonian waters. Amazoniana 8, 2:255–264

Volkmer-Ribeiro C, Traveset A (1987) Annotated catalog of the type specimens of Potts' species of freshwater sponges. Proc Acad Nat Sci Phila 139:223–242

# A Systematic Revision of the Central-Atlantic Halichondrida (Demospongiae, Porifera). Part I: Evaluation of Characters and Diagnosis of Genera

M.C. Diaz, R.W.M. van Soest, and S.A. Pomponi

## Introduction

The order Halichondrida has been erected to separate demosponges that possess a confused skeletal arrangement, a simple spicule complement of oxea, and/or styles and a differentiated ectosomal skeleton (Levi 1973; Bergquist 1978; Hartman 1982). Although at present, they are considered a separate order, the problem is recognized of establishing the relationships within the group and with other Demospongiae orders.

The simplicity of its skeletal elements has been claimed as a basic problem for the taxonomy of the Halichondrida (Levi 1957). The classification of the group was heavily affected by the major change in the taxonomy of Porifera introduced by Levi (1956), and based on consideration of reproductive strategies as main systematic characters. Before this, the group was considered to be part of a varying assemblage of families always including Axinellidae, Ridley and Dendy 1887 (see Topsent 1928; De Laubenfels 1936).

The occurrence of viviparous larvae in some species of the typical genera *Hymeniacidon* and *Halichondrida* as well as in all Haplosclerida, Poecilosclerida, and most Keratosa studied, resulted in the placement of these orders in the subclass Ceractinomorpha.

The order Halichondrida was redefined to include only the families Halichondriidae and Hymeniacidonidae. The other families previously grouped with the Halichondrids because of morphological similarities, were placed in the Tetractinomorpha, based on the assumed oviparous reproduction.

The existence of a number of exceptions to the oviparity-viviparity rule (Reiswig 1976; Bergquist 1978, 1980; Hoppe and Reichert 1987), the low number of genera per order upon which these important phylogenetic conclusions were based, and the recognition of the value of morphological characters in the ecology, evolution and systematics of the Porifera lead us to recommend that the present scheme of classification be revised. The order Halichondrida, families Halichondriidae and Hymeniacidonidae sensu Levi 1973, can be associated with some Axinellida sensu Levi 1973 (excepting Raspailiidae, Euryponidae, Sigmaxinellidae, and Hemiasterellidae), by a convincing series of intermediate genera and families. The major features of Axinellida (Levi 1973) i.e., axial condensation of choanosomal skeleton and indiscriminate occurrence of styles and oxea, and the major features of Halichondrida, i.e., ectosomal tangential

J. Reitner and H. Keupp (Eds.)
Fossil and Recent Sponges

skeleton and confused choanosomal skeleton, intergrade nicely in the series of genera: *Axinella* (Axinellida) – *Phakellia* (Axinellida) – *Myrmekioderma* (?) – *Acanthella* (?) – *Dictyonella* (?) – *Axynissa* (?) – *Topsentia* (Halichondrida) – *Ciocalypta* (Halichondrida). Also, the genus *Didiscus*, presently located in the order Hadromerida or Poecilosclerida, is morphologically more similar to some Halichondrida genera than to hadromerid or poecilosclerid genera.

Our first approach to evaluate the classification of the group was to examine representative genera from the Central Atlantic presently included in the order or related to it by strong morphological similarities.

We analyzed specimens included under a new concept of the Halichondrids, viz, Demospongiae with a confused to plumoreticulate skeletal architecture built of interchangeable styles and oxea and intermediate spicules of widely diverging sizes and not functionally localized, and which usually develop some kind of ectosomal skeleton.

The genera included in this concept are: *Spongosorites*, Topsent 1896; *Halichondria*, Fleming 1828; *Hymeniacidon*, Bowerbank 1862; *Didiscus*, Dendy 1922; *Topsentia*, Berg 1899 (including *Coelocalypta*); *Myrmekioderma*, Ehlers 1870; *Epipolasis*, De Laubenfels 1936; *Petromica*, Topsent 1898; *Ciocalypta, Bowerbank 1862; Axinyssa*, Lendenfeld 1897; *Amorphinopsis*, Carter 1887; *Collocalypta*, Dendy 1922. (The later two genera are not represented in our collections in the Central Atlantic).

Our main objective in this chapter is to evaluate the taxonomic value of the characters presently used in the classification of the group and to produce clear diagnoses of the most common genera from the Central Atlantic. Possible phylogenetic relationships within the Halichondrida are discussed. This study is the first part of a revision of the Halichondrida undertaken by van Soest, Pomponi, and Diaz.

## Materials and Methods

Specimens studied include material collected by the authors and museum specimens from the Central Atlantic area. Specimens from the Bahama Islands, Gulf of Mexico, southeast Florida, Caribbean Colombia and Venezuela are part of the collections of the Division of Biomedical Marine Research, Harbor Branch Oceanographic Institution (HBOI). Barbados and Curacao specimens are part of the collections of the Institute of Taxonomic Zoology, University of Amsterdam (ZMA) (Fig. 1).

Shallow-water samples (from depths less than 40 m) were collected by scuba diving and snorkelling. Deep water samples (40 to 1000 m) were collected by trawling or by the Johnson-Sea-Link manned submersibles.

Museum material was studied from the Zoologisch Museum, University of Amsterdam (ZMA), the National Museum of Natural History, Washington D.C. (NMNH), Yale Peabody Museum, Yale University (YPM), the British Museum of Natural History, London (BMNH) and the Museum National d'Histoire Naturelle, Paris (MNHN).

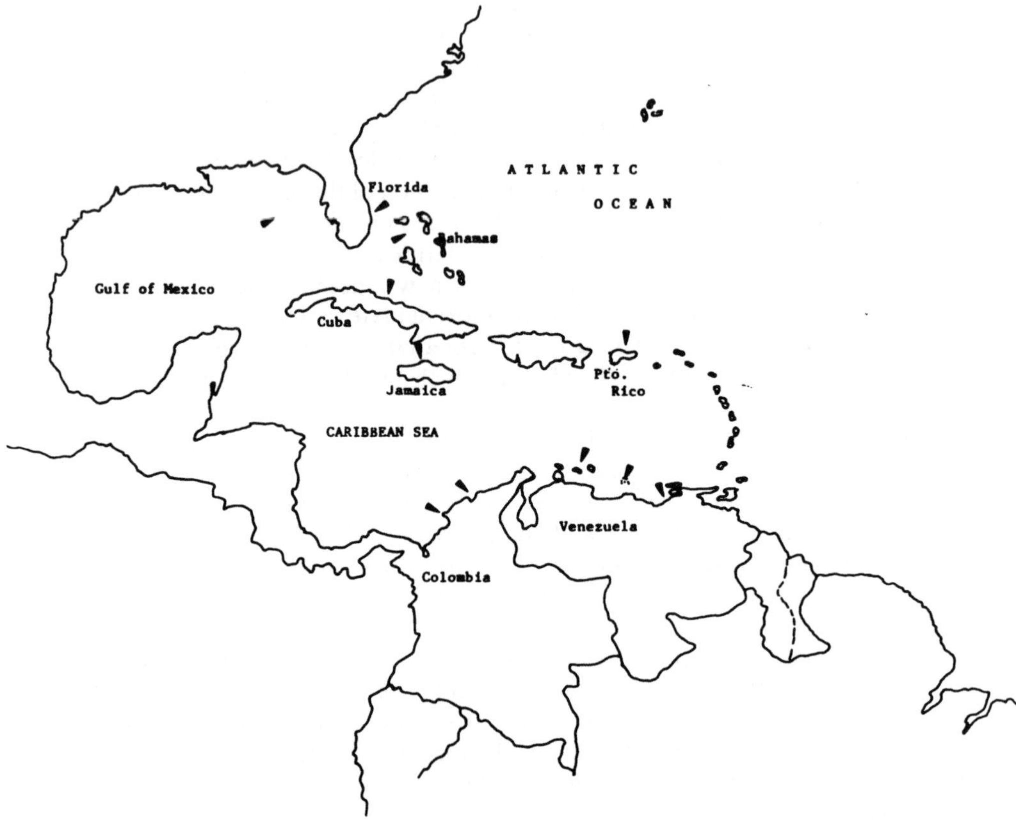


**Fig. 1.** Localities where sponge material has been collected by the Division of Biomedical Marine Research, Habor Branch Oceanographic Institution

Spicule preparations were made by dissolving small fragments of the sponges in concentrated nitric acid, rinsing with water, dehydrating in ethanol, and mounting with Permount. Thick sections were made by hand or microtome on unembedded or paraffin-embedded samples to study skeletal architecture. Approximately 100 specimens in ten genera were studied. Spicule measurements and skeletal architecture were the primary basis for comparisons. Data on morphology, habitat, and ecology were analyzed for all the material collected by the authors and part of the museum material.

## Results

Characters previously considered as diagnostic for a genus were found to vary intragenerically or sometimes intraspecifically. Analysis of each morphological character and diagnoses of the genera studied are presented.

***Morphological Characters***

*Growth forms.* Halichondrids in general are massive amorphous to lobate in shape. Few genera are characterized by a particular habit. Exceptions are *Ciocalypta* and *Collocalypta*, which consist of a massive base buried in the sand, from which fistules project. The fistules of *Ciocalypta* have an tangential ectosomal crust, while this is absent in *Collocalypta*. Shape can vary intraspecifically, so it is not a valid character for species or generic distinctions.

*Surface.* The surface texture is related to the ectosomal and/or choanosomal skeleton, which is characteristic for most of the genera.

A paratangential or tangential ectosomal arrangement results in a smooth surface as in *Halichondria* and *Spongosorites* or a microhispid surface as in *Topsentia*, and *Myrmekioderma*. The conulose surface of *Axinyssa* is formed by radial tracts of spicules which project to the surface.

A grooved surface as in *Myrmekioderma* and *Didiscus* occurs as thick and sinuous depressions (e.g., *M. styx*), or as thinly channeled surface depressions (e.g., *M. rea* and *D. oxeata*). Observation of such features is often difficult in museum specimens. Nevertheless, we conclude that the surface in this group is a valid character, both for generic and species diagnoses.

*Consistency.* This character is related to the density of spicules, the quantity of spongin present, and the arrangement of skeletal elements. It can have generic taxonomic value, e.g., the brittle siliceous consistency of the genus *Topsentia*. It is a consistent character among the species of a genus, and only rarely will it have diagnostic value at the species level.

*Color.* The taxonomic value of this character is low for most genera. *Spongosorites* species are characterized by a color change from yellow to brown or black in alcohol. Some specimens of *S. ruetzleri* are aerophobic, which may be correlated with their association with a vermetid gastropod.

*Choanosomal Skeleton.* The tendency of spicules to be strewn or grouped in tracts, the proportion of spicules to organic elements, and the orientation, if any, of the skeleton, are characteristic of the generic level. *Topsentia*, with a high density of spicules and brittle consistency and low spongin content represents an extreme case of utterly confused spicules. *Halichondria*, *Hymeniacidon*, *Myrmekioderma*, *Didiscus*, and *Spongosorites* represent a medium condition in which spicules, spongin and tissue are present in about equal amounts.

In *Hymeniacidon* and *Halichondria* the spicules tend to form directionless tracts, while in both *Myrmekioderma* and *Didiscus*, strewn spicules and tracts are in equal proportions, with a tendency to a radial orientation of tracts.

*Ectosomal Skeleton.* This character has strong taxonomic value at the generic level. The ectosome can be formed by small or large spicules, arranged tangentially, paratangentially or perpendicularly (palisade), forming organized tracts or strewn in confussion. Some genera have a consistent ectosomal architecture: *Spongosorites* (paratangential crust), *Myrmekioderma-Didiscus* (palisade and tangential), *Petromica-Ciocalypta* (radial columns supporting a tangential organized skeleton in the fistules), *Hymeniacidon-Halichondria* (detachable tangential layer).

In other genera, however, the ectosomal skeleton can vary among species (e.g. *Topsentia*) or even within a single individual (*Epipolasis suluensis*). Therefore we can not assign a particular ectosomal type with a single genus or genus-group, as can be done in other Demospongiae (e.g., Haplosclerida).

*Spiculation.* The spicule complement of the group has always been referred to as a simple one, of oxea and styles. It is necessary to add that in general the spicules are present in more than one size category, and that the predominance of oxea or styles, or the existence of both shapes without discrimination in a genus is a stable taxonomic character.

The smaller category of spicules usually occurs in the ectosomal skeleton, forming for example, a perpendicular palisade in *Myrmekioderma-Didiscus*, or a crust in *Spongosorites*.

Spicule malformations have been used repeatedly in the past to characterize some genera, but were found in our study to vary intragenerically or intraspecifically, thus losing their diagnostic value. The possession of double bent spicules, previously considered diagnostic for *Spongosorites*, was found to vary intraspecifically in *S. ruetzleri*, and was not common in the other two species studied. Another striking example is the case of *Myrmekioderma*, characterized in the past by the possession of acanthose microxeas. After studying approximately 30 specimens of *M. styx* from different areas and depths in the Caribbean and Bahamas, specimens were found with acanthose or smooth oxeas or a combination of both. As a consequence, two genera will have to be synonymized (*Myrmekioderma* and *Anacanthea*) and several species will be included now in *Myrmekioderma* (*Epipolasis reiswigi* Diaz et al. 1987 = *M. styx*, *Anacanthea rea* De Laubenfels 1934 = *M. rea*).

The existence of true diactinal microscleres, filling the main skeleton is recognized. Trichodragmata, which are widespread among Axinellidae, Poecilosclerida and even occur in Spirophorida probably represent the retention of an ancestral character.

## Diagnosis of the Halichondriidae Genera

*Halichondria* Fleming 1828
Type: *Spongia panicea* Pallas 1766.

Species studied: *H. panicea* (Pallas 1766) (YPM 1925, 8492; ZMA from Europe); *H. bowerbanki* Burton 1930 (ZMA from Europe); *H. melanadocia* De Laubenfels 1936 (types YPM 5218, NMNH 22463, 30247; HBOI and ZMA specimens from the Central Atlantic); *H. magniconulosa* (holotype YPM 5039; HBOI and ZMA from the Caribbean); *H. bibula* (NMNH 1017); *H. modesta* (Pulitzer-Finali 1986 as *Aponastra modesta*) (Holotype from Mus. Genova); *H. lutea* Alcolado 1984 (fragment in the NMNH, ZMA, and HBOI from the South Caribbean).

*Definition (Fig. 2).* Massive amorphous Demospongiae with a basic skeleton formed by ill-defined, directionless spicule tracts, and a clearly detachable tangential ectosomal skeleton. The spicules, in a wide size range, are predominantly oxea, although styles or styloids are sometimes present. Tufted larvae have been found for *H. panicea* and *H. bowerbanki* (Wapstra and van Soest 1987). It is assumed that they occur in all *Halichondria* species.

*Remarks.* The density of the spicule tracts and their orientation vary among the species studied. An oblique to parallel direction of the tracts is most common. The development of the ectosomal skeleton also varied intragenerically. It ranges from a packed tangential arrangement of spicule bundles as in *H. panicea*, to a dermal organic layer with a low density of scattered spicules, as in *H. lutea*.

*Distribution.* Cosmopolitan, more diverse in cold waters. Common in shallow coastal environments as rocky shores and mangroves (*H. melanadocia*, *H. panicea*, *H. bowerbanki*). *Halichondria lutea* is restricted to deep areas (20–30 m) of coral reefs or steep shores, where coral rubble and sand are the main substrates.

*Hymeniacidon* Bowerbank 1864.
Type: *H. caruncula* Bowerbank 1864.

Species studied: *H. perlevis* (Montagu 1818) (ZMA specimens, from Europe), *H. heliophila* (Parker 1910) (NMNH 23341, 23581; HBOI from Florida); *H. caerulea* Pulitzer-Finali, 1986 (NMH & ZMA from the Caribbean).

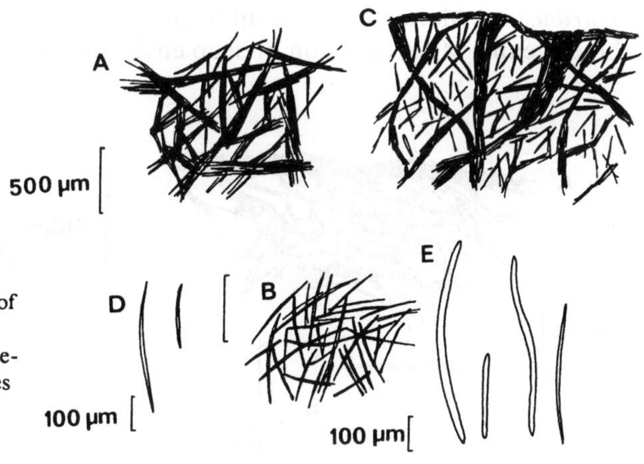


**Fig. 2.** *Halichondria* spp. **A** Tangential view of ectosome of *H. melanadocia*. **B** Tangential view of ectosome of *H. lutea*. **C** Perpendicular section of peripheral skeleton of *H. melanadocia*. **D** Spicules of *H. melanadocia*. **E** Spicules of *H. lutea*

*Definition (Fig. 3).* Massive encrusting to fistulose Demospongiae with a skeletal plan similar to that of *Halichondria*. The main difference is that oxeote spicules have been lost, retaining styles and stylotes.

*Remarks.* The tangential ectosome is normally less well developed than for *Halichondria*, and there is a decidely fleshly consistency in the type and many other species.

*Distribution.* Cosmopolitan, but not Antartic or Artic. Available information points to shallow coastal environments, including intertidal areas.

*Spongosorites* Topsent 1896.
Type: *S. placenta* Topsent 1896.

Species studied: *S. placenta* (NMNH Slide DT. 904, 1895 and topotypical specimen); *S. ruetzleri* (ZMA and HBOI specimens from Bahamas and Barbados); *S. siliquaria* (ZMA specimens from Barbados); *Spongosorites* sp. (ZMA and HBOI from Bahamas).

*Definition (Fig. 4).* Massive amorphous to subspherical Demospongiae with a smooth, thick ectosomal crust (50–400 μm) of relatively thin spicules, paratangentially arranged. The choanosomal skeleton of spicules strewn in confusion and occasionally grouped in spongin enforced tracts, directed mainly parallel and oblique to the surface.

*Remarks.* Main changes in the definition have been a consequence of the invalidity of crooked oxea as a generic character, the consistency of the ectosomal crust, and the choanosomal skeletal arrangement. The taxonomy of *Spongosorites* spp. has been confused in the past (Bergquist 1968; De Laubenfels 1936); however, the characters in the revised definition are clearly represented in all the material studied.

*Distribution.* Mainly tropical, subtropical and occasionally temperate in all the three oceans. More common in deep environments, 150–600 m deep.

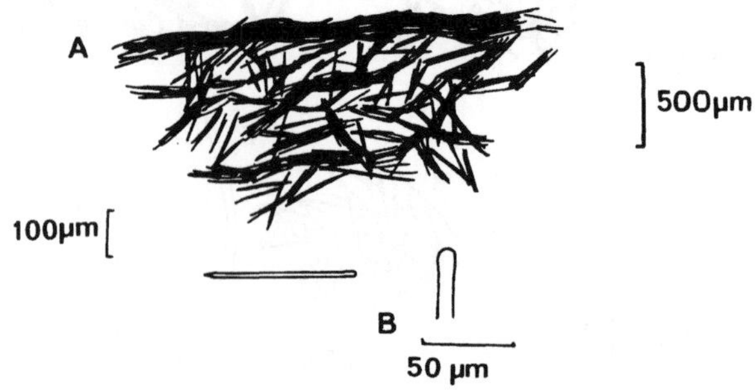


**Fig. 3.** *Hymeniacidon heliophila*. **A** Perpendicular section of peripheral skeleton. **B** Style

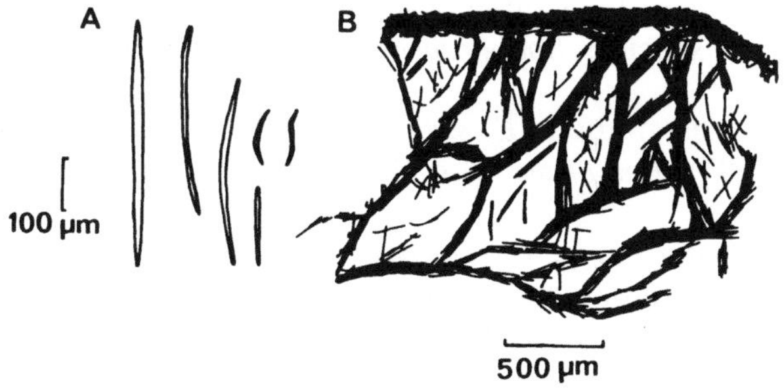


**Fig. 4.** *Spongosorites* sp.
**A** Spicules. **B** Perpendicular section of peripheral skeleton

*Topsentia* Berg 1899
Type: *Anisoxya glabra* Topsent 1898

Species studied: *T. glabra* (Topsent 1898) (MNHN type slide DT. 1169 and topotypical specimen DT. 1389); *T. sinuata* (Pulitzer-Finali 1986 as *Spongosorites*) (holotype Mus. Genova and ZMA specimens); *T. roquensis* (holotype ZMA. 5408 and specimens from Central Atlantic).

*Definition (Fig. 5).* Massive amorphous to lobate, brittle Demospongiae with a skeleton apparently completely lacking spongin; as a consequence spicules show a confused, directionless and packed arrangement around canals, subdermal spaces, etc. Smaller spicules usually arranged without any organization produce a compact, paratangential, ectosomal layer, creating a microhispid surface. Spiculation consists of oxeas in a wide size range, with two or three size-classes distinguished. Twisted, bent and double-bent spicules sometimes present. No microscleres.

*Remarks.* Species in the genus lack bright colors; dirty white, greenish drab or brownish tones are predominant. Characters for specific diagnosis are ectosomal characteristics, spicule shape, habit, and color.

*Distribution.* Tropical and subtropical, in all three oceans. More common in deep environments (> 15 m).

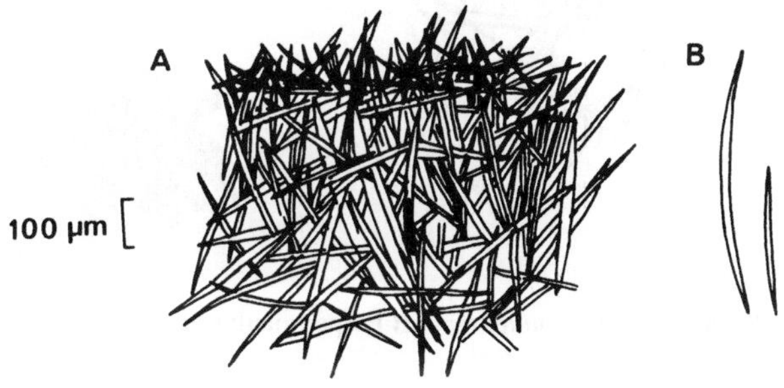


**Fig. 5.** *Topsentia* sp.
**A** Perpendicular section of peripheral skeleton.
**B** Spicules

*Epipolasis* De Laubenfels 1936.
Type species: *Spongosorites suluensis* Wilson 1925.

Species studied: *E. suluensis* (Wilson 1925 as *Spongosorites*) (type NMNH 21297); *Epipolasis* spec. (HBOI from Bahamas).

*Definition (Fig. 6).* Massive-amorphous to flabellate Demospongiae with a *Topsentia*-like choanosome of oxea in a wide size range and an ectosomal palisade of smaller spicules, tightly packed. Trichodragmata almost always present.

*Remarks.* The original definition is largely modified, adding detailed information about the ectosomal specialization and trichodragmata. The previous characterization emphasizing the radial spicule tracts and the wide size range of oxea would include *Myrmekioderma* and *Anacanthea*.

The holotype is flabellate in shape with a different ectosomal type on each side of the body: a parchment-like feltwork of tangential oxea in the smooth side and a palisade in the oscular side. In spite of the discussion of the taxonomy of this group (Bergquist 1968; De Laubenfels 1936) the combination of a compact spicule skeleton lacking spongin, with an ectosomal palisade, differentiates the two studied species from the other genera.

*Distribution.* Philippines, Bahamas. Records restricted to two specimens, one from shallow and the other from deep water.

*Ciocalypta* Bowerbank 1862.
Type: *C. penicillus* Bowerbank 1862.

Species studied: *C. penicillus* Bowerbank 1862 (holotype in BMNH and topotypical specimens in BMNH and ZMA); *C. massalis* (Carter 1886 as *Leucoph-*

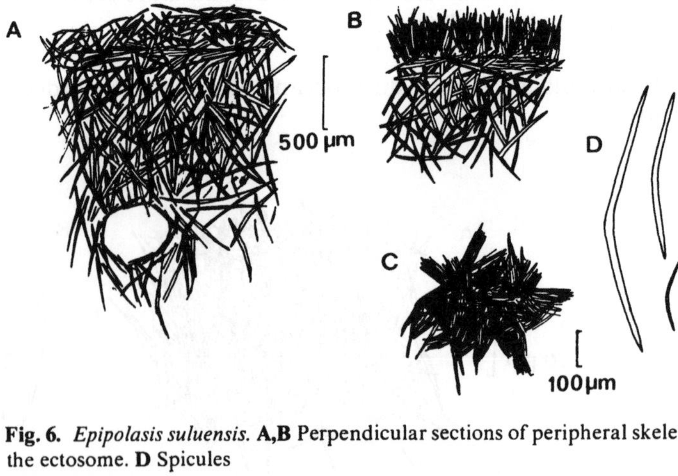


**Fig. 6.** *Epipolasis suluensis.* **A,B** Perpendicular sections of peripheral skeleton. **C** Tangential view of the ectosome. **D** Spicules

*loeus*) (Topotypical specimens in BMNH), *Ciocalypta* sp. (ZMA specimen from Indonesia).

*Definition (Fig. 7).* Massive fistulose Demospongiae with a *Topsentia*-like choanosomal skeleton, a tangential ectosomal layer of spicule bundles and fistule projections supported by a characteristic skeletal architecture. Fistule skeleton consists of a central column of spongin-enforced anastomosing longitudinal spicule tracts, from which issue at regular distances perpendicular spicule bundles, which cross extensive subdermal lacunae and support the ectosomal crust. Spiculation mainly of styles although oxea can occur.

*Remarks.* Variations among species are found in the spicule type, in the development of ectosomal skeleton and in external characteristics.

*Distribution.* Temperate, tropical and subtropical regions of all oceans. Typically in sandy bottoms, buried and with projecting fistules.

*Petromica* Topsent 1898
Type: *P. grimaldi* Topsent 1898.

Species studied: *P. ciocalyptoides* (Van Soest and Zea 1986, as *Monanthus*) (holotype in the Leiden Museum, ZMA and HBOI specimens from the Caribbean).

*Definition (Fig. 8).* Massively encrusting to fistulose Demospongiae with a basic *Topsentia*-like choanosomal skeleton which shows some radiality towards the surface, a tangential, well-developed ectosomal reticulation of spicule bundles, and a sublithistid spiculation of styles or oxeas and desmas. Fistulose hollow projections with an architecture very similar to that of *Ciocalypta*, but with the supporting skeleton situated in the wall of the fistule.

*Remarks.* Species differ in spicule modifications and size and in external characters. Two species representing the genus have been previously reported, *P.*

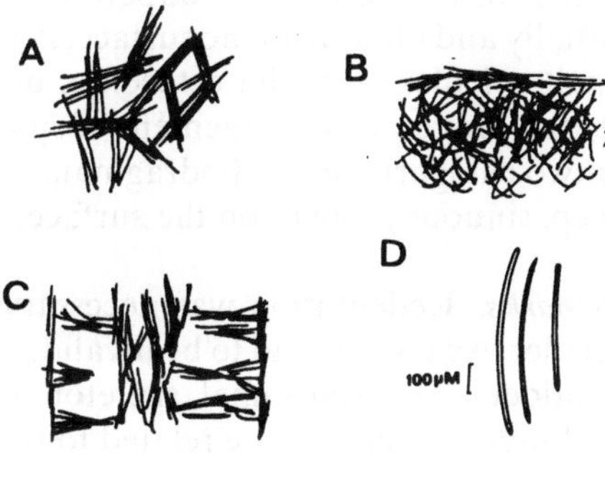


**Fig. 7.** *Ciocalypta* sp. **A** Tangential view of ectosome. **B** Perpendicular section of basal body. **C** Perpendicular section of fistule. **D** Spicules

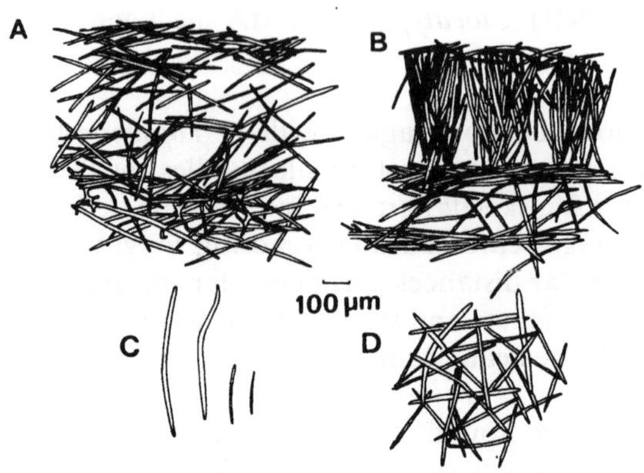


**Fig. 8.** *Petromica ciocalyptoides.* **A** Perpendicular section of the peripheral skeleton of the basal mass. **B** Perpendicular section of fistule. **C** Spicules. **D** Tangential view of ectosome

*plumosus* from South Africa and *P. ciocalyptoides* from the West Indian region. *Petromica ciocalyptoides* shows strong similarities in skeleton and habit to *Ciocalypta* species, which suggests a close relationship between the two genera. Desmas are of low taxonomic value because they might have been lost or independently evolved in several, unrelated lineages.

*Distribution.* Caribbean and southeast Africa. Restricted to sandy, rubble bottoms at depths of 10–40 m.

*Myrmekioderma* Ehlers 1870
Type: *Alcyonium granulatum* Esper 1794.

Species studied: *M. granulata* (Esper 1794) (ZMA specimens, holotype of *Neoprosypa atina*); *M. styx* De Laubenfels 1953 (holotype NMNH 23400; YPM 6281, 7701, 4727; HBOI and ZMA from the Carribean); *M. rea* De Laubenfels 1934 (as *Anacanthea*) (holotype NMNH 22 301; HBOI and ZMA specimens from the Caribbean); *M. spelea* Pulitzer-Finali 1983 (as *Raphisia*); *Epipolasis reiswigi* (= *M. styx*).

*Definition (Fig. 9).* Massive-amorphous to lobate Demospongiae, with spicules strewn in confusion or grouped in spongin-enforced plumose tracts, running radially and oblique to the surface. A palisade, vertically packed arrangement of small spicules from the ectosome, usually continued internally by a paratangential-tangential arrangement of spicules. Two size categories of oxea and one or two categories of trichodragmata are present. Most species have a pattern of deep, sinuous grooves on the surface.

*Remarks.* Redefinition was necessary because the acanthose condition of the smaller oxea was found to be invalid. A more detailed description of the skeleton is added. The ectosomal skeleton ranges from tangential to perpendicular (palisade) which may be related to the surface structure (i.e. grooves).

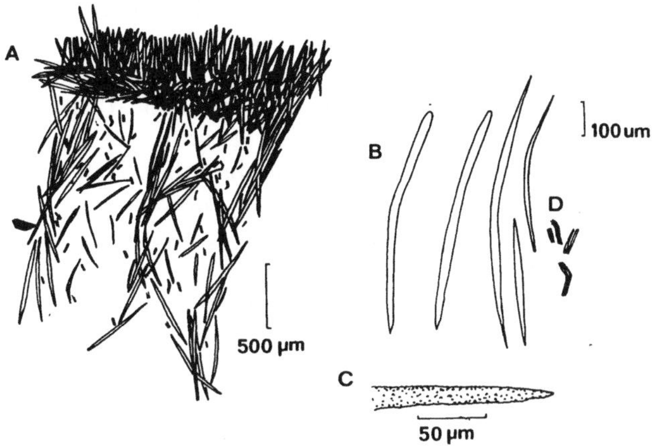


**Fig. 9.** *Myrmekioderma styx*. **A** Perpendicular section of peripheral skeleton. **B** Megascleres. **C** Acanthoxea. **D** Trichodragmata

*Distribution.* Tropical and subtropical, three oceans. Hard substrates, sand and rubble bottoms in subtidal waters, 20–100 m depth.

*Didiscus* Dendy 1922
Type: *D. placospongioides* Dendy 1922.

*Species studied: D. oxeata* Hechtel 1983 (holotype YPM 8968 from Brasil; YPM 4874, 5327, 4631; HBOI and ZMA specimens from the Carribean); *D. habanensis* (= *D. oxeata*) (fragment in the NMNH).

*Definition (Fig. 10).* Massive, amorphous to lobate Demospongiae with spicules strewn in confusion and sporadically grouped in spongin-enforced directionless tracts. The ectosome is a gradation from a perpendicular palisade to a tangential or paratangential arrangement of spicules. The skeleton is very similar to *Myrmekioderma*, except for the presence of discorabds microscleres which are concentrated in the ectosome and scattered throughout the choanosome. Most species exhibit a system of grooves in the surface.

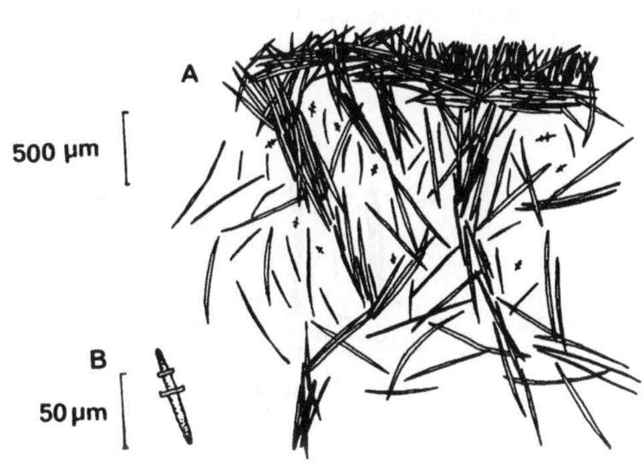


**Fig. 10.** *Didiscus oxeata*. **A** Perpendicular section of peripheral skeleton. **B** Discorhabd

*Remarks.* The main modification with respect to previous definitions is the observed relationship with *Myrmekioderma*. *Didiscus oxeata* and *M. styx* are indistinguishable in the field, without microscopic evaluation of the spicules.

The striking similarity in all morphological aspects decreases the value of the presence of discorhabds as a basis for grouping *Didiscus* with the Hadromerida.

It is considered here that the discorhabds of *Latrunculia* and *Didiscus* are not homologous. The discorhabds of *Latrunculia* have discs with rays and spines, similar to asterose microscleres of the Hadromerida. *Didiscus* discorhabds are microxea with smooth or frayed discs. Considering all the similarities that the *Myrmekioderma* and *Didiscus* share with the basic concept of the Halichondrida, it is proposed here to include them within the group.

*Distribution.* Tropical, subtropical three oceans. Ecology similar to *Myrmekioderma* spp.

*Axinyssa* Lendenfeld 1897.
Type: *Axinyssa topsenti* Lendenfeld 1897.

Species studied: *A. topsenti*, Lendenfeld 1897 (holotype ZMB); *A. tethyoides* Kirkpatrick 1903 (holotype BMNH); *A. ambrosia* (De Laubenfels 1934, as *Raphisia*) (holotype NMNH 22445, ZMA from Curacao); *A. lewisi* (Van Soest and Stentoft 1988, as *Leucophloeus*) (holotype ZMA 5401; HBOI specimens from Bahamas).

*Definition (Fig. 11).* Massive-amorphous Demospongiae with spicules strewn in confusion or forming lose bundles in the interior of the body; radial tracts of spicules separated at regular intervals give rise to conules at the surface. The ectosome is a thick organic skin with sparsely scattered spicules. Strongly collagenous mesohyl. Spicules are oxea, strongyloxea or styles usually of one size class.

*Remarks.* Burton's definition: "Axinellidae of massive form in which the skeleton is composed of vertically ascending fibers of oxea of one sort only, the fibers branching and anastomosing throughout their length and ending at the surface in small tufts which project to a greater or lesser extent".

The holotype specimen of Kirkpatrick is quite unusual in many ways: it is black, it has a dendritic reticulation of spongin bound tracts, a thick organic skin,

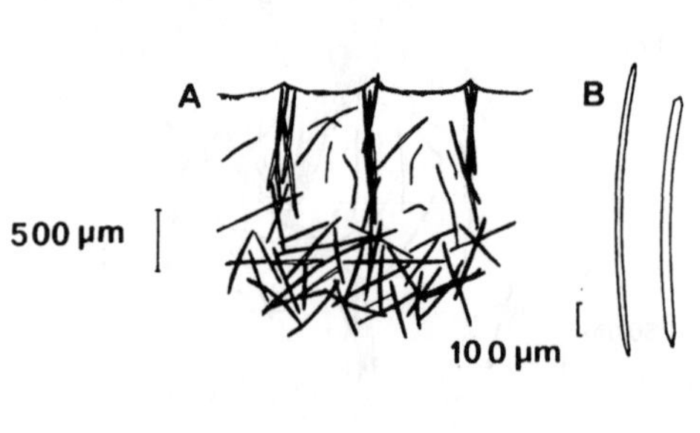


**Fig. 11.** *Axinyssa lewisi*. **A** Perpendicular section of peripheral skeleton. **B** Spicules

and large quantities of choanosomal organic foreign material; the spicule tracts are quite discrete and seem to be unattached to the organic parts. Possibly the specimen is not a healthy representative of the species as no other similar specimens have ever been described. Burton's choice of it as the type of an otherwise quite common group of mostly tropical Halichondrids lacking ectosomal skeletal protective cover is most unfortunate. However, Burton makes it clear that this genus was erected to accommodate species without distinct size categories of oxeotes and which do not have a reticulated skeleton, but rather a confused one.

*Distribution.* Tropical regions of three oceans. Shallow reef to deeper water environment.

## Discussion and Conclusions

Although it is acknowledged that the difficult taxonomy of the Halichondrids is due mainly to their simple spicule complement, we have found that the combination of various skeletal characteristics such as spicule type, tendency to form tracts, orientation of skeleton and ectosomal specializations, allows us to propose revised definitions of halichondrid genera.

The necessary changes in generic diagnostic characters have decreased the taxonomic value of certain spicule morphologies and malformations (e.g., spined, bent). Trichodragmata and discorhabds are regarded as useful generic diagnostic characters. The diactinal nature of all the microscleres found in the group supports their close relationship. The skeletal arrangement for the genera studied is considered valuable in characterizing related genera.

We propose to include within the Halichondrids genera previously placed in other orders (*Myrmekioderma*, *Didiscus*, *Epipolasis*, *Spongosorites*). Their skeletal arrangement, spicule complement, and habit show more similarity to typical halichondrid genera (*Topsentia*, *Halichondria*, *Hymeniacidon*) than to other Demospongiae.

The intergradation in the occurrence of oxea, and styles found in some genera (e.g., *Ciocalypta*, *Spongosorites* and *Halichondria*) strongly suggests that the division of families based on the predominance of oxea or styles can no longer be upheld. The halichondrid genera defined above can be included in the family Halichondriidae Vosmaer (1887), which would be restricted to "Demospongiae with a choanosomal skeleton consisting of spicules in high density, strewn in confusion or arranged in ill-defined tracts, and usually with ectosomal skeletal specializations".

On the basis of morphological similarities discussed above, there are genera which can be clearly associated:

- *Halichondria* and *Hymeniacidon*, with their main difference being the absence of oxeas in the later. The separation of Halichondrida into the families Hymeniacidonidae and Halichondriidae on the basis of predominance of oxea or styles can no longer be upheld. Styles, styloids and oxeas are found to intergrade in some genera (*Ciocalypta, Halichondria, Axinyssa, Epipolasis*).

- *Myrmekioderma* and *Didiscus* share a particular surface structure, but are differentiated by their distinct and different microsclere complement.
- *Topsentia, Ciocalypta, Petromica* and *Epipolasis* are all characterized by a brittle choanosome with a disorganized arrangement of spicules. In this group, there is a striking similarity in morphology between *Ciocalypta* and *Petromica*.

Some genera such as *Spongosorites* and *Axinyssa* were found difficult to relate to the other genera studied. *Axinyssa*, with its radial tracts and low developed ectosomal skeleton, may represent an intermediate form between the halichondrids and axinellids. *Spongosorites* could be intermediate between the *Myrmekioderma-Didiscus* group (tracts and confused spicules) and *Topsentia* (paratangential ectosome, wide size range of spicules and spicule deformations).

With respect to the relationship of the Halichondrids with other Demospongiae, a close relationship with certain Axinellids must be recognized. We consider that the evidence used to separate these close morphological groups is based in too few genera per order. Furthermore, the systematic value of the morphology of these organisms, must be emphasized. Morphological characters have been useful for diagnosis of most of the well-defined sponge taxa (e.g., Astrophorida, Haplosclerida, Poecilosclerida, Hexactinellida).

Further studies, including a cladistic analysis and biochemical characterization of halichondrid and axinellid genera are in progress as part of a revision of the Halichondrida by Van Soest, Pomponi, and Diaz.

## References


Alcolado P (1984) Nuevas especies de esponjas encontradas en Cuba. Poeyana 271:1-22

Berg C (1899) Substitucion de nombres genericos III. Commun Mus Nac Buenos Aires 1:77-80

Bergquist PR (1968) The marine fauna of New Zealand: Porifera, Demospongiae Part I. NZ Oceanogr Inst Mem 37:1-105

Bergquist PR (1978) Sponges. University of California Press, Berkeley, pp 1-268, Pls 1-12

Bergquist PR (1980) The ordinal and subclass classification of the Demospongiae (Porifera); appraisal of the present arrangement, and proposal of a new order. NZ J Zool 7:1-6

Bowerbank JS (1862) On the anatomy and physiology of the Spongiadae Part III. Philos Trans R Soc Lond 152:1087-1135, pls 72-74

Bowerbank JS (1864) A monograph of the British Spongiadae I. Ray Society London I-XX, pp 1-290, pls I-XXXVII

Burton M (1930) Additions to the sponge fauna at Plymouth. J Mar Biol Assoc UK 16(2):489-507

Burton M (1931a) Studies on Norwegian sponges I Det Kong. Nor Vidensk Selsk Forh 4(38):136-139

Burton M (1931b) A collection of marine sponges mostly from the Natal coast. Ann Natal Mus 6(3):337-358 pl XXIII

Carter HJ (1886) Descriptions of sponges from neighborhood of Port Philip Heads, South Australia, continued. Ann Mag Nat Hist (5) 17:40-43, 112-127, 431-441, 502-516 & (5) 18:34-55, 126-149

Carter HJ (1887) Report on the marine sponges, chiefly from King Island in the Mergui Archipelago. J Linn Soc London 21:61-84, pls 5-7

De Laubenfels MW (1934) New sponges from the Puerto Rican deep. Smithson Misc Colt 91 (17):1-28

De Laubenfels MW (1936) A discussion of the sponge fauna of the Dry Tortugas in particular and the West Indies in general, with material for a revision of the families and orders of the Porifera. Publ Carnegie Inst Wash 467 Pap Tortugas Lab 30:1-225

De Laubenfels MW (1953) Sponges from the Gulf of Mexico. Bull Mar Sci Gulf Caribbean 2:511-557

Dendy A (1922) Report on the Sigmatotetraxonidae collected by HMS "Sealark" in the Indian Ocean. Trans Linn Soc Lond Zool 18(1):1-164, pls 1-18

Diaz MC, Alvarez B, van Soest RWM (1987) New species of Demospongiae, (Porifera) from the National Park Archipielago Los Roques, Venezuela. Bijdr Dierkd 57(1):31-41

Ehlers EH (1870) Die Esperschen Spongien in der Zoologischen Sammlung der K. Universität Erlangen. E Th Jacob, Erlangen, pp 36

Esper EJC (1791-94) Die Pflanzentieren in Abbildungen nach der Natur mit Farben erleuchtet nebst Beschreibungen. Raspe, Nürnberg Zweyter Theil, pp 1-303, pls 1-109

Fleming J (1828) A history of British Animals. Edinburgh, pp 1-156

Hartman WD (1982) Porifera. In: Parker SP (ed) Synopsis and classification of living organisms, vol 1. McGraw-Hill, New York, pp 640-666

Hechtel G (1983) New species of marine Demospongiae from Brazil. Theringia Ser Zool Porto Alegre (63):58-89

Hoppe WF, Reichert MJM (1987) Predictable annual mass release of gametes by the coral reef sponge *Neofibularia nolitangere* (Porifera, Demospongiae). Mar Biol 94:277-285

Kirkpatrick R (1903) Descriptions of South African sponges Part III. Mar Invest South Africa 2:233-264, pls 5-6

Lendenfeld R von (1897) Spongien von Sansibar. Abh Senckenb Naturforsch Ges 21:93-133

Levi C (1956) Etude des *Halisarca* de Roscoff, Embryologie et systematique des Demosponges. Arch Zool Exp Gen 93(1):1-181

Levi C (1957) Ontogeny and Systematics in Sponges. Syst Zool 6:174-183

Levi C (1973) Systematique de la classe Demospongia (Demosponges). Traite Zool 3(1):577-631

Montagu G (1818) An essay on sponges with descriptions of all the species that have been discovered on the coast of Great Britain. Mem Wern Nat Hist Soc 2(1):67-122, pls 3-16

Pallas PS (1766) Elenchus Zoophytorum. P Van Cleef The Hague, pp 1-451

Parker GH (1910) The reactions of sponges with a consideration of the origin of nervous system. J Exp Zool 8:765-805

Pulitzer-Finali G (1983) A collection of the Mediterranean Demospongiae (Porifera) with, in appendix, a list of the Demospongiae hitherto recorded from the Mediterranean sea. Estratt Dagli Ann Mus Civ Stor Nat Genova 84:445-621

Pulitzer-Finali G (1986) A collection of Demospongiae from the West Indies with, in appendix, a list of Demospongiae hitherto recorded from the West Indies. Ann Mus Civ Stor Nat Giacomo Doria 86:1-216

Reiswig HE (1976) Natural gamete release and oviparity in Caribbean Demospongiae. In: Harrison FW, Cowden RR (eds) Aspects of sponge biology. Academic Press, London, pp 99-112

Topsent E (1896) Materiaux pour servir a l'etude de la faune des spongiaires de France 11:225-255

Topsent E (1898) Eponges nouvelles des Agores. Mem Soc Zool-Fr (11):225-255

Topsent E (1928) Spongiaires de l'Atlantique et de la Mediterranee provenant de croisieres du Prince Albert ler de Monaco. Res Camp Sci Prince Monaco 74:1-376, pls I-XI

Van Soest RWM, Stentoft N (1988) Barbados deep water sponges. Stud Fauna Curacao Other Caribb Isl 70:1-175, 12 pls

Van Soest RWM, Zea S (1986) A new sublithistid sponge *Monanthus ciocalyptoides* n.sp. (Porifera, Halichondrida) from the West Indian region. Bull Zool Mus Univ Amst 10(4):201-205

Vosmaer GCJ (1887) Porifera. In: Bronn HG (ed) Die Klassen und Ordnungen des Thierreichs 2:369-496, pls XXVI-XXXIV

Wapstra M, van Soest RMW (1987) Sexual reproduction, larval morphology and behaviour in Demospongiae from the south west of the Netherlands. In: Vacelet J, Boury-Esnault N (eds) Taxonomy of Porifera from the NE Atlantic and Mediterranean Sea. NATO ASI Ser 13:283-307

Wilson HV (1925) Siliceous and horny sponges collected by the US Fisheries Steamer 'Albatros' during the Philippine Expedition 1907-10. Bull US Nat Mus 100(2,4):273-506, pls 37-52

# A Systematic Revision of the Central Atlantic Halichondrida (Demospongiae, Porifera). Part II. Patterns of Distribution of Secondary Metabolites

S.A. Pomponi, A.E. Wright, M.C. Diaz, and R.W.M. van Soest

## Introduction

The order Halichondrida consists of poorly defined genera among which the systematic relationships are not well understood. Based on a reevaluation of morphological characters presently used to define halichondrid genera, a new concept of the Halichondrida has been proposed by Diaz et al. (this Vol.). However, many morphological characters previously considered diagnostic at various taxonomic levels within the Halichondrida not only vary within taxa, but also occur in nonhalichondrid taxa (Diaz et al., this Vol.).

Biochemical characters provide an additional data set which can be used to either corroborate or refute existing classifications based on morphological criteria. Both primary metabolites, i.e., metabolic products essential to growth or by-products of catabolism, and secondary metabolites, i.e., products synthesized from intermediate and end products of primary metabolism and which have no obvious role in cell metabolism (Stanbury 1988), have been used to elucidate taxonomic relationships in the Porifera. These include sterols, nucleosides, amino acids, carotenoids, fatty acids, terpenes, acid mucopolysaccharides, and brominated compounds (Bergquist and Wells 1983; Lawson et al. 1984; Liaaen-Jensen et al. 1982).

Certain halichondrid genera exhibit unusual patterns of distribution of secondary metabolites with antifungal, antitumor, or antiviral activity (Kashman et al. 1987; Wright et al. 1987), which we hypothesize could be useful as taxonomic characters in our revision of the Halichondrida. The purpose of this study is to examine patterns of distribution of secondary metabolites in representative samples of halichondrid genera from the Central Atlantic and to evaluate their usefulness as taxonomic characters.

## Methods

Samples were collected by the authors from the Bahama Islands, Belize, Colombia, Venezuela, Mauritania, and the Netherlands (Table 1). Shallow-water samples (less than 40 m) were collected by wading, snorkelling, and scuba. Deep-water samples (40–500 m) were collected by trawling or by the Harbor Branch Oceanographic Institution (HBOI) Johnson-Sea-Link manned sub-

J. Reitner and H. Keupp (Eds.)
Fossil and Recent Sponges

**Table 1.** Halichondrid species and samples analyzed by TLC

| Species | Sample Number[a] | TLC Group[b] | Depth (m) | Site[c] | VG/AC/BT[d] |
|---|---|---|---|---|---|
| *Halichondria melanadocia* | 10-V-88-1-6 | 1 | 1 | C | |
| | 10-V-88-2-25 | 1 | 5 | C | |
| | 3-VI-88-3-4 | 1 | 1 | V | |
| | 7-VI-88-3-1 | 1 | 9 | V | |
| | 29-V-88-3-8 | 1 | 3 | V | |
| | 29-V-88-3-24 | 1 | 3 | V | |
| | 18-XI-85-2-7 | 1 | 1 | Z | |
| | 6-VI-88-3-14 | 2 | 2 | V | |
| | 29-V-88-3-21 | 2 | 3 | V | |
| *Halichondria panicea* | H-2 | 1 | 1 | N | |
| *Halichondria bowerbanki* | H-3 | 2 | 1 | N | |
| *Hymeniacidon perlevis* | H-4 | 3 | 1 | N | |
| *Spongosorites ruetzleri* | 22-III-87-1-8 | 1 | 218 | B | Y/Y/Y |
| | 13-XI-87-1-1 | 1 | 157 | B | Y/Y/Y |
| | 15-III-87-1-8 | 2 | 461 | B | Y/Y/Y |
| | 24-VIII-85-1-13 | 3 | 172 | B | N/N/Y |
| | 23-VIII-85-1-45 | 3 | 213 | B | N/N/Y |
| | 11-XI-87-1-14 | 3 | 148 | B | N/N/Y |
| *Spongosorites* aff. *S. ruetzleri* | 14-XI-87-1-8 | 4 | 165 | B | Y/Y/N |
| *Spongosorites* sp. A | 8-XI-87-2-3 | 5 | 480 | B | Y/N/N |
| | 22-III-87-1-3 | 6 | 232 | B | Y/N/N |
| *Topsentia roquensis* | 28-V-88-2-14 | 1 | 43 | V | |
| *Topsentia* sp. A | 24-VIII-85-1-31 | 1 | 160 | B | |
| | 2-VII-87-5-21 | 1 | 37 | B | |
| | 5-VII-87-3-8 | 1 | 30 | B | |
| | 10-VII-87-5-3 | 1 | 37 | B | |
| *Topsentia* sp. B | 6-VII-87-2-8 | 1 | 37 | B | |
| *Topsentia* aff. *Halichondria* | 14-XI-87-1-27 | 1 | 58 | B | |
| *Topsentia* aff. *C. porrecta* | 9-XI-87-1-6 | 1 | 46 | B | |
| *Coelocalypta porrecta* | C-21 | 1 | 20 | M | |
| *Petromica ciocalyptoides* | 4-VI-88-1-3 | 2 | 15 | V | |
| | 4-VI-88-2-9 | 2 | 24 | V | |
| *Epipolasis* sp. A | 9-XI-87-1-2 | 1 | 357 | B | |
| | 9-XI-87-1-17 | 1 | 464 | B | |

**Table 1.** (*Continued*)

| Species | Sample Number[a] | TLC Group[b] | Depth (m) | Site[c] | VG/AC/BT[d] |
|---|---|---|---|---|---|
| *Epipolasis* | | | | | |
| sp. A | 9-XI-87-1-18 | 1 | 352 | B | |
| | 16-XI-87-1-8 | 1 | 337 | B | |
| *Myrmekioderma* | | | | | |
| *styx* | 18-III-87-1-16 | 1 | 76 | B | |
| | 16-III-87-1-18 | 1 | 82 | B | |
| | 18-III-87-3-14 | 1 | 55 | B | |
| | 18-III-87-1-24 | 1 | 51 | B | |
| | 22-VI-87-2-2 | 1 | 37 | B | |
| | 28-V-88-2-17 | 1 | 43 | V | |
| | 28-V-88-4-10 | 1 | 43 | V | |
| | 29-V-88-1-14 | 1 | 30 | V | |
| | 17-III-87-1-14 | 1 | 62 | B | |
| | 15-V-88-3-1 | 2 | 18 | C | |
| | 27-VI-87-2-3 | 2 | 21 | B | |
| | 11-III-87-3-8 | 2 | 20 | B | |
| *Myrmekioderma* | | | | | |
| *rea* | 4-VI-88-1-9 | 3 | 15 | V | |
| | 31-V-88-1-12 | 3 | 18 | V | |
| | 22-III-87-1-19 | 3 | 76 | B | |
| | 12-XI-87-3-13 | 3 | 65 | B | |
| *Didiscus* | | | | | |
| *oxeata* | 8-VI-85-1-14 | 4 | 4 | B | |
| | 27-V-88-1-3 | 4 | 21 | V | |

[a] First three components of sample number are Day-Month-Year of collection. Samples H-1 to H-4 were collected in winter 1987. Sample C-1 was collected in July 1988.
[b] TLC Group refers to similarity of TLC patterns within the taxonomic grouping as presented in Results.
[c] Site codes: B-Bahamas, C-Colombia, M-Mauritania, N-Netherlands, V-Venezuela, Z-Belize.
[d] For *Spongosorites* species, the presence (Y) or absence (N) of vermetid gastropods (VG), aerophobic color change from yellow to black (AC), the compound bromotopsentin (BT).

mersibles. Voucher specimens are deposited at the Indian River Coastal Zone Museum (HBOI).

Immediately after collection, 2 g subsamples were homogenized and extracted in 100% ethanol (Pharmco) and the extracts stored at −20° C. Samples were analyzed using thin layer chromatography, a technique for separating a mixture of chemical compounds into various components. A solvent, or mobile phase, carries the sample mixture through the sorbent, or stationary phase. Separation occurs due to selective adsorption of the components of the mixture onto the stationary phase. Polar compounds are retained on silica gel stationary phases longer than nonpolar compounds. Polar mobile phases move compounds more quickly than nonpolar mobile phases. The result is a series of separated bands, spots, or zones on the sorbent which may be detected visually (in the case of colored compounds), by physical methods (e.g., UV absorbance) or by chemical methods (e.g., spraying with specific reagents). Results can be quantified for comparison by calculating a retention factor or Rf, which is defined as the ratio of the distance a compound moves from the origin to the distance the solvent front moves from the origin. For a complete explanation of TLC methods, see Touchstone and Dobbins (1978).

Prior to TLC analysis, extracts were dried and resuspended in 5% heptane-methanol at a concentration of 100 mg/ml. The sorbent was DC-Plastikfollen Kieselgel 60 F254, 0.2 mm thinkness (EM Science). Plate size was 10×20 cm or 10×10 cm. Four solvent systems (of increasing polarity) were used: (A) ethyl acetate/heptane (1:1); (B) ethyl acetate/methanol (9:1); (C) ethyl acetate/methanol (99:1); (D) n-propanol/ethyl acetate/water (7:2:1). A 1-$\mu$l sample was spotted by capillary tube at the origin. Plates were developed in rectangular glass chambers. The length of development for solvent systems A-C was 7 cm while that of solvent D was 6 cm. Both physical and chemical detection methods were used. Plates were viewed with ultraviolet light at 254 and 365 nm and subsequently sprayed with a solution of 2 g vanillin (Aldrich) in 100 ml concentrated sulfuric acid (CMS) followed by heating of the plate. For the *Spongosorites* group, no heat was necessary to detect the components after spraying with vanillin/$H_2SO_4$.

## Results

### *Halichondria* and *Hymeniacidon*

Eleven specimens (three species) of *Halichondria* collected from shallow temperate and tropical waters could be divided into two groups with similar TLC patterns (Table 1). Group 1, which consists of *H. melanadocia* and *H. panicea*, contains components at Rf's of 0.8, 0.55, 0.0, and a minor component at 0.7 (solvent system A). Group 2, which consists of *H. melanadocia* and *H. bowerbanki*, is missing the component at Rf = 0.8, has a larger concentration at Rf = 0.7, and an additional trace component at Rf = 0.5. Neither group has a suite of compounds which could characterize the genus. The patterns of the major metabolites are reminiscent of TLC patterns for sterols, which are ubiquitous in the Porifera (Bergquist et al. 1980). There are very subtle differences between the two groups in the patterns of trace metabolites which cannot, however, be correlated with species, depth, location, or time of collection of the samples (Table 1).

A single sample of *Hymeniacidon perlevis* was compared with the three species of *Halichondria*. It exhibited an additional minor component in solvent system D (detected by vanillin/$H_2SO_4$, heat) at an Rf of 0.05 but was otherwise very similar to the *Halichondria* patterns. It is difficult to draw any conclusions about relationships between the two genera based on this study as none of the *Halichondria* or *Hymeniacidon* samples examined contained significant quantities of secondary metabolites in their crude extracts.

### *Spongosorites*

*Spongosorites* species from the Mediterranean and West Indian regions have been characterized biochemically by the presence of bromotopsentin or topsentin, indole alkaloids (Bartik et al. 1987) with antitumor and antiviral activities (Tsujii et al., 1988). Within the *Spongosorites* samples examined, there were six different TLC groups/patterns (Table 1). *Spongosorites ruetzleri* exhibits three very similar, but not identical, patterns of secondary metabolites. The major metabolite in all of the *S. ruetzleri* is bromotopsentin which, in solvent system C, is observed at an Rf of 0.78 as a bright yellow compound which fluoresces yellow under long wavelength UV light (365 nm). Group 1 contains bromotopsentin and a component observed at an Rf of 0.64 (system C) which fluoresces blue under long wavelength UV. Group 2 contains bromotopsentin and a component which fluoresces blue under long wavelength UV observed at an Rf of 0.51 (system C). Group 3 contains bromotopsentin, lacks the blue fluorescing compounds, but has a short wavelength UV absorbing compound observed at an Rf of 0.65 (system C).

A single specimen of *Spongosorites* aff. *S. ruetzleri* is quite different from the TLC patterns of *S. ruetzleri* in the lack of bromotopsentin. In solvent system C, this sample contains a yellow fluorescing (long wavelength UV) compound at the origin and two blue fluorescing components (long wavelength UV) observed at Rf's of 0.53 and 0.05 which slowly air oxidize to purple. This sample differed from the typical *S. ruetzleri* in size (massive), color (brown-yellow), thickness of the ectosome (thinner than in *S. ruetzleri*), and spicule morphology (crooked oxea).

Two specimens identified as *Spongosorites* sp. A and distinguished from *S. ruetzleri* on the basis of morphology (spherical), surface (papillate), and choanosomal skeleton (which incorporates sand grains) differ biochemically from *S. ruetzleri* as well as from each other. Neither sample contains bromotopsentin. Sample 8-XI-87-2-3 contains very polar components which, in solvent system D after visualization with vanillin/$H_2SO_4$ and heat, appear as a large brown component with an RF of 0.69, a smaller brown component at 0.48, and a yellow-green component with an Rf of 0.63. Sample 22-III-87-1-3, under the same conditions, has two small brown components with Rf's of 0.75 and 0.55, and an orange component of Rf = 0.62. The two samples are morphologically similar. They differ in depth and season of collection (Table 1) but with the small sample size (n = 2) it is not possible to correlate the different TLC patterns with environmental variables.

TLC patterns may be useful at both the specific and generic levels within the *Spongosorites* group, particularly if the minor intraspecific differences in TLC patterns can be correlated with environmental (e.g., depth), biological (e.g., symbiotic associations), and/or morphological (e.g., color) factors. For example, samples of *S. ruetzleri* in TLC groups 1 and 2 have vermetid gastropods and an aerophobic color change from yellow to black. Group 3 samples have neither vermetids nor aerophobic properties (Table 1).

### *Topsentia, Coelocalypta, and Petromica*

All specimens of *Topsentia* showed similar or identical TLC patterns. The secondary metabolites are quite polar and do not migrate in the less polar solvent systems A-C. In the most polar solvent system D, there is a characteristic pattern of metabolites: a purple component at an Rf of 0.83, maroon at 0.62, yellow at 0.48, and purple at 0.39 (visualized by vanillin/$H_2SO_4$, heat). *Coelocalypta porrecta* is quite similar to the *Topsentia* species except that it has a lower concentration of all the metabolites. Since this genus is monotypical, a merging with *Topsentia* may be considered. *Petromica* (a senior synonym of *Monanthus*) *ciocalyptoides* has some similarities to *Topsentia* but also contains non-polar metabolites at Rf's of 0.74, 0.64, and 0.5 in solvent system A (visualized by vanillin/$H_2SO_4$, heat). TLC patterns are useful at the generic level within this group.

### *Epipolasis*

Four specimens of *Epipolasis* sp. A were compared. The samples, all collected from depths of 335–450 m in the Bahamas (Table 1), show the same TLC pattern of purple compounds at Rf = 0.68 and at the origin (solvent system A), but the metabolites are not present in sufficient quantity to enable comparisons with other genera using TLC.

### *Myrmekioderma and Didiscus*

Perhaps the most interesting TLC patterns were found in *Myrmekioderma styx* and *Didiscus oxeata*. Two distinct TLC patterns emerged in *Myrmekioderma styx*, corresponding quite clearly with depth (Table 1). All samples collected from depths greater than 30 m had four major components which were separated in solvent system A: a purple component at Rf = 0.8, a blue component at 0.7, an aqua component at 0.67, and a purple component at 0.44 (visualized with vanillin/$H_2SO_4$, heat). All samples from depths shallower than 20 m, regardless of geographic location, under the same chromatographic conditions, had one major purple component at Rf = 0.67 and two minor fluorescent components at Rf's of 0.79 and 0.54. There were no samples analyzed from depths between 20 and 30 m. Deep-water samples of *M. styx* contain reiswigin, a diterpene with antiviral activity (Kashman et al. 1987) but it is not known at this time if reiswigin is one of the compounds detected by TLC.

*Myrmekioderma rea* contains only trace metabolites at Rf = 0.63 and at the origin in solvent system A, so we cannot draw any conclusions about the relationship of this species with *M. styx* or with other halichondrids based on secondary metabolites. The TLC patterns were identical in both shallow-water (15–18 m) and deep-water (65–76 m) specimens.

*Didiscus oxeata*, although indistinguishable from *M. styx* in the field, is quite different chemically. It has two major metabolites which, after visualization with vanillin/$H_2SO_4$ and heat, are observed as characteristic red-pink spots at Rf = 0.45 and 0.77 (in solvent system A). *Didiscus oxeata* contains the unique sesquiterpene phenols curcuphenol and curcudiol which have antifungal activity (Wright et al. 1987). These compounds are not known to occur in any other sponges and are diagnostic for this species. We have not examined other species of *Didiscus*.

## Discussion

Of the samples analyzed, the *Halichondria* species, *Hymeniacidon* species, *Epipolasis* species, and *Myrmekioderma rea* produce only trace amounts of secondary metabolites and are, therefore, not amenable to chemotaxonomic analysis by thin layer chromatography. Novel isocyanide-containing compounds have been reported for *Hymeniacidon amphilecta* and the axinellid *Acanthella* sp. (Faulkner et al. 1979; Wratten et al. 1978). Oroidin-type compounds have been reported for *Hymeniacidon aldis* and three other axinellid species (Cimino et al. 1982; Kitagawa et al. 1983). Careful screening for these compounds (as suggested by Bergquist and Wells 1983) may clarify phylogenetic relationships between the Halichondrida and some Axinellida genera.

A suspected relationship between *Topsentia*, *Coelocalypta*, and *Petromica* based on morphological analyses of this group (Diaz et al., this Vol.; Van Soest et al. 1990) is supported by TLC patterns of secondary metabolites.

The two unique TLC patterns of *Myrmekioderma styx* are useful to distinguish deep-water from shallow-water populations of the species. This was the only case in which biochemical variation could be clearly correlated with an environmental variable (depth). The secondary metabolites found in *M. styx* could not be shown in *M. rea*, so they may only be useful at the species or subspecies level. *Didiscus oxeata*, although very similar morphologically to *M. styx* except for microsclere complement, is quite distinct in its production of secondary metabolites. More species of *Didiscus* need to be examined to determine if the unique phenols are diagnostic at the generic or specific level.

Caution must be exercised when trying to analyze taxonomic relationships based on patterns of secondary metabolites. The occurrence or patterns of distribution of secondary metabolites may be produced by epibionts, symbionts, or other associated organisms, or as a result of the association (Bergquist and Wells 1983). They could be affected by environmental variables, or may represent different phenotypic expression of the same genotype (Sole-Cava and Thorpe 1987). Careful subsampling to eliminate epibionts and other associated organisms, as well as sampling from a wide temporal, spatial, and depth range will reduce ecological artifacts. Enzyme electrophoresis has been used to determine levels of genetic variation in sponge populations (Sole-Cava and Thorpe 1987). However, if the secondary metabolites are produced by the sponge and show a consistent distribution over a taxon, they can serve as a good synapomorphy and thus determine that the taxon is monophyletic.

Although not of equivalent weight, patterns of secondary metabolites may provide useful taxonomic criteria to supplement morphological character analyses in the halichondrid genera *Spongosorites*, *Topsentia*, *Coelocalypta*, *Petromica*, *Didiscus*, and to a certain extent, *Myrmekioderma*. Although outside the scope of the present study, we acknowledge the necessity for chemical characterization of the compounds detected by TLC analysis before they can be reliably used as taxonomic characters. Such qualitative analyses are currently in progress (by AEW) as part of a revision of the Halichondrida and reevaluation of the relationships between this group and the Axinellida by Diaz, Pomponi, and Van Soest.

*Acknowledgments.* We thank the crews of the Research Vessels SEA DIVER, EDWIN LINK, SEWARD JOHNSON, and JOHNSON-SEA-LINK *I* and *II* for their assistance in collection of sponge samples. We also acknowledge the technical assistance of D. Nadine Neil in preparation of extracts and TLC analyses. We thank Dr. Frank Koehn (HBOI) and Dr. Richard L. Raesly (the Pennsylvania State University) for constructive discussions about halichondrid biochemistry and systematics.

## References

Bartik KJ, Braekman JC, Daloze D, Stoller C, Huysecom J, Van de Vyver G, Ottinger R (1987) Topsentins, new toxic bis-indole alkaloids from the marine sponge *Topsentia genitrix*. Can J Chem 65:2118–2121

Bergquist PR, Wells RJ (1983) Chemotaxonomy of the Porifera: the development and current status of the field. In: Scheuer PJ (ed) Marine natural products, vol 5. Academic Press, New York, p 1

Bergquist PR, Hofheinz W, Oesterhelt G (1980) Sterol composition and the classification of the Demospongiae. Biochem Syst Ecol 8:423–435

Cimino G, De Rosa S, De Stefano S, Mazzarella L, Puliti R, Sodano G (1982) Isolation and X-ray crystal structure of a novel bromo-compound from two marine sponges. Tetrahedron Lett 23:767–768

Diaz MC, Van Soest RWM, Pomponi SA (1991) A systematic revision of the Central Atlantic Halichondrida (Demospongiae, Porifera). Part I. Evaluation of characters and diagnosis of genera (this Vol)

Faulkner DJ, Armstrong RW, Djura P, Higgs MD, Ravi BN, Stierle DB, Wratten SJ (1979) Some chemical constituents of Caribbean sponges. Colloq Int CNRS 291:401–406

Kashman Y, Hirsch S, Koehn F, Cross S (1987) Reiswigins A and B, novel antiviral diterpenes from a deepwater sponge.Tetrahedron Lett 28:5461–5464

Kitagawa J, Kobayashi M, Kitanaka K, Kido M, Kyogoku Y (1983) Marine natural products. XII. On the chemical constituents of the Okinawan marine sponge *Hymeniacidon aldis*. Chem Pharm Bull 31:3321–2328

Lawson MP, Bergquist PR, Cambie RC (1984) Fatty acid composition and the classification of the Porifera. Biochem Syst Ecol 12:375–393

Liaaen-Jensen S, Renstrom B, Ramdahl T, Hallenstvet M (1982) Carotenoids of marine sponges. Biochem Syst Ecol 10:167–174

Sole-Cava AM, Thorpe JP (1987) The use of electrophoresis in sponge taxonomy. In: Vacelet J, Boury-Esnault N (eds) Taxonomy of Porifera. NATO ASI Series, vol G13. Springer, Berlin Heidelberg New York, p 243

Stanbury PF (1988) Fermentation technology. In: Walker JM, Gingold EB (eds) Molecular biology and biotechnology. The Royal Society of Chemistry, London, p 1

Touchstone JC, Dobbins MF (1978) Practice of thin layer chromatography. Wiley, New York

Tsujii S, Rinehart KL, Gunasekera S, Kashman Y, Cross SS, Lui MS, Pomponi SA, Diaz MC (1988) Topsentin, bromotopsentin, and dihydrodeoxybromotopsentin, antiviral and antitumor bis(indolyl) imidazoles from Caribbean deep-sea sponges of the family Halichondriidae. J Org Chem 53:5446–5453

Van Soest RWM, Diaz MC, Pomponi SA (1990) Phylogenetic classification of the halichondrids (Porifera, Demospongiae). Beaufortia 40:15–62

Wratten SJ, Faulkner DJ, Hirotsu K, Clardy J (1978) Diterpenoid isocyanides from the marine sponge *Hymeniacidon amphilecta*. Tetrahedron Lett 45:4345–4348

Wright AE, Pomponi SA, McConnell OJ, Kohmoto S, McCarthy PJ (1987) (+)-Curcuphenol and (+-curcudiol, sesquiterpene phenols from shallow and deep water collections of the marine sponge *Didiscus flavus*. J Nat Prod 50:976–978

# Taxonomic Value of Microstructural Features in Calcified Tissue from Recent and Fossil Demospongiae and Calcarea

J.-P. CUIF and P. GAUTRET

## Introduction

Since the recent discoveries of living Demospongiae with calcified nonspicular skeletons, there is a need for a common classification including both recent and fossil sponges. More precisely, it is now obvious that the classical paleontological system, in which major taxa are based on the morphological characteristics of the calcified structures (i.e., Sphinctozoa, Inozoa, chaetetids, etc.), does not fit with the new biological data (Vacelet 1983).

This is currently a much debated question between paleontologists. A recent classification of the microstructures from living forms (Wendt 1984) including in the same section skeletal tissues built in obviously distinct classes (i.e., "spherulitic" from *Astrosclera* and *Petrobiona*) suggests that phylogenic lineages cannot be based on data provided by calcareous skeletons.

However, calcareous skeletons are virtually the only source of information concerning fossil species: the various molds of siliceous spicules are uncommon, and rarely usable. Thus, in recent years, much attention has been paid to the characteristics of the microstructure of skeletons, to estimate their potential value (Gautret 1986).

Examining more closely the previous works in which microstructural terminology is employed, it appears that the terms utilized extend over very different levels of analysis. Frequently, the microstructure is expressed only in very general terms, sometimes by a single word.

Among the different microstructural terms employed for the description of the skeletal parts of calcified Demospongiae or Calcaraea, two, "irregular" and "spherulitic", demonstrate the trends of this work. In the living species, the character "irregular" is used (Wendt 1984) for the skeletal tissue of *Vaceletia* and *Acanthochaetetes*. Typical aspects of these tissues are illustrated by this author (Wendt 1984; Fig. 1c,d). The second term ("spherulitic") is applied to the arrangement of the fibrous units in the skeletons of *Petrobiona* and also *Astrosclera*.

Obviously, among the Demospongiae, *Vaceletia* and *Acanthochaetetes* are phylogenetically distinct. On the other hand, *Petrobiona* and *Astrosclera* belong to separate classes. If we use the same term to describe their calcified skeletons, these structures become, of course, "nonphylogenetic" (Wendt 1979, Fig. 2; 1984, Text-Fig. 1).

J. Reitner and H. Keupp (Eds.)
Fossil and Recent Sponges

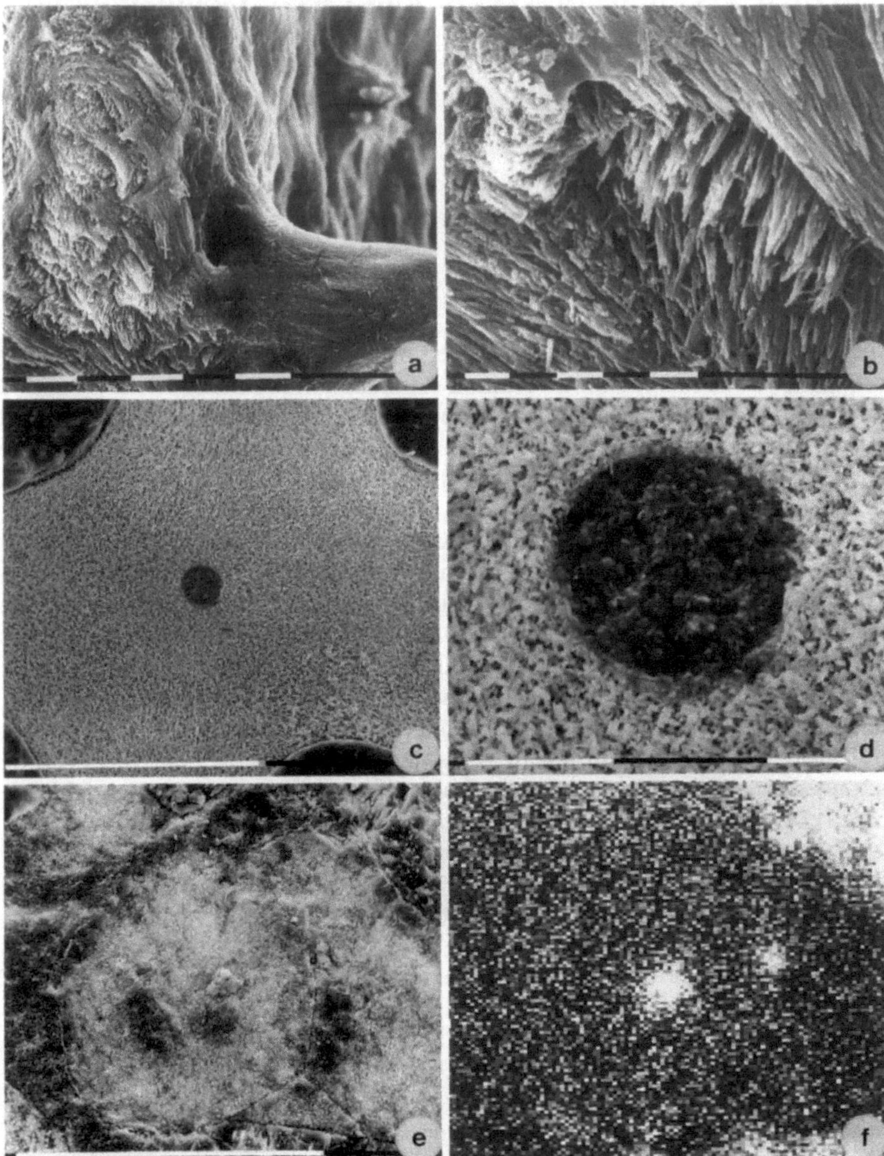

**Fig. 1. a,b** *Acanthochaetetes wellsi* Hartman and Goreau. The walls are built by successive layers (**a** *right*). In each layer, calcitic fibers are parallel, but their direction is different from one layer to another. **a** Bar = 100 μm; **b** 10 μm. Etching: dilute formic acid 5%, 5 s. **c,d** *Vaceletia crypta* (Vacelet). The centers of a biomineralization units are not mineralized. **c** Bar = 100 μm, **d** 10 μm. Polished and etched surface (formaldehyde and acetic acid mixture). **e,f** Vaceletid-like biomineralization units of a Triassic Inozoa. **e** Polished and etched section (bar = 100 μm). *Arrows* center of the biomineralization units. **f** Iron X-ray mapping of the same sector: concentration of Fe in the calcitized center of the biomineralization units

But, before accepting this point of view, we should answer two usual questions:

- Are the skeletons joined together in these microstructural groups actually similar? In other words, is it possible to obtain significant data concerning characteristics of the various calcified tissues, *avoiding confusion between them*, even in fossil material having suffered some diagenetic alterations?
- what is the probability of a *repeated production of the same type of skeletal secretion* in separate groups?

This work is an attempt to improve our knowledge of carbonate structures of sponges, with special attention to the traces of skeletal growth process which can be observed in the skeletons of both living and fossil species.

## Method

Various experimental approaches have shown that organic materials remain included in the skeletal units, after their formation, in specimens belonging to every taxa. By using appropriate reagents, it is possible to reveal the successive growth stages of the microstructural units, and thus the skeletal growth modalities become accessible to analysis.

Selective destruction of organic matter is obtained either by usual oxydizing compounds, or by enzymatic solutions. Some experiments utilize the proteolytic action of bacterial cultures (see in Cuif et al. 1987).

On the other hand, decalcifying solution including fixative reagents enables one to observe organic remnants within the microstructural units. However, one should keep in mind that the main part of the organic phase is composed of soluble matter and the precise topographic data are rather difficult to obtain.

However, the soluble nature of the organic intracrystalline matrix makes possible a biochemical approach, after complete dissolution of the mineral phase. To date, the most significant results have been obtained by gel filtration chromatography (cf. results and some technical data in Fig. 3a).

Concerning the mineral component of the skeleton, the usual technique of microprobe analysis has been extended by the use of the X-ray mapping system, which provides pictures which may express the diagenetic state of the fossil materials.

## Case Studies of Microstructures: Comparison of Skeletal Growth Processes in Four Living Species

### *Vaceletia and Acanthochaetetes*

In the typical sphinctozoan *Vaceletia*, the successive convex (more or less hemispheric) cupolae of which the skeleton is built are prepared in the initial step of the biomineralization process, within a whole organic structure (Vacelet 1977,

1979). In spite of its homogenous appearance, this primary organic skeleton is already differentiated into mineralization units. These consist of spherical nodules with well-defined centers which obviously are the basis for the next step in skeletal formation (Gautret 1985). As a matter of fact, these biomineralization units become more clearly visible at the beginning of the mineralization itself, when microcystalline units are produced.

These nodular microstructural units can readily be demonstrated with the SEM, on polished surfaces etched by dilute formic acid or fixative/decalcifying mixtures. They are limited by a thin organic envelope and, curiously, the center of the units does not mineralize (Fig. 1c,d).

Mineral components appear as cylindrical aragonitic crystals, either without any particular orientation, or sometimes with fasciculate or more or less parallel disposition.

Examining briefly the crystalline units of the *Acanthochaetetes* skeleton, one may observe that the very minute and elongated elements bear some resemblance to the skeletal tissue of *Vaceletia*. But the analysis of the growth process expressed by the calcified skeleton alone always provides very clear and effective criteria for the distinction of these two species.

In the *Acanthochaetetes*, skeletal development is effected by adjunction of successive growth layers, each about 5 to 10 μm thick. This is a first important difference with the growth process of *Vaceletia*. Moreover, the SEM pictures show clearly that, in each secretion stage, the crystalline units are arranged in the same direction, which is different from one layer to another (Fig. 1a,b). This is another clear difference with the random disposition of the *Vaceletia* skeletal crystals.

### *Comparison of the Skeletal Growth Process in Astrosclera and Petrobiona*

The astrosclerids seem to be accepted as being typical representatives of the spherulitic growth process, having isodiametric and perfectly radiating microstructural units. Previous studies (Gautret 1986) have shown the sequence of spherulite formation. An initial intracellular stage is followed by a migrating phase by which the initial spherulite is transported to the upper part of the growing skeleton. During a second phase an epithelial secretion is produced by which the existing spherulitic skeleton is reinforced. Finally, in a third step, there is a more or less complete in-filling of functional cavities.

This evolution can be demonstrated by specific preparations on the skeleton of astroclerids: limit of the initial spherulite (Fig. 2c), or growth lines in the epithelial secretion stage (Fig. 2d). The abundance of organic matter within the initial spherulite is probably important during diagenesis of these structures because, in most fossil materials, the central spherulites are generally more strongly recrystallized than other parts of the skeleton (Fig. 2e).

Compared to this very precise and regular sequence, the skeletal structure of *Petrobiona* is obviously distinct, even if, in some parts, fan systems of fibers exhibit a radiating aspect (Gautret 1986). In *Petrobiona*, the basal skeleton is produced by addition of successive layers and the mode of secretion, even from the beginning, appears to be entirely epithelial. There is no indication of an initial

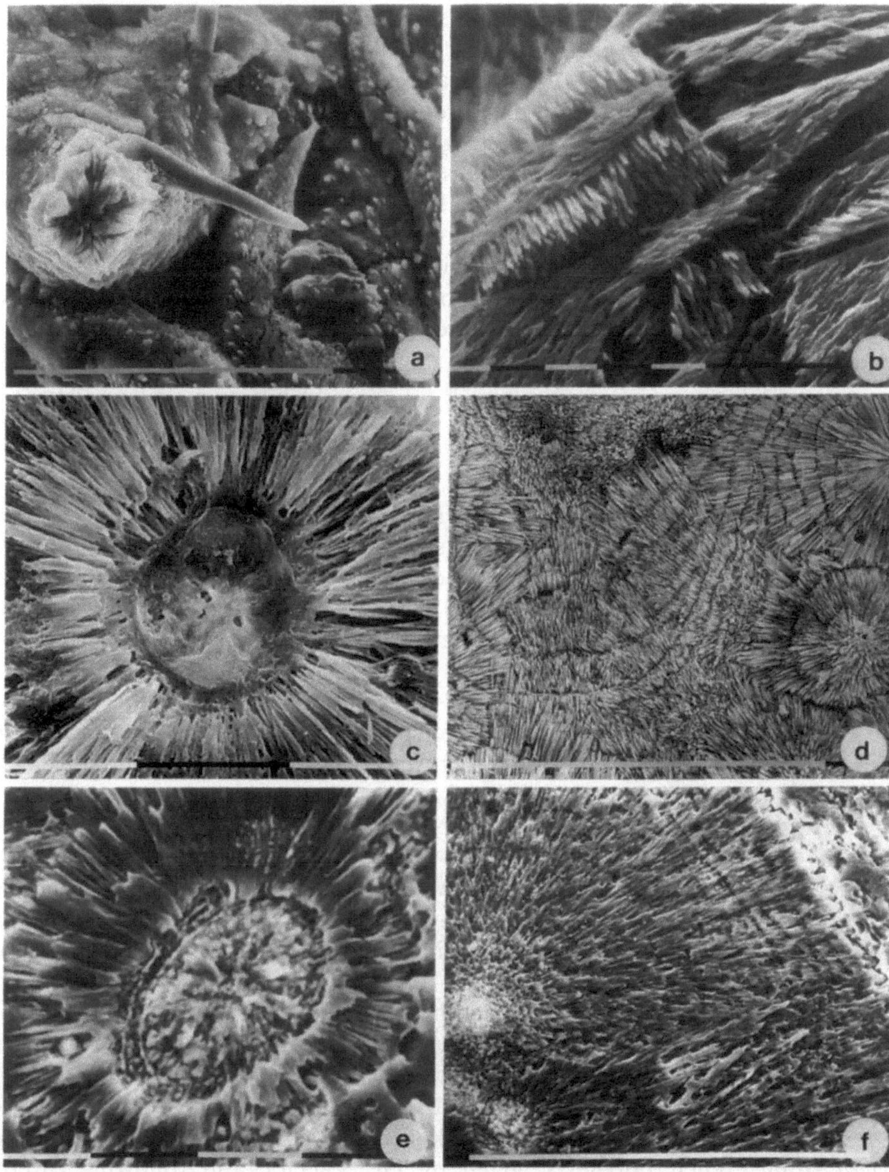


**Fig. 2.** **a,b** *Petrobiona massiliana* Vacelet and Levi. **a** radiating but not truly spherulitic disposition of fibers in the basal skeleton (bar = 1 mm). **b** Feather-like disposition of elementary units in fibers of the basal skeleton, visible after action of enzymatic solution (bar = 1 μm). **c,d** *Astrosclera willeyana* Lister. Specific preparation demonstrating the successive stages of the growth process (formaldehyde and acetic acid mixture). **c** Aspect of the central part of a microstructural unit showing the limit of the initial spherulite (bar = 10 μm). **d** Epithelial secretion in the late part of the fibers' growth (bar = 100 μm). **e,f** Astrosclerid organization of spherulites in Triassic specimens. **e** Initial stage of spherulite which has been calcitized whereas fibers of the second stage are well preserved (bar = 10 μm). **f** Epithelial secretion of the last stage demonstrated by growth line, after light acid etching (bar = 100 μm)

intracellular stage. The very important spinose processes (but not truly spicular in nature), visible on the external surface of the skeleton, and also observable in thin section, are also typical for the basal skeleton of *Petrobiona* (Fig. 2a).

More precise observation of the fibrous units also enables an easy distinction between specimens belonging to the genus *Astrosclera* or *Petrobiona*. In *Astrosclera*, the main part of the spherulites are constructed of rectilinear and simple fibro-crystals. In *Petrobiona* the crystalline units exhibit very unusual organization, with a feather-like disposition: on each side of a median crystal, lateral units are branched, slightly oblique with respect to the direction of the main crystal (Fig. 2b). The taxonomic value of this pennular-shaped crystallization is demonstrated by the occurrence of the same disposition in another Calcarea *Murrayona*.

## *Conclusion*

As for the "irregular" tissues of *Vaceletia* and *Acanthochaetetes*, there is no possible confusion between the skeletal structures of *Astrosclera* and *Petrobiona*. The only term, "spherulitic", cannot be used as a common designation of the skeletal structures in these two forms. These two living species, obviously separated from the taxonomic point of view, are also distinct with respect to the microstructure of their skeletons.

The differences between these four types of skeleton are obvious at various levels of observation: from the general disposition of biomineralization units to the most precise observation of microcrystals. These are confirmed in the comparison of the associated organic matrix.

Using gel-filtration chromatography, after careful acid decalcification of the various skeletal structures (pH always upper 4), it appears that the respective organic components exhibit significant differences, even under the general examination of their molecular masses: the elution curves of *Acanthochaetetes*, *Petrobiona*, *Astrosclera* are clearly distinct (Fig. 3a) [not enough material was available to analyze *Vaceletia*].

Chromatographic analysis has also been carried out on the skeletons of living Ceratoporellids. These confirm the particular combination of organic components which is characteristic of each skeletal type.

It may be noted that when subfossil specimens have been studied [*Acanthochaetetes* from Palau Islands, kindly provided by Dr. Kato, or *Astrosclera* from French Polynesia, given by Dr. Vacelet, and following careful decontamination of skeletal parts, each skeleton gave the typical elution profile, compared with the living reference (Fig. 3b,c).

Thus it appears that each type of biomineralization process involves specific organic material. These results have to be refined by more precise biochemical analysis, but, even in their present states, they provide new evidence for the precise correlation between taxonomy and skeletal characteristics.

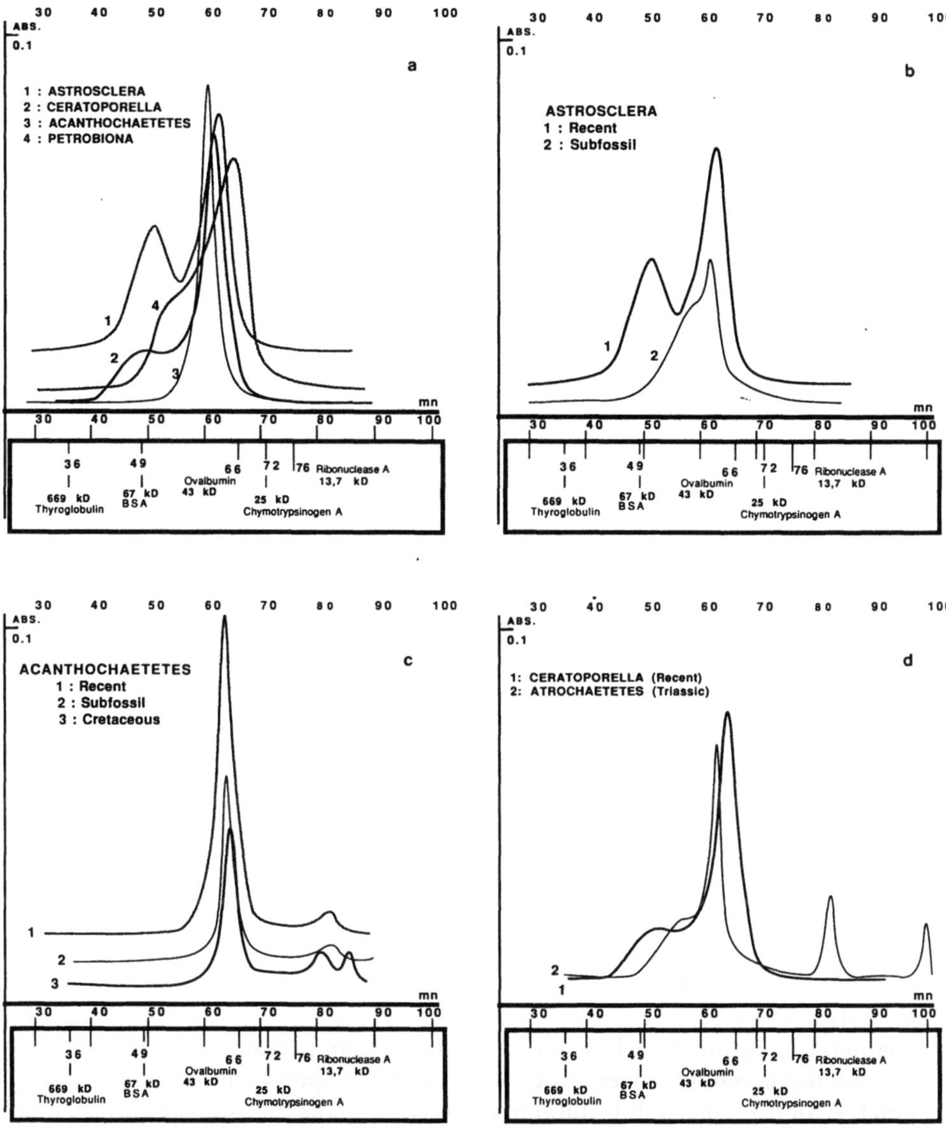


**Fig. 3.** In the intraskeletal organic matter of living species, the molecular weights are distributed in a few continuous ensembles, showing the characteristic aspect of polymer chromatographic curves. In the fossil skeletons, these bio-polymers are transformed: during diagenesis there is a reduction of the mean molecular weight (cf. *Astrosclera*). **a** Elution curves of soluble intraskeletal matrix from four living calcified Sponges (Demospongiae: *Astrosclera*, *Ceratoporella* and *Acanthochaetetes*; Calcarea: *Petrobiona*). **b** Soluble matrix of calcareous skeleton in living and subfossil *Astrosclera*. **c** Comparison of soluble matrix in living *Acanthochaetetes*, recent and very well preserved specimen from Cretaceous of Provence (France). **d** Example of intraskeletal matrix from *Ceratoporella* and a triassic ceratoporellid: *Atrochaetetes* sp. All these results have been obtained in the same experimental conditions: Gel-filtration chromatographic column: Pharmacia High resolution column SUPEROSE S12, elution buffer: ammonium acetate 33mM, flow rate: 0.25 ml/mm; U.V. Detection: 280 nm

## Low Probability of Repeated Occurrences of Skeletal Growth Processes

As may be expected, it is not possible to offer a totally convincing argument concerning this question. We can only estimate whether these complex skeletal growth processes, each of them involving specific assemblages of organic molecules, could appear twice or more through time, in different biological lineages.

However, a summary of data provided by faunas from Permo-Triassic formations in the Dolomites, Djebel Tebaga (Tunisia) and Turkey, and by French Cenomanian from the Pyrenees and Provence, help in replying to this question.

### *Evidence for Presence of Astrosclerid-, Vaceletid- and Acanthochaetetid-Like Skeletons in Ancient Formations*

1. In Permian and specially the Triassic formations, the spherulitic astrosclerid-like skeletal growth process seems to be the most abundant. Very close resemblances can be observed in numerous skeletons belonging to various architectural types: Sphinctozoa; chaetetids, or others (Gautret Razgallah 1987; Dieci et al. 1974; Cuif 1979). All the recognized steps of skeletal formation in *Astrosclera* are present, demonstrating in most cases very precise similarities (Fig. 2e,f: size of initial stages, epithelial secretion in the late part of the skeletal growth, size and shape of fibers, etc.). A previous analysis of the intraskeletal organic matter from three spherulitic forms (Recent, Triassic, and Permian) has already been attempted and has shown very comparable molecular weight curves (Cuif and Gautret 1987).

If, by hypothesis, a polyphyletic development of an "astrosclerid-like" skeletal process is pressumed for all these architectural types (believed, in this case, to belong to separate lineages), synchronous occurrence of such phenomena, for these various phyla, appears improbable.

2. In the same reef fauna existed many sphinctozoan (*Stylothalamia*, for example) exhibiting in their skeleton the closest resemblances with the vaceletid skeletal growth process: biomineralization units of same size, with differentiated unmineralized centers, each with an external envelope (Gautret 1985).

But these resemblances occur also in other specimens, which are not spinctozoan from the architectural point of view. In the illustrated example, the center and the limits of biomineralization units have been accentuated by diagenesis (Fig. 1e,f).

If these resemblances are not caused by phylogenetic relationships, we need another "convergent evolutive phenomena" to explain the close similarities between the "vaceletid-like" Triassic species and the living *Vaceletia*. This is, of course, highly improbable, if we take into account the precise synchronism with an identical phenomenon concerning the astrosclerid-like species.

3. The similarity between the Cretaceous *Acanthochaetetes* from France (or from Spain: see Reitner and Engeser 1987), and the originally called *Tabulo-*

*spongia* from Palau Island provides the third example of this interpretative choice. In the fossil forms, similarities of histological and microstructural criteria confirm entirely the resemblance demonstrated by the general architecture of the skeleton.

Summarizing these three examples, it appears that to deny a phylogenetical relationship between recent and fossil forms, in spite of their close resemblance, would be more difficult that to accept these relationships.

### *Presence of Typical Ceratoporellids in the Triassic of Dolomites and Turkey*

It is also possible to find in the Triassic formations of the Dolomites or Turkey specimens with typical ceratoporellid characteristics (Cuif 1973), which are now confirmed by study of organic intraskeletal matrix (Fig. 3d). These already appear completely distinct with respect to their skeletal features from the contemporaneous astrosclerids.

This suggests two remarks:

- If such a clear distinction occurs between skeletal structures of living ceratoporellid and astrosclerid (two groups with a well-established taxonomic proximity), then the appearance of an "astrosclerid-like" skeletal growth process, similar in all its stages, is improbable in other, more distant, biological groups.
- Again, if the similarity between Triassic "ceratoporellid-like" forms and the living form is also interpreted as a "convergent evolution case", its synchronism with the same phenomenon postulated for the astrosclerid becomes more improbable again.

N.B. It should be noted that sometimes skeletal fibers of *Ceratoporella* can exhibit a completely radiating disposition (Cuif 1974, Pl. 29, Fig. 2). From a purely descriptive point of view, this skeletal aspect can also be named a "spherulitic" structure. But we found in these "geometrical spherulites" no indication of a true astrosclerid skeletal secretory process (initial intracellular stage, etc.).

### *Hypothesis of a Possible "Environmental Action" in the Production of the Nonspicular Calcareous Skeleton of Sponges*

Acceptance of a phylogenetical relationship between Triassic and living forms of ceratoporellids, vaceletids, and astrosclerids, or between Cretaceous and recent acanthochaetetids, forces one to explain the very important chronological gap between the fossil data and the recent documents.

We consider that the main part of the Jurassic sponge fauna is not the descendant of the mineralized Triassic species: the biomineralization criteria do not allow such a phylogenetic succession (Gautret and Cuif in press). However, the synchronism in the disappearance of the Triassic fauna and the reappearance, in recent times, of the same biomineralization types forces one to admit that some external causes have an influence on the biological mechanisms.

It seems clear that this external action cannot influence the whole biological sequence by which skeletons are built. We do not see how the oceanic parameters (i.e., temperature, pH, or others) could determine the production, in various lineages, of the complex biological sequences involving specific organic matter in each biomineralization type.

This hypothetical external action can influence only the last part of the sequence, but is essential for the production of fossil data: the reaction by which calcium carbonate crystals are formed. It may be suggested that, during the Jurassic and Cretaceous times, astrosclerids and vaceletids had entirely organic skeletons, possibly because the water chemistry was unfavorable for mineral precipitation. Moreover, external conditions being different, it is also possible that other skeletal types become observable, produced by sponges which were without skeleton until then.

This hypothesis, based on the comparison of four important biomineralization processes, has to be verified by the same approach, in which a complete description of the skeleton production replaces the usual microstructural terminology. Thus, the important demosponge fauna described, for example, by Hudson (1960) and Wood and Reitner (1986) could help us in describing the faunal change between the Triassic and Jurassic periods, and thus enable us to correlate this evolution with the equivalent phenomena currently studied in other Invertebrate groups.

## References

Cuif JP (1973) Mise en évidence des premières Sclérosponges fossiles dans le Trias des Dolomites. CR Acad Sci Paris 277-Ser D:2333–2336

Cuif JP (1974) Rôle des Sclérosponges dans la faune récifale du Trias des Dolomites (Italie du Nord). Geobios 7-2:139–153

Cuif JP (1979) Caractères morphologiques et microstructuraux de trois Sclérosponges triasiques. Association avec des Chaetetida. Coll Int CNRS Biol Spong Paris 291:475–481

Cuif JP, Gautret P (1987) Comparaison des modalités de diagnèse du squellette de Spongiaires carbonatés dans le Trias de Turquie et le Permien de Tunisie. Geobios 20-6:757–773

Cuif JP, Dauphin Y, Denis A, Gautret P, Lawniczak A, Raguideau A (1987) Résultats récents concernant l'analyse des biocristaux carbonatés; implications biologiques et sédimentologiques. Bull Soc Géol Fr 8-3-2:269–288

Dieci G, Russo A, Russo F (1974) Nota preliminare sulla microstruttura di spugne aragonitiche del Trias medio-superiore. Boll Soc Paleontol Ital 13-1-2:99–107

Gautret P (1985) Organisation de la phase minérale chez *Vaceletia crypta* (Vacelet), Démosponge, Sphinctozoaire actuelle. Comparaison avec des formes aragonitiques du Trias de Turquie. Geobios 18-5:553–562

Gautret P (1986) Utilisation taxonomique des caractères microstructuraux du squellette aspiculaire des Spongiaires. Etude du mode de formation des microstructures attribuées au type sphérolitique. Ann Paléontol (Vert-Invert) 72-2:75–110

Gautret P, Cuif JP (1990) Les Démosponges calcifiées des bioherrmes du Jurassique supérieur du Sud-Tunisien (Oxfordien de la région de Tataouine). Geobios (in press)

Gautret P, Razgallah S (1987) Architecture et microstructure des Chaetetides du Permien du Djebel Tebaga (Sud Tunisie). Ann Paléontol 73-2:59–82

Hudson RGS (1960) The tethyan Jurassic stromatoporoids *Stromatoporina*, *Dehornella*, and *Astroporina*. Palaeontology 2:180–199

Reitner J, Engeser TS (1987) Skeletal structures and habitats of Recent and fossil *Acanthochaetetes* (subclass Tetractinomorpha, Demospongiae, Porifera). Coral Reefs 6:13–18

Vacelet J (1977) Une relique du Secondiare: un représentant actuel des éponges fossiles Sphinctozoaires. CR Acad Sci Paris 285 Sér D:509–510

Vacelet J (1979) Description et affinités d'une éponge Sphinctozoaire actuelle. Coll Int CNRS Biol Spong Paris 291:483–493

Vacelet J (1983) Les éponges calcifiées et les récifs anciens. Pour la Science 6:14–22

Wendt J (1979) Development of skeletal formation, microstructure, and mineralogy of rigid calcareous Sponges from the late Paleozoic to Recent. Coll Int CNRS Biol Spong Paris 291:449–457

Wendt J (1984) Skeletal and spicular mineralogy, microstructure and diagenesis of coralline calcareous sponges. Paleontogr Am 54:326–336

Wood RA, Reitner J (1986) Poriferan affinities of mesozoic stromatoporoids. Palaeontology 29-3:469–473

# Deficient Spiculation in a New Species of *Merlia* (Merliida, Demospongiae) from the Balearic Islands

J. VACELET and M.J. URIZ

## Introduction

Among living calcified demosponges (Sclerospongiae), the genus *Merlia* Kirkpatrick 1908, is highly remarkable by the facultative character of the calcareous skeleton which is present in the genotype, *M. normani* Kirkpatrick 1908, but absent in *M. deficiens* Vacelet 1980, and in another species described by Hoshino (in press).

In addition to a tabulate, aspicular calcareous skeleton *Merlia normani* is characterized by unique siliceous spiculation composed of fascicles of small tylostyles and of a microsclere complement including a highly distinctive clavidisc. This sponge is very widespread since it has been found in cryptic habitats of tropical and subtropical regions of the Atlantic Ocean, Mediterranean, Red Sea, and Indian Ocean, with remarkable constancy in its characters in all these areas. *Merlia deficiens* is similar with respect to its spicules and living tissue, but has no trace of a calcareous skeleton. It also displays a wide tropical and subtropical distribution, being found in dark caves of the Caribbean Sea, Bermuda, Mediterranean Sea, Indian Ocean and Pacific Ocean, again with a high degree of constancy in spiculation. The clavidiscs of *Merlia deficiens* differ slightly in size from those of *M. normani*. Hoshino's new species, which does not have a calcareous skeleton, differs from *M. deficiens* with respect to spicule size and appears to be restricted to Japan, where it lives in non-cryptic habitats. The distribution of these species is shown in Fig. 1.

The two species without a rigid skeleton were classified in the genus *Merlia* mainly because of their spiculation, which is highly distinctive in the class Demospongiae. This interpretation supposes that in *Merlia* the calcareous skeleton is an unstable character that may disappear in the course of speciation without greatly affecting other characteristics. This phenomenon could be more than anecdotic and represent a major trend in the evolution of the Porifera. Indeed, it appears that rigid, calcified skeletons, that were common in several different groups of the phylum in Paleozoic and Mesozoic times, *have been replaced in modern sponge fauna by spicule and spongin skeletons* (Vacelet 1985).

A sponge collected in a cave of the Balearic Islands demonstrates that in the genus spiculation including the highly diagnostic clavidiscs, can also disappear, at least partly. These specimens are described here as a new species of *Merlia, M. lipoclavidisca*.

J. Reitner and H. Keupp (Eds.)
Fossil and Recent Sponges

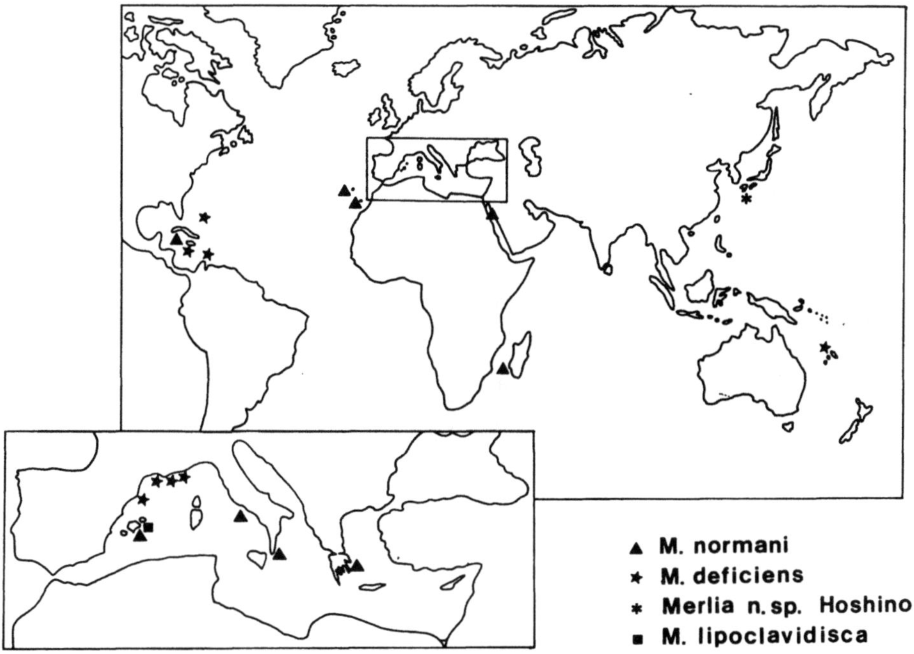


**Fig. 1.** Distribution of the genus *Merlia*

## Material and Methods

Specimens were collected by Scuba diving in La Catedral cave, 12 m deep, in the northeastern part of Mallorca (Balearic Islands, Mediterranean Sea). They were living on a rocky substratum partially covered with dead Bryozoa at the mouth of the cave. They were preserved in formalin.

## Description of *Merlia Lipoclavidisca* n.sp.

Holotype: NO. LBIM.D.J.V. 48, Museum National d'Histoire Naturelle, Paris.
Paratypes: CEAB-MC-2 and CEAB-MC-3, Centro de Estudios Avanzados de Blanes.
*Derivatio nominis*: from the Greek leipein = to lack: lacking clavidisc.
The external features of the three specimens as well as the characteristics of their calcareous skeleton and living tissue (Fig. 2a-d, Fig. 4) are exactly the same as *M. normani* as described by Kirkpatrick (1911). They are encrusting sponges, 1-1.5 mm thick and measured 60/30 mm (holotype), 25/25 mm and 45/30 mm. Living tissue covers and fills a basal calcareous crust made of vertical tubes ("pseudocalicles"), divided up to horizontal tabulae. They are grayish blue in the living state and gray in alcohol. The surface is finely granulous due to the terminal

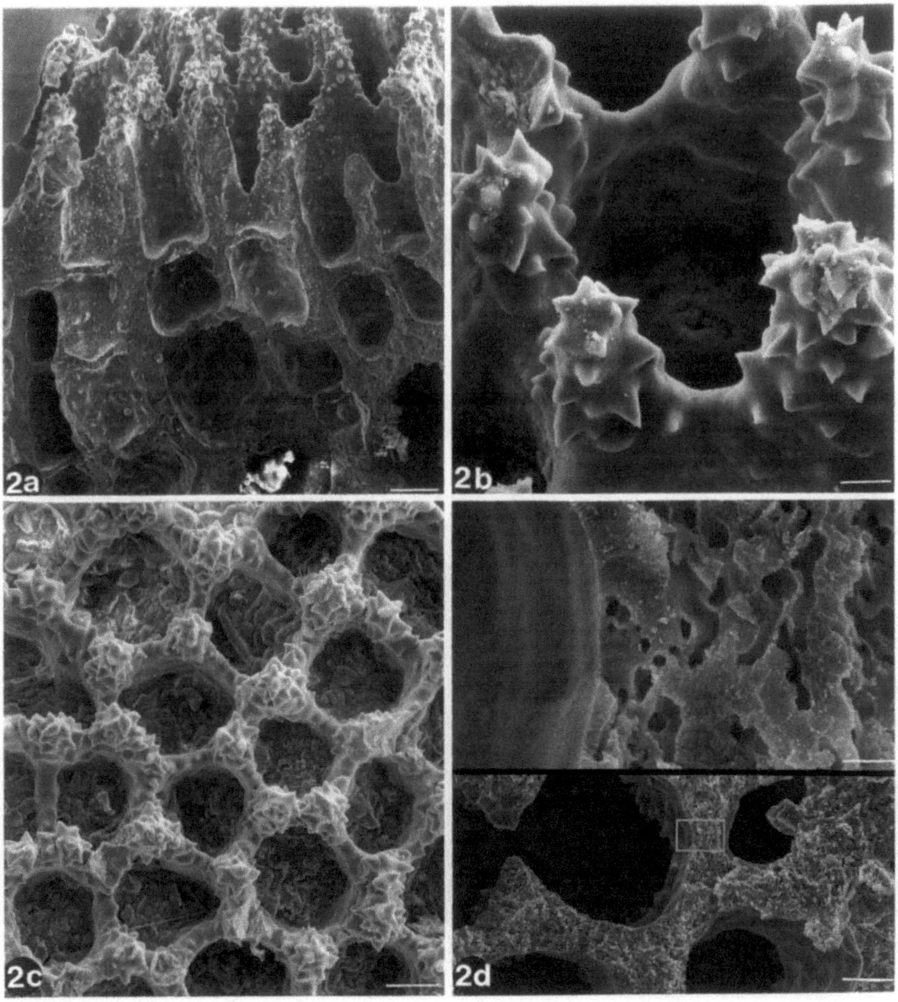


**Fig. 2.** **a** Vertical section through the skeleton of *Merlia lipoclavidisca*, showing the pseudocalicles and the tabulae forming a roof over basal crypts. × 70. Bar = 100 μm. **b** Surface view of pseudocalicle closed by a perforate tabula and surrounded by pillars with nodal tubercles. × 300. Bar = 25 μm. **c** Surface view of the skeleton. × 100. Bar = 66 μm. **d** Undersurface view of a fracture in the skeleton. Note the large number of pits made by boring-alga, *Ostreobium*. × 160 (bar = 50 μm) and × 1600 (bar = 5 μm)

superficial tubercles of the pillars of the rigid skeleton. No large apertures are clearly visible, but small holes, 60–100 μm in diameter, can be observed in the central part of some pseudocalicles.

The ectosome is composed of a pinacoderm and of a thin fibrillar layer. The choanosome is somewhat contracted in its upper layers, where the choanocyte chambers are elongated. In tissue within the tubes of the skeleton, which is less contracted, the choanocyte chambers are spherical (maximum diameter: 25 μm). A fascicle of spicules surrounded by elongated cells is observed in the central area of each pseudocalicle. One to three layers of cell masses are entrapped within the basal cavities of the skeleton. After decalcification, this tissue appears as moniliform cylinders. As in *M. normani* (Vacelet, in press), their cells are large and packed with reserve granules. No reproductive stages were observed.

The calcareous skeleton displays the same organization and microstructure as *M. normani* (Kirkpatrick 1911; Cuif et al. 1979). Its basal and middle parts contain many boring organisms, mainly filamentous algae (Figs. 2d, 4). No such organisms are found on the upper parts of the skeleton.

Spicules (Fig. 3a-d): Only tylostyles were observed. They are thin and straight, with an oval or, most often, roughly ovoid terminal or subterminal swelling. A secondary swelling is often present at some distance from the head. The length of these spicules was 135–180 μm (160.9 μm ± 10.48) for the holotype and 146–175 μm (159.44 μm ± 7.9) and 138–182 μm (157.28 μm ± 8.5) for the paratypes (mean ± standard deviation obtained for 50 spicule measurements). Their diameters were 0.9–1.1, 0.9–1.5 and 0.7–1.5 μm, respectively.

Most of these tylostyles are grouped in diverging bundles, their sharp ends generally pointing outwards. These megascleres are less abundant than in the living tissue of *M. normani* or *M. deficiens*.

## Discussion

The first problem in describing a sponge with a supplementary skeleton is to be certain that it is not a composite organism. This objection was raised for *M. normani* by Weltner (1909), who claimed that the specimens studied by Kirkpatrick were calcareous skeletons of an unknown organism parasited by a boring sponge, a hypothesis that was ruled out by Kirkpatrick (1911). Although the possibility is remote, these specimens with incomplete spiculation could conceivably correspond to dead skeletons of *M. normani* secondarily inhabited by another sponge. While the abundance of boring microorganisms in the older parts of the skeleton could be interpreted as an indication that they have not

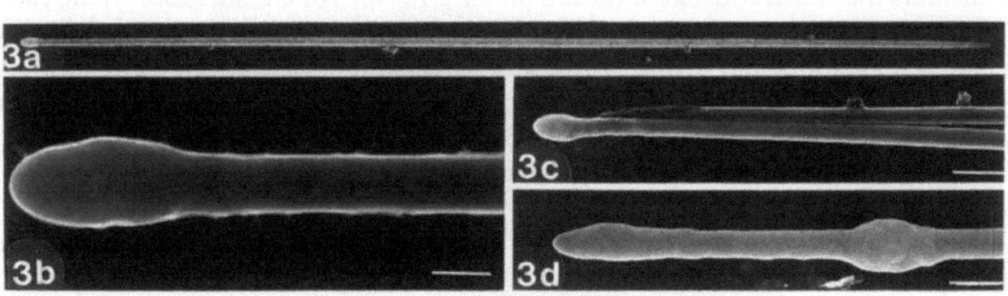


**Fig. 3. a** Siliceous tylostyle. × 700. Bar = 10 μm. **b** Head and tip of a tylostyle. × 2300. Bar = 3.25 μm. **c** Head of tylostyle. × 7800. Bar = 1 μm. **d** Head and secondary swelling of tylostyle. × 5000. Bar = 1.5 μm

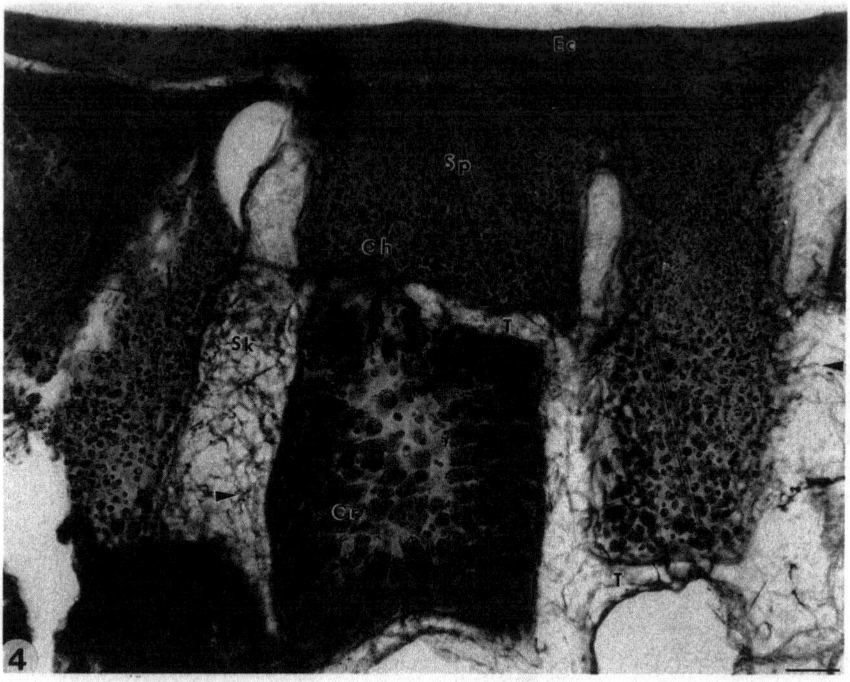


**Fig. 4.** Ground section of living tissue and skeleton. *Ec* ectosome; *Ch* choanosome; *Cr* mass of crypt cells; *Sk* calcareous skeleton, the basal and middle layers of which are perforated by *Ostreobium* (*arrowheads*). *Sp* fascicle of siliceous tylostyles. *T* tabula. × 270. Bar = 27 μm

contained living tissue for a while, such algae are absent from the superficial, younger areas of the skeleton which would also have been exposed to parasites. More importantly, the tylostyles, the living tissue, and the general organization resemble *M. normani* so closely that there is no doubt the skeleton is proper to the sponge.

The status of the species described under the genus *Merlia* is still controversial. According to the system of Hartman and Goreau (1972), which considers the calcareous skeleton as a major criterion for classification at the class level, specimens without skeleton should be classified in a completely different class (Demospongiae) from those with a skeleton (Sclerospongiae), despite striking detailed similarities in spicule and living tissue characteristics. Another radical position is that of van Soest (1984) who considers the calcareous skeleton to be a facultative, nonspecific character and thus that *M. normani* and *M. deficiens* are synonymous.

Vacelet (1980) proposed that forms with and without a skeleton should be grouped under the same genus defined on the basis of spicule and living tissue characteristics but separated at the species level. This position is corroborated by a slight difference in clavidisc size between *M. normani* and *M. deficiens* that appears to be constant in the wide and somewhat overlapping geographic distribution of the two species. Hoshino (in press) followed the same reasoning when describing a new *Merlia* devoid of calcareous skeleton. The variation described here demonstrates that spicules are not as reliable a feature as thought for defining the genus *Merlia* since the highly diagnostic clavidiscs may be absent. Overall, however, the similarities between these specimens and *M. normani* are so obvious that classification in the same genus is inevitable.

Another problem is posed by the three specific differences between these specimens and *M. normani*. The first difference involves living color, our specimens being grayish blue as compared with bright red to orange in *M. normani*. The second difference is that our specimens have no microscleres of any kind (raphides, comates and clavidiscs). The third difference which is slight but significant (Table 1) consists in longer and thinner megascleres. Whereas the mean length of tylostyles in *M. normani* from diverse localities is 130.7 μm (with a confidence interval of ± 3.2), it is 159.3 μm ± 0.33 in *M. lipoclavidisca*. However, their diameter (0.7–1.5 μm) is smaller than in *M. normani* (usually 1.5–2 μm). In *M. deficiens*, tylostyles are similar to those in *M. normani* (133 μm ± 3.5/1.5–2 μm). In an abstract to the Third Conference on the Biology of Sponges, Hoshino reports that his new species also has long tylostyles.

Although these differences may, as supposed here, be species-specific, the possibility that they represent ecomorphic or individual variations cannot be

**Table 1.** Length in μm [minimum-maximum (mean ± standard deviation)] of the tylostyles of the species of *Merlia*, based on ten measurements, except for *M. lipoclavidisca* (50 measurements) and Hoshino's species (unknown)

| | |
|---|---|
| *Merlia normani* | |
| Madeira (Atlantic) | 130–150 (137.0 ± 9.48) |
| Madeira (Atlantic) | 130–150 (140.0 ± 6.66) |
| Lindos (Medit., Greece) | 120–135 (128.5 ± 5.79) |
| Matapan (Medit., Greece) | 140–170 (152.0 ± 10.60) |
| Sorrente (Medit., Italy) | 105–125 (119.2 ± 4.80) |
| Naples (Medit., Italy) | 120–145 (133.0 ± 6.40) |
| Balearic Is. (Medit., Spain) | 112–150 (130.9 ± 10.36) |
| Europa (Indian Ocean) | 115–132 (123.3 ± 5.53) |
| Jamaica (Caribbean) | 100–120 (112.7 ± 6.40) |
| *Merlia deficiens* | |
| Marseille (Medit., France) | 135–165 (145.0 ± 11.05) |
| Marseille (Medit., France) | 120–150 (137.5 ± 8.83) |
| Medas Islands (Medit., Spain) | 120–152 (132.0 ± 8.10) |
| Jamaica (Caribbean) | 120–140 (130.0 ± 7.45) |
| New Caledonia (Pacific) | 127–147 (139.5 ± 5.37) |
| New Caledonia (Pacific) | 110–120 (114.9 ± 4.25) |
| *Merlia* n. sp. (Hoshino) | |
| Japan (Pacific) | 146–180 (168 ± ?) |
| *Merlia lipoclavidisca* | |
| Holotype | 135–180 (160.9 ± 10.48) |
| Paratype | 146–175 (159.4 ± 7.90) |
| Paratype | 138–182 (157.3 ± 8.50) |

ruled out, especially since sponges living in dark caves frequently display spicule abnormalities and color changes. In this case the distinctive characteristics noted here could be ecologically determined in individuals of *M. normani* living under special conditions. However, the conditions prevailing in the areas of the cave where *M. lipoclavidisca* was collected were not the extreme ones described in the darker recesses of this habitat (Harmelin et al. 1985), but rather were similar to those under which *M. normani* is found.

The main distinctive characteristic of this new species is the absence of microscleres. There are many examples in which specimens of sponges presumably belonging to the same species differ with respect to the absence or relative abundance of one or more types of microscleres. The most accurate data on this phenomenon were given by Bergquist and Sinclair (1973), in a study of spiculation in larvae or newly settled demosponges. They concluded that "absence of microsclere types as a sole criterion for the separation of two species should be viewed with suspicion". Although the absence of microscleres is not the sole criterion here, some caution is called for in classifying our specimens as a new species, since we have found only three specimens and since normal specimens of *M. normani* are present in the cave.

Whether these specimens correspond to an intraspecific variation of the same species or to a new species, they demonstrate that either of the two features of the genus, i.e., the clavidisc microsclere and calcareous skeleton, may be dropped out. Conceivably specimens may exist in which both features are absent thus making their identification as *Merlia* impossible. Such specimens may have already been encountered and mistakenly classified under other genera of demosponges, e.g., as *Terpios*. Bergquist and Fromont (pers. comm.) have under study a new species of *Desmacella* which may correspond to such incomplete *Merlia* specimens.

In the interpretation of many important groups of fossils, the absence of spicules has been, and is still sometimes, used as one of the main criteria against considering them as Porifera. Interestingly siliceous spicules are somewhat inconstant features in existing calcified demosponges. They are entirely missing in the living Sphinctozoa, *Vaceletia crypta* (Vacelet). They are also absent in Central Pacific populations of *Astrosclera willeyana* Lister. Conversely, they are well developed in Indian Ocean populations presumably of the same species. Intermediary population in which spicules are ill-formed and inconstant (Vacelet 1981) or present only in small specimens (Ayling 1982) can be found in the West Pacific. It now appears that the highly distinctive microsclere complement of the genus *Merlia* may be absent. Although species of this relic genus apparently have a high genetic stability, as shown by the constancy of the features over very wide geographic areas, some characteristics appear to be instable. Porifera apparently is the only phylum in which the same organism is able to secrete both a calcium carbonate and a siliceous skeleton. In the present fauna this possibility is now restricted to a few relic forms. In some of these relics, this unique possibility is seemingly labile and one mineral may be partially or totally absent.

Another problem raised by the genus *Merlia* is that of its affinities within Demospongiae. Most authors (Kirkpatrick 1911; Hartman and Goreau 1970; van Soest 1984) have considered that the characteristic clavidisc indicates affinities with the order Poecilosclerida, subclass Ceractinomorpha, since its shape is reminiscent of the diancistra of the Hamacanthidae, which is a spicule at an intermediary stage between sigma and clavidisc. This observation led van Soest (1984) to classify *Merlia* within Poecilosclerida Biemnidae, in which he included the family Hamacanthidae. Vacelet (1979, 1981) proposed on the contrary to place the genus in a special order Merliida, classified in the subclass Tetractinomorpha which includes another sponge with a tabulate calcareous skeleton: *Acanthochaetetes wellsi* Hartman and Goreau. This proposal is based on collagen characteristics (smooth-type microfibrils (Garrone 1978) which appear to be distinctive for Tetractinomorpha), and the ability to accumulate masses of reserve cells which is unknown in Poecilosclerida but well developed in Hadromerida. This position is adhered to here, since the Hamacanthidae-Biemnidae affinities appear, in fact, questionable. The unusually small tylostyle megascleres in *Merlia* are arranged differently from in *Hamacantha*, and the resemblance between diancistra and clavidisc may be a convergent similarity. Young diancistra are clearly related to sigma, whereas young clavidics most often form a complete ring in which the two ends may largely overlap. Unfortunately, the reproduction has not been observed to date in the genus and its subclass assignment still relies on inconclusive evidence.

## References


Ayling AL (1982) A redescription of *Astrosclera willeyana* Lister, 1900 (Ceratoporellida, Demospongiae), a new record from the Great Barrier Reef. Mem Natl Mus Victoria 43:99–103

Bergquist PR, Sinclair ME (1973) Seasonal variation in settlement and spiculation of sponge larvae. Mar Biol 20:35–44

Cuif JP, Debrenne F, Lafuste JG, Vacelet J (1979) Comparaison de la microstructure du squelette carbonaté non spiculaire d'Eponges actuelles et fossiles. Coll Int CNRS n°291: Biol Spong Paris, Déc 1978: 459–465

Garrone R (1978) Phylogenesis of connective tissue. Morphological aspects and biosynthesis of sponge intercellular matrix. Karger, Basel, pp 250

Harmelin JG, Vacelet J, Vasseur P (1985) Les grottes sousmarines obscures: un milieu extrême et un remarquable biotope refuge. Téthys 11(3–4):214–229

Hartman WD, Goreau TF (1970) Jamaican coralline sponges: their morphology, ecology and fossil relatives. Symp Zool Soc Lond 25:205–243

Hartman WD, Goreau TF (1972) *Ceratoporella* (Porifera: Sclerospongiae) and the chaetetid "corals". Trans Conn Acad Arts Sci 44:133–148

Hoshino T A new *Merlia* (Poecilosclerida) from japanese waters. In: Hartman W, Rützler K (eds) Proc III Int Conf on the Biology of Sponges, Woods Hole (in press)

Kirkpatrick R (1908) On two new genera of recent pharetronid sponges. Ann Mag Nat Hist 8(2):503–514

Kirkpatrick R (1911) On *Merlia normani*, a sponge with a siliceous and calcareous skeleton. Q J Microsc Sci 56:657–702

Vacelet J (1979) Description et affinités d'une Eponge Sphinctozoaire actuelle. Coll Int CNRS n°291: Biol Spong Paris Déc 1978:483–493

Vacelet J (1980) Squelette calcaire facultatif et corps de régénération dans le genre *Merlia*, Eponges apparentées aux Chaetétidés fossiles. CR Acad Sci D 290:227–230
Vacelet J (1981) Eponges hypercalcifiées ("Pharétronides", "Sclérosponges") des cavités des récifs coralliens de Nouvelle-Calédonie. Bull Mus Nat Hist 3A:313–351
Vacelet J (1985) Coralline sponges and the evolution of the Porifera. In: Conway Morris S, Gibson R, Platt HM (eds) The origins and relationships of lower invertebrates. Syst Assoc Spec Vol 28:1–13
Vacelet J The storage cells of calcified relict sponges. In: Hartman W, Rützler K (eds) Proc III Int Conf on the Biology of Sponges, Woods Hole (in press)
van Soest RWM (1984) Deficient *Merlia normani* Kirkpatrick, 1908, from the Curaçao Reefs, with a discussion on the phylogenetic interpretation of sclerosponges. Bijdr Dierk 54(2):211–219
Weltner W (1909) 1st *Merlia normani* Kirk. ein Schwamm? Arch Naturgesch 75:139–141

# Phylogenetic Aspects and New Descriptions of Spicule-Bearing Hadromerid Sponges with a Secondary Calcareous Skeleton (Tetractinomorpha, Demospongiae)

J. Reitner

## Introduction

In contrast to the nonrigid modern demosponges and Calcarea, the sponges with an additional calcareous skeleton are extremely rare and restricted to dark and cryptic habitats within tropical reefs and submarine caves of the Mediterranean Sea. Only 14 recent genera of the so-called sclerosponges or coralline sponges and pharetronids are known (nine demosponges and five calcisponges) (Kirkpatrick 1908, 1910, 1911; Lister 1900; Döderlein 1897; Vacelet 1964, 1981; Hartman and Goreau 1970, 1975; Hartman 1979, etc.). Coralline sponges are quite common in the fossil record. During the Paleozoic and Cretaceous they were important reefbuilders or members of reefal communities. Traditionally, most of these sponges are classified as Stromatoporoidea, Chaetetida, Sphinctozoa, or Pharetronida. These terms describe only the type of soft tissue arrangement within these particular sponge skeletons and have no taxonomic value (Reitner 1987a,b,c; Wood 1987). Hartman and Goreau (1972) have established a new class of sponges which incorporates all sponges with siliceous spicules and an additional calcareous basal skeleton. However, based on the similarities of the spicules and soft tissue with nonrigid demosponges and Calcarea, van Soest (1984), Vacelet (1985), Reitner (1987a,b,c) and Wood (1987) have proposed a polyphyletic origin of the class "Sclerospongiae". It is possible to recognize different phylogenetic lineages of fossil and modern calcified demosponges which demonstrate in some cases extremely long-ranging clades (Reitner and Engeser 1985, 1987; Wood et al. 1989).

The purpose of this chapter is to demonstrate the polyphyletic nature of calcified hadromerida and to establish a phylogenetic model for this group.

## Material and Methods

This study is based on the type material of the species *Boswellia mortoni* Gray (1980) (British Museum of Natural History (BMNH) no. R4429), *Cassianothalamia zardini* Reitner (1987b) (Institut für Paläontologie Freie Universität Berlin (IPFUB) no. 87/1/2) and other newly collected fossil material deposited within the IPFUB. The Recent counterparts were loaned from the BMNH, Institute of Taxonomic Zoology Amsterdam (ITA), and the marine lab Endoume, University

J. Reitner and H. Keupp (Eds.)
Fossil and Recent Sponges

of Marseille. Some were collected in Bermuda and the western part of the Mediterranean Sea near Banyuls sur Mer.

From all sponges thin- or histological sections were made and studied under a light microscope. The histological sections were stained with acid fuchsin and toluidine blue to study the soft tissue arrangement. In order to separate the spicules from the soft tissue, the sponges were treated with $H_2O_2$. To expose the spicules from the calcareous basal skeletons, polished surfaces were etched with 5% Tritriplex-III solution ($C_{10}H_{14}N_{20}18 \times 2H_2O$).

After 1 day of treatment the samples were cleaned with distilled water and then dried. Subsequently the etched surfaces were examined with a scanning electron microscope (SEM). Micro- and ultrastructures were studied on freshly broken surfaces. The chemical analyses of the calcitic skeletons were performed with atomic absorption spectroscopy (AAS), energy dispersive X-ray analysis (EDAX), and electron microprobe analysis.

## Systematic Position of the Hadromerida

The Hadromerida are sponges which possess monactine megascleres as tylostyles or subtylostyles. The microscleres are always asters or modified asters. The megascleres are generally organized in a radial pattern but, in most cases, this radial pattern is ill developed. The microscleres, if present, are concentrated within the cortical zone of the sponge. Within the Hadromerida, spongin is present but never as fibers.

The hadromerid sponges are oviparous, wherever they have been observed. Based on spicule morphology, oviparity, and simple spongin, these particular sponges are classified within the subclass Tetractinomorpha (Levi 1973; Bergquist 1978). The Hadromerida are one of the best-characterized sponge groups.

### *Skeletal Types of Extant Hadromerid Sponges*

Within the Hadromerida a reduction of the spicular skeleton is observed in various families. This reducted part of the primary skeleton is substituted by strong collagenous skeletal elements. It is possible to distinguish different stages of sclere reduction:

a) reduction of microscleres (Suberitidae, Polymastidae),
b) reduction of megascleres (*Chondrilla*) or a partial reduction, as seen in *Spirastrella*,
c) reduction of micro- and megascleres (*Chondrosia*)

A special type of indirect skeletal formation is observed within some species of *Spirastrella* and the genus *Cliona*. *Cliona* excavates burrows in any calcareous material and the entire sponge, except the marginal inhalant and exhalant systems, lives inside the burrows (Rützler 1974; Keupp and Reitner this Vol.).

Calcareous basal skeletons, the main topic of this chapter, are observed only within the extant genus *Acanthochaetetes*, which possesses a spicular skeleton similar to *Spirastrella* (Reitner and Engeser 1987). Within the fossil record, calcareous basal skeletons are observed in several taxa.

## Fossil and Modern Hadromerida with a Secondary Calcareous Skeleton

Remarks:
The author has decided to erect new genera for new fossil taxa. The distinct types of calcareous skeletons of these particular fossils should be recognizable as morphotypes for all fieldworking paleontologists, since spicules are not seen or preserved in all cases. Using the principles of phylogenetic-systematics new genera must not be erected in all cases of calcareous skeletons.

### *Family Suberitidae Schmidt*

The Suberitidae are characterized by tylostyle and subtylostyle megascleres and rare microstrongyle microscleres. The typical hadromerid radial arrangement of the scleres is present only at the surface of the sponge. In the center part of the sponge the orientation of the spicules is confused. The spicules which cover the exhalant canals often show a loose plumose orientation (Fig. 3a,b).

*Chaetetes* (*Boswellia*) *mortoni* Gray (Figs. 1,2)

1980 *Chaetetes* (*Boswellia*) *mortoni* Gray, p. 808, Pl. 103

This species from the Lower Carboniferous (Lower Asbian) from Northern Wales is the oldest known coralline sponge with hadromerid character.

Primary skeleton:
*Chaetetes mortoni* possesses tylostyle megascleres (length 170–210 $\mu$m) which are arranged in bundles as observed in *Suberites* (Fig. 3a,b). These bundles are relicts of a radial pattern, characteristic of the Hadromerida.

Microscleres are not observed. The spicules are preserved as diagenetic silica or calcite (Fig. 2c).

Secondary skeleton:
The spicules are covered by a clinogonal ("water jet" structure, or fasicular fibrous) calcite (probably Mg-calcite) (Fig. 1c), which is common in coralline sponges. In this case the secondary skeleton is analogous to the primary collageous/spicular skeleton, which is substituted by Mg-calcite (Fig. 1b). The entire basal skeleton exhibits a chaetetid structure: that means the calicles are separated by tabulae which mark the boundary between the living sponge and the dead part of the basal skeleton (for measurements see Gray 1980, p.809).

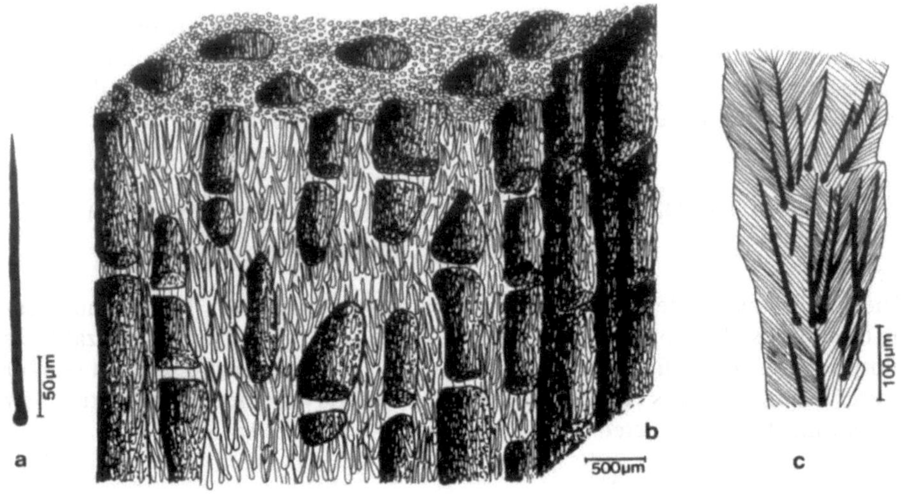


**Fig. 1.** *Chaetetes* (*Boswellia*) *mortoni* Gray. **a** Tylostyle megasclere. **b** Reconstruction of the spicular arrangement which is similar to the extant genus *Suberites*. **c** Primary and secondary skeleton of a single wall

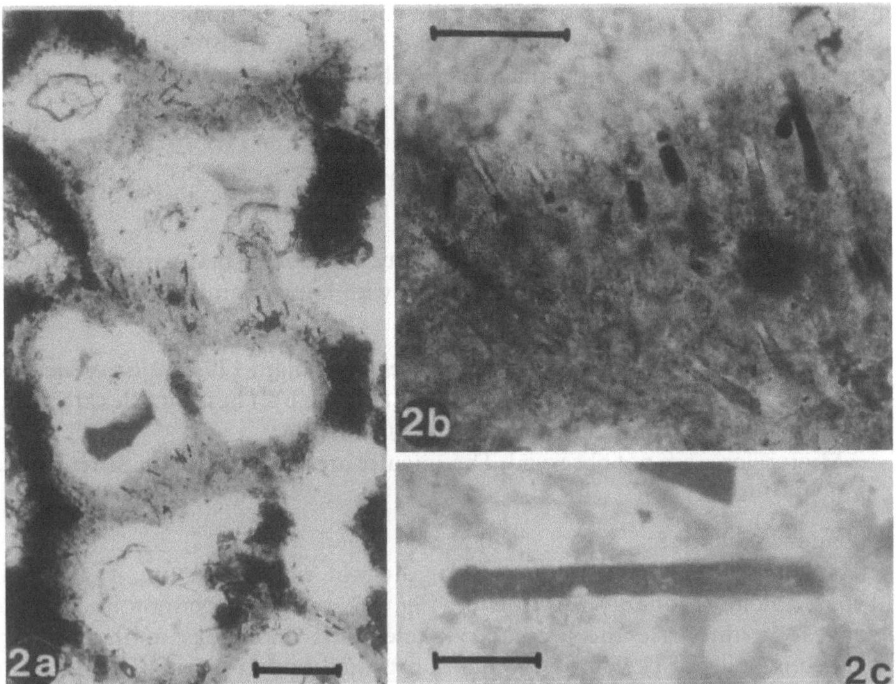


**Fig. 2.** **a** Horizontal section of *Ch.*(*B.*) *mortoni* Gray demonstrating the spicule bundles. Gray's paratype (British Museum Nat. Hist. no R 4429). Scale = 200 µm. **b** Detail of **a** Spicules partly preserved in pyrite. Scale = 100 µm. **c** Tylostyle megasclere of *Ch.*(*B.*) *mortoni*. Scale = 50 µm

**Fig. 3.** **a** *Suberites carnosus*; view on the surface of the dermal layer. Corallingene, Banyuls sur Mer, marine station Arago. **b** Detail of dermal layer of the specimen of **a**. The spicules are arranged in a subradiate way

Remarks:

The spicular skeleton of *Chaetetes mortoni* is comparable with the dermal layers of the families Suberitidae and Polymastiidae. Both groups have only rare microscleres and tylostyle megascleres. Microscleres are not known from *C. mortoni*. Significant are the bundles of megascleres, arranged in a loose perpendicular orientation, which is characteristic for the Suberitidae (Fig. 3a,b). This species is the oldest known hadromerid sponge.

*Chaetetopsis favrei* (Deninger 1906) (Figs. 4,5)

1906 *Monotrypa favrei* Deninger, p. 64, Pl. VI, Fig. 5a,b
1930 *Chaetetopsis favrei* (Deninger); Peterhans, p. 35, Pl. I,II
1970 *Chaetetopsis favrei* (Deninger); Fischer, p. 190
1979 *Chaetetopsis favrei* (Deninger); Kazmierczak, Figs. 2–4

This early Cretaceous coralline sponge species is very common in Barremian and Aptian reefs. Kazmierczak (1979) reported the first monaxonic spicule remains from a specimen from the Crimea. The spicules are preserved in rows of framboid pyrite. The exact shape of the spicules is not shown in this particular specimen. Newly collected material from the Albian of Greece allows one to distinguish the spicule type and arrangement, as well as the microstructure of the calcareous skeleton.

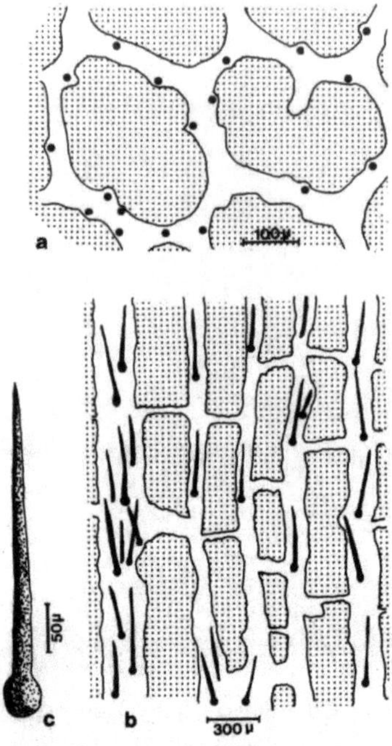


**Fig. 4.** *Chaetetopsis favrei* (Deninger). **a** Horizontal section exhibiting the spicule arrangement (*black dots*). Redrawn from Kazmierczak 1979. **b** Vertical cut of a specimen from Arachova (Aptian? from Greece). **c** Tylostyle megasclere of the specimen from Arachova

Primary skeleton:
The tylostyle spicules (120–450 μm length, 5–10 μm thick) (Figs. 4c, 5c-e) are arranged in loose bundles (Figs. 4,5c). The specimen from the Barremian of Crimea covers partially the inner wall of the calicles (Fig. 4a). Horizontal spicule elements and microscleres are not observed. The loose arrangement of the spicules indicates a prominent collagenous skeleton which has a matrix function for the secondary calcareous skeleton.

Secondary skeleton:
The entire basal skeleton demonstrates a chaetetid structure (Figs. 4c, 5a). The calicles are subdivided by tabulae. The tabulae are regular and thin (10–20 μm) and composed of minimicrite. The vertical wall exhibits a fasicular fibrous microstructure, as seen in Paleozoic *Chaetetes* (Fig. 5b). The thickness of the wall varies only a little (300–700 μm). The spicules are intramural. Some of the spicules protrude from the calcitic wall. In this case the spicules were primarily not covered

**Fig. 5.** *Chaetetopsis favrei* (Deninger). **a** Vertical section of the specimen of Arachova (near Delfi, Greece; Aptian? component in an Eocene "Qphiolithe Conglomerate"). Scale = 200 μm. **b** Fascicular/hemispherulitic microstructure of the calicle wall of the Arachova specimen. Scale = 75 μm. **c** Tylostyle megasclere of the Arachova specimen. The spicule is partly preserved by pyrite spherules (framboid pyrite, probable formed by bacteria). Scale = 50 μm. **d** Tylostyle megasclere of the Arachova specimen. The spicule is preserved by a granular calcite. Scale = 50 μm. **e** Intramural tylostyle megascleres of the Arachova specimen. Scale = 75 μm

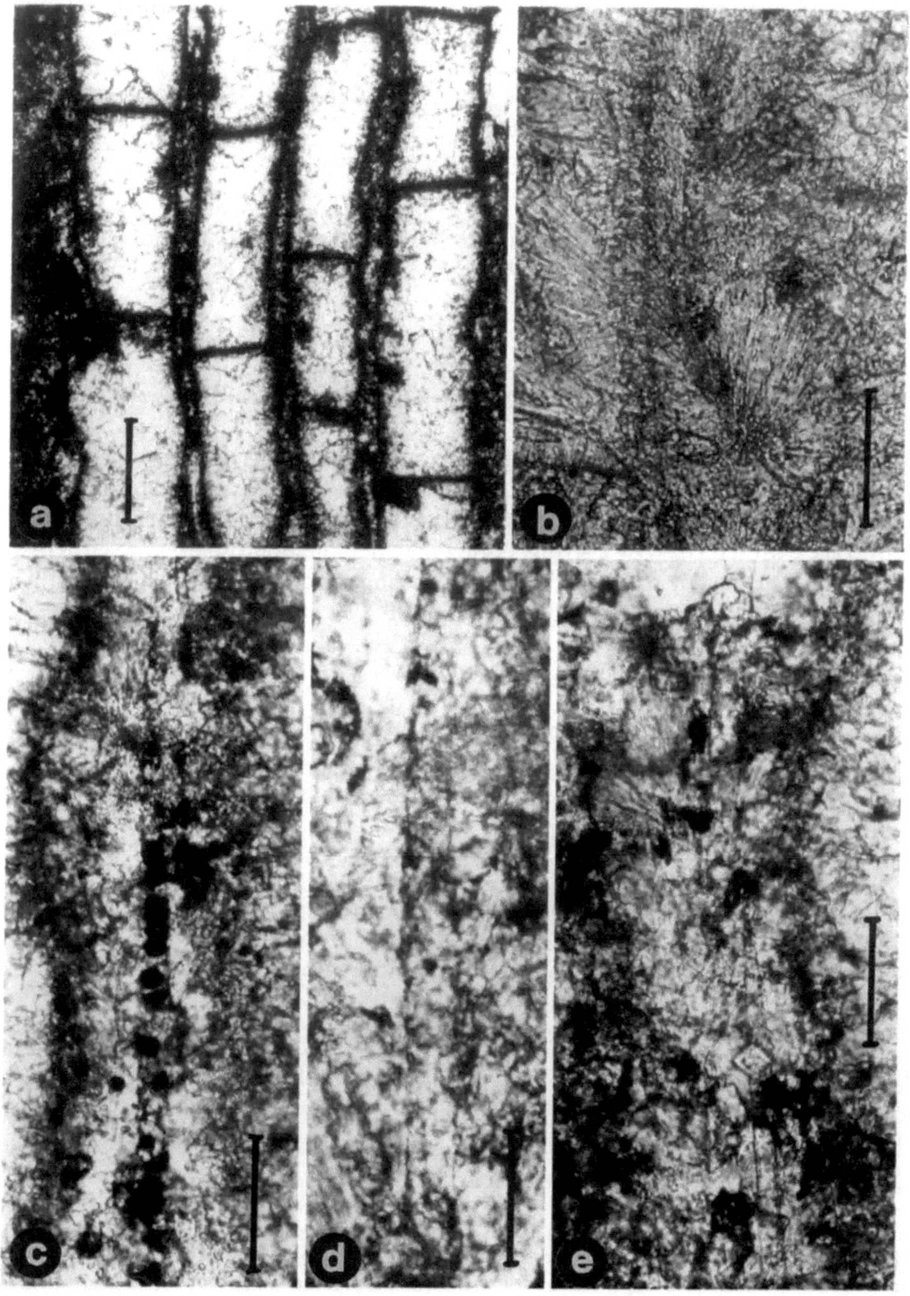
a
b
c
d
e

by collagenous soft tissue. The basal skeleton reflects the primary sponge skeleton. The primary microstructure is, in most cases, not recognizable. The skeleton exhibits a diagenetic granular or blocky calcite. Only in a few examples are relicts of the original structure observed. Based on this observation, an aragonitic primary mineralogy is very probable.

Remarks:
Spicule type and the relict radially oriented spicules point toward a classification within the hadromerids. The axially oriented intramural tylostyle scleres indicate that *Chaetetopsis* as well as *Chaetetes mortoni* Gray is also a member of the family Suberitidae.

### *Family Chondrosiidae Schulze*

The Chondrosiidae are sponges with a strong collagenous skeleton lacking, in most cases, the spicular skeleton. Only aster microscleres, mostly euasters, are present in the genus *Chondrilla*. Megascleres are unknown. All members of this group are characterized by a strong cortex in which the microscleres are enriched if present. The cortex is formed by fibrillar collagen. The sponges are encrusting or massive.

Genus *Chondrochaetetes* n.gen.

Type species: *Chondrochaetetes longitubus* n.gen. n.sp.

Derivatio nominis: Combined from the hadromerid family Chondrosiidae "Chondro-" and the term "Chaetetes" for basal skeletons with subdivided by tabulae into calicles.

Diagnosis:
Hemispherical sponges with a chaetetid secondary calcareous basal skeleton. The microstructure of the vertical walls is fascicular fibrous. The tabulae are micritized. Intramural aster microscleres are common within the vertical walls.

*Chondrochaetetes longitubus* n.gen., n.sp. (Fig. 6a-e)

Derivatio nominis: After the Latin word "longitubus" since this species possesses very long calicles.

**Fig. 6. a** *Chondrochaetetes longitubus* n.gen. n.sp., holotype from the Carboniferous of the USSR. Horizontal cut exhibiting asterose microscleres (*arrows*) (BMNH no. R 27318). Scale = 500 μm. **b** Vertical section of the holotype of *Chondrochaetetes*. The very thin tabulae and the calicle wall are covered by a granular calcitic cement. Scale = 500 μm. **c** Detail of the calicle wall of *Chondrochaetetes* demonstrating the fascicular fibrous wallstructure and intramural asterose microscleres. Scale = 200 μm. **d** Intramural asterose microsclere of *Chondrochaetetes*. Preservation is in granular calcite. Scale = 50 μm. **e** Intramural asterose microscleres (euasters?) of *Chondrochaetetes*. Black cores of the spicules are pyrite. Some of them exhibiting relicts of the star rays (*arrow*). Scale = 50 μm

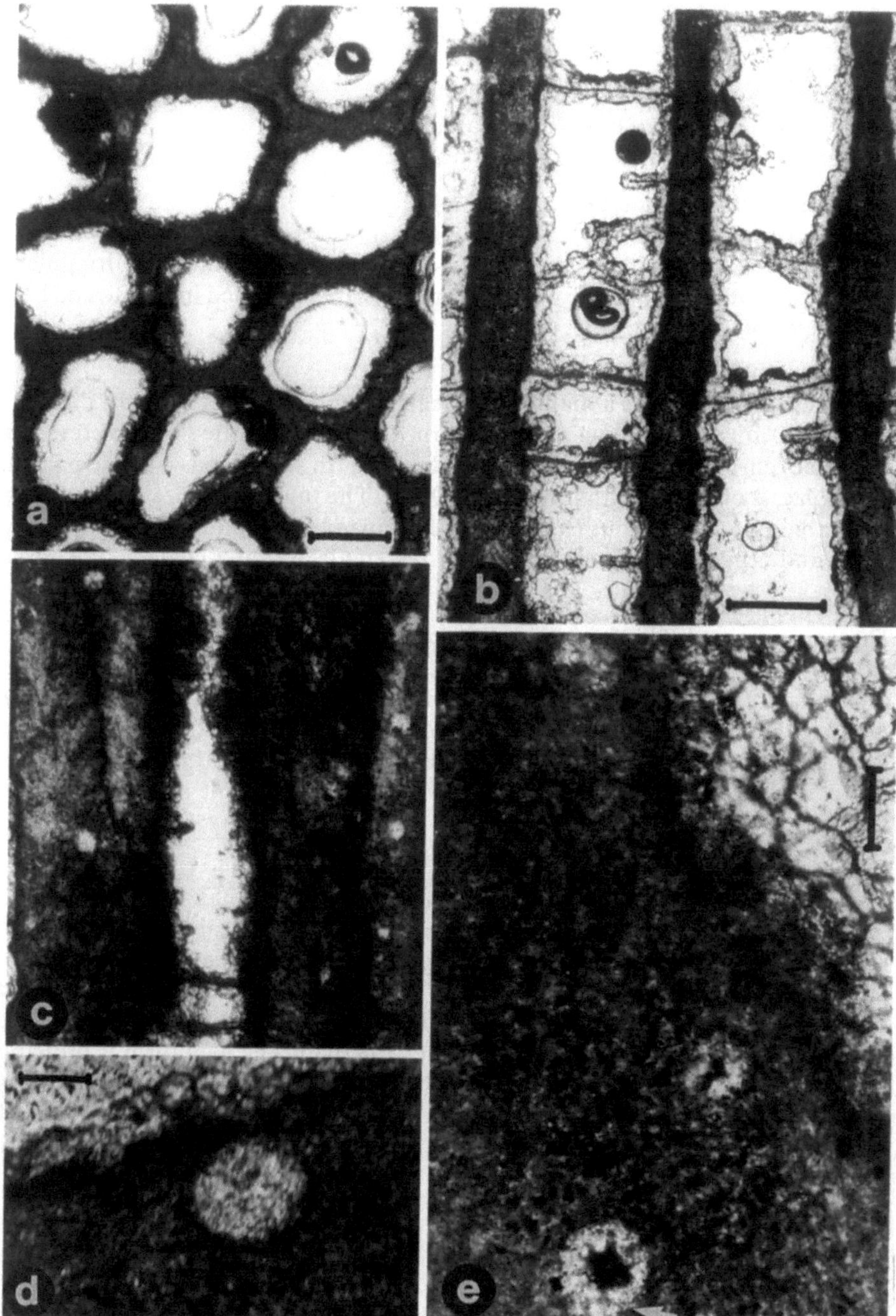
a
b
c
d
e

Holotype: British Museum Natural History no. R 27318 (labeled as "*Chaetetes radians*" Fischer).
Diagnosis: See genus diagnosis
Locus typicus: Kaluga Borovitch, Yaldai, Russia
Stratum typicum: Carboniferous Limestone

Description of the holotype:
Primary skeleton:
The observed spicules are asters, probably euaster, and they are irregularly distributed within the calcitic wall. The spicules are preserved in calcite and have a diameter of 25–35 μm (Fig. 6c-e).

Secondary skeleton:
The uncut specimen has a diameter of nearly 5 cm. The calicles have a diameter of 200–300 μm and exhibit a radial pattern. They are separated only by few (3 or 4 tabulae/cm) thin (ca. 30 μm) tabulae (Fig. 6b). In some areas a dense packing of tabulae is observed with more than 10/cm. The microstructure of the vertical walls are fascicular fibrous (Fig. 6c). The tabulae have a minimicritic structure. Walls and tabulae are covered by a late diagenetic granular cement (Fig. 6b).

Differential diagnosis:
The new species differs from *Chaetetes mortoni* in possessing aster microscleres only. Other differences of the new species *Calcistella tabulata* include a missing canal system within the calicle wall. The microstructure of *Chondrochaetetes* is fascicular fibrous in contrast to the dense micritic structure of *Calcistella*. The microscleres are significantly smaller with a mean diameter of 30 μm in comparison with microscleres of *Calcistella*, which exhibit a mean diameter of 55 μm.

Remarks:
This Carboniferous spicule-bearing calcified demosponge has, based on the aster (euaster?) scleres, close affinities to the family Chondrosiidae. However, an enrichment of microscleres in any zone of the chaetetid skeleton is not observed and there are no traces of a dermal layer. The spicules are irregularly distributed, probably caused by early diagenetic dissolution of the scleres and rapid epitactical cementation of the molds (cf. Reitner 1987b,c). The new form is probably related to the genus *Chondrilla*, which contains euaster microscleres only. In contrast to the tylostyle megasclere-bearing species, the chaetetid basal skeleton of this particular form is not analogous to the primary spicular/organic skeleton.

Genus *Calcistella* n.gen. (Fig. 7a-d)

Type species: *Calcistella tabulata* n.gen., n.sp.
Derivatio nominis: After the Latin words "calcium" and "stella" = star.

Diagnosis:
The aster microscleres-bearing sponge possesses a modified chaetetid secondary micritic skeleton. Diameter of the calicles is exceptionally large. The calicle walls

are structured by numerous pores and canals. The asters are restricted to the mineralized micritic part of the vertical walls.
Differential diagnosis: See species description

*Calcistella tabulata* n.gen., n.sp. (Fig. 7a-d)

Derivatio nominis: After the Latin word "tabula" = plate
Holotype: Institut für Paläontologie der Freien Universität Berlin (IPFUB, JR1/89).
Locus typicus: Arachova near Delfi, Greece; sampled by H. Keupp
Stratum typicum: Albian component within the "Ophiolithe-Conglomerate" of Eocene age.

Diagnosis: See generic diagnosis

Description of the holotype:
Primary skeleton:
The spicular skeleton is composed of aster microscleres only (Fig. 7c,d). The asters are preserved as spherules and exhibit only, in a few cases, relicts of spines. Based on this observation, the asters were probably euasters. The diameter of the spherules varies little (50–65 μm with a mean value of 55 μm). The asters are restricted to dark micritic areas within the calicle wall (Fig. 7c). The dark areas form a more or less irregular network within the vertical skeletal elements, interrupted by pores which mark complicated canal systems inside the calicle wall. The canal pores are separated from the dark micritic areas by thin granular sparitic layers. The diameter of the canal systems are 140–400 μm. The enrichment of spicules in the micritic parts only indicates a thick fibrillar primary collagenous skeleton. The canal systems are probably excurrent canals. Spicules are never observed within the tabulae.

Secondary skeleton:
The entire sponge exhibits a hemispherical chaetetid structured basal skeleton (6 cm diameter measured in thin section) with extraordinarily big tubes (Fig. 7a,b). The diameter of the tubes varies from 1.2–3 mm (mean 1.8 mm). The microstructure of tabulae and the vertical wall is micritic. Sparitic lamellae are common above the tabulae, at the base of the tubes, and within the excurrent canals (Fig. 7b). These ca. 20 μm thick lamellae are interrupted by thin micritic layers and mark the upward migration of the living sponge tissue. The sparitic layers mark the gap between the secondary skeleton and pinacoderm of the living sponge. The primary mineralogy of the basal skeleton was a high Mg-calcite. All associated aragonitic organisms are recrystallized. EDAX analysis show Mg anomalies within the wall and support this assumption.

Remarks:
This type of chaetetid basal skeleton is analogous to the primary collagenous skeleton of the living sponge. The collagenous skeleton was probably very strong and became calcified synvivo, as indicated by the sparitic growth of lamellae. The microscleres are concentrated within the thick fibers only. The calicle wall was

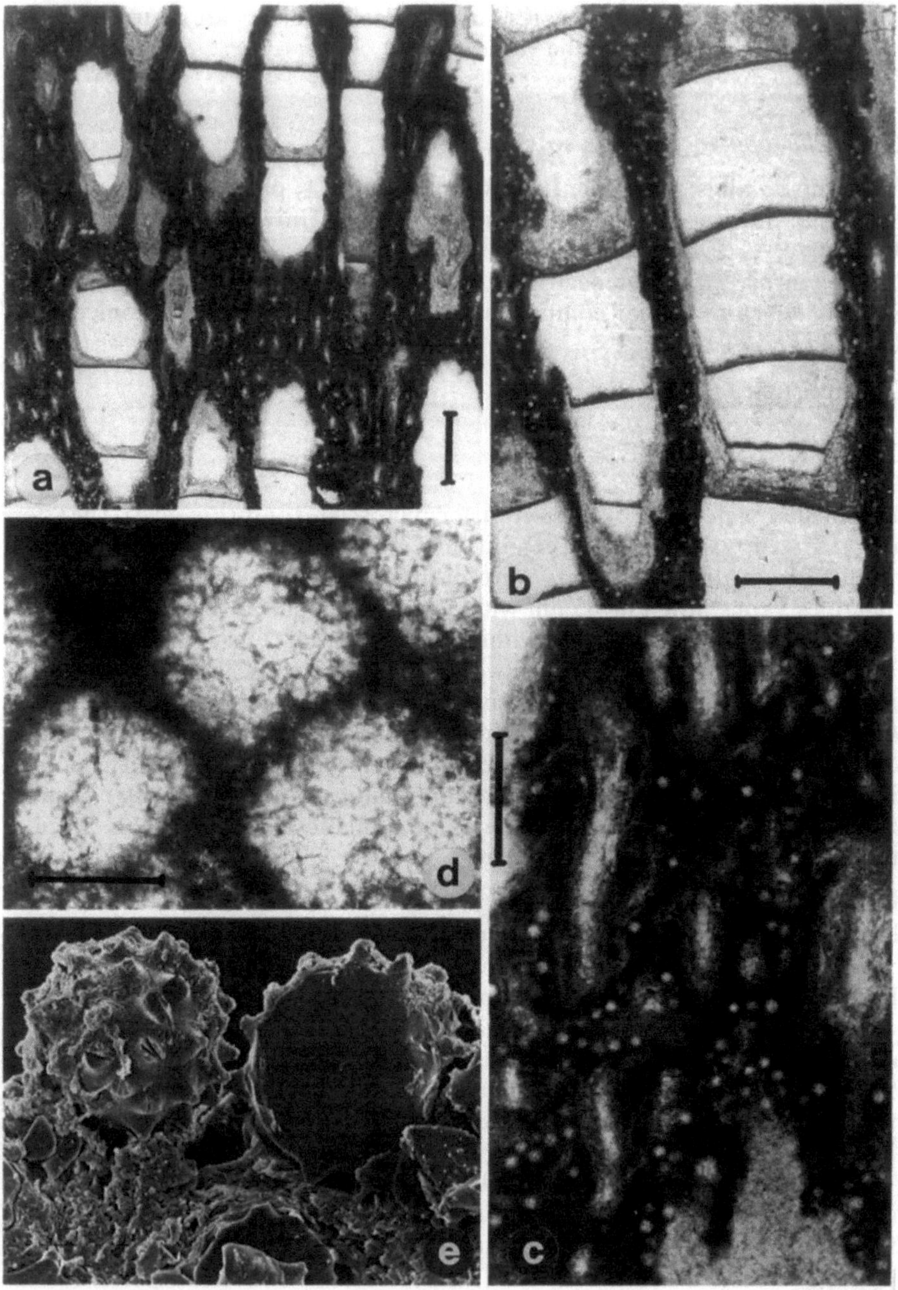
a
b
d
e
c

also penetrated by choanosomal soft tissue with prominent excurrent canals. This is in contrast to all other observed chaetetid sponge skeletons.

The very big microscleres, the strong collagenous skeleton, and the canal systems are comparable with the family Chondrosiidae, particularly with the genus *Chondrilla* (Figs. 7e,10b). The euasters of *Chondrilla grandistellata* show dimensions as measured in *Calcistella* (Figs. 7e,10b). The mineralized internal canal systems were probably formed by the endopinacoderm layer. The calical walls became mineralised by calcification of the collagenous fibers, as observed in the extant coralline sponge *Vaceletia* (Reitner 1987b). The tabulae were probably formed by the exopinacoderm layer, as observed in the extant hadromerid sponge *Spirastrella* (*Acanthochaetetes*) *wellsi*. This may explain the different observed structures of the calcitic secondary rigid body.

Differential diagnosis:
This new form differs from the genus *Chondrochaetetes* n.gen. by having a complicated excurrent canal system within the calicle walls, micritic microstructure, and aster microscleres 55 μm in diameter in contrast to 30 μm measured in *Chondrochaetetes*. The chaetetid species from the early Cretaceous of China, *Pseudomillestroma reticulata* Deng (Deng 1982), differs from the new species in having smaller asters and missing a canal system inside the calicle wall.

Genus *Calcichondrilla* n.gen. (Fig. 8a-d)

Type species: *Calcichondrilla crustans* n.gen., n.sp.
Derivatio nominis: After the Latin word "Calcium" and the genus name "Chondrilla" for hadromerid sponges which have only euaster microscleres.

Diagnosis:
Crustose sponges which have large aster microscleres only. The scleres are enriched in the primary strong collagenous parts of the skeleton. Prominent excurrent canals do exist. The entire primary collagenous skeleton of the sponge is calcified.

Differential diagnosis: See species diagnosis

*Calcichondrilla crustans* n.gen., n.sp. (Fig. 8a-d)

Derivatio nominis: After the Latin word "crustae" for crusts.
Holotype: Deposited in the IPFUB, JR2/89 (Fig. 8a-d)

**Fig. 7.** **a** Tangential cut of the basal skeleton of *Calcistella tabulata* n.gen. n.sp. (holotype) from the Aptian? of Arachova (Greece). Within the thick calicle walls the characteristic canal systems are visible. Scale = 1 mm. **b** Calicles of *Calcistella* exhibiting the tabulae and the growing lamellae of the upward moving soft tissue preserved in a granular calcite. The calicle wall exhibiting numerous asterose microscleres. Scale = 1 mm. **c** Detail of **a** showing the internal canal systems and the spicules. Scale = 800 μm. **d** Calcite filled scleres molds (euasters?) of *Calcistella*. Scale = 50 μm. **e** Euaster microscleres of *Chondrilla grandistellata* (SEM microgaph). Compare **d**

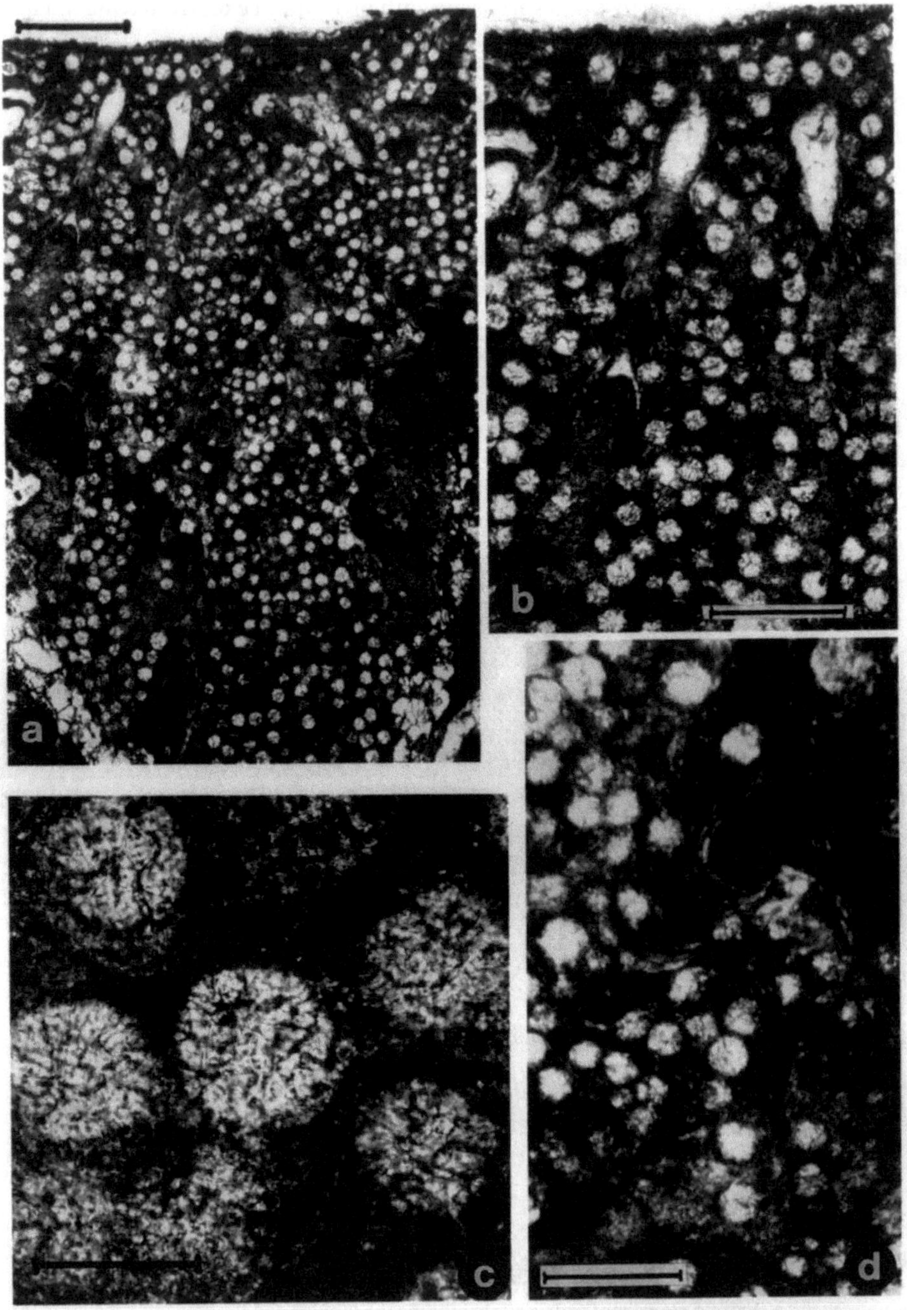


**Fig. 8.** **a** Total view (oblique cut) of the holotype of *Calcichondrilla crustans* n.gen. n.sp. from the Middle Albian of Saturarran (northern Spain). Scale = 1 mm. **b** Detail of **a** exhibiting the excurrent canals and the spicular skeleton. Scale = 400 μm. **c** Probably euasters of *Calcichondrilla*. Molds preserved in a polycrystalline calcite. Scale = 100 μm. **d** Lamellar structure of the calcified primary strong collagenous skeleton of *Calcichondrilla*. Scale = 200 μm

Paratypes: Deposited in the IPFUB, JR3/89 (Fig. 9a-c)
Locus typicus: Beach of Saturarran, Basque Lands in Northern Spain
Stratum typicum: Middle Albian clast from a debris flow of Late Albian age.

Description of the holotype:
Primary skeleton:
The incomplete specimen is found in a thin section and is 5 mm in diameter. The specimen is cut obliquely and shows different niveaus of the excurrent canals (Fig. 8a,b). The diameter of the excurrent canals varies between 100 and 400 μm. The excurrent canal systems are radially arranged. The spicules are enriched in areas between the canals. The inner surface of the canals are free of spicules (Fig. 8b,d). The spicules are big euasters (75–100 μm diameter, mean 80 μm). The asters are preserved as calcitic spherules which exhibit, in some cases, relicts of short rays (Fig. 8c). The entire skeleton demonstrates irregular lamellae which reflects the fibrillar character of the strong primary collagenous structures (Fig. 8d). No separate dermal layer enriched in microscleres is present.

Secondary skeleton:
The whole primary skeleton of the sponge is calcified. The entire skeleton is becoming calcified and is analogous to the primary collagenous skeleton. The primary mineralogy was a high Mg calcite. The microstructure is irregular. In most cases, the calcite is brownish in color.

Description of paratype I (Fig. 9a,b)
The specimen was found in a thin section of a Lower Albian rock of the *Euzkadiella*-facies of the red lagoonal rudist limestone of Ereño, 30 km west of Saturraran. *Calcicondrilla* forms a thin crust (mean 200 μm) on the surface of an already dead specimen of the coralline haplosclerid sponge *Euzkadiella erenoensis*. In some areas the sponge exhibits small buds (200–250 μm). The diameter of the scleres varies from 75–115 μm. The secondary skeleton has the same irregular brownish calcite. Because of the thin crust, no excurrent canals are observed.

Description of paratype II (Fig. 9c)
The specimen was cut horizontally and is located in a thin section of the Lower Albian gray reef limestone of Ereño. *Calcichondrilla* crusts on microsolenid corals of the *Acanthochaetetes* community. The thick part of the crust exhibits horizontal cuts of the excurrent canals. The very thin crusts overgrow biogenes, similar to bryozoans and algae. On the sponge a thecidellinid brachiopod is cemented with its fixed valve. This indicates the rigid character of the living sponge. The spicules are very densely arranged. Relict structures of the collagenous skeleton, as seen in the holotype, are not observed. The entire sponge represents first-order framebuilders of the *Acanthochaetetes* community.

Remarks:
The new genus *Calcichondrilla* exhibits many similarities with the genus *Chondrilla* (Figs. 7e, 10a,b) especially the type of microscleres and the internal structures. The species *Chondrilla grandistellata* bears the same large spicules (euaster diameter 75–150 μm, mean 100 μm), the same type and dimensions of

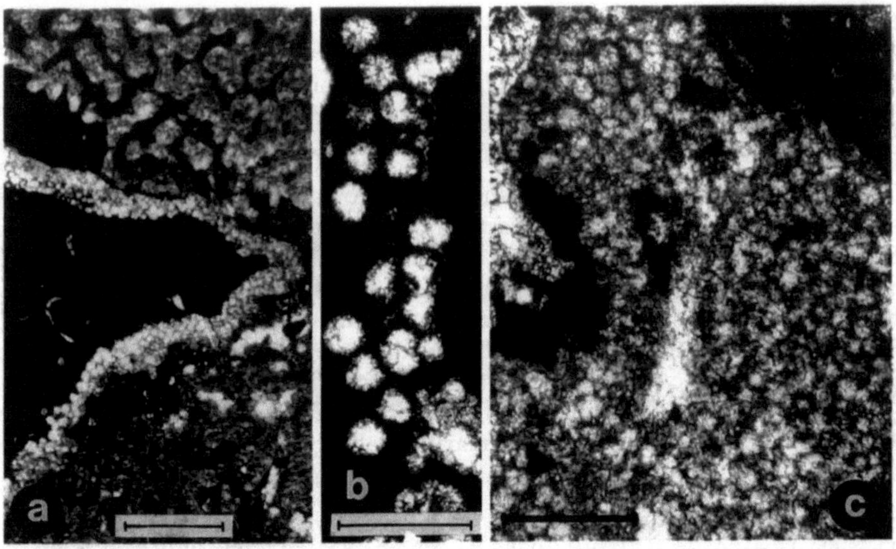


**Fig. 9. a** Paratype I of *Calcichondrilla crustans* n.gen. n.sp. from the Late Aptian "Red Rudist Limestone" of Ereño (northern Spain). The specimen is growing on the haplosclerid stromatoporoid *Euzkadiella erenoensis* Reitner. Scale = 1 mm. **b** Detail of the spicular skeleton of paratype I. Scale = 300 μm. **c** Paratype II from the late Aptian gray reef limestone of Ereño (northern Spain). Scale = 500 μm

excurrent canals, and the same lamellae in the collagenous skeleton. All species of *Chondrilla* exhibit a characteristic dermal layer with an enrichment of spicules. Within the holotype no dermal layer is observed, but the thin crusts seen in paratype I and II may be interpreted as dermal layers. The systematic position within the family Chondrosiidae and a close relationship with the genus *Chondrilla* is very probable.

This species is part of the *Acanthochaetetes* community in the Aptian/Albian reefs (Reitner 1989a). *Calcichondrilla* is observed in northern Spain and within the Mural Limestone in Arizona (Reitner et al., in press).

Differential diagnosis:
The new species differs from *Calcistella tabulata* in having larger microscleres, the crustose growing pattern, and the lack of chaetetid basal skeleton. The closely related genus *Chondrilla* is missing a calcareous secondary skeleton.

***Family Spirastrellidae Ridley & Dendy 1886, Sensu Hentschel 1909***

Emend. diagnosis
The Spirastrellidae are sponges with tylostyle megascleres and spiraster microscleres with great variability. The sponges encrust or form massive bodies.

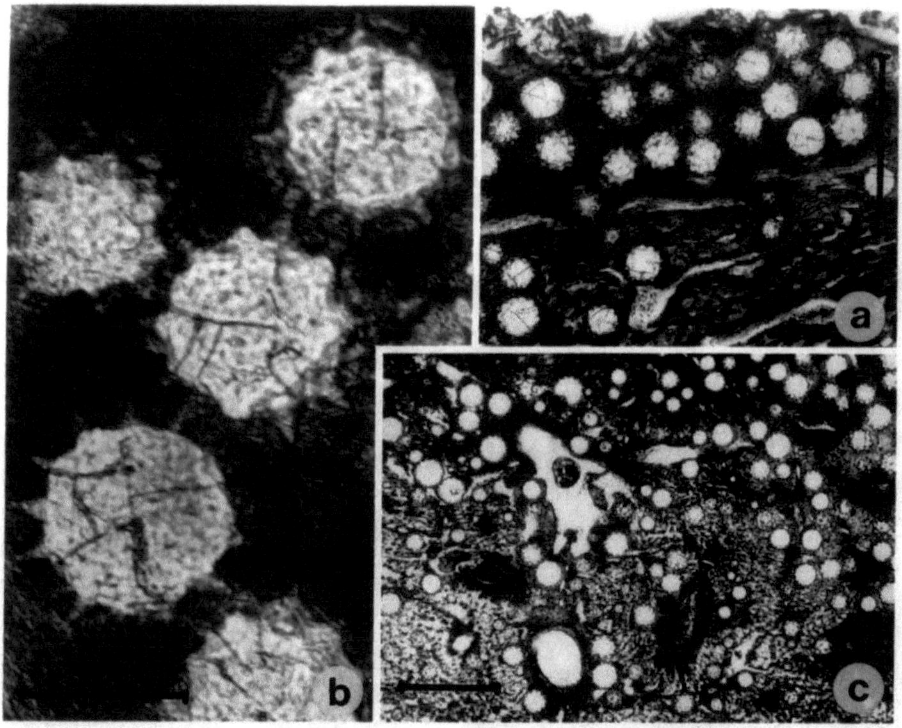

**Fig. 10.** **a** Vertical section through the dermal layer of Recent *Chondrilla grandistellata* exhibiting the enrichment of euasters. Scale = 500 μm. **b** Euasters of **a**. Scale = 100 μm. **c** Choanosomal area of *Ch. grandistellata* exhibiting the excurrent canals (compare Fig. 8b). Scale = 500 μm

In some cases prominent astrorhizal excurrent canals are visible on the surfaces of encrusting specimens (Figs. 11e-g, 13e).

In very few cases, excavating species are observed. High Mg calcite chaetetid-structured secondary skeletons are present in the subgenus *Acanthochaetetes* since the early Cretaceous (Jurassic?).

Genus *Spirastrella* Schmidt 1868
Synom. list: See Wiedenmayer 1977
Type species: *Spirastrella cunctatrix* Schmidt 1868

Subgenus *Acanthochaetetes* Fischer 1979 n.stat.
1970 – *Acanthochaetetes* n.gen. – Fischer, p. 199
1976 – *Tabulospongia* n.gen. – Mori, p. 5
Type species: *Acanthochaetetes seunesi* Fischer 1970

Remarks:
The validity of the taxon "*Acanthochaetetes*" is questionable. Rios and Almela (1944) have described a chaetetid sponge, *Septachaetetes eocenus*, from the Eocene of the southern Pyrenees, which exhibits a basal skeleton similar to *Acanthochaetetes*. New collections of this taxon do not show any spicule remains.

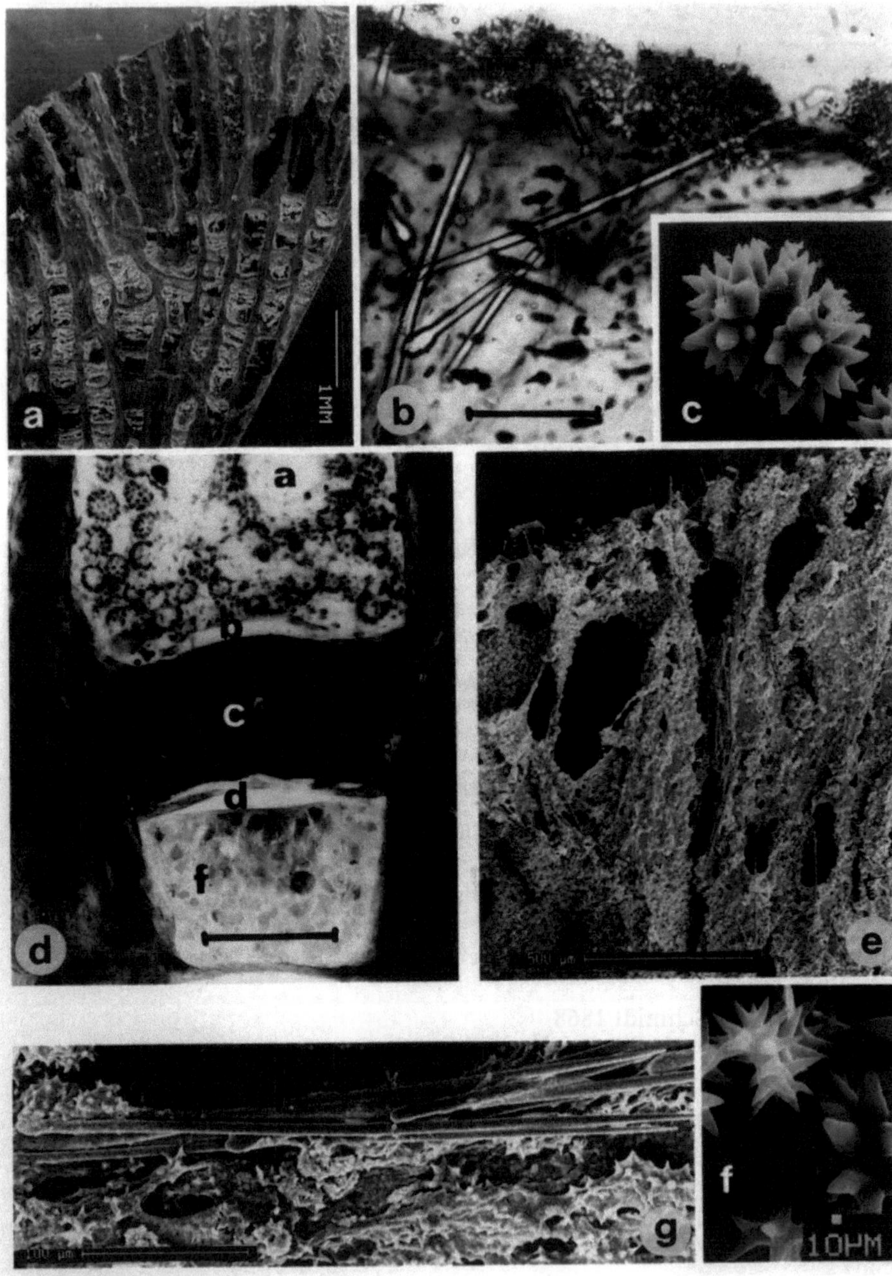
a
1MM
b
c
a
b
c
d
f
d
e
500 µm
f
10µM
100 µm
g

The calicles of this species exhibits only few spines, or in most cases, the spines are missing. *Septachaetetes* is the older synonym of *Acanthochaetetes*. If both taxons are similar, *Septachaetetes* is the valid name. But this question remains unanswered, because spicules are missing.

Diagnosis:
Sponges with tylostyle megascleres and spiraster microscleres are similar to *Spirastrella* with an additional high Mg-calcite chaetetid skeleton. Within the chaetetid secondary skeleton modified archaeocytes could be enclosed (modified gemmules).

*Spirastrella* (*Acanthochaetetes*) *wellsi* (Hartman and Goreau) n.comb. (Fig 11a-d)
Selected synonym list:
1975 – *Acanthochaetetes wellsi* n.sp. – Hartman and Goreau, pp. 2, Figs. 1-14.
1976 – *Tabulospongia horiguchii* n.gen., n.sp. – Mori, pp. 2, Pl. 1-5
1977 – *Tabulospongia japonica* n.sp. – Mori, pp. 2, Pl. 1-2
1987 – *Acanthochaetetes wellsi* – Reitner and Engeser, Figs. 1,3,5,7,8

Diagnosis: See subgenus diagnosis

Description of the primary skeleton and soft tissue arrangement:
The spicules are straight or slightly curved tylostyle megascleres (200–360 μm length) (Fig. 11b). In contrast to the extremely abundant microscleres the tylostyles are rare and demonstrate no preferred orientation. The microscleres are spirasters with extreme variation in shape (Fig. 11b,c).

Simple star-shaped scleres and all variations up to elongated amphiaster types are observed. The microscleres are enriched within the dermal layer (Fig. 11b). Collagen fibrils are present to anchor the soft tissue with the calcitic basal skeleton. These fibrils are seen only inside of the calicle wall and are never observed within the upper part of the spines and tabulae. Vacelet and Garrone (1985) describe additional intracellularly collagen fibrils and large bundles of fibrils between cells at the bottom of the calicles forming a matrix for the tabulae. Spongin is not observed.

---

**Fig. 11. a** SEM micrograph of *Spirastrella* (*Acanthochaetetes*) *wellsi* (Hartman & Goreau) from Guam (Pacific) demonstrating the chaetetid character of the basal skeleton. Only the youngest part is free of early diagenetic cements. **b** Histological section of the dermal layer *S.*(*A.*) *wellsi* exhibiting the microscleres coat and the tylostyle megascleres (Vacelet coll.) Scale = 100 μm. **c** Spiraster microsclere of *S.*(*A.*) *wellsi* (SEM micrograph). **d** Lower part of the youngest part of a calicle in *S.*(*A.*) *wellsi*. The base of the tube is covered by choanosomal tissue exhibiting the choanocyte chambers (*a*). The bottom of the calicle is closed by an organic tabula (*b*). The black part (*c*) are stained totipotent archaeocytes enclosed by tabulae. The lower tabula (*d*) is becoming mineralized. The archaeocytes below the first archaeocyte chamber are totally enclosed by mineralized tabulae (*f*). (Vacelet coll.) Scale = 300 μm. **e** *Spirastrella cunctatrix* Schmidt from the Mediterranean Sea near Banyuls sur Mer. The specimen exhibiting strong vertical canal systems comparable with the calicle systems in *S.* (*Acanthochaetetes*) *wellsi*. (SEM micrograph). **f** Spirasters of *Spirastrella cunctatrix* (SEM micrograph). **g** Tylostyles of *Spirastrella cunctatrix* (SEM micrograph)

The soft tissue covers only the youngest calicle (Fig. 11a). The choanosome covers the calicle wall. The biggest space of the calicle is needed for the aquiferous excurrent system. The choanosome is anchored with the endopinacoderm at the basal skeleton. The choanocyte chambers are small and have a mean diameter of 20 μm (Hartman and Goreau 1975; Vacelet 1981) (Fig. 11d). The surface of the sponge is covered by a dermal layer which is penetrated by incurrent pores and canals as well as superficial excurrent systems demonstrating an astrorhizal pattern. For further detailed descriptions, see Hartman and Goreau (1975). Very important is the occurrence of totipotent archaeocytes which are captured by tabulae in older parts of the calicles (Fig. 11d).

Secondary skeleton:
The sponge has an unique high Mg-calcite skeleton. The entire skeleton exhibits a classical chaetetid structure (Fig. 11a). The microstructure of the skeleton is characterized by irregular lamellae of elongated Mg-calcite crystals (for detailed descriptions and chemical compositions see Reitner and Engeser 1987). The calcite lamellae are present within a thin mucus layer between the exopinacoderm and the rigid calcitic skeleton. The thickness of the mucus layer is ca. 200 nm, measured by one discrete, newly formed crystal layer. The newly formed crystals are needle-like and show a mean length of 2–4 μm and diameter of less than 250 nm. These primary crystals increase size by rapid epitactical crystal growth. The separate crystals grew together and formed a rigid calcareous body. Within the tabulae, the crystals are oriented more or less parallel. In some cases the crystals exhibit a random orientation. Within the empty calicles a randomly oriented high Mg-calcite cement ("back-fill") is observed (Fig. 11a).

Very common are asexual buds. The buds have their roots in one single calicle of the parent sponge. This budding feature stems from escaped, mobile, totipotent archaeocytes, forming a genetically identical sponge. The size of the calicles are mostly smaller than those seen in the parent sponge.

These simple procedure for skeletal formation is supported by stable isotope geochemical data. An equilibrium with seawater (high positive delta $C^{13}$ (+4) and delta $O^{18}$ (-0.9 data) is indicated (Reitner 1989b; Reitner and Grothe 1988).

*Spirastrella* (*Acanthochaetetes*) *seunesi* (Fischer) n.comb.

This species is very common in Late Aptian to Late Albian Urgonian reefs with a worldwide distribution (Reitner 1989a).

Selected synonymy list:

1970 – *Acanthochaetetes seunesi* n.gen., n.sp. – Fischer, p. 201, Fig. 32, Pl. F, Figs. 3–5
1972 – *Acanthochaetetes seunesi* – Fischer and Lafuste, Figs. 1–7, Pl. 7
non 1973 – *Acanthochaetetes seunesi* – Cuif et al. Pl. 1, Fig. 6
1983 – *Acanthochaetetes seunesi* – Reitner and Engeser, Pl. 1, Figs. 4–8
1987 – *Acanthochaetetes seunesi* – Reitner and Engeser, Figs. 2,4,6,8,10.

Description of the skeleton:
This sponge has straight or slightly curved tylostype megascleres and star-shaped spiraster microscleres. The microscleres do not show the extreme variation as seen

in *S.(A.) wellsi*. The spicules are embedded within the calcitic skeleton. The megascleres have no preferred orientation within the secondary skeleton. The microscleres are quite often enriched within the tabulae. Megascleres are never entrapped within the tabulae. Within the calicle walls, microscleres are rare (further details see Reitner 1982, 1987d; Reitner and Engeser 1983, 1987). The structure and chemistry of the chaetetid basal skeleton is similar to *S.(A.) wellsi*. Some specimens show the same budding features as seen in *S.(A.) wellsi*.

*Spirastrella (Acanthochaetetes) dendroformis* n.sp. (Figs. 12b, c; 13d)

1983 – *Acanthochaetetes ramulosus*. – Reitner and Engeser, pp. 774–775, Pl. 1, Figs. 1–3.
1987d – *Acanthochaetetes ramulosus*. – Reitner, Pl. 18, Fig. 2 (not *A. ramulosus* Michelin)

Derivatio nominis: After the Creek word "dendr-" = tree shaped branches
Holotype: Depository IPFUB, JR4/89 (Figs. 12a-c, 13a-d).
Material: Over 50 specimens
Locus typicus: Peña Albeniz near the village Araya, northern Spain.
Stratum typicum: Late Albian fore reef sediments of the Albeniz reef mound.

Diagnosis:
This sponge exhibits a ramified chaetetid basal skeleton. Structure and chemistry is similar to other *S. (Acanthochaetetes)*. Only the axial part of the calcareous skeleton is irregularly arranged. In this part only few calicles separated by tabulae are visible. The intramural tylostyle megascleres are arranged radially within the

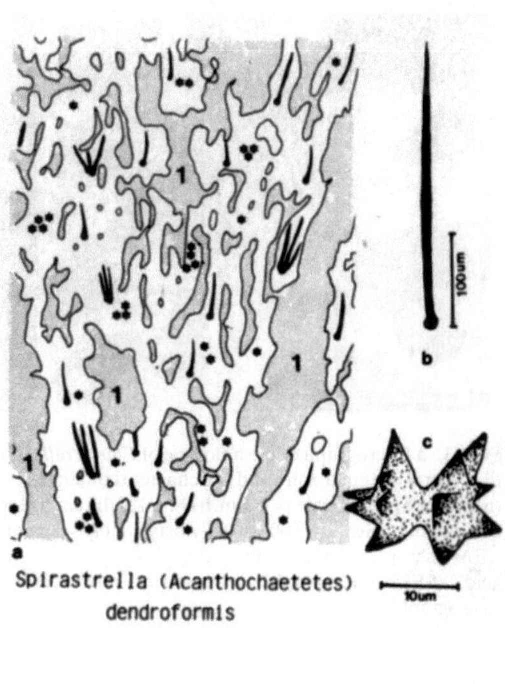


**Fig. 12.** *Spirastrella (Acanthochaetetes) dendroformis* n.sub. gen., n.sp. **a** Drawing of the ontogenetic older parts of the central zone of *S.(A.)* dendroformis exhibiting an irregular stromatoporoid structure of the basal skeleton and the spicule arrangement (aquiferous systems dotted). **b** Tylostyle megasclere. **c** Spiraster ("amphiaster")

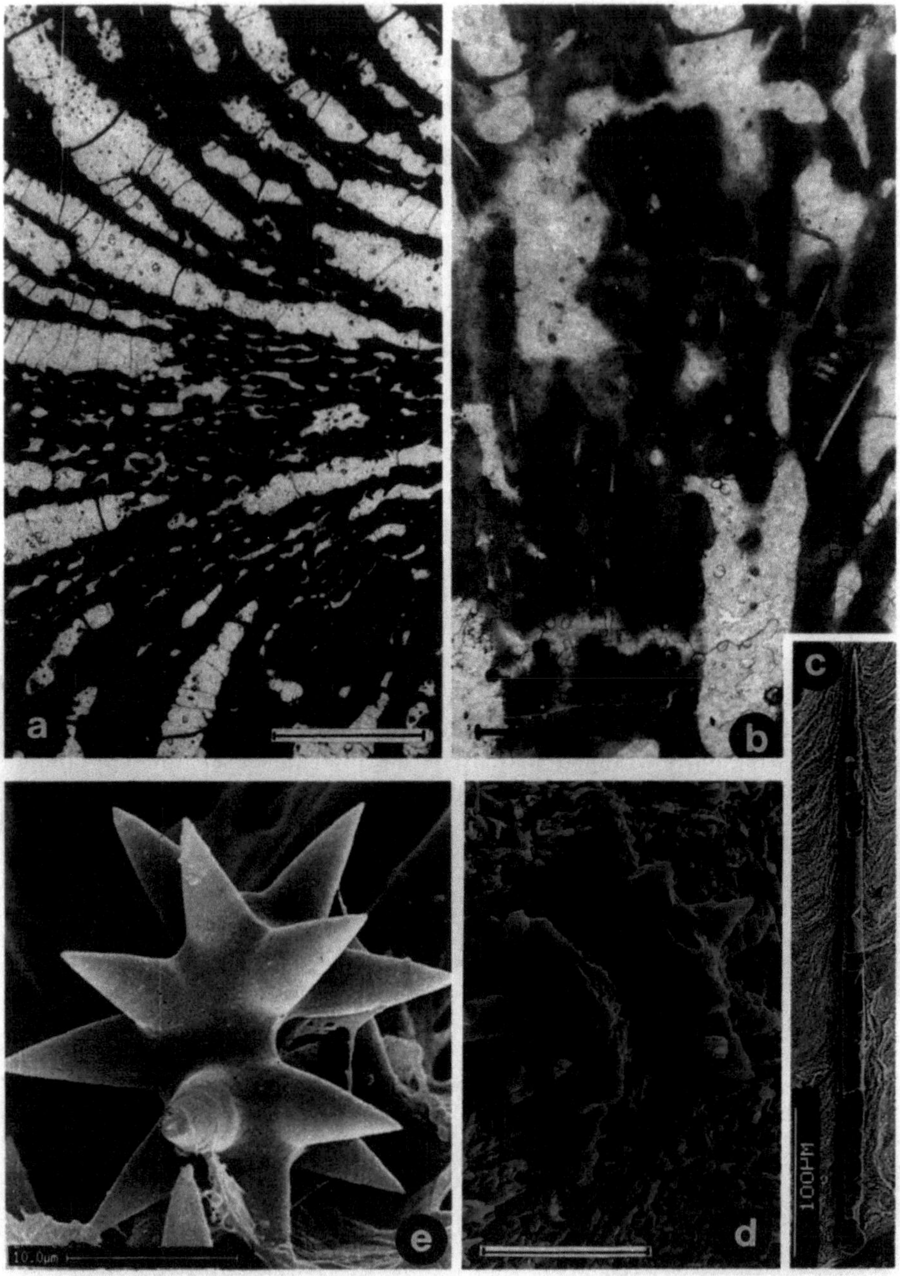


**Fig. 13.** **a** Micrograph of the holotype of *Spirastrella* (*Acanthochaetetes*) *dendroformis* demonstrating the central irregular zone and the chaetetid outer zone. Specimen from Late Albian reefs near Araya (northern Spain). Scale = 1 mm. **b** Detail of the central zone of the holotype exhibiting the megascleres arrangement. Scale = 500 µm. **c** Tylostyle megasclere of *S*.(*A*.) *dendroformis* etched with titriplex acid. *SEM* micrograph. **d** Spiraster ("amphiaster") microsclere of *S*.(*A*.) *dendroformis* etched with titriplex acid. (SEM micrograph). **e** Spiraster microsclere of *Spirastrella cunctatrix* Schmidt (SEM micrograph)

axial part. Microscleres are modified ("amphiaster") spirasters, which are enriched within the axial part and tabulae of the outer part of the sponge.

Description of the holotype:
Spicular skeleton:
The spicular skeleton is composed of tylostyle megascleres with a length of 325–450 μm (mean 365 μm) (Figs. 12a-c, 13b,c) and spiraster microscleres mostly of amphiaster shape (Figs. 12a-c, 13d). The diameter varies from 25–35 μm (mean 28 μm). The tylostyles are common in the central part and arranged in the same fan shaped pattern as observed by the tube systems (Fig. 13a,b). In some cases the spicules are arranged in small bundles which reflects the original radial pattern (Fig. 12a-c).

Some of the megascleres protrude into the open space of the calicle. Microscleres are common and mostly irregularly distributed (Fig. 12a-c). An enrichment is observed within the tube walls of the central part and the tabulae of the outer part.

Secondary skeleton:
From the holotype, three thin sections were made: the horizontally, vertically, and oblique section. The vertical median section has a length of 7.5 cm. The stem has a thickness of 1.7 cm. The irregular center part of the sponge varies from 5 mm–10 mm. The center part is characterized by irregularly ascending tubes. The whole skeletal structure is fan shaped (Fig. 13a,b). Only a few tubes are separated by thin tabulae. The boundary between the center and the outer part is marked by a prominent tabulae layer. The outer part is characterized by a true chaetetid structure as seen in other acanthochaetetids. Microstructure and chemistry of the calcitic skeleton is similar to other *Spirastrella* (*Acanthochaetetes*) species. The inner calicle diameter varies from 320–520 μm (mean 433 μm). The outer calicle diameter varies from 480–1000 μm (mean 703 μm). Buds as seen in hemispherical species of *S.*(*A.*) are not present.

Differential diagnosis:
The new species exhibits close similarities with the secondary skeleton of *Spirastrella* (*Acanthochaetetes*) *ramulosus* (Michelin). But, in *S.*(*A.*) *ramulosus*, only few megascleres are observed and never spiraster microscleres. In *S.*(*A.*) *seunesi* (Fischer), no "amphiaster"-type spirasters are observed. Tylostyles are rare, never arranged in bundles, and never exhibit any radial pattern.

Remarks:
The spicular skeletons of all basal skeleton bearing spirastellids are similar to the species *Spirastrella cunctatrix* Schmidt (Figs. 11e-g, 13e). The extant *S.*(*A.*) *wellsi* (Hartman and Goreau) exhibits a wide variability of spiraster microscleres which is never observed within fossil forms. In *S. cunctatrix* microscleres are very abundant. The tylostyles are arranged in small bundles with a maximum of 20 scleres. Most bundles possess five to six spicules (Fig. 11g). Within the extant *S.*(*A.*) *wellsi* this feature is also observed. Within all fossil species, only *S.*(*A.*) *dendroformis* exhibits this feature.

An unique feature is the presence of modified archaeocytes within the older portions of the calicles. These archaeocytes are comparable with gemmule bodies of freshwater sponges (Fig. 11d).

In some investigated a prominent vertical exhaltant tube system is observed, perhaps comparable with the calicles in *Spirastrella* (*Acanthochaetetes*) (Fig. 11e).

### *Family Cassianothalamiidae Reitner (1987b)*

The affinities of this extinct coralline sponge group from the Late Triassic (Lower Carnian) with the order Hadromerida is based on the presence of spiraster microscleres. Besides spirasters, sterrasters and rare monaxonic megascleres are also observed. Atypical are the fusiform spicules.

*Cassianothalamia zardinii* Reitner (1987b)

1985 – ? *Stylothalamia* n.sp. – Reitner and Engeser, p. 170; Pl. 4, Figs. 8–12
1987 – *Cassianothalamia zardinii* n.gen., n.sp. – Reitner, p. 573, Pl. 1–3; Pl. 4, Fig. 1

Primary and secondary skeleton:
The entire sponge exhibits a thalamid structure, convergent with the modern Vaceletidae. The thalamid collagenous skeleton was subsequently mineralised by an irregular micritic high Mg-calcite. A prominent spongocoel is present. Intramural spicules are rare. Mostly present are small spirasters, some bigger sterrasters, and very few monaxonic megascleres which rarely protrude into the open space of the chambers. The primary organic skeleton is marked by meniscus-shaped structures which cover the megascleres in the outer part of the vertical pillars (for further details see Reitner 1987b).

Remarks:
*Cassianothalamia* is probably a hadromerid sponge, based on the spirasters. The relationship to any extant Hadromerida is not known. The problem lies with the sterrasters, which are observed only in the Geodiidae, order Choristida, while the characteristic triaene megascleres are missing in *Cassianothalamia*. The hadromerid sponge *Placospongia* exhibits sterraster-like scleres (selenasters) beside small spirasters. Any further relationships are not known.

The thalamid or sphinctozoid rigid calcitic skeleton exhibits the same shape as the extant ceractinomorph sphinctozan sponge *Vaceletia crypta* (Vacelet).

## Functional Interpretation of the Observed Basal Skeleton Types

Seven different secondary skeletal types are present within the Hadromerida. Five of them exhibit a chaetetid structure, one a thalamid, and one a modified crust, which is the simplest basal skeleton type.

Chaetetid skeletons are known from two extant demosponges, *Spirastrella* (*A*.) *wellsi* Hartman and Goreau, linked to the Hadromerida, and *Merlia normani* Kirkpatrick, linked to the Poecilosclerida. One species from the Calcarea, *Petrobiona massiliana* Vacelet, linked to the Calcaronea, is known, which contains a modified chaetetid skeleton (Reitner 1987a). In all three examples, the soft tissue of the active living sponge covers only the top of the skeleton and occupies only the youngest calicle. These sponges also produce totipotent archaeocytes enclosed in the older parts of the calicles. In *S.*(*A*.) *wellsi* the archaeocytes are separated from the active living sponge by organic or mineralized tabulae. In *Merlia normani* separation is also made by mineralized tabulae with a central hole and in *Petrobiona* the archaeocytes are located at the end of narrow canals within the basal skeleton (Vacelet 1988; Reitner 1987a; Reitner and Keupp 1989).

Besides the rigid *Merlia normani* a "soft" form is known which does not have any calcareous skeleton (*Merlia deficiens* sensu Vacelet 1980). Van Soest (1984) pointed out that both *Merlia* species are the same. This is important because the soft form does not construct any gemmule bodies. Both forms are probably ecotypes but the problem is, they occur together in same ecological niches. This feature is also observed in *Spirastrella*. The spicular skeleton of the subgenus *Acanthochaetetes* is closely related to *Spirastrella*. *Acanthochaetetes* has been restricted to deep fore reef areas and cryptic niches of coral reefs since the Lower Cretaceous and *Spirastrella* is not part of the *Acanthochaetetes* community (Reitner 1989a). An ecological controlling factor in the formation of the basal skeleton is proposed for *Acanthochaetetes*. This may be true for the calcisponge *Petrobiona*, which is located only in submarine caves and under huge rocks (Vacelet 1964; Reitner 1989a). The formation of a chaetetid or modified chaetetid skeleton is linked, in these cases, with ecological parameters. All modern demosponge chaetetid skeletons are secreted by the exopinacoderm and are not analogous to the primary skeleton of the sponge.

Within *Chaetetes* (*Boswellia*) *mortoni*, *Chaetetopsis favrei*, and *Chondrochaetetes longitubus*, the secondary skeleton is analogous to the primary skeleton. The living sponge occupied in these cases the ontogenetic youngest parts of the secondary skeleton. The thickness of the soft tissue is marked by the last tabulum. Archaeocytes may have been present within the extinct taxa. Besides the classic chaetetid skeletons, the ceratoporellid type of chaetetid skeleton is known, which is characterized by a rapid, early diagenetic, epitactical cementation of the older calicles. In this case, no totipotent archaeocytes can be protected. Probably all open chaetetid skeletons had the function of huge modified gemmule bodies. This idea is supported by the observation of small buds, which are connected to certain calicles of the parent sponge, as seen in *S.*(*A*.) *wellsi*.

The primary collagenous skeleton of the thalamid hadromerid sponge *Cassianothalamia* became calcified immediately after new chamber formation (Reitner 1987b). Common small asexual buds linked with the parent skeleton are observed. A formation from modified gemmules is probable.

Two further noncalcified hadromerids form gemmules, *Cliona lobata* and several types of the *Suberites* taxon. Gemmulue formation is therefore not restricted to haplosclerids.

In *S.(A.) dendroformis* two ontogenetic stages of secondary skeletal formation are observed. In the early stage calcitic skeleton is analogous to the primary skeleton. The irregular tubes probably reflect the primary aquiferous system. Megascleres are common, regularly oriented, and exhibit, in some cases, the primary aggregates as seen in *Spirastrella*. The ontogenetic late stage is characterized by a normal acanthochaetetid skeleton which is not analogous to the primary sponge skeleton.

In *Calcichondrilla* the entire primary sponge skeleton became calcified. No chaetetid structure or special cavities for totipotent archaeocytes are observed. This basal crust is occasionally an important reef-frame stabilizer.

## Phylogenetic Theories (Figs. 14,15)

To demonstrate possible phylogenetic relationships, the method of phylogenetic systematics was used. The Hadromerida and Tetractinellida are a monophylum, based on a number of well-defined synapomorphies (Hartman 1982; van Soest 1987). Van Soest (1987) has discussed three different cladograms based on the classification of Hartman (1982), his own ideas, and a cladogram based on a computer program developed by W.N. Ellis. In all three cladograms the Hadromerida and Tetractinellida are classified as a monophylum. Within the computer cladogram, the Chondrosiidae are grouped with the Vaceletidae, based on the lack of any spicules. Within the genus *Chondrilla*, asters are present. A close relationship between *Chondrilla* and *Chondrosia* is based on the similarity of the soft tissue and collagenous skeleton. *Chondrosia* is a chondrillid which lost its spicules. The Chondrosiidae must be classified within the monophylum Tetractinellida/Hadromerida.

There are three views on the significance of the secondary skeletons. Hartman and Goreau (1972) proposed that all calcareous basal skeleton bearing sponges are a monophylum and have created a separate class of sponges (Sclerospongiae). Van Soest (1984) and Vacelet (1985) pointed out the similarities of certain spicule skeletons with noncalcified demosponges. Van Soest (1984) suggested that calcified basal skeletons were developed only once and therefore an important plesiomorphy of the demosponges. Vacelet (1985) showed a polyphyletic origin of the calcified demosponges based on spicules similarities with noncalcified taxa. This idea was supported by Reitner (1987a,b), Reitner and Engeser (1985, 1987), Reitner and Keupp (1989), and Wood (1987), Wood and Reitner (1988), Wood et al. (1989) based on findings of many different spicule-bearing groups of calcified demosponges. These authors also pointed out the convergent nature of chaetetid, stromatoporoid, crust, and thalamid type of basal skeletons.

The two cladograms discussed here are based mainly on skeletal features to derive the phylogenetic significance of the secondary calcareous skeletons.

The analysis is based on the following features:

1. Radial pattern of megascleres
2. Asterose microscleres

3. Tetractinellid spicules
4. Loss of tetractinellid spicules
5. Tylostyle or modified tylostyle spicules
6. Reduction of radiate pattern (subradiate)
7. Spiraster microscleres
8. Calcareous secondary skeleton (css)
8a. Chaetetid high Mg-calcite basal skeleton combined with internal stromatopororid basal skeleton and a remaining radial pattern of tylostyles.
8b. Chaetetid high Mg-calcite microlamellar-structured secondary skeleton without any similarities to the primary sponge skeleton.
8c. Chaetetid high Mg-calcite irregularly structured secondary skeleton analogous to the primary collagenous/spicular skeleton.
8d. Chaetetid high Mg-calcite fascicular fibrous basal skeleton analogous to the primary collagenous/spicular skeleton.
8e. Chaetetid aragonitic fascicular fibrous basal skeleton analogous the primary skeleton.
8f. Thalamid high Mg-calcite secondary skeleton analogous to the primary skeleton.
8g. Crust type or modified stromatoporoid Mg-calcite secondary skeleton analogous to the primary skeleton.
9. Modified totipotent archaeocytes
10. Loss of asterose scleres
11. Reduction of megascleres
11a. Strong collagenous primary skeleton
12. No intramural spicules
13. Loss of megascleres
14. Loss of the entire spicular skeleton
15. Loss of the calcareous secondary skeleton
16. Burrowing ability in calcareous material
17. Clionid-type spirasters

*Theory I (Fig. 14)*
This theory is based on a close relationship between the Tetractinellida (Choristida) and Hadromerida. Two strong synapomorphies of these groups are the radiate or subradiate architecture of the megascleres (1) and the presence of asterose microscleres (2). The Choristida are characterized by four-rayed spicules (3) which are not present or lost within the Hadromerida (4). An important apomorphy of the Hadromerida are tylostyle or modified tylostyle scleres. Van Soest (1987) has placed the Tethyidae as a sister group of the Astrophorida (Choristida) because both groups exhibit a strong radiate pattern never observed in other Hadromerida and do not possess tylostyle spicules. The Cassianothalamidae (8f) may show some affinities with the Tethyidae and must separate from the Hadromerida s.st.

A typical hadromerid group are the Spirastrellidae, which are characterized by spirasters (7) and some excavating species (16). The Clionidae based on spicule morphology and burrowing ability are a possible sister group of the Spirastrel-

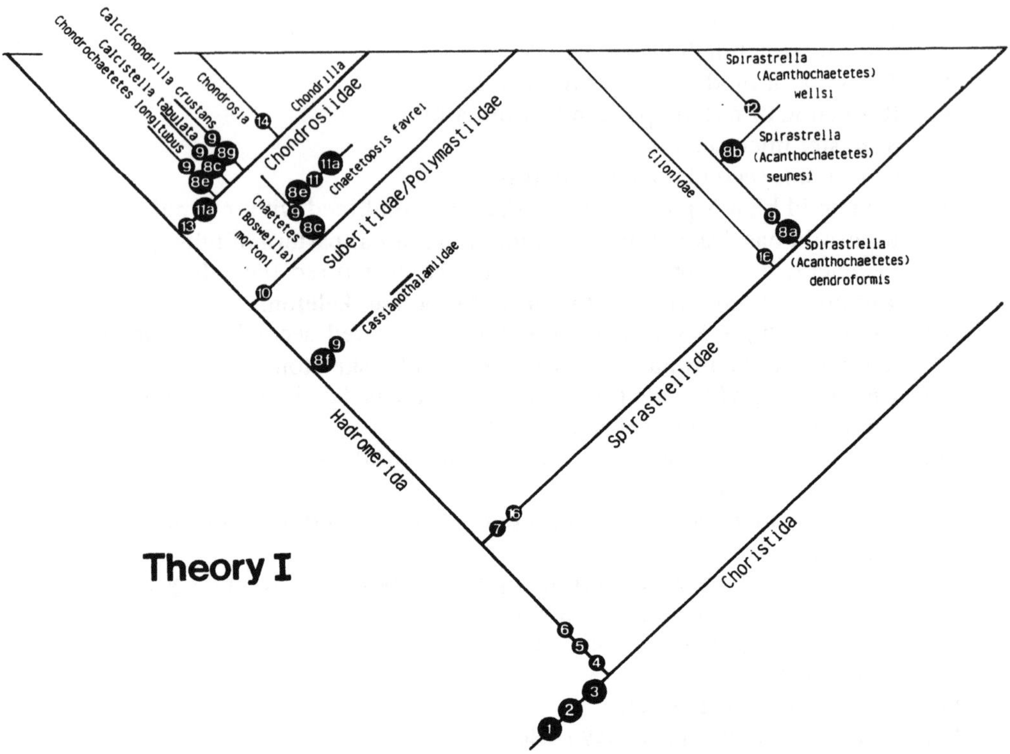


**Fig. 14.** Theory I. Cladogram demonstrating possible phylogenetic relationships based on a polyphyletic origin of the calcareous secondary skeletons. Character numbers see text

lidae. Important is the occurrence of a particularly high Mg-calcite chaetetid basal skeleton (8b) (*Acanthochaetetes*); a strong apomorphy of this sublineage. Related to this feature are modified totipotent archaeocytes. The oldest known representatives of this extant lineage were found in fore-reef and cryptic environments of Lower Cretaceous reef carbonates (occurrences from the Late Jurassic are doubtful) (Fischer 1970; Reitner and Engeser 1983, 1987). The specimens of the Lower Cretaceous exhibit a relatively high diversity (ca. six species) and some of them have intramural spicules. The younger forms have lost the intramural spicules (12). Besides the characteristic "adapted" chaetetid skeleton, one form is present which exhibits a combined chaetetid/stromatoporoid skeleton (8b). Probably this form is the primary one.

The Suberitidae/Polymastiidae are a sponge group in which the asterose microscleres are lost (10). The affinities to the Hadromerida is based on the relict radial pattern of the scleres. Boury-Esnault (1987) discussed this problem and concluded that the systematic position of this particular is open at the moment. For this present study the Suberitidae/Polymastiidae are placed within the Hadromerida. The oldest known (Lower Carboniferous) hadromerid sponge, is

*Chaetetes* (*Boswellia*) *mortoni*. *Boswellia* possesses a chaetetid skeleton analogous to the primary skeleton, which is similar to the genus *Suberites* (8d). The Lower Cretaceous form *Chaetetopsis favrei* exhibits nearly the same type of chaetetid skeleton, and is probably evolved from *Boswellia*. In contrast to *Boswellia*, *Chaetetopsis* exhibits an aragonitic chaetetid skeleton (8e) and a reduced density of intramural spicules (11). The primary skeleton is probably replaced by a strong collagenous skeleton (11a).

The third group of Hadromerida which exhibits calcareous secondary skeletons are the Chondrillidae within the Chondrosida. The genus *Chondrilla* has lost the tylostyle megascleres (13) and is characterized by asterose microscleres enriched in an ectosomal crust. An important apomorphy of the Chondrosida is a strong collagenous skeleton (11a). Three different basal skeletons are observed. A classic chaetetid skeleton with a fascicular fibrous microstructure and rare intramural aster microscleres from the Carboniferous (*Chondrochaetetes longitubus*) (8d), a chaetetid skeleton with an irregular micritic microstructure with preserved remains of the aquiferous systems within the calicle walls and numerous asterose spicules (*Calcistella tabulata*) (8c), and the crust type or modified stromatoporoid calcitic basal skeleton of the genus *Calcichondrilla* (8g) is observed. This form demonstrates a very close relationship with the extant noncalcified species *Chondrilla grandistellata*. All three basal skeletons are analogous to the primary skeleton. Within the genus *Chondrosia*, the whole spicular skeleton is lost (14). Based on this theory, the secondary calcareous basal skeletons have developed independently seven times from noncalcified ancestors and are therefore convergent. The entire structure and function of the secondary skeleton are different. The chaetetid type skeleton is probably linked with the occurrence of totipotent archaeocytes and therefore these skeletal types are modified gemmule bodies. Basal skeleton formation is probably related to ecological parameters.

*Theory II (Fig. 15)*

The main problem with theory I is the independent development of secondary skeletons. According to van Soest (1984), calcareous secondary skeletons are a plesiomorphic feature of the Demospongiae. This eliminates the problem of a repeated origin for calcareous basal skeletons. Within Calcarea, both subclasses, the Calcinea and Calcaronea, evolved from rigid basal skeleton-bearing forms. The Murrayonida are probable ancestors of the modern Calcinea and the Minchinellida of the nonrigid modern Calcaronea (Reitner 1987a). Interpreting the basal skeletons as a plesiomorphic feature of the Hadromerida clade, it is much easier to explain the common occurrence of calcareous secondary skeletons. Based on this theory, the modern noncalcified forms are evolved from basal skeleton-bearing forms. This view is supported by the fossil record. Most fossil Hadromerida have only formed a basal skeleton, excepting *Cliona*. In this clade *Cliona* is related to Spirastrella as a sister group. Both have lost the calcareous skeleton (15) and have a burrowing ability (16). All clionids burrow and are characterized by elongated spirasters (17), a good autapomorphy of this taxon. Isolated asterose and tylostyle spicules which are common in Mesozoic sediments

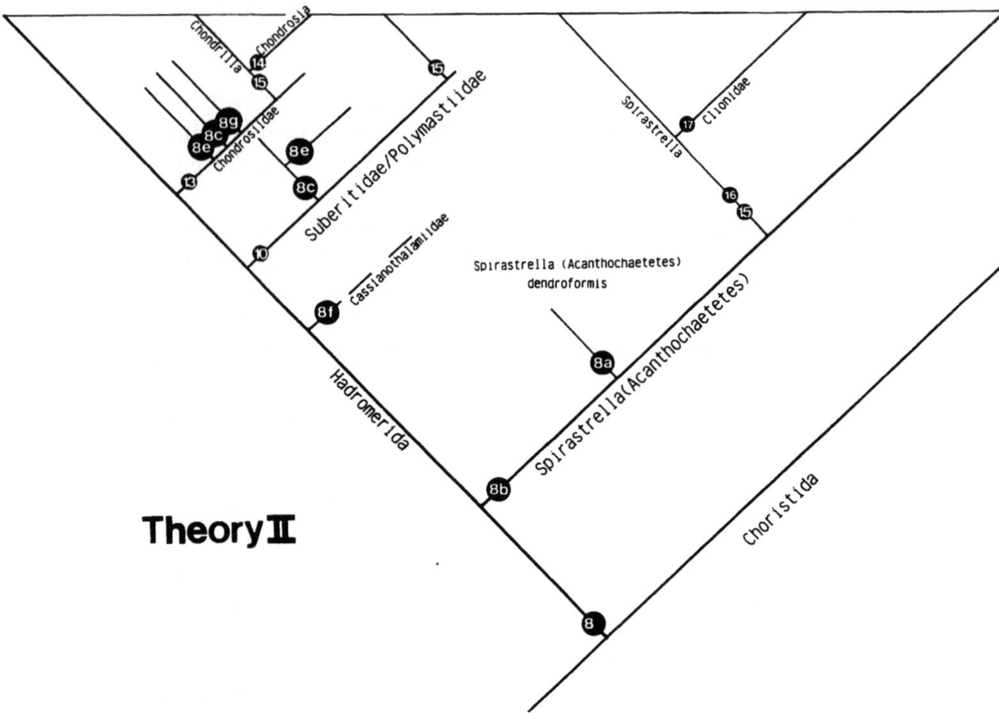


**Fig. 15.** Theory II. Cladogram based on the monophyletic origin of the calcareous basal skeleton. Character numbers see text

may be of hadromerid origin. But both spicule types occur also in other sponge groups.

The main criticism about this theory is the different micro- and ultrastructures as well as different mineralogies of the basal skeletons. To decide which theory is correct, we must know more about the ontogeny of young sponges after settlement of the larva. No data are available at this time from these particular sponges. Basal skeletons are definitely not a plesiomorphic character of the entire Demospongiae, but single clades can evolve from calcareous rigid skeleton bearing ancestors.

A monophyletic origin of the calcareous skeletons of the Hadromerida cannot be fully excluded, but this opinion must be proven by studies of the early ontogeny of these sponges (e.g., *Acanthochaetetes*). A further problem are the preservation potentials of nonrigid sponges. We do not know much about "soft" demosponges in the past. Only isolated scleres are common and in most cases the entire spicula sponges skeleton is not known. It is very probable that "soft" Hadromerida have existed since early Mesozoic time.

Based on this current knowledge a polyphyletic/convergent nature of the basal skeletons of the Hadromerida is most probable.

## Conclusions

a) The Hadromerida s.str. are a monophyletic sponge group in which three types of skeletal formations are observed: (1) primary skeleton composed of collagenous skeletal material and siliceous micro- and megascleres which can be reduced or lost; (2) secondary calcareous basal skeletons; and (3) indirect basal skeleton formation by excavating burrows.

b) Five different chaetetid skeletons are observed. Four of them are analogous to the primary sponge skeleton. Only one does not exhibit any similarities with the primary skeleton (*Acanthochaetetes*). In all cases the chaetetid skeletons can be explained as modified gemmule bodies in which totipotent archaeocytes are present. Besides the chaetetid skeletons, one crust type or modified stromatoporoid skeleton (*Calcichondrilla*), and one thalamid skeleton are observed (*Cassianothalamia*).

c) Only three new asterose spicules-bearing taxa are described: *Calcistella tabulata* n.gen. n.sp. from the Lower Cretaceous of Greece, *Chondrochaetetes longitubus* n.gen. n.sp. from the Carboniferous of Russia, and *Calcichondrilla crustans* n.gen. n.sp. from the Lower Cretaceous of northern Spain. All three new genera are classified within the chondrosiidae.

   The chaetetid genus *Acanthochaetetes* is included as a subgenus within the genus *Spirastrella* based on the similarity of the spicular skeleton. One new species is described: *Spirastrella* (*Acanthochaetetes*) *dendroformis* n.sp. It is characterized by an internal stromatoporoid skeleton with radially oriented spicules and a normal chaetetid adult structure.

d) The Cassianothalamiidae are probably linked with the Tethyidae.

e) The fossil species *Chaetetes* (*Boswellia*) *mortoni* Gray and *Chaetetopsis favrei* (Deninger) are classified within the tylostyle-bearing Hadromeria (Suberitidae/Polymastiidae).

f) Two phylogenetic theories in which the origin and significance of the different basal skeletons are discussed. A polyphyletic origin is, based on the current knowledge, probable. A monophyletic origin cannot be fully excluded, if the young adult sponges do not possess any calcareous skeletons.

*Acknowledgments.* I am indebted to Prof. J. Vacelet (Station Marine Endoume Marseille), Prof. H. Keupp, Dipl. Biol. F. Grothe (Freie Universität Berlin), Dr. R. van Soest (Zoological Institute Amsterdam), and Miss S. Stone, curator of the sponge section of the British Museum N.H. in London, for the permission to study specimens in their collections and some donations. I thank Dr. B. Gierlowski-Kordesch (Freie Universität Berlin) for translational assistance. The Deutsche Forschungsgemeinschaft is acknowledged for financing this investigation (Re 665 1-1).

## References

Bergquist PR (1978) Sponges. Hutchinson, University of California Press, pp 268

Boury-Esnault N (1987) The *Polymastia* species (Demosponges, Hadromerida) of the Atlantic Area. In: Vacelet J, Boury-Esnault N (eds) Taxonomy of Porifera. Nato Asi Ser G13, Springer, Berlin Heidelberg New York Tokyo, pp 29–66

Cuif J-P, Feuillee P, Fischer J-C, Pascal A (1973) Prèsence d'Astrorhizes chez les Chaetetida Mèsozoiques. CR Acad Sci Paris 277:2473–2476
Deng Z (1982) Mesozoic Milleporina and tabulatomorphic corals from Xizang. Paleontol Xizang 4:184–188
Deninger K (1906) Einige neue Tabulaten und Hydrozoen aus Mesozoischen Ablagerungen. N Jahrb Min Geol Paläontol 1:61–70
Döderlein L (1897) Über die Lithonina, eine neue Gruppe von Kalkschwämmen. Zool Jahrb 10:15–32
Fischer J-C (1970) Rèvision et essai de classification des Chaetetida (Cnidaria) post-paleozoiques. Ann Palèontol (Invert) 56:151–220
Fischer J-C, Lafuste J (1972) Nouvelles observations sur la palèohistologie du genre *Acanthochaetetes* (Hydrozoa, Chaetetida). Bull Soc Geol Fr 14(7):320-324
Gray D (1980) Spicule pseudomorphs in a new Paleozoic chaetetid, and its sclerosponge affinities. Palaeontology 23:803–820
Hartman WD (1979) A new sclerosponge from the Bahamas and its relationship to Mesozoic stromatoporoids. Coll Int CNRS 291:467–474
Hartman WD (1982) Porifera. In: Parker SP (ed) Synopsis and classification of living organisms. McGraw-Hill, New York, pp 640–666
Hartman WD, Goreau TF (1970) Jamaican coralline sponges: their morphology, ecology, and fossil relatives. In: Fry WG (ed) The biology of Porifera. Symp Zool Soc Lond 25:205–243
Hartman WD, Goreau TF (1972) *Ceratoporella* (Porifera: Sclerospongiae) and the chaetetid "corals". Trans Conn Acad Arts Sci 44:133–148
Hartman WD, Goreau TF (1975) A Pacific tabulate sponge, living representative of a new order of sclerosponges. Postilla 167:1–14
Kazmierczak J (1979) Sclerosponge nature of chaetetids evidenced by spiculated *Chaetetopsis favrei* (Deninger 1906) from the Barremian of Crimea. N Jahrb Geol Paläontol Mh 1979:98–108
Kirkpatrick R (1908) On two new genera of recent pharetronid sponges. Ann Mag Nat Hist 8:503–514
Kirkpatrick R (1910) On a remarkable pharetronid sponge from Christmas Island. Proc R Soc 83:124–133
Kirkpatrick R (1911) On *Merlia normani*, a sponge with a siliceous and calcareous skeleton. QJ Microsc Sci 56:657-702
Levi C (1973) Systematique de la classe Demospongiaria (demosponges). In: Grasse PP (ed) Traitè de Zoologie III(1) Masson, Paris, pp 577–631
Lister JJ (1900) *Astrosclera willeyana*, the type of a new family of sponges. In: Willey A (ed) Zool Results Part 4:461–482
Mori K (1976) A new sclerosponge from Ngargol, Palau Island and its fossil relatives. Tohoku Univ Sci Rep 2 Ser Geol 46:1–9
Mori K (1977) A calcitic sclerosponge from the Ishigaki-shima coast, Ryukyu Islands, Japan. Tohoku Univ Sci Rep 2 Ser Geol 47:1–5
Peterhans E (1930) Etude du genre *Chaetetopsis* Neumayr et classification nouvelle des Chaetètidès. Eclogae Geol Helv 22:81–85
Reitner J (1982) Die Entwicklung von Inselplattformen und Diapir-Atollen im Alb des Basko-Kantabrikums (Nordspanien). N Jahrb Geol Paläontol Abh 165:87–101
Reitner J (1987a) Phylogenie und Konvergenzen bei rezenten und fossilen Calcarea (Porifera) mit einem kalkigen Basalskelett ("Inozoa, Pharetronida"). Berl Geowiss Abh 86:87–125
Reitner J (1987b) A new calcitic sphinctozoan sponge belonging to the Demospongiae from the Cassian Formation (Lower Carnian; Dolomites, Northern Italy) and its phylogenetic relationship. Geobios 20:571–589
Reitner J (1987c) *Euzkadiella erenoensis* n.gen. n.sp. ein Stromatopore mit spikulärem Skelett aus dem Oberapt von Ere.o (Prov Guipuzcoa, Nordspanien) und die systematische Stellung der Stromatoporen. Paläontol Z 61:203–222
Reitner J (1987d) Mikrofazielle, palökologische und paläogeographische Analyse ausgewählter Vorkommen flachmariner Karbonate im Basko-Kantabrischen Strike Slip Fault-Becken-System (Nordspanien) an der Wende von der Unterkreide zur Oberkreide. Doc Nat 40:1–248
Reitner J (1989) Lower and Mid-Cretaceous Coralline Sponge Communities of the Boreal and Tethyan Realms in Comparison with the Modern Ones – Palaeoecological and Palaeogeographic

Implications. – In: Wiedmann J (ed) Cretaceous of the Western Tethys. Proceedings 3rd International Cretaceous Symposium, Tübingen 1987, pp 851–878. E Schweizerbart'sche Verlagsbuchhandlung, Stuttgart

Reitner J (1989b) Struktur, Bildung und Diagenese der Basalskelette bei rezenten Pharetroniden unter besonderer Berücksichtigung von petrobiona massiliana Vacelet & Levi 1958 (Minchinellida, Porifera). Berl Geowiss Abh (A) 106:343–384

Reitner J, Engeser TS (1983) Contributions to the systematics and the paleoecology of the family Acanthochaetetidae (Fischer 1970) Order Tabulospongida, Class Sclerospongiae. Geobios 16:773–779

Reitner J, Engeser TS (1985) Revision der Demospongier mit einem thalamiden, aragonitischen Basalskelett und trabekulärer Internstruktur ("Sphinctozoa" pars). Berl Geowiss Abh 60:151–193

Reitner J, Engeser TS (1987) Skeletal structures and habitats of Recent and fossil *Acanthochaetetes* (subclass Tetractinomropha, Demospongiae, Porifera). Coral Reefs 6:151–157

Reitner J, Grothe F (1988) Origin of the basal skeleton of *Petrobiona massiliana* Vacelet (Calcaronea, "Pharetronida"). Berl Geowiss Abh 100:29

Reitner J, Keupp H (1989) Basalskelette bei Schwämmen – Beispiel einer polyphyletischen Entwicklung. Geowissenschaften 1989:71–78

Reitner J, Cuffey RJ, Scott RW Albian coralline sponge, Caribbean Province, Arizona. N Jahrb Geol Paläont Mh (in press)

Rios JM, Almela YA (1944) Un Chaetetido del Eoceno español. Notas Comun Inst Geol Min Esp 12:19–37

Rützler K (1974) The burrowing sponges of Bermuda. Smithsonian Contrib Zool 165:203–216

Vacelet J (1964) Etude monographique de l'eponge calcaire Pharetronide de Mediterranèe, *Petrobiona massiliana* Vacelet & Levi. Les Pharetronides actuelles et fossiles. Recl Trav St Mar Endoume 50:1–125

Vacelet J (1980) Squelette calcaire facultatif et corps de regènèration dans le genre *Merlia*, éponges apparentèes aux Chaetètidès fossiles. CR Hebd Seances Acad Sci Paris 290:227–230

Vacelet J (1981) Eponges hypercalcifièes ("Pharetronides", "Sclerosponges") des cavites recifs coralliens de Nouvelle-Caledonie. Bull Mus Natl Hist Paris 4 Ser 3 1981 sec a 2:313–351

Vacelet J (1985) Coralline sponges and the evolution of Porifera. In: Conway Morris S, George JD, Gibson R, Platt HM (eds) The origins and relationships of lower invertebrates. Syst Assoc Spec Vol 28:1–13

Vacelet J (1988) The storage cells of calcified relict sponges. In: Proceedings Third Internat conf Sponge Biol. Smithsonian Inst Press, Washington, DC

Vacelet J, Garrone R (1985) Two distinct populations of collagen fibrils in a "sclerosponge" (Porifera). In: Bairati A, Garrone R (eds) Biology of invertebrate and lower vertebrate collagens. NATO ASI Ser ser A: Life Sciences 93. Springer, Berlin Heidelberg New York, pp 183–189

Van Soest RWM (1984) Deficient *Merlia normani* Kirkpatrick, 1908, from the Curacao reefs, with a discussion on the phylogenetic interpretation of sclerosponges. Bijdr Dierkd 54:211–219

Van Soest RWM (1987) Phylogenetic exercises with monophyletic groups of sponges. In: Vacelet J, Boury-Esnault N (eds) Taxonomy of Porifera. NATO ASI Ser, ser G: Ecol Sci 13. Springer, Berlin Heidelberg New York, pp 227–241

Wiedenmayer F (1977) Shallow-water sponges of the western Bahamas. Exper Suppl 28:287

Wood R (1987) Biology and revised systematics of some Late Mesozoic stromatoporoids. Spec Pap Paleontol 37:1–89

Wood R, Reitner J (1988) The "chaetetid" demosponge *Stromatoaxinella irregularis* (Michelin) and its systematic implications. N Jahrb Geol Paläontol Mh 1988:213–224

Wood R, Reitner J, West R (1989) Systematics and phylogenetic implications of the haplosclerid stromatoporoid *Newellia mira* nov. gen. Lethaia 22

# Further Evidence for Poriferan Affinities of Favositids

J. Kaźmierczak

## Introduction

Rediscoveries of peculiar modern sponges with mixed siliceous and calcareous hard parts called coralline sponges or sclerosponges (Hartman and Goreau 1970, 1975; Mori 1976, 1977) revived Kirkpatrick's (1912) early suggestion of a possible sponge nature of the favositids, a large and widespread group of Paleozoic marine calcareous fossils commonly attributed to corals. Due to the lack of crucial evidence, the cnidarian affinity of favositids was advocated on the basis of their skeletal traits such as assumed coloniality, calycal and septal structures connecting structures between the alleged corallites, etc. (for review see, e.g., von Schouppé and Oekentorp 1974; Flügel 1976; Scrutton 1987). Their sponge nature, in turn, was argued on the basis of a similarity of the favositid skeletal architecture to the basal calcareous skeletons of such modern sclerosponges as *Merlia normani* Kirkpatrick, *Acanthochaetetes wellsi* Hartman et Goreau or *Tabulospongia japonica* Mori (for review see Flügel 1976; Scrutton 1987). Since favositids are similar also to the fossil Chaetetida, the finding of unquestionable pseudomorphs of siliceous spicules entrapped in the chaetetid calcareous skeletal material (Kaźmierczak 1979; Gray 1980) strengthened the position of those specialists who opted for a poriferan origin of favositids. Discussion revealed, however (Hartman and Goreau 1972; Flügel 1976; Oliver 1979, 1986; Scrutton 1987), that a poriferan affinity of favositids will not be accepted unless the presence of poriferan spicules entrapped in their calcareous skeletons can be conclusively demonstrated. In fact, such spicules have been recently shown in the Devonian favositid *Thamnopora*, where calcitic pseudomorphs of desma-like sclerites occurring within the calcareous skeletal substance have been described (Kaźmierczak 1984). The single thin section of *Thamnopora* precluded a detailed analysis of the character and mode of distribution of the spicules. Others suggested that these may rather be traces of boring microorganisms (Oekentorp 1985; Finks 1986).

As of almost the same time Copper (1985) reported on the findings of "unmistakable calcified polyps" preserved on the skeletal surface of Silurian favositids, opinions have been expressed (Oekentorp and Stel 1985; Oliver 1986; Scrutton 1987) that the weight of evidence is now strongly in favor of favositids being cnidarians. Copper's "polyps", however, are an apparent misinterpretation, as the structures he described do not represent, in the author's opinion, a

J. Reitner and H. Keupp (Eds.)
Fossil and Recent Sponges

sensational fossilization of coelenterate soft parts but simply well-preserved skeletal elements called pseudopercula known for a long time in favositids (e.g., Dunbar 1927; Swann 1947; Flügel 1986). It has nonetheless become obvious that only further discoveries of convincing spicule-bearing favositids may bring a new impetus to the current discussion on favositid affinities.

Reexamination by the author of a large collection of tabulate thin sections collected by the late Dr. Anna Stasińska and housed in the Institute of Paleobiology of the Polish Academy of Sciences resulted in findings of a number of spicule-bearing specimens assigned to several favositid genera. Specimens belonging to the most representative favositid genera such as *Palaeofavosites* Twenhofel, *Mesofavosites* Sokolov, and *Favosites* Lamarck have been selected for the present work. The purpose of this chapter is to describe these new spicule findings in favositids and to compare them with spicules occurring in modern and fossil sponges. It is the author's hope that this chapter, along with the author's report on a spicule-bearing halysitid tabulate (Kaźmierczak 1989), will stimulate further research on tabulate affinities.

## Material

A large collection of Tabulata, comprising ca. 2000 thin sections stored in the Institute of Paleobiology of the Polish Academy of Sciences in Warsaw, have been examined for the presence of traces of spiculation in their calcareous skeletons. As a result, distinct pyritic and calcitic pseudomorphs of primary siliceous spicules embedded in the calcareous skeletons or laying loosely in the inter-skeletal spaces have been discovered in specimens belonging to various taxa. For the purpose of the present chapter three representatives of spiculated favositids are chosen; they are exemplified by specimens traditionally classified to such favositid genera *Palaeofavosites*, *Mesofavosites*, and *Favosites*.

Abbreviations used: PMO – Paleontological Museum Oslo, ZPAL – Institute of Paleobiology of the Polish Academy of Sciences, Warsaw.

*Example 1 (Figs. 1A-E and 4D,E):*
Undescribed specimen, designated originally as *Palaeofavosites hystrix* Sokolov 1951, represented by three thin sections labeled PMO 40788a-c. The specimen comes from the locality Malmøy is Oslo Region (Norway) and has been collected from the Silurian sequence, series 7a (*Pentamerus*) belonging to the Llandovery (Strand and Henningsmoen 1960). Topotypic specimens of the spicule-bearing *P. hystrix* have been described and illustrated by Stasińska (1967: 69, Pl. XVII, Fig. 3).

*Example 2 (Figs. 2A-C and 4C):*
Undescribed specimen, designated originally as *Mesofavosites* sp. cf. *M. imbelis* Klaamann, represented by only one longitudinal thin section labeled ZPAL T/III-31 E. The specimen was found as a small erratic (glacial) boulder of Baltic

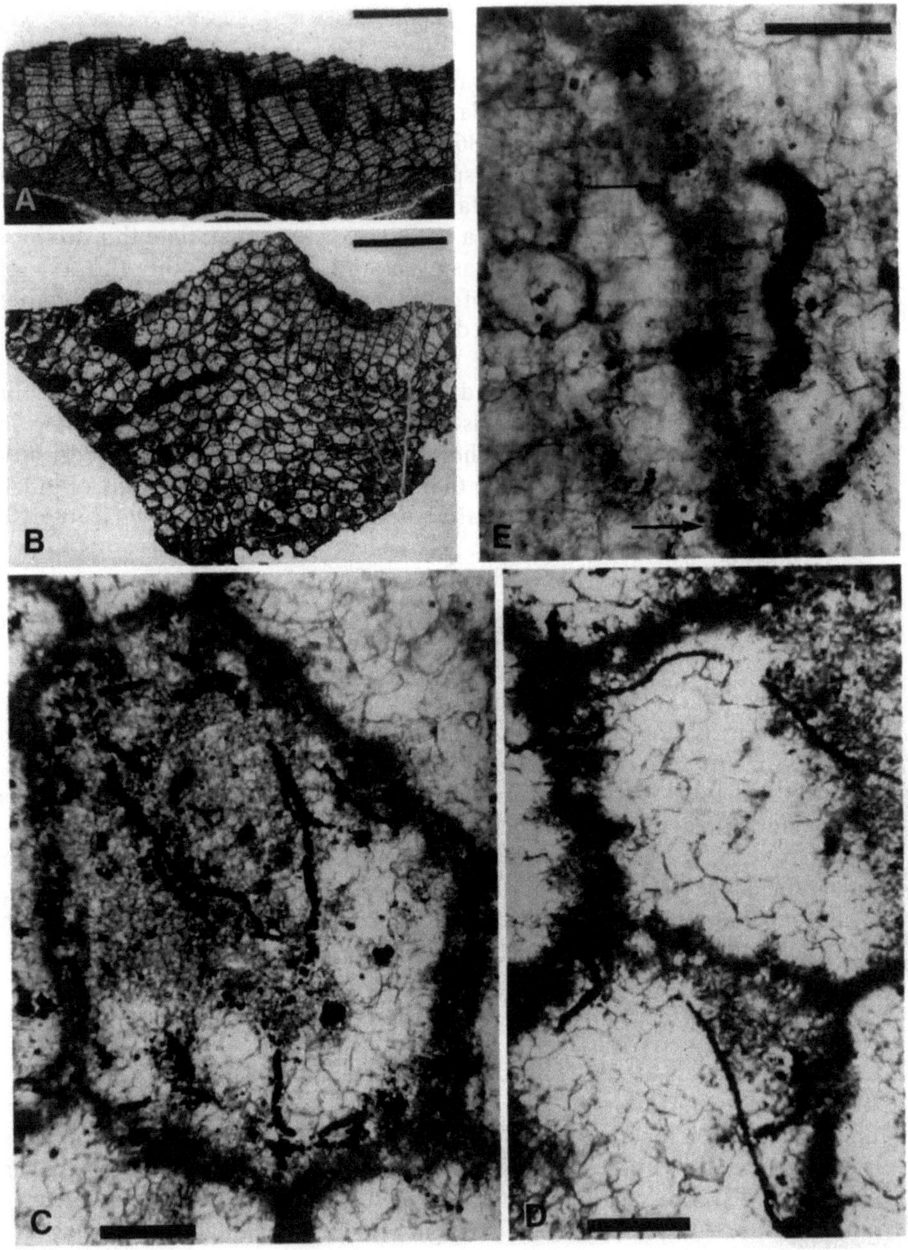


**Fig. 1.** *Palaeofavosites hystrix* Sokolov, Malmøy, Oslo Region (Norway); *Pentamerus* series (7a), Llandovery, specimen PMO 40788. **A,B** Longitudinal and transverse sections through the calcareous skeleton; bar = 5 mm. **C,D** Transverse sections through the skeletal tubes show the mode of preservation and distribution of the ophirhabd-like pyritic pseudomorphs of the originally siliceous spicules. Note that only some spicules are embedded in the calcareous skeletal substance, whereas many others are loosely distributed in the lumen of the tubes. The light spots within the tube walls represent oblique and cross-sections of calcitized spicules which have thus far been interpreted as traces of commensal worm-like organisms (e.g., *Chaetosalpinx*); bar = 200 μm. **E** Cross-section of the skeletal tube wall with cross-sections of pyritic and calcitic spicule pseudomorphs (larger *arrows*). Note also indistinct longitudinal section of a calcitized spicule (*small arrows*). In the right portion of the picture two poorly preserved pyritic pseudomorphs of ?ophirhabd-like spicules are visible in the interskeletal space; bar = 100 μm

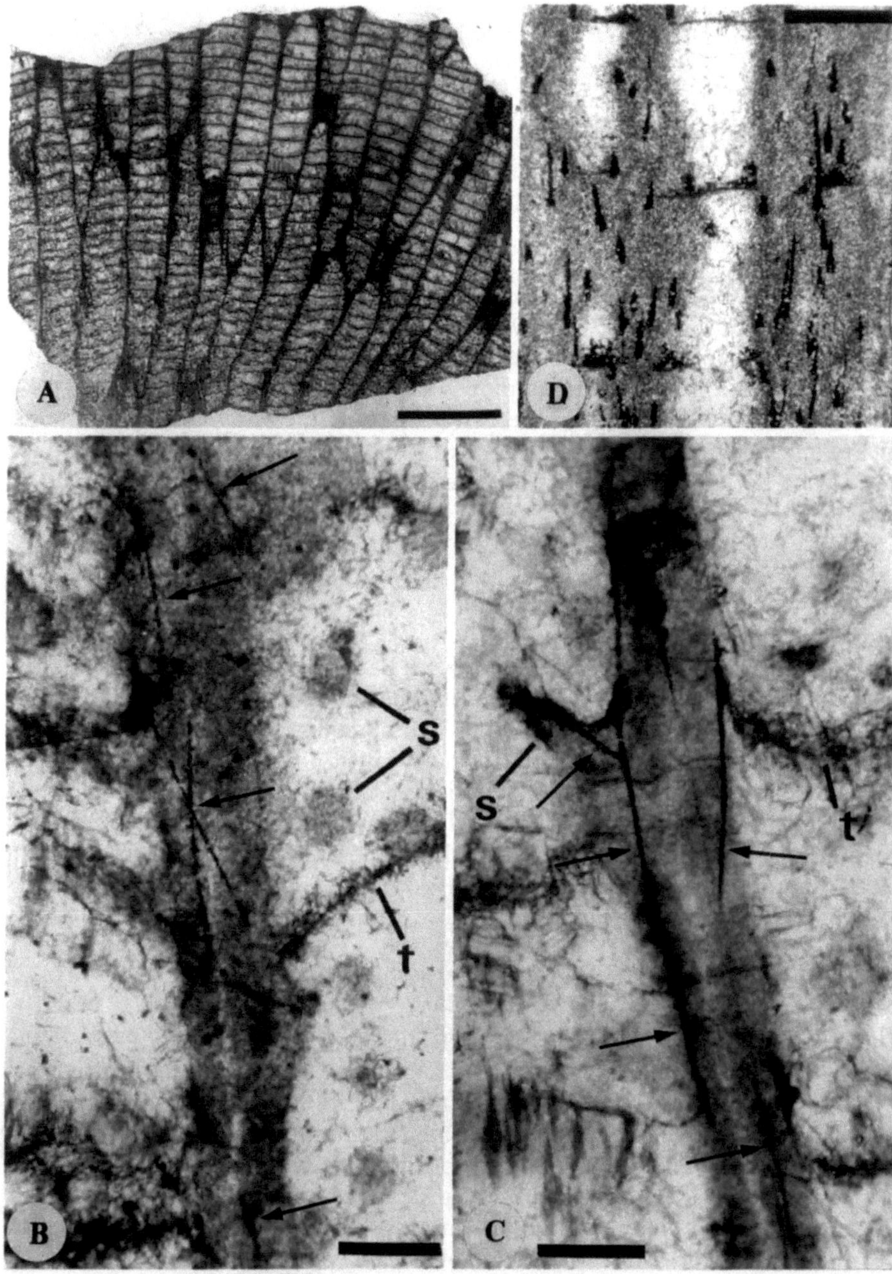


**Fig. 2.** *Mesofavosites* cf. *imbelis* Klaamann, Wyszogród (Poland); ?Wenlockian, glacial erratic boulder of Baltic origin, specimen ZPAL T/III-31E. **A** Longitudinal section through the calcareous skeleton; bar = 5 mm. **B,C** Enlarged fragments of **A** show numerous pyritic pseudomorphs of long oxea-like spicules (some are *arrowed*) entrapped subvertically within the walls of the skeletal tubes; *s* septal spines; *t* tabulae; bar = 200 μm. **D** Chaetetid sclerosponge *Chaetetopsis favrei* (Deninger), Belaya Mountain (Crimea); Barremian, specimen ZPAL Pf. III/1a. Longitudinal section of an enlarged portion of the calcareous skeletal tubes enclosing numerous, subvertically arranged pyritic pseudomorphs of siliceous style-like spicules (cf. Kaźmierczak 1979). Note the striking similarity in spicule preservation and distribution in this specimen with the favositid oxea-like spicules shown in **B** and **C**; bar = 200 μm

origin on the Vistula shore near Wyszogród (central Poland). By comparison with typical *Mesofavosites imbelis* occurring in the upper part of the Silurian Jaani Stage of Estonia (Klaamann 1961), the age of the material can be evaluated as Wenlockian. The specimens ascribed by Stasińska (1967) to *M. imbelis* come from Wenlockian deposits of the Oslo Region (Norway).

*Example 3 (Figs. 3 and 4A,B):*
Specimens described and illustrated by Stasińska (1958: 189, Pl. V, Fig. 2; Pl. VI, Figs. 1–3) as *Favosites goldfussi* D'Orbigny 1850, represented by two thin sections labeled ZPAL Nos. 7 and 16. The material comes from marly siltstones correlating with the lithological complex XVI of the Skalska Series exposed by trenching near Skały village in the eastern part of the Holy Cross Mountains, Poland (cf. Pajchlowa 1957: 215, Fig. 1). Conodont stratigraphy established for the Skalska Series (Malec 1986) indicates a late Eifelian age for the *F. goldfussi*-bearing horizon.

## Spicules: Observations and Comparison

In all three cases, the traces of spicules have been studied in thin sections only. Acid-etched residues of the spicule-bearing specimens have not provided spicule materials satisfactory for examination. All spicules are preserved as pyritic and/or calcitic pseudomorphs after primary silica. Space limitation precludes, however, a detailed discussion of the basic pseudomorphosic processes known in fossil siliceous sponge spicules. Interested readers are referred to Rauff (1893–94), Kaźmierczak (1974, 1979), Rigby and Dixon (1979), Gray (1980), Van Kempen (1983), Van Kempen and Hellingwerf (1983).

### *Case 1: Palaeofavosites Twenhofel (Figs. 1A-E and 4D,E)*

Under standard magnification (Fig. 1A,B), the thin sections prepared from the spicule-bearing specimen PMO 40788, designated originally as *P. hystrix* Sokolov, reveal features corresponding to the definition of the genus *Palaeofavosites*, i.e., the presence of mural pores located at the extreme edges of the faces of prisms (cf. Klaamann 1964; Oekentorp and von Schouppé 1969). Traces of primary siliceous spicules are preserved as numerous pyritic pseudomorphs. They occur both within the walls of calcareous skeletal tubes and in the lumen of the tubes (Fig. 1C,D). The replacement of silica by pyrite is not perfect and many pseudomorphs are preserved as discontinuous alignments of pyrite grains and framboids which only approximately resemble the original shape of spicules. Pyrite often replaces only partially the primary volume of the spicules, and the pseudomorphs are therefore somewhat thinner than the original spicules. A considerable proportion of the pyritic pseudomorphs have been partially or wholly replaced by calcite (Fig. 1E), probably due to late diagenetic or epigenetic weathering (oxidation) of pyrite. Cross-sections of such secondarily calcitized

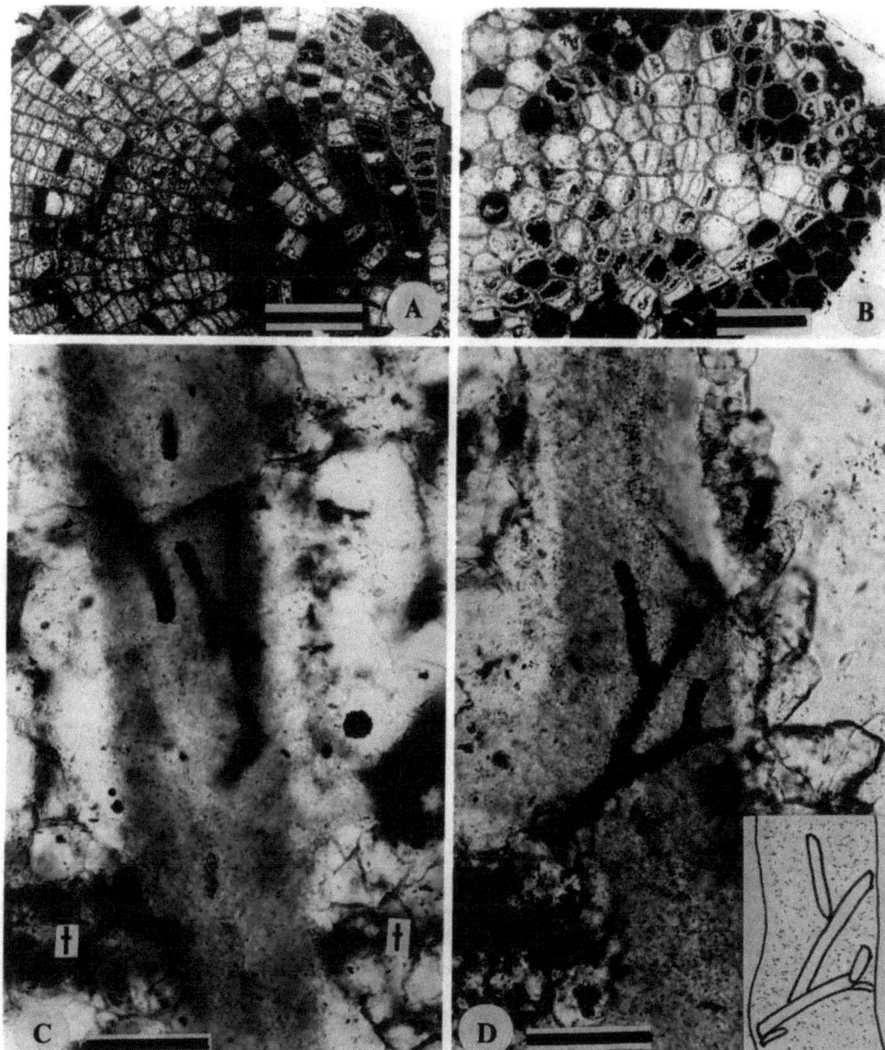


**Fig. 3.** *Favosites goldfussi* d'Orbigny. Skały/Miłoszów, Holy Cross Mountains (Poland); Skalska Formation, Eifelian, specimens ZPAL Nos.7 and 16. **A,B** Longitudinal and transverse sections through the calcareous skeleton; bar = 5 mm. **C,D** Enlarged longitudinal sections with traces of pyritic pseudomorphs of monaxonic siliceous spicules embedded within the calcareous walls of skeletal tubes; *t* tabulae. The spicules in **D**, making an impression of branching represent in fact a monaxon meshwork as shown schematically in the insert; bar = 100 μm

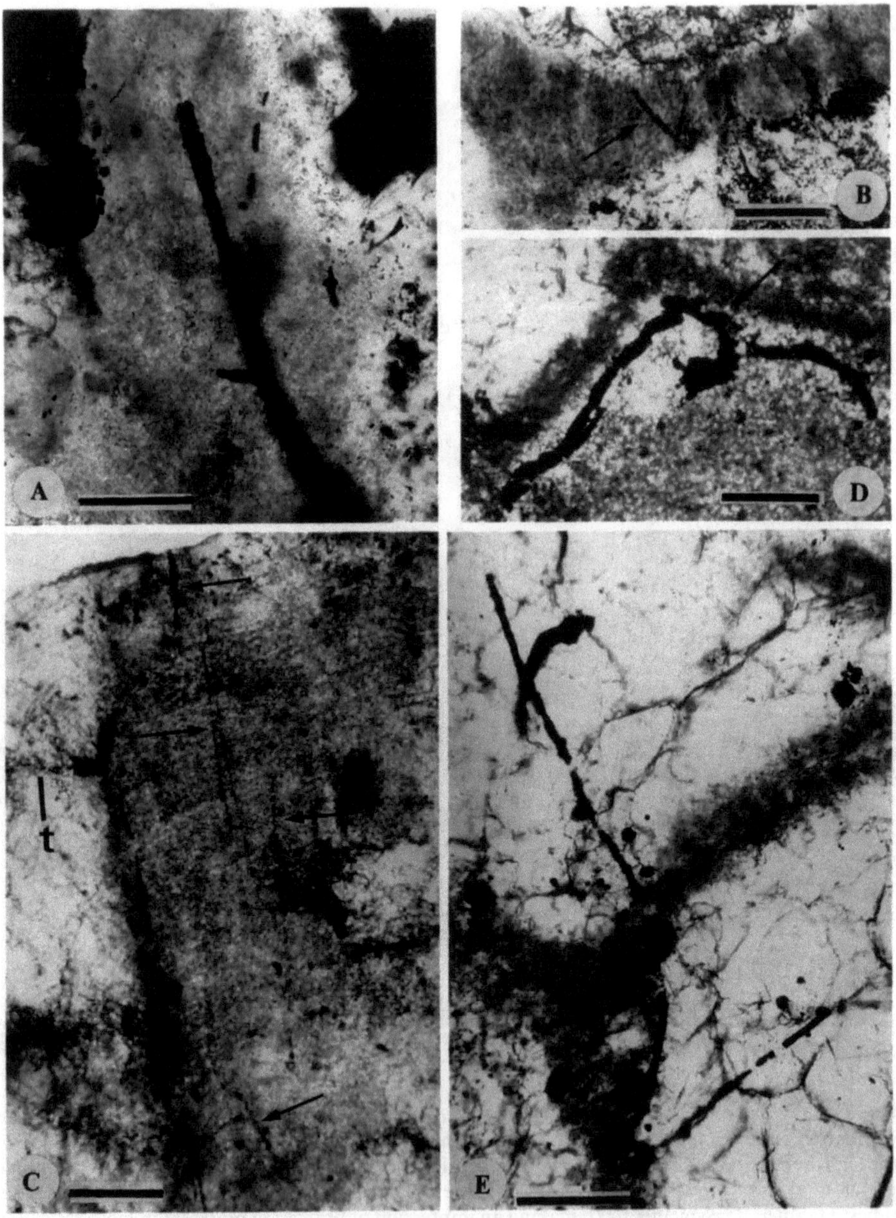

**Fig. 4.** Examples of spicule preservation in various favositids. **A,B** *Favosites goldfussi* d'Orbigny, Skały/Miłoszów, Holy Cross Mountains (Poland); Skalska Formation, Eifelian, specimen ZPAL No. 7. **A** Longitudinal section of the skeletal wall with traces of pyritic spicule pseudomorphs variable in size; bar = 100 μm. **B** Cross-section of the skeletal wall with pyritic pseudomorphs of monaxonic spicules (*arrowed*); bar = 200 μm. **C** *Mesofavosites* cf. *imbelis* Klaamann, Wyszogród (Poland); ?Wenlockian, glacial erratic boulder of Baltic origin, specimen ZPAL T/III-31E: longitudinal section of the calcareous skeletal tube with traces of strongly obliterated pyritic pseudomorphs of oxea-like

spicules are visible as white, roundish spots in transversal sections (Fig. 1C,D) or as sinuous, tube-like structures in longitudinal sections of skeletal walls. It is interesting to note that identical structures have long been known in favositids and interpreted as traces of symbiotic (commensal) worms named *Chaetosalpinx* or *Helicosalpinx* (cf. Sokolov 1948; Oekentorp 1969; Stel 1976). Remnants of the earlier pyrite are often discernible in the calcitized spicules (Fig. 1E). The spicules seem to be randomly distributed but a closer analysis shows their preference for subvertical and subhorizontal location. Judging by the abundance of spicules in the calcareous skeletal tubes, the primary spicular network was quite dense, and calcium carbonate was originally only a tiny fraction of the siliceous spicular material building the skeletal tubes. The quasi-regular character of the spicular lattice is accentuated by the approximately equal distances between cross-sections of vertically oriented spicules in tube walls (Fig. 1D,E). The distribution density of spicules in the lumen of skeletal tubes is lower than in the calcareous walls and rather variable. It may be concluded that the primary density of spicules within the lumen of the skeletal tubes was similar to that entrapped in skeletal walls but the preservation and/or pseudomorphosis conditions were there less favorable. For instance, the formation of the neospar, which fills most of the interskeletal spaces, must have been destructive for both the pyritized and calcified spicules.

The spicular meshwork is composed of two spicule types. Long, thin, thread-like spicules are dominant. They are straight, slightly bent or even a little wavy, equidimensional, smooth or irregularly tuberculated, up to 450 μm long and usually 10–12 μm in diameter (Figs. 1C,D and 4E). They are well comparable to the modified oxeas called ophirhabds (cf. Schrammen 1936; de Laubenfels 1955; Rigby 1981) which occur within the members of the family Cephaloraphiditidae Reid 1968 of the demospongean order Choristida. Ophirhabds or sinuous oxeas occur in some cephaloraphiditid sponges as the only type of spicules (e.g., the Recent *Ophiraphidites tuberosus* Carter 1876 – fide Laubenfels 1955) or they may be associated with triaenes, styles, tylostyles, tylotes and oxeas (e.g., the Eocene *Ophiraphidites* Schrammen – see Rigby 1981).

Less common are thicker, shorter, and strongly bent spicules reaching 30–60 μm in diameter and up to 200 μm in length (Fig. 4D,E). They are covered with large, irregular tubercles and rods which, in the arc-shaped forms, occur only on the external side (cf. Fig. 4D). Such spicules are very similar to imperfect rhizoclonal desmas called heloclones which are characterized by rather poorly developed zygomes and the lack of branches (e.g., Schrammen 1936; Reid 1970). They represent sponge sclerites transitional between typical ophirhabds and

spicules (some are *arrowed*) arranged subvertically within the calcareous wall. Well visible is also the lamellar microstructure of the calcareous skeleton; *t* tabula; bar = 200 μm. **D,E** *Palaeofavosites hystrix* Sokolov, Malmøy, Oslo Region (Norway); Llandovery specimen PMO 40788. **D** Transverse section showing a few pyritic pseudomorphs of ophirhabd- and heloclone-like (the latter *arrowed*) spicules located in the lumen of the skeletal tube; bar = 200 μm. **E** Same as **D** but with pyritic pseudomorphs of spicules partially embedded in the calcareous skeleton; bar = 100 μm rhizoclones. In some Paleozoic siliceous sponges ophirhabds and heloclones occur together with other spicules (e.g., Rigby and Dixon 1979). Sponges having such spicules are classified within the order Lithistida as the so-called sublithistids. Based upon the observed spicule types, the studied *Palaefavosites* specimen can be tentatively classified as a choristid sponge with some features of sublithistid spicular organization.

### *Case 2: Mesofavosites Sokolov (Figs. 2A-C and 4C)*

The assignment of the specimen ZPAL T/III-31E to *Mesofavosites* is authenticated by the presence of mural pores on the faces, as well as at the edges of the faces, of prisms diagnostic of this genus (for generic diagnosis see, e.g., Sokolov 1951; Hill 1981).

The spicules are preserved as either pyrite- or limonite-stained calcite pseudomorphs after primary silica. They occur within the walls of the calcareous skeletal tubes and in septal spines, where they are vertically or subvertically arranged (Figs. 2B,C and 4C). In contrast to the above-described specimen of *Palaeofavosites*, the spicules of *Mesofavosites* have not been observed in the interskeletal spaces. The distribution density of spicules is rather low and their occurrence very irregular. They are satisfactorily preserved in only a few places in the area of longitudinal thin section available for study. Generally, they are either lacking, or visible as indistinct linear accumulations of pyrite grains, or as lighter than the background striation, running perpendicularly to the horizontally arranged calcareous growth lamellae typical of favositids (Lafuste and Plusquellec 1987).

The spicule pseudomorphs can be easily identified as remnants of smooth, long, fusiform oxeas (diactines). They have 5–10 $\mu$m in diameter in their thickest portion and are up to 450 $\mu$m long. Since the pyritization of siliceous spicules is usually associated with thickness reduction of the resulting spicule pseudomorphs (e.g., Rauff 1893–1894, Pl. 17:3; van Kempen and Hellingwerf 1983; Pl. 4), the original oxeas were probably somewhat thicker.

The presence of one single kind of fusiform oxeas permits assignment of *Mesofavosites* to the demospongean order Monactinellida sensu Levi 1973 or Ceractinellida Levi. More detailed taxonomy of the specimen is not possible without knowing details of the whole architecture of the spicular skeleton. Sponges with megascleres consisting exclusively of fusiform oxeas are known from the Cambrian (*Leptomitus* Walcott 1886) through Silurian (*Climacospongia radiata* Hinde 1884), Devonian (*Lasiocladia compressa* Hinde 1884), Pleistocene (*Tethya logani* Dawson 1857) to modern times (*Halichondria* Fleming 1828, *Trachyopsis* Dendy 1905). In terms of spicule arrangements, morphology, and preservation, it is instructive to compare the described specimen with the Cretaceous chaetetid *Chaetetopsis favrei* which encloses numerous pyritic pseudomorphs of monaxonic sclerites entrapped vertically in the calcareous skeleton (see Fig. 2D, and Kaźmierczak 1979).

### *Case 3: Favosites Lamarck (Figs. 3 and 4A,B)*

The assignment of the spicule-bearing specimens ZPAL Nos. 7 and 16 to the genus *Favosites* as *F. goldfussi* by Stasińska (1958) is justified by the presence of mural pores on prism faces (for *Favosites* diagnosis see Hill 1981). The structure of the calcareous skeleton of the examined specimen in thin sections under standard magnification is shown in Fig. 3A,B.

The spicules, which are preserved as pyrite pseudomorphs after primary silica, are visible at only a few points in the thin sections. They occur mostly within the walls of the calcareous skeletal tubes but some have also been observed in the tube lumen or extending from the calcareous walls into the interskeletal spaces. Basically, two kinds of slightly upwards-radiating monaxonic spicules can be recognized: (1) poorly preserved, very thin, thread-like, smooth sclerites, mostly 4–6 μm in diameter and up to a few hundred micrometers long, usually visible as discontinuous alignments of pyrite blobs (Fig. 4A); and (2) better preserved and much thicker, thread-like, straight or slightly bent, indistinctly tuberculated sclerites, 12–17 μm in diameter and up to 400 μm long. These two types of spicules are comparable to the ophirhabds occurring in the above described specimen of *Palaeofavosites*, thus pointing to a choristid affinity of the specimens under consideration. Remnants of rare and poorly preserved heloclone-like sclerites visible in the lumen of some skeletal tubes of *F. goldfussi* make the similarity between the two forms even closer, and indicate, as in the case of *Palaeofavosites*, a possible sublithistid kinship.

## Conclusions

The new findings of spicule-bearing specimens of typical favositids confirm previous suggestions of a sponge (sclerosponge) nature of these common Paleozoic fossils. Similar to the calcareous skeletons of other modern and fossil sclerosponges (coralline sponges), favositid fossils are basally excreted hard parts of various demosponges and, as such carry little information about the biological organization of the organisms producing them. The kinds of spicules occurring in favositids, though generally monaxonid in character, indicate that the forms assigned to *Palaefavosites* and *Favosites*, which have ophirhabds and ?heloclones, are related to choristid and/or sublithistid demosponges, whereas those assigned to *Mesofavosites*, enclosing fusiform oxeas in their calcareous skeleton, are related to monactinellid or ceractinellid demosponges. Further examination of favositids is needed before a new taxonomy of this enormously splintered group can be established, which will take into account its poriferan affiliation. Nevertheless, terms such as "corallum", "corallites", "colony", "coenosteum", and others stemming from the long-accepted cnidarian attribution of the favositids should be abandoned. The poriferan affinity of favositids requires also a considerable reevaluation of their paleoecology, particularly concerning the role they played in ancient communities.

*Acknowledgments.* Warm thanks are extended to Antoni Hoffman (Institute of Paleobiology, Polish Acad. Sci., Warsaw) for reviewing the paper. Supported by the Polish Academy of Sciences (project CPBP 04.03).

## References

Copper P (1985) Fossilized polyps in 430-yr-old *Favosites* corals. Nature 316:142–144
De Laubenfels MW (1955) Porifera. In: Moore RC (ed) Treatise on invertebrate paleontology, part E. Geol Soc Amer and Univ Kansas Press, New York, Lawrence, pp 21
Dunbar CO (1927) Pseudopercula in the tabulate coral *Favosites*. Am J Sci 13:101–114
Finks RM (1986) "Spicules" in *Thamnopora*. Fossil Cnidaria 15(1.2):22
Flügel HW (1976) Ein Spongienmodell für die Favositidae. Lethaia 9:405–419
Flügel HW (1986) *Imilce* Flügel 1975 (Khmeriidae, Demospongea) aus der Yabeina-Zone (Perm) von Tunis. Mitt Österr Geol Ges 78 (for 1985):267–289
Gray DI (1980) Spicule pseudomorphs in a new Palaeozoic chaetetid, and its sclerosponge affinities. Palaeontology 23:803–820
Hartman WD, Goreau TF (1970) Jamaican coralline sponges: their morphology, ecology and fossil relatives. Symp Zool Soc Lond 25:205–243
Hartman WD, Goreau TF (1972) *Ceratoporella* (Porifera: Sclerospongiae) and the chaetetid "corals". Trans Conn Acad Arts Sci 44:133–148
Hartman WD, Goreau TF (1975) A Pacific tabulate sponge, living representative of a new order of sclerosponges. Postilla 167:1–14
Hill D (1981) Rugosa and Tabulata. In: Teichert C (ed) Treatise on invertebrate paleontology, part F (suppl 1). Geol Soc Amer and University of Kansas Press, Boulder, Lawrence, pp 762
Kaźmierczak J (1974) Lower Cretaceous sclerosponge from the Slovakian Tatra Mountains. Palaeontology 17:341-347
Kaźmierczak J (1979) Sclerosponge nature of chaetetids evidenced by spiculated *Chaetetopsis favrei* (Deninger 1906) from the Barremian of Crimea. Neues Jahrb Geol Paläontol Monatsh 2:97–108
Kaźmierczak J (1984) Favositid tabulates: evidence for poriferan affinity. Science 225:835–837
Kaźmierczak J (1989) Halysitid tabulates: sponges in corals' clothing. Lethaia 22:195–205
Kirkpatrick R (1912) *Merlia normani* and its relation to certain Palaeozoic fossils. Nature 89:502–503
Klaamann E (1961) Tabulyaty i geliolitidy venloka Estonii. Tr Inst Geol AN Est SSR 6:69–110
Klaamann E (1964) Pozdneordovikskiye i rannesiluriyiskiye Favositida Estonii. Inst Geol AN Est SSR, Tallin, pp 118
Lafuste J, Plusquellec Y (1987) Structure et microstructure de *Favosites cylindrica* Michelin 1847, espece-type de *Ohiopora* n. gen. (Tabulata, Dévonien). Can J Earth Sci 24:1465–1477
Malec J (1986) Stratygrafia dewonu w profilu Grzegorzowice. CXII Sesja Naukowa "Badania IG nad dewonem Gór Swiętokrzyskich (unpublished lecture)
Mori K (1976) A new recent sclerosponge from Ngargol, Palau Islands and its fossil relatives. Sci Rep Tohoku Univ 2nd Ser 46:1–9
Mori K (1977) A calcitic sclerosponge from the Ishigaki-shima Coast, Ryukyu Islands, Japan. Sci Rep Tohoku Univ 2nd Ser 47:1–5
Oekentorp K (1969) Kommensalismus bei Favositiden. Münster Forsch Geol Paläontol 12:165–217
Oekentorp K (1985) Spicules in favositid Tabulata – remarks to J Kaźmierczak's interpretation. Fossil Cnidaria 14(1):34–35
Oekentorp K, Stel J (1985) *Favosites* – a true coral. Remarks to P Copper's discoveries of fossilized polyps. Fossil Cnidaria 14(2):28–29
Oekentorp K, von Schouppé A (1969) Kritische Betrachtungen über die Anordnung der Poren bei *Palaeofavosites* Twenhofel, 1914. Neues Jahrb Geol Paläontol Abh 133:89–100
Oliver WA (1979) Sponges, they are not. Paleobiology 5:188–190
Oliver WA (1986) Favositids are corals – further remarks. Fossil Cnidaria 15(1.2):19–21
Pajchlowa M (1957) Dewon w profilu Grzegorzowice-Skały. Inst Geol Biul 122:145–254
Rauff H (1893–1894) Palaeospongiologie. Palaeontographica 40:1–346
Reid REH (1970) Tetraxons and demosponge phylogeny. Symp Zool Soc Lond 25:63–89

Rigby JK (1981) The sponge fauna of the Eocene Castle Hayne Limestone from East-Central North Carolina. Tulane Stud Geol Paleontol 16:123–144
Rigby JK, Dixon OA (1979) Sponge fauna of the Upper Silurian Read Bay Formation, Somerset Island, District of Franklin, Arctic Canada. J Paleontol 53:587–627
Schrammen A (1936) Die Kieselspongien des oberen Jura von Süddeutschland. Palaeontographica Abt A 85:1–114
Scrutton CT (1987) A review of favositid affinities. Palaeontology 30:485–492
Sokolov BS (1948) Kommensalizm u favozitid. Izv Akad Nauk SSSR, Ser Biol 1:101–110
Sokolov BS (1951) Tabulyaty paleozoya evropeyskoy chasti SSSR, chast 2: Silur Pribaltiki (Favozitidy llandoverskogo yarusa. Vses neft nauchno-issled geol-razved Inst (VNIGRI) Tr n s 52:1–124
Stasińska A (1958) Tabulata, Heliolitida et Chaetetida du Dévonien moyen des Monts de Sainte-Croix. Acta Palaeontol Pol 3:161–282
Stasińska A (1967) Tabulata from Norway, Sweden and from the erratic boulders of Poland. Palaeontol Pol 18:5–112
Stel J (1976) The Paleozoic hard substrate trace fossils *Helicosalpinx*, *Chaetosalpinx* and *Torquaysalpinx*. Neues Jahrb Geol Paläontol Monatsh 12:726–744
Strand T, Henningsmoen G (1960) Cambro-silurian stratigraphy. Nor Geol Unders 268:128–169
Swann DH (1947) The *Favosites alpensis* lineage in the Middle Devonian Traverse Group of Michigan. Contrib Mus Paleontol Univ Mich 6:235–318
Van Kempen THMG (1977) *Roepella solanensis*, a new sponge from the mid-Tertiary of S.E. Spain. First fossil record of a hymeniacidonid, with remarks on other fossilized sponges with monaxonid megascleres. I, II. Proc K Ned Akad Wet B 80:106–131
Van Kempen THMG (1983) The biology of aulocopid lower parts (Porifera-Lithistida). J Paleontol 57:363–376
Van Kempen THMG, Hellingwerf RH (1983) Notes on the occurrence of coring monaxons in *Archaeoscyphia*, and the preservation and mineralogy of such spicules in an aulacopiid lower part (Porifera-Lithistida). Proc K Ned Akad Wet B 86:69–77
Von Schouppé A, Oekentorp K (1974) Morphogenese und Bau der Tabulata. Palaeontographica Abt A 145:79–194

# “Sphinctozoa”: An Overview

B. SENOWBARI-DARYAN

## Introduction

The term “sphinctozoa” (= “Thalamida” De Laubenfels 1955) was used by Steinmann (1882) for those “pharetronid” sponges (sponges with a rigid calcareous skeleton) which show an internal segmentation (external segmentation may not always be developed). Because their spicular skeletons are rarely preserved, even if originally developed, the systematic position of most sphinctozoan genera is uncertain (Hartman et al. 1980) or they are classified as Calcispongea (de Laubenfels 1955; Seilacher 1962; Ott 1967; Ziegler and Rietschel 1970, etc.). The discovery of the modern segmented sponge – *Vaceletia crypta* (Vacelet) – by Vacelet (1977) and its affiliation to the Demospongea (as indicated by its histological and cytological features, as well as its sexual reproduction; *Vaceletia* does not have any spicules) prompted reevaluation of the systematic position of sphinctozoan sponges within the Porifera and made their certain affiliation to the Calcispongea questionable. Some workers (e.g., Vacelet 1979, 1985; Pickett 1982; Pickett and Jell 1983) placed all or most of the “sphinctozoa” Steinmann, even those without any spicular skeleton, into the Demospongea. Others (e.g., Rigby and Webby 1988) still include such sponges in the Calcispongea (see Senowbari-Daryan and Rigby 1988).

Vacelet (1979, 1985) suggested placing the Jurassic-Cretaceous genera *Barroisia*, *Thalamopora*, and *Sphaerocoelia* in a separate group, the Sphaerocoelida, and retaining them in the Calcispongea because they possess calcite spicules. All other “spinctozoa” he would place in the group Sphinctozoida within the Demospongea.

Drawing on the model of *Vaceletia crypta* (Vacelet), Reitner and Engeser (1985) placed some Triassic genera and all Jurassic to modern thalamid sponges with a trabecular internal structure (pillar filling tissue) and an aragonitic rigid skeleton with irregular microstructure in the Demospongea. *Vaceletia* possesses such a trabecular internal structure and an irregular aragonitic skeleton. Most of genera included by Reitner and Engeser (1985) do not have proven spicules.

In addition to presence or absence of spicules and composition and microstructure of the rigid skeleton (aragonite or high-Mg-calcite), there are several additional morphological features that should be used for the description and classification of fossil sphinctozoan sponges. These criteria are discussed briefly below:

J. Reitner and H. Keupp (Eds.)
Fossil and Recent Sponges

## Structural Features of Segmented Sponges

### *Spicular Skeleton*

Within the thalamid sponges spicules may be missing (not preserved?) or rare. In addition to some Jurassic and Cretaceous genera (e.g., *Barroisia*, Reid 1967, 1968; Debrenne and Lafuste 1972; Reitner 1987a; Senowbari-Daryan 1989) unquestioned primary spicules have been found in a few Upper Triassic genera, including *Colospongia* Laube, *Thaumastocoelia* Steinmann, and *Celyphia* Pomel from Cassian beds (Carnian) in the Southern Alps, and in *Stylothalamia* sp. and other species from lowermost Norian Cipit-boulders of the Antalya area in Turkey (Figs. 1, 2, see Senowbari-Daryan 1989). These originally siliceous spicules are now replaced by calcite. Most of them are monaxones, but rarely other shapes also occur (Fig. 2g). However, sphinctozoans from other Triassic or Permian localities lack spicules (preservation?) or spicules occur rarely (e.g., *Pisothalamia* Senowbari-Daryan and Rigby 1988; Senowbari-Daryan 1989, 1990). Other structures in some sphinctozoans (e.g., *Stylothalamia* Ott and *Cassianothalamia* Reitner), here called "pseudospicules" (Fig. 1e-f, 2a-g), are interpreted as cavities (originally filled by organic substances filled by calcite during diagenesis). Similar structures observed in Cretaceous segmented sponges have been interpreted as spicules (replaced by pyrite) by Reitner and Engeser (1985). The primary nature of these structures as spicules is also in question (Senowbari-Daryan 1989).

### *Composition of the Rigid Skeleton*

The rigid skeleton of sphinctozoan sponges is composed of aragonite or high-Mg-calcite. Because of recrystallization or replacement the original skeletal mineralogy of Carboniferous and older sphinctozoans is not known (Rigby and Potter 1986). An aragonitic skeleton was secreted by most genera of Permian, Triassic, some Jurassic and Cretaceous sphinctozoans, as well as by modern *Vaceletia* (Wendt 1979, 1984; Reitner and Engeser 1985; Gautret 1985). High-Mg-calcite skeletons have been found in some Triassic genera (e.g., *Uvanella* Ott, *Cassianothalamia* Reitner, etc.). Skeleton of some Jurassic and Creta-

**Fig. 1. a** Monaxon spicules within the rigid skeleton of *Thaumastocoelia cassiana* Steinmann from the Cassian beds (Carnian) in the Dolomites, Italy. The octahedral crystal is pyrite. **b** Monaxons of *Celyphia submarginata* Pomel arranged in a bundle. Sample 10, Cassian beds (Carnian). **c** Spicules embedded within the wall of *Celyphia submarginata* Pomel. Same sample as **b**. **d** Few spicules of *Celyphia submarginata* Pomel. Same sample as **b**. **e** *Stylothalamia* sp. from the Lower Norian of the Antalya area (Turkey) showing the "pseudospicules" (*large arrowheads*) and the true spicules (*small arrowheads*) within the rigid skeleton. The "pseudospicular" structures are three times larger than the true spicules. **f** Enlargement of **e** showing the "pseudospicules" (*large arrowheads*) and the spicules (*small arrowheads*) in detail

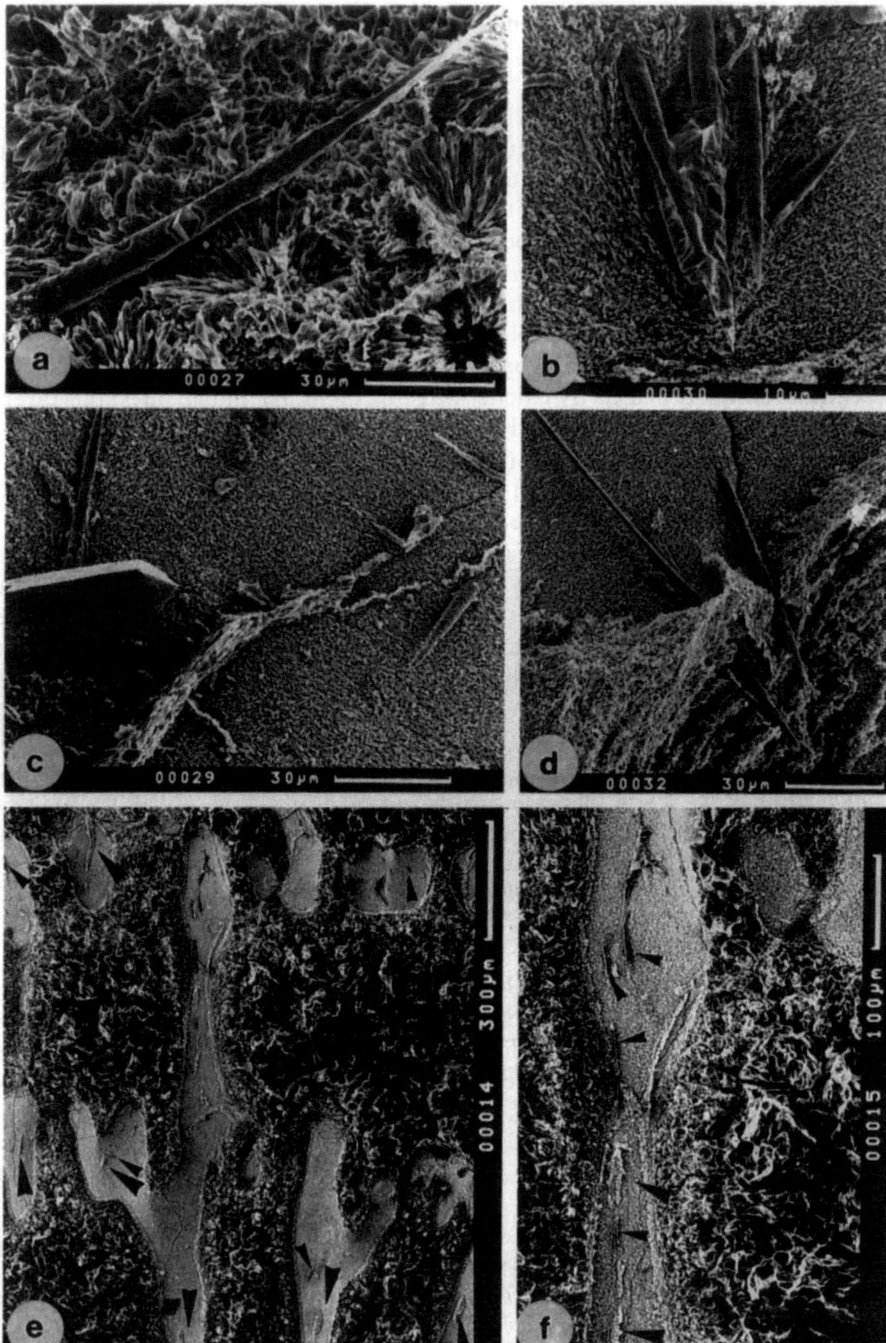


**Fig. 1.** (see caption p. 225)

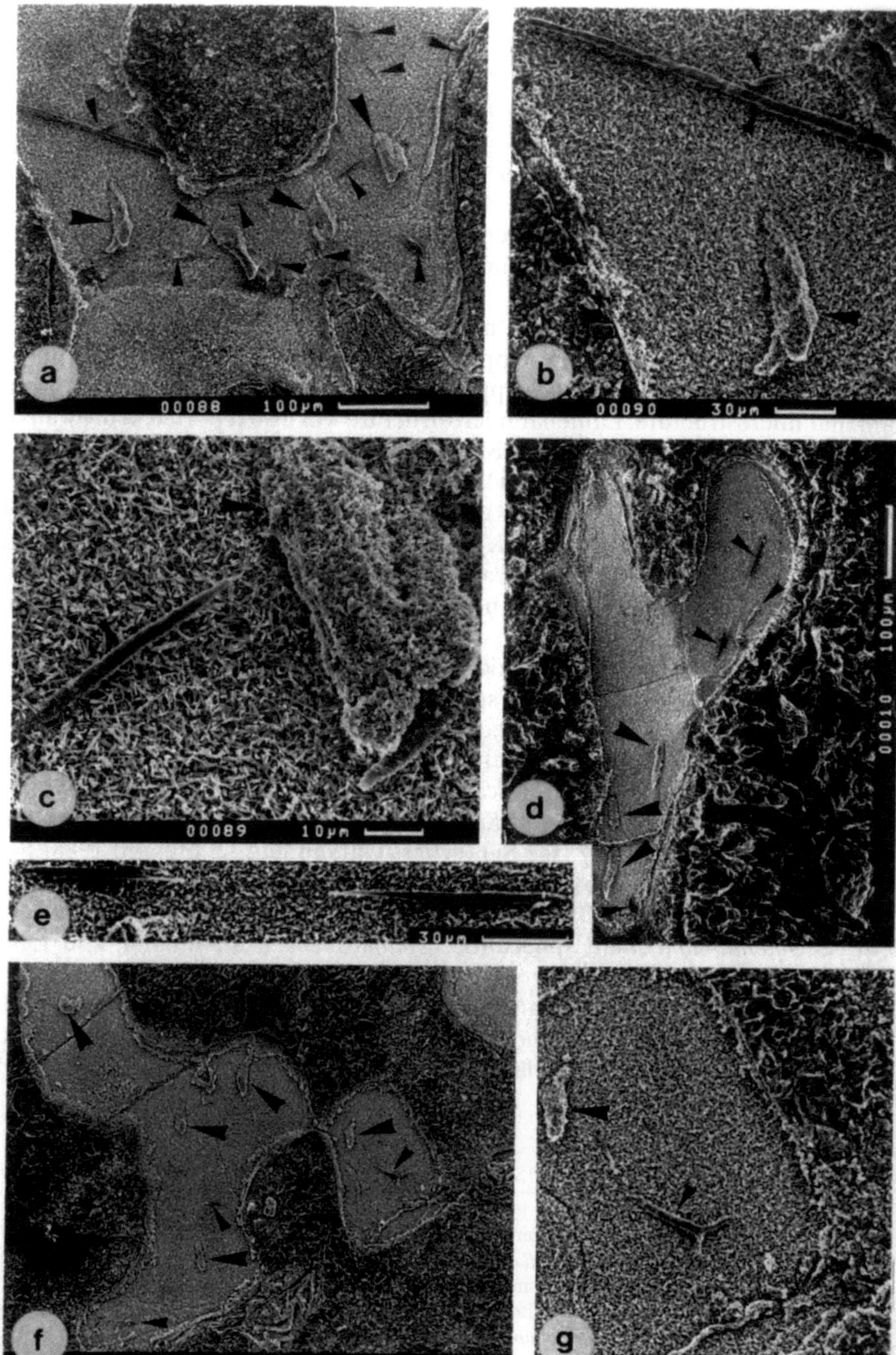


**Fig. 2.** (see caption p. 228)

ceous "hydrozoans", which may be placed into the sphinctozoan sponges (e.g., some species of *Actinostromaria*), were also most probably composed of high-Mg-calcite.

### *Microstructure of the Rigid Skeleton*

Wendt (1979, 1984) observed different types of microstructures in rigid skeletons of segmented sponges. These types include spherulitic, irregular, clinogonal, and orthogonal.

In addition to these types of microstructures, we also found a lamellar structure in the skeleton of the Upper Triassic sphinctozoan *Celyphia* Pomel. According to Russo (1981), the Upper Triassic *Montanaroa* Russo also has a lamellar microstructure. Lamellar microstructure was also reported in the wall of the Ordovician *Angullongia* by Webby and Rigby (1985) and Rigby and Potter (1986). Jones (1979) concluded that lamellar microstructure is a primary feature within the sponges, but Wendt (1984) believes that this type of microstructure was produced by diagenetic processes. Kruse (1987: 551), perhaps following Wendt, also interpreted lamination in the wall of *Angulloniga* to be of secondary origin. The lamellar structure in *Celyphia*, however, represents a primary structure (see Senowbari-Daryan 1989, 1990).

Clinogonal and orthogonal microstructures have not been found within Permian or Triassic sphinctozoan sponges. These types of microstructure are known only in Jurassic and Cretaceous segmented sponges (Wendt 1984).

The oldest known microstructure of segmented sponges is spherulitic (lamellar: *Angullongia*?). Spherulitic microstructure has been found in the typical Carboniferous and Permian genus *Sollasia* Steinmann, as well as in the Permian genus *Crystothalamia* Girty (see Senowbari-Daryan and Di Stefano 1988; Senowbari-Daryan and Rigby 1988). Spherulitic and irregular types of microstructure are common in Triassic sphinctozoans (e.g., irregular: *Colospongia* Laube, *Solenolmia* Pomel; spherulitic: *Thamastocoelia* Steinmann, and *Enoplocoelia* Steinmann).

Some Triassic sphinctozoan genera, such as *Cassianothalamia zardinii* Reitner, *"Cystothalamia" slovenica* Senowbari-Daryan, *Uvanella irregularis* Ott, and *Zardinia* Dieci et al. display skeletons composed of homogeneous and microgranular high-Mg-calcite.

**Fig. 2.** **a** *Stylothalamia* sp. from the Lower Norian of the Antalya area (Turkey). The *large arrowheads* show the "pseudospicules" and the *small arrowheads* show the true spicules within the rigid skeleton. **b** Enlargement of part of **a** showing the small spicules (*small arrowheads*) and large "pseudospicules" (*large arrowheads*). **c** Enlargement of **a** shows the size and structural differences between spicules and "pseudospicules" in detail. **d** *Stylothalamia* sp. from the same locality as the specimen in **a**. The "pseudospicules" (*large arrowheads*) and the spicules (*small arrowheads*) embedded in the rigid skeleton. **e** Enlargement of **d** showing two (or only one?) monaxon spicules. **f** Same sample as in **a** showing the segment wall with spicules (*small arrowheads*) and "pseudospicules" (*large arrowheads*). **g** Enlargement of **f**. A triaene spicule and "pseudospicules" in detail

## Morphological Features of Segmented Sponges

Other morphologic features of segmented sponges are also considered to be important for classification.

a) Perforation of external and internal segment walls by numerous small openings, called pores by Seilacher (1962, Porata Seilacher), or by single or rare larger openings, called ostia by the same author (Aporata Seilacher).
b) The absence or presence of central tube (spongocoel) and its formation.
c) The absence or presence of internal structures (filling tissue) in the interior of the chambers and the different types of such internal structures.
d) The growth form of the sponges, arrangement, and shapes of the chambers and other features.

These morphologic features should also be used for description and classification of segmented sponges.

In practice, subdivision of sphinctozoan sponges into Porata and Aporata, as introduced by Seilacher (1962), is feasible. The determination whether taxa are porate or aporate is difficult or impossible in only rare cases. Perforations of external and internal walls generally are similar but perforations also may be different (see Fig. 3). Pores of the chamber or segment walls are usually simple openings, but branched pores may also occur in some genera (see Senowbari-Daryan 1990).

Six types of internal structures (filling tissue) are known in segmented sponges (Figs. 4, 5) and include: reticular, tubular, trabecular, septate, spore-like, and pisolitic. Different types of internal structures generally do not occur in the same sponge. A combination of septate and reticular filling tissue does occur in one Triassic genus *Ceotinella* Pantic (see Flügel et al. 1978). Vesicular structures may occur with other internal structures within the same sponge and are not considered to be true filling tissue.

Most Cambrian and Ordovician sphinctozoans do not display internal filling structures. Some tubular structures, though not well developed, occur in the

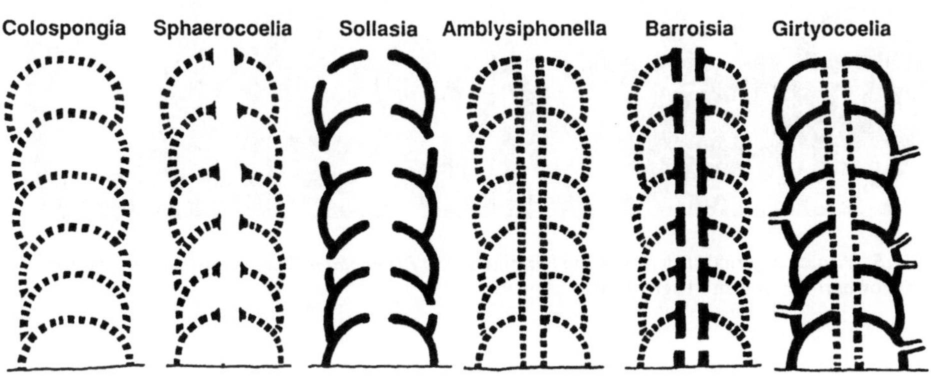


**Fig. 3.** Skeletal structures of chamber walls and the nature or absence of a spongocoel in segmented sponges, displaying different patterns

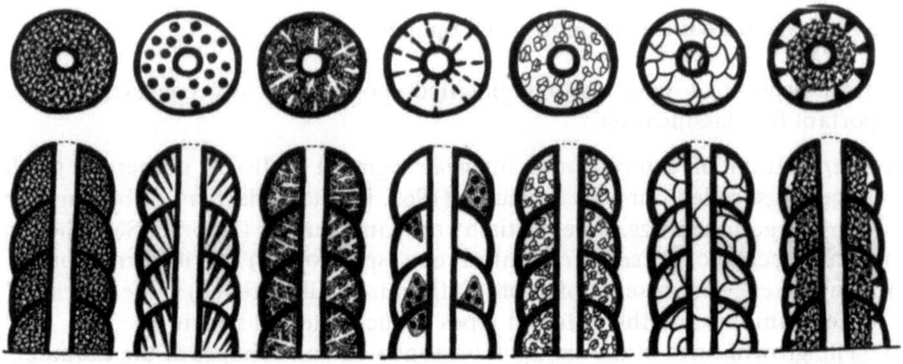


**Fig. 4.** Different kinds of internal filling structures within chambers of segmented sponges

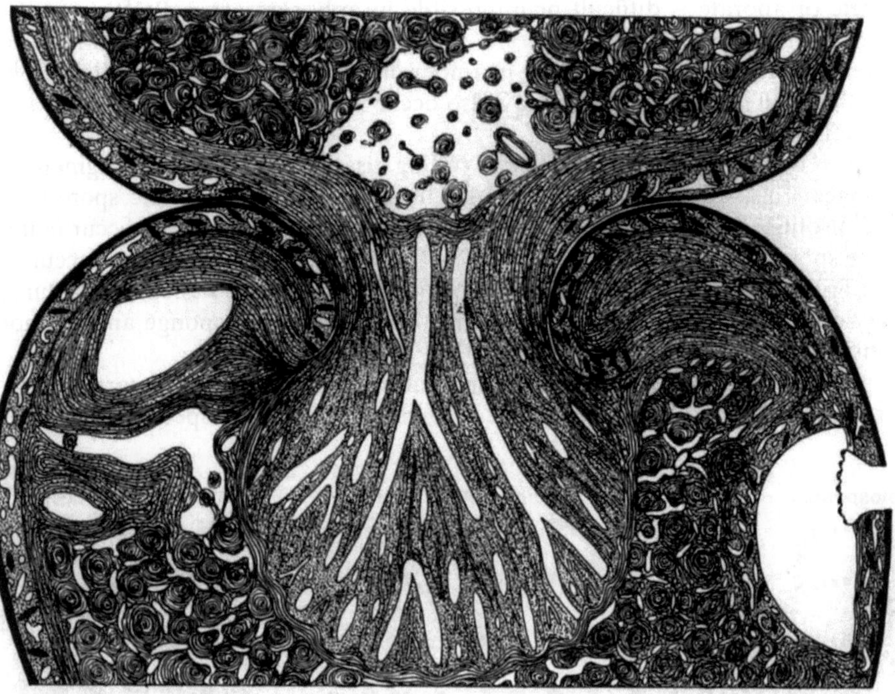

**Fig. 5.** Pisolitic internal structure in the Permian genus *Pisothalamia* Senowbari-Daryan and Rigby (Senowbari-Daryan and Rigby 1988)

Ordovician species *Amblysiphonella tubulara* Rigby and Potter and in *Girtyocoelia cana* Rigby and Potter. The only known Silurian sphinctozoan genus – *Rigbyspongia* Freitas – possesses a trabecular internal structure. This type of internal structure is common in Permian and Triassic sphinctozoans, and in all thalamid sponges with internal structures from Jurassic up to the modern *Vaceletia*. The septate-type structure is developed in the Devonian *Radiothalamos* Pickett and Rigby and the Triassic *Phragmocoelia* Ott. This type of internal structure is not known from Carboniferous and Permian genera.

Reticular internal structure may occur in the Ordovician species *Amblysiphonella reticulata* Rigby and Potter (1986); however, it certainly occurs in the Devonian *Hormospongia* Rigby and Blodgett, in some Permian genera, e.g., *Preverticillites* Parona, and is abundant in Triassic genera, e.g., *Solenolmia* Pomel and *Senowbaridaryana* Engeser and Neumann. Reticular internal structure is known in only one Jurassic genus, *Muellerithalamia* Reitner.

Tubular internal structure is known from the Ordovician (see above), Permian (*Pseudoamblysiphonella* Senowbari-Daryan and Rigby), and Triassic (*Polytholosia* Rauff). In these genera tubes of filling tissue are developed between the exo- and endowall, almost perpendicular to the axis of the sponge.

The Ordovician genus *Cliefdenella* Webby was initially placed into the stromatoporoids (Webby 1969); however, Rigby and Potter (1986) later included it in the segmented sponges. *Cliefdenella* and the Triassic *Madonia* Senowbari-Daryan and Schäfer have tubes oriented parallel to the axis of the central tube of the sponges.

Spore-like internal structure is known only from the Permian genus (*Intrasporeocoelia* Fan and Zhang, see Rigby et al. 1988) from Sichuan, Southwestern China. A pisolitic-like internal structure is known only in the Permian genus *Pisothalamia* Senowbari-Daryan and Rigby, from southern Tunisia (Fig. 5). The lack of internal structures in Carboniferous genera is remarkable.

The spongocoel or central tube is an important feature of most genera of segmented sponges (Senowbari-Daryan 1990). However, it is not always possible to determine its nature, whether retro-, pro-, ambi- or pseudosiphonat, as defined by Seilacher (1962).

Usually there is only one central tube within a segmented sponge, but tubes may number up to more than 100 in a single sponge (Senowbari-Daryan and Schäfer 1983). The spongocoel is generally cylindrical. The development of funnel-shaped spongocoel is questionable. Such openings were used in branched stems for systematic classification of Cretaceous thalamid sponges by Reitner and Engeser (1985).

Chambers of segmented sponges are usually spherical to hemispherical in longitudinal sections, although tube-like chambers are developed in some Permian and Triassic genera, such as in *Subascosymplegma* Deng. Cup- or funnel-like chambers are known from the Ordovician *Pseudoimperatoria* Senowbari-Daryan and Rigby and the Triassic *Panormida* Senowbari-Daryan. Crescent-like chambers occur in Permian and Triassic genera and are common in younger segmented sponges.

With the exception of some Middle Cambrian single-chambered genera such as *Blastulospongia* Pickett and Jell (the affiliation of *Blastulospongia* Pickett and

Jell to the sponges is questionable), *Jawonya* Kruse, and *Wagima* Kruse, all other sphinctozoan sponges are multi-chambered organisms. The chambers are usually arranged in a line, one above the other (moniliform: without spongocoel; or cateniform, catenulate: with spongocoel). In some sponges the chambers are glomerate. If the chambers are arranged in one layer the sponge is monoglomerate but if more than one layer it is polyglomerate. A monoglomerate sponge is similar to kernels around a cob, like in *Discosiphonella* Inai. *Cystothalamia* Girty is polyglomerate. Stratiforme (Finks 1983) chambers are developed in some species of *Guadalupia*, e.g., *G. zitteliana* Girty, or in *Neoguadalupia* Zhang. The Ordovician *Corymbospongia* Rigby and Potter, the Permian *Parauvanella* Senowbari-Daryan and Di Stefano and the Triassic *Uvanella* Ott chambers are arranged above and beside each other like a bunch of grapes (Fig. 6).

The external growth habit of segmented sponges is usually cylindrical (e.g., *Amblysiphonella* Steinmann), but conical or sheet-like types may also occur (Fig. 6). Moniliform linear individuals are common, but branching stems are also known.

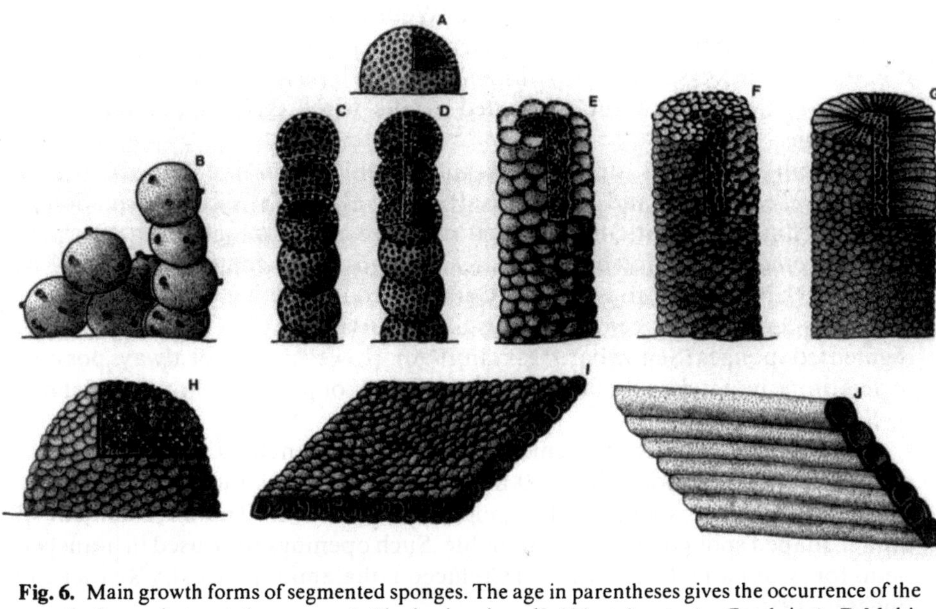


**Fig. 6.** Main growth forms of segmented sponges. The age in parentheses gives the occurrence of the growth forms, but not the genus. **A** Single-chambered? (*Blastulospongia*, Cambrian). **B** Multi-chambered (e.g., *Celyphia*, Cambrian-Triassic, Cretaceous?). **C** Moniliform (e.g., *Colospongia*, Cambrian-Ordovician, Carboniferous-Cretaceous). **D** Cateniform (e.g., *Amblysiphonella*, Cambrian-Recent). **E,G** Monoglomerate (**E**: e.g., *Discosiphonella*, Ordovician, Carboniferous-Cretaceous. **G**: e.g., *Guadalupia* Girty, Permian). **F** Polyglomerate (e.g., *Cystothalamia*, Permian-Triassic). **H** Tuberous (e.g., *Uvanella*, Ordovician, Permian Triassic). **I,J** Sheet-like (e.g., *Ascosymplegma*: sheet-like with tubiform chambers, Permian-Triassic, *Guadalupia zitteliana* Girty: sheet-like with stratiforme chambers, Ordovician, Carboniferous-Triassic)

## Main Groups of Segmented Sponges

Using the mineralogical composition of the rigid skeleton (aragonite, calcite), the presence or absence of a spicular skeleton, and mineralogical composition of spicule (calcite or siliceous), the sphinctozoans have been subdivided into various groups (orders?) by different workers (see Senowbari-Daryan 1990).

a) Sphaerocoelida Vacelet 1979: This term was proposed by Vacelet (1979) for segmented sponges whose skeleton is composed of aragonite or calcite (primary aragonite?) with calcite spicules. Jurassic-Cretaceous. Calcispongea.

Only one family, the Sphaerocoeliidae Steinmann, is included here and contains three Jurassic-Cretaceous genera: *Barroisia* Steinmann, *Sphaerocoelia* Steinmann, and *Thalamopora* Roemer. All three genera have no internal filling structures and no vesiculae. Reitner (1987a) established a new genus – *Muellerithalamia* – whose spicular (monactine, triactine, and caltropse) and rigid skeleton is composed of calcite. According to Reitner, the probable affiliation of this sponge to the Minchinellida (Calcaronea) is based on the type and arrangement of the scleres.

b) Verticillitida Termier and Termier 1977 (including "Murguiathalamida" Reitner and Engeser 1985): Skeleton composed of aragonite with siliceous spicules (Reitner and Engeser 1985) and trabecular internal structure. Some representatives of the group do not have any spicular skeleton. Triassic-Recent (*Vaceletia*: without spicules). Demospongea.

The "Murguiathalamida" are included in the Verticillitida because the nature of the "triaxone" spicules of this group, as described by Reitner and Engeser (1985), cannot be confirmed (Senowbari-Daryan 1989, 1990).

c) Permosphincta Termier and Termier 1974: This term was used by Termier and Termier (1974) for those fossil sphinctozoan sponges whose skeleton is composed of aragonite (if known) without any spicular skeleton. Middle Cambrian-Upper Triassic, Cretaceous? Demospongea?, Calcispongea?.

In fact spicular skeletons are not known (preservation?) from most Cambrian to Tertiary sphinctozoan genera. The occurrence of probably siliceous spicules in several Upper Permian and Upper Triassic "typical" sphinctozoan genera (Senowbari-Daryan 1989) strongly suggests a demospongean nature for the majority of the Permosphincta. However, more information is needed.

d) Pisothalamida Senowbari-Daryan and Rigby 1988: Skeleton composed of aragonite, with monaxon siliceous spicules and pisolitic internal structure. Permian. Demospongea.

No siliceous spicules were known from these types of Paleozoic and Triassic sphinctozoans. Therefore, the group was established by the authors to differentiate the demospongid sphinctozoans with a pisolitic type of internal structure from other sphinctozoan sponges whose systematic positions remain uncertain. This "order" might be included in the Verticillitida, once the pisothalamids are better known.

e) "Hadromerida" (pars): Reitner (1987b) described the sphinctozoan genus *Cassianothalamia* from the Upper Triassic Cassian formation. Its skeleton is composed of high-Mg-calcite with microscleres and (uncertain) macroscleres. According to Reitner, the scleres were originally siliceous and the type of microscleres (asters) is typical of the Order Hadromerida. Triassic-Cretaceous? Demospongea.

We could not find undoubted microscleres in *Cassianothalamia*, furthermore the interpretation of some structures as microscleres and macroscleres by Reitner (1987b) remains fairly uncertain (see Senowbari-Daryan 1989, 1990). *Cassianothalamia* differs from representatives of the Verticillitida in the composition of the rigid skeleton (high-Mg-calcite) and in possession of microscleres, which are not known from the Verticillitida (preservation?). An affiliation of *Cassianothalamia* with the Verticillitida cannot be excluded.

In addition to *Cassianothalamia*, there are numerous mainly Upper Triassic sphinctozoan sponges whose skeletons are composed of high-Mg-calcite with a granular microstructure. No spicules have been found in any of them. Representatives of this group may or may not have internal filling structures, for example *Ceotinella* Pantic and *Zardinia* Dieci et al. do and *Uvanella* Ott and *"Colospongia" andrusovi* Jabloňsky do not. Using the composition of the rigid skeleton it is not possible to place these sponges certainly into either the Demospongea or the Calcispongea at the present time.

f) Guadalupiida Termier and Termier 1977: Skeleton composed of calcite (primarily aragonite, according to Finks 1983) without spicules, astrorhizoid system developed (e.g., *Guadalupia zitteliana* Girty). Ordovician and Permian. Sclerosponges (Demospongea).

An astrorhizoid system is not supposed to occur in sphinctozoan sponges. Such a canal system is typical of stromatoporoids and sclerosponges. The type of growth and astrorhizoid system of *Guadalupia* is similar to the Chaetetida. Most probably *Guadalupia* belongs to the sclerosponges.

The "sphinctozoans" are a highly polyphyletic group of sponges; thus the term "sphinctozoa" does not have any systematic significance and cannot be used as a systematic category. The systematic position of most sphinctozoan genera still remains uncertain. Siliceous spicules in some "typical" sphinctozoan sponges (see Senowbari-Daryan 1989, 1990) suggest that these genera should be placed into the Demospongea (Ceractinomorpha). By inference the affiliation of the majority of Paleozoic and Triassic sphinctozoan genera is with the Demospongea.

## Occurrence

Until a few years ago, sphinctozoans were known only from rocks of Carboniferous to Cretaceous ages. Discovery of new sphinctozoan genera in the Cambrian (Pickett and Jell 1983; Kruse 1987, Debrenne and Wood 1990), Ordovician (Rigby and Potter 1986; Rigby and Webby 1988; Webby and Rigby 1984), Silurian (de Freitas 1987), Devonian (Pickett and Rigby 1983; Rigby and

Blodgett 1983), and Tertiary rocks (Clausen 1982; Pickett 1982), as well as the occurrence of the thalamid sponge *Vaceletia crypta* (Vacelet 1977) in modern oceans, proves the existence of segmented sponges from the Cambrian up to modern times. Sphinctozoans are known from all continents, with the exception of Antarctica. Sphinctozoans were important frame builders, especially in the Permian and Triassic. Of the about 108 genera reported in the literature, more than 70 genera are known from Permian and Triassic rocks (Fig. 7).

Several evolutionary crises and changes in generic composition of segmented sponge assemblages have been recognized in the geological record. There were several occasions when nearly all or at least the majority of genera became extinct and new genera arose. The most important crises are at the Ordovician/Silurian, the Permian/Triassic, and the Carnian/Norian boundaries or in Lower Norian time (see Riedel and Senowbari-Daryan this Vol.), at the Triassic/Jurassic and the Cretaceous/Tertiary-boundaries (Fig. 7).

## Paleoecology

The modern sphinctozoan sponge – *Vaceletia crypta* (Vacelet) – lives in cryptic habitats in the outer slope (15–38 m depth) of coral reefs of the Indopacific (Vacelet 1979; Basile et al. 1984). Fossil sphinctozoans inhabited similar biotopes. For example, a zone with sponges (sphinctozoans and inozoans) associated with

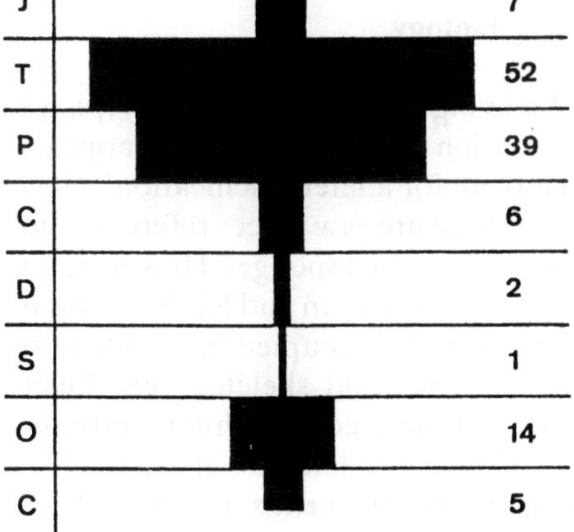


**Fig. 7.** Frequency of described genera segmented sponges in the geological record. The *numbers* indicate the generic diversity from the Cambrian up to Recent

"hydrozoans", "tabulozoans", bryozoans, thamnasteroid corals, and miliolid foraminifers has been described in Upper Triassic carbonate platform deposits in Sicily (Senowbari-Daryan et al. 1982; Di Stefano and Senowbari-Daryan 1985). Sedimentological (mostly micrite) and paleontological data indicate that rocks of this zone accumulated in a low-energy and deeper-water (below wave-base) environment. These rocks contrast sharply to the "coral-algal-biolithite" developed on the platform. The porostromate algae (solenoporaceans and diverse "*Cayeuxia*") and the dasycladaceans associated with high-growing corals ("*Thecosmilia*"), megalodonts and agglutinated foraminifers, are very rare within the "sponge-bryozoan-zone" (see Senowbari-Daryan et al. 1982).

A similar biozonation is reported from Upper Rhaetian reefs of the Northern Calcareous Alps (Schäfer 1979; Senowbari-Daryan 1980; Piller 1981). Sphinctozoans are rare or missing in the lagoonal facies. Reitner and Engeser (1985) described a similar environment for thalamid sponges in the Cretaceous of Spain. Comparable associations of organisms have been also reported from Permian reefs of southern Europe and northern Africa (Flügel et al. 1984; Senowbari-Daryan and Di Stefano 1988; Senowbari-Daryan 1990). The association of sphinctozoan sponges and algae (*Archaeolithoporella*, dasycladaceans) in the Permian reefs indicates that these sponges lived within the photic zone (Senowbari-Daryan and Rigby 1988). Thalamid sponges were shallow water organisms (normally about 15–50 m deep) living associated with other organisms in the low energy zone of reefs, usually at the reef slope.

Kruse (1987, p. 550) concluded that Cambrian sphinctozoans did not live in reef environments, as was typical for late Paleozoic and Mesozoic forms. Sedimentological and paleontological data indicate that during the uppermost Lower Permian to the Middle Permian sphinctozoans changed biotopes. Sphinctozoans of the Ordovician up to Carboniferous are found in shelf sediments, but Permian and Mesozoic genera are restricted to reef limestones. These distribution document migrations of sphinctozoan sponges from open shelf regions to reef biotopes during earth history.

## Paleobiology

The living tissue of *Vaceletia crypta* (Vacelet) fills a series of chambers and the formation "of new chambers appears first as an organic matrix, which is the template for a later calcification" (Vacelet 1979, p. 483).

There are few direct references to observations of living tissue evidence in fossil segmented sponges. However, some evidence points out (see Reitner 1987a; Senowbari-Daryan and Rigby 1988) that the soft body in fossil representatives of the group also occupied the inside of several chambers, like in modern *Vaceletia crypta*. The rigid skeleton (exoskeleton) of fossil sphinctozoan sponges was probably covered by a thin membrane, as in modern *Vaceletia*.

Concentric laminae of the rigid skeleton inclose the spicular skeleton in the Upper Triassic genus *Celyphia* Pomel. This relationship indicates that the spicular skeleton was secreted first. The rigid skeleton developed second.

Water circulation through different shapes of segmented sponges was probably also different. Three types of construction may be recognized (Fig. 8):

### 1. *Amblysiphonella type (Fig. 8a)*

In *Amblysiphonella* (representative of the porate group) and *Girtyocoelia* (representative of the aporate group) an axial tube or a bundle of tubes runs through the whole sponge. Water apparently passed through inhalant canals (exopores or ostia) into chamber interiors and continued (after circulation in choanocyte chambers) through the chambers to exit at the apopores (endopores, Finks 1983) into the axial channel (Spongocoel). Water left the sponge through the osculum, in a pattern like that in the living *Vaceletia crypta* (Vacelet). This type of water pathway occurs in Cambrian?/Ordovician up to the Recent thalamid sponges.

### 2. *Sphaerocoelia type (Fig. 8b)*

In the Jurassic-Cretaceous genus *Sphaerocoelia* Steinmann and in the Cambrian genus *Nucha* Pickett and Jell there is no walled axial tube (spongocoel), such as is well developed in *Amblysiphonella*. Rather chamber roofs are pierced by a large

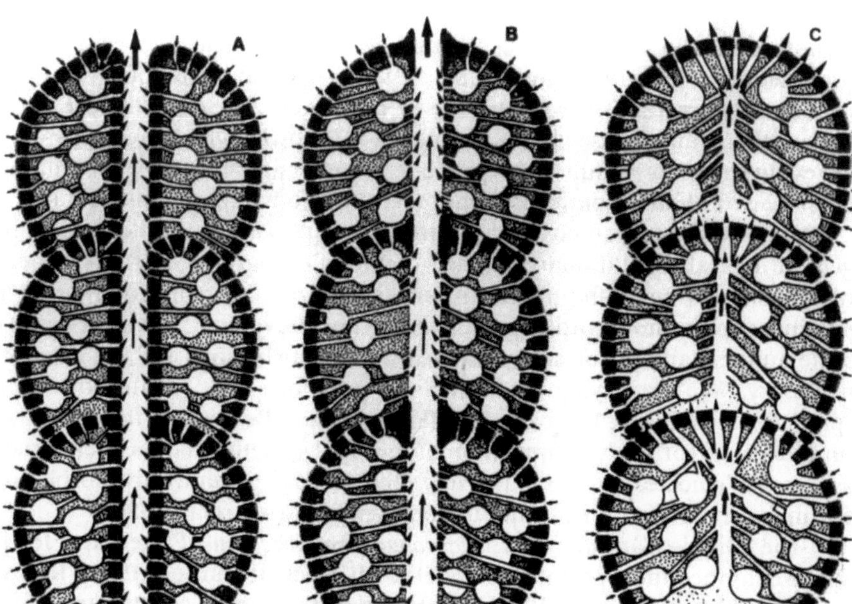


**Fig. 8.** Possible water current patterns within segmented sponges. **A** *Amblysiphonella* type. **B** *Sphaerocoelia* type. **C** *Colospongia* type

opening or osculum (cryptosiphonate sensu Seilacher 1962). The rim around the osculum (see Hillmer and Senowbari-Daryan 1986, Pl. 6, Fig. 4) points to the probable existence of a central opening surrounded by the soft sponge body through the whole sponge, but that opening lacked a calcified wall (Fig. 8b). This type of construction is developed within the Cambrian porate *Nucha* and probably also in *Jawonya*, and *Wagima*, as well as in the Jurassic-Cretaceous genus *Sphaerocoelia*. Some Ordovician and Triassic genera such as *Porefieldia?* or *Thaumastocoelia?* might have had this type of circulatory pattern too. Aporate *Sollasia* Steinmann, and probably *Follicatena* Ott, and *Henricellum* Wilckens, might have had this type of canal system.

### *3. Colospongia type (Fig. 8c)*

*Colospongia*, a porate form, and *Celyphia*, an aporate form, have chambers that are pierced either by numerous small pores (*Colospongia*) or by a few ostia (*Celyphia*). There is no indication of a spongocoel or osculum in either. It is difficult to understand the manner of water circulation in these genera. We believe that circulation in *Colospongia* was generally similar to that in *Amblysiphonella* or *Sphaerocoelia* (Fig. 8c). Circulation in *Celyphia* and similar genera must have been essentially the same as in *Colospongia* (see Ziegler and Rietschel 1970, p. 32).

## Phylogeny of Segmented Sponges

Hypothetical phylogenetic reconstruction for sphinctozoan sponges were proposed by Ott (1967) and Finks (1970). Possible relationships between Paleozoic sphinctozoans were discussed by Pickett and Jell, (1983) Rigby and Potter (1986), and Kruse (1987). Zhang (1987) showed what he considered to be the evolution of the family Sebargasiidae. Termier and Termier (1977) discussed possible evolutionary patterns within the calcareous sponges in general.

Because of the polyphyletic nature of the sphinctozoans it is impossible to give an overall evolutionary pattern of the whole group until the systematic relationships of the different groups are recognized. More information to fill in possible gaps in the record (e.g., in the Silurian, Devonian, Lower Triassic), and better understanding of structural features, distribution, and occurrence are necessary. We are trying to discuss the possible relationships of "typical" sphinctozoans during Paleozoic and Triassic time. Because of the lack of structural features (primary composition of the rigid skeleton and its microstructure, and lack of spicules) in most Paleozoic sphinctozoan genera, only morphological features can be used for this discussion (Senowbari-Daryan 1990).

The Middle Cambrian *Blastulospongia monothalamus* Pickett and Jell has a single chamber and a perforated skeletal wall. Pickett and Jell (1983), Rigby and Potter (1986), and Kruse (1987) reported that the simply constructed *Blastulospongia* might represent the ancestor of at least most Paleozoic porate sphinctozoans. The assignment of this organism to the sponges is not certain. If

*Blastulospongia* is a true sponge, it is the oldest and simplest sebargasid sponge that we know. All other Cambrian sphinctozoan sponges described by Kruse (1987) are also single-chambered sponges and display sebargasiid construction.

*Amblysiphonella* has a spongocoel and perforated chamber walls and may have derived from the Cambrian *Nucha* Pickett and Jell. *Amblysiphonella* may represent the ancestral form of most Paleozoic and Triassic sphinctozoan sponges. Evidently the cryptosiphonate construction in *Nucha* was not a favorable feature, because this kind of skeletal architecture has not been recognized within sphinctozoan sponges of Ordovician age. However, such architecture reappears independently in the Jurassic-Cretaceous *Sphaerocoelia* (Calcispongea).

One line of evolution from *Amblysiphonella* may have led to the Ordovician genus *Amblysiphonelloides* Rigby and Potter, by developing tubular internal structures, as well as to those younger genera that display similar internal structures (Polytholosiidae). The Silurian genus *Rigbyspongia*, with trabecular filling structures and the Devonian genera *Hormospongia* and *Radiothaolamos* with reticular or septal structures could have derived also from the generalized *Amblysiphonella*. Thus sebargasiid sphinctozoans might be the ancestoral group of most later porate sphinctozoans at least up to the Triassic.

Aporate sphinctozoans appear first in the Ordovician (*Cliefdenella*, *Cystothalamiellea*, *Girtyocoelia*, etc., see Rigby and Potter 1986). Rigby and Potter (1986) concluded that the aporate wall of early representatives of the group (*Cliefdenella*, *Girtyocoelia*) developed by a loss of pores within a porate wall sometime in the Middle Ordovician.

It seems unlikely that *Cliefdenella* which has a complex skeleton, was an ancestral genus of the later aporate sponges. All other presently known Ordovician aporate genera (*Girtyocoelia*, *Cystothalamiella*, *Porefieldia*) have more complex skeletal structure than many later genera (e.g., the Carboniferous-Permian *Sollasia* or the Triassic *Celyphia*). It is probable that the ancestral aporate genus (perhaps a form like *Sollasia*) might yet be found in Lower Ordovician or even in Cambrian rocks.

*Acknowledgments.* This research is part of the project Evolution of reefs, F142/49-2 at the Institute of Paleontology in Erlangen and was supported by the Deutsche Forschungsgemeinschaft. Thanks to Prof. Dr. E. Flügel, Dr. W. Ch. Dullo, Dipl. Geol. H.W. Kügel, Dipl. Geol. P. Riedel (all Erlangen), Prof. Dr. G.F. Tietz (Hamburg) and Prof. Dr. J.K. Rigby (Provo/USA) for their help and discussions.

## References

Basile LL, Cuffey RJ, Kosich DF (1984) Sclerosponges, pharetronids and sphinctozoans (relict cryptic hard-bodied Porifera) in the modern reefs of Enewetak Atoll. J Paleontol 58(3):636–650

Clausen CK (1982) *Wienbergia*, new genus for *Barroisia faxensis* (Porifera, Demospongea) from the Middle Danian of Denmark. Bull Geol Soc Den 29:111–115

Debrenne F, Lafuste J (1972) Microstructure du squelette de quelques Sphinctozoaires (Exemples pris dans le Permien du Jbel Tebaga et le Crétacé d'Angleterre. Bull Soc Géol Fr 14(7):325–330

Debrenne F, Wood R (1990) A new Cambrian sphinctozoan sponge from North America, its relationship to archaeocyaths and the nature of early sphinctozoans. Geol Mag 125(5):435–443
De Freitas T (1987) A Silurian sphinctozoan sponge from east-central Cornwallis Island, Canadian Arctic. Can J Earth Sci 24:840–844
De Laubenfels MW (1955) Porifera. Treatise Invertebr Paleontol Part *E*: Geological Society of America, University of Kansas, pp 21–112
Di Stefano P, Senowbari-Daryan B (1985) Upper Triassic dasycladaceans (green algae) from the Palermo Mountains (Sicily, Italy). Geol Rom 24:189–220
Finks RM (1970) The evolution and ecologic history of sponges during Palaeozoic times. In: Fry WG (ed) The biology of the Porifera. Symp Zool Soc Lond 25:3–22
Finks RM (1983) Pharetronida: Inozoa and Sphinctozoa. In: Broadhead TW (ed) Sponges and spongiomorphs. University of Tennessee, pp 55–59
Flügel E, Lein R, Senowbari-Daryan B (1978) Kalkschwämme, Hydrozoen, Algen und Mikroproblematika aus den Cidarisschichten (Karn, Ober-Trias) der Mürztaler (Alpen (Steiermark) und des Gosaukammes (Oberösterreich). Mitt Ges Geol Bergbaustud Österr 25:153–195
Flügel E, Kochansky-Devidé V, Ramovš A (1984) A Middle Permian calcisponge/algal/cement reef: Straza near Bled, Slovenia. Facies 10:179–256
Gautret P (1985) Organisation de la phase minérale chez *Vaceletia crypta* (Vacelet) Démosponge, Sphinctozoaire Actuelle. Comparaison avec de formes aragonitiques du Trias de Turquie. Geobios 18(5):553–562
Hartman WD, Wendt J, Wiedenmayer F (1980) Living and fossil sponges. Sedimentata VIII:1–274
Hillmer G, Senowbari-Daryan B (1986) Sphinctozoa aus dem Cenoman von Mühlheim-Broich, SW-Westfalen. Mitt Geol Paläontol Inst Univ Hamb 61:161–187
Jones WC (1979) The microstructure and genesis of sponge biominerals. Colloq Int Cent Natl Rech Sci 291:425–447
Kruse PD (1987) Further Australian Cambrian sphinctozoans. Geol Mag 124(6):543–553
Ott E (1967) Segmentierte Kalkschwämme (Sphinctozoa) aus der alpinen Mitteltrias und ihre Bedeutung als Riffbildner im Wettersteinkalk. Bayer Akad Wiss Math Naturwiss Kl Abh N F 131:1–96
Pickett J (1982) *Vaceletia progenitor*, the first Tertiary sphinctozoan (Porifera). Alcheringa 6:241–247
Pickett J, Jell PA (1983) Middle Cambrian Sphinctozoa (Porifera) from New South Wales. Mem Assoc Aust Paleontol 1:85–92
Pickett J, Rigby JK (1983) Sponges from the early Devonian Garra Formation, New South Wales. J Paleont 57(4):720–741
Piller W (1981) The Steinplatte reef complex, part of an Upper Triassic carbonate platform near Salzburg, Austria. SEPM Spec Publ 30:261–290
Reid REH (1967) Spicules in the Skeleton of *Tremacystia d'orbigny* (HINDE). Nature 15:875–876
Reid REH (1968) *Tremacystia*, *Barroisia* and the status of Sphinctozoida (Thalamida) as Porifera. Paleontol Contrib Univ Kansas 34:1–10
Reitner J (1987a) Phylogenie und Konvergenzen bei rezenten und fossilen Calcarea (Porifera) mit einem kalkigen Basalskelett ("Inozoa", "Pharetronida"). Berl Geowiss Abh A 86:87–125
Reitner J (1987b) A new calcitic sphinctozoan sponge belonging to the Demospongiae from the Cassian Formation (Lower Carnian; Dolomites, northern Italy) and its phylogenetic relationship. Geobios 20(5):571–589
Reitner J, Engeser TH (1985) Revision der Demospongier mit einem thalamiden, aragonitischen Basalskelett und trabeculärer Internstruktur ("Sphinctozoa" pars). Berl Geowiss Abh A 60:151–193
Rigby JK, Blodgett B (1983) Early Middle Devonian sponges from the McGrath quadrangle of west-central Alaska. J Paleontol 57(4):773–786
Rigby JK, Potter AW (1986) Ordovician sphinctozoan sponges from the eastern Klamath Mountains, northern California. J Paleontol Suppl 60(4):1–47
Rigby JK, Webby BD (1988) Late Ordovician Sponges from the Malongulli Formation of central New South Wales, Australia. Palaeontogr Am 56:1–147
Rigby JK, Fan J, Zhang W (1988) The sphinctozoan sponge *Intrasporeocoelia* from the Middle and Late Permian of China: re-examination of its filling structures. J Paleontol 62(5):747–753

Russo F (1981) Nouve spugne calcaree triassiche di campo (Cortina d'Ampezzo, Belluno). Boll Soc Paleontol Ital 20:3–17

Schäfer P (1979) Fazielle Entwicklung und palökologische Zonierung zweier obertriadischer Riffstrukturen in den Nördlichen Kalkalpen ("Oberrhät"-Riff-Kalke, Salzburg). Facies 1:3–245

Seilacher A (1962) Die Sphinctozoa, eine Gruppe fossiler Kalkschwämme. Akad Wiss Lit Abh Math Naturwiss Kl 1961 (10):720–790

Senowbari-Daryan B (1980) Fazielle und paläontologische Untersuchungen in oberrhätischen Riffen (Feichtenstein- und Gruber-Riff bei Hintersee, Salzburg, Nördliche Kalkalpen). Facies 3:1–237

Senowbari-Daryan B (1989) Spicula in segmentierten Schwämmen. Berl Geowiss Abh A 106:473–515

Senowbari-Daryan B (1990) Die systematische Stellung der thalamiden Schwämme und ihre Bedeutung in der Erdgeschichte. Münchner Geowiss. Abh 21:1–324

Senowbari-Daryan B, Di Stefano P (1988) Mikrofacies and sphinctozoan assemblage of some Lower Permian breccias from the Lercara Formation (Sicily). Rev Ital Paleont Stratigr 94(1):3–34

Senowbari-Daryan B, Rigby JK (1988) Upper Permian segmented sponges from Djebel Tebaga, Tunisia. Facies 9:171–250

Senowbari-Daryan B, Schäfer P (1983) Zur Sphinctozoen-Fauna der obertriadischen Riffkalke ("Pantokratorkalke") von Hydra, Griechenland. Geol Paleontol 17:179–205

Senowbari-Daryan B, Schäfer P, Abate B (1982) Obertriadische Riffe und Rifforganismen in Sizilien. Facies 6:165–184

Steinmann G (1882) Pharetronen-Studien. N Jb Miner etc 1882 II:139–191

Termier H, Termier G (1974) Spongiaires Permiens du Djebeltebaga (Sud Tunisien). CR Acad Sci Paris 279:247–274

Termier H, Termier G (1977) Structure et evolution des Spongiaires Hypercalcifies du Paleozoique superior. Mém Inst Géol Univ Louv 2:57–109

Vacelet J (1977) Une nouvelle relique du Secondaire: un représentant actuel de Éponges fossiles Sphinctozoaires. CR Acad Sci Paris 285 (Ser D): 509–511

Vacelet J (1979) Description and affinités d'une éponge sphinctozoaire actuelle. Colloq Int Cent Natl Rech Sci 291:483–493

Vacelet J (1985) Coralline sponges and the evolution of Porifera. In: Morris SC, George JD, Gibson R, Platt HM (eds) The origin and relationship of lower invertebrates. Syst Assoc 28:1–13

Webby BD (1969) Ordovician stromatoporoids from New South Wales. Palaeontology 12(4):125–135

Webby BD, Rigby JK (1985) Ordovician sphinctozoan sponges from central New South Wales. Alcheringa 9:209–220

Wendt J (1979) Development of skeletal formation, microstructure, and mineralogy of rigid calcareous sponges from the late Palaeozoic to Recent. Colloq Int Cent Natl Rech Sci 291:449–457

Wendt J (1984) Skeletal and spicular mineralogy, microstructure and diagenesis of coralline calcareous sponges. Palaeontogr Am New York 54:326–336

Zhang W (1987) A new genus *Neoguadalupia* with notes on connections of interrelated genera in Sebargassiidae, Sphinctozoa. Sci Geol Sin 7(3):231–238

Ziegler B, Rietschel S (1970) Phylogenetic relationships of fossil calcisponges. In: Fry WG (ed) The biology of the Porifera. Symp Zool Soc Lond 25:23–40

# A Cautionary Tale: Difficulties in Inferring Living Morphology from Post-Depositional Appearance

J. PICKETT

## Introduction

During fieldwork on One Tree Island in 1973 a specimen taken to be a calcareous sponge with a conspicuous array of spicules was discovered among coral being broken up for an analysis of the invertebrates living among it, particularly those causing degradation of the coral skeleton. This specimen, initially interpreted as a sclerosponge, was set aside because of its small size and the lack of any supplementary material. Because of the circumstances on One Tree Island at the time, the specimen was not preserved in a manner suitable for soft tissue examination. Much later, further specimens were collected on Lady Elliott Island, but these were dead and eroded skeletons collected on the beach.

The cryptic habit of the species, ultimately recognised as the foraminiferan *Sporadotrema mesentericum* (Carter 1880), means that it is capable of being preserved in cavities of reefal limestones. The association with siliceous spicules is likely to render interpretation of this and similar forms in the fossil record less than straightforward. The skeleton of the species and its associated spicules are described in some detail, and their implications for interpretation of fossil forms are discussed. The description presents a process of secondary thickening of the wall not previously described in *Sporadotrema*, although Hickson (1911, p. 449) implied a progressive thickening in *S. cylindricum* (Carter 1880).

## Description

The species forms small, deep rose-pink, irregular masses, reaching a maximum diameter of at least 33 mm. The main skeleton is composed of calcium carbonate in the form of calcite, and is made up of anastomosing palisades which are 1–2 mm wide at their distal margins, and possibly exceeding 3 mm in the proximal parts. The palisades of smaller specimens are fairly open (Fig. 1), while those of larger

**Fig. 1.** Appearance of *Sporadotrema mesentericum* in ordinary light. Specimen AM Z5002, Long Bank, One Tree Island, 8 m. **A-C** Lateral and vertical (**C**) views, × 3. **D** Unbroken oxeote spicules projecting from crestal furrow (even these few have now been broken off with repeated handling), × 5. **E** Apical view of palisade showing crestal furrow and lateral branches. Spicules are in situ, but have been broken off with handling, × 8. **F** fractured surface of palisade showing internal chambers and thickened, porous outer wall, × 8

J. Reitner and H. Keupp (Eds.)
Fossil and Recent Sponges

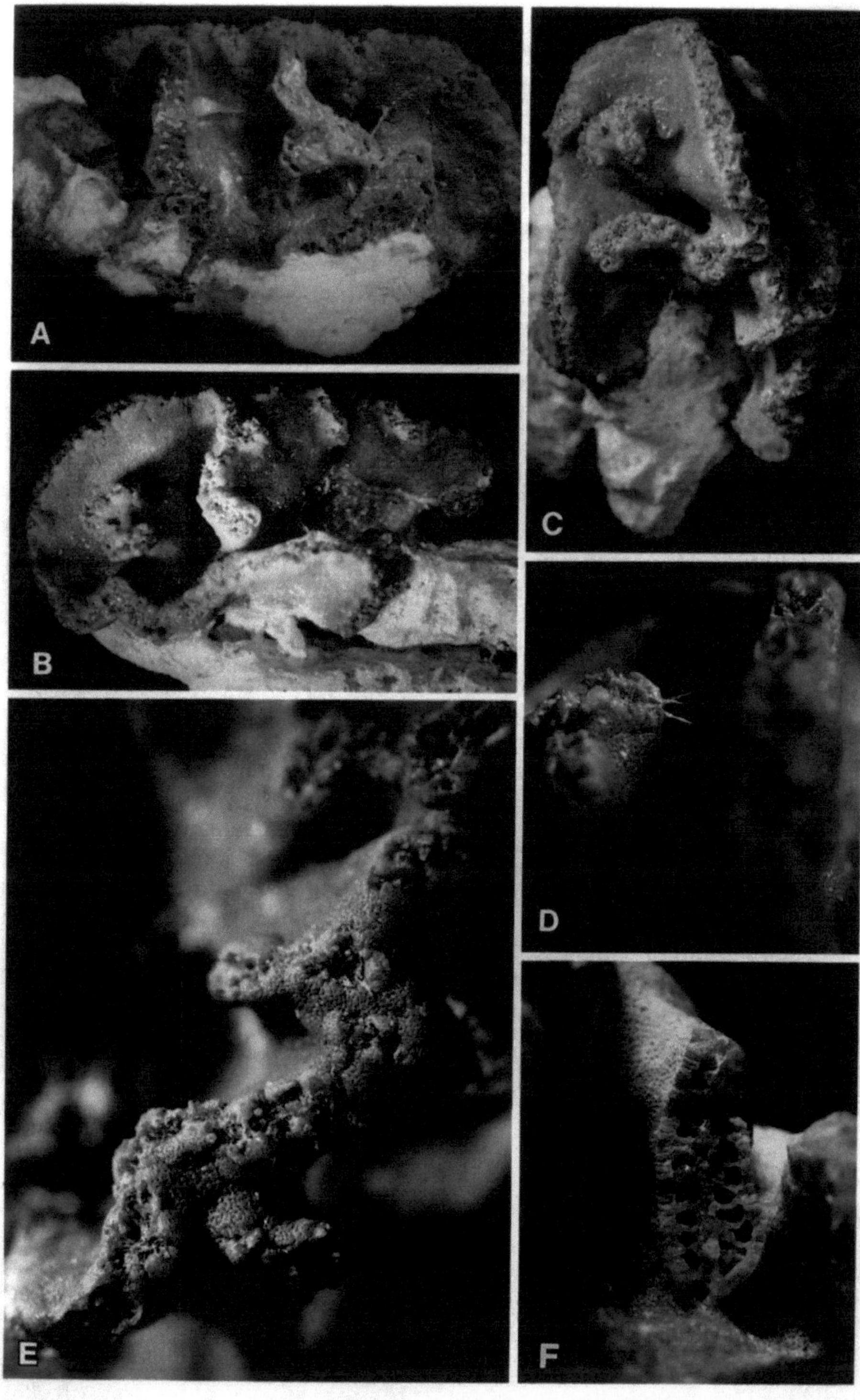
A
B
C
D
E
F

(and presumably older) specimens are much more closely spaced (Fig. 2C). The sides of the palisades are smooth, somewhat undulose, and entirely covered by round or oval pores 0.02–0.08 mm in diameter and separated by a similar distance. Along the crests of the palisades is a more or less continuous but irregular furrow, protected by short, finger-like projections 0.1–0.2 mm in diameter and less than 0.5 mm long, directed roughly inwards and upwards (Figs. 1E, 3B). At the base of the furrow are apertures which are obscured by the spicules and the finger-like projections; measurements made on sections of the skeleton suggest diameters of ca. 0.1 mm. Protruding from the distal furrow are relatively long, acutely terminated siliceous spicules,extending up to 1 mm beyond the distal margin of the palisade (Fig. 1D).

Internally the palisade is divided into a series of irregular chamberlets with convex-upward walls. In longitudinal section (Fig. 2D) the details of the convex chamber wall are clear on the outer side, but at the centre the relations are rather confused because of the inclusion of the finger-like projections into the internal structure with the additions of new chamberlets.

Growth takes place by addition of blister-like chamberlets, generally at the top of the palisades. The undulose surfaces of the palisades reflect the shape of each addition. However, new chambers can be added at the base of a wall, presumably to strengthen it, and new palisades may be produced by lateral deposition of new chambers on an older palisade. These latter two processes account for the denser nature of older parts of the skeleton.

In thin section there is a clear distinction in structure between those parts of the skeleton which either presently form or earlier formed part of the outer wall, and those which are entirely internal, or lay in the crestal furrow. The former are penetrated by pores normal to the surface, while the latter are imperforate. This is not to say that there is no internal communication between chambers; that is achieved by relatively large pores 0.4–0.6 mm in diameter, but which are relatively rare (one or two per chamber), in contrast to the closely spaced pores of the external wall. The outer wall is much thicker than the internal ones, and is divided into two distinct zones. The outer, thicker zone is traversed by the pores visible on the exterior, 0.032 to 0.064 mm in diameter; this zone may reach 0.36 mm in thickness. In the inner zone the pores divide into much finer canals, 0.012 to 0.014 mm in diameter, so that the internal openings of the pores are much finer than those on the exterior. This zone is of more constant thickness, 0.056 to 0.064 mm, and remains so, whereas the outer zone shows distinct variations in density of pigmentation, suggesting addition of layers of calcite during growth. Within

→

**Fig. 2.** Appearance of *Sporadotrema mesentericum* in thin section. Specimens from beach shingle, Lady Elliott Island. **A** Section of outer wall, showing variation in pigmentation of secondary wall, and internal dark line of primary wall continuous with that of non-porous internal walls. Specimen AM Z5005, × 30. **B** Section of part of palisade showing dark line of primary wall and blister-like shape of chambers. Specimen AM Z5006, × 30. **C** Section of slightly eroded specimen AM Z5006 and part of coral substrate, × 7. **D** Palisade of same specimen with eroded apex, showing shape of chambers, × 30

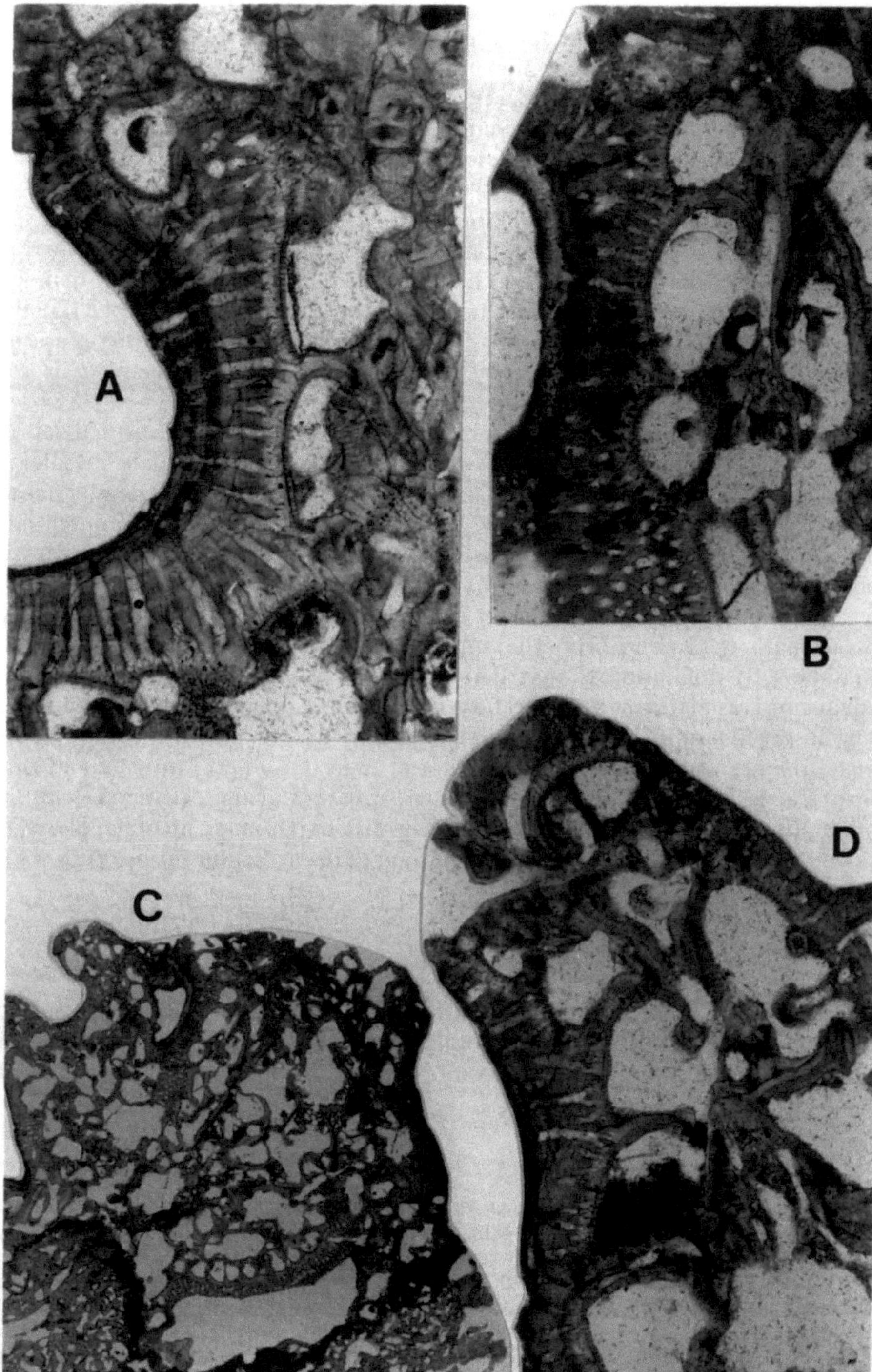
A
B
C
D

the inner zone there is a dark line (Fig. 2B), which lies nearer to the outer zone than to the inner edge of the wall, and which, by analogy with the septa of corals, is taken to represent the locus of initial mineralisation, implying that the inner zone of the wall is deposited first, mineralisation occurring on both sides of the dark line. This dark line can be traced continuously from the porous outer wall to the imperforate inner parts of the skeleton (Fig. 2A,B). The youngest chambers, as shown clearly on the broken wall of a chamber on the top of a palisade (Fig. 3C), consist only of the inner layer with fine pores. This confirms that the inner layer is formed prior to deposition of the more coarsely porous outer wall. A careful examination of the very youngest chambers on the original specimen demonstrates that the fine pores of the inner wall are initially exposed, but rapidly covered by the more coarsely porous outer wall. In thin section some of the apical siliceous spicules have occasionally been observed, either embedded in the calcitic skeleton, or lying free in the chambers.

Most of the siliceous spicules become separated when the soft tissue is removed by treatment (or by decay). Spicules occur in two sites. Firstly, the lateral walls carry a thin investment of soft tissue, in which there are very slender oxeas lying parallel to the surface (Fig. 4B,C), running between the pores. The outer wall of dead specimens may bear an anastomosis of fine grooves showing the course of the spicules, but these impressions do not appear to be left on all specimens. Secondly, the largest spicules occur in the crestal furrow. These spicules only occasionally become incorporated into the basal skeleton or remain within the chambers. It thus appears that they are either resorbed, or retained in the advancing crestal furrow and merely pushed forward as new chambers are added. The observed terminations of all long spicules are oxeote (most are now broken); dichotriaenes with a directional ray shorter than those that bifurcate and calthrops at least are also present. The microstructure of the calcitic skeleton is minutely and rather irregularly fibrous (Fig. 5B), or it is made up of flat plates of calcite which appear to be orientated normal to the accretion surface (Fig. 5A).

## Discussion

The above description has been presented avoiding morphological terms specific to foraminifera. In this way the description becomes applicable to a hypothetical sponge, in which the pores of the wall become prosopores, with an excurrent

**Fig. 3.** Appearance of *Sporadotrema mesentericum* in SEM. Specimen AM Z5002, Long Bank, One Tree Island. **A** Section of mature part of palisade, showing imperforate inner skeleton and primary (finely perforate) and secondary (coarsely perforate) layers of outer wall, uncoated, × 40. **B** Apical view of palisade showing crestal furrow with in situ spicules (all protruding rays broken off), finger-like processes and porous outer wall. The fine pores of the primary layer are visible only on the most apical portions of the youngest chambers. Pt-coated, × 80. **E** Spicules in crestal furrow, calthrops near centre. Pt-coated, × 150. **C** Horizontal section with coarsely perforate wall, × 40. **D** Agglutinated foreign sponge spicules; × 80

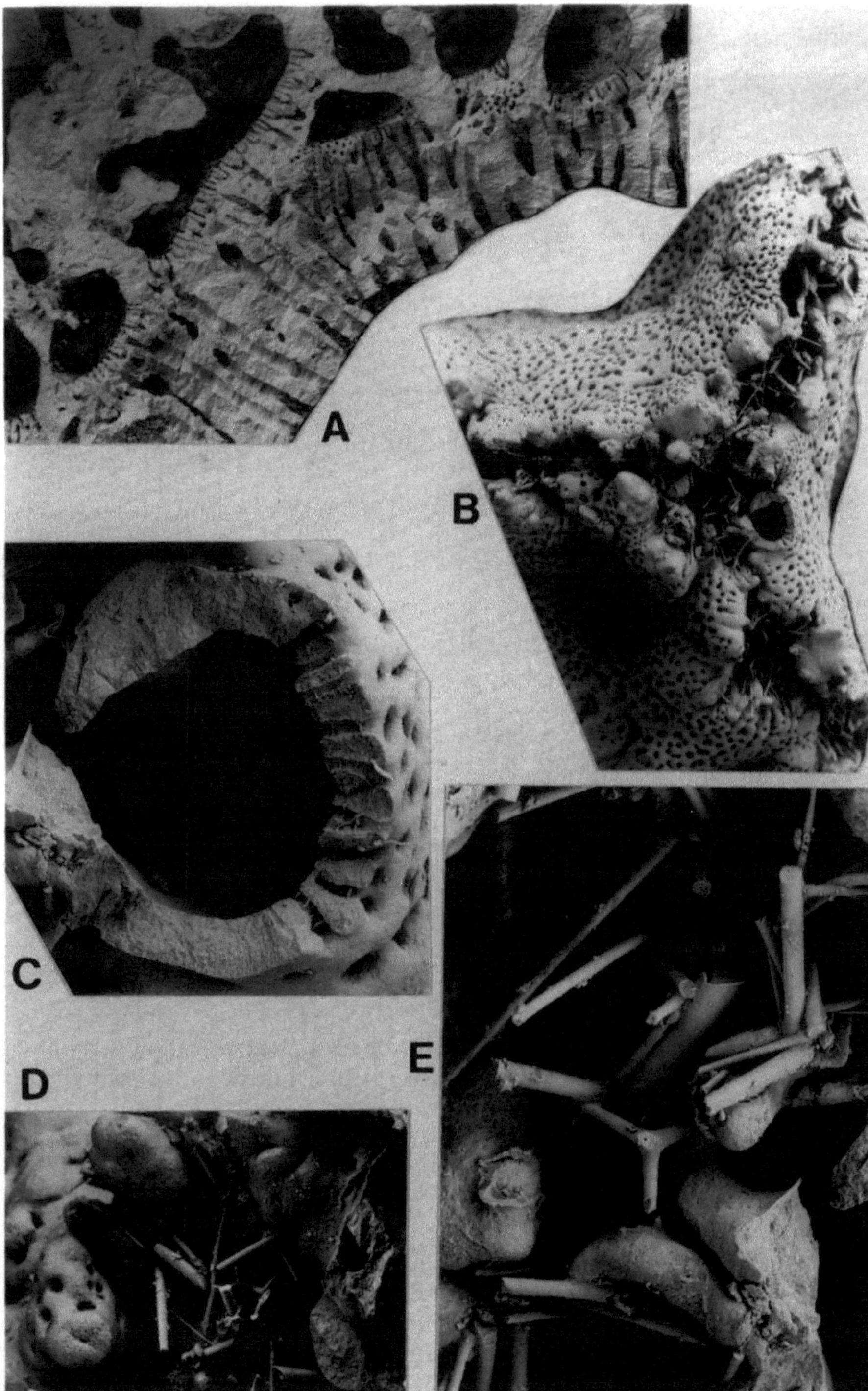
A
B
C
D
E

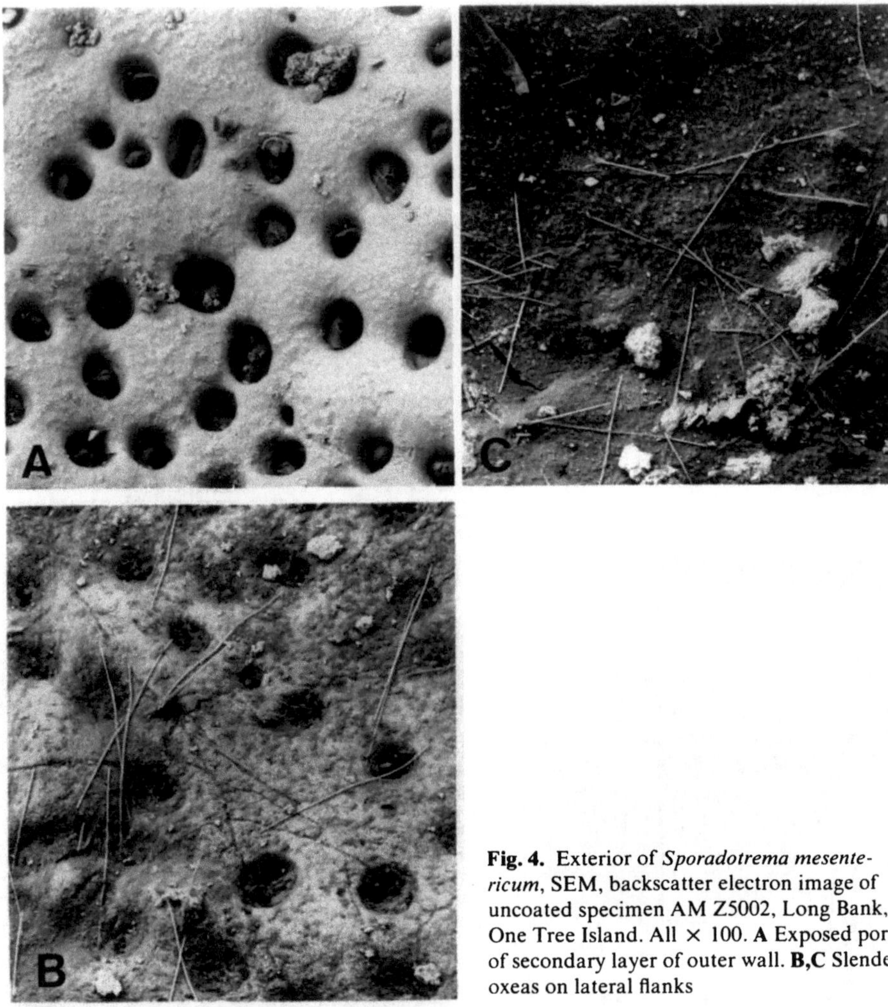


**Fig. 4.** Exterior of *Sporadotrema mesentericum*, SEM, backscatter electron image of uncoated specimen AM Z5002, Long Bank, One Tree Island. All × 100. **A** Exposed pores of secondary layer of outer wall. **B,C** Slender oxeas on lateral flanks

system emptying through the crestal furrow. Such an interpretation is favoured by the general presence of spicules, and their partition into two areas, with coarser spicules in the crestal furrow.

When such a form is considered in a geological context, in which there is no possibility of examining the soft tissue, the difficulties of interpretation become more real. The size of the pores is exceptionally great (0.02–0.08 mm) for a foraminiferan, whereas the normal range lies an order of magnitude lower.

Within a sponge context, the association of a calcareous skeleton and siliceous spicules, particularly styles, is immediately reminiscent of the sclerosponges. In addition, both calcitic (*Acanthochaetetes*) and aragonitic (most others) basal skeletons are known from the group; *Merlia* even has a skeleton including both minerals (Hartman 1983). However, the basal skeleton of scle-

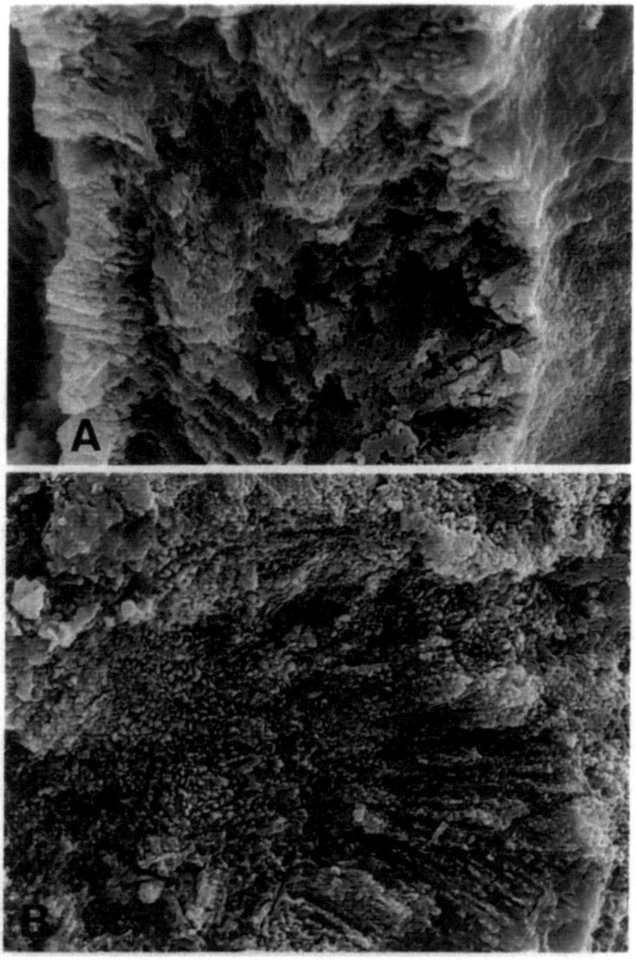


**Fig. 5.** Microstructure of *Sporadotrema mesentericum*. Both fields are enlarged sections of Fig. 3C, and are of the primary wall layer. Both × 4000 approx. **A** Area between two prosopores, showing platy fabric of calcite. **B** Imperforate section of wall at top of Fig. 3C, showing apparently fibrous fabric

rosponges is caliculate, in some cases also tabulate, and its manner of growth is fundamentally continuous. The functional parts of the sponge are more or less external to the skeleton, lying in the calicles. The largest canals of the excurrent system are external, producing the astrorhizae characteristic of a number of genera.

The order Sphinctozoa is characterized by its distinctive manner of growth: the periodic addition of chambers, generally at the distal end of the skeleton. In the more typical members of the group, such as the extant species *Vaceletia crypta* (Vacelet 1977), this addition occurs in a linear series, producing cylindrical sponges which may or may not branch, and which may or may not have an axial spongocoel. However, there are fossil forms which achieve much greater complexity, notably genera such as *Cystauletes*, *Cystothalamia*, *Uvanella* and *Girtycoelia*, in which the incremental units are substantially different from the typical torus-shaped form. In these genera they are basically blister-like, just as in *Sporadotrema*, and are added in regular orientation.

These genera are all included in the family Guadalupiidae by Finks (1983); the cylindrical forms (*Cystauletes*, *Cystothalamia*, included in the family Cystothalamiidae by Ott 1967) Finks regards as "enrolled" versions of the unilateral *Guadalupia* morphology. If the structure of *Sporadotrema* were regarded as a "folded over" or bilateral version of the same morphology, and its finger-like processes in the crestal furrow as a relic of the *Guadalupia* trabecularium, then a false interpretation of the *Sporadotrema* morphology becomes easily possible.

The siliceous megascleres of all known modern sclerosponges are basically styles, most of them spinose. No polyactinal spicules have reliably been reported from the group so far. Mori (1977) records dichotriaenes in *Acanthochaetetes japonica*: all known specimens of this species are eroded fragments collected from beach shingle, so there is some probability that the spicules are foreign.

The presence of spicules within the wall of the sphinctozoan genera *Sphaerocoelia* and *Barroisia* has been known since the last century (Hinde 1883–1888), and has been used by workers such as Rauff (1911, 1913) and Reid (1968) to reinforce the placement of the Sphinctozoa as a whole with the Calcarea.

In their new genera, *Murguiathalamia*, *Boikothalamia* and *Vascothalamia*, Reitner and Engeser (1985) have reported structures which they consider to be traces of siliceous spicules that have undergone diagenetic solution and replacement by either low magnesium calcite or pyrite (p. 173).

The skeletons of the two specimens from Lady Elliott Island from which thin sections were prepared contain an anastomosis of fine, probably fungal borings ca. 0.004 mm in diameter. These are absent from the section of the specimen collected live, indicating that the invasion took place after death of the organism. Care should be taken in working with fossil material that such borings are not interpreted as ghosts of spicules within the skeleton.

In establishing the genus *Sporadotrema*, Hickson (1911) remarked (p. 448) "this is truly a giant foraminifer". The sheer size, coupled with the abundance of spicules and their remarkable separation according to location/size mean that recognition of the true nature of fossil representatives of such forms is likely to be extremely difficult.

*Acknowledgments.* Thanks to Joachim Reitner for constructive and helpful discussions during the symposium on Fossil and Recent Sponges, Berlin 1988.

## References

Carter HJ (1880) Report on specimens dredged up from the Gulf of Manaar. Ann Mag Nat Hist Ser 1880, 5:437

Finks RM (1983) Pharetronida: Inozoa and Sphinctozoa. In: Broadhead TW (ed) Sponges and spongiomorphs: Notes for a short course. University of Tennessee, Department of Geological Sciences, Studies in Geology 7, pp 55–69

Hartman WD (1983) Modern and ancient Sclerospongiae. In: Broadhead TW (ed) Sponges and spongiomorphs: Notes for a short course. University of Tennessee, Department of Geological Sciences, Studies in Geology 7, pp 116–129

Hickson SJ (1911) On Polytrema and some allied genera. A study of some sedentary Foraminifera based mainly on a collection made by Prof. Stanley Gardiner in the Indian Ocean. Linn Soc Lond Trans Zool Ser 2, 14:443–462, pl. 30–32

Hinde GJ (1883–1888) Catalogue of the fossil sponges in the department of the British Museum (Natural History) with description of new and little known species. London, Brit Mus Nat His, 248 p

Mori K (1977) A calcitic sclerosponge from the Ishigaki-shima Coast, Ryukyu Islands, Japan. Sci Rep Tohoku Univ Sendai Second Ser (Geol) 47(1): 1–5, pl. 1–2

Ott E (1967) Segmentierte Kalkschwämme (Sphinctozoa) aus der alpinen Mitteltrias und ihre Bedeutung als Riffbildner im Wettersteinkalk. Abh k bayer Akad Wiss Math naturwiss Kl Neue Folge 131:1–96

Rauff H (1911) Pharetronen-Studien, Neues Jahrb Mineral Geol Paläontol 2:139–191

Rauff H (1913) Barroisia und die Pharetronenfrage. Paläontol Z 1:75–144

Reid REH (1968) Tremacystia, Barroisia, and the status of Sphinctozoida (Thalamida) as Porifera. Univ Kans Palaeontol Contrib 34:1–10

Reitner J, Engeser T (1985) Revision der Demospongier mit einem thalamiden, aragonitischen Basalskelett und trabekulärer Internstruktur ("Sphinctozoa" pars). Berl Geowiss Abh (A) 60:151–193, pl. 1–6

Vacelet J (1977) Une nouvelle relique du Secondaire: un représentant actuel des éponges fossiles Sphinctozoaires. CR Acad Sci Paris 285:509–511

# Recent Calcarea with a Reinforced Skeleton ("Pharetronids")

J. VACELET

## Introduction

The fossil record of the Porifera is obviously restricted to those forms with stiffened skeletons. In Recent fauna, a number of reinforcement mechanisms are observed in highly diverse evolutionary lines. However, rigid skeleton are not predominant in Recent sponges and, as far as can be judged from the incomplete fossil record, have been widely replaced by flexible spongin skeletons.

In Recents seas, rigid skeletons have been observed in all major groups of siliceous sponges except the subclass Homoscleromorpha. Sponges of the class Hexactinellida have the dictyonine type of skeleton, while, in the class Demospongiae, the lithistid type with fused or interlocked spicules (desmas) is found mainly in Tetractinomorpha (Tetractinellida), but also in some Ceractinomorpha. Calcareous, rigid skeletons of the sclerosponge type occur in "coralline" sponges belonging to diverse groups of both Tetractinomorpha and Ceractinomorpha Demospongiae (Vacelet 1985). For a long time the fossil remains of these calcareous organisms were wrongly classified either in the phylum Cnidaria or in the class Calcarea of the phylum Porifera. Thanks to a few remnants surviving in Recent seas, these misinterpretations have been corrected and most fossil stromatoporoids, chaetetids, and sphinctozoids are now assumed to be sponges belonging to the class Demospongiae.

That is not to say that rigid skeletons able to fossilize cannot be observed in Calcarea. Indeed, some fossil remains widely known as "pharetronids" must be classified as Calcarea. The distinction between Calcarea and Demospongiae is easy to make in living animals but is difficult or impossible to make in fossils in which spicules and microstructures have deteriorated. Despite recent claims that the demosponges of the subclass Homoscleromorpha and sponges of the class Calcarea share certain characters (amphiblastula larvae or shape of triactine spicules) and thus may come from the same evolutionary line (van Soest 1984; Reitner 1987), the long-standing consensus is that Demospongiae and Calcarea constitute two widely separated lines. Apparently this claim is based on the fact that the same word is used to describe highly different characters, i.e., "calthropes" for similarly shaped spicules of completely different composition and biogenesis or "amphiblastula" for larvae of widely different embryogenesis.

Given the difficulties involved in interpreting fossil remains, it is of prime importance that the characters used for classification of Recent forms be me-

J. Reitner and H. Keupp (Eds.)
Fossil and Recent Sponges

ticulously defined. A few years ago a review was published on the Recent Pharetronids (Vacelet 1970). Since then, new data have been provided which are summarized by Hartman (1982) and Reitner (1987), whereas the structure and genesis of biominerals have been excellently reviewed by Jones (1979). With these new observations, a critical re-examination was needed. The aim of this new review is to catalog the characters shared by living "pharetronids" and to discuss their evolutionary relationships with other Recent Calcarea.

## Classification of the Recent Calcarea

Presently the most widely accepted system of classification for Calcarea is that of Bidder (1898) revised by Hartman (1958) and Borojević (1979). This sytem recognizes two evolutionary lines forming two subclasses: Calcaronea Bidder 1898, and Calcinea Bidder 1898.

In the calcaronean line, the choanocyte has an apical nucleus from which the flagellum arises. The larva is an amphiblastula obtained by eversion of a stomoblastula. Triradiate spicules are mostly sagittal. Diactines are the first spicules to appear during ontogeny. The carrier-cell system of fertilization is well documented.

In the calcinean line, the choanocyte has a basal nucleus without connection with the flagellum. The larva is a coeloblastula which becomes a solid blastula during its free swimming life. Spicules includes regular triactines, i.e., triradiates with rays of equal length at equal angles. Triactines are the first spicules to appear during ontogeny. Fertilization is known in only one species.

The most coherent scheme of evolution is that of Borojević (1979). According to this scheme, the water current system developed in parallel in the two lines from the simple asconoid grade of organization through syconoid to leuconoid, together with diverse specializations of the free spicules. A remarkable fact emphasized by Borojević is that the diverse taxa of both subclasses exhibit a wide range of degrees of evolutionary sophistication with numerous coexisting intermediary forms. This suggests that these sponges represent a recent burst of evolution, although some of them are the simplest stages known in the Porifera and as such may be wrongly considered as the most primitive. Genera *Clathrina* and *Leucosolenia*, despite the fact that they are the simplest sponges imaginable, are neither Precambrian fossils nor hypothetical stages of the early evolution of the Porifera, but rather dynamic living forms that differentiated a large number of species in highly competitive coastal environments.

Most members of class Calcarea have skeletons composed of calcium carbonate spicules supported by organic material and living tissue. Since such skeletons are not able to fossilize, fossil records for these sponges are doubtful. On the other hand, a few living forms ("pharetronids") have rigid skeletons obtained by diverse means of reinforcement which may be the same as those found in fossil remains.

**Recent Calcarea with Reinforced Skeletons**

Three types of skeleton reinforcement are known in Recent Calcarea. The first consists in an intertwining or fusion of spicules which form a rigid skeleton. In the second the rigid structure is obtained by secretion and fusion of aspicular calcareous bodies (sclerodermites). The third mode of reinforcement involves spicule tracts linked by organic material. These differences in skeletal structures are important for classification. However, they are not the only characters that should be taken into account and it is necessary to examine whether the three groups they allow to distinguish are consistent with the other characters used in the general classification of Calcarea. Most of the characters of these sponges are summarized in Table 1.

1. The first group, in which spicules are fused, constitutes the family Minchinellidae. Despite the relatively high number of living and fossil members, this family appears to be fairly homogeneous. Its principal features have been described by Vacelet (1970) and by Hartman (1982). In this family, genera are mainly defined by the organization of the rigid skeleton. Interestingly, their calcareous skeleton is somewhat similar in morphology to that of the lithistid siliceous skeleton of the demosponges and this is a remarkable homeomorphy between the two classes.

In genus *Minchinella* Kirkpatrick 1900 (two known species), the skeleton is made up of a primary network of tetractines whose basal rays are fused and progressively embedded in calcareous cement (Fig. 1a,b). The calcitic needles of the cement lie perpendicular to the surface of the spicule. The final result is a solid skeleton with an orthogonal microstructure, in which the primary spicules are completely enclosed and only visible at fractures. Such cement reinforcements have never been observed in lithistid siliceous skeletons but are common in hexactinellid frameworks.

Most Recent species have been classified under the genus *Plectroninia* Hinde 1900, which was created for miocene sponges, and similarity between them and fossil material needs to be checked. Recent *Plectroninia* (with 11 known species) are small encrusting sponges with an ectosomal layer of tangential, regularly placed free spicules. The underlying choanosomal skeleton is made up of two networks: a primary one consisting of large tetractines and a secondary one consisting of small tetractines (Fig. 1c). In both networks the basal rays of tetractines are linked, while the apical ray remains free and pointed outward. The basal actines are welded by their thickened ends forming a structure like the zygomes of the lithistid desmas. Although it may be reinforced by a secondary deposit of calcareous cement, the articulation would appear to be mainly mechanical but little information is available on the fine structure and mode of formation of the "zygosis".

A similar structure is observed in *Monoplectroninia hispida* Pouliquen & Vacelet 1970, except that there is only one basal network, the large fused tetractines being absent. In *Petrostroma schulzei* Döderlein 1892, which is a more

**Table 1.** Characters of Recent Calcarea with a reinforced skeleton. Calcinean characters are underlined

| | Nucleus of choanocyte | Larva | Regular triactine | Perioscular circlet | Tuning-fork or tract spicule | Reinforcement of skeleton | Habitat | Distribution |
|---|---|---|---|---|---|---|---|---|
| *Lelapia* 3 sp. | Apical | Amphiblastula[a] | No | No? | Tuning-fork tracts | Spicule tracts + diactines | Open littoral | Indopacific |
| *Kebira* 1 sp. | Apical | Amphiblastula[b] | No | No? | Modified tri-actine tracts | Spicule tracts + diactines | Open littoral | Red Sea |
| *Lepidoleucon* 1 sp. | Apical | Amphiblastula | Yes (scales) | Tetractines | No | Scales | Cryptic | Indopacific |
| *Minchinella* 2 sp. | Apical | Unidentifiable embryos | No | Tetractines | Tuning-fork, free | Fused spicules and cement | Cryptic | Pacific Ocean |
| *Plectroninia* 11 sp. | Apical | Amphiblastula[c] | In 3 species | Tetractines, rarely triacts | Tuning-fork in tracts or free | Fused spicules ± tracts | Cryptic and bathyal | Indopacific, Atlanto-Medit. |
| *Monoplectroninia* 1 sp. | Apical | ? | No | Tetractines | Tuning-fork, free | Fused spicules | Cryptic | Mediterranean |
| *Petrostroma* 1 sp. | ? | ? | No | ? | Tuning-fork, in tracts? | Fused spicules + tracts? | Bathyal | Pacific Ocean |
| *Tulearinia* 1 sp. | Apical | ? | No | Triactines and tetractines | No | Interlocked spicules | Cryptic | Indopacific |
| *Petrobiona* 1 sp. | Apical | Amphiblastula | No | Tetractines | Tuning-fork, free | Solid mass | Cryptic | Mediterranean |
| *Lelapiella* 1 sp. | Basal | ? | Yes | Triactines | Diactine tracts | Spicule tracts tripods | Cryptic | Indopacific |
| *Murrayona* 1 sp. | Basal | Blastula | Yes (scales) | Triactines | Tuning-fork, free | Reticulate rigid skeleton | Cryptic | Pacific and E. Indian oceans |
| *Paramurrayona* 1 sp. | Basal | Blastula | Yes (scales) | Tetractines | Tuning-fork in tracts | Spicule tracts scales, plates | Cryptic | Indopacific, Caribbean |

[a] Unpublished observations on *Lelapia australis*, specimen 1887-7-12-30 of the British Museum.
[b] Personal communication by M. Ilan.
[c] Known in one species (*P. tecta*).

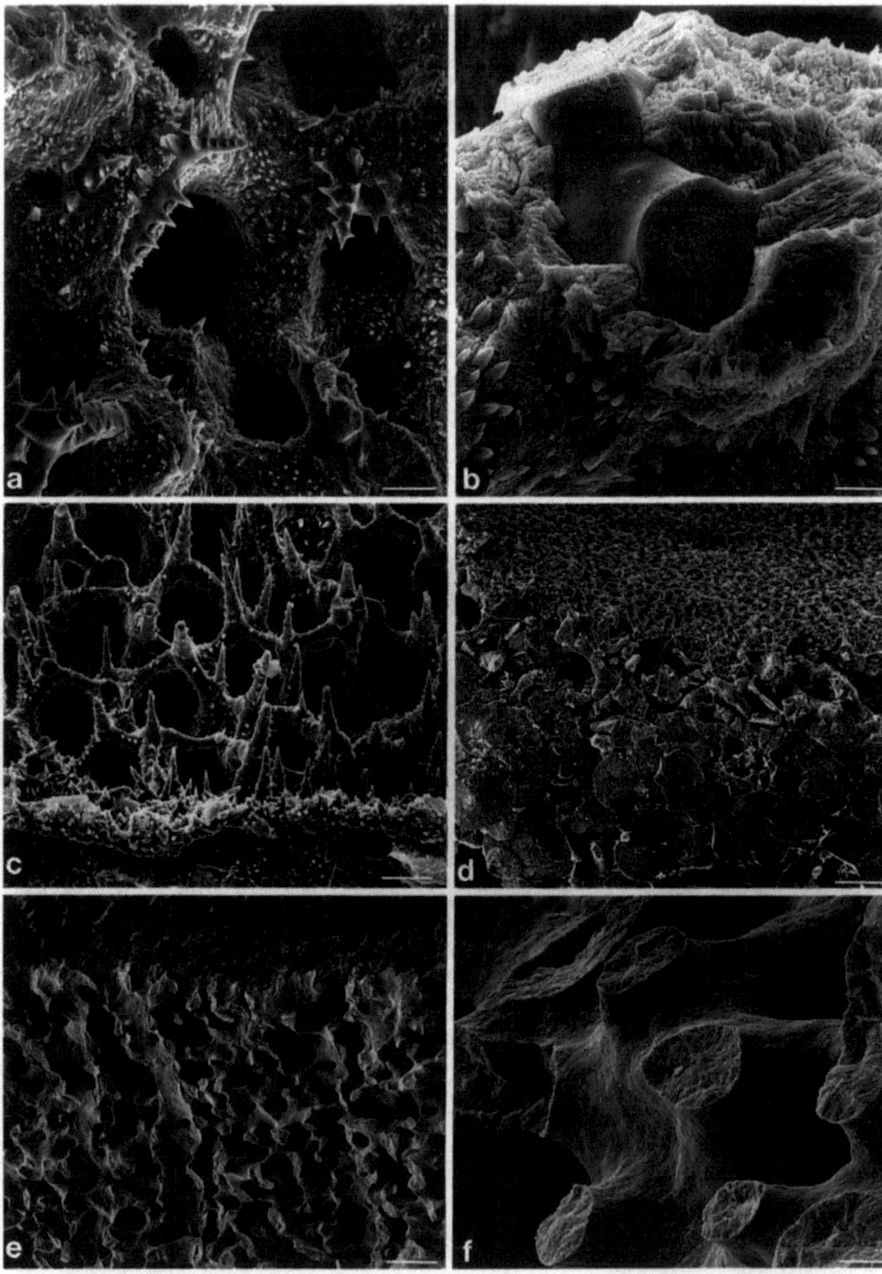
a
b
c
d
e
f

massive sponge, large fused tetractines form radiate lines which are in turn joined by a network of small fused tetractines.

Genera such as *Bactronella*, *Porosphaera*, *Tretocalia*, etc., are known only through Mesozoic or Cenozoic fossils. On the basis of present evidence, it is difficult to distinguish these sponge from *Plectroninia*. Inclusion of Recent *Plectroninia* species in a genus described for fossils, as was first done by Kirkpatrick (1900), is controversial. According to Pickett (pers. commun.), the type of the genus, *P. halli* Hinde, which dates from the mid-Miocene, was nonencrusting and probably lived free on a muddy sand bottom. If so, this sponge would be quite different from the Recent forms of "*Plectroninia*" and would rather resemble the Recent genus *Petrostroma*.

As stated earlier, Minchinellidae is a family with many living and fossil members. Nevertheless, and although the genera remain to be defined, this group is highly homogeneous and can be considered as monophyletic. Assignment of Minchinellidae to the subclass Calcaronea is unquestionable (Table 1). However, the lack of knowledge about larvae as well as the fact that regular triactines are dominant in some species of *Plectroninia* should be noted. Most species present a perioscular circlet of tetractines but a circlet of triactines may occur in a few species. Tuning-fork spicules, a consistent feature, are not always grouped in tracts. Thus this family interestingly illustrates the range of "characteristic" features.

Another sponge which deserves mention here is *Tulearinia stylifera* Vacelet 1977, for it may represent a first step in the process of spicule linkage that characterizes Minchinellidae. Indeed while this encrusting sponge exhibits an entanglement of large and small tetractines in the basal layer, its inclusion in the Minchinellidae is questionable because actine tips are only slightly modified without true zygoses and tuning-fork spicules are absent. Thus this mode of reinforcement may be seen as a transitional phase either to the Minchinellid configuration or to a convergent spicule linkage observed in a sponge adapted to high energy environments like the tunnels found in coral reef fronts.

The fossil record shows that other grades of skeletal organization based on calcareous spicules and in particular on tetractine fusion have existed at one time or another but are now extinct. Evidence strongly suggests that the Sphinctozoans Sphaerocoeliida were Calcarea (Reid 1968) and, according to Wood (1987), some Stromatoporoids such as "*Burgundia*" *tutcheri*, from the Middle Jurassic, may be also a Calcarea. The spicules in these sponges were fused as in the Minchinellidae

**Fig. 1. a** Surface view of skeleton in *Minchinella*, showing the tetractine framework progressively embedded in calcareous cement. × 240. Bar = 31 μm. **b** Fracture in the skeleton of *Minchinella*, showing a spicule embedded in cement. The microstructure is of the orthogonal type. × 800. Bar = 10 μm. **c** Main and underlying secondary network of *Plectroninia*. In both, the three basal rays of the tetractines are fused while the apical ray remains free and outwardly pointed. × 150. Bar = 50 μm. **d** Surface view of *Murrayona*, showing inhalant area (*top*, with a network of free triactines) and exhalant area (*bottom*, with a covering of scales). × 25. Bar = 300 μm. **e** Fracture and surface (*top*) of skeleton in *Murrayona*. × 20. Bar = 400 μm. **f** Skeleton of *Murrayona*. × 130. Bar = 60 μm in "*Burgundia*" *tutcheri* and *Muellerithalamia*; however their arrangement was different in Sphaerocoeliida (Reitner 1987).

Minchinellidae have been found in coastal cryptic habitats of the Indopacific and Mediterranean. This area of distribution is larger than first thought (Vacelet et al. 1989). About 30 specimens of diverse species of *Plectroninia* have been found in bathyal zone at depths up to 1600 m in the Mediterranean Sea, Atlantic Ocean (Açores, Brazil, South Africa), Indian Ocean (including the subantarctic Crozet Islands, 525 m deep) and Pacific Ocean. Thus the genus *Plectroninia* actually has a wide distribution in Recent seas, but its representatives have been generally overlooked due to their habitat and their small size.

2. In the second group the reinforced skeleton is obtained by fusion of calcitic bodies (sclerodermites) which are very different from spicules. This type of skeleton is surprisingly similar to the sclerosponge calcareous skeleton described in several demosponges. However, contrary to sclerosponge skeleton, an exact equivalent of its microstructure has never been observed in the fossil record. This skeleton is found in two Recent calcareans which differ greatly with respect to their other characters.

The first is *Murrayona phanolepis* Kirkpatrick 1910. Its reticulate skeleton is made of calcitic sclerodermites (Fig. 1d-g). This skeleton does not entrap spicules. Living tissue and reproduction are of the calcinean type. Most triactines are sagittal. However, special superficial scales are derived from regular triactines and free tuning-fork triactines are present. *Murrayona* live in cryptic habitats of the Pacific Ocean.

The second species is *Petrobiona massiliana* Vacelet and Lévi 1958. Its skeleton is formed by a solid mass of calcite, composed of elongate sclerodermites, forming a series of crests between which the living tissue (Vacelet 1964). The skeleton also presents narrow canals in which numerous reserve cells are stored. This survival structure is unique in the Calcarea, but is reminiscent of the "pseudogemmulae" enclosed in the calcareous skeleton of two calcified demosponges (Vacelet 1990). How this massive skeleton is formed is still unclear. Unpublished observations show that sclerodermites appear first either as erect spines (containing no spicule) or as sinuous, elongate bodies (Fig. 2a,b). Contrary to my first description calcareous spicules including free tuning-fork triactines are frequently entrapped within this skeleton (Fig. 2a). These entrapped spicules are

**Fig. 2.** **a** Surface view of skeleton in *Petrobiona*, showing sclerodermites with their terminal spines and numerous entrapped spicules. × 150. Bar = 50 μm. **b** Surface view of skeleton in *Petrobiona* showing an area with rare entrapped spicules and numerous sinuous sclerodermites. × 115. Bar = 65 μm. **c** A serpulid tube lying tangentially on the skeleton of *Petrobiona*, in the process of being overgrown. Patches of newly formed skeleton can be seen on its surface. × 20. Bar = 400 μm. **d** Enlarged area of **c**. The new skeleton appears as elongated sclerodermites which entrap a triactine spicule. × 130. Bar = 60 μm. **e** Newly formed skeleton on the surface of a serpulid tube in *Petrobiona*. × 130. Bar = 60 μm. **f** An overgrown serpulid tube lying vertically in the skeleton of *Petrobiona*. × 215. Bar = 35 μm

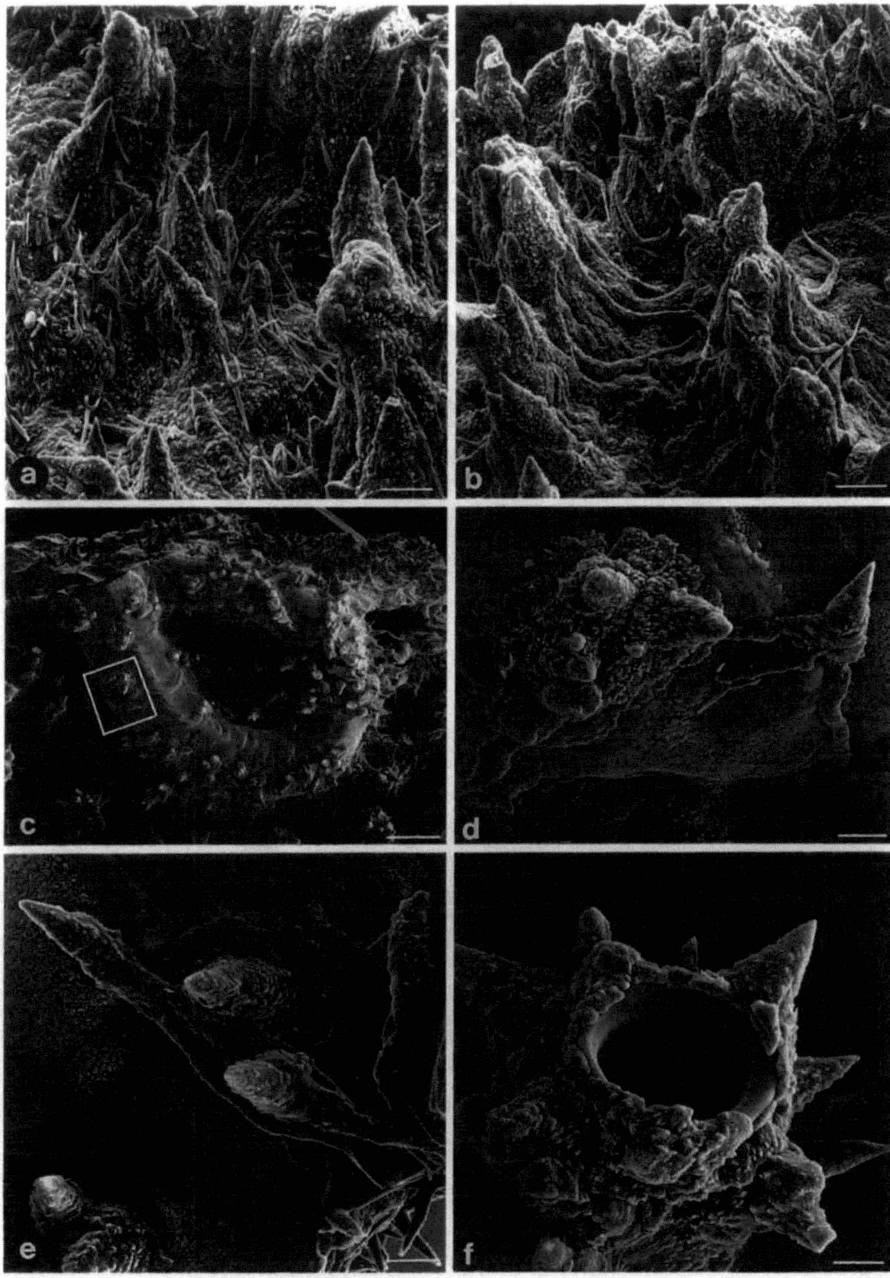
a
b
c
d
e
f

especially numerous in the basal layers, and, unlike the siliceous spicules entrapped in the skeleton of calcified demosponges, do not dissolve in the depths of the skeleton. They are randomly arranged and apparently do not determine the way sclerodermites are deposited. In actively growing areas, like the surface of serpulid tubes being entrapped in the skeleton, sinuous sclerodermites are laid down before spicule secretion begins (Fig. 2c-f). Different observation have been made on this point by Reitner (pers. commun.). Thus this skeleton differs widely from Minchinellidae, since it is not a secondary deposit upon a primary spicule framework. Soft tissue and spicule characters are typical of the Calcaronea. Reproduction is also of a calcaronean type, but nourishment of both oocytes and embryos is unusually complex (Vacelet 1964; Gallissian and Vacelet 1990). This complexity may stem from the fact that this sponge, a long-lived K-strategist unlike most other littoral calcareans, produces only a few larvae but very rich in reserves. *Petrobiona massiliana* is known to live only in cryptic habitats in the Mediterranean at depths between 1 and 25 m. Its presence in this sea is ancient since subfossils have been found in a cave in Crete that emerged some 1500 years ago (Vacelet 1980) and fossils have been noted in South Italy in Pleistocene strata dating back about 30,000 years (unpubl.).

Thus, despite similarities in skeletal structure, *Murrayona* and *Petrobiona* belong to highly different evolutionary lines. In fact, their skeleton shows differences in fine structure (Cuif et al. 1979) and may correspond to different biomineralization mechanisms.

A near relative of *Murrayona* is *Paramurrayona corticata* Vacelet 1967. The main difference between these two sponges is that, instead of a rigid choanosomal skeleton, *Paramurrayona* has only a cortical skeleton made of calcareous scales similar to those of *Murrayona* and underlying free calcareous plates of unknown origin. Tuning-fork spicules are present, but unlike in *Petrobiona* and *Murrayona*, they are arranged in rows. The distribution of *Paramurrayona* is remarkable: it is common in the Indopacific area and one specimen similar to the Indopacific ones has been found in a cryptic habitat of the Caribbean (Vacelet 1981). In animals with low dispersal abilities this distribution pattern is indicative of a very ancient origin and great genetic stability. *Paramurrayona* is the only "pharetronid" known from the Caribbean, where calcified demosponges are relatively well represented.

3. The third group of pharetronids, with spicule tracts, displays less distinctive characters and its classification apart from Calcarea with nonreinforced skeleton is debatable. Indeed in genera *Lelapia* Gray 1867 and *Kebira* Row 1909, the skeleton is reinforced only by tracts of spicules formed either by tuning-fork triactines or triactines with vestigial paired rays ("nail" spicules). Assembly is insured only by organic material and these tracts do not fossilize. Unlike other calcareans with reinforced skeletons which are cryptic, these sponges live in open coastal habitats of the Indopacific area. *Lelapiella incrustans* Vacelet 1977, has been classified in the same family on the ground that it has tracts made of special diactines. However, two calcinean characters suggest that this little-known sponge may have other affinities. The first is that its choanocytes appear to be basinucleated and the second is the presence of regular triactines (Table 1).

In the Lepidoleuconidae, with one known species, *Lepidoleucon inflatum* Vacelet 1967, skeletal reinforcement consists only in the presence of a coat of scales formed by modified triactines. Tuning-fork triactines are absent. The affinities of the family are unclear.

## Discussion

The place of "pharetronids" in the evolution of the Calcarea is a debated topic. As the classification proposed by Bidder and by Hartman for Calcarea has gained greater acceptance, the general consensus has been in favor of a polyphyletic interpretation of pharetronids, similar to that of "sclerosponges". Indeed, Murrayonidae, and Paramurrayonidae exhibit affinities with subclass Calcinea, whereas Minchinellidae, Petrobionidae, Lelapiidae, and Lepidoleuconidae exhibit affinities with subclass Calcaronea. Accordingly, pharetronids have thus been divided between the two lines of evolution in Calcarea, with the order Murrayonida Vacelet 1981 being classed in the Calcinean line and the others (order Lithonida Vacelet 1981) in the Calcaronean line. Lelapiidae and Lepidoleuconidae have even been included in the order Sycettida of the "normal" Calcaronea by Hartman (1982).

This diphyletic interpretation of the evolution of pharetronids is based on their great diversity with respect to the main classification criteria. However, they share a number of characters which have not been considered as homeomorphies (Vacelet 1970) and which accordingly might indicate that the group is actually monophyletic. These characters, namely hypercalcification, tuning-fork triactine, cortical specializations, perioscular circlet of spicules, and habitat are critically reexamined here.

As previously shown, hypercalcification can result from a variety of processes. Three convergent structures are observed in Minchinellidae, in Petrobionidae, and in Murrayonidae. As it is also observed in diverse demosponges, hypercalcification cannot be considered as a truly shared character. However, it may indicate affinities with Paleozoic and Mesozoic groups in which it was a more common occurrence.

Emphasis has been placed on tuning-fork triactines because they have long been considered as the main component of the "fibers" that allow fossilization of pharetronids. Actually in Recent forms tuning-fork spicules are never rigidly linked and the tracts they form are not fossilizable. Thus it appears that this feature has been overstressed. Since they are secreted in intercellular spaces instead of intracellularly, calcareous spicules are more adaptable to their position and function than siliceous spicules. Phyletically these spicules are thus less informative than siliceous spicules. Tuning-fork spicules may correspond to a convergent adaptation to the constitution of spicular tracts and as such may have been more frequent in ancient Calcarea when rigid skeleton were more common. An example of this convergent mutation of triactines is found in *Guancha lacunosa*, a calcinean sponge that forms a peduncle with special spicules. In contrast with the regular triactines that predominate in this sponge, peduncle spicules include triactines with short, nearly parallel rays reminiscent of tuning forks. However, tuning-fork triactines may be present in the absence of tracts, as demonstrated in the genus *Plectroninia*, which includes species with tuning forks in tracts as well as species with free tuning forks. Similarly, *Murrayona phanolepis*, which displays a rigid skeleton, has loose tuning forks while *Paramurrayona corticata*, which does not have a rigid skeleton, has tuning forks in tracts.

Cortical spicule specializations such as scales or tripods are probably related to high energy environments like the ones frequently encountered in cryptic habitats in coral reefs. Thus some caution is needed in interpreting the phyletic implications of these features. In *Paramurrayona* and *Murrayona*, which have many shared characters, the scales may indicate a common origin. However, in *Lepidoleucon*, which is a highly different sponge, scales are probably a convergent adaptation to reef front tunnels.

Another shared character in the pharetronids is the frequent presence of a circlet of modified tetractines, or more rarely triactines, around the oscule. This structure, which is perhaps a remnant of papillae, is of a doubtful phyletic importance, since perioscular specializations are common in calcareans.

Ecological and biogeographical similarities are also observed between pharetronids. With the exception of the Lelapiidae, which live in coastal environments similar to that of nonpharetronid Calcarea, all pharetronids live in coastal cryptic habitats or on steep walls on upper levels of the bathyal zone. Most have a large and seemingly very ancient geographic distribution and belong to monotypic genera or families, the only exception being the genus *Plectroninia* which has 11 known Recent species. Obviously neither of these two features has phyletic significance, since both have been observed in several unrelated calcified or lithistid demosponges. A simple explanation for the presence of these strongly skeletonized calcareans and demosponges in this habitat would be that these relicts from various nearly extinct groups have found a common refuge in this special environment either because the level of competition is lower than in open habitats or because conditions are somewhat similar to those in ancient reef environments.

In conclusion, my present opinion is that the similarities between pharetronids, some of which are shared with the calcified demosponges, have no phyletic significance. These features are more likely homeomorphies or convergent characters, that were more common in Calcarea in ancient times when more heavily skeletonized sponges were important reef-builders. These characters have survived in a few relicts which contrast with Recent Calcarea presenting free-spicule and loose skeletons that result from a remarkable recent burst of evolutionary activity. It is important to note that these Recent Calcarea did not descend directly from the calcified, conservative Calcarea but rather evolved in parallel from the same two main stocks, i.e., Calcinea and Calcaronea. It is likely that this evolution took place recently perhaps in relation with the loss of hypercalcification. The outline in Fig. 3 sums up this viewpoint which admittedly does not answer all questions about the evolution of the Calcarea. For example, how can one explain the surprising coexistence in Recent seas of so many stages of the progressive complication of the canal system and skeleton ranging from a highly "primitive" simple asconoid structure to the complicated

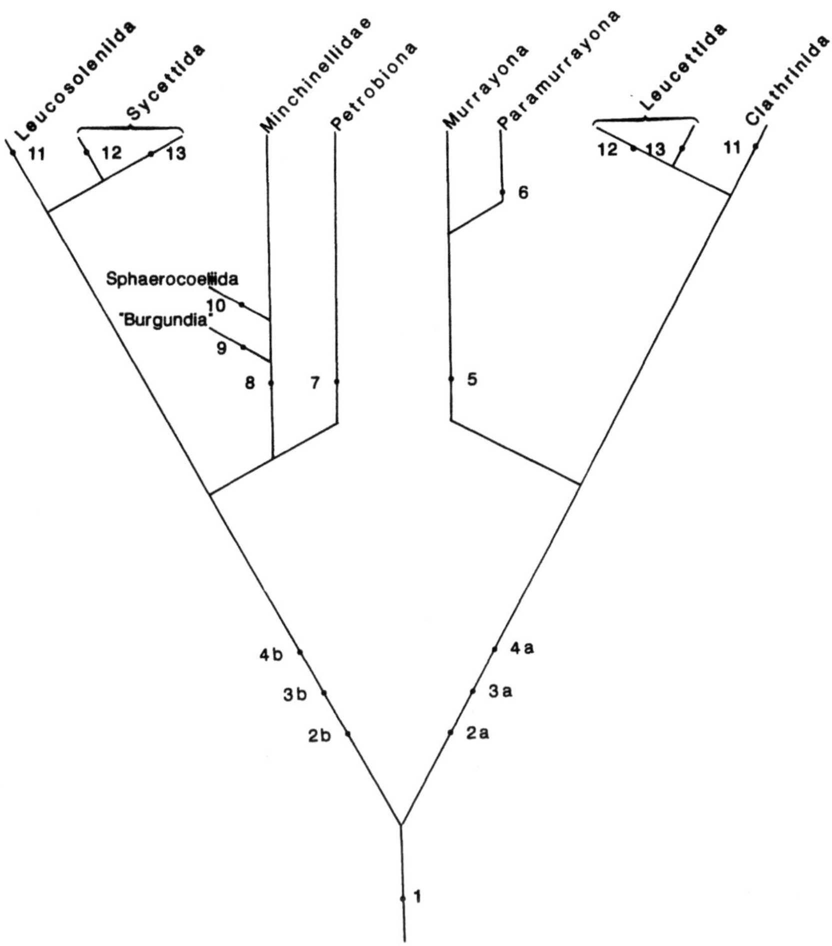


**Fig. 3.** Tentative outline of the evolution of Calcarea. The lines for Leucosoleniida, Sycettida, Leucettida and Clathrinida do not correspond to natural groups, but, for simplification, are based on the three types of canal system organization. *1* calcium carbonate spicules; *2a* basinucleate choanocyte; *2b* apinucleate choanocyte; *3a* blastula larva; *3b* amphiblastula larva; *4a* mostly regular triactines; *4b* mostly sagittal triactines; *5* aspicular calcitic skeleton of the *Murrayona* type; *6* loss of aspicular skeleton; *7* massive calcitic skeleton of the *Petrobiona* type; *8* skeleton made of fused tetractines; *9* stromatoporoid grade of skeletal organization; *10* sphinctozoid grade of skeletal organization; *11* asconoid grade of canal system organization; *12* syconoid grade of canal system organization; *13* leuconoid grade of canal system organization

hypercalcified forms which are isolated, mainly monotypic taxa? Based on the present hypothesis this could be explained by the fact that pharetronids are relicts of several ancient specializations, while nonhypercalcified Calcarea result from a recent burst of evolutionary activity. Conceivably, this burst occurred by neoteny from the simple olynthus, the earliest stage after larva metamorphosis.

Obviously, however, this explanation remains in the realm of pure conjecture since nothing is known on the ancient fauna of nonhypercalcified Calcarea. In particular there is no way of knowing if simple forms similar to the Recent *Clathrina* or *Leucosolenia* coexisted with the "pharetronids" in Paleozoic and Mesozoic times. However, a canal system simpler than the leuconoid one has never been demonstrated in the fossil record.

## References

Bidder GP (1898) The skeleton and classification of calcareous sponges. Proc R Soc London 64:61–76

Borojevic R (1979) Evolution des spongiaires Calcarea. Colloq Int CNRS n° 291: Biol Spong: 527–530

Cuif JP, Debrenne F, Lafuste JG, Vacelet J (1979) Comparaison de la microstructure du squelette carbonaté non spiculaire d'Eponges actuelles et fossiles. Colloq Int CNRS n° 291: Biol Spong: 459–465

Döderlein L (1892) Ueber *Petrostroma schulzei*, n.g. n.sp. der Kalkschwämme. Verh Dtsch Zool Ges 2:143–145

Gallissian MF, Vacelet J (1990) Fertilization and nutrition of the oocyte in the hypercalcified sponge *Petrobiona massiliana* (Calcarea, Calcaronea). Proc III Int Conf on the Biology of Sponges, Woods Hole (in press)

Gray JE (1867) Notes on the arrangement of sponges, with the description of some new genera. Proc Zool Soc Lond 492–558, pl 17–18

Hartman WD (1958) A re-examination of Bidder's classification of the Calcarea. System Zool 7(3):97–110

Hartman WD (1982) Porifera. In: Parker SP (ed) Synopsis and classification of living organisms. McGraw-Hill, New York, pp 641–666

Hinde GJ (1900) On some remarkable calcisponges from the Eocene strata of Victoria (Australia). Q J Geol Soc 56:50–66

Jones WJ (1979) The microstructure and genesis of sponge biominerals. Colloq Int CNRS n° 291: Biol Spong: 425–447

Kirkpatrick R (1900) Description of Sponges from Funafuti. Ann Mag Nat Hist 7(6):345–362

Kirkpatrick R (1908) On two new genera of recent Pharetronid Sponges. Ann Mag Nat Hist 8(2):503–514

Kirkpatrick R (1910) On a remarkable pharetronid sponge from Christmas Island. Proc R Soc 83:124–133

Pouliquen L, Vacelet J (1970) Nouvelles observations sur des éponges Pharétronides Minchinellidae de Méditerranée. Téthys 2(2):437–442

Reid REH (1968) *Tremacystia, Barroisia* and the status of sphinctozoida (Thalamida) as Porifera. Univ Kans Paleontol Contrib 34:1–10

Reitner J (1987) Phylogenie und Konvergenzen bei rezenten und fossilen Calcarea (Porifera) mit einem kalkigen Basalskelett ("Inozoa", "Pharetronida"). Berl Geowiss Abh 86(A):87–125

Row RWH (1909) Report on the Sponges collected by Mr Cyril Crossland in 1904–5. Part I. Calcarea (XIII-th report on the marine biology of the Sudanese Red Sea). J Linn Soc Lond 31:182–214

Vacelet J (1964) Etude monographique de l'Eponge Calcaire Pharétronide de Méditerranée, *Petrobiona massiliana* Vacelet et Lévi. Les Pharétronides actuelles et fossiles. Recl Trav Stn Mar Endoume 34(50):1–125

Vacelet J (1967) Descriptions d'Eponges Pharétronides actuelles des tunnels obscurs sous-récifaux de Tuléar (Madagascar). Recl Trav Stn Mar Endoume Suppl 6:37–62

Vacelet J (1970) Les Eponges Pharétronides actuelles. Symp Zool Soc Lond 25:189–204

Vacelet J (1977) Eponges Pharétronides actuelles et Sclérosponges de Polynésie française, de Madagascar et de la Réunion. Bull Mus Natl Hist Nat 444:345–366

Vacelet J (1980) Les affinités du peuplement de Spongiaires de la Méditerranée. Journ Etud Syst Biogéogr Médit Cagliari CIESM, pp 29–30

Vacelet J (1981) Eponges hypercalcifiées ("Pharétronides", "Sclérosponges") des cavités des récifs coralliens de Nouvelle-Calédonie. Bull Mus Natl Hist Nat 3A:313–351

Vacelet J (1985) Coralline sponges and the evolution of the Porifera. In: Conway Morris S, George JO, Gibson R, Platt HM (eds) The origins and relationships of lower invertebrates. Syst Assoc Spec Vol 28:1–13

Vacelet J (1990) The storage cells of calcified relict sponges. Proc III Int Conf on the Biology of Sponges, Woods Hole (in press)

Vacelet J, Lévi C (1958) Un cas de survivance en Méditerranée du groupe d'Eponges fossiles des Pharétronides. CR Acad Sci 246:318–320

Vacelet J, Boury-Esnault N, Zibrowius H (1989) Unexpected deepwater record of calcareous sponges (Calcarea). Deep Sea Newsletter (in press)

Van Soest RWM (1984) Deficient *Merlia normani* Kirkpatrick, 1908, from the Curaçao reefs, with a discussion on the phylogenetic interpretation of sclerosponges. Bijdr Dierk 54(2):211–219

Wood R (1987) Biology and revised systematics of some late Mesozoic stromatoporoids. Palaeontol Assoc Lond Spec Pap 37:1–89

# Aspects of Sponge Biology

# Introduction

The conference in Berlin dealt mainly with phylogenetic, taxonomic and ecological aspects. However, biological problems of sponges are often the key to e.g. ecological or taxonomic problems. Most talks at the "Third International Conference on the Biology of Sponges" in Woods Hole (USA) 1985 dealt mainly with biological problems and presented new investigations and goals (Rützler and Hartman 1985). An important, recent contribution to the biology of sponges is the book of N. Weissenfels (1989), *Biologie und mikroskopische Anatomie der* Süß*wasserschwämme*, in which, in an excellent way, new scientific results from freshwater sponges are presented.

The few presented papers on the biology of sponges represent only part of the current research on the biology of sponges, but they exhibit many important new scientific developments which are shortly reviewed. The papers presented here deal mainly with morphological and structural characters in different sponges, with respect to classification problems. In one chapter biochemical differences are used for identification (Hooper et al., this Vol.) and the importance of this method is elucidated.

A further trend in the better understanding of population dynamics and intraspecific variation of species is shown in Gaino et al. (this Vol.) whose results are based on long-term in situ investigations combined with underwater photographic documentations.

The formation of skeletal elements in sponges is always of principal interest in sponge research. Mainly the paleospongiologists have great interest in these investigations. The formation of siliceous spicules in demosponges has been known for a long time but details are still unknown (review in Simpson 1984). The spicule formation in Hexactinellida is in most respects unclear. Within the Calcarea the papers of Ledger (1976) and Ledger and Jones (1977) help in understanding the formation of calcareous spicules. Many open questions still remain, e.g. organic components within the spicules. Calcareous spicules are formed extracellularly and it seems that characteristic axial filaments and organic sheaths are present. The chapter by Ledger and Jones demonstrates in an excellent way that decalcification artifacts should not be misinterpreted as organic components. Calcareous spicules are free of detectable organic material. This result supports the idea of the convergent phylogenetic development of spicules. The formation of non-spicular calcareous skeletons is mainly investigated by paleontologists (e.g. Jones 1979; Wendt 1984; Reitner 1989; Cuif and

Gautred, this Vol.; Wood, this Vol.). Many fossil sponges exhibit aragonitic or Mg-calcitic skeletons and play an enormous role within the fossil record. Historically, these particular sponges are called "stromatoporoids, chaetetids, or sclerosponges". The non-spicular skeletal formation of this polyphyletic group of sponges (see Part I) has not been well investigated. It is possible to determine an intracellular formation of small aragonitic spherules in *Astrosclera willeyana* Lister and a common extradermal and/or pinacodermal formation within a polysaccharide mucus layer. An important role is played by synvivo-diagenetic processess (Reitner 1987; 1989), which form solid, reinforced calcareous skeletons. The chapter by Wood (this Vol.) gives an overview of current knowledge and some new data on the formation of secondary calcareous skeletons. The formation of secondary skeletons is initiated by the organism itself, but the growing process is uncontrolled which is evidenced by stable isotope data (delta $O^{18}$, delta $C^{13}$). The carbonates are in equilibrium with the seawater (Reitner 1989).

Further important questions pertain to the uptake of bacteria, cyanobacteria and algae within sponge tissue. Many freshwater and marine sponges bear these organisms within the soft tissue but the role and function, weather symbiosis, commensalism, or parasitism are mostly unclear (e.g. Hartman 1969; Vacelet 1971; review Simpson 1984). The contribution by U. Saller on the role of the unicellular green algae *Chlorella* within the soft tissue of *Spongilla lacustris* demonstrates the positive symbiotic effect of this companionship.

The contributions presented at the conference represent reports of current research on the biology of sponges and demonstrate directions of further investigation.

## References

Hartman WD (1969) New genera and species of coralline sponges (Porifera) from Jamaica. Postilla 137:1-39

Jones WC (1979) The microstructure and genesis of sponge minerals. Coll Int Cent Natl Rech Sci 291:425-447

Ledger PW (1976) Aspects of the secretion and structure of calcareous sponge spicules. PhD Thesis, Univ College N Wales, 125 pp

Ledger PW, Jones WC (1977) Spicule formation in the calcareous sponge *Sycon ciliatum*. Cell Tissue Res 181:553-567

Reitner J (1987) *Euzkadiella erenoensis* n.gen.n.sp. ein Stromatopore mit spiculärem Skelett aus dem Oberapt von Ereño (Prov Guipuzcoa, Nordspanien) und die systematische Stellung der Stromatoporen. Paläont Z 61:203-222

Reitner J (1989) Struktur, Bildung und Diagenese der Basalskelette bei rezenten Pharetroniden unter besonderer Berücksichtigung von *Petrobiona massiliana* Vacelet & Levi 1958 (Minchinellida, Porifera). Berliner Geowiss Abh A 106:343-383

Rützler K, Hartman WD (1985) Third International Conference on the Biology of Sponges. Abstracts 76 pp (Woods Hole)

Simpson TL (1984) The cell biology of sponges. Springer, Berlin Heidelberg New York, 662 pp

Vacelet J (1971) Etude en microscopie electronique de l'association entre une cyanophycée chroococcale et une époge du genre *Verongia*. J Micros 12:363-380

Weissenfels N (1989) Biologie und mikroskopische Anatomie der Süßwasserschwämme (Spongillidae). Gustav Fischer, 110 pp

Wendt J (1984) Skeletal and spicular mineralogy, microstructure and diagenesis of coralline calcareous sponges. Palaeontol Am 54:326-336

# Morphometric and Biochemical Differences Between Sympatric Populations of the *Clathria* "Spicata" Species Complex (Demospongiae: Poecilosclerida: Microcionidae) from Northern Australia

J.N.A. HOOPER, R.J. CAPON, C.P. KEENAN, and D.L. PARRY

## Introduction

Morphological systematics of Recent Porifera are relatively difficult in comparison with other invertebrate phyla. On the one hand, morphological differences between supposedly distantly related higher taxa may appear to be minute or trivial, which is apparently a product of a long but conservative evolutionary history which was not produced by any radical morphological changes (Wiedenmayer 1977), but has produced an extremely diverse biochemical constitution (Bergquist 1978). Conversely, species may show a high degree of intraspecific morphological variability (e.g., Pansini 1982). In some cases that variability may be correlated directly with measurable differences in environmental factors (e.g., Palumbi 1984, 1986; Thompson and Murphy 1989), whereas in other cases species appear to be much more morphologically conservative across wide geographical ranges (e.g., Hooper 1987). Our knowledge of the effects of various environmental factors on intraspecific variability in sponges is still very limited (e.g., review in Simpson 1984), and it is not presently possible to construct models which may predict the behaviour of species (and their characters) under environmental changes (e.g., Simpson 1978; Palumbi 1984). These uncertainties have allowed authors to make quite different interpretations of the systematic importance of morphometric characters and levels of taxonomic divergence within the classification (cf. Hallmann 1912; Wilson 1921; Hooper 1991b). The present study examines a case in which only cryptic morphometric characters were able to consistently differentiate between two sympatric sibling species of *Clathria*, but species were otherwise clearly differentiated into two chemical populations. Methodology, descriptions of taxa and more complete results are presented elsewhere (Hooper et al. 1990).

## Results

During the course of a larger study on the Microcionidae (Hooper 1991b), approximately 60 specimens of a soft, red-brown, branching or reticulate *Clathria* Schmidt were collected from various depths and localities along approximately 2300 km of the northwestern Australian coast (from Northwest Cape, Western Australia, to the Wessel Islands, Northern Territory). The "species" appeared to be relatively polymorphic in growth form, displaying up to four distinctive

J. Reitner and H. Keupp (Eds.)
Fossil and Recent Sponges

morphotypes (Fig. 1). It also showed a few other apparently minor variations in morphological characters, such as the presence or absence of basal spination on megascleres and toxa thickness and curvature, whereas other features such as skeletal architecture, geometry of most spicules, spicule size, and live pigmentation were much more consistent, regardless of growth form type. Through a combination of those relatively consistent features, the "species" was certainly adequately defined and relatively easily differentiated from other Australasian *Clathria*. Intraspecific variability between specimens was remarkable, but certainly no greater than known to occur in other sponge populations recorded in the literature (e.g., Pansini 1982). The "species" was characterized in particular by its skeletal architecture and localization of spicules within and around fibers. It was an obvious member of Hallmann's (1912) "spicata" species complex, which contained ten nominal species (Hooper 1991a), and the earliest available name for the taxon was *Clathria lendenfeldi* Ridley and Dendy (Hooper et al. 1990).

Subsequent biochemical analyses (general protein, carotenoid pigment and free amino acid profiles) and screening for antimicrobial activity showed that the "species" was in fact distinctly heterogeneous. Furthermore, two populations (chemotypes) were clearly differentiated, but those chemical populations did not appear to be correlated with any significant or obvious morphological key characters. Similarly, no correlations were found between those "chemical populations" and ecological, geographical or bathymetric distributions of samples. The null hypothesis initially considered that the two sympatric populations were members of the same species (viz. *C. lendenfeldi* sensu lato), but this was tentatively rejected on the basis that electrophoretic "fingerprints" could consistently differentiate populations, and corroborated by other biochemical data. Much more rigorous systematics methods eventually provided some empirical evidence to support that conclusion, showing that there were in fact consistent (albeit relatively cryptic) morphometric differences between populations, and on that basis the two chemotypes were subsequently treated as sympatric sibling species. These were *C. lendenfeldi* Ridley and Dendy (1886), with six nominal species or subspecies as junior synonyms, and *C. major* Hentschel (1912) which was originally described as a subspecies of *C. frondifera* (Bowerbank) (Hooper 1991b). Morphometric and biochemical data which lead to those conclusions are summarized as follows.

### *Morphometrics*

*Shape* (Fig. 1). Four distinctive morphological forms were discovered for both species, showing an approximately similar range of gross morphological types: (1) branching (Fig. 1B,F), comprising 62% of specimens, with gross morphology of nominal species *C. lendenfeldi*, *C. spicata* Hallmann, and *R. bispinosus* Whitelegge; (2) clathrous, *Echinoclathria*-like (Fig. 1D,G), comprising 28% of specimens, and resembling the growth form of *C. whiteleggei* Dendy; (3) bushy, planar or multiplanar, lamellated fans (Fig. 1C,H); and (4) low growing irregularly bushy sponges (Fig. 1E), together comprising 10% of specimens.

**Fig. 1.** Morphological variability in *Clathria lendenfeldi* (**A-E**) and *C. major* (**F-I**). (**A** NTM Z1339; **B** NTM Z0875; **C** NTM Z2701; **D** NTM Z1011; **E** NTM Z1870; **F** NTM Z1409; **G** NTM Z2091; **H** NTM Z2096; **I** NTM Z1755)

*Colour.* Pigmentation of both species was relatively consistent and stable, with the same range of hues: vivid red (Munsell 2.5R 4–5/10), bright red (2.5R 5/12–5R 4/10), bright red-orange (5R 7/8), mauve-red (2.5R 4/8–5R 5/6) or pale red-brown (5 R 6/6) in shallow-water specimens, whereas typical pigmentation was absent in deeper water specimens, varying from light brown (2.5YR 6/6), gray-brown (2.5YR 7/4) or light gray-brown (7.5YR 6–7/4–6). Several deeper-water specimens were gray-brown with a turquoise (oxidized copper) ectosomal tinge (2.5B 7/4). Preserved material was beige, brown, gray-brown or muddy gray. Pigment was alcohol soluble. The presence of reddish colouration in shallow-water populations of both species (intertidal-20 m) may be indicative of the functional photoprotective role of the carotenoid pigments, whereas those pigments would not be expected to be as predominant in specimens from deeper coastal waters (39–108 m).

*Skeletal Structure* (Fig. 2). The development of the *ectosomal skeleton* (consisting of brushes of small, basally spined auxiliary subtylostyles) partially obscured by a paratangential subectosomal skeleton (brushes of larger, basally smooth or spined auxiliary subtylostyles) appeared to vary significantly in both species, and spicimens of both species spanned the nominal genera *Thalysias* (or *Rhaphidophlus*) and *Clathria*. These data and similar evidence derived from population studies on other northern Australian microcionids (Hooper 1991b) provided sufficient reason to abandon the taxonomic differentiation between the genera *Clathria* and *Thalysias*.

a) 13% of specimens lacked an ectosomal skeleton entirely (true *Clathria* condition);
b) where present, the major proportion of dermal brushes were of subectosomal origin, and in many cases those structures masked the presence of smaller ectosomal subtylostyles, characteristic of the *Thalysias* condition;
c) other specimens exhibited a gradual progression from *Clathria* to *Thalysias* states, with a poorly developed paratangential ectosome (64%), or with occasional erect brushes of both ectosomal and subectosomal spicules, mainly in thicker regions (18%), to obvious well-developed *Thalysias* structures (5% of those specimens);
d) 6% of specimens had significant quantities of inorganic detritus on the ectosome and scattered throughout the mesohyl, but never incorporated into fibers (cf. the *Clathriopsamma* condition).

In the *subectosomal region:*

a) 87% of specimens had characteristically poorly developed subectosomal regions, identical in both species;
b) 5% showed marked structural differences between ectosomal and subectosomal regions (these differences were most obvious in thicker regions of branches (e.g., surface projections);
c) 8% had a well-developed subectosomal region, characteristic of the *Thalysias* condition.

    a) 26% of specimens had an undifferentiated and open extra-axial reticulation of fibers and spicule tracts;
    b) 14% had markedly plumose and dendritic extra-axial primary tracts, connected by pauci- or aspicular (reticulate) fibers;

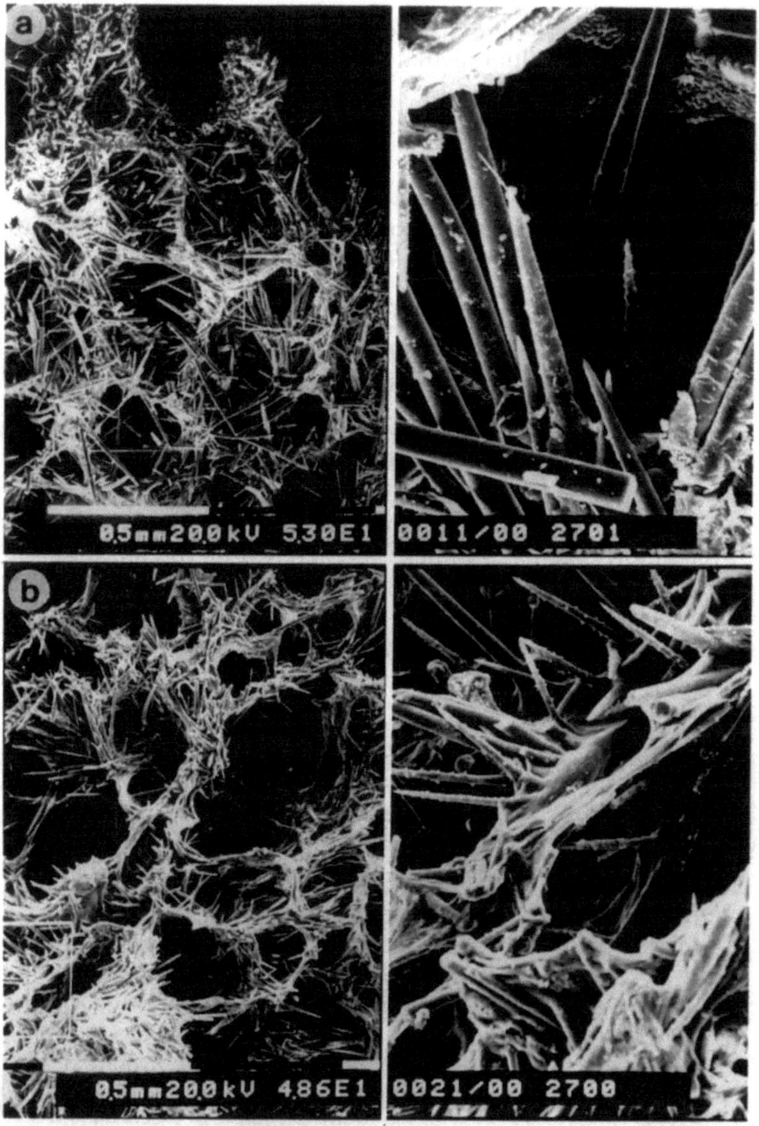


**Fig. 2.** Scanning electron micrographs of skeletal architecture in (**a**) *C. lendenfeldi* (NTM Z2701; *left view* magnified 53 times, *right view* magnified 424 times) and (**b**) *C. major* (NTM Z2700; *left view* magnified 48.6 times, *right view* magnified 238 times). Bar = 0.9 mm

c) 60% exhibited an extra-axial structure somewhere between those two extremes, with reticulate fiber and plumo-reticulate spicule skeletons.

The *choanosomal architecture* of both species was identical and equally as variable (Fig. 2):

a) 46% of specimens had an openly reticulate fiber and spicule skeleton, without axial condensation or obvious axial and extra-axial differentiation (95% of that material had more-or-less regular reticulate architecture);
b) 13% had very slightly axially condensed fibers with plumose spicule tracts;
c) 23% had an obvious axial condensation;
d) 18% had multiple axial cores in the choanosome, which produced several series of radiating plumose or plumo-reticulate extra-axial tracts.

The presence of heavily echinated fibers by both acanthostyles and principal megascleres, particularly at fiber nodes, was a consistent feature of both species, and characteristic for Hallmann's (1912) "spicata" group of *Clathria* in general. The inclusion of spherical pigment granules dispersed throughout the mesohyl was common to many specimens, but that feature was not always present: 61% of specimens had abundant pigmented bodies, mostly confined to the periphery, whereas in 8% those bodies were scattered evenly throughout the mesohyl, in 13% they were rare, and in 18% of specimens they were absent entirely.

*Megasclere Geometry* (Fig. 3).

1. The bases of *choanosomal principal megascleres* in both species were typically spined, although fewer examples of entirely smooth spicules were also found in most (85%) specimens. The proportion of entirely smooth choanosomal megascleres ranged from 0–4% of spicules (44% of specimens), 5–10% (25%), 11–20% (27%), up to 40% of megascleres sampled (6% of specimens). In only one specimen smooth principal subtylostyles were predominant. Rare examples of strongylote principal megasclere modifications were observed in few specimens (8%). Similarly, there was a tendency for acanthostyles and choanosomal subtylostyles to intergrade, with intermediate forms relatively common (Fig. 3). Both categories of spicules were found echinating fibers. The proportion of choanosomal principal megascleres with partly spined shafts ranged from 0–4% of spicules (48% of specimens), 5–10% (22%), 11–20% (25%), up to 32% of megascleres sampled (5% of specimens).
2. The geometry and distribution of larger subectosomal subtylostyles was relatively consistent in *lendenfeldi*. The majority of those samples had microspined subtylote bases, whereas fewer had smooth bases. The proportion of subectosomal subtylostyles with smooth bases ranged from 0–4% of spicules (46% of specimens), 5–10% (25%), 11–20% (26%), up to 40% of megascleres sampled (3% of specimens). In contrast, in specimens of *major*, up to 80% of larger auxiliary megascleres had spinose and hastate apices, the remainder being the usual fusiform and smooth forms.

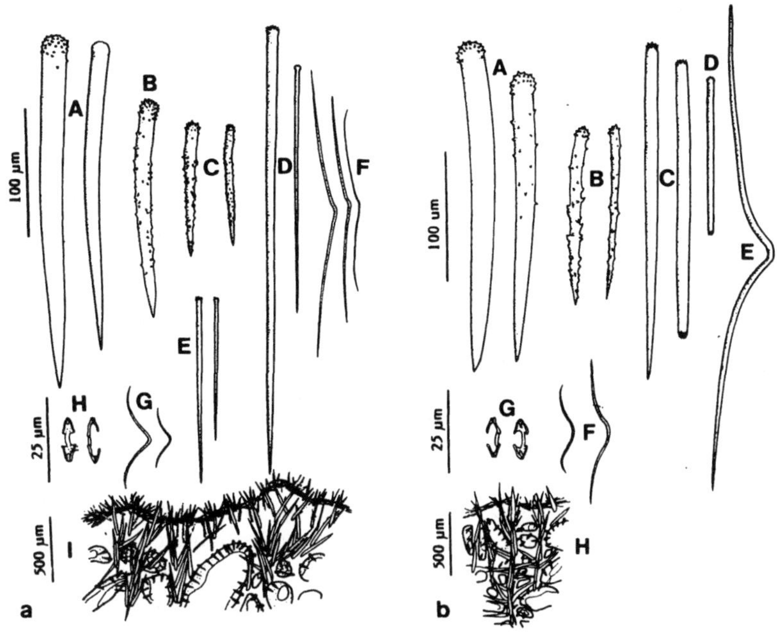


**Fig. 3.** Megascleres microscleres and skeletal structures of (**a**) *C. lendenfeldi* (NTM Z2095) and (**b**) *C. major* (NTM Z0858)

3. The bases of ectosomal auxiliary subtylostyles were invariably spined. They were morphologically similar to subectosomal spicules, but differed in size and distribution in the skeleton. Specimens of *major* had a variable proportion (30–80%) of apically spined and hastate spicules, and that feature was one of very few consistent characters differentiating the two species. In one specimen of *major* ectosomal megascleres were greatly reduced, and the dermal skeleton was composed mainly of larger auxiliary subtylostyles, and in another, both classes of auxiliary spicule were practically absent, replaced almost entirely by acanthostyles and choanosomal styles protruding through the surface.
4. The form and even spination of acanthostyles did not vary between specimens, regardless of seasonal or geographical distribution of samples. Acanthostyles of *major* were also significantly larger than those of *lendenfeldi* ($F = 5.38$, $P < 0.0005$), which appeared to be due to the greater elongation of the aspinose neck region on spicules.

*Microsclere Geometry* (Fig. 3).

1. The geometry of isochelae was entirely homogeneous within and between species. Isochelae were always present in samples but they often varied in abundance, ranging from abundant (3%), common (86%), uncommon (8%), to rare (3% of specimens). In addition, 18% of specimens had a few contort

isochelae (between 4–16% of spicules sampled), but those forms were not considered to be characteristic for the species. Isochelae of *major* were slightly smaller than those of *lendenfeldi* ($F = 1.56$, $P < 0.0025$), although their geometry was identical.

2. Both species had toxa microscleres which varied from long thin straight morphs to short thin recurved morphs (Fig. 3aG-F; 3bE-F), although the thickness and extent of curvature differed between species. The separation of toxas into two distinct morphological categories in either species was artificial, since many intermediate forms were present. Toxas varied greatly in length and abundance between samples of both species: in 36% of specimens they were rare, 23% uncommon, 26% moderately common, and in only 15% of specimens they were noticeably abundant. Toxas of *major* had a distinctly different geometry (Fig. 3bE) and thickness ($F = 2.66$, $P < 0.0005$) from those of *lendenfeldi*, and that feature appeared to be a consistent cryptic morphological feature differentiating the two taxa.

*Variability in Spicule Dimensions* (Fig. 4). In general, *lendenfeldi* had a high intraspecific variability in spicule size, for all spicule categories (Fig. 4A-F), but that variability was as consistent within groups as it was between groups (subdivided on the basis of locality, seasonal, bathymetric, and growth form classes). In fact, only the width of choanosomal styles showed any statistically significant differences between seasonal distributions of samples (analyzed for samples from the Darwin region only, $P < 0.05$) (Hooper et al. 1990). In other words, the species was equally as variable throughout Western Australian, Northern Territory, Queenslands, and Indonesian populations. In contrast, most megascleres in most specimens of *lendenfeldi* were generally (statistically) smaller than those in *major*. That trend was clearly indicated in visual plots of mean spicule lengths and widths ($\pm 2$ standard errors) (Fig. 4), but it was not an absolute difference and the two species showed some overlap. Statistical differences between the species were significant for acanthostyles only (Hooper et al. 1990), but those low significance levels are probably due to high levels of within-group variability.

## *Reproduction*

Only three specimens of *lendenfeldi* and one of *major* were recorded with reproductive products, over 4 years observations. Viviparous larvae were oval to elongate, measuring 210–313 × 190–212 μm, and 290–330 × 380–397 μm, respectively, and fecund individuals were collected from both the Darwin and Cobourg Peninsula regions of the Northern Territory (April-May, September-October). Larvae contained well-developed juvenile styles, isochelae and toxas located predominantly near the posterior pole and forming radial tracts, in addition to abundant spongin, and scattered pigment granules, identical with those observed in the adult. Patterns of ciliation on larvae were not observed (preserved material).

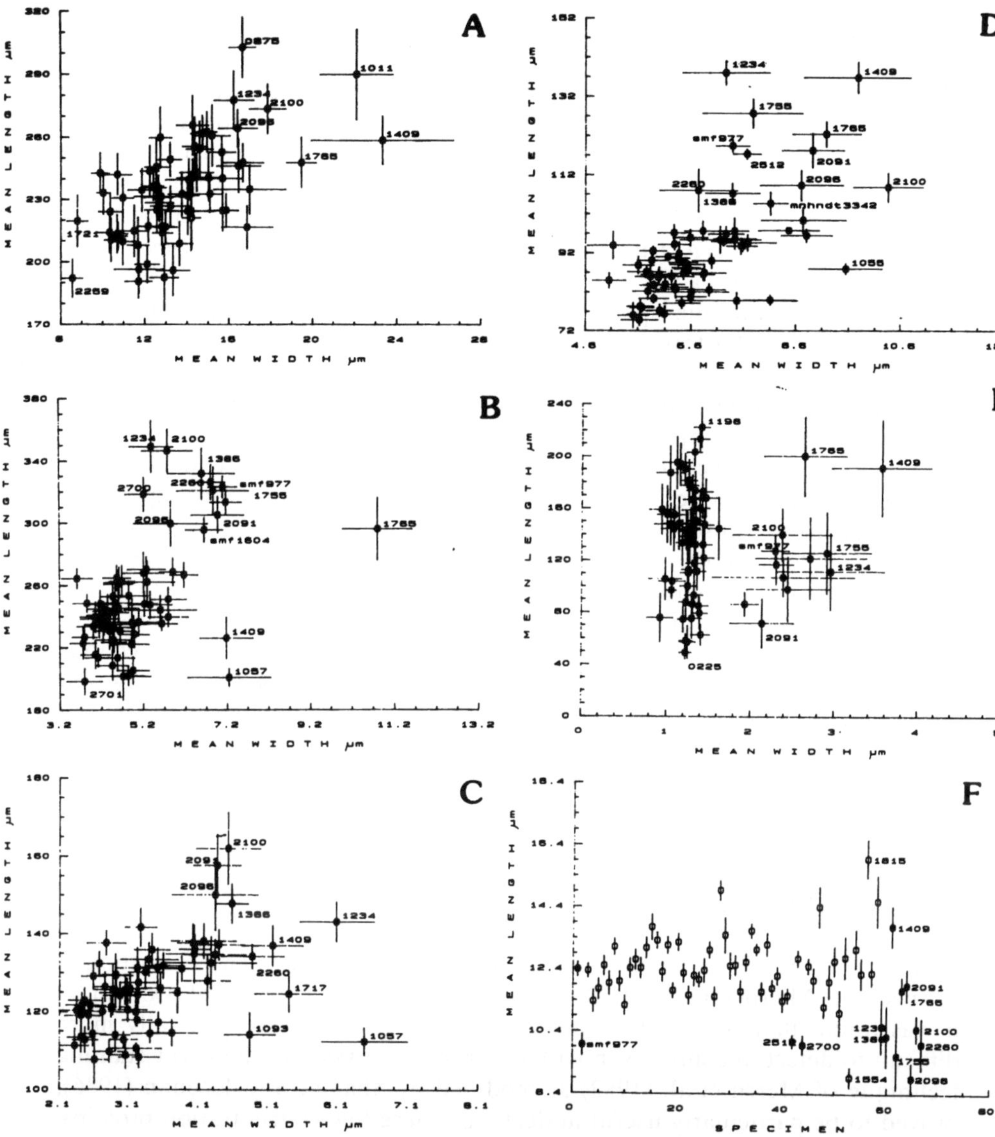


**Fig. 4.** Variability in spicule dimensions of *C. lendenfeldi* (*open circles*) and *C. major* (*solid circles*). Plots of mean spicule lengths and widths (± 2 standard errors of the mean) are based on 25 measurements of each spicule category for each specimen. **A** Choanosomal principal styles. **B** Subectosomal auxiliary styles. **C** Ectosomal auxiliary styles. **D** Echinating acanthostyles. **E** Toxas. **F** Palmate isochelae (length only)

### *Ecology and Biogeography*

*Clathria lendenfeldi* (sensu stricto) has been recorded previously from the Pacific coast of Australia and the Gulf of Manaar, India. Other species synonymized with *lendenfeldi* (Hooper et al. 1989) are also known to have a widespread distribution throughout (a) the Indian Ocean (*C. spicata*-Western Australia, Cargados Carajos, and Gulf of Aden; *C. whiteleggei*-Saya de Malha, Natal coast, Gulf of Aden and Arabian coast); (b) Arafura and Timor Seas (*C. coppingeri* var. *aculeata*-Aru I., Indonesia), and (c) Pacific coast of Australia (*R. bispinosus* -NSW coast). The present study found the species to be relatively abundant throughout the northwest coast of Australia. *C. major* has been previously recorded only from Aru I., Indonesia (Hentschel 1912), but in the present study it was found to have a relatively widespread distribution throughout northwest Australian waters, comparable to (and sympatric with) *C. lendenfeldi*.

Shallow-water specimens of both species (0–19 m depth) were found on the sides, tops, and in crevices of rock and dead coral heads, invariably exposed to currents (personal observations). Deeper-water specimens (39–108 m depth) were exposed to currents, attached to flat rock or coral substrate, with at least the basal portion of stems covered by silt or shell-grit (unpublished CSIRO in situ trawl photographs, P Young pers. commun.). Polychaete infestations were common in both species, and 41% (*lendenfeldi*) and 36% (*major*) of sections examined had at least one polychaete worm in the mesohyl [*Typosyllis spongicola* (Scyllidae)]. One specimen of *lendenfeldi* contained numerous isopods on the external surface and burrowing into the sponge, and another was infested with ophiuroids, whereas one specimen of *major* was covered entirely with epiphytic algae.

## Biochemical Systematics

*Proteins* (Fig. 5). From an electrophoretic study of 276 specimens, consisting of 13 species of Microcionidae, 15 Raspailiidae, and 23 other species of Axinellida, it was found that protein banding patterns of sponges were generally more difficult to detect in comparison with other marine taxa, but the silver-staining technique of Merril et al. (1982) as modified by Shaklee and Keenan (1986), proved to be particularly useful in detecting these low-concentration proteins. This technique also produced color differences in many bands, and these differences were found to be useful in determining whether or not proteins with the same relative mobilities (Rm) were in fact homologous. Banding patterns of all species examined were shown to be relatively consistent regardless of geographic, seasonal, or other differences between samples (Hooper 1991b). Visual interpretation of "fingerprints" from ten specimens of *C. lendenfeldi* and seven of *C. major* showed that each species had quite different general protein banding patterns (Fig. 5). Intraspecific variability was relatively low, with mean levels of similarity between samples [$\pm$ 1 standard error of the mean, (and coefficient of variation)] being 85.86% $\pm$1.31 (CoV = 4.58) and 85.90% $\pm$0.95 (CoV = 2.71) for

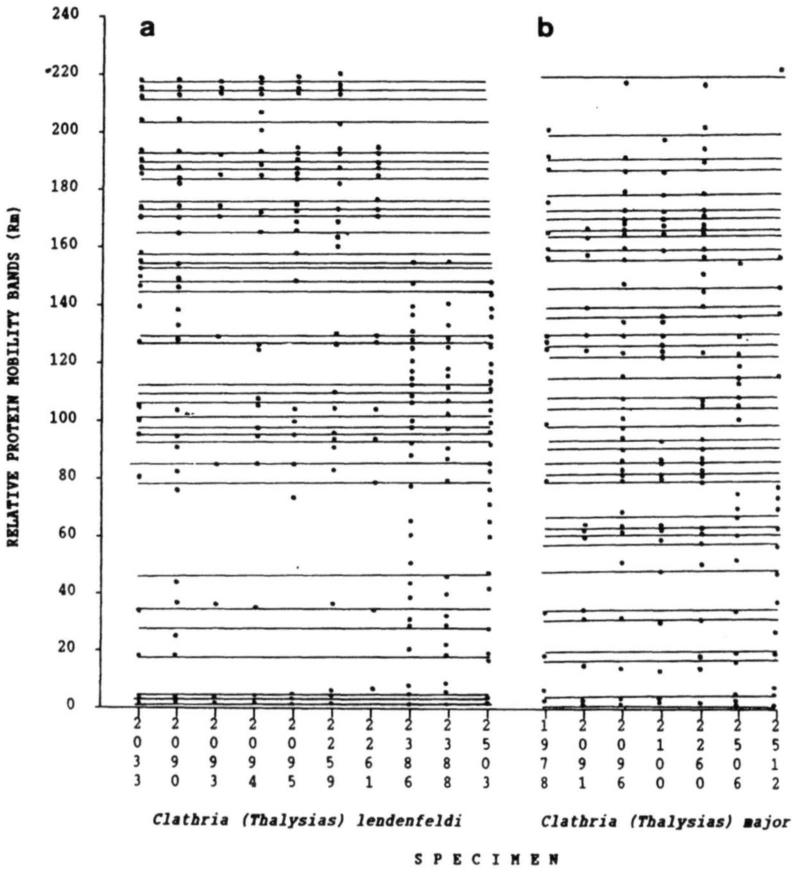


**Fig. 5a,b.** Intra- and interspecific comparisons in electrophoretically detectible protein profiles, showing the major relative mobility (Rm, millimeters) banding pattern of proteins in samples of *C. lendenfeldi* and *C. major*, collected from different localities and in different seasons

*lendenfeldi* and *major*, respectively. An interspecific comparison in protein banding patterns between *lendenfeldi* and *major* produced and average Gower metric similarity index of 80.4% between samples. This index included all relative mobility bands, regardless of consistency or possible origin from symbiotic organisms. A further comparison was made between all 13 microcionid species from northern Australia (Fig. 6a), in which similarity indices were averaged for each species, and in which only consistent relative mobility bands were considered. From those data it was noted that for electrophoretic proteins at least, *lendenfeldi* and *major* were not necessarily the most closely related species, as would be otherwise expected from their postulated close morphological affinities.

*Carotenoid Pigments* (Table 1, Fig. 6b). The utilization of HPLC in the present study, over methods based on TLC (Lee and Gilchrist 1985) enabled much more

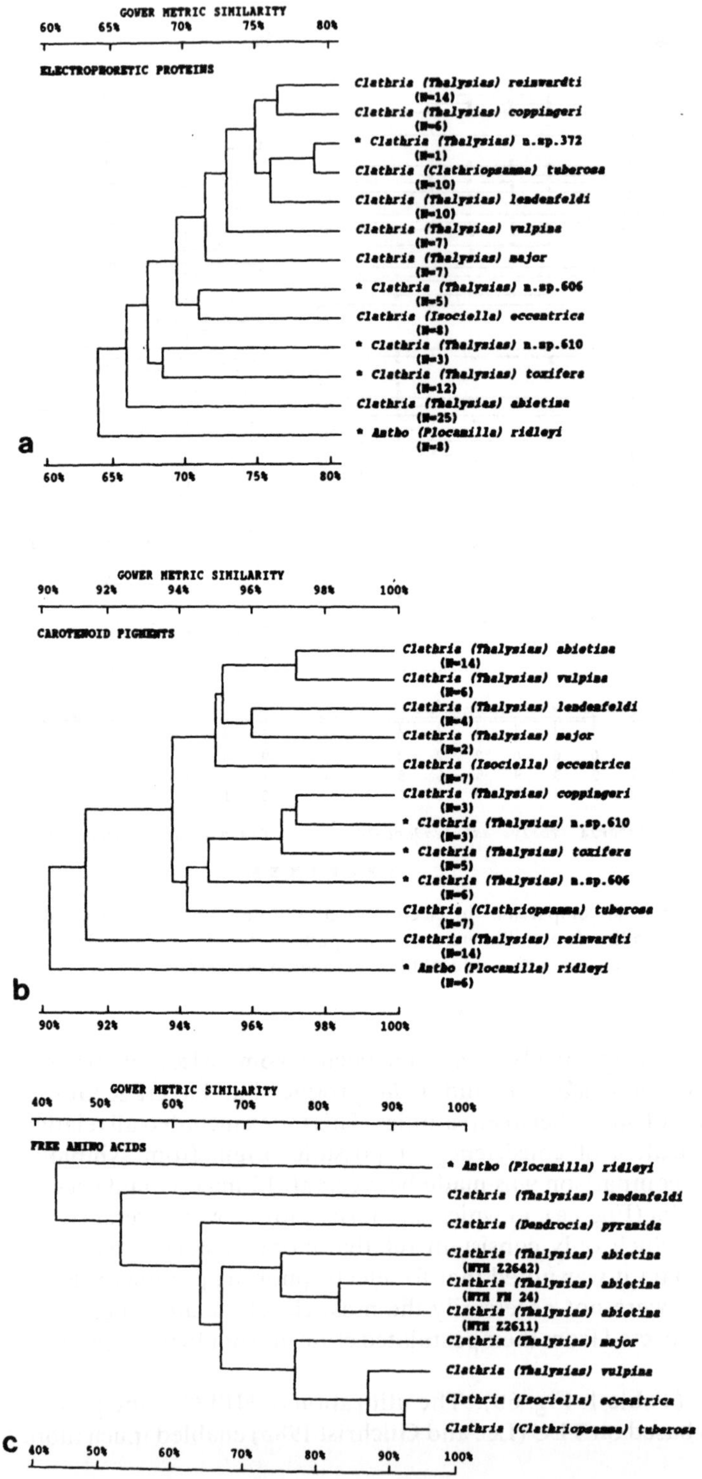
GOWER METRIC SIMILARITY
60% 65% 70% 75% 80%
ELECTROPHORETIC PROTEINS
Clathria (Thalysias) reinwardti (N=14)
Clathria (Thalysias) coppingeri (N=6)
* Clathria (Thalysias) n.sp.372 (N=1)
Clathria (Clathriopsamma) tuberosa (N=10)
Clathria (Thalysias) lendenfeldi (N=10)
Clathria (Thalysias) vulpina (N=7)
Clathria (Thalysias) major (N=7)
* Clathria (Thalysias) n.sp.606 (N=5)
Clathria (Isociella) eccentrica (N=8)
* Clathria (Thalysias) n.sp.610 (N=3)
* Clathria (Thalysias) toxifera (N=12)
Clathria (Thalysias) abietina (N=25)
* Antho (Plocamilla) ridleyi (N=8)
a
60% 65% 70% 75% 80%
GOWER METRIC SIMILARITY
90% 92% 94% 96% 98% 100%
CAROTENOID PIGMENTS
Clathria (Thalysias) abietina (N=14)
Clathria (Thalysias) vulpina (N=6)
Clathria (Thalysias) lendenfeldi (N=4)
Clathria (Thalysias) major (N=2)
Clathria (Isociella) eccentrica (N=7)
Clathria (Thalysias) coppingeri (N=3)
* Clathria (Thalysias) n.sp.610 (N=3)
* Clathria (Thalysias) toxifera (N=5)
* Clathria (Thalysias) n.sp.606 (N=6)
Clathria (Clathriopsamma) tuberosa (N=7)
Clathria (Thalysias) reinwardti (N=14)
* Antho (Plocamilla) ridleyi (N=6)
b
90% 92% 94% 96% 98% 100%
GOWER METRIC SIMILARITY
40% 50% 60% 70% 80% 90% 100%
FREE AMINO ACIDS
* Antho (Plocamilla) ridleyi
Clathria (Thalysias) lendenfeldi
Clathria (Dendrocia) pyramida
Clathria (Thalysias) abietina (NTM Z2642)
Clathria (Thalysias) abietina (NTM PH 24)
Clathria (Thalysias) abietina (NTM Z2611)
Clathria (Thalysias) major
Clathria (Thalysias) vulpina
Clathria (Isociella) eccentrica
Clathria (Clathriopsamma) tuberosa
c
40% 50% 60% 70% 80% 90% 100%

**Table 1.** Carotenoid pigment profiles, showing the distribution (retention times; Rt minutes) and relative abundance (%) of carotenoids in samples of *C. lendenfeldi* and *C. major*, collected from different localities and in different seasons. (+ = trace detected only)

| Retention Time (Rt) (min) | Sample[a] *C. lendenfeldi* 1 | 2 | 3 | 4 | *C. major* 1 | 2 |
|---|---|---|---|---|---|---|
| 1.74–1.78 | 26.1 | 11.8 | 2.9 | 4.1 | + | + |
| 1.94–1.98 | 51.3 | 20.1 | 7.1 | 29.6 | 6.3 | 13.9 |
| 2.43–2.58 | | | + | + | 1.1 | 1.3 |
| 2.82–2.97 | | | | | | |
| 3.04–3.17 | 7.6 | | | | | |
| 3.58–3.77 | | | | | + | + |
| 4.23–4.41 | | | | | 2.5 | 1.1 |
| 4.82–5.36 | 3.0 | 33.9 | 8.9 | 35.5 | 5.7 | 5.5 |
| 5.40 | | | | | | |
| 5.61–5.75 | | | | | | |
| 5.90–6.51 | + | + | 7.6 | 14.6 | 13.3 | 10.6 |
| 6.80–7.18 | + | | + | + | | |
| 7.24–7.55 | | + | | | + | + |
| 7.70–8.06 | + | 2.5 | 6.0 | 8.7 | 6.9 | 5.9 |
| 8.47–8.88 | | + | 5.0 | 1.6 | + | + |
| 8.94 | | | | | | |
| 9.62 | | | | | | |
| 9.90–10.53 | 12.0 | 21.5 | 44.4 | 4.2 | 42.3 | 61.8 |
| 10.84–11.07 | | | | | 17.4 | + |
| 11.32–11.65 | | + | | | | |
| 12.08–12.12 | | | | | | |
| 12.34–12.43 | | | | | | |
| 12.72–12.97 | | 5.1 | 7.1 | 0.8 | 2.3 | 3.4 |
| 14.61–14.82 | + | 5.0 | 11.0 | 1.0 | 2.1 | 2.2 |

[a] Samples:

1. *C. lendenfeldi*, NTM Z2093, Darwin region, dry season.
2. *C. lendenfeldi*, NTM Z2259, Darwin region, wet season.
3. *C. lendenfeldi*, NTM Z2386, Darwin region, pre-dry season.
4. *C. lendenfeldi*, NTM Z2503, Cobourg Peninsula region, dry season.

1. *C. major*, NTM Z2512, Cobourg Peninsula region, dry season.
2. *C. major*, NTM Z2506, Cobourg Peninsula, dry season.

**Fig. 6.** Numerical similarity (%) in general protein electrophoresis (**a**), carotenoid pigment profiles (**b**) and free amino acids (**c**) of representative Australasian *Clathria* species, based on a cluster analysis using the Gower metric similarity index and UPGMA (group-average) sorting (* incrusting species; *N* number of specimens analyzed) accurate detection of even minute differences in both the composition and concentration of carotenoids between samples. However, conclusions derived from both these studies were based on phenetic comparisons between samples and species, which must assume that retention time (RT) peaks with the same or similar numerical values represent homologous proteins, whereas more sophisticated analyses are capable of molecular recognition of carotenoids (e.g., Liaaen-Jensen et al. 1982). Consequently, reliable conclusions from these data can only be made for comparisons based on multiple samples of individual species. From an examination of 178 samples, consisting of 12 species of Microcionidae, 13 Raspailiidae, and 12 other species of Axinellida, it was found that levels of intraspecific variability between samples were generally much lower than those recorded for electrophoretically detectable proteins (Hooper 1991b). Four specimens of *C. lendenfeldi* and two of *C. major* were analyzed (Table 1), and mean levels of similarity between samples were 95.53% ±0.73 (CoV = 1.32) and 95.30% respectively. The average level of similarity between the two species was 92.7%, and on the basis of their carotenoid pigment characteristics at least, the two taxa appeared to be more closely related to each other than they were to any of the other 10 species of Microcionidae examined (Fig. 6b). Another important difference found between *C. lendenfeldi* and *C. major* was that the water-soluble fraction of sample extracts in *major* only showed antibiotic activity against *B. subtilis*. The biological activity observed was not pronounced, nor have analyses proceeded further to enable the identification of active molecules, but activity was consistent for both samples of the species and may be an accurate representation.

*Free Amino Acids* (Table 2, Fig. 6c). Free amino acid (FAA) analyses provided accurate molecular comparison between samples, and these data are also directly comparable with results of earlier studies (Bergquist and Hogg 1969; Bergquist and Hartman 1969). Thirty-nine specimens, consisting of nine species of Microcionidae, nine Raspailiidae, and 12 other species of Axinellida were examined. All species were clearly differentiated from each other by the distribution and concentrations of FAA's, and these profiles did not appear to vary significantly between conspecific populations (checked for multiple samples of *Clathria abietinus*) (Fig. 6c). Substantial differences were noted to occur between FAA's in *C. lendenfeldi* and *C. major* (Table 2). *C. lendenfeldi* had a greater diversity of amino acids, although many of those occurred only in relatively low concentrations (and which less sophisticated methods of detection would probably record as merely trace quantities), and FAA's such as serine, alanine and leucine were relatively more common. By comparison, *C. major* had fewer FAA's but significantly higher concentrations of proline, DOPA, and glycine, in particular (per given unit weight of sample); the glycine:proline ration was 11:1 in *C. lendenfeldi* and 5:1 in *C. major*. In general, the FAA profiles of the 2 species of Microcionidae examined here, the other 7 species examined (Fig. 6c), and the three species known from previous studies (Bergquist and Hogg 1969; Bergquist and Hartman 1969) were in relatively close agreement. A significant difference between those studies was the low concentrations of taurine in present samples, whereas earlier results showed that the amino acid occurred at relatively high and

**Table 2.** Free amino acid (FAA) profiles of *C. lendenfeldi* and *C. major* from the Darwin region, showing the relative abundance (% area of major peaks) of each FAA and range of retention times (Rt minutes)

| FAA | Rt (min) | % Concentration *C. lendenfeldi* | *C. major* |
|---|---|---|---|
| Aspartic acid | 2.07 | 0.63 | – |
| Glutamic acid | 2.37 | 1.53 | – |
| Serine | 3.39–3.53 | 4.27 | 1.40 |
| Glycine | 3.79–3.80 | 54.29 | 72.52 |
| Histidine | 4.23 | 1.19 | – |
| (Unknown) | 4.74 | 0.75 | – |
| Taurine | 5.03–5.04 | 2.39 | 1.14 |
| Arginine | 5.39–5.40 | 2.82 | 0.76 |
| Threonine | 5.65–5.66 | 2.63 | 1.12 |
| Alanine | 5.99 | 6.13 | – |
| Proline | 6.27–6.34 | 4.59 | 13.82 |
| DOPA | 8.22–8.23 | 1.07 | 3.59 |
| (Unknown) | 8.95 | 0.67 | – |
| (Unknown) | 9.44–9.62 | 0.65 | 1.60 |
| Valine | 10.32–10.36 | 4.37 | 1.24 |
| Tyrosine | 10.78–10.82 | 0.85 | 0.75 |
| (Unknown) | 11.62 | – | 1.17 |
| Isoleucine | 13.27 | 2.78 | – |
| Leucine | 13.54–13.58 | 3.41 | 0.89 |
| (Unknown) | 14.12 | 1.02 | – |
| Lysine | 15.14 | 1.57 | – |
| (Unknown) | 15.74 | 2.39 | – |

consistent levels. Similarly, both species of *Clathria*, and all other microcionids examined here had relatively high levels of proline, whereas only small concentrations were previously known to occur in the family. High levels of glycine and marginally lower but significant levels of glutamine and serine were found in all samples, and that profile is apparently characteristic for the Poecilosclerida in general (Bergquist and Hogg 1969; Bergquist and Hartman 1969).

## Discussion

Taxonomic differentiation between these closely related sympatric species, *C. major* and *C. lendenfeldi*, was clearly supported by heterogeneity between groups in electrophoretic proteins, free amino acids, carotenoid pigment profiles, and the presence of antibiotic activity in *C. major*, and there was also an equally clear within-group biochemical homogeneity for both species. However, morphological features that correlated with those biochemical data were less obvious, to the extent that when used alone morphological evidence may be insufficient to enable clear and consistent differentiation between the taxa. Both species belonged to Hallmann's (1912) "spicata" group of *Clathria*, which is characterized in having mostly aspiculose (uncored) fibers heavily echinated by both acanthostyles and choanosomal principal subtylostyles, an ordered extra-fiber skeleton consisting of more or less plumose ascending spicule tracts composed of both choanosomal and subectosomal megascleres (producing the "spicate" arrangement from which the group derives its name), and the presence of choanosomal and subectosomal subtylostyles which project through (and mostly mask) the ectosomal skeleton.

Shape and surface features were found to be the least consistent of all characters examined, although it was possible to define variability to within finite ranges, and several nominal *Clathria* species which had been previously differentiated from other members of the "spicata" group on the basis of growth form (e.g., *C. whiteleggei, C. spicata*) were found to be clear synonyms of *C. lendenfeldi* (Hooper 1991a). Similarly, many features which have been used previously in the traditional identification of species (and even genera) of Microcionidae were not found to be particularly reliable in indicating new taxa. Those features were shared between these sibling species, in numerous other microcionid species complexes (Hooper 1991b), and it was concluded that for this group only combinations of characters appear to be reliable indicators.

Of the 148 nominal Australasian Microcionidae species examined, including 16 species which occurred relatively commonly in the northwestern Australian region, *C. lendenfeldi* and *C. major* were clearly the most morphologically diverse sponge species of that group, so much so that some specimens were assigned to either taxon only with hesitation. Those anomalous examples differed from typical forms of both species only in details of questionable systematic importance (megasclere basal spination, isochelae size, degree of axial condensation of the skeleton). Specimens referred here to *major* differed from the remainder only by the geometry and thickness of toxas (cf. Fig. 3aF-G; 3bE-F), in the generally marginally, statistically) larger size of most megascleres (Fig. 4A-F), and in the presence of apically microspined auxiliary megascleres, similar to those found in a small group of other *Clathria* [viz. *C. (Clathriopsamma) australiensis* (Carter) from southern Australia, *C. (Clathria) parthena* (de Laubenfels) from California, and *C. haplotoxa* (Topsent) from north Africa]. In other details, *C. major* was equally as variable as *C. lendenfeldi*, and both species had sympatric geographical, ecological, and bathymetric distributions, similar ranges of growth forms, skeletal architecture, and spiculation. As shown above, it is highly improbable that those morphological and biochemical differences merely represent phenotypic variability between populations of a single species, whereas it is quite likely that they do represent fixed differences in the sponge genotype. Despite the impracticalities associated with "field identifications" of that species complex, it is considered here that a combination of cryptic morphological characters, statistically different morphometric features, and biochemical differences between the taxa clearly differentiate them at the species level of classification.

*Acknowledgements.* We wish to thank Daniel Low Choy, Stephen Brand, Diane Barton, Peter Scammels, and Paul Lynden for technical assistance during various aspects of this study, and Russell Hanley (NT Museum) for identifying polychaete worms. Shirley Stone (BMNH), Klaus Rützler (USNM), Dietrich Kühlmann (ZMB), Manfred Grasshoff (SMF), Claude Lévi (MNHN), Lester Cannon (QMU), and Loisette Marsh (WAM) kindly provided numerous type specimens and other material used in this study. Peter Young (CSIRO Fisheries) generously provided access to his in situ trawl photographs of sponges from the Northwest Shelf. The Northern Territory University Planning Authority, Northern Territory Museum, University of Queensland and Australian National University provided funds and equipment for biochemical analyses. The Australian Heritage Commission and CSIRO Fisheries are thanked for providing funds and shiptime, enabling collections to be made from the Cobourg Peninsula and the Northwest Shelf, respectively.

## References

Bergquist PR (1978) Sponges. Hutchinson, London, pp 268

Bergquist PR, Hartman WD (1969) Free amino acid patterns and the classification of the Demospongiae. Mar Biol 3(3):247–268

Bergquist PR, Hogg JJ (1969) Free amino acid pattern in Demospongiae: a biochemical approach to sponge classification. Cah Biol Mar 10:205–220

Hallmann EF (1912) Report on the Sponges obtained by the FIS "Endeavour" on the Coasts of New South Wales, South Australia, Queensland, and Tasmania, 1909–10. Part 1. Zool Res Fish Exp FIS "Endeavour" 1909–10, (2):117–300

Hentschel E (1912) Kiesel- und Hornschwämme der Aru und Kei-Inseln. Abh Senckenb Naturforsch Ges (1912):295–448

Hooper JNA (1987) New Records of *Acarnus* Gray (Porifera: Demospongiae: Poecilosclerida) from Australia, with a synopsis of the genus. Mem Queensl Mus 25(1):71–105

Hooper JNA (1991a) Character stability, systematics and affinities between Microcionidae (Poecilosclerida) and Axinellida. Proc 3rd Int Conf Biol Sponges. Smithson Contrib Zool: (in press)

Hooper JNA (1991b) Revision of the Australasian Microcionidae (Porifera: Demospongiae). Invert Tax (in preparation)

Hooper JNA, Capon RJ, Keenan CP, Parry DL (1990) Biochemical and morphometric differences of two sympatric sibling species of *Clathria* (Porifera: Demospongiae: Microcionidae) from northern Australia. Invert Tax 4(1):123–148

Lee WL, Gilchrist BM (1985) Carotenoid patterns in twenty-nine species of sponges in the order Poecilosclerida (Porifera: Demospongiae): a possible tool for chemosystematics. Mar Biol 86:21–35

Liaaen-Jensen S, Renstrom B, Ramdahl T, Hallenstvet M, Bergquist P (1982) Carotenoids of Marine Sponges. Biochem Syst Ecol 10(2):167–174

Merril CR, Goldman G, van Keuren ML (1982) Simplified silver protein detection and image enhancement methods in polyacrylamide gels. Electrophoresis 3:17–23

Palumbi SR (1984) Tactics of acclimation: morphological changes of sponges in an unpredictable environment. Science 225 (4669): 1478–1480

Palumbi SR (1986) How body plans limit acclimation: responses of a demosponge to wave force. Ecology 67(1):208–214

Pansini M (1982) Notes on some Mediterranean Axinella with description of two species. Boll Mus Ist Biol Univ Genova 50–51:79–98

Ridley SO, Dendy A (1886) Preliminary report on the monaxonida collected by the HMS "Challenger". Ann Mag Nat Hist (5)18:325–351, 470–493

Shaklee JB, Keenan CP (1986) A practical laboratory guide to the techniques and methodology of electrophoresis and its application to fish fillet identification. CSIRO Mar Lab Aust Rep (177):1–59

Simpson TL (1978) The biology of the marine sponge *Microciona prolifera* (Ellis and Solander). 3. Spicule secretion and the effect of temperature on spicule size. J Exp Mar Biol Ecol 35(1):31–42

Simpson TL (1984) The cell biology of sponges. Springer, Berlin Heidelberg New York Tokyo, pp 662
Thompson JE, Murphy PT (1989) Environmentally induced variation in the diterpene composition of a Great Barrier Reef sponge. Proc 3rd Int Conf Biol Sponges. Smithson Contrib Zool (in press)
Wiedenmayer F (1977) The Nepheliospongiidae Clarke, 1900 (Demospongea, Upper Devonian to Recent) an ultraconservative, chiefly shallow-marine sponge family. Eclogae Geol Helv 70(3):885–918
Wilson HV (1921) The genus *Raspailia* and the independent variability of diagnostic features. J Elisha Mitchell Sci Soc Univ N Carolina Chapel Hill 37:54–60

# Histological Indications of the Phylogenesis of the Haplosclerida (Demospongiae, Porifera)

P.-F. LANGENBRUCH

## Introduction

In spite of the rather homogeneous skeletal appearance of the Haplosclerida, histological features have recently been found (Langenbruch 1983, 1985, 1988; Langenbruch et al. 1985; Langenbruch and Scalera-Liaci 1986; Langenbruch and Weissenfels 1987), which call into question whether this group of sponges is really monophyletic. The focus of these investigations has been the structure of the choanocyte chambers, their relation to the mesenchymal tissue, and their arrangement between the canal systems.

Meanwhile, some additional haplosclerid sponge species have been investigated in this regard. The results obtained to date by the author till now in this field are summarized in the present chapter and discussed with regard to their phylogenetic and systematic implications.

## Material and Methods

Specimens of 19 different marine haplosclerid sponge species were collected from the Mediterranean (Bari, Italy; Marseille, France), the North Atlantic (Helgoland, FRG; Anglesey, Wales, UK) and the Caribbean (Discovery Bay, Jamaica). Most of them [*Haliclona elegans* (Bowerbank 1866); *Haiclona mediterranea* (Griessinger 1971); *Haliclona rosea* (Bowerbank 1866); *Haliclona oculata* (Pallas 1766); *Haliclona limbata* (Montagu 1818); *Haliclona indistincta* (Bowerbank 1866); *Adocia simulans* (Johnston 1842); *Reniera mucosa* (Griessinger 1971); *Reniera fulva* (Topsent 1893); *Reniera sarai* (Pulitzer 1969); *Reniera* sp.; *Gellius ravus* (Stephens 1912); *Pellina fistulosa* (Bowerbank 1866); *Dendroxea lenis* (Topsent 1892); *Petrosia ficiformis* (Poiret 1789)] were cut into small pieces with razor blades immediately after collection or after a stay of several days in aquaria supplied with running seawater and fixed for 2 h at 4°C in 2% glutaraldehyde and 2% $OsO_4$ in 0.2 M Na-cacodylate buffer (pH 6.9) containing 5.5% sucrose, desilicified for 1 h in 5% hydrofluoric acid (HF) in 0.2 M Na-cacodylate buffer containing 6.9% sucrose, dehydrated in ethanol, block-stained for 2 h at 4°C with 1% phosphotungstic acid and 1% uranyl acetate in the 70% ethanol of the dehydration series, and embedded in styrol-methacrylate.

J. Reitner and H. Keupp (Eds.)
Fossil and Recent Sponges

Another part of the specimens [*Callyspongia* sp.; *Niphates digitalis* (Lamarck 1814); *Niphates* sp.; *Amphimedon compressa* (Duchassaing and Michelotti 1864)] was fixed in situ by a method developed by Willenz (modified after Eisenman and Alfert 1981) for 5–10 min in a mixture (combined in situ immediately before use) of 10 parts 4% glutaraldehyde in 0.2 M Na-cacodylate buffer (pH 7.4) supplemented with 0.35 M sucrose and 0.1 M NaCl, and 1 part 1% $OsO_4$ in 0.2 M Na-cacodylate supplemented with 0.3 M NaCl. Fixation was then continued in the same fixative without the added $OsO_4$ at 0°C to 4°C for 48 h. The material was postfixed for 3 h in 1% $OsO_4$ in 0.2 M Na-cacodylate supplemented with 0.3 M NaCl, dehydrated in ethanol, desilicified for 2 h in 5% hydrofluoric acid (HF) in the 70% ethanol of the dehydration series and embedded in styrol-methacrylate.

In addition, specimens of four freshwater sponge species [*Ephydatia fluviatiis* (Linné 1758); *Ephydatia muelleri* (Lieberkühn 1856); *Spongila lacustris* (Linné 1759); *Eunapius fragilis* (Leidy 1851)] were cultivated in laboratory aquaria, and after a growth period of 3 to 11 weeks have been fixed for 1 h in 1% $OsO_4$ and 1% $K_2Cr_2O_7$ adjusted by 2.5 N KOH to pH 7.3, desilicified for 30 min in 5% hydrofluoric acid (HF) in 0.025 M phosphate buffer (pH 7.3) containing 6.9% sucrose, dehydrated in ethanol, blockstained for 2 h at 4°C with 1% phosphotungstic acid and 1% uranyl acetate in the 70% ethanol of the dehydration series and embedded in styrol-methacrylate.

All fixation techniques used were well suited for the intended investigations, with regard to phase-contrast microscopy as well as for transmission and scanning electron microscopy.

Preparation for scanning electron microscopy was carried out by a method developed by Weissenfels (1982), described already several times (Langenbruch 1983, 1984, 1985; Langenbruch et al. 1985; Langenbruch and Scalera-Liaci 1986).

For phase-contrast microscopy a Leitz Dialux microscope with Wild MPS 45/51 automatic photography attachment and for transmission electron microscopy a Zeiss EM 9 S-2 was used. Scanning electron microscopy was carried out on a Siemens Autoscan (Fig. 1b), the Field Emission SEM Hitachi S800 (Fig. 4a) and a Leitz AMR 1000 (Fig. 4b).

## Results

With regard to the relation of the choanocyte chambers to the mesenchyme, two fundamentally different types of choanocyte chamber arrangement were found in different species of the investigated Haplosclerida. They are exemplified in the following.

### *Choanocytes in Contact with the Mesenchyme*

Examples of this kind of choanocyte chamber arrangement are shown in Fig. 1a-c. The section through a tissue region of *Spongilla lacustris* in Fig. 1a contains four choanocyte chambers (ChC), pinacocyte-bounded spaces of the incurrent

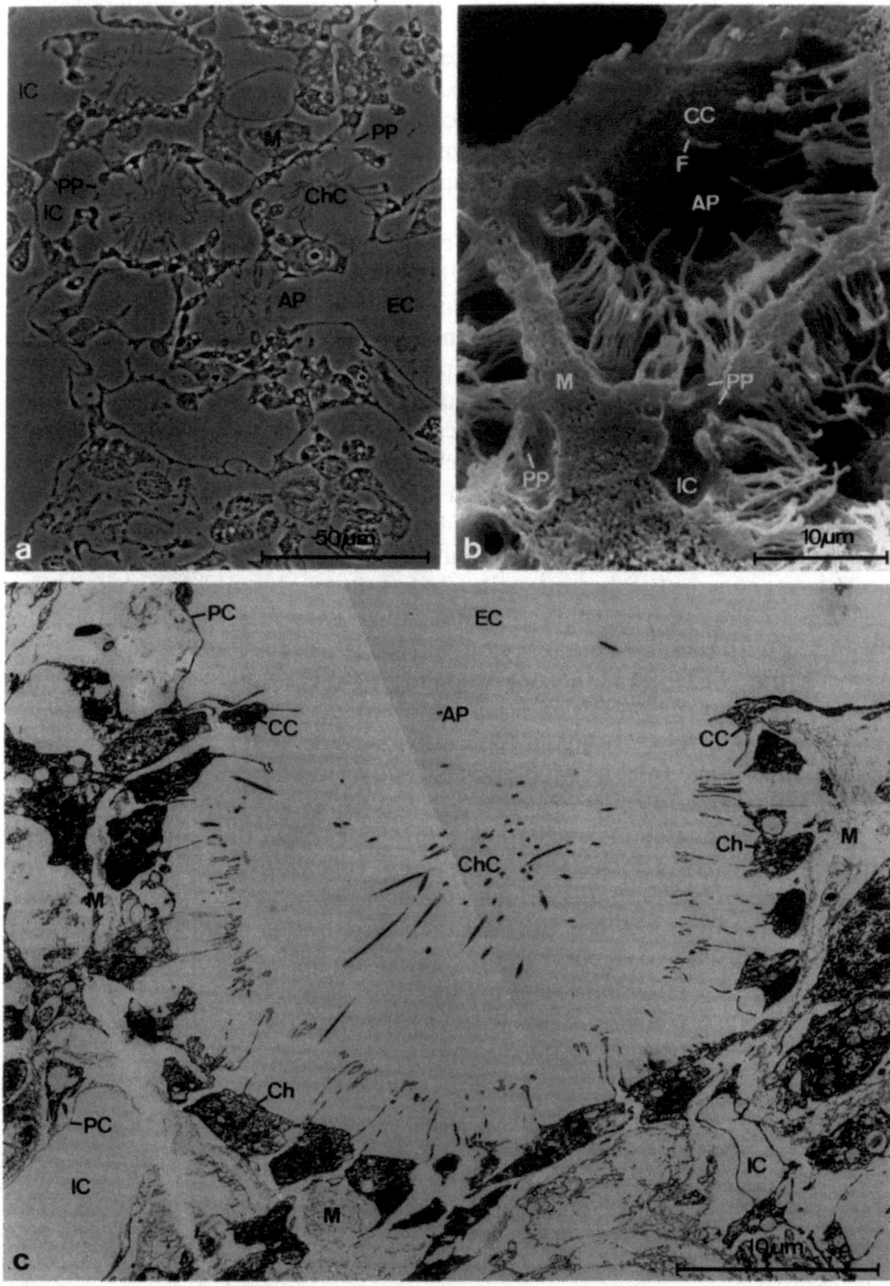


**Fig. 1.** **a** Tissue region of *Spongilla lacustris* (phase-contrast micrograph). *IC* incurrent canal; *EC* excurrent canal; *ChC* choanocyte chamber; *PP* prosopyle; *AP* apopyle; *M* mesenchyme. **b** Sectioned tissue piece of *Dendroxea lenis* (scanning electron micrograph). *IC* incurrent canal; *PP* prosopyle; *CC* cone cell; *AP* apopyle; *F* flagellum; *M* mesenchyme. **c** Median section through a choanocyte chamber (*ChC*) of *Haliclona indistincta* (transmission electron micrograph). The choanocytes (*Ch*) are at their outer surface in contact with the mesenchyme (*M*). *IC* incurrent canal; *EC* excurrent canal; *PC* pinacocyte; *CC* cone cell; *AP* apopyle

(IC) and of the excurrent canal (EC) systems and mesenchyme (M) regions filled with different kinds of cells. The choanocyte chambers of *Spongilla lacustris* are of the normal eurypylous type, which is common to many other Demospongiae. They are by prosopyles (PP) of about 5 μm diameter connected with the incurrent canals (IC) and open each by one wide apopyle (AP) to the excurrent canal (EC) system.

The arrangement of this choanocyte chamber type between the canal systems is demonstrated in Fig. 1b in a scanning electron micrograph of a sectioned tissue piece of the marine sponge *Dendroxea lenis* with one half sectioned and two partly visible choanocyte chambers. Small branches of the incurrent canal (IC) system approach the chambers and, where the pinacocytes of the canal wall come into contact with the choanocytes, small holes, prosopyles (PP), are seen, which connect the chambers with the incurrent canal system. The apopyle (AP), through which the water is expelled to the excurrent canal system, is surrounded by a ring of cone cells (CC). These flagellated (F) cells are modified choanocytes (Weissenfels 1981), which are inserted at the seam between the choanocyte epithelium and the pinacocyte epithelium of the excurrent canal wall.

The arrangement of the cone cells (CC) at the border of the apopyle (AP) is also seen in Fig. 1c in a median section through a choanocyte chamber of *Haliclona indistincta*. Two pinacocyte-bounded (PC) small incurrent canals (IC) approach the chamber at the opposite side of the apopyle, but in this case prosopyles are not sectioned. Of special importance is the fact that in this chamber type the choanocytes (Ch) are, at their outer surface, in direct contact with the mesenchymal tissue (M), which consists of different kinds of cells and collagen fibrils in the intercellular spaces.

All the investigated marine haplosclerid species with choanocytes in contact with the mesenchyme possess a special kind of cells with large inclusions (CI in Fig. 2a,b) in the mesenchyme, which appear in none of the other investigated Haplosclerida. Figure 2a demonstrates their distribution in the mesenchyme of *Reniera sarai*. These cells are often found in higher concentrations near the excurrent canals. Their inclusions (I in Fig. 2b) are membrane-bound and contain an obviously fibrillar material in a variable density. The cytoplasma around the vacuoles is in some regions (→) extremely dense.

### *Choanocytes Separated from the Mesenchyme*

Examples of this second type of choanocyte chamber arrangement in haplosclerid sponges are shown in Figs. 3a,b and 4a,b). The phase-contrast aspect (Fig. 3a), demonstrated here in *Haliclona oculata*, is already remarkably different. The mesenchyme (M), is only poorly developed and many of the choanocyte chambers (ChC) are standing far out into large incurrent canals (IC).

In Fig. 3b the construction of this kind of choanocyte chamber is shown in a transmission electron micrograph of a section through a choanocyte chamber of *Callyspongia* sp. The chamber is situated between an incurrent (IC) and an

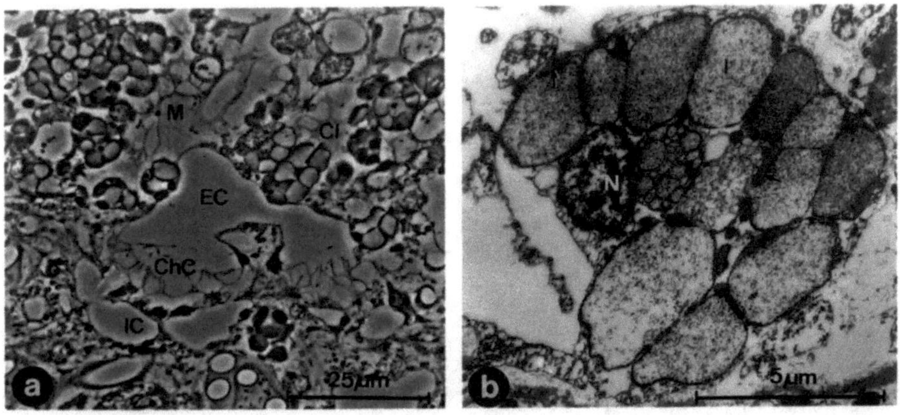


**Fig. 2.** **a** Tissue region of *Reniera sarai* (phase-contrast micrograph). *IC* incurrent canal; *EC* excurrent canal; *ChC* choanocyte chamber; *CI* cell with large inclusions; *M* mesenchyme. **b** Cell with large inclusions (*I*) in the mesenchyme of *Dendroxea lenis* (transmission electron micrograph). *N* nucleus; → dense cytoplasm

excurrent canal (EC). On both sides the mesenchymal tissue (M) approaches the chamber, but in this case the choanocytes (Ch) are completely separated from the mesenchyme (M) by a continuous pinacocyte cover (→), which extends in prolongation of the pinacocyte epithelium (PC) of the incurrent canal (IC) wall all over the outer chamber surface to the apopyle (AP). At the border of the apopyle the pinacocyte cover (→) is in contact with a cone cell (CC) ring which surrounds the apopyle. Consequently in this case none of the choanocytes gets in contact with the mesenchymal cells.

As already mentioned, many of the choanocyte chambers of this special organizational type are standing far out into the incurrent canals (Fig. 3a). By scanning electron microscopy (Fig. 4a,b) they are seen at the incurrent canal (IC) walls covered to the canal lumen by a nearly complete pinacocyte cover (PCC in Fig. 4a) or only by some processes (PCP in Fig. 4b) of the canal wall pinacocytes. In some cases the chambers may even be uncovered to the incurrent canal lumen. Where a pinacocyte cover is missing, the cell bodies of the choanocytes (Ch) are seen from the incurrent canal side. They are connected to one another by small processes (→). Thus a sieve-like structure results. There are no distinctive prosopyles in these only scarcely covered chambers, but the water flows in by all spaces between the choanocytes.

### *Synopsis of the Investigated Species*

The investigated haplosclerid sponge species are listed in Table 1 according to their type of choanocyte chamber arrangement. The particular chamber types are schematically drawn at the bottom of the table.

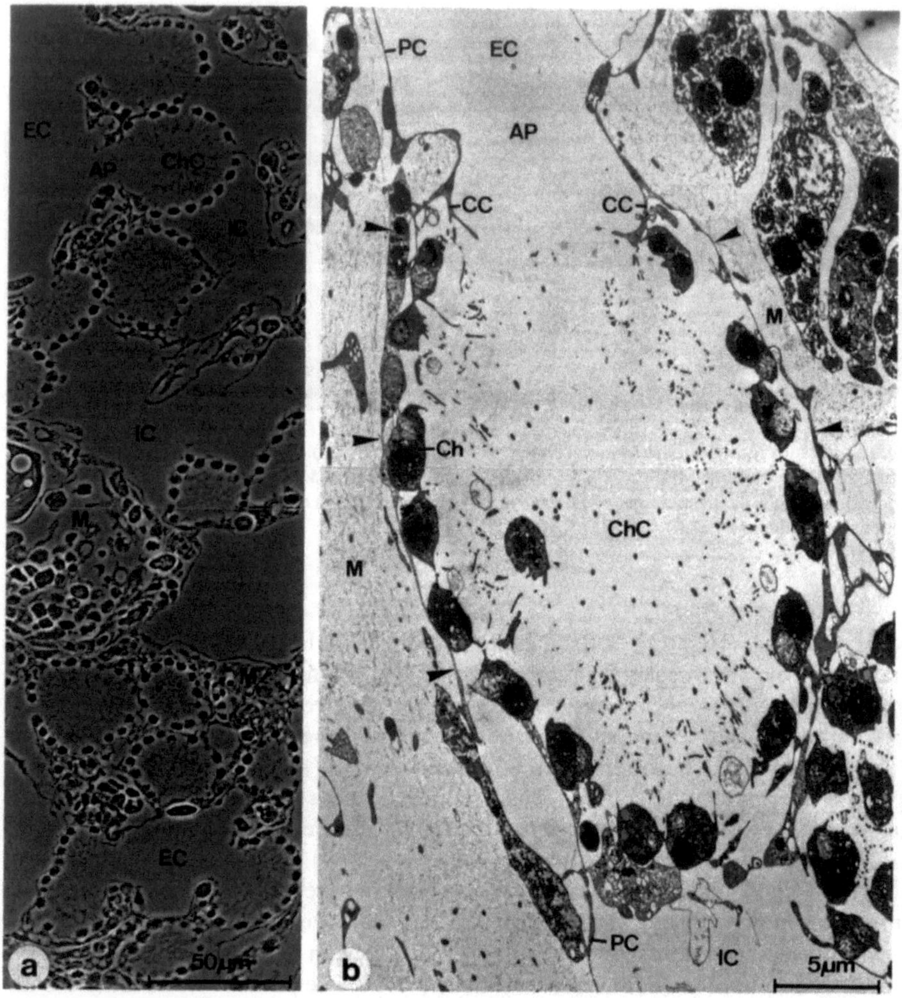


**Fig. 3.** **a** Tissue region of *Haliclona oculata* (phase-contrast micrograph). *IC* incurrent canal; *EC* excurrent canal; *ChC* choanocyte chamber; *AP* apopyle; *M* mesenchyme. **b** Choanocyte chamber (*ChC*) of *Callyspongia* sp. (transmission electron micrograph). The choanocytes (*Ch*) are separated from the mesenchyme (*M*) by a pinacocyte (*PC*) cover (→). *IC* incurrent canal; *EC* excurrent canal; *CC* cone cell; *AP* apopyle

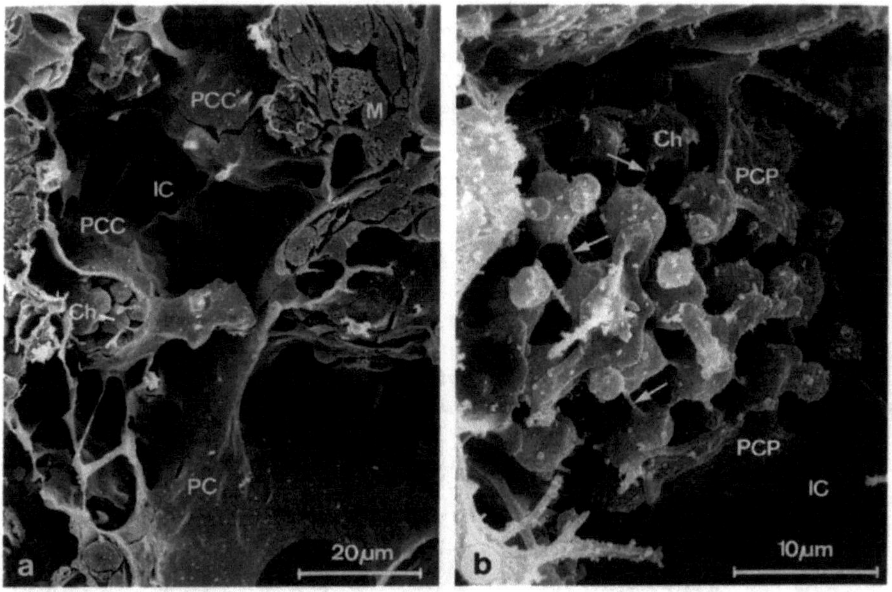


**Fig. 4. a** Sectioned tissue piece of *Gellius ravus* (scanning electron micrograph). View to an incurrent canal (*IC*) with choanocyte (*Ch*) chambers standing out from the canal wall. *PCC* pinacocyte cover on the outer surface of the chambers; *PC* pinacocyte; *M* mesenchyme. **b** View from the incurrent canal (*IC*) side to the outer surface of a choanocyte chamber of *Niphates* sp. (scanning electron micrograph). The choanocytes (*Ch*) are connected by small processes (→). *PCP* pinacocyte process

Choanocytes, which are in contact with the mesenchyme, are found in four of the investigated marine Haplosclerida and in all four investigated freshwater sponges. In the remaining 15 marine species the choanocytes are separated from the mesenchymal tissue.

## Discussion

The feature "choanocytes separated from the mesenchyme" has until now been found only in haplosclerid sponges and by Hartman and Willenz (1989) also in the calcified sponge *Calcifibrospongia actinostromarioides*, which also by spicule morphology is similar to the Haplosclerida (Vacelet 1983, 1985) and may actually be a haplosclerid sponge. Choanocytes in contact with the mesenchyme on the other hand are found not only in the Haplosclerida listed on the left side of Table 1, but as far as is known, in all other Demospongiae (Brien 1973; Simpson 1984).

On the basis of these results the out-group comparison leads to the conclusion that the considerably more widespread feature "choanocytes in contact with the mesenchyme" represents the ancestral state, while the feature "choanocytes separated from the mesenchyme", which is restricted to a rather small group of

**Table 1.** Distribution of the two different kinds of choanocyte chamber arrangement in the investigated haplosclerid sponge species. The schematic drawings show sections through choanocyte chambers (*ChC*) between incurrent (*IC*) and excurrent canals (*EC*) and the intervening mesenchyme (*stippling*). *Arrows* direction of water flow

| Choanocytes in contact with the mesenchyme | Choanocytes separated from the mesenchyme | |
|---|---|---|
| Haliclona indistincta | Haliclona elegans | Niphates sp. |
| Amphimedon compressa | Haliclona mediterr. | Reniera mucosa |
| Reniera sarai | Haliclona rosea | Reniera fulva |
| Dendroxea lenis | Haliclona oculata | Reniera sp. |
| Ephydatia fluviatilis | Haliclona limbata | Gellius ravus |
| Ephydatia muelleri | Adocia simulans | Pellina fistulosa |
| Spongilla lacustris | Callyspongia sp. | Petrosia ficiformis |
| Eunapius fragilis | Niphates digitalis | |

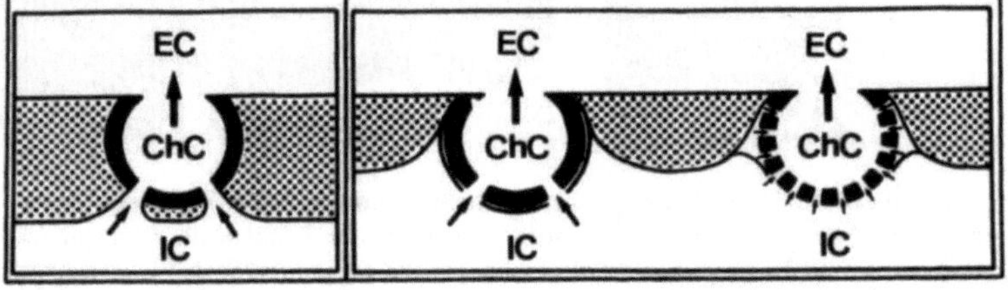


sponge species, is the derived one. Consequently the feature "choanocytes separated from the mesenchyme" would be a synapomorphy of the sponge species on the right side of Table 1, which by this would be proved to be monophyletic.

With regard to the eight species listed on the left side of Table 1, no synapomorphy has been found till now. The four marine species listed above the dotted line have the described cells with large inclusions in common, which in future may help to clear up their descent.

The absence of these cells in the investigated freshwater sponges, however, renders a statement on the descent of these sponillid species rather difficult. They may have developed under loss of this special kind of cell from the same ancestor like the other sponge species of this group, but this remains uncertain until a clear synapomorphy for the whole group has been found.

By the present result, that there are two groups of sponges in the present Haplosclerida with fundamentally different organizational types, the question arises, whether the Haplosclerida are really monophyletic. A regular ectosomal unispicular reticulum, which has been considered by Van Soest (1987) as a synapomorphy of the Haplosclerida and Petrosida, is not common to all Haplosclerida and thus in my opinion cannot show the monophyletic status of this sponge group. The only feature which is common to all Haplosclerida is an overall reticular skeletal structure, generally with only oxeas; but they share this character with the Desmacidonidae, which for this reason were transferred by De

Laubenfels (1936) into the Haplosclerida, but because of their characteristic microscleres are traditionally thought to be Poecilosclerida (Levi 1973).

After all, in my opinion the only clear synapomorphy that remains is the separation of the choanocyte chambers from the mesenchyme in the great majority of the present Haplosclerida. Thus it seems to be reasonable to restrict the order Haplosclerida to sponges with this character.

The species with reticulate skeleton and choanocytes in contact with the mesenchyme (left side of Table 1), which are at present included in the Haplosclerida, may actually be Desmacidonidae and thus Poecilosclerida, which lost the ability to produce the characteristic microscleres. This may be proven by further histological investigations, which should include desmacidonid sponge species. Volkmer-Ribeiro (1986) recently also assumed a poecilosclerid origin for the genus *Metania* and some other neotropical freshwater sponges.

The original order Haplosclerida has recently been split by Berquist (1980) into the two separate orders, Haplosclerida and Nepheliospongida. This view has been adopted by Hartman (1982), who proposed the name "Petrosiida" instead of "Nepheliospongida". The histological investigation of *Petrosia ficiformis*, however, showed that this species is connected to other haplosclerid sponges by the possession of choanocyte chambers which are separated from the mesenchyme. In the light of these results, it seems to be more convenient in future to separate the two groups in question on the subordinal level.

In both groups listed in Table 1 there are species of the genera *Haliclona* and *Reniera*. In consequence of the different kind of choanocyte chamber arrangement a revision of the generic names of at least *Haliclona indistincta* and *Reniera sarai* seems to be necessary.

*Acknowledgments.* I wish to express my thanks to Prof. Dr. N. Weissenfels (Zoologisches Institut der Universität Bonn, FRG) for helpful support. I thank Prof. Dr. L. Scalera-Liaci (Università di Bari, Italy), Prof. Dr. J. Vacelet and Dr. N. Boury-Esnault (Station Marine d'Endoume, Marseille, France), Prof. Dr. W.D. Hartman, and Dr. Ph. Willenz (Yale University, New Haven, CT/USA), and Dr. W.C. Jones (University College of North Wales, Bangor, Wales/U.K.) for their help in collecting and identifying the sponges. I should also like to thanks Mrs. I. Nüssle, Mrs. U. Müller, and Mrs. B. Zarbock for technical assistance. The scanning electron micrographs were prepared in the Institute für Angewandte Zoologie, Universität Bonn, and in the Max-Planck-Institut für Systemphysiologie, Dortmund. I thank Prof. Dr. W.J. Kloft and Mrs. K Ulmen (IAZ Bonn), and Prof. Dr. R. Kinne, Dr. D. Schäfer, and Dr. J. Hentschel (MPI Dortmund). This work was supported by the Deutsche Forschungsgemeinschaft.

## References

Bergquist PR (1980) The ordinal and subclass classification of the Demospongiae (Porifera); appraisal of the present arrangement and proposal of a new order. NZJ Zool 7:1–6

Brien P (1973) Les Démosponges. In: Grassé P-P (ed) Traité de Zoologie, vol III: Spongiaires. Masson, Paris, pp 133–461

De Laubenfels MW (1936) A discussion of the sponge fauna of the Dry Tortugas in particular and the West Indies in general, with material for a revision of the families and orders of the Porifera. Pap Tortugas Lab 30:1–225

Eisenman EA, Alfert M (1981) A new fixation procedure for preserving the ultrastructure of marine invertebrate tissues. J Microsc 125:117–120

Hartman WD (1982) Porifera. In: Parker SP (ed) Synopsis and classification of living organisms, vol I. McGraw-Hill, New York, pp 640–666

Hartman WD, Willenz Ph (1989) Organization of the choanosome of three Caribbean sclerosponges. Proc III Int Conf Biol Sponges, Woods Hole (in press)

Langenbruch P-F (1983) Body structure of marine sponges. I. Arrangement of the flagellated chambers in the canal system of *Reniera* sp. Mar Biol 75:319–325

Langenbruch P-F (1984) Vergleichende rasterelektronenmikroskopische Darstellung der Gemmulaschalen von *Ephydatia fluviatilis, E. muelleri* und *Spongilla fragilis* (Porifera) Zoomorphology 104:79–85

Langenbruch P-F (1985) Die Aufnahme partikulärer Nahrung bei *Reniera* sp. (Porifera). Helgol Meeresunters 39:263–272

Langenbruch P-F (1988) Body structure of marine sponges. V. Structure of choanocyte chambers in some Mediterranean and Caribbean haplosclerid sponges (Porifera). Zoomorphology 108:13–21

Langenbruch P-F, Scalera-Liaci L (1986) Body structure of marine sponges. IV. Aquiferous system and choanocyte chambers in *Haliclona elegans* (Porifera, Demospongiae). Zoomorphology 106:205–211

Langenbruch P-F, Weissenfels N (1987) Canal systems and choanocyte chambers in freshwater sponges (Porifera, Spongillidae). Zoomorphology 107:11–16

Langenbruch P-F, Simpson TL, Scalera-Liaci L (1985) Body structure of marine sponges. III. The structure of choanocyte chambers in *Petrosia ficiformis* (Porifera, Demospongiae) Zoomorphology 105:383–387

Lévi C (1973) Systématique de la classe des Démospongiaria (Démosponges). In: Grassé PP (ed) Traité de Zoologie, vol III. Spongiaires. Masson, Paris, pp 577–631

Simpson TL (1984) The cell biology of sponges. Springer, Berlin Heidelberg New York Tokyo

Vacelet J (1983) Les éponges hypercalcifiées, reliques des organismes constructeurs de récifs du Paléozoique et du Mésozoique. Bull Soc Zool Fr 108:547–557

Vacelet J (1985) Coralline sponges and the evolution of Porifera. In: Morris SC, George JD, Gibson R, Platt HM (eds) The origins and relationships of lower invertebrates. Syst Assoc Spec vol 28:1–13

Van Soest RWM (1987) Phylogenetic exercises with monophyletic groups of sponges. In: Vacelet J, Boury-Esnault N (eds) Taxonomy of Porifera. NATO ASI Series, vol G13. Springer, Berlin Heidelberg New York Tokyo, pp 227–241

Volkmer-Ribeiro C (1986) Evolutionary study of the freshwater sponge genus *Metania* Gray, 1867: III. Metaniidae, new family. Amazoniana 9:493–509

Weissenfels N (1981) Bau und Funktion des Süßwasserschwamms *Ephydatia fluviatilis* L. (Porifera). VIII. Die Entstehung und Entwicklung der Kragengeißelkammern und ihre Verbindung mit dem ausführenden Kanalsystem. Zoomorphology 98:35–45

Weissenfels N (1982) Rasterelektronenmikroskopische Histologie von spongiösem Material. Microsc Acta 85:345–350

# Symbiosis of *Spongilla lacustris* (Spongillidae) and Green Algae. Algae Uptake, Distribution and Final Whereabouts

U. Saller

## Introduction

It has long been well known that freshwater sponges contain unicellular algae (Brandt 1882). According to Goetsch and Scheuring (1927), these algae belong to the genus *Chlorella*. In Protozoa, marine Porifera, and other vertebrates a companionship with algae is also known too (Goetsch and Scheuring 1927; Karakashian et al. 1968; Sarà and Liaci 1964; Sarà 1971; Vacelet 1971).

First detailed examinations of the chlorellae in freshwater sponges were performed in the middle of this century. They dealt predominantly with the ultrastructure and chemistry of the algae (Agnes and Brøndsted 1953; Gilbert and Allen 1973a,b; Williamson 1979), but little is known about the uptake of the algae by the sponge, their distribution, and their final whereabouts (Williamson 1979).

## Material and Methods

For in vivo observation and algae cultivation, well-grown samples of *Spongilla lacustris* were collected in the river Sieg near Rosbach and transferred to an aquarium. Small pieces were cut out of the sponges and placed on cover slides, to which they attached after a few days. With the sponge downwards, the cover slide was put upon a microscopic slide, which was prepared with a water-filled ring. Thus it was possible to observe the sponges for many hours in vivo by light and phase-contrast microscopy.

For *Chlorella* cultivation an earth-extract cultivation medium was used. Small sponge pieces, cut out of the field specimens, were cleaned carefully, minced, and added to the culture medium. Within 1 week the infected solution was colored green by the reproducing algae. Cold light, warm light, and fluorescent neon tubes lighted up the culture solution for 12 h a day.

To analyze the algae uptake by phase-contrast and electron microscopy, algae were added for 30 min to 4-week-old white sponges hatched from gemmules. The sponges were fixed after ten different times of incubation. After feeding, the sponges were cleaned carefully. The fixation was made with 1% $OsO_4$ and 1% $K_2Cr_2O_7$ in 0.025 M sodium cacodylate buffer solution, pH 7.3, the embedding in styrol-metacrylate (see Saller and Weissenfels 1985). Postcontrasting (Fig. 4a) with lead citrate after Reynolds (1963).

J. Reitner and H. Keupp (Eds.)
Fossil and Recent Sponges

## Results

### *In Vivo Microscopy*

The pinacocytes (P in Fig. 1a) of the peripheral rim region are very suitable for cytological in vivo observations. In sponges originating from the field and attached on cover slides, and also in sponges hatched from green gemmules, the pinacocytes contain few, up to many intracellular chlorellae (C in Fig. 1a). These are stationary and are displayed only by the pulsation of the contractile vacuoles (CV). Many more algae are found in the ameboid mesenchymatic cells.

When the algae cultivation solution was added to sponges in sandwich cultures, the uptake of the chlorellae could be observed directly by inverse phase-contrast microscopy. With the inflowing water the algae pass through the dermal pores of the pinacoderm and reach the subdermal space. By incurrent canals they reach the choanocyte chambers and are finally caught in the apical and in the basal region of the choanocytes. Twenty minutes after addition, the first chlorellae appear in archeocytes in the vicinity of the choanocyte chambers.

Six hours later most of the sponge cells contain chlorellae except the granulocytes, microscleroblasts, and the pinacocytes of the peripheral rim region and of the pinacoderm.

### *Phase-Contrast and Electron Microscopy*

The evaluation of the semi- and ultrathin sections supports the results of the in vivo observations. After passing the prosopyles, the algae reach the outer region inside the choanocyte chamber (C in Fig. 1b). By a rather dense, web-like net of filamentous material (←) which connects the collars of the choanocytes, the algae are prevented from leaving the choanocyte chambers and finally are taken in by the choanocytes (C in Fig. 1c). Electron microscopice examination shows that the 1–1.5 μm large chlorellae are taken in by the choanocytes at their apical collar region and at their base by cytoplasmatic cell processes. In Fig. 2a an approximately longitudinally sectioned choanocyte (Ch) is shown. Cytoplasmatic cell processes of the apical collar region (Cl) surround the *Chlorella* (C) nearly totally. The algae in the cross-sectioned choanocyte collar (*Cl*) in Fig. 2b is totally enclosed. In both figures the original position of the collar microvilli involved in the incorporation of the alga can be identified (→). The algae in Fig. 2c (C) is nearly surrounded by cell processes of the choanocyte body (Ch) and next to its incorporation. Algae are taken up also by endopinacocytes (P in Fig. 3a) near to the prosopyles (PP) of the choanocyte chambers (ChC). Figure 3b shows a *Chlorella* (C) in higher magnification inside of an endopinacocyte (P). The dish shaped chloroplast (CP) is distinctly recognizable.

While algae first occur in choanocytes and in endopinacocytes of the incurrent canals, in the following hours they are found more and more in mesenchymatic cells.

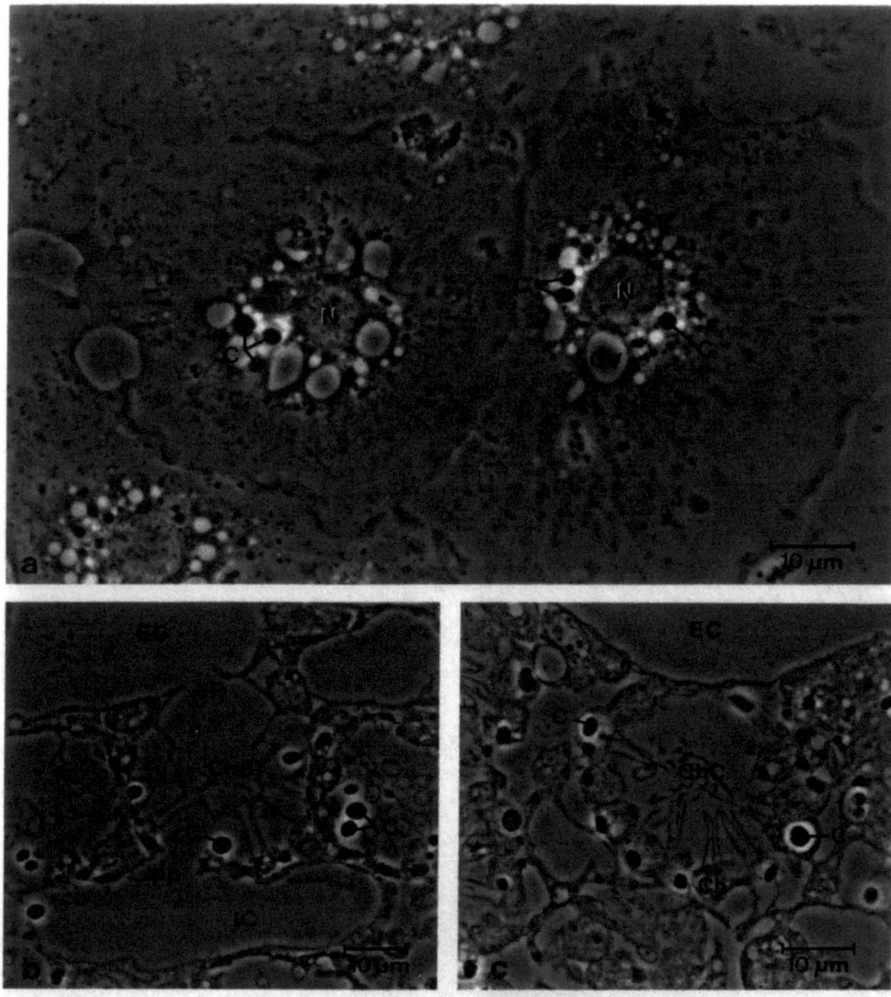

**Fig. 1. a** Detail from the peripheral rim region of a *Spongilla lacustris* which hatched from a green gemmule and was grown on a cover slide. The pinacocytes (*P*) of the peripheral rim region contain few unicellular chlorellae (*C*). *N* nucleus, *CV* contractile vacuole. Phase-contrast microscope photograph (PhC) in vivo. **b,c** Choanocyte chambers (*ChC*) of 4-week-old sponges. **b** 2 min after algae addition; **c** 10 min after algae addition. *IC* incurrent canal; *PP* prosophyle; *C Chlorella*; *Ch* choanocyte; ← fibrillar material; *AP* apopyle; *EC* excurrent canal (PhC)

The transmission of the chlorellae is very specific. There is no release of the algae into the intracellular mesenchyme. After the donor cell and the receiver cell have approached one another, the vacuole with the *Chlorella* inside bulges outward and is surrounded by cell processes of the receiving cell. The vacuole opens, the donor cell retracts, and the receiver cell closes around the alga. Finally, the alga is incorporated into the receiver cell. The electron microscopical aspects

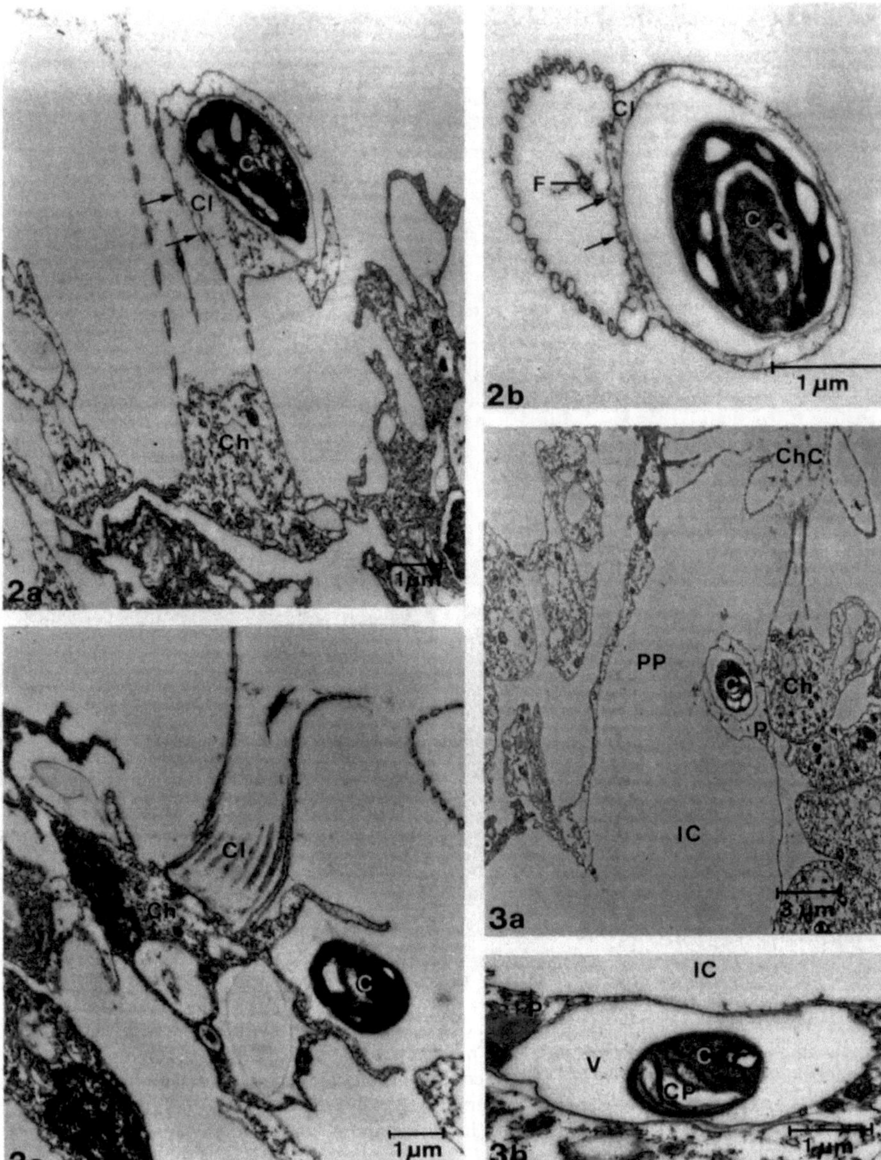


**Fig. 2.** Choanocytes (*Ch*) 2 min after algae addition. **a** Cell processes of the collar (*Cl*) envelop a *Chlorella* (*C*) extensively (EM). **b** Cross-section through the apical collar region (*Cl*) with totally enveloped alga (*C*). *F* flagellum; → original microvilli (EM). **c** Cell processes of a choanocyte (*Ch*) extend to a *Chlorella* (*C*. *Cl* collar (EM)

**Fig. 3.** **a** Prosopyle region (*PP*) of a choanocyte chamber (*ChC*) with an alga (*C*) taken up by an endopinacocyte (*P*). *IC* incurrent canal; *Ch* choanocyte (EM). **b** Detail of an endopinacocyte (*P*) with a *chlorella* (*C*) with dish-shaped chloroplast (*CP*). *V* vacuole (EM) of the alga transmission are shown in Fig. 4a and b. Small cell processes of the receiver cell (RC in Fig. 4a) are close to the bulged out alga vacuole of the donor cell (DC). In Fig. 4b the transmission of the alga is more advanced. The *Chlorella* at the end of the cell process of the donor cell (DC) is surrounded very closely by the receiver cell. In this late stage of transmission the alga is covered only by the cell membrane and the vacuole membrane of the donor cell. In this magnification these two membranes are not clearly distinct. In front to the receiving cell the vacuole is open (*).

One to two days after algae feeding, chlorellae can be found in archeocytes, amoebocytes, endopinacocytes, choanocytes, spongioblasts, and scleroblasts of the sponge.

The chlorellae are able to divide inside of the sponge cells. Fig. 4c shows a vacuole (V) of an archeocyte (AC) with a divided *Chlorella* 2 days after algae addition. Both daughter cells (C′) are surrounded by one common cell wall (W).

The sponges taken from the field show a bright green color. In well-grown, 1-cm or thicker specimens, the green color is often restricted to the apical zone. Their cells contain large numbers of algae (C in Fig. 4d). In electron microscopice sections as many as three dozen chlorellae were counted in one cell.

## Discussion

The chlorellae which occur under natural conditions in the cells of green *Spongilla lacustris* are not absolutely dependent on their host. Specimens of the genus *Chlorella* also live together with other invertebrates, like Protozoa, Hydrozoa, Mollusca, and Turbellaria. In Mollusca the algae are parasites rather than commensals or even symbionts because they damage their host (Goetsch and Scheuring 1927). In *Paramecium bursaria* the algae are referred to as true symbionts (Karakashian 1963; Karakashian et al. 1968).

The mutual profit in the companionship has been reported to be protection of the algae inside the sponge and the receiving of products of algae photosynthesis by the sponge (Agnes and Brøndsted 1953). The latter has been demonstrated by autoradiographical observations by Gilbert and Allen (1973b). According to Muscatin et al. (1967), this product is glucose. It results in better growth of green sponges compared with sponges without algae (Frost 1978; Frost and Williamson 1980). This becomes evident also in laboratory cultures of sponges. Also the observed faster hatching of green gemmules indicates a positive physiological effect.

The occurrence of brown and green gemmules in *Spongilla lacustris* is interpreted as an adaptation to the unsteady conditions in spring. The green gemmules hatch early in spring and thus settle the substratum previous to the brown ones. If these young sponges which hatched from green gemmules are killed by low temperature, the brown gemmules guarantee protection of the species by hatching later than the green ones (Agnes and Brøndsted 1953). A division of the chlorellae inside the gemmule thesocytes does not occur (Williamson 1979).

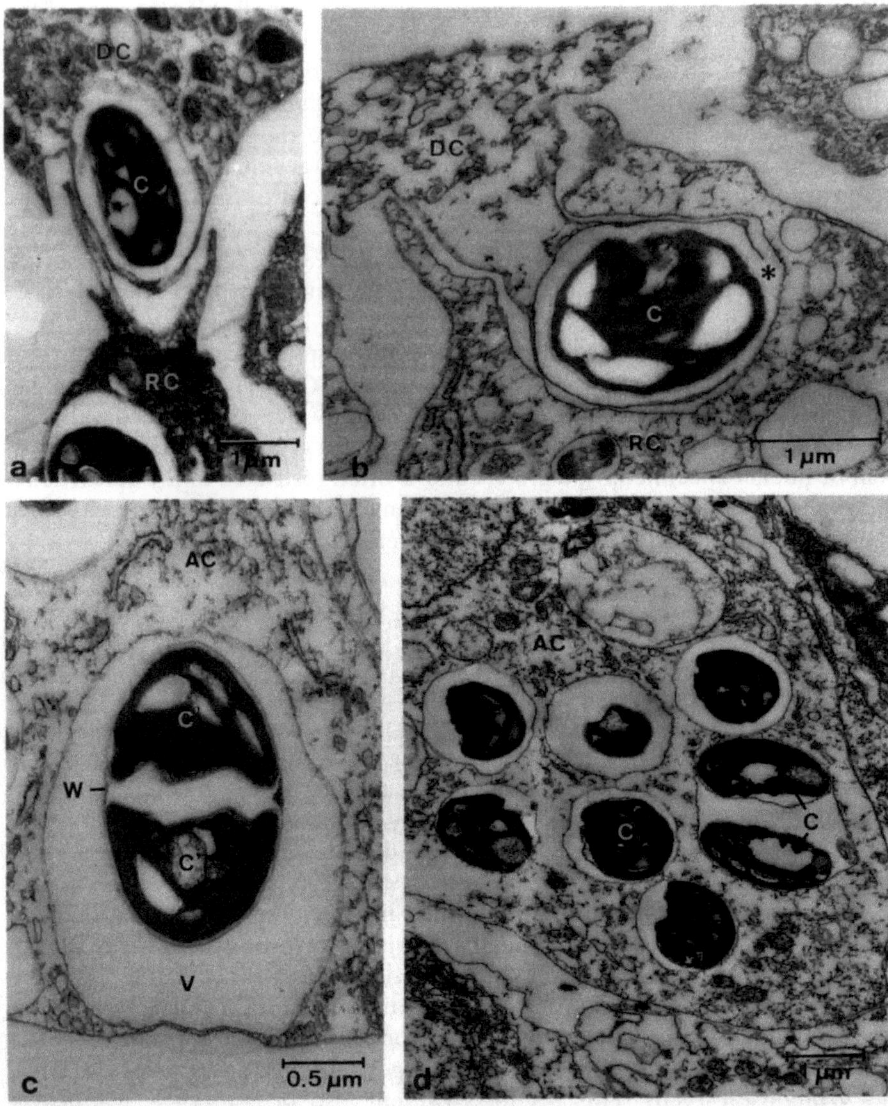


**Fig. 4.** **a,b** Two stages of *Chlorella* transmission. **a** Early stage of alga transmission. Cell processes of the receiver cell (*RC*) extend to the bulged-out alga vacuole (*C*) of the donor cell (*DC*) (EM). **b** Late stage of alga transmission. A handle-like cell process of the donor cell (*DC*) with an open alga vacuole (*) is covered closely by the receiving cell (*RC*). *C Chlorella* (EM). **c** Division of a *Chlorella*. Both daughter cells of the alga (*C'*) lie in one vacuole (*V*) and are covered by a common cell wall (*W*). *AC* archeocyte (EM). **d** Archeocyte (*AC*) of a green sponge from the field showing numerous chlorellae (*C*). Two algae are located in a common host vacuole. Postcontrasted with lead citrate (EM)

The appearance of algae in pinacocytes of the peripheral rim region in sponges which have been grown from green sponge pieces from the field and their lack in the pinacocytes of the peripheral rim region of the infected sponges can be explained by the differentiation of algae-containing archeocytes to pinacocytes in the course of the reorganization process of the sponge pieces.

The algae are transmitted among the sponge cells in a very specific way. A similar manner of transfer has been described by Willenz (1980) after experimental feeding of latex beads to *Ephydatia fluviatilis*. In this case the transfer takes place by a very large contact region between the involved cells and the content of different vacuoles is often transferred simultaneously.

## References

Agnes B, Brøndsted HV (1953) The effect of symbiotic zoochlorellae on the germination rate of gemmules of *Spongilla lacustris* (L.). Viddensk Medd Dan Naturhist Foren 115:133–144

Brandt K (1882) Über die morphologische und physiologische Bedeutung des Chlorophylls bei Tieren. Arch Anat Phys Abt 125

Frost TM (1978) The ecology of the freshwater sponge *Spongilla lacustris*. Thesis Dartmouth College Hannover N H: pp 1–196

Frost TM, Williamson CE (1980) In situ determination of the effect of symbiotic algae on the growth of the freshwater sponge *Spongilla lacustris*. Ecology 61:1361–1370

Gilbert JJ, Allen HL (1973a) Studies on the physiology of the green freshwater sponge *Spongilla lacustris*: primary productivity, organic matter, and chlorophyll content. Verh Int Ver Limnol 18:1413–1420

Gilbert JJ, Allen HL (1973b) Chlorophyll and primary productivity of some green, freshwater sponges. Int Rev Ges Hydrobiol 58:633–658

Goetsch W, Scheuring L (1927) Parasitismus und Symbiose der Algengattung *Chlorella*. Z Morphol Ökol 7:220–253

Karakashian SJ (1963) Growth of *Paramecium bursaria* as influenced by the presence of algal symbionts. Physiol Zool 36:52–68

Karakashian SJ, Karakashian MW, Rudzinska MA (1968) Electron microscopic observations on the symbiosis of *Paramecium bursaria* and its intracellulae algae. J Protozool 15(1):113–128

Muscatin LM, Karakashian SJ, Karakashian MW (1967) Soluble extracellular products of algae symbiosis with a ciliate, a sponge, and a mutant hydra. Comp Physiol Biochem 20:1–12

Reynolds ES (1963) The use of lead citrate at high pH as an electron-opaque stain. J Cell Biol 17:208–212

Saller U, Weissenfels N (1985) The development of *Spongilla lacustris* from the oocyte to the free larva (Porifera, Spongillidae). Zoomorphology 105:367–374

Sarà M (1971) Ultrastructural aspects of the symbiosis between two species of the genus *Aphanocapsa* (Cyanophyceae) and *Ircinia* (Demospongiae). Mar Biol 11:214–221

Sarà M, Liaci L (1964) Symbiontic association between zooxanthellae and two marine sponges of the genus *Cliona*. Nature (Lond) 203:321

Vacelet J (1971) Étude en microscopie électronique de l'association entre une cyanophycée chroococcale et une éponge du genre *Verongia*. J Microsc 12:363–380

Willenz P (1980) Kinetic and morphological aspects of particle ingestion by the freshwater sponge *Ephydatia fluviatilis* L. In: Smith DC, Tiffon Y (eds) Nutrition in the Lower Metazoa. Pergamon, Oxford, pp 163–178

Williamson CE (1979) An ultrastructural investigation of algal symbiosis in white and green *Spongilla lacustris* (L.) (Porifera, Spongillidae). Trans Am Microsc Soc 98(1):59–77

# Cortical Structure and Adaptation in the Genus *Tethya* (Porifera, Demospongiae)

M. SARÀ and E. MANARA

## Introduction

The taxonomy of Porifera is traditionally based upon the types of spicules, their shape and dimensions, which help to distinguish the different species. A taxonomy based on spicules shows evident shortcomings especially when the taxonomist is faced with sponges as morphologically complex as those of the genus *Tethya*. Until now, the taxonomy of *Tethya*, based essentially on spicular characters, has remained confused, as shown by the two very contradictory revisions made by Topsent (1918) and by Burton (1924), despite the quantity of available data and an extensive literature. In fact, in a genus such as *Tethya*, specific differentiation in the genus need not be restricted to spicular differences but can be extended to include the whole morpho-functional organization of the sponge, especially its cortical structure. Spicular divergence should be considered only in this broader context. In any case, for the recognition of species, the skeletal arrangement of the spicules is as important as their shape and dimensional details.

A strong divergence in the cortical structure has been found in *T. aurantium* (Pallas) and *T. citrina* Sarà and Melone which live sympatrically along the European shores. Morphological differences in the cortex of *T. aurantium* and *T. citrina* depend both on the skeletal structure as well as the inhalant aquiferous system. These structures appear to be coordinated to act in a functional way.

The anatomical characteristics also influence some external features of the sponges, such as surface texture and body plasticity. The different morpho-functional characteristics are apparently adaptations to the different ecological requirements of the two species. Some general hypotheses on the origin of these species and on the mechanisms of speciation in *Tethya* may be advanced on the basis of the adaptive significance of their morpho-functional divergence.

## Material and Methods

Morphological data on the skeletal arrangement and the inhalant cortical structures were obtained from several specimens of sympatric populations of *T. aurantium* and *T. citrina*. Samplings were made by skin diving in the coastal

J. Reitner and H. Keupp (Eds.)
Fossil and Recent Sponges

lagoon of Stagnone (Marsala, Sicily), at 0–2 m depth in partially overlapping habitats. Sponges were fixed in formol 4% or in Bouin. Paraffin sections were stained by iron hematoxylin-eosin or Masson's technique. In order to check the uniformity of skeletal and aquiferous system traits, specimens obtained from different populations (Rovigno, Palese, Porto Cesareo) were also investigated by examination of stained and unstained sections.

## Results

### *General Remarks on the Cortical Structure of Tethya Species*

*Tethya* species typically show a massive spherical or subspherical shape with a radiate structure produced by bundles of megascleres. The body is divided into two well-marked concentric zones: (a) periferal cortex or ectosome; (b) a central medulla, endosome, or choanosome, which contains the choanocyte chambers. Many other sponges have a cortical structure but that of *Tethya* is certainly one of the more specialized and complex. The cortex may accomplish two functions: (1) protection against the physico-chemical factors, predators and microorganisms; (2) regulation of the aquiferous system and, consequently, of sponge feeding.

The main differences in the cortex of *Tethya* species concern the skeletal arrangement and the inhalant aquiferous system. In the cortex there is a well-developed mesohyl with layers of collagenous fibers, various cell types, skeletal elements, lacunae, and canals. It is also possible to distinguish a more hyaline external region (ectocortex) and a denser, fibrous inner region (endocortex). External characters, such as surface protuberances and depressions, consistency and plasticity, are related to the morphological structure of the cortex but may also vary according to the physiological state of the sponge.

The megasclere bundles, which hispidate the surface of the sponge, can regulate shape, size, and height of the tubercles (papillae or conules): *Tethya* shows a very high degree of contractibility and in some species its volume may be reduced almost to a 1/2–2/3; in *T. crypta* Reiswig (1971) has shown the occurrence of daily exogenous and longer endogenous rhythms of expansion and contraction. How the different components of the cortex are involved in this contractibility is not presently known. However, the degree of plasticity does differ among the species and it is certainly related to their different cortical structure.

### *Cortical Structure in T. aurantium and T. citrina*

*T. aurantium* and *T. citrina* were hitherto regarded as belonging to the same species, before the discovery was made that they are clearly separated species (Sarà and Melone 1965). They are sympatric along the Mediterranean shores with distinct contiguous and partially overlapping habitats. *T. aurantium* has a wider geographic distribution, which is anfiatlantic, including records from the Caribbean Sea.

The two species are separated by many characteristics (Sarà and Melone 1965; Sarà 1987; Sarà and Gaino 1987). They can be easily distinguished in the field by color, consistency, dimensions and surface structure.

The main anatomical differences concern the cortical skeleton and the inhalant aquiferous system (Fig. 1). *T. aurantium* has a more developed and thicker cortex than *T. citrina*, which is softer and more contractile. This fact could be related to a different pumping activity and feeding behavior in the two sponges. However, an analysis of the mechanism and physiological consequences of this divergence has not been made yet.

*Cortical Skeleton*

As in other species of *Tethya*, the cortical spicules are composed of micrasters, megasters, and monaxon megascleres. The cortical micrasters of *T. aurantium* and *T. citrina* are slightly different. A greater difference between them has been observed in the choanosome (Sarà and Gaino 1987).

In both species the megasters occur only in the cortex and at the boundary between the cortex and the choanosome. They show a remarkable and constant difference between the two species in their shape and arrangement. The megasters of *T. aurantium* are typical spherasters with $R/C < 1$ (the ratio between the length of a ray and the diameter of the center) while those of *T. citrina* are oxyspherasters with $R/C > 1$. In *T. aurantium* the spherasters are densely packed in many layers inside the endocortical mesohyl without having any close relationship with the aquiferous cavities. On the contrary, the oxyspherasters of *T. citrina* are placed in monolayers along the walls of lacunae and canals in the sponge ectocortex and endocortex. Consequently, the functions of the megasters seem to differ in the two sponges, acting protectively in the cortex or as a reinforcement of the walls of the aquiferous system, respectively. It may be suggested also that the different spicular shape is an adaptive feature for these different functional roles.

The bundles of megascleres in the cortex of *T. aurantium* are less numerous and more terminally divergent than in *T. citrina*. Protuberances in the cortex are more marked in *T. aurantium*, producing in this species a more papillate surface than in *T. citrina*. On the other hand, *T. citrina* is provided with a set of slender styles (strongyloxeas) in the subcortical region, probably serving as a reinforcement of the more feeble cortical structure found in this species.

*Collagen Matrix*

In *T. aurantium* the ectocortex is thicker and more compact than in *T. citrina*, where it is largely excavated into superficial lacunae. The endocortex is fibrous in both species, but the collagenous matrix is more extended in *T. aurantium*, partially involving the ectocortex. In *T. citrina* the collagen fibers of the endocortex form a clearcut ring around the choanosome.

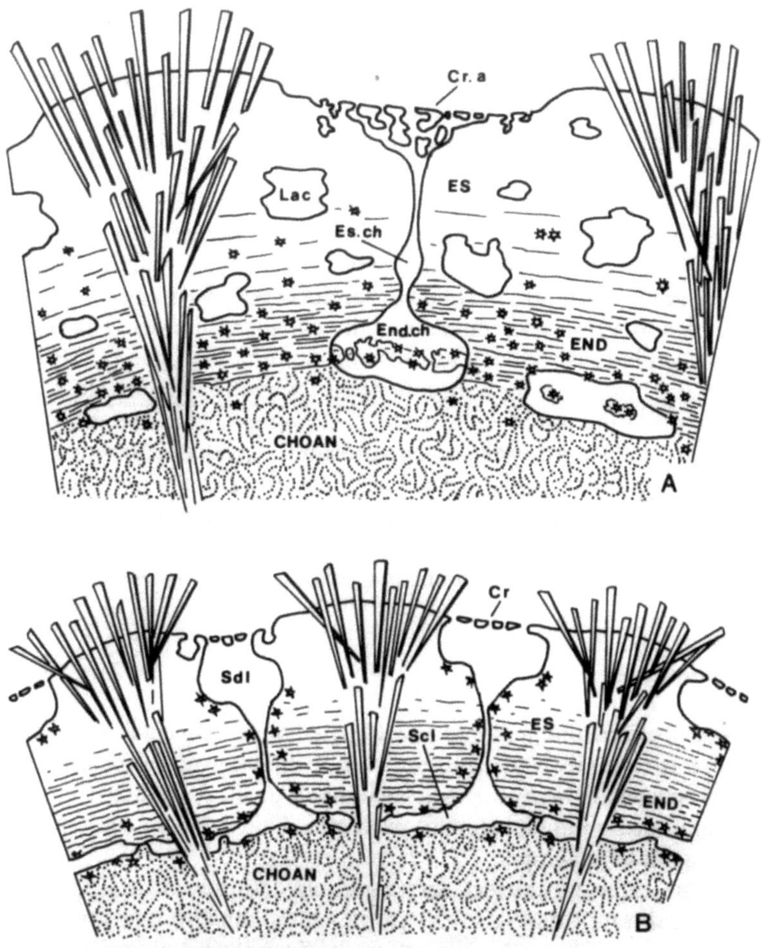


**Fig. 1.** Cortical structure in *T. aurantium* (**A**) and *T. citrina* (**B**). *CHOAN* Choanosome; *Cr* cribre; *Cr.a* cribral area; *END* endocortex; *End.ch* endochone; *ES* ectocortex; *Es.ch* esochone; *Lac* lacuna; *Sdl* subdermal lacuna; *Scl* subcortical lacuna. In the figure the bundles of styles and the megasters are included but the micrasters are omitted

*Aquiferous System*

The overall arrangement of the *T. aurantium* aquiferous system was described by Delage et Herouard (1899) and Topsent (1900). In the depressions between the papillae, cribriporal chones formed by a dense system of small canals lead to a larger canal in which it is possible to detect an outer esochone and an inner endochone. The canal shows some lateral enlargements into the ectocortex. The myocytes of the endocortex form a sort of sphincter around the inhalant canal in its endocortical tract. In the inner part of the cortex the canal widens into an endochone which is transversely crossed by a filter-like structure, sustained by a sheet of megasters. The endochone opens into subcortical lacunae which do not constitute a continuous system as in *T. citrina*.

In *T. citrina* the ostia are grouped in depressions around the small papillae, forming cribres as described for *Polymastia* by Boury Esnault (1972). They open directly into wide superficial lacunae (vestibules) occupying most of the ectocortex. Each vestibule opens into an inhalant canal which is not subdivided, as in *T. aurantium*, into an esochone and endochone. The inhalant canal runs in a simple and direct way (euchone) without enlargements through the endocortex until it opens in the subcortical crypts. These crypts form a continuous lacunary system at the boundary between the endocortex and the choanosome, interrupted only by some thin trabeculae and the spicular bundles. The inner walls of the superficial lacunae as well as those of the inhalant canal and of the subcortical crypts are lined by a more or less dense layer of oxyspherasters, apparently with a structural function. In contrast to *T. aurantium*, *T. citrina* shows restricted zones of contact between cortex and choanosome. Another difference in the aquiferous system of the two species is seen in the number of the cribriform structures; the plagiochones or cribriporal chones of *T. aurantium* are very frequent and scattered over the whole surface of the sponge, while in *T. citrina* cribres are less numerous and restricted to more specialized zones around the papillae.

## Discussion

Literature data and personal unpublished observations show that other species of *Tethya* are equally well differentiated by their cortical structure. Further study of other species of *Tethya* is thus also likely to show different types of aquiferous systems and related skeletal patterns. This has an ecological as well as a phylogenetic relevance.

The different cortical structures may represent adaptations to particular ecological conditions. It is very probable that the different cortical structures of *T. aurantium* and *T. citrina*, based on a different type of aquiferous system, have a functional meaning in relation to the pumping activity of the sponge and therefore to its feeding mechanism. In the surveyed Mediterranean stations, the two species were generally found in different, neighboring habitats. Occasionally there are overlapping zones. The ecological niche of *T. aurantium* is different from that of *T. citrina*. *T. citrina* is more tolerant to calm waters and sedimentation. *T. aurantium* is more eurytopic and tolerant of higher values of hydrodynamism and light. The water turbulence with its associated factors seems the main limiting factor. The morpho-functional features distinguishing the two species are likely to behave as adaptations to their different ecological requirements. *T. aurantium*, which lives in more exposed places, has a more rigid and thicker cortex, reinforced by the packing in the endocortex of stout spherasters. *T. citrina*, which lives in more sheltered places, has a thinner and feebler cortex. On the other hand, the more plastic structure of the cortex of *T. citrina* favors a larger contractibility and may be an adaptive device of this species for increasing its pumping activity in calmer waters, since it does not require strong protection against the physical factors. The arrangement of the oxyspherasters on the walls of the aquiferous system, as well as their shape, may be considered a further adaptation to the specialized niche of *T. citrina*.

Phylogenetically the different cortical structures may be considered as evolutionary steps reflecting not only adaptive requirements but also morphological and developmental constraints. The cortex has evolved in *Tethya* according to Burton (1928) as an adaptation to shallow waters in a sponge which shows its derivation from deep-sea ancestors. However, the thinner cortex of *T. citrina* is apparently a very specialized one considering the structure of the aquiferous system, with stomia grouped in cribres (Boury Esnault 1972) and the shape and arrangement of the megasters. Therefore, it is likely that the reduction of the cortex in *T. citrina* represents a secondary process determined by the specialized niche of this species.

Some suggestions about the origin of *T. citrina* from *T. aurantium* or from a common ancestor similar to *T. aurantium* have been previously advanced (Sarà 1989). Following former considerations on a sympatric or micro-allopatric (parapatric) speciation in sponges (Sarà 1956, 1957, 1959) based on features of species distribution, life histories, and restricted dispersal, it has been hypothesized that an evolutionary scenario for *T. citrina* is one of parapatric origin (White 1978, Murray and Clarke 1980). The occurrence of asexual reproduction besides the sexual one may have facilitated isolating mechanisms after the founder event. Genetic data obtained by electrophoretic techniques (Sarà 1989) indicate for *T. aurantium* and *T. citrina* a high genetic distance with a divergence time or the order of 10 million years. The two species show a different population genetic structure. Mediterranean populations of *T. aurantium* are surprisingly uniform in their allozyme complement while *T. citrina* populations are variously divergent (Sarà et al. 1989).

The strong morphofunctional divergence, apparently adaptive, between the two species is consistent with a strong impact from ecological pressures on their speciation and not with stochastically based mechanisms as the genetic revolutions associated to the peripatric speciation (Mayr 1954, 1982). Founderflush mechanisms (Carson and Templeton 1984) contrast with the high genetic uniformity, intra- and interpopulational, shown by the suggested ancestor of *T. citrina*, *T. aurantium*. The two species occupy two distinctive adaptive peaks. Therefore, even if one considers distruptive selection, their separation should have been produced by large macromutational events due to some change in major, developmental regulating genes and to the occupation by the mutated founder of a new niche. Alternatively, the genetic assimilation theory (Waddington 1961), also in the new interpretations of Ho (1984), could be taken into consideration.

The distribution range of the two species, in addition to a phylogenetically important morphological feature such as the microsclere shape, agrees with the hypothesis of an origin of *T. aurantium* and *T. citrina* from a common ancestor. However, the alternative view that the two species came from different ancestors, speciated in allopatry, and only secondarily became sympatric, considering the genetic and morphofunctional divergence, cannot be presently discarded. A wider knowledge about the systematics of the genus *Tethya* is needed for a choice between these two contrasting hypotheses, as well as on the question of whether it should be considered as a supra-generic entity with the "morpho-species" as genera split into less differentiated species.

## References


Boury-Esnault N (1972) Une structure inhalante remarquable des Spongiaires, le crible: étude morphologique et citologique. Arch Zool Exp Gén 113:7–24

Burton M (1924) A revision of the sponge family *Donatiidae*. Proc Zool Soc Lond 4:1033–1046

Burton M (1928) Comparative study of shallow-water and deep-sea sponges, with notes on their external form and reproduction. J Queckett Microsc Club 16:44–70

Carson HL, Templeton AR (1984) Genetic revolutions in relation to speciation phenomena: the founding of new populations. Annu Rev Syst Ecol 15:97–132

Delage Y, Herouard E (1899) Traité de Zoologie concrète: Mesozaires-Spongiaires. Schleicher, Paris, pp 1–244

Ho MW (1984) Environment and heredity in development and evolution. In: Ho MW, Saunders PT (eds) Beyond neo-darwinism. Academic Press, London, pp 267

Mayr E (1954) Change of genetic environment and evolution. In: Huxley J, Hardy AC, Ford EB (eds) Evolution as a process. Allen und Unwin, London, pp 157

Mayr E (1982) Processes of speciation in animals. In: Barigozzi C (ed) Mechanisms of speciation: Alan Liss, New York, p 1

Murray J, Clarke B (1980) The genus *Partula* in Moorea: speciation in progress. Proc R Soc Lond B Biol Sci 211:83–117

Reiswig HM (1971) In situ pumping activities of tropical Demospongiae. Mar Biol (Berl) 9:38–50

Sarà M (1956) Variabilità della specie ed ecologia nei Poriferi. Boll Zool 23:65–78

Sarà M (1957) Problemi di speciazione negli animali marini. Arch Zool Ital 42:273–379

Sarà M (1959) Sulla coesistenza di specie strettamente affini di Poriferi in biotopi litorali. Boll Zool 26:83–96

Sarà M (1987) A study of the genus *Tethya* (Porifera Demospongiae) and new perspectives in sponge systematics. In: Vacelet J, Boury-Esnault N (eds) Taxonomy of Porifera of the NE Atlantic and the Mediterranean sea. Springer, Berlin Heidelberg New York Tokyo, pp 205

Sarà M (1989) Divergence between the sympatric species *Tethya aurantium* and *T. citrina* and speciation in sponges. In: Ruetzler K, Hartman W (eds) New perspectives in sponge biology. Smithsonian Inst Press, Washington DC (in press)

Sarà M, Gaino E (1987) Interspecific variation in arrangement and morphology of micrasters of *Tethya* species (Porifera Demospongiae). Zoomorphology 107:313–317

Sarà M, Melone N (1965) Una nuova specie del genere *Tethya*, *T. citrina* sp.n. dal Mediterraneo (Porifera Demospongiae). Atti Soc Peloritana Sci Fis Mat Nat Suppl 11:123–138

Sarà M, Mensi P, Manconi R, Bavestrello G, Balletto E (1989) Genetic variability in Mediterranean populations of *Tethya* sp. (Porifera Demospongiae). In: Ryland JS, Tyler PA (eds) Reproduction, Genetics and Distribution of marine organisms. Olsen and Olsen, pp 293–298

Topsent E (1900) Etude monographique des Spongiaires de'France. III Monaxonida Hadromerina. Arch Zool Exp Gén (3) 8:1–331

Topsent E (1918) Eponges de San Thomé. Essai sur le genres *Spirastrella*, *Donatia* et *Chondrilla*. Arch Zool Exp Gén 57:535–618

Waddington CH (1961) Genetic assimilation. Adv Genet 10:257–294

White MJD (1978) Modes of speciation. Freeman, San Francisco

# A New Mediterranean Species of *Axinella* Detected by Biochemical Genetic Methods

A.M. Solé-Cava, J.P. Thorpe, and R. Manconi

## Introduction

There has been a recent revival of interest in the use of marine sponges as models in molecular biology (Akiyama and Johnson 1983; Garrone 1984; Muller et al. 1985; Dams 1986), immunology (Smith and Hildemann 1984; Buscema and van de Vyver 1985; Kaye and Reiswig 1985; Smith and Hildemann 1986), evolutionary genetics (Balakirev and Manchenko 1985; Solé-Cava and Thorpe 1986, 1987, 1990), and chemical ecology (Amade et al. 1987; Braekman and Daloze 1986; Dyrynda 1985). This interest can be related to the particular position sponges occupy in zoological classification. Work on the evolution of any given cellular or biochemical process often needs to take into account "how far down" the zoological scale that process is found (Scudder 1987), and the Porifera, as an ancient group, provide an invaluable subject for such studies.

However, for these studies to have any true biological meaning, the identities of the experimental organisms must be well established. This is necessary for the reproducibility of experiments and to avoid the use of mixtures of species treated as if they belonged to just one biological species (Avise 1974). Moreover, a good classification is essential if relationships are to be inferred and predictions made within the Porifera. To this end, new classification techniques and approaches, such as the chemical (Bergquist and Wells 1983; Bergquist et al. 1986) and biochemical (Solé-Cava and Thorpe 1986, 1987; Sara 1988) taxonomy, and cladistics (van Soest 1987) have been used with varying degrees of success.

The genus *Axinella* is common worldwide, and is well represented in the Mediterranean (Pansini 1983). There has been a fair amount of confusion in relation to the identity of two common Mediterranean species of this genus, *A. damicornis* (Esper) and *A. verrucosa* (Esper). This is probably due to the paucity of knowledge on the extent of true intraspecific variation displayed by each species. In phenotypically similar species, finding an objective way of determining species boundaries morphologically becomes difficult, and often one is faced with the problem of deciding whether individual samples belong to a single, highly polymorphic species, or to a set of distinct species, with narrow and overlapping ranges of morphological characteristics. Solutions to this kind of problem can be found through detailed morphological analysis of a series of samples or by the use of genetic markers to verify the reproductive isolation of each putative morphological group (Solé-Cava and Thorpe 1987).

J. Reitner and H. Keupp (Eds.)
Fossil and Recent Sponges

In the case of Mediterranean *Axinella*, the matter is further complicated by the presence of a third "morph", of more or less intermediate appearance between *A. damicornis* and *A. verrucosa* (though closer to *A. damicornis*). This morph was only detected through the extensive ecological and taxonomic work which is currently being carried out on *Axinella* sponges from the Mediterranean (Pansini 1983; Pansini, pers. commun.). If this new morph should belong to a different species, this could account, for example, for the very low level of "intraspecific" allogenetic graft acceptance found in *A. damicornis* (Buscema and van de Vyver 1984). It may be that some of the grafts involved individuals from this third morph, which would thus be interspecific rather than intraspecific. In the same way, other works on *A. damicornis*, such as those on chemical defence (see, e.g., the review by Braekmann and Daloze 1986), could involve mixtures of species.

There are four possible explanations for the existence of this third morph: (a) It represents an intermediate form of a single, highly variable species, which includes *A. damicornis* and *A. verrucosa*; (b) it belongs to one of those two species, extending its range of known variation; (c) it is the product of hybridization between *A. damicornis* and *A. verrucosa*; and (d) it belongs to a third biological species.

The aim of this work was to test these hypotheses using population genetic techniques. Samples from the gene pools of each population were studied, and comparisons on gene frequencies at specific loci were used to address the problem of possible gene flow between them. Hypotheses (a) and (b) could be ruled out by the extremely low probability of panmixia occurring between the three populations, whereas hypothesis (c) was eliminated by the kind of genotypes observed for the samples of the new *Axinella* morph. We concluded, therefore, that the third morph belonged to a different biological species.

## Materials and Methods

The specimens studied were collected in November 1986, by Scuba diving in the locality of Paraggi (Golfo Marconi, Ligurian Sea). All individuals chosen were sympatric, at a depth of 10–20 m on a rocky vertical cliff. The usual precaution of collecting individual sponges separated by a few meters, in order to avoid the collection of clonemates (Solé-Cava and Thorpe 1986, 1987), was not taken in the case of these sponges, since they are known to reproduce exclusively sexually (Buscema and Van de Vyver 1984). In any case, the banding patterns of each individual could be compared throughout the enzyme loci for absolute genetic identity (the results show that no possible clonemates were found in any of the morphotypes).

After collection, the specimens were stored at approximately 4°C and brought to the laboratory for electrophoretic and morphological examination. The specimens could be separated into three morphotypes, solely on the basis of external morphology. *A. damicornis* was typically characterized by flattened and coalescing branches, forming palmate ramifications, whereas *A. verrucosa* had thinner, digitated branches, which rarely fused. The third morphotype showed greater similarity to *A. damicornis*, but the branches tended to be more fused, forming an irregularly shaped mass. This morphotype will be referred to as "unknown" for the moment, until further taxonomic work is done and a holotype is named.

Seven individuals of each morphotype were analyzed using horizontal starch gel electrophoresis. Twenty enzyme systems were tried, but only eight enzyme loci could be recorded consistently for all morphotypes. This was probably due to the low concentration of enzymes generally present in sponge tissue (Solé-Cava and Thorpe 1987), or to the inadequacy of the methods used here to study the enzymes of these particular sponges. The electrophoretic methods and the recipes for the staining solutions employed are described elsewhere (Solé-Cava 1986; Solé-Cava and Thorpe 1986).

For each of the individuals analyzed genetically, 15 to 20 spicules of each type were selected and their length and width measured. Because of the small number of specimens sampled for each species and the large standard deviations found between individuals, the results were compared by a non-parametric (Kolmogorov-Smirnov) test (Sokal and Rohlf 1981).

## Results

Both styles and oxeas from *Axinella verrucosa* were generally bigger than those of the other two species (Table 1 and Fig. 1), but the difference was not statistically significant (Table 2).

The eight loci observed show a moderate amount of gene variation (Table 3). The heterozygosity values observed were only a little higher than those found by Sara (1988) for *Tethya citrina* and *T. aurantium*, but were much smaller than the values reported for other demosponges (in the range 0.137–0.335; Balakirev and Manchenko 1985; Solé-Cava and Thorpe 1986, 1987, 1990). This difference probably due to the small number of loci analyzed for *Tethya* and for *Axinella*

**Table 1.** Mean, standard deviation and sample sizes (total number of spicules measured in the seven individuals of each species) for the dimensions of the two spicule types of *A. verrucosa*, *A. damicornis*, and *A.* sp.("Unknown"). All measurements are in micrometers

| | | Oxea | | Style | |
|---|---|---|---|---|---|
| | | Length | Width | Length | Width |
| *A. damicornis* | Mean | 348.5 | 7.8 | 485.4 | 7.1 |
| | S.D. | 137.3 | 2.9 | 318.9 | 2.4 |
| | N | 115 | | 110 | |
| "Unknown' | Mean | 336.7 | 7.0 | 423.4 | 8.3 |
| | S.D. | 63.5 | 2.8 | 261.9 | 2.2 |
| | N | 128 | | 109 | |
| *A. verrucosa* | Mean | 488.1 | 6.2 | 540.5 | 6.0 |
| | S.D. | 88.4 | 2.2 | 94.4 | 2.3 |
| | N | 95 | | 96 | |

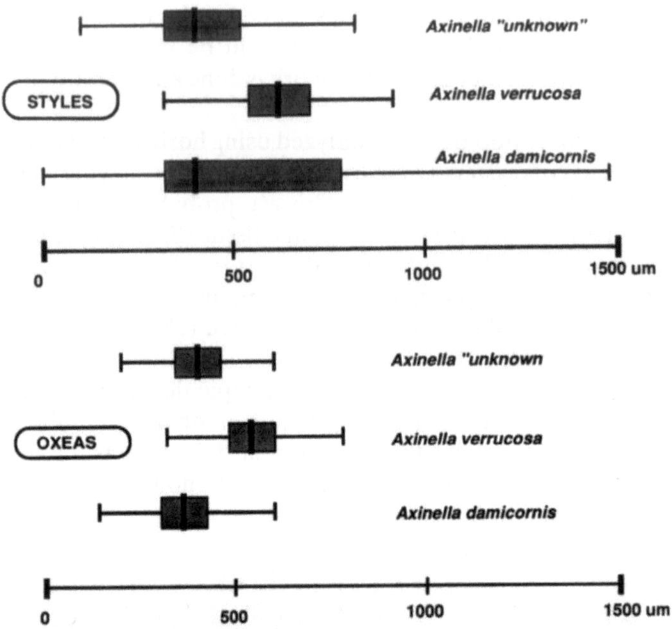


**Fig. 1.** Box-and-whiskers plot for spicule lengths of three *Axinella* morphotypes

(eight in both cases) as compared to the 16 to 28 loci assayed for the other species of demosponges.

The genetic identity (I, Nei 1972) between the three *Axinella* morphotypes studied here is small; I = 0.700 between *A. damicornis* and "unknown" and I = 0.125 between either of these and *A. verrucosa*. This indicates a large gene divergence between their populations (Nei 1972; Thorpe 1979) and possible reproductive isolation.

## Discussion

The genetic identities found between the morphotypes of *Axinella* are in the range of typical values found for interspecific comparisons (Thorpe 1982, 1983; Avise 1983), which indicates that they belong to different species. Additionally, a number of diagnostic loci (sensu Ayala 1983) were found for each morphotype. The probability of finding X loci fixed for different alleles in two samples from the same population, in the absence of selection and in sexually reproducing organisms is smaller than

$$n1^{-(2*n2*X)},$$

where n1 and n2 are the sample sizes for the samples 1 and 2 respectively, arranged in a way that n1 ≧ n2 (as deduced from Carter and Thorpe 1979 and

**Table 2.** Significances of the difference found between the ranked values of the oxea (above the diagonal) and style (below diagonal lengths, for pairwise Kolmogorov-Smirnov test comparisons of the samples from the three morphotypes. Observe that none of the values is under 0.05

| | *A. damicornis* | "Unknown" | *A. verrucosa* |
|---|---|---|---|
| *A. damicornis* | – | 0.215 | 0.500 |
| "Unknown" | 0.194 | – | 0.500 |
| *A. verrucosa* | 0.500 | 0.500 | – |

**Table 3.** *Axinella damicornis, A. verrucosa,* and *A.* sp. ("unknown") Gene frequencies for eight enzyme loci. *Ca* Carbonic anydrase; *Est* Esterases; *Np* Nucleoside phosphorylase; *Pep* Peptidases; *Pgd* Phosphogluconate dehydrogenase; *Sod* Superoxide dismutase; $H^e$ mean expected heterozygosity; *n* number of alleles analyzed in each population

| Locus | Allele | *Damicornis* | Unknown | *Verrucosa* |
|---|---|---|---|---|
| Ca | 1 | 0.786 | 0.714 | 0 |
| | 2 | 0.214 | 0.286 | 0 |
| | 3 | 0 | 0 | 1.000 |
| Est-1 | 1 | 0 | 0 | 0.500 |
| | 2 | 0 | 0 | 0.500 |
| | 3 | 1.000 | 1.000 | 0 |
| Est-2 | 1 | 0.143 | 0 | 0 |
| | 2 | 0.786 | 0.786 | 0 |
| | 3 | 0.071 | 0.214 | 0 |
| | 4 | 0 | 0 | 0.500 |
| | 5 | 0 | 0 | 0.500 |
| Np | 1 | 0 | 0 | 1.000 |
| | 2 | 1.000 | 0 | 0 |
| | 3 | 0 | 1.000 | 0 |
| Pep-1 | 1 | 1.000 | 1.000 | 1.000 |
| Pep-2 | 1 | 0 | 0 | 1.000 |
| | 2 | 1.00 | 0 | 0 |
| | 3 | 0 | 1.000 | 0 |
| Pgd | 1 | 1.000 | 1.000 | 0 |
| | 2 | 0 | 0 | 1.000 |
| Sod | 1 | 1.000 | 1.000 | 0 |
| | 2 | 0 | 0 | 1.000 |
| H | | 0.087 | 0.093 | 0.125 |
| | n | 14 | 14 | 14 |

Solé-Cava and Levy 1987). In the case of the *Axinella* species, this probability is lower than $7^{-(2*7*2)}$ or $< 2.2 \times 10^{-24}$ for the comparison of *A. damicornis* and the "unknown" *Axinella* species, and lower than $7^{-(2*7*7)}$ or $< 1.5 \times 10^{-83}$ for the comparison of either of those with *A. verrucosa*. It is thus extremely unlikely that the morphotypes belong to one large, panmictic population. This certifies the reproductive isolation between these sympatric populations and, therefore, confirms their specific status.

As suggested by the morphological work done prior to the electrophoretical analysis (Fig. 2), the new species of *Axinella* was genetically closer to *A. damicornis* than to *A. verrucosa*. This agreement between overall morphological and genetic similarities is often found in biochemical taxonomy studies, indicating a general parallelism between morphological and molecular evolution, if not in absolute rates (Lewontin 1984), at least in direction.

One could speculate about the origins of these three sympatric, morphologically similar, species. Speciation is generally considered to come about as a result of the evolutionary divergence of isolated populations (allopatric speciation) or possibly through reproductive isolation brought about by niche

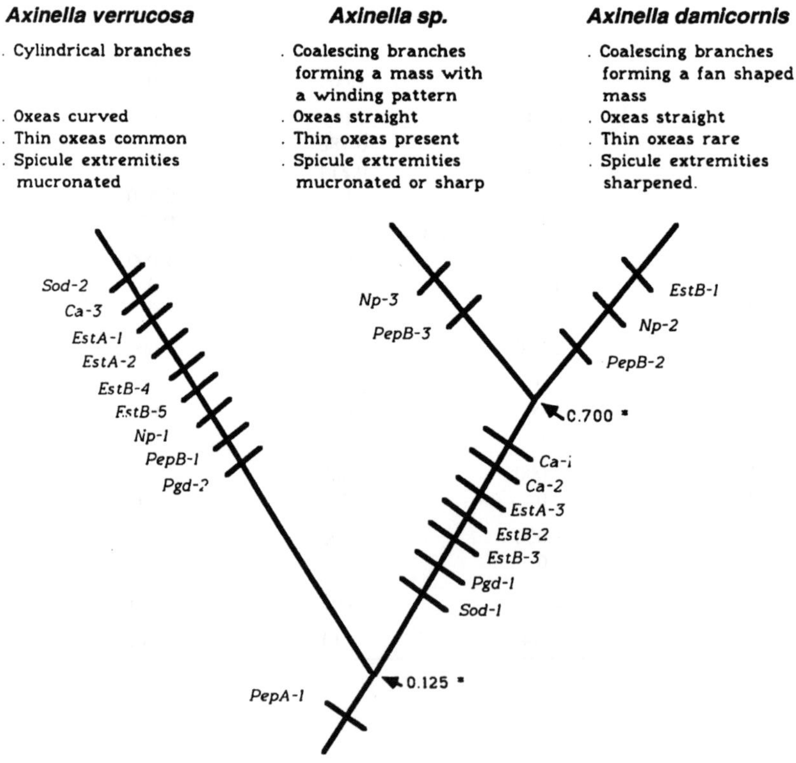


**Fig. 2.** "Consensus" tree, based on morphology, genetic similarity indices (*) and derived electromorphs for the three species of *Axinella*

differentiation in sympatric populations. The latter model has been used to explain the existence of three morphologically similar species of calcareous sponges of the genus *Leucosolenia* living sympatrically in the Mediterranean (Sara 1956, 1957). Similarly, *Axinella damicornis* and *A. verrucosa* have important ecological differences, such as distinct and widely spaced reproductive periods: spermatogenesis occurs during a 3–4-week period in April/May for *A. verrucosa* and August/September for *A. damicornis* (Siribelli 1962). This anachronic reproductive isolation certainly plays a role in maintaining the identity of these species, but does not necessarily indicate sympatric speciation.

A pre-requisite for the theory of sympatric speciation is a high level of local gene polymorphism (Sara 1957), which is indeed found in many sponge species (Sole-Cava and Thorpe 1987, 1990). However, this model has been contested on paleontological (Wiedenmayer 1974) and ecological (Vethaak et al. 1982) grounds, and a new picture emerges, where allopatric, micro-allopatric and parapatric speciation processes best explain the occurrence of sympatric sibling species in sponges (Sara 1987, 1988). In the light of our findings, it would now be extremely interesting to investigate the reproductive period of this new *Axinella* species, and study the possible vicariant or ecological processes associated to its origin.

The lack of significant difference in spicule size between the three species indicates that spicule dimensions should be used with extreme caution in sponge taxonomy, particularly because of the high intraspecific variation which may be encountered (Siribelli 1961; Pansini 1983; Jones 1984, 1987). In the genus *Suberites* from the Isle of Man (Irish Sea), for example, three species could be well distinguished genetically, but no differences in spicule dimensions were observed between them (Solé-Cava and Thorpe 1986). It is imperative, therefore, that studies on the amplitude of the intraspecific spicule length variation, like those of Jones (1984, 1987) be undertaken on a larger number of species, principally on groups with simple spicule sets, such as some of the haplosclerids and axinellids.

Biochemical taxonomy has been used, to date, to help clarify species problems in three unrelated genera of demosponges (*Suberites*, *Tethya*, and *Axinella*), thus confirming the usefulness of the technique in sponge taxonomy (Solé-Cava and Thorpe 1987). The collaboration of classical taxonomists with geneticists seems to be both necessary and promising, as it is in the similar field of chemical taxonomy (Bergquist 1978; Bergquist and Wells 1983). If a revolution in sponge taxonomy is to happen, it will require the joint effort of ecologists, immunologists, chemists, cladists, and geneticists, all led by classical taxonomists, who will show the way and indicate the obstacles to be faced.

*Acknowledgments.* The authors are thankful to J. Hotchkiss for helpful comments on the manuscript, to T. Norton for the provision of facilities, and to J. Reitner for Logistic support during the Symposium. A. Solé-Cava was supported by grants from CAPES and FAPERJ (Brazilian Government).

## References


Akiyama SK, Johnson MD (1983) Fibronectin in evolution: presence in invertebrates and isolation from *Microciona prolifera*. Comp Biochem Physiol 76B:687–694

Amade P, Charroin C, Baby C, Vacelet J (1987) Antimicrobial activities of marine sponges from the Mediterranean Sea. Mar Biol 94:271–275

Avise JC (1974) The systematic value of electrophoretic data. Syst Zool 23:465–481

Avise JC (1983) Protein variation and phylogenetic reconstruction. In: Oxford GS, Rollinson D (eds) Protein polymorphism: adaptive and taxonomic significance. Academic Press, London. pp 103–130

Ayala FJ (1983) Enzymes as taxonomic characters. In: Oxford GS, Rollinson D (eds) Protein polymorphism: adaptive and taxonomic significance. Academic Press, London, pp 3–26

Balakirev ES, Manchenko GP (1985) Intraspecies genetic variation in the sponge *Suberites domuncula*. Biol Morya 5:36–40

Bergquist PR (1978) Sponges. Hutchinson, London, pp 268

Bergquist PR, Wells RJ (1983) Chemotaxonomy of the Porifera: the developments and current status of the field. In: Scheurer PJ (ed) Marine Natural Products. Academic Press, London, pp 1–50

Bergquist PR, Lavis A, Cambie RC (1986) Sterol composition and classification of the Porifera. Biochem Syst Ecol 14:105–112

Braekman JC, Daloze D (1986) Chemical defence in the sponges. Pure Appl Chem 58:357–364

Buscema M, van de Vyver G (1984) Cellular aspects of alloimmune reactions in sponges of the genus *Axinella* II. *Axinella verrucosa* and *Axinella damicornis*. J Exp Zool 229:19–32

Buscema M, van de Vyver G (1985) Cytotoxic rejection of xenografts between marine sponges. J Exp Zool 235:297–308

Dams E (1986) Sequence analysis of the 5S rRNA from 19 species proves the validity of a universal secondary structure model of this molecule. Dissert Abstr Int 47C:882

Dyrynda PEJ (1985) Chemical differences and the structure of subtidal epibenthic communities. Eur Mar Biol Symp 19:411–421

Garrone R (1984) Formation and involvement of extracellular matrix in the development of sponges, a primitive multicellular system. In: Trelstad RL (ed) The role of extracellular matrix in development. Alan Liss, New York, pp 461–477

Jones WC (1984) Spicule dimensions as taxonomic criteria in the identification of the haplosclerid sponges from the shores of Anglesey. Biol J Linn Soc 80:239–259

Jones WC (1987) Skeletal variation in embryo-containing specimens of *Haliclona rosea* (Bowerbank) from Anglesey. North Wales. In: Vacelet J, Boury-Esnault N (eds) Taxonomy of Porifera. Springer, Berlin Heidelberg New York Tokyo, pp 101–124

Kaye HR, Reiswig HM (1985) Histocompatibility responses in *Verongia* species (Demospongiae): implications of immunological studies. Biol Bull 168:183–188

Lewontin RC (1984) Detecting population differences in quantitative characters as opposed to gene frequencies. Am Nat 123:115–124

Muller WEG, Zahn RK, Zahn G, Rijavec M, Batel R, Kurelec B, Dorn A, Muller I (1985) Sponge cell aggregation: biochemical characterization of the system. Thalassia Yugosl 19:285–296

Nei M (1972) Genetic distance between populations. Am Nat 106:283–292

Pansini M (1983) Notes on some Mediterranean *Axinella* with description of two new species. Boll Mus Ist Biol Univ Genova 50–51:79–98

Sara M (1956) Aspetti genetici ed ecologici dell'ibridazione naturale fra differentie specie di Leucosolenia (Calcispongiae) a Roscoff. Boll Zool 23:1–3

Sara M (1957) Problemi di speciazione negli animali marini. Arch Zool Ital 42:273–379

Sara M (1987) A study of the genus *Tethya* (Porifera Demospongiae) and new perspectives in sponge systematics. In: Vacelet J, Boury-Esnault N (eds) Taxonomy of Porifera of the NE Atlantic and Mediterranean Sea. Springer, Berlin Heidelberg New York, pp 205–225

Sara M (1989) Divergence between the sympatric species *Tethya aurantium* and *T. citrina*, an example of speciation in sponges. In: Hartman WD, Rützler K (eds) Proceedings of the third International Conference on the biology of sponges. Smithsonian Institution Press, Washington (in press)

Scudder GGE (1987) The next 25 years: invertebrate systematics. Can J Zool 65:786–793

Siribelli L (1961) Differenze nell-aspetto esterno e nello scheletro fra *Axinella verrucosa* O.S. e *Axinella damicornis* (Esper) OS (Demospongiae). Annu Ist Mus Zool Napoli 13:1–24

Siribelli L (1962) Differenze nel ciclo sessuale di popolazioni conviventi di *Axinella damicornis* (Esper) ed *Axinella verrucosa* O.S. (Demospongiae). Annu Inst Mus Zool Univ Napoli 14:1–8

Smith LC, Hildemann WH (1984) Alloimmune memory is absent in *Hymeniacidon sinapium*, a marine sponge. J Immunol 133:2351–2355

Smith LC, Hildemann WH (1986) Allograft rejection, autograft fusion and inflammatory responses to injury in *Callyspongia diffusa* (Porifera: Demospongiae). Proc R Soc Lond 226:445–464

Sokal RR, Rohlf FJ (1981) Biometry. The principles and practice of statistics in biological research. Freeman, San Francisco, pp 859

Solé-Cava AM (1986) Studies of biochemical genetics and taxonomy in coelenterates and sponges. PhD thesis. The University of Liverpool, pp 227

Solé-Cava AM, Levy J (1987) Biochemical evidence for a third species of angel shark off the East coast of South America. Biochem Syst Ecol 15:139–144

Solé-Cava AM, Thorpe JP (1986) Genetic differentiation between morphotypes of the marine sponge *Suberites ficus* (Demospongiae: Hadromerida). Mar Biol 93:247–253

Solé-Cava AM, Thorpe JP (1987) The uses of electrophoresis in sponge taxonomy. In: Vacelet J, Boury-Esnault N (eds) Taxonomy of Porifera. Springer, Berlin Heidelberg New York Tokyo, pp 243–258

Solé-Cava AM, Thorpe JP (1990) High levels of gene variation in marine sponges. In: Hartman WD, Rutzler K (eds) Proceedings of the third International Conference on the biology of sponges. Smithsonian Institution Press, Washington (in press)

Thorpe JP (1979) Enzyme variation and taxonomy: the estimation of sampling errors in measurements of interspecific genetic similarity. Biol J Linn Soc 11:369–386

Thorpe JP (1982) The molecular clock hypothesis: biochemical evolution, genetic differentiation and systematics. Annu Rev Ecol Syst 13:139–168

Thorpe JP (1983) Enzyme variation, genetic distance and evolutionary divergence in relation to levels of taxonomic separation. In: Oxford GS, Rollinson D (eds) Protein polymorphism: adaptive and taxonomic significance. Academic Press, London, pp 131–152

Van Soest RWM (1987) Phylogenetic exercises with monophyletic groups of sponges. In: Vacelet J, Boury-Esnault N (eds) Taxonomy of Porifera. Springer, Berlin Heidelberg New York, pp 227–242

Vethaak AD, Cronie RJA, Van Soest RWM (1982) Ecology and distribution of two sympatric, closely related sponge species, *Halichondria panicea* (Pallas, 1766) and *H bowerbanki* Burton, 1983 (Porifera, Demospongiae), with remarks on their speciation. Bijdr Dierk 52:82–102

Wiedenmayer F (1974) Recent marine shallow-water sponges of the West Indies and the problem of speciation. Verh Naturforsch Ges Basel 84:361–375

# Non-Spicular Biomineralization in Calcified Demosponges

R. Wood

## Introduction

The sponges occupy an unusual evolutionary position as the most primitive of metazoans. However, despite their relatively small number of cell types they have evolved a considerable array of morphological forms and have evolved into a variety of ecological habitats. Although they possess no tissue sensu stricto, their cellular activities attain a significant degree of integration and they are able to secrete both calcium carbonate (as aragonite or high- or low-Mg calcite) and amorphous or hydrated silica. These minerals may be precipitated in the form of spicules, scales, plaques, cements, desmas, massive skeletons and granules. In the absence of autogenic precipitations, foreign mineral grains may even be incorporated and so it would appear that some mineralized component is necessary to the functioning of the sponge, presumably to maintain structural rigidity or for protection. Sponges therefore offer considerable opportunities to study mechanisms of biomineralization.

The rediscovery of a diverse living fauna of calcified sponges which possess a calcareous skeleton in addition to a siliceous spicule framework has provoked renewed interest in the biology, phylogeny and ecology of related fossil reef-building forms and the fossil record has since been realized to contain an extremely rich record of sponges which combine both these skeletal elements. In fact, some Recent forms have been found to be "living fossils" with very conservative histories extending back as far as the Upper Permian (Wood 1987). These living and fossil forms are known as calcified demosponges, and their calcareous skeletons have been found to be convergent in many features, including mineralogy, grade of organization and microstructural type (Vacelet 1985; Wood 1987; Reitner 1987). Many of the fossil forms were previously allocated to the Cnidaria due to the convergence shown by their skeleton to that of corals and the living forms were also initially classified together using similar criteria. Therefore, intriguing questions are raised as to the nature of the controls (extrinsic or intrinsic) which caused the formation of massive calcareous skeletons in such a taxonomically diverse group of sponges.

In this review, I will attempt to assess the status and role of calcareous skeletons in sponges by a compilation of what is known of their biomineralization mechanisms, resultant microstructures and systematic distribution, and also by a discussion of any unusual features shared by these sponges which may point to a broader understanding.

J. Reitner and H. Keupp (Eds.)
Fossil and Recent Sponges

## Modes of Biomineralization

Most work on the calcareous skeletons of sponges has concerned the detailed description of microstructural types, with little attention being given to their mechanisms of precipitation. Sponges not only produce a wide variety of microstructural fabrics (Fig. 1) but form them in a number of ways. These mechanisms can be classed under a series of genetic sub-divisions which are described below. This has the advantage of separating microstructures which have previously been classified together under the same descriptive heading, but which clearly have different modes of genesis (see Gautret 1986). This could lead to a more appropriate method of classification which would describe living calcified forms by their organizational grade, biomineralization mechanism and resultant microstructure type, in conjunction with the systematic placing of their non-calcified close-relatives. *Ceratoporella* could be described as a calcified agelasid of chaetetid grade with a modified spherulitic microstructure formed by a secretory pinacoderm and *Astrosclera*, a calcified agelasid of stromatoporoid grade, with a compound spherulitic microstructure of intracellular origin.

Table 1 summarises information concerning the supposed non-spicular biomineralization mechanisms found in Recent calcified demosponges, their products, systematic distribution and the inferred placing of fossil forms.

### *Mineralization of a Collagenous Matrix*

Collagenous frameworks are commonly found in a wide variety of Recent non-calcified demosponges (e.g. *Esperiopsis*, Fig. 2g). In this mode of calcareous skeleton formation, the calcareous skeleton is wholly defined by the extent of this pre-formed collagenous framework. This would seem to indicate an "organic matrix-mediated" mechanism, in which the cell manipulates an organic framework upon which regulated mineralization occurs (Lowenstam 1981). Spicules may or may not play a role, but where present they appear to act as a framework around which the organic matrix is precipitated and, in the case of

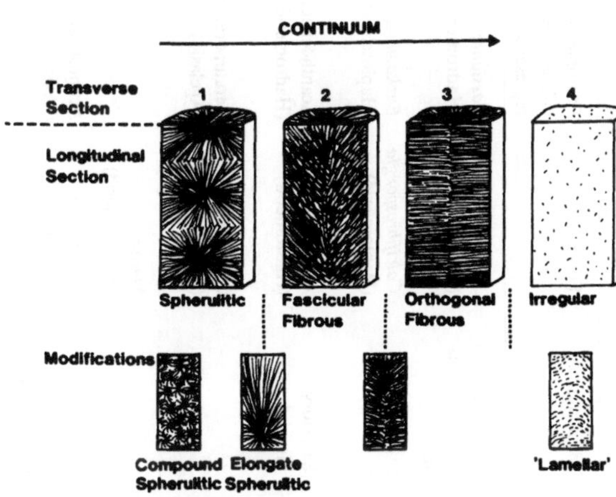


**Fig. 1.** Four main calcareous skeleton microstructural types of the calcareous skeleton found in demosponges, illustrated schematically as they appear in sections through a radial element. Modifications of these types are shown below (Wood 1987)

**Table 1.** Summary table of characteristics of living sponges with a calcareous skeleton

| Mechanism | Form | Examples of inferred fossil representatives | Micro-structure | Mineralogy Recent/Fossil | Skeletal parts | Role of spicules | Rich symbiotic fauna | Remarks |
|---|---|---|---|---|---|---|---|---|
| Mineralization of collagenous matrix | *Vaceletia* (Ceractinomorph) | *Cassianothalamia* (Hadromerid) | Irregular | A/MgC | 1° Skeleton<br>Backfill<br>No tabulae present | Aspicular | Yes | Periodic calcification |
| | *Calcifibrospongia* (Haplosclerid) | *Euzkadiella* (Haplosclerid) | Spherulitic | A/MgC | 1° Skeleton<br>No tabulae<br>or backfill present | Spicule framework | Yes | |
| Secretory pinacoderm | *Acanthochaetetes* (Hadromerid) | Acanthochaetetids (Hadromerid) | Irregular (aligned in one plane) | High MgC | 1° Skeleton<br>1° and 2° Tabulae | Chance incorporation | No | Collagenous anchoring strands |
| | Ceratoporellids (Agelasid) | Ceratoporellids (Agelasid) | Elongate spherulitic | A | 1° Skeleton<br>?2° Tabulae<br>backfill | Partial incorporation | Yes | Continual calcification |
| | *Merlia* (Poecilosclerid) | | Fascicular fibrous | MgC | 1° Skeleton<br>1° Tabulae | No incorporation | No | Thin encruster |
| ?Passive secretory pinacoderm (cement) | *Minchinella* (Calcinea) | Actinostromaridae (?Axinellid) | Orthogonal | MgC | 1° Skeleton | Spicule framework | No | |
| | *Murrayona* (Calcinea) | | Irregular | MgC | 1° Skeleton | No incorporation | No | Aggregated to form skeleton |
| | *Petrobiona* (Calcaronea) | | Flake-spherulitic | MgC | 1° Skeleton | Spicule framework | No | Solid mass |
| Intracellular origin | *Astrosclera* (Agelasid) | *Follicatena* (?) | Compound spherulitic | A | 1° Skeleton<br>backfill | Chance incorporation | Yes | |

A: Aragonite; MgC: Magnesium calcite; 1°, primary, 2°, secondary.

fibrous microstructures, as nucleation sites for the acicular crystal growth. Two forms are known which calcify by this method, but they are systematically unrelated and produce totally different microstructures. However, both forms bear dense bacterial populations within their tissues.

*Vaceletia* (a Recent aspiculate ceractinomorph sphinctozoan genus) has a small (1–3 cm long) chambered skeleton with a central exhalant siphon and is found in the Mediterranean and Pacific. It bears a thin ectosomal layer and a leuconoid, dense choanosome with small choanocyte chambers. The living tissue is mainly internal within the last few hemispherical chambers, but a pinacoderm layer covers the surface of the uppermost chamber. The chamber walls are perforated by ostia and are composed of a feltwork of aragonitic microfibrils (Fig. 2a). New chambers are added periodically, and originate by the formation of a collagenous template below the upper pinacoderm layer. Mineralization first occurs within isolated spherical regions, extending along the length of the new chamber wall (Gautret 1985). Crystals first form as bundles of disorientated acicular rods (Fig. 2c, compact in the terminology of Gautret 1985), which later become more granular (Fig. 2b, composite in the terminology of Gautret 1985) during a further precipitation phase. In some parts of the skeleton, the acicular bundles may show a preferred orientation parallel to the collagenous fibres between which they grow (Fig. 2c). The pillars which support the hemispherical chambers form during the first mineralization phase by mineralization of organic strands (Figs. 2d,e). Older, abandoned chambers become filled with layered lenses of irregular microstructure and although it is not clear whether these are the result of inorganic or organic mineralization, their form does suggest an organic origin.

Non-spicular fossil sphinctozoans which bear an irregular calcareous skeleton are inferred, by suggested affinity to *Vaceletia*, to have possessed this mode of biomineralization. In addition, spiculate fossil forms are known e.g. *Newellia mira* (a Carboniferous haplosclerid stromatoporoid) and *Actinostromarianina lecompti* (a Jurassic axinellid stromatoporoid, Fig. 2f) whose irregular calcareous skeleton appears to drape around a primary spicular skeleton, also implying direct mineralization of a collagenous template (Wood 1987; Wood et al. 1989).

*Calcifibrospongia* (a Recent haplosclerid stromatoporoid genus) possesses a massive reticulate skeleton of sheets of aragonite. These are pierced by holes allowing a continuity of living tissue to occupy the interskeletal spaces up to a depth of 3 cm beyond the vestibule floor, with the mesohyl and choanoderm occurring in the interstices of the calcareous skeleton. The entire skeletal meshwork encloses a lattice of siliceous strongyles some of which protrude beyond the surface of the skeleton (Fig. 3c). In addition, the exopinacoderm bears a reticulum of tracts of thin strongyles covering the entire sponge surface. The calcareous skeleton is formed of elongate sclerodermites, 60–110 $\mu$m long (Fig. 3b). Hartman (1979) suggests that the centres of calcification may form simultaneously along or within a particular length of fibre, the acicular crystals of aragonite growing out in all directions from each centre until they reach the boundary of the organic fibre or are stopped by the crystals of a neighbouring sclerodermite. Spicules frequently serve as the centres of the sclerodermites, but

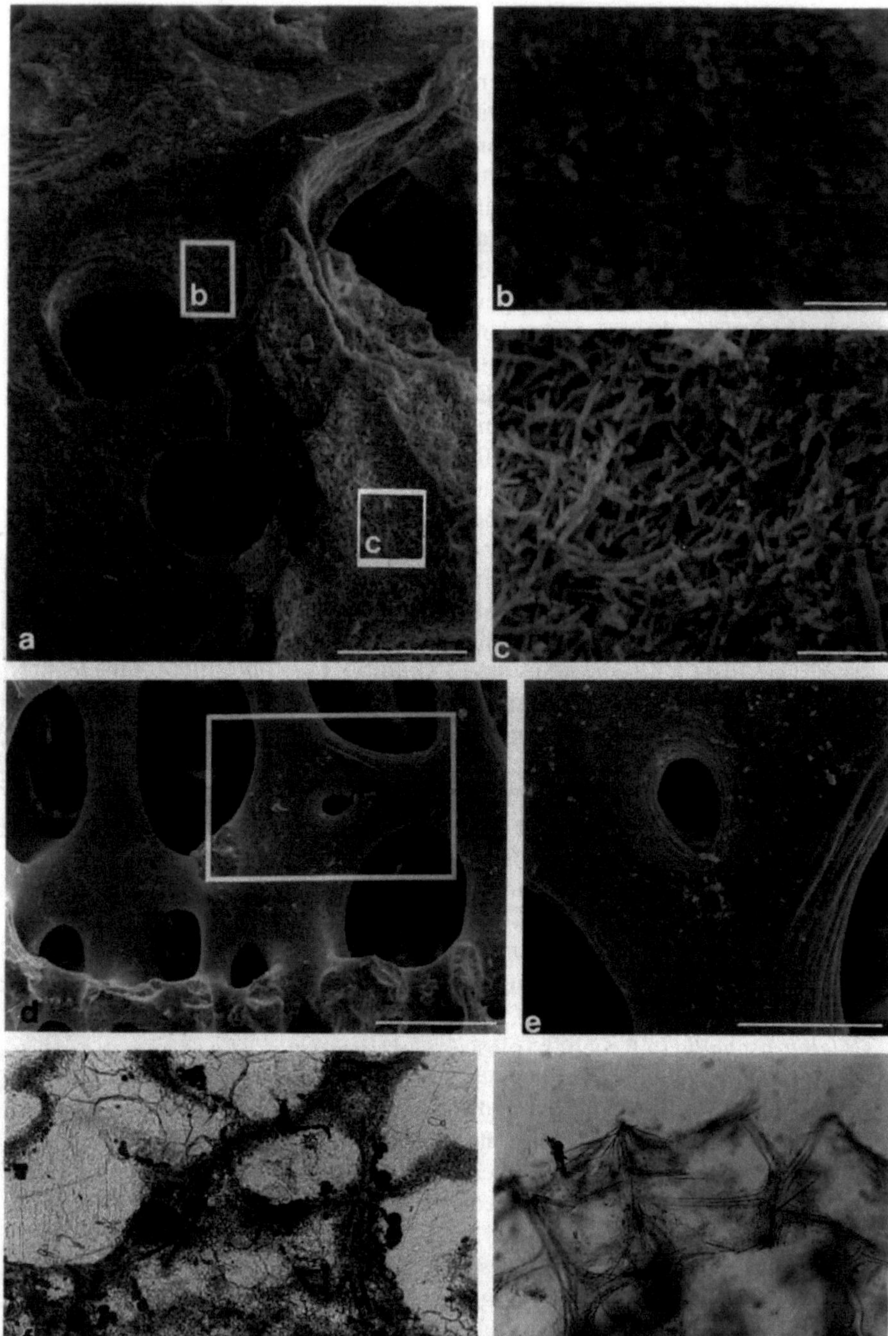
b
c
a
b
c
d
e
f
g

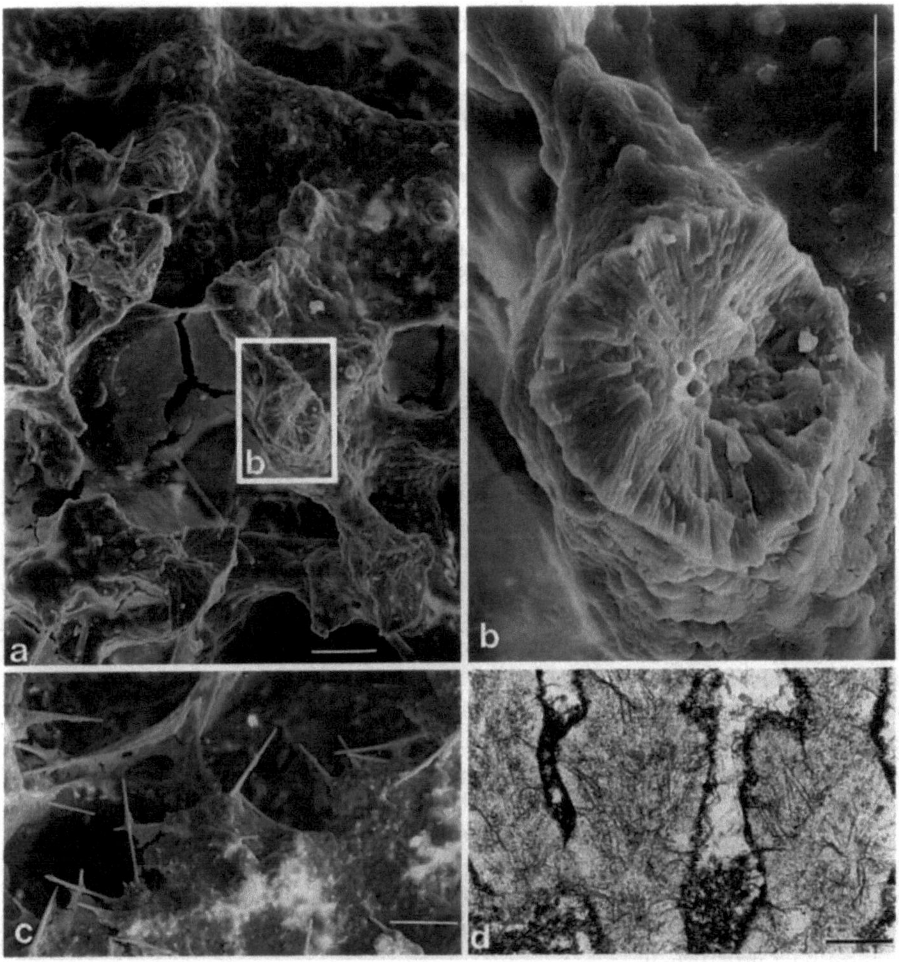


**Fig. 3.** Mineralization of a collagenous matrix to produce a spherulitic microstructure. **a-c** SEM micrographs of *Calcifibrospongia actinostromariodes*. Recent, Jamaica. **a** Broken surface of element showing aragonitic spherulitic sclerodermites. **b** Enlargement showing sclerodermite seemingly initiated from spicule shafts. Bar = 50 μm. **c** Strongyle spicules projecting from the surface of the calcareous skeletal elements. Bar = 100 μm. **d** Longitudinal view of *Euzkadiella erenoensis* (Cretaceous haplosclerid stromatoporoid) showing a similar microstructure. Bar = 100 μm

**Fig. 2.** Mineralization of a collagenous matrix to produce an irregular microstructure. **a-e** SEM micrographs of *Vaceletia crypta*. Recent, Mediterranean. **a** Tangential broken surface of an abandoned chamber showing aragonitic irregular microstructure. Bar = 300 μm. **b** Enlargement showing granular (composite) tissue. Bar = 50 μm. **c** Enlargement showing acicular (compact) tissue. Bar = 50 μm. **d** Pillars. Bar = 300 μm. **e** Enlargement of **d** showing calcified collagenous fibres. Bar = 100 μm. **f** *Actinostromarianina lecompti* (Jurassic axinellid stromatoporoid) showing Mg-calcite skeleton of irregular microstructure seemingly draped around a spicular framework of pyrite pseudomorphs. Bar = 100 μm. **g** *Esperiopsis anomala* (Recent non-calcified axinellid) showing collagenous fibres draped around spicular framework. Bar = 100 μm it is not clear whether the crystals in fact initiate from these points (Fig. 3c).

*Calcifibrospongia* can show deposits of epithecal material varying from 40–250 μm, in thickness, representing planes of the successive dieback and regrowth of the skeleton, but with no other filling tissue or secondary skeleton present. The choanocytes chambers are exceptionally small, 17–18 μm, in diameter, and are densely arranged within the mesohyle, with a relatively large number of choanocyte chambers compared to other calcified demosponges (Vacelet 1988).

*Euzkadiella* (a Cretaceous haplosclerid stromatoporoid genus, Fig. 3d) possesses a similar microstructure as *Calcifibrospongia* and this has been inferred to have been formed by the same process (Reitner 1987).

## *Secretory Pinacoderm*

This mode of biomineralization is probably mediated by a secretory pinacoderm via a thin mucus- or fluid-filled layer (possible polysaccharide) like that of calcareous algae. This would indicate a "biologically controlled" mechanism of precipitation, in which the cell appears to act as a causative agent and in some way controls the precipitation of minerals (Mann 1983). The muco-polysaccharide layer might provide a migration path and medium for joining $Ca^{2+}$ and $CO_3^{2-}$ ions, as known in hermatypic corals. This layer may also to a lesser extent serve as a template controlling the configuration of crystals. This mechanism is diagramatically illustrated in Fig. 4 showing calicle and tabula formation. Several forms are at present collected under this sub-heading, until more details allow us to refine the scheme.

Of the four Caribbean genera comprising the *Ceratoporellidae* (agelasid chaetetids), *Ceratoporella* and *Hispidopetra* are encrusting and massive forms, *Stromatospongia* bears an irregular gross morphology and is an obligate commensal of serpulid worms, and *Goreauiella* is saucer-shaped and attached to the substrate by a peduncle.

The living tissue forms a thin veneer, no more than 1 mm thick over the calcareous mass, dipping 1–1.2 mm into calicles (0.2–0.5 mm diameter) or meandroid furrows in the surface of the aragonite. The siliceous spicules

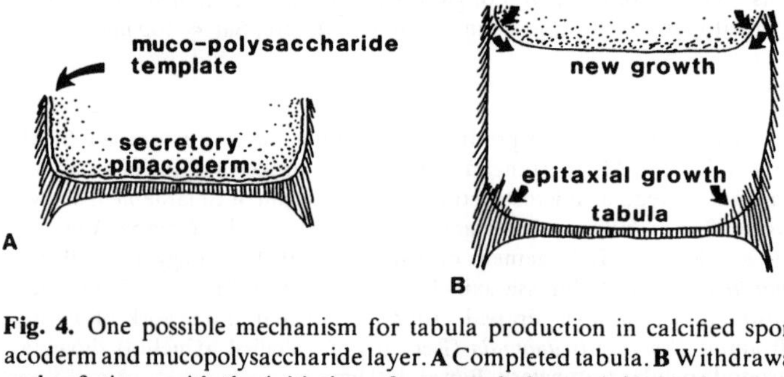


**Fig. 4.** One possible mechanism for tabula production in calcified sponges, via a secretory pinacoderm and mucopolysaccharide layer. **A** Completed tabula. **B** Withdrawal of secretory pinacoderm and soft-tissue with the initiation of a new tabula precipitation by growth from the calicle walls. Epitaxial growth may begin in abandoned areas. Not to scale

(acanthostyles, acanthostrongyles) and spongin become trapped in the aragonite as the calcareous skeleton grows upwards. The calicles increase by longitudinal fission and are filled with an epitaxial growth of aragonite as the living tissue moves upward, although tabulae have also been noted (Fig. 5a). These appear as outgrowths from the calicle walls, which meet in the centre (Fig. 5b). The calcareous skeleton is formed of acicular or fibrous crystals arranged in a modified spherulitic form, which Hartman and Goreau (1975) observed to form around blobs of organic matrix (?spongin) surrounding the heads of some spicules. In *Ceratoporella*, the spicules project upwards and outwards from the calicle walls. However, in *Hispidopetra* dense "bushes" of stylote siliceous spicules are enclosed. Thus, to some extent, the positioning of the calcareous skeleton appears to the determined by the placing of the spicule and spongin frameworks.

One ostium, or sometimes two, open out into the tissue filling a single calicle. These lead to inhalant canals which give off branches that communicate with the choanocyte chambers, which are small, 12–23 μm in diameter. The mesohylar layer is thick containing few choanocytes but large numbers of symbiotic bacteria.

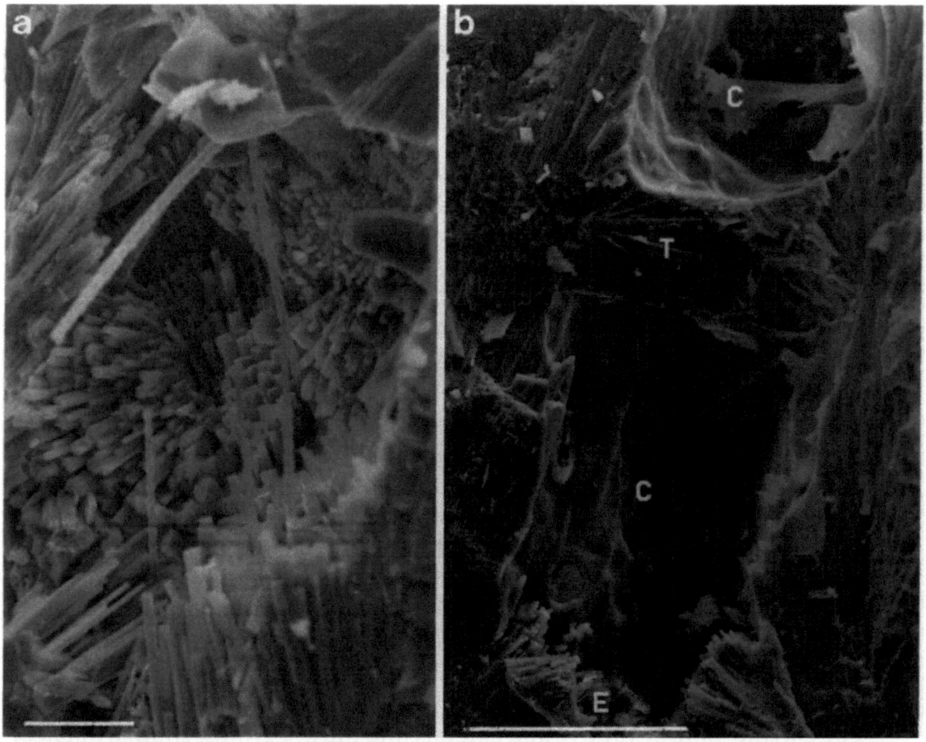


**Fig. 5.** Mineralization via a secretory pinacoderm. SEM micrographs of *Ceratoporella nicholsoni*. Recent, Jamaica. **a** Aragonitic epitaxial backfill of presumed inorganic origin. Note multiple orientation of crystal bundles. Bar = 100 μm. **b** Longitudinal section of a calicle showing: *T* tabula; *C* calicle; *E* epitaxial backfill. Tabulae are produced as secondary outgrowths from the skeletal walls. See Fig. 4 for explanation of their formation. Bar = 300 μm

In contrast, the chambers of *Goreauiella* are surrounded by a much thinner mesohylar layer containing scarce, dispersed bacteria. In *Hispidopetra*, large ovoid cells with long processes are found clustered at the base of the living tissue in contact with the skeleton.

The basal calcareous skeleton of *Acanthochaetetes* (hadromerid chaetetid) is of high-Mg calcite, with an irregular microstructure of crystals (1–8 μm long) orientated in one plane only, giving the appearance of a lamellar microstructure in longitudinal section. The living tissue forms a thin veneer. The surface of the skeleton is pitted with calicles reminiscent of *Ceratoporella*, but spines which are irregularily clumped or arranged in vertical rows project from the walls of the calicles. The calicles are 1.2–2 mm deep and 0.3–0.6 μm in diameter. Instead of the abandoned parts of the skeleton being filled with a secondary epitaxial backfill, the portions below the living tissue are sectioned off by a series of tabulae or horizontal partitions.

Growth of the calicle walls is incremental and layered, and the spines and tabulae grow in an integrated way as outgrowths from the calcicle walls. However, secondarily precipited filling tissue in the form of tabulae is also known. The microscleres form a dense layer at the surface of the living tissue while the tylostyles are aligned parallel to the calicle walls. In living forms, the spicules are not incorporated into the calcareous skeleton, but in fossil representatives they are, although by chance (Reitner and Engeser 1983). The microscleres often occur within the tabulae, perhaps indicating their formation near the living tissue surface.

Nanometre-sized organic fibres act as the matrix for calcification and the calicle wall centres are richest in organic material. Collagenous strands extend into the calcareous skeleton and these seem to act as anchorage points for the soft tissue as noted in some hermatypic corals. The excurrent canals sometimes leave impressions on the skeletal surface as in ceratoporellids.

Both ceratoporellids and acanthochaetetids have considerable fossil records, extending to the Upper Permian and Upper Cretaceous respectively.

*Merlia* (a Recent poecilosclerid chaetetid genus) is an extremely thin encrusting sponge, usually found on gastropod shells. Of the four species known, only two bear a calcareous skeleton and there is even debate as to whether these two are really separate species. The basal calcareous skeleton is of low-Mg calcite, composed of contiguous rows of calicles with openings 0.12–0.15 μm in diameter. Tabulae, each perforated with a central hole, are laid down as the skeleton is built up. No more than four tabulae occur, dividing the calicle into a maximum of five compartments. The pillars are stout fibre fascicles which form the calicle corners and from these form three vertical flanges at 120°, which join with adjacent pillars to form hexagonal spaces. Kirkpatrick (1912) suggested that the tabulae are secreted by the basal pinacoderm.

The choanosome lies above and extends into the uppermost compartment of the calicles. Deeper compartments house crypt cells which accumulate reserves. They may function as gemmules. The canals and choanocyte chambers occupy the depressions between the growing flanges at the tip of this structure. There are no traces of the aquiferous system on the skeleton, the tissue being relatively thick compared with that of the ceratoporellids.

### *Passive Secretory Pinacoderm (Cement)*

Orthogonal microstructure is common in many fossil calcified sponges, including fossil demosponges (Fig. 6) and also in some living calcarean sponges, e.g. *Minchinella*. This simple microstructural type could be explained by the cement-like precipitation of acicular crystals with their c-axis perpendicular to a secretory membrane, and would as such constitute a "biologically induced" precipitation mechanism, in which the cell appears to act as a causative agent in the precipitation of minerals, but has little control (Lowenstam 1981). The Upper Jurassic axinellid stromatoporoid *Actinostromarianina lecompti* shows a primary irregular calcareous skeleton and a secondary orthogonal fibrous one, which has a banded distribution forming latilaminae throughout the growth of the individual. This would seem to point to the mediation of a periodic environmental effect, such as warmer temperatures, causing an intermittent "biologically induced" precipitation.

### *Intracellular Origin*

*Astrosclera* (a Recent agelasid stromatoporoid genus) is the only known calcified demosponge whose calcareous skeleton has an intracellular origin.

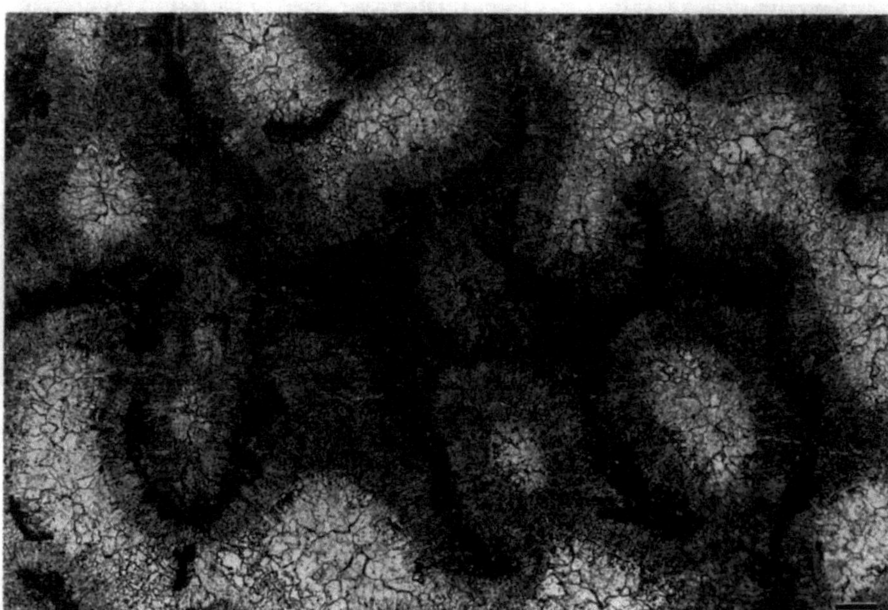

**Fig. 6.** Mineralization via a passive secretory pinacoderm. *Actinostromaria* sp. Jurassic ?axinellid stromatoporoid. Longitudinal section showing an orthogonal fibrous-like cement around a primary siliceous spicule framework (now composed of calcareous pseudomorphs). Bar = 100 μm

The basal calcareous skeleton of *Astrosclera* consists of a fine reticulum of aragonite with no differentiation into pillars and lamellae (Fig. 7a). Calcareous spherules are formed intracellularly as small nuclei (15 $\mu$m) by amoebocytes near the surface of the sponge and then transported to a position where they grow larger by epitaxial growth and add to the general calcareous skeletal network (Fig. 7b). They are known to possess an organic calcification centre of some 7 $\mu$m diameter (Gautret 1986), and the spherules show a periodicity of organic-rich layers, one of which corresponds to the attachment of the spherules to the skeletal wall by a covering organic envelope. The living tissue penetrates the mesh of the skeleton to a depth of approximately 1 cm, and a secondary deposit of acicular crystals partially fills the abandoned inner skeletal reticulum. In the Indian ocean, *Astrosclera* secretes siliceous acanthostyles, but Pacific populations largely lack them, although when present, the spicules are incorporated into the calcareous skeleton. There appears to be no discernible direct relationship between the spicules and the positioning of the calcareous skeleton; they are incorporated by chance. Below the pinacoderm are large exhalant canals, 60–80 $\mu$m in diameter, which branch downwards and outwards into the skeletal tissue to form astro-rhiza-like structures. No tabulae are found. A secondary epitaxial backfill occurs as penicillate tufts in the abandoned parts of the skeleton.

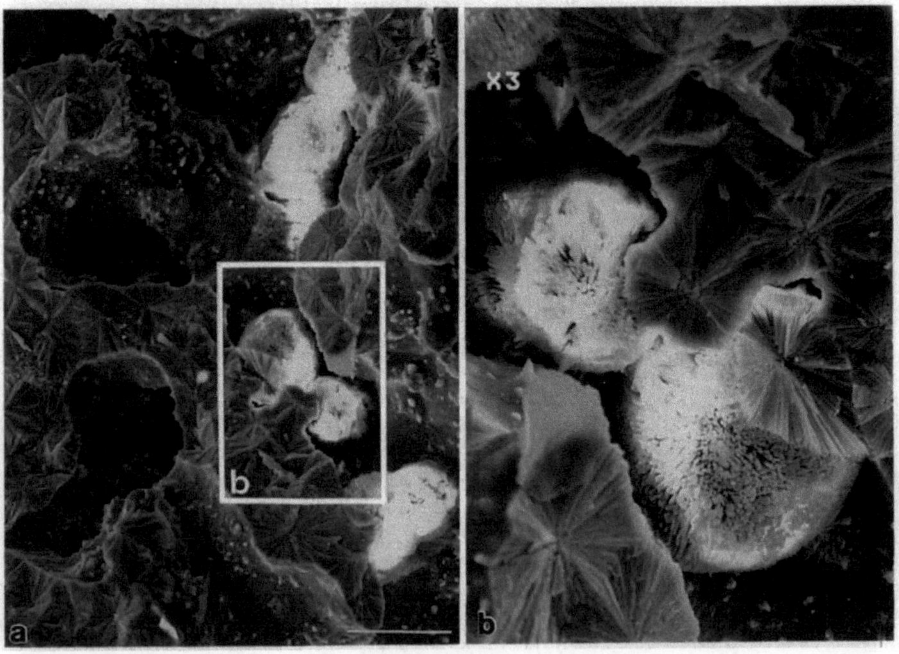


**Fig. 7.** Mineralization via an intracellular origin. SEM micrographs of *Astrosclera willeyana*. Recent, Jamaica. **a** Broken tangential surface showing aragonitic compound spherulitic microstructure. Note spiny outgrowths from the smooth surfaces. Bar = 100 $\mu$m. **b** Enlargement ($\times$3) showing individual spherules growing in situ by epitaxial growth to form the skeletal walls. Note the organic rich centres of the spherules

A wide variety of fossil forms are known which show this microstructural type, especially from the Upper Triassic of Italy and Turkey. However, although the biomineralization mechanism was probably the same, taxonomic affinity cannot be assumed until we have spicule confirmation.

## *Discussion*

When all three skeletal elements are present, the sequence of precipitation is always the same: (a) spicule formation, transportation and orientation, (b) spongin/collagen fibres to bind the spicules in place, (c) calcareous skeleton precipitation.

All biomineralization mechanisms appear to precipitate crystals from beneath an organic layer, which to varying extents acts as template. The distribution of nucleation sites, presumably determined by organic tissue organization, also appears to be an important factor in determining the resultant microstructural type. In most cases, with the exception of irregular microstructures, the crystals are precipitated with their c-axes orientated at right angles to the pinacoderm.

Lowenstam (1981) suggested the phrases "biologically induced" and "organic matrix-mediated" for describing the varying degrees of control of the organic component. Mann (1983) added the additional intermediate status of "biologically controlled". Calcified demosponges appear to show representatives of all these types.

It is often difficult to differentiate between seemingly organically mediated and inorganic precipitations within demosponges, which may be an indication that different biomineralization mechanisms are able to operate within one individual. Fossil calcified demosponges frequently show the phenomenon of latilaminae, which are large-scale concentric bands within the calcareous skeleton indicating some sort of periodicity. These are usually formed by a different spacing, thickness, or even an additional microstructural type-coating of the skeletal elements. This phenomena may be explained as the result of additional "biologically induced" precipitations, caused by periodic environmental fluctuations, or by an environmental influence acting upon those precipitated by a normally more biologically controlled mechanism. Another common deposit found in calcified demosponges in an epitaxial growth of acicular crystals, which initiates after the soft tissue has abandoned an area. This too can be explained by inorganic precipitation mechanisms.

In most Recent sponges the aquiferous system shows great mobility; oscular positions are often rearranged on a daily basis to adjust the fluctuating environmental conditions or adjust to soft tissue growth (S. Stone pers. commun.). Enclosing the aquiferous system within skeletal substances, e.g. spongin or $CaCO_3$, would therefore greatly restrict aquiferous system mobility, as the aquiferous system would antedate all skeletogenic products. We could conclude either that the environments in which calcified demosponges lived were relatively stable, producing only minor amounts of clogging sediment, dispensing with the frequent need to alter aquiferous system arrangement, or that the risks of tissue abrasion were more important for them than aquiferous system flexibility (Wood 1987).

The possession of a rigid calcareous skeleton enclosing the aquiferous system might also account for the incremental growth pattern of many calcified demosponges. The rigid enclosure of the aquiferous system might require the wholesale upward movement of soft tissue into a new layer of skeletogenesis in order for the individual to grow.

All living forms show a smooth calcareous covering over the interskeletal areas occupied by soft tissue, indicating some type of final precipitation after the bulk of skeleton-building calcium carbonate. Spiny outgrowths from these thin linings, consisting of crystals clusters, are also common.

*Ceratoporella*, *Merlia* [and *Petrobiona* (Calcaronea)] contain accumulations of cells with numerous storage granules and food vacuoles. In the two demosponges they resemble the thesocytes of the Suberitidae and are found enclosed within the calcareous skeleton at the bottom of the pseudocalicles where they form monoliniform cylinders (Vacelet, in press). These cells communicate with the choanosome by a small hole or via collagenous partitions within the calcareous tabulae and so are effectively placed in protected areas sectioned off by tabulae. Their development from normal cells seems likely to be due to isolation from well integrated endosomal tissue, like that of gemmule formation, although these masses of storage cells are not true gemmules as they possess no organic envelope and cannot function as free reproductive bodies. Vacelet notes that although these cells do not build the calcareous skeleton, they may have a role in hibernation or regeneration processes, which may give some explanation for the discontinuous growth found in many calcified demosponges.

It is known that the marine *Pseudomonas* bacteria can produce aragonite by releasing ammonia, and sulphate-reducing bacteria can also produce spherulitic aragonite in sediments (Jones 1979). Therefore the presence of dense bacterial populations within many of these calcified demosponges may be significant and they may have a parallel function to that of zooxanthellae in hermatypic corals.

## Taxonomic Status of the Calcareous Skeleton

Considering the systematically widespread presence of calcareous skeletons in demosponges, it is clear that the mere possession of a calcareous skeleton has little higher taxonomic significance. In addition, they are convergent in many characteristics. Table 2 lists those factors which appear to influence and possibly determine the form of calcareous skeletons in demosponges and Fig. 8 suggests the taxonomic status of the various calcareous skeletal characteristics.

### *Grade*

The various grades of organization must to some exent reflect non-systematically related soft tissue and aquiferous system arrangements. Forms which have a thin veneer of tissue and therefore show traces of the excurrent canal system as astrorhizal grooves or ridges in the surface of the calcareous skeleton, produce a calcareous skeleton with a calicular chaetetid grade of organisation. Reitner (this

**Table 2.** Factors which affect four sets of demosponge calcareous skeletal characters. Only the detailed internal arrangement of the skeleton may be correlated with systematic position. All other characters can be convergent

| | |
|---|---|
| 1. Mineralogy<br>$pCO_2$ fluctuations, i.e. origination during Aragonite/high-Mg-calcite or low-Mg-calcite facilitating phases | 3. Microstructure<br>a) Organic template + biomineralization mechanism |
| 2. Grade of organization<br>a) Soft-tissue arrangement (folding of pinacoderm)<br>b) Soft-tissue thickness<br>c) Organization of aquiferous system | 4. Internal organization<br>a) Systematic position<br>b) Environmental fluctuations |

Vol.) has suggested that the formation of tabulated calicles in demosponges may be to produce protected areas to contain the totipotent archaeocytes, which are commonly found at the bases of calicles in living "chaetetids". In contrast, forms which have a greater thickness of living tissue have penetrating aquiferous systems thus producing an 'open' calcareous skeleton, i.e. one of stromatoporoid grade.

## *Mineralogy*

Figure 9 shows the distribution of mineralogical type in known demosponges through time. It is clear that aragonitic forms predominated in Permian/Triassic and Recent times, which correspond to the proposed aragonite facilitating phases of Sandberg (1983). Mineralogy in demosponges appears to be a response of physiochemical precipitates relative to water chemistry and may be detemined by the appearance of calcification relative to Sandberg's scheme within a particular clade. Since different mineralogies are found within members of the same order, mineralogical type may not be of very high taxonomic significance.

| TAXONOMIC CATEGORIES > / TAXONOMIC FEATURES V | CLASS | ORDER | FAMILY | GENUS | SPECIES |
|---|---|---|---|---|---|
| SPICULES: | | | | | |
| TYPE | X | X | X | X | X |
| ARRANGEMENT | X | X | X | X | X |
| CALCAREOUS SKELETON: | | | | | |
| GRADE | | | ? | X | |
| MODE OF BIOMINERALIZATION | | | ? | X | |
| MICROSTRUCTURE | | | | X | |
| MINERALOGY | | | ? | X | X |
| INTERNAL ORGANISATION | | | | | X |
| FILLING TISSUE | | | ? | X | X |

**Fig. 8.** Diagrammatic representation of the taxonomic use of biomineralogical features within sponges

### *Microstructure*

Microstructure seems to be the result of the form of organic template and the biomineralization mechanism or mechanisms employed by the sponge. The type of biomineralization mechanism, when higher taxonomic placing has been achieved by spicule means, is of considerable significance in lower taxonomic categories. However, too little is known about the systematic distribution of mechanisms and mineralogy in sponges to assess the relative status of differing microstructures, but it would seem probable that "biologically controlled" precipitations, e.g. intracellular spherulites, would have greater higher systematic use than "biologically induced" ones, e.g. orthogonal cement.

## Role of Biomineralization

The active scavenging of sediment grains falling upon the surface of aspicular sponges suggests a need for some form of mineralized reinforcement, but existing lines of evidence suggest that non-spicular skeleton production is a simple process and that sponges can calcify with relative ease as:

a) Calcareous skeletons have a wide systematic distribution. Calcified demosponges are known from the Orders Haplosclerida, Axinellida, Hadromerida and Poecilosclerida (see Fig. 9).

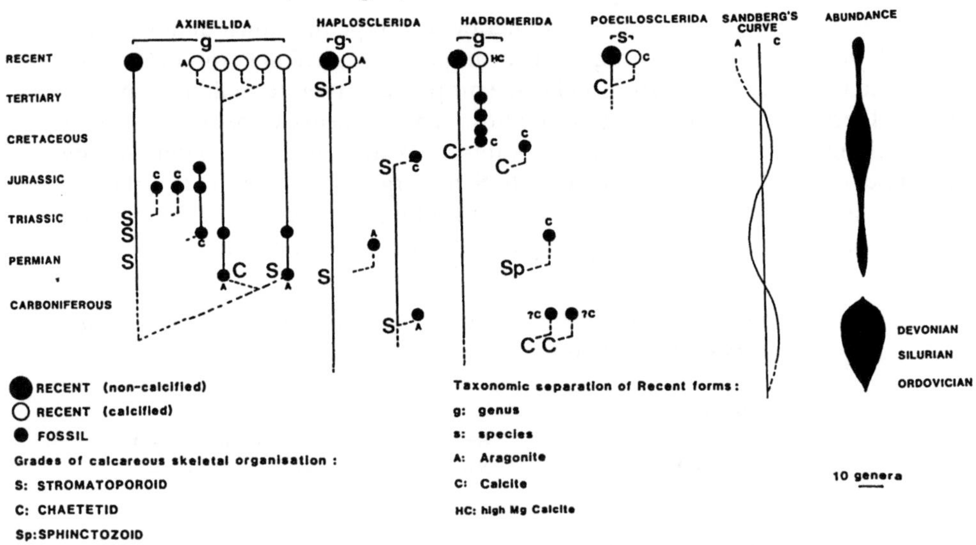


**Fig. 9.** The temporal distribution of known calcified demosponges, including some determined clades, within four Recent demosponge orders. The occurrence and grade of calcareous skeleton development is shown within these orders. Spicule clades are defined by Recent non-calcified forms which show a close relationship to Recent calcified forms. The close affinity between calcified and non-calcified forms is shown by *connected dotted lines* and an indication of taxonomic separation. *Unconnected dotted lines* indicate affinity to order level only. Sandberg's proposed oscillating mineral trend (Sandberg 1983) and the total estimated temporal abundance of calcified sponges are given for comparison. (After Wood et al. 1989)

b) Different microstructures and seemingly different biomineralization mechanisms are present within closely related forms, e.g. both *Astrosclera* and *Ceratoporella* are agelasids, but one produces a spherulitic microstructure with an intracellular origin and the other an elongate spherulitic one via a secretory pinacoderm. Only future research will clarify the degree of taxonomic separation between these two forms. In addition, members of the same clade appear to have independently produced different calcareous mineralogies and microstructures at different times, e.g. *Newellia mira* and *Euzkadiella erenoensis* (Wood et al. 1989).
c) Closest relatives of Recent calcified forms are non-calcified, e.g. *Spirastrella* to *Acanthochaetetes*, and *Agelas* to *Astrosclera* and *Ceratoporella* (see Fig. 9).
d) Some species appear to have a facultative ability. There are four species of the genus *Merlia*, but only two produce a calcareous skeleton, and
e) $\partial\ ^{18}O$ data suggest that in demosponges vital fractionation of sea water is minimal. The crystals precipitated by sponges do not have a unique crystal habit, trace element or isotopic composition which points to a minimal intervention of biological processes. The $\partial\ ^{18}O$ signature of calcified demosponges all fall within the field of normal sea water (J. Reitner pers. commun.), indicating that little fractionation occurs. This is in marked constrast to calcareous sponges which show heavy $\partial\ ^{18}O$ signatures when compared to sea water.

Calcarean sponges which bear an additional calcareous skeleton may be temperate whereas calcified demosponges are sub-tropical/tropical both in their Recent and inferred fossil distribution. It would therefore seem that calcified demosponge mineralization is temperature-related.

It remains to discern the factors and mechanisms responsible for the development of calcareous skeletons within demosponges, for the independent development at different times within the same lineage points to the mediation of ecological or environmental pressures. Changes in ocean chemistry have been linked to biological developments, in particular to the development of hard parts. Riding (1982) has suggested that changes in $pCO_2$ have facilitated the production of calcium carbonate skeletons in calcareous algae. Among metazoans the formation of hard parts might appear to be more independent of extrinsic events, but the apparent ease with which sponges gain a calcareous skeleton would appear to point just such an environmental factor or factors. Figure 9 represents a preliminary gathering of data showing the systematic and stratigraphic distribution of known spiculate forms with a note of their grade and mineralogy. Relative abundance of calcified sponges through time and Sandberg's proposed oscillating mineral trend are also given for comparison.

Kazmierczak et al. (1985) suggest that biomineralization is an adaptation to deleterious amounts of dissolved $Ca^{2+}$ in the cell surroundings and that Recent calcified demosponges have formed calcareous skeletons as detoxification products, as the cryptic habitats in which they live are highly supersaturated with calcium and are subject to intense lithification. However, since these forms are relicts of once-diverse reef-building faunas, a more probable explanation might be that global chemical changes allowed the production of wave-resistant calcareous skeletons which in turn enabled sponges to become reef frame-builders in tropical and sub-tropical environments.

We know that at least some living calcified demosponges grow extremely slowly (0.2 mm/year Willems and Hartman 1985) and, since individuals can grow up to 1 m in diameter, are also very old (up to 5000 years). The mechanical strength of their skeletons is also considerable (Shuhmacher and Plewka 1981) and there is no reason why many extinct forms did not share these characteristics. In fact, the mechanical strength of calcified sponges exceeds that of corals, but it appears they were out-competed by these faster-growing reef-builders when they appeared in the Lower Jurassic. The reef-building niche of calcified demosponges was then subsequently "squeezed" and the forms which survived where those able to colonise habitats were corals could not live, i.e. poorly lit habitats or very shallow ones, or forms which already inhabited such niches. Recent calcified demosponge faunas are of low diversity and abundance.

Another possible contributory explanation for calcification might be the formation of gemmulae, which are found within a taxonomically diverse range of calcified demosponges but are otherwise not common in non-calcified forms. These gemmulae allow the regeneration of the sponge individual after the death of living tissue, and in calcified demosponges they are found within protected areas of the skeleton. Could calcareous skeletons be produced as a protected environment in which to harbour gemmulae through times of environmental stress? Their presence may also, as suggested by Vacelet (1985) explain the seemingly disjunct growth patterns show by may calcified demosponges.

The facultative calcareous skeleton of the living genus *Merlia* is likely to hold the key to these problems. Although calcified and non-calcified species are found in the same macro-environments (J. Vacelet pers. commun.), a detailed study of the differing micro-environmental and micro-ecological requirements of the forms might reveal those biological, ecological or environmental factors responsible for the formation of calcareous skeletons.

Calcified demosponges are the only organisms known which simultaneously secrete hydrated silica and calcium carbonate. This was previously thought to be physiologically impossible, so how have these sponges overcome the problem of the thermo-dynamic instability of these two minerals? Spicules are secreted within special cells and during life the spicules bear a protective coating. This coating breaks down when they become embedded within the calcareous skeleton, and eventually the spicules corrode away. It is apparent that these two biominerals are formed by distinct cell types within the sponge, but this might be further evidence for the passive role taken by the sponge in the formation of its calcareous skeleton.

## Conclusions

Intriguing questions remain as to why demosponges have gained a calcareous skeleton in addition to a primary siliceous spicular one throughout significant periods of their geological history. To appraise the significance of a calcareous skeleton to demosponges it is important to understand the mechanisms of non-spicular calcite formation and its systematic distribution in Recent representatives.

Apparently, Recent demosponges show three main non-spicular precipitation mechanisms and within these a variety of microstructures may be formed. A fourth type can also be inferred for some fossil forms by analogy to Recent calcareans. (1) Direct mineralization of an extensive organic framework, e.g. *Calcifibrospongia* (spherulitic microstructure) and *Vaceletia* (irregular microstructure); (2) via a thin secretory pinacoderm and a possibly mucupolysaccharide filled layer, e.g., *Acanthochaetetes* (irregular microstructure) and *Ceratoporella* (elongate spherulitic microstructure); (3) a "passive" almost cement-like formation of orthogonal fibres, e.g., *Minchinella* (calcarean); and (4) an intra-cellular mechanism of spherule production, transportation to the skeletal wall and then in situ epitaxial crystal growth to form the skeletal wall, e.g. *Astrosclera*. From microstructural characteristics, fossil representatives can be inferred for all four mechanisms. The distribution and form of organic tissue appears, not surprisingly, to play an important role. However, the influence of the organic material is variable, with calcified demosponges showing a range of seemingly "biologically controlled", "biologically-induced" through to "organic matrix-mediated" mechanisms. In contrast, the role of spicules is only of subsidiary importance, but they may in some cases act as nucleation sites.

Considering the sub-tropical or tropical distribution of both Recent and fossil calcified demosponges, biomineralization mechanisms appear to be temperature dependent. The presence of gemmules and thesocyte-like cells in some calcified demosponges, although not responsible for calcification, may explain the considerable storage and regenerative capacity of these forms.

Existing lines of evidence suggest that non-spicular skeleton production is a simple process and that sponges can calcify with relative case, as (1) calcareous skeletons have a wide systematic distribution; (2) different microstructures and seemingly different biomineralization mechanisms are present within closely related forms; (3) closest relatives of Recent calcified forms are non-calcified; (4) some species appear to have a facultative ability; (5) it is difficult to differentiate between areas of organic and inorganic precipitation; and (6) $\partial\ ^{18}O$ data suggests that vital fractionation of sea water is minimal.

Many features of the demosponge calcareous skeleton are convergent, including grade of organization and microstructure. Only details of the internal organization may be systematically determined. There appears to be a good correlation between Sandberg's proposed oscillating mineral trend and the mineralogical type of the calcareous skeleton in demosponges through time, with aragonitic forms predominating in the Permian-Triassic and Recent.

The relative ease at which demosponges are able to calcify and the convergent formation of these skeletons in a taxonomically wide range of forms, suggests the mediation of extrinsic controls to promote calcareous skeleton development. It therefore remains to discern the ecological and environmental pressures which encourage or facilitate the production of calcareous skeletons in sponges. The formation of calcareous skeletons may have been in response to global changes in ocean chemistry, such as rises in $pCO_2$, which enabled tropical or sub-tropical forms to calcitize and exploit reef-building niches.

*Acknowledgments.* Thanks to John Whiteman (Sorby laboratory, University of Sheffield) and Dave Newling (Department of Earth Sciences, University of Cambridge) for technical support. This work was carried out while under tenure of a N.E.R.C. Research Fellowship (GT5/F/87/GS/1), which is gratefully acknowledged. This is Earth Sciences contribution no. ES. 1289.

## References

Gautret P (1985) Organisation de la phase minérale chez *Vaceletia crypta* (Vacelet) demosponge, sphinctozoaire actuelle. Comparison avec des formes aragonitiques du Trias de Turquie. Geobios 18:553–562

Gautret P (1986) Utilisation taxonomique des caracteres microstructuraux du skelett aspiculaire des spongiaires: Etude du mode de formation des microstructures attribuées au type sphérolitique. Ann Paléontol (Vert Invert) 72:75–110

Hartman WD (1979) A new sclerosponge from the Bahamas and its relationship to Mesozoic stromatoporoids. Collog Int CNRS 291, Biol Spong 467–475

Hartman WD, Goreau TE (1975) A Pacific tabulate sponge, living representative of a new order of sclerosponges. Postilla 167:1–21

Jones WC (1979) The microstructure and biogenesis of sponge biominerals. Collog Int CNRS 291, Biol Spong: 425–449

Kazmierczak J, Ittekko TV, Degens ET (1985) Biocalcification through time: environmental challenge and cellular response. Paläontol Z 59:15–33

Kirkpatrick R (1912) *Merlia normani* and its relation to certain Palaeozoic fossils. Nature (Lond) 87:502–503

Lowenstam HA (1981) Minerals formed by organisms. Science 211:1126–31

Mann S (1983) Mineralization in biological systems. Struct Bonding 54:125–74

Reitner J (1987) *Euzkadiella erenoensis* n.gen. n.sp. ein Stromatopore mit spikularem Skelett aus dem Oberapt von Eveño (Prov. Guipuzcoa, Nordspanien) und die systematische Stellung der Stromatoporen. Paläontol Z 61:203–222

Reitner J, Engeser T (1983) Contributions to the systematics and palaeoecology of the family Acanthochaetetidae Fischer 1970 (Order Tabulospongida, Class Sclerospongiae) Geobios 16:773–779

Riding R (1982) Cyanophyte calcification and changes in ocean chemistry. Nature 299:814–815

Sandberg PA (1983) An oscillating trend in Phanerozoic non-skeletal carbonate mineralogy. Nature 305:19–22

Shuhmacher H, Plewka M (1981) Mechanical resistance of reefbuilders through time. Oecologia 49:279–282

Vacelet J (1985) Coralline sponges and the evolution of the Porifera. In: Conway Morris S et al. (eds) The origins and relationships of lower invertebrates. Syst Assoc Spec Vol 28:1–13 Oxford

Vacelet J (in press) The storage cells of relict calcified sponges. In: Ruetzler K, Martmauw D (eds) Proc Third Int Conf on the Biology of Sponges, Woods Hole

Willems P, Hartman WD (1985) Calcification rate of *Ceratoporella nicholsoni* (Porifera: Sclerospongiae): an in situ study with calcein: Proceedings Fifth Internat Coral Reef Congress 5:113–118

Wood R (1987) Biology and revised systematic of some late Mesozoic stromatoporoids. Spec Pap Palaeontol 37:1–89

Wood R, Reitner J, West R (1989) Systematics and phylogenetic implications of the haplosclerid stromatoporoid *Newellia mira* nov. gen. Lethaia 22:85–93

# On the Structure of Calcareous Sponge Spicules

P.W. Ledger and W.C. Jones

## Introduction

Calcareous sponge spicules are composed of calcite (Sollas 1885) admixed with magnesium, sodium, strontium and sulphate. The magnesium content is species-specific and has been found to vary from 1.25–3.15%, or 5.2–12.9 M% $MgCO_3$ (Jones and Jenkins 1970). It also increased with spicule size when different types of spicules were compared in two of these species. The spicules in addition contained traces of barium, manganese, lithium and silicon, the latter probably as contaminating quartz. Water and iron oxide have also been detected, at 3.14% and 0.26% respectively, by Bütschli (1908), in spicules of *Leucandra aspera*. Calcareous sponge spicules are thus composed of impure magnesian calcite. In addition, each is a single crystal with crystalline continuity throughout. There is also a precise relationship between the directions of the spicule rays and the crystallographic axes of the spicule calcite (von Ebner 1887; Jones 1954a, 1954b, 1955a, 1970, 1984).

Towe and Cifelli (1967) have pointed out that optical evidence alone may not distinguish between a single crystal unit and an aggregate of highly oriented microcrystals. However, examination of large triacts of *Leucandra nivea* with the scanning electron microscope has failed to reveal any suggestion of the latter on fractured or etched spicule surfaces at magnifications of up to × 43,000 (Jones and James 1972). Nevertheless, the spicules are not entirely homogeneous. Under the light microscope, longitudinal striations can be seen in the rays of large monact and triact spicules (Jones 1970). Their appearance is enhanced by fracturing and etching, whereupon it becomes clear that they represent concentric laminations with a spacing of 3–5 μm (von Ebner 1887). Under the scanning electron microscope it is apparent that the broad bands are composed of finer lamellae, 0.29–0.53 μm thick (Jones and James 1972). Electron probe microanalysis of *Leucandra* spicules has failed to reveal any compositional (Mg, Ca, Sr, S) variations across the laminations (Jones and James 1969), but possibly the resolution or sensitivity of the probe might have been inadequate.

The extent and nature of organic material associated with the mineral phase pose key questions when one is attempting to understand processes of biomineralization. However, the concept of a coherent organic phase permeating throughout the spicule calcite is alien to its acceptance as a single crystal (Towe 1972). Maas (1904) and Weinschenk (1905) believed that an organic phase existed

J. Reitner and H. Keupp (Eds.)
Fossil and Recent Sponges

in spicule calcite, but their evidence was convincingly refuted by Bütschli (1907) and the concept met with little approval (Jones 1967), until it was revived by Travis (1970). More controversial has been the existence of a discrete central thread of organic material within each spicule ray. Minchin and Reid (1908) reviewed previous work on the axial filament and presented their own results, which seemed to confirm the presence of a true axial structure stainable with picronigrosin. Jones (1967) demonstrated how in the small spicules of *Clathrina coriacea* and *Leucosolenia complicata* this structure was an artefact produced by the contraction of the elastic spicule sheath. Also, in the first electron micrographs of calcareous spicules to be published, he noted that the space left by a decalcified spicule contained no structure whatsoever, certainly not of the order of size and integrity as described by Minchin and Reid. This work was criticized by Fjerdingstad (1970) who considered that Jones's Fig. 5 did, in fact, demonstrate the presence of an axial filament about 0.3 μm in diameter. However, one of the so-called filaments of that figure is on a knife mark and looks as if it could be irrelevant debris, another possible filament is in an obliquely sectioned ray and thus would be expected to be elongated within the ray section, but it is not, and finally, one would have expected that Jones's conclusions would have been based on more than just one photograph, and indeed the rays in Fig. 6 of the same paper reveal no such "filaments".

That there is some peculiarity of the axial region is not doubted; it corrodes relatively easily, due to the presence of variable amounts of inorganic impurities and occluded water (von Ebner 1887; Bütschli 1908; Jones 1967).

Another aspect of the association of organic material with the spicule calcite is that of the sheath. However, when prepared by some methods the sheath that is revealed is in fact an inorganic artefact (Bütschli 1901; Jones 1955b). Jones (1967) was also able to show that the true sheath first appears as a delicate membrane between the spicule and its enveloping sclerocytes and that subsequently, where the spicules are exposed to the mesohyle, a dense layer of fibrils is formed immediately around it.

The work of Travis et al. (1967) and Travis (1970) presented a completely different picture of spicule organization. Travis (1970), working on *Sycon* (= *Scypha*) sp., described a complex organic matrix dividing up the largest spicules into compartments and subcompartments. Small spicules (rays $< 0.1$ μm thick) were considered to be only poorly crystalline and showed an axial filament that could be interpreted as being composed of impure mineral. Calcite deposition in medium-sized spicules was said to occur in the sheath, sometimes leaving a central non-mineralized portion.

Clearly the structure of the spicules is still a matter of controversy, particularly as regards the presence of any organic material. In view of the great relevance of this to the secretory and morphogenetic mechanisms at work in spicule formation, we have reinvestigated the ultrastructure of sponge spicules by a number of methods.

## Materials and Methods

Specimens of *Sycon ciliatum* (Fabricius) [= *Scypha ciliata* (Fabricius) Burton 1963], collected at Church Island, Menai Bridge, Anglesey, or from the floating pier at Menai Bridge, and specimens of *Leucandra* (= *Leuconia*) *nivea* (Grant) from Church Island were used in the investigations reported here. Specimens for electron microscopy were treated immediately or after limited maintenance (2 days) in a sea water aquarium.

### *Preparation for Transmission Electron Microscopy*

Pieces were fixed in 2.5% glutaraldehyde in 80% filtered seawater adjusted to pH 7.8 (1 to 1½ h), rinsed in 80% filtered sea water at pH 7.8 ($3 \times 5$ min and $3 \times 20$ min) and then post-fixed in 1% osmium tetroxide in sea water at pH 7.8 (1 h). After fixation, the material was rinsed in the appropriate buffer and dehydrated in 25%, 50%, 75% ethanol (5–10 min each) and absolute ethanol (three changes, 20 min each). The embedding resin used was either araldite, or Spurr's low viscosity resin (Spurr 1969). Thin sections were cut on diamond knives and double-stained with uranyl acetate and lead citrate. Some material was block-stained with 2% uranyl acetate for 30 min after the last rinse and before dehydration. In this case the sections were stained only with lead citrate. When accidental dissolution of spicule calcite was to be avoided, all aqueous media, from the fixative to the final ethanol dilution, were saturated with calcium and magnesium carbonates.

### *Spicule Isolation*

Spicules were isolated by shaking fragments of the sponges for a few minutes in 5% sodium hypochlorite solution. After a brief centrifugation the supernatant was removed and the spicules rinsed three times in distilled water. They were allowed to settle, the water was decanted off and absolute alcohol was added. The spicules were stored in this until required.

### *Replicas*

Isolated spicules were dried from alcohol and embedded either in araldite made to a hard formula, or metallurgical embedding resin. The former was found more suitable for the large *Leucandra* spicules, the latter for the much smaller *Sycon* spicules. After rough grinding to expose the spicules, the block face was polished with 800 grade abrasive paper followed by 6 μm, 1 μm and 0.25 μm diamond paste. Etching of the calcite was effected with either 10% D-tartaric acid, or 2.5% EDTA in 10% formaldehyde at pH 7.5, or 5% aqueous uranyl acetate. The optimum time for etching was determined by exposing the blocks to the etchants for short (30-s) periods, rinsing in distilled water and alcohol, drying and then taking a first stage (plastic) replica, which was checked under the light microscope to see if sufficient etching had occurred. If necessary, further etching was carried out and the block face checked again. Optimum times varied with spicule types and etchants, but were between 30 s and 3 min. For electron microscopy a two-stage replicating method was used. The first stage involved the use of 0.034 mm thick acetylcellulose film (Bioden RFA, Polaron Equipment Ltd) which was softened (not dissolved) in ethyl acetate for 1½ min and applied to the block in the region of the spicules. Solvent evaporation was carried out in a covered Petri dish to slow down the process and prevent premature peeling of partially hardened film. When dry. the plastic was peeled off and mounted (replica upwards) on pieces of glass slides. The final replica was of vacuum-evaporated carbon with gold/palladium shadow. Shadow was applied first at 45° to a thickness of about 100–200 Å, followed by a film of carbon about 100 Å thick. Suitable areas were selected and cut out, and the carbon was scored into 1-mm squares. Removal (dissolution) of the plastic first stage tended to destroy the evaporated replica due to swelling of the plastic before it dissolved. This was overcome by supporting the carbon with wax. The following schedule proved to be the least harmful.

The replica was placed carbon side down on a chip of solid paraffin wax on a glass slide. The slide was warmed under a lamp until the wax melted, whereupon the heat was removed to allow the wax to set. The slide was then immersed in methyl acetate in a shallow dish at room temperature for 3 to 4 min, after which it was again placed under a lamp. During this time the film softened and partially dissolved, but the wax held the replica securely. After a few minutes the slide was transferred to warm acetone to remove the rest of the film, dissolve the wax and free the carbon replica. The free-floating final replicas were then transferred to clean acetone and picked up on uncoated electron microscope grids.

### *Investigation of the Axial Filament*

Some particular methods were used for the investigation of the so-called axial filament. Isolated spicules were decalcified with 0.2% aqueous uranyl acetate, or 0.1 M HCl, or 0.01 M HCl. For optical studies spicule suspensions were allowed to settle on slides, reagents were added and the decalcification process watched and photographed using interference contrast. For electron microscopy spicules were allowed to dry on celloidin-coated EM grids and the grids were immersed in the reagents until decalcification was complete. They were then carefully rinsed in distilled water and blotted dry.

### *Grid Decalcification of Sections*

Thin sections of material that had not been decalcified were immersed in drops of 1% phosphotungstic acid or 0.5% uranyl acetate for 10–30 mins. This was done after examination of the mineral in the electron microscope, as suggested by Watabe (1963), but also sections were decalcified without prior examination.

Reagent-grade microcrystalline calcium carbonate was embedded in araldite and sections of this were also treated as above.

## Observations

### *Transmission Electron Microscopy of Sectioned Spicules and Replicas*

Most of the spicules in sections of non-decalcified sponge tissue are seen to be made up of numerous polygonal sub-units. The long axes of these units are invariably perpendicular to the cutting direction and they bear no constant relationship to the morphological or crystalline axes of the spicules. Moreover, as similar units are also formed when chemically precipitated calcium carbonate is sectioned, there can be no doubt that such units are sectioning artefacts like those described by Towe and Hamilton (1968) for molluscan shell. Garrone (1969), Simpson and Vaccaro (1974) and Fjerdingstad (1970) obtained similar artefacts in sections of siliceous spicules, although the last author considered them to be basic structural units. Very occasionally an unshattered spicule is encountered (Fig. 1a) and, by comparison with the results of Newesely and Helmcke (1971), who sectioned calcite of geological origin, one can suggest that these spicules were fortuitously oriented in the plane of sectioning in such a way that they were cleaved rather than sectioned.

While observing thin spicule sections, extinction contours caused by electron diffraction are seen (Fig. 1a). With changes of beam intensity they move across the calcite and there is no justification for regarding them as representing variations in the composition or density of the mineral. Apart from the extinction contours and knife marks the calcite appears quite homogeneous. There are no indications of lamellae or differentiation of the axial region. However, after a short exposure to the electron beam, a vague substructure becomes visible that apparently consists of units 10 nm or less in size (Fig. 1e). Thin sections of chemically precipitated calcium carbonate reveal similar structures (Fig. 1f), thus demonstrating that they are not evidence of the presence of organic compartments or of a lack of crystallinity in the spicule. Since this type of structure is not apparent when sections are first viewed, it is probably caused by electron beam damage, possibly involving calcination of the calcite to calcium oxide, as described by Towe and Thompson (1972) for ion beam thinned, non-biological aragonite. A different effect was obtained by these authors after beam damage to similar sections of molluscan nacre. Here the single crystal units developed a frothy appearance which they considered to be due to water and organic material trapped within the crystal, but not present as a discrete organic matrix.

Decalcification of thin sections has been used to demonstrate crystal/matrix relationships of molluscan shell (Watabe 1963) and was used by Travis (1970) on *Sycon* spicules. Ideally, part of a section is photographed before and after decalcification, so that a direct comparison can be made between the organic and inorganic phases. When this was performed on sections of spicules in the present study, an almost perfect replica of the mineral was left (Fig. 1b). This figure does

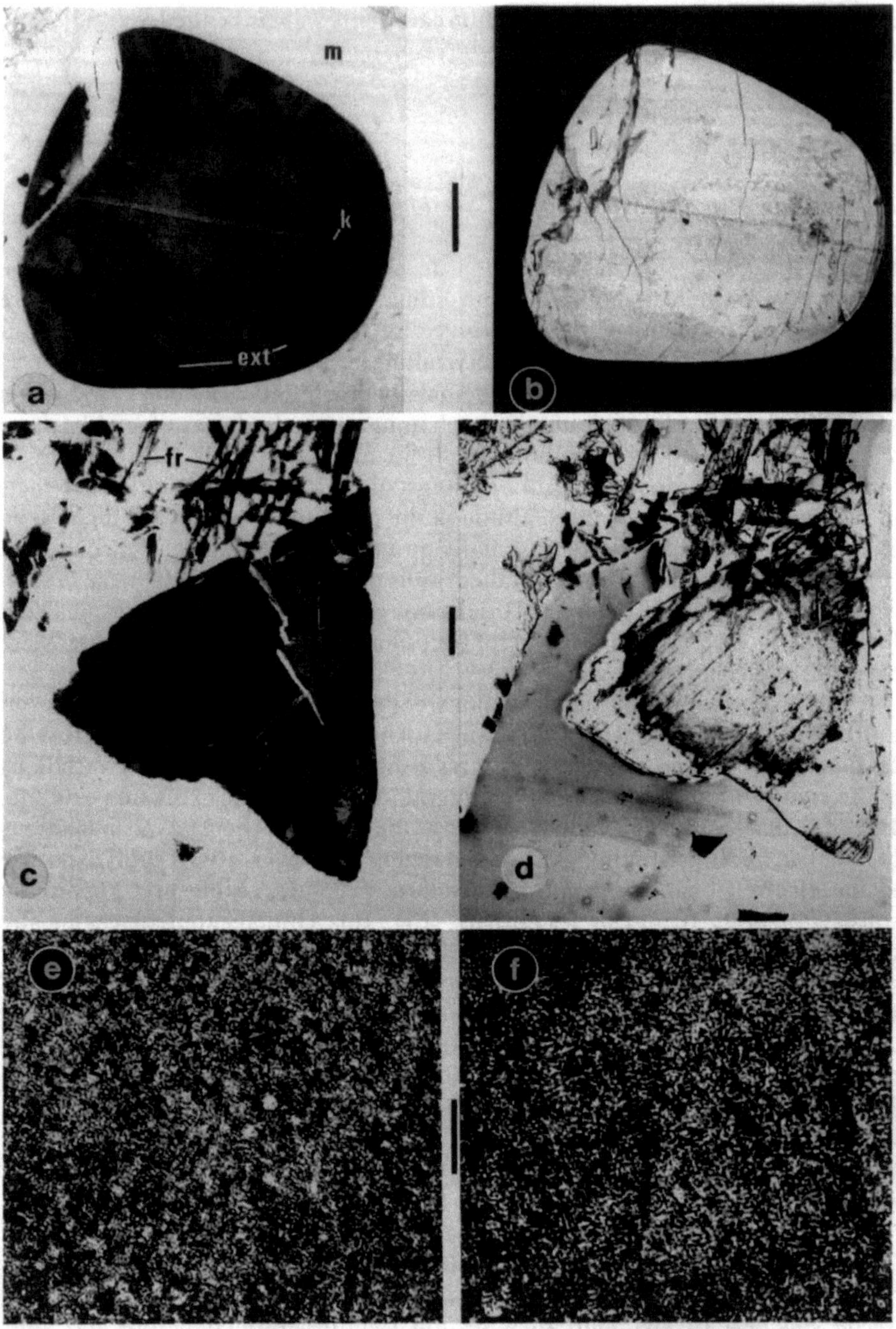
m
k
ext
a
b
fr
c
d
e
f

not show exactly the same section as Fig. 1a. There was one section in between, but the spicule in Fig. 1b was likewise exposed to the electron beam before decalcification. Fragments of shattered spicules left an exact copy of each fragment. When the procedure was repeated on sections of chemically precipitated calcium carbonate, they behaved in exactly the same way as the spicules and left a similar deposit (Fig. 1d). Sections of either spicules or calcium carbonate that had not been exposed to the electron beam left no detectable residues after decalcification. Thus one can conclude that the apparent organic matrix shown in Fig. 1b is really an artefact caused by the exposure of the mineral to the electron beam before decalcification. The artefact could arise by partial transformation of the calcite to an insoluble mineral, or perhaps by a replica of the mineral being left in the plastic support film.

Replicas of *Leucandra* spicules that have been ground, polished and etched with 10% D-tartaric acid exhibited concentric lamellae similar to those described by Jones and James (1972) (Fig. 2a-e). Examination of carbon replicas at high magnification showed that the smallest discernible lamellae are only tenths of a micron wide (Fig. 2c). The broad bands were not easily seen on carbon replicas but can be distinguished on the optical micrograph of the first stage plastic replica (Fig. 2e). The surface exposed in this figure is in a plane which includes the axis of one of the three rays, the other two being symmetrically directed out of this plane. The centre of transverse sections of the rays was not usually etched back as a circular pit in the way Jones and James (1972) found, but was undifferentiated (Fig. 2d) or sometimes sunk into an etch pit (Fig. 2a). This figure also shows other etch marks in parallel (arrowed). The granular microtexture visible at high magnifications is caused by the preparatory technique and was present even on the replicated surfaces of the polished resin around the spicules.

The investigation of lamellae in *Sycon* was more difficult because instead of differentiating the bands, the etchants tended to corrode the whole spicule, so making replication impossible. However, lamellae were revealed by using EDTA (Fig. 2f). The bands appeared to be narrower than those of Fig. 2c, although this could have been due to the apparently finer "grain" and possibly better resolution of the *Sycon* preparations. Etching with 10% D-tartaric acid often resulted in the formation of pits that were poorly regular, but were sufficiently oriented to suggest that they were etch pits.

**Fig. 1.** Spicule structure: thin sections. **a** An almost complete transverse section through a ray of a triact. It has an irregular shape because the section is near the junction of the three rays. The oblique parallel lines are knife marks (*k*) and the irregular bands are extinction contours (*ext*). Mesohyle (*m*). Bar = 1.0 μm. **b** A section adjacent to the above which has been decalcified with 0.5% UA after examination in the electron microscope. The material left has copied the knife marks and the fractured parts of the spicule. **c** A section of a microcrystal of inorganic calcium carbonate. Fragments of an adjacent particle (fr) are also present. Bar = 1.0 μm. **d** The same section decalcified with 0.5% UA. **e** Granular substructure of spicule calcite. Bar = 0.1 μm. **f** Granular substructure of inorganic crystalline calcium carbonate. Bar = 0.1 μm

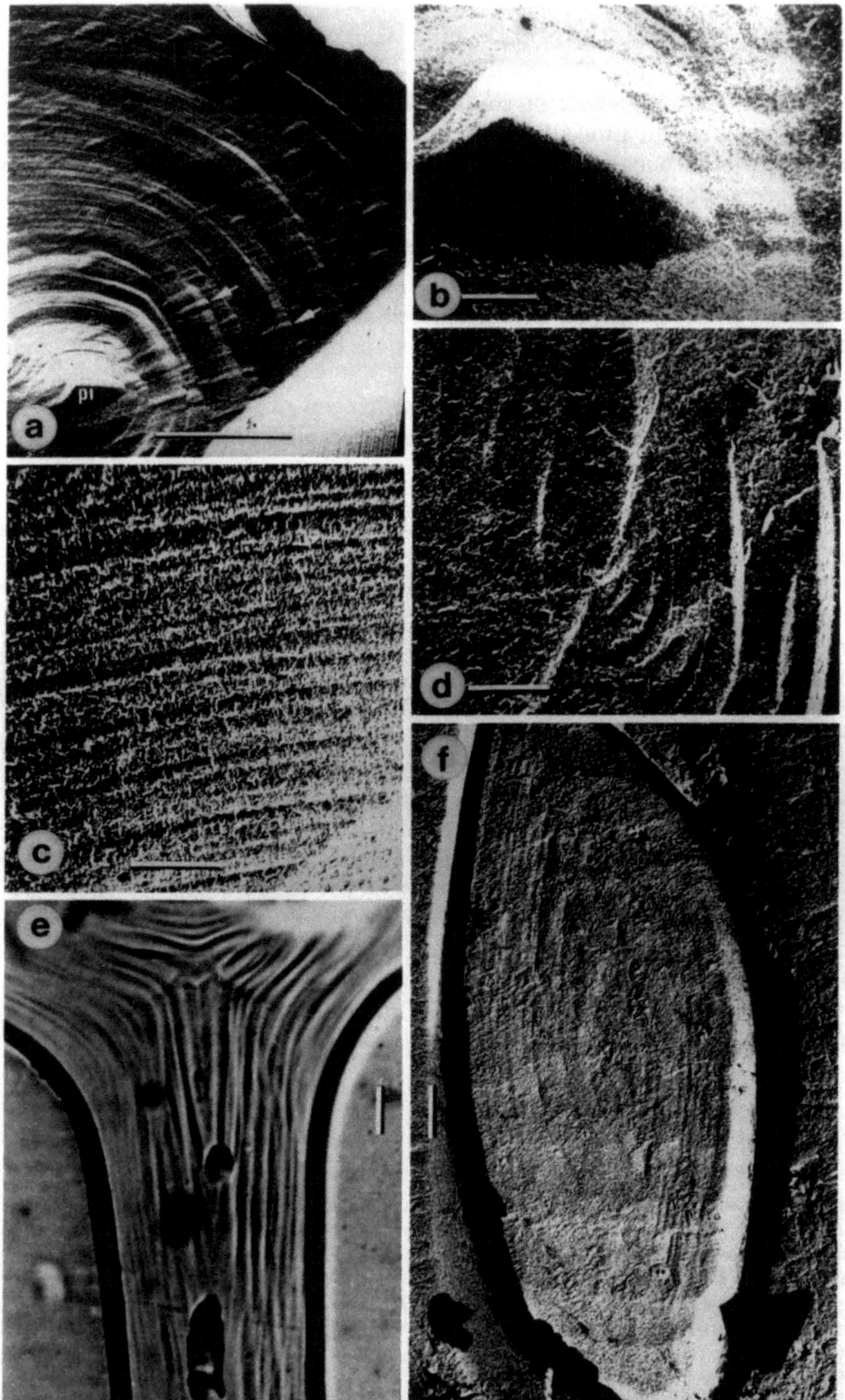
a
pi
b
c
d
e
f

### *The Axial Filament (So-Called)*

Examination of material decalcified fortuitously during fixation reinforced the observations of Jones (1967) that no axial filament is left in a spicule cavity after dissolution of the calcite. However, in material block-stained by 2% uranyl acetate (UA) after fixation, a discrete, electron-dense, tubular structure, about 0.1 to 0.2 μm in diameter was found in practically every spicule space (Fig 3a,b). Longitudinal sections in particular (Fig. 3b) suggested a coherence and rigidity of the tube that initially seemed inconsistent with its interpretation as an artefact. For several reasons, however, its original existence within the spicule was doubted. For example, despite its discrete appearance in some sections (Fig. 3a,b), it could also appear more diffuse, or even widely dilated. Also, in non-UA-treated material, even the many sections passing transversely or sagittally through the growing tips of the spicules gave no hint of the apparent filament (e.g. Fig. 6 of Ledger and Jones 1977), yet here it would have been in process of formation. Finally, replicas and thin sections of spicules have yielded no suggestion of the presence of such a filament. In order to clarify the above observations, isolated spicules were treated with various decalcifying agents under interference contrast and on electron microscope grids.

### *Decalcification by HCl*

The effect of HCl on *Sycon* spicules is similar to that observed by Jones (1955b) on those of *Leucosolenia*. The sheath is revealed by dissolution of the calcite and it contracts a certain amount depending on the concentration of the acid and on the particular ray. In strong HCl it immediately shrivels to a thin filament. Figure 3c shows a spicule undergoing decalcification by dilute HCl. The collapse of the sheath of the unpaired ray occurs almost immediately beyond the dissolving tip, but the sheath of the paired rays does not collapse so much and no "axial filament" is visible within it. When spicules on electron microscope grids are decalcified with HCl, a similar contraction of the sheath occurs (Fig. 3e), which can be confirmed by continuous observation of the process of dissolution. Figure 3c is so

**Fig. 2.** Spicule structure: concentric lamellae as revealed by the replication technique. **a-e** are replicas from large triacts of *Leucandra*. **a** A replica taken from a plane transverse to the long axis of a ray. Note the numerous concentric bands, the parallel etch marks (*arrowed*) and the etch pit (*pi*) at the centre of the ray. This pit appears as a mound thanks to the reversal of the topography caused by the replication technique. Bar = 10.0 μm. **b** Detail of the etch pit. Its faces correspond to crystallographic planes of the calcite. Bar = 1.0 μm. **c** Detail of the concentric bands. Bar = 1.0 μm. **d** The centre of a ray which has not formed an etch pit. There is no indication of the presence of an axial filament. Bar = 1.0 μm. **e** A first stage (plastic) replica. The plane exposed contains the axis of one of the rays. The lines correspond to successive surfaces during formation of the ray and confirm that bands such as those in **a** and **c** are formed where concentric lamellae meet the exposed surface. Bar = 10.0 μm. **f** Lamellae of a *Sycon* spicule exposed by etching with EDTA. The exposed surface is in a plane oblique to the long axis of the ray; hence the oval shape. Bar = 1.0 μm

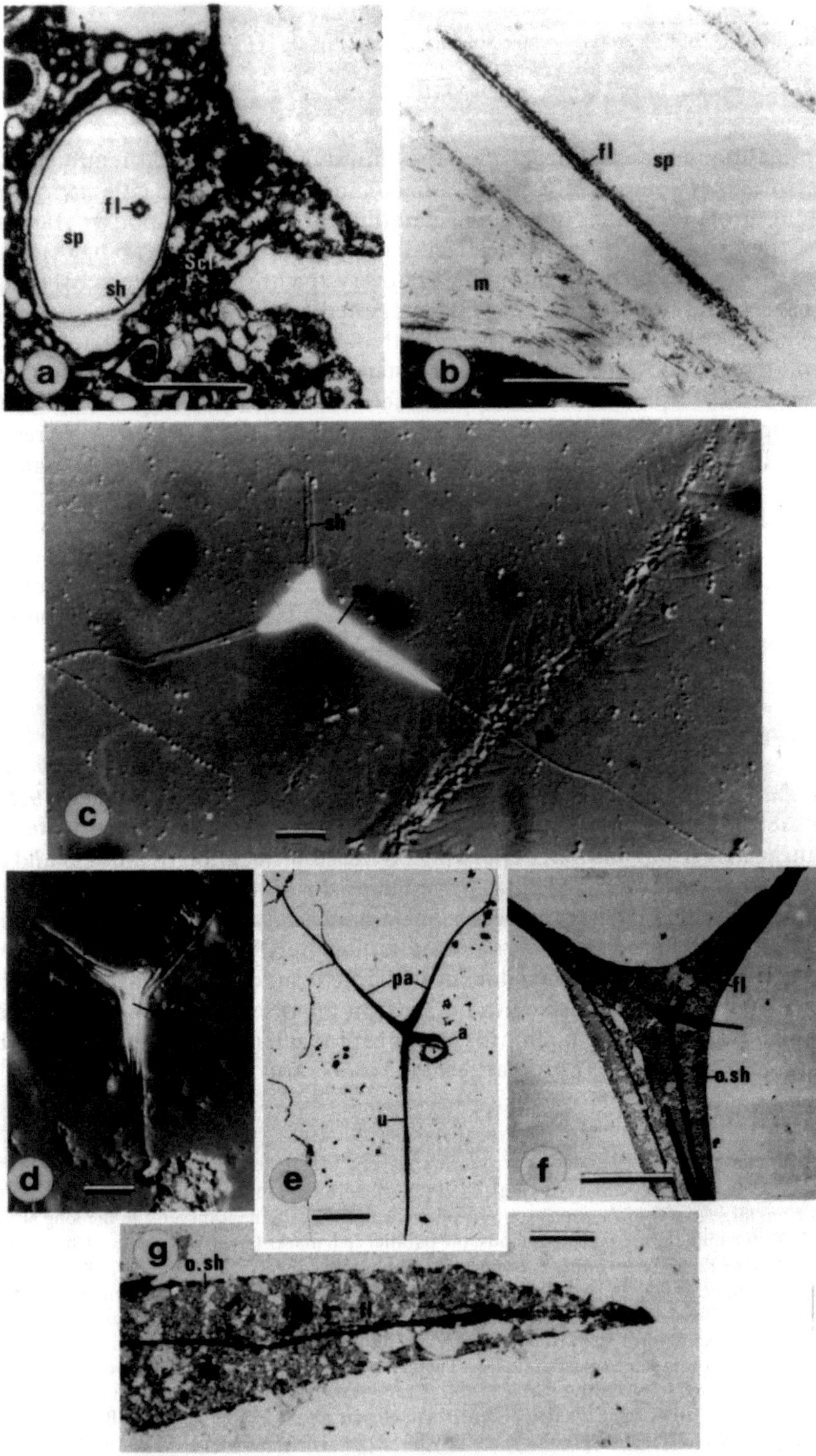
a
fl
sp
sh
Scl
b
fl
sp
m
c
sh
d
e
pa
a
u
f
fl
o.sh
g
o.sh
fl

similar to Figs. 2–4 of Travis et al. (1967) that there can be little doubt that their so-called axial filaments were likewise the contracted spicule sheaths.

### *Decalcification by UA: The Composite Sheath*

When UA is the decalcifying agent there is a different result in that the outline of the spicule is maintained by the formation of a thick sheath (Fig. 3d). This sheath partially dissolves when HCl is added to the slide and the remnant then collapses to form a structure like the contracted sheath revealed by HCl-decalcification. Bearing in mind the similar action of potash on sponge spicules (Jones 1955b), it is probable that the thick UA-produced sheath consists largely of a relatively insoluble reaction product of UA and $CaCO_3$ (perhaps uranyl carbonate). This is dissolved by the HCl, whereupon the organic sheath is released and contracts.

### *Decalcification by UA: Filament and HCl-Insoluble Sheath*

Sometimes within the UA-produced composite sheath one can distinguish a thin, often irregular filament running along the axis of the rays and splaying out to form a triangle where the rays meet at the spicule centre. When UA-decalcified spicules are examined under the electron microscope, similar filaments are seen, surrounded by material that more or less maintains the outline of the original spicule (Fig. 3f,g). This outer layer does not dissolve in HCl and thus is not the composite sheath. Its origin was explained by the observation that additional UA solution dissolves the inorganic component of the composite sheath, probably because it is acidic and, when fresh, unsaturated in respect of the inorganic component. Whereas spicules and composite sheaths under a coverslip require fresh UA for dissolution of the inorganic component, in electron microscope preparations the

◄

**Fig. 3.** Spicule structure: the apparent axial filament and its origin. All except **c** and **d** are electron micrographs. **a** Transverse section of a spicule ray from material block-stained with UA. The sheath (*sh*) outlines the space left by the dissolved spicule (*sp*), within which there is a tubular axial filament (*fl*). Mesohyle (*m*). Bar = 1.0 μm. **b** A spicule space sectioned obliquely and containing an almost longitudinal section through the axial filament (*fl*). Material had been block-stained with UA. Bar = 1.0 μm. **c** A triact (*sp*) partially decalcified with dilute HCl and photographed using interference contrast. The sheath of the unpaired ray (*sh″*) has contracted to a thin filament, but those of the paired rays (*sh′*) remain sheath-like. Bar = 10.0 μm. **d** A triact partially decalcified with 0.5% UA (interference contrast). The original outline of the spicule is maintained by the composite sheath (*c. sh*). Bar = 10.0 μm. **e** A whole tetract completely decalcified by HCl on an electron microscope grid. The sheath is filamentous towards the extremities of the unpaired (u), paired (pa) and apical (a) rays, but widens at their bases. Bar = 10.0 μm. **f** The junction between the rays of a triact decalcified on an E.M. grid with 0.5% UA. The approximate outline of the spicule is maintained by the organic component of the composite sheath (o. sh). What appears to be an axial filament can be seen within the rays. The arrow is referred in the text. Part of a monaxon lies besides the triact. Bar = 10.0 μm. **g** The tip of a monaxon decalcified by 0.5% UA. The filament (*fl*) extends to the tip of the organic component of the composite sheath. Bar = 1.0 μm technique is such that there is a large excess of solution per spicule. Consequently after the initial formation of a composite sheath, dissolution of the inorganic component could occur by diffusion of unsaturated solvent into the vicinity of the sheath.

Under the light microscope it can be seen that the organic material freed by UA-dissolution of the composite sheath does not contract as much as when it is freed by the action of HCl. In addition, when the less contracted sheaths are picked up on electron microscope grids, they flatten and dry against the support film. This increases their apparent width and presumably modifies their contractility in the presence of HCl. However, one has still to explain the origin of the filament seen within the outline of the rays. The previously mentioned objections to regarding this as a true axial filament still apply. In addition, at the junction of the thin filaments (arrow, Fig. 3f) the width of the material is greater than the size of spicule primordia that have been observed in thin sections (see Fig. 3, Ledger and Jones 1977) and by optical microscopy of intact spicules. Furthermore, no similar dense material was visible in a primordium seen in thin sections of UA-decalcified material.

## Serial Transverse Sections Through the Tip of a Spicule Ray

The filament described by Minchin and Reid (1908) was the result of contraction and staining of the spicule sheath (Jones 1967), and the apparent axial filament demonstrated above can be similarly explained. Well-preserved "filaments" are always tubular, as would be expected if they were contracted sheaths. Ultrastructural evidence for contraction of the sheath is presented in Fig. 4. Figure 4a-e shows a series of transverse sections through the tip of an apical ray. Figure 4a,b and c are truly serial, whereas between this group and Fig. 4d, and between Fig. 4d and 4e there were several intermediate sections. Figure 4a is nearest the tip and shows a spicule cavity about 0.1 μm in diameter which is lined by particularly dense material. In Fig. 4b the cavity has widened out, a process that continues into Fig. 4c, in which the diameter is about 0.15 μm. However, in this section the darkly staining material appears to be somewhat disrupted. It is even more so in Fig. 4d, in which the cavity has widened to 0.4 μm. At the level of Fig. 4e an "axial filament" has appeared. Several similar examples of such serial sections have

**Fig. 4.** Spicule structure: the apparent axial filament and its origin. These figures are explained in the text. Bar = 1.0 μm. **a-e** Adjacent sections through the apical ray of a tetract of *Sycon ciliatum*. Sclerocyte of apical ray (*a.scl*); sea water (*sw*); sheath (*sh*); filament (*fl*); converging spaces (*co*). **f** An oblique section through a spicule ray. Space left after spicule decalcification (*sp*); sheath (*sh*); filament (*fl*). **g** A section of the same ray passing almost longitudinally through the ray tip. The organic material appears to consist of a thin, dense layer (*a*) which is pulling away from the surrounding, more diffuse, material (*b*). Converging spaces (*co*); dense cup (*d*). **h** An axial filament (*fl*) in the space (*sp*) left by a partially decalcified spicule. The remaining calcite (*cal*) is sheathed by material (*s*) similar to, and continuous with, that which makes up the filament

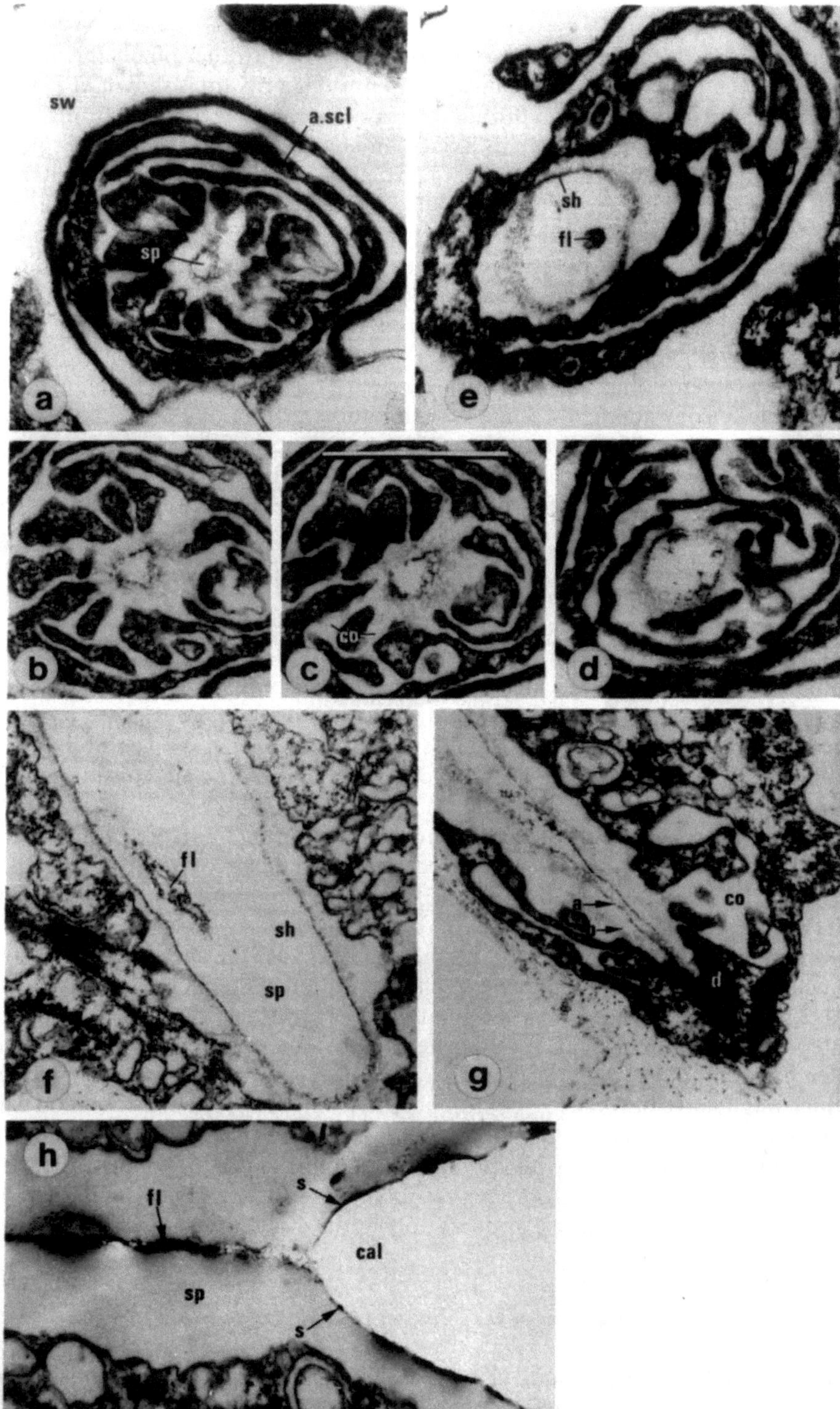

sw
a.scl
sp
a
sh
fl
e
b
co
c
d
fl
sh
sp
f
a
co
g
h
fl
s
cal
sp
s

been seen, this one being in no way an exception. In the other plane, the nearest that has been obtained to a sagittal section through a ray tip is shown in Fig. 4g. Figure 4f is an oblique section through the same ray at some distance from the tip and shows an obliquely sectioned "axial filament". At the tip (Fig. 4g) the spicule cavity contains diffuse material around a layer of more darkly staining material. This arrangement is interpreted as showing the dense layer pulling away from the surrounding diffuse material, particularly at a distance of about 1.0 μm from the tip. Final evidence for a contracting sheath comes from Fig. 4h which shows material that has been partly decalcified. Extending from the remains of the spicule is an apparent axial filament, but material of a similar density and appearance can also be seen around the spicule. Whilst this does not prove that all of the filament comes from the sheathing material, it does show that the sheath is capable of contracting to form a very thin tube. Combining evidence from all these micrographs, it seems that at the tip the tendency for the contractile sheath to separate from the outer material is less, possibly because it is less stretched. Further from the tip, where the radius of the spicule ray is greater, the contact between the two materials is insufficient to hold them together and the inner material contracts more or less uniformly to form a tube. However, in intermediate regions there is a tendency for the contractile material to remain attached to the outer substance, so causing the disrupted appearance seen in Fig. 4d.

An explanation is required of why the production of a spurious axial filament should be limited to UA-decalcified material and also why it involves separation of the sheath into layers that are not apparent in material decalcified by other methods. The significance of UA could stem from the properties it combines. The 2.0% and 0.5% UA solutions used have a pH of about 4.2 and, considering the contraction caused by HCl, would be expected to have a particularly strong contractile effect on the sheath. This, and the fixing and staining properties of UA could together result in the very thin, coherent and densely stained tube seen in UA-decalcified material. Other decalcification methods used (e.g. acid alcohol, EDTA, fortuitous decalcification in buffered fixatives) presumably do not combine these properties of UA and thus do not cause the sheath to contract so much, or do not preserve it if they do. The separation of the sheath into layers can be explained by assuming that originally it was infiltrated only partially by the inorganic component of the composite sheath. The non-infiltrated part could then have contracted and torn itself away from the remaining organic material, which might have become less contractile as a result of its former association with the inorganic component. On the other hand, the layers might be of chemically different organic material, with varying contractility, or they might be of the same substance, with gradation from a highly polymerized layer that lies tightly against the spicule, to a more diffuse and less polymerized outermost layer.

### *Variation in the Composition of the Mineral Component*

The possibility that the mineral in the axes of spicule rays may have a composition different from that in other parts is suggested by differences in the rate of dissolution (Jones 1970; Jones and James 1972) and, in the present study, by differences in the resistance to accidental dissolution. The latter has only been seen once, in partially decalcified material, where part of the spicule had unevenly corroded back to leave a thin mineral rod. Possibly the size and apparent coherence of the organic axial tube could have been caused by its initial moulding around this slightly less soluble structure which, in most cases, is dissolved by the subsequent processing. The occurrence of a different mineral in the centre of the rays would not necessarily presuppose a process of specialized secretion because the ray axis is nothing but an extension of the ray tip and the high surface to volume ratio at the tip could cause crystallographic or electrostatic phenomena resulting in discrimination for or against particular ions, and thus confer a particular pattern of corrodability on the ray axes.

The occurrence of an axial filament has been highly disputed in calcisponge biology. Comparison with siliceous sponge spicules would suggest that one ought to be present. However, the constituent mineral is so different in the two types of spicules, that there is really no valid reason why both should possess an axial filament. The organic axial filament of the demosponge *Haliclona elegans* has a regularity of structure suggestive of crystalline properties (Garrone 1969); bearing in mind the non-crystalline state of the hydrated silica, it could play a role in morphogenesis similar to that of the crystalline calcite (Jones 1984).

## Discussion

The regularity of etch pits and other etch marks, their continuity across several bands and the continuity of the extinction contours across the whole spicule are all consistent with the concept of spicules being single crystals.

The absence of an organic matrix is supported by the decalcification test which produces either negative or undoubtedly artefactual results. In addition, the lack of "frothy" structures arising when the material is examined using an electron beam suggests that diffuse, trapped, organic material is absent from the spicule calcite.

The demonstration of growth banding in *Sycon* spicules brings the structure of small spicules into conformity with that of the large triacts of *Leucandra*. The appearance of these bands after etching may indicate variations in the concentration of emergent dislocations (Jones and James 1972); if lamellae are laid down discontinuously, slight lattice disorientations between adjacent layers could arise. Another possibility is that the lamellae appear because of variations in the concentrations of ions or molecules that have been incorporated in the calcite lattice. Water is certainly present, in amount about 3.5%, and, according to Bütschli (1901, 1908), this causes the appearance of concentric layers of vesicles in the large monaxons of *Leucandra aspera* when these are subject to mild heating up to about 400°C. These vesicles became progressively coarser as the heating was continued, and because there was negligible change in spicule volume or specific gravity Bütschli concluded that normal, unheated spicule calcite contained an alveolar structure consisting of vesicles too fine to be visible by light microscopy. Heating would vapourize the water and the pressure and heat would bring about "plastic softening" or "viscous mobility" of the crystalline material, whereby the tiny alveoli could fuse to form larger, irregular cavities. Mineral calcspar contains only about 0.5% water and a layered alveolar microstructure does not develop with mild heating, but even the unheated mineral can exhibit fine laminations in places at times (Bütschli 1908). It thus seems clear that the spicule lamination is caused by the presence in varying concentration of tiny droplets of water, of size probably no bigger than a few molecules, within the crystal lattice. There may also be variations in the concentrations of impurities in the spicule calcite. The fact that electron-probe microanalysis of *Leucandra* spicules by Jones and James (1969) did not reveal such compositional differences is not a conclusive objection, because the beam width of their instrument was 1.0 $\mu$m and it was not known then that the 3–4 $\mu$m bands are made up of much finer lamellae. Such compositional variations can produce growth bands parallel to the faces of crystals grown from solution (Authier 1973), but this process is not directly comparable to spicule growth because the surfaces of spicule rays are not crystal faces. However, crystals grown from the melt can exhibit similar growth bands which are parallel to the crystal-melt interface, and independent of crystal faces (Morizane et al. 1966). The possibility that growth bands might be able to form parallel to the growth surface on minerals grown from solution cannot therefore be discounted. The cause of the banding, however, is still a matter for speculation. In their discussion of the bands of *Leucandra* spicules, Jones and James (1972) were hindered by lack of knowledge concerning the number of cells involved in the formation of these spicules and the length of time required for completion of growth. However, for *Sycon* spicules it is known that each ray is formed by two cells and that the spicules take no more than 2 days to grow (Jones 1979; Jones and Ledger 1986). Thus tidal or circadian rhythms and/or sequential secretion by different cells can be discounted as factors contributing to spicule lamination, in view of the number of bands visible in Fig. 2f. An alternative possibility is that the secretory process is intrinsically cyclical or discontinuous and thereby somehow causes the compositional variations or slight dislocations between layers. At yet another level, the origin of the laminations may lie in some feature of the actual growth of the calcite. For example, even if secretion of ions were a continuous process, growth of the spicule could occur discontinuously by successive nucleation on existing calcite, followed by layer formation; each layer would deplete the crystallization liquor of ions so that growth would cease and not recur until their concentrations had risen again. Variation in the rate of deposition would in turn result in variation in the degree of incorporation of different ions, water molecules, or lattice dislocations, such that lamellae would become apparent after etching.

This study has provided no evidence to support the opinion of Travis (1970) that calcareous sponge spicules "have organic matrices of varying degrees of complexity which consist of amorphous-appearing material in which collagen fibrils may or may not be embedded . . . , and in which the mineral is deposited as poorly crystalline, or amorphous phase, or as well-ordered and oriented polycrystalline aggregates of calcite". The different arrangements were dependent on spicule size. In some of the largest spicules of *Scypha*, calcite was claimed to occur in compartments enclosed by walls made up partly of large collagen fibrils, described by Travis et al. (1967) and, in the mesohyle, by Ledger (1974). However, the walls were stated to be only 400 to 800 Å thick whereas the fibres were quoted as being 0.2 to 0.3 μm wide. Clearly this is physically impossible. Ledger (1976) never saw these fibres within the spicules. Other observations of Travis also cannot be related to the evidence given above. The discrete crystallites apparently occupying compartments in an organic matrix were probably artefacts resulting from spicule fracture during sectioning or incomplete spicule corrosion. The technique used by Travis (1970) had required 3 months for the infiltration of the spicules by epon or methacrylate prior to sectioning, during which time partial corrosion could well have taken place, particularly as the spicule rays depicted (Travis 1970, Fig. 1) have widths which are far too narrow in relation to length (length/width = 63) for such small spicules. Travis also stated that "... the same or alternate consecutive sections were demineralized on the grid ...", which suggests that at least some aspects of the so-called organic matrix may have arisen as a result of the effects of the electron beam, as described above. Fractured spicules leave replicas of their fragments which could be erroneously interpreted as organic matrix. She also referred to the deposition of calcite "in the matrix of the sheath", a notion that is quite alien to the idea of a sheath, for as soon as impregnation occurred the "sheath" would no longer be ensheathing the spicule. Also this sheath was claimed to be the origin of the thin (150 Å) fibrils, even though their extra-spicular origin had previously been observed by Jones (1967, Fig. 4).

The only methods that seem erroneously to have demonstrated the presence of an axial filament and intraspicular organic matrix have involved the removal of calcite and thereby the production of artefacts, as demonstrated above. Definite proof of the presence of organic matter within the spicule calcite would better come from non-decalcified spicules. Until such is forthcoming, we consider that the evidence strongly supports the view that organic matter is absent from within the spicule calcite, or at least not present in amounts or degree of organization detectable by conventional electron microscopy.

## References

Authier A (1973) X-ray topography as a tool in crystal growth studies. J Cryst Growth 13/14:34–38

Burton M (1963) A revision of the classification of the calcareous sponges. British Museum (Nat Hist), London, pp 693

Bütschli O (1901) Einige Beobachtungen über Kiesel- und Kalknadeln von Spongien. Z Wiss Zool 69:235

Bütschli O (1907) Über die Einwirkung von konzentrierter Kalilauge und konzentrierte Lösung von kohlensaurem Kali auf kohlensauren Kalk, sowie über dabei entstehende Doppelsalze von kohlensaurem Kali und kohlensaurem Kalk. Verh Naturhist-Medicin Vereins Heidelberg 8:277

Bütschli O (1908) Untersuchungen über organische Kalkgebilde, nebst Bemerkung über organische Kieselbilde. Abh K Ges Wiss Göttingen 60:1–177

Fjerdingstad EJ (1970) Ultrastructure of the spicules of *Spongilla lacustris* (L.). Symp Zool Soc Lond 25:125–133

Garrone R (1969) Collagène, spongine et squellette minéral chez l'éponge *Haliclona rosea* (O.S.) (Démosponge, Haplosclérìdae). J Microsc 8:581–598

Jones WC (1954a) The orientation of the optic axis of spicules of *Leucosolenia complicata*. Q J Microsc Sci 95:33–48

Jones WC (1954b) Spicule form in *Leucosolenia complicata*. Q J Microsc Sci 95:191–203

Jones WC (1955a) Crystalline properties of spicules of *Leucosolenia complicata*. Q J Microsc Sci 96:129–149

Jones WC (1955b) The sheath of spicules of *Leucosolenia complicata*. Q J Microsc Sci 96:411–412

Jones WC (1967) Sheath and axial filament of calcareous sponge spicules. Nature (Lond) 214:365–368

Jones WC (1970) The composition, development, form and orientation of calcareous sponge spicules. Symp Zool Soc Lond 25:91–123

Jones WC (1979) Spicule growth and production in juvenile *Sycon ciliatum*. In: Lévi C, Boury-Esnault N (eds) Biologie des Spongiaires. Colloq Int CNRS No 291:67–77

Jones WC (1984) Spicule form in calcareous sponges (Porifera:Calcarea). The principle of uniplanar curvature. J Zool Lond 204:571–584

Jones WC, James DWF (1969) An investigation of some calcareous sponge spicules by means of electron probe microanalysis. Micron 1:34–39

Jones WC, James DWF (1972) Examination of the large triacts of the calcareous sponge *Leuconia nivea* Grant by scanning electron microscopy. Micron 3:196–210

Jones WC, Jenkins DA (1970) Calcareous sponge spicules: a study of magnesian calcites. Calcif Tissue Res 4:314–329

Jones WC, Ledger PW (1986) The effect of diamox and various concentrations of calcium on spicule secretion in the calcareous sponge *Sycon ciliatum*. Comp Biochem Physiol 84A:149–158

Ledger PW (1974) Types of collagen fibres in the calcareous sponges *Sycon* and *Leucandra*. Tissue Cell 6:385–389

Ledger PW (1976) Aspects of the secretion and structure of calcareous sponge spicules. Ph D thesis, University of Wales

Ledger PW, Jones WC (1977) Spicule formation in the calcareous sponge *Sycon ciliatum*. Cell Tissue Res 181:553–567

Maas O (1904) Ueber die Wirkung der Kalkentziehung auf die Entwicklung der Kalkschwämme. Sber Ges Morph Physiol München 50:4–21

Minchin EA, Reid DJ (1908) Observations on the minute structure of the spicules of calcareous sponges. Proc Zool Soc Lond 2:661–677

Morizane K, Witt AF, Gatos HC (1966) Impurity distributions in single crystals. I. Impurity striation in nonrotated InSb crystals. J Electrochem Soc 113:51–54

Newesley H, Helmcke JG (1971) Dünnschnittpräparation kompakter Kristalle und mineralisierter Hartgewebe für Röntgenfluoreszenz-Analysen mit dem Elektronenmikroskop (Elmiskopsonde). Biomineralization 3:39–50

Simpson TL, Vaccaro CA (1974) An ultrastructural study of silica deposition in the freshwater sponge *Spongilla lacustris*. J Ultrastruct Res 47:296–309

Sollas WJ (1885) On the physical characters of calcareous and siliceous sponge-spicules and other structures. Sci Proc R Dubl Soc 4:374–392

Spurr AR (1969) A low-viscosity epoxy resin embedding medium for electron microscopy. J Ultrastruct Res 26:31–43

Towe KM (1972) Invertebrate shell structure and the organic matrix concept. Biomineralization 4:1–14

Towe KM, Cifelli R (1967) Wall ultrastructure in the calcareous foraminifera: crystallographic aspects and a model for calcification. J Paleontol 41:742–762

Towe KM, Hamilton GH (1968) Ultramicrotome-induced deformation artifacts in densely calcified material. J Ultrastruct Res 22:274–281

Towe KM, Thompson GR (1972). The structure of some bivalve shell carbonates prepared by ion-beam thinning. A comparison study. Calcif Tissue Res 10:38–48

Travis DF (1970) The comparative ultrastructure and organization of five calcified tissues. In: Schraer H (ed) Biological Calcification: cellular and molecular aspects. Appleton-Century-Crofts, New York, pp 203

Travis DF, François CJ, Bonar LC, Glimcher MJ (1967) Comparative studies of the organic matrices of invertebrate mineralized tissues. J Ultrastruct Res 18:519-550
von Ebner V (1887) Über den feineren Bau der Skelettheile der Kalkschwämme nebst Bemerkung über Kalkskelete überhaupt. Sber Akad Wiss Wien Abt 1 95:55-149
Watabe N (1963) Decalcification of thin sections for electron microscope studies of crystal-matrix relationships in mollusc shells. J Cell Biol 18:701-703
Weinschenk E (1905) Ueber die Skeletteile der Kalkschwämme. Zentralbl Miner Geol Paläont 19:581-588

# Morphological and Structural Variations in *Clathrina clathrus* (Porifera, Calcispongiae)

E. Gaino, M. Pansini, R. Pronzato, and F. Cicogna

## Introduction

Several studies on calcareous sponges of the genus *Clathrina* deal with habitat preferences of the species (Sarà 1960), gametogenesis (Sarà 1955, 1974; Tuzet 1964), and reproductive cycles (Johnson 1978). However, very few data about their temporal evolution are available.

The life history of two closely related species *Clathrina coriacea* and *C. blanca* was followed by Johnson (1979) in a field work that represents the first attempt to quantify temporal fluctuations in calcareous sponges. The monitoring of several individuals by monthly photographic records allowed Johnson to point out that both the species are annual, increase in winter and spring, and die in summer. Less recent observations on the development and behaviour of some littoral sponges, including two species of the genus *Clathrina*, were made by Burton (1949).

However, the total time span covered by these investigations does not exceed 2 years. A recent study by Pansini and Pronzato (1990) provided data about growth, recruitment, mortality, and seasonal variations of several Demospongiae species over a period of 6 years.

In the present research the same technique of monitoring through underwater photography was used in order (a) to study the dynamics of a group of individuals in a population of *Clathrina clathrus* over a long period and (b) to investigate the mechanism of contraction and dilation which characterizes its constitutive network of tubules. Structural rearrangements related to this behaviour were observed also by scanning electron microscopy (SEM).

## Material and Methods

Plastic frames (30 × 30 cm) fixed to the bottom at 28 m depth on a rocky cliff of the Portofino Promontory (north-western Mediterranean) were periodically monitored by underwater photography. Further details may be found in Pansini and Pronzato (1990). A Hasselblad camera in a waterproof box was used for the photographic surveys, which were first performed monthly and later at longer intervals. A series of daily photographs was taken too. Maps were drawn after the color transparencies by the camera lucida of a Wild stereo-microscope. The

J. Reitner and H. Keupp (Eds.)
Fossil and Recent Sponges

projections of the sponge borders were considered; perimeters and areas covered by each specimen were measured by a graphic digitizer coupled to an IBM S30 PC. The two morphological forms (contracted and expanded) of *Clathrina* were always distinguished, but since they can show relatively rapid changes, an automatic time-lapse photographic device (Cicogna and Pronzato 1985) was used (at 30-min intervals over a 37-h period) to ascertain the characteristics of these short-lasting phenomena. The population was left undisturbed.

Histological examination of specimens periodically collected was carried out on paraffin-embedded material, in order to investigate reproductive periods. Ultrastructural analyses by scanning electron microscopy (SEM), were performed on selected material after fixation underwater in 1.2% glutaraldehyde in Gibbons at pH 7.6. Specimens, stored at 4°C for 1 h, were then transferred to 2.5% glutaraldehyde in Gibbons for a further hour. After rinsing in Gibbons, specimens were dehydrated in a graded ethanol series, critical point dried in $CO_2$, covered with gold-palladium using coating unit E 5000 (Polaron Equipment Ltd. Watford, Herts, England) and observed in an ISI DS 130 scanning electron microscope.

## Results

Since fragmentation of *Clathrina clathrus* specimens occurs all year round without any particular peak of activity, it is rather difficult to verify the actual recruitment, in terms of sttlement of new specimens deriving from larvae. On the other hand, the period of sexual maturity of this species has not been identified yet, notwithstanding the surveys we are performing.

In the frame monitored monthly throughout 8 years, the total number of specimens of *Clathrina clathrus* shows variations in the long run, ranging from 1 to 12 specimens over more than 100 months of observations (Fig. 1). After about 70 months of observations a sharp decrease in the number of specimens was noted, due to a high mortality in the whole population (Fig. 1). The total area covered shows some oscillations from July 1979 to November 1983, followed by a decrease after that date (Fig. 2). Owing to fragmentation and merging phenomena, the growth of single specimens is unverifiable, since they usually maintain their individuality only for short periods. Indeed, a comparison of a series of 5 daily monitored photographs shows the number of specimens ranging from 16 (on the 3rd day) to 9 (on the 5th day) (Fig. 3). In addition, the areas covered by 3 specimens of *Clathrina* (a,b,c) display a three- or fourfold increase or decrease (Fig. 4).

The breaking up of specimens is a common phenomenon that takes place rather rapidly, in not more than 24 hours (arrows in Fig. 3).

Time-lapse photographic records with one frame every 30 min for 37 h (from 17.30 h on 6/26 to 06.30 h on 6/28/1985), revealed that *Clathrina clathrus* of the typical expanded form (Fig. 5a) may change its size and shape by a gradual contraction (Fig. 5b) of its constitutive network of tubules.

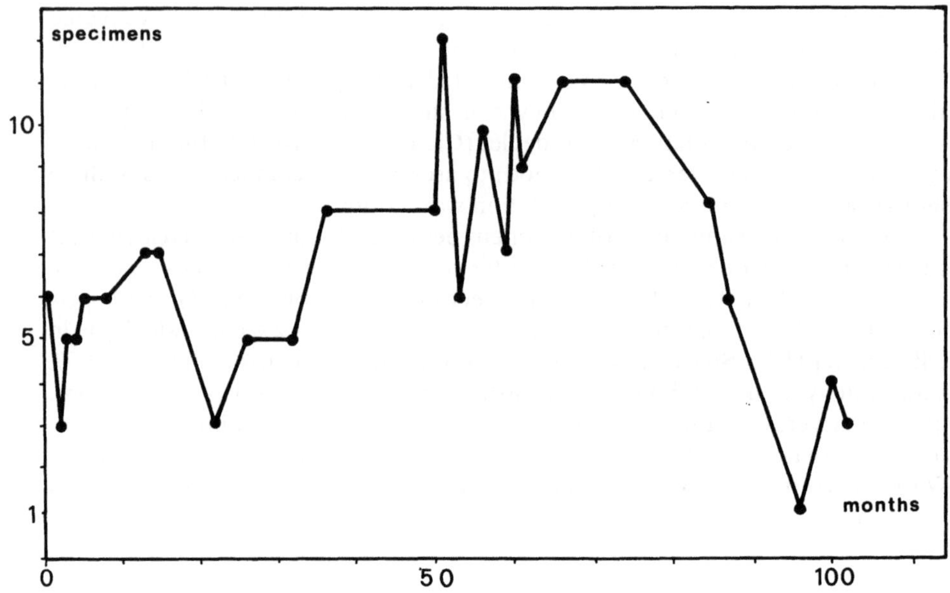


**Fig. 1.** Total number of *Clathrina* specimens monitored over 8 years

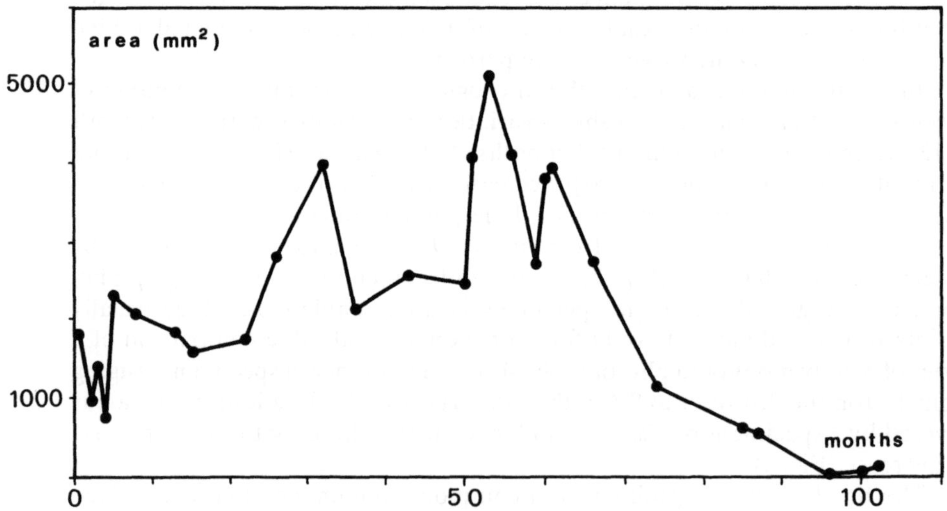


**Fig. 2.** Values of the area covered by *Clathrina* specimens over 8 years

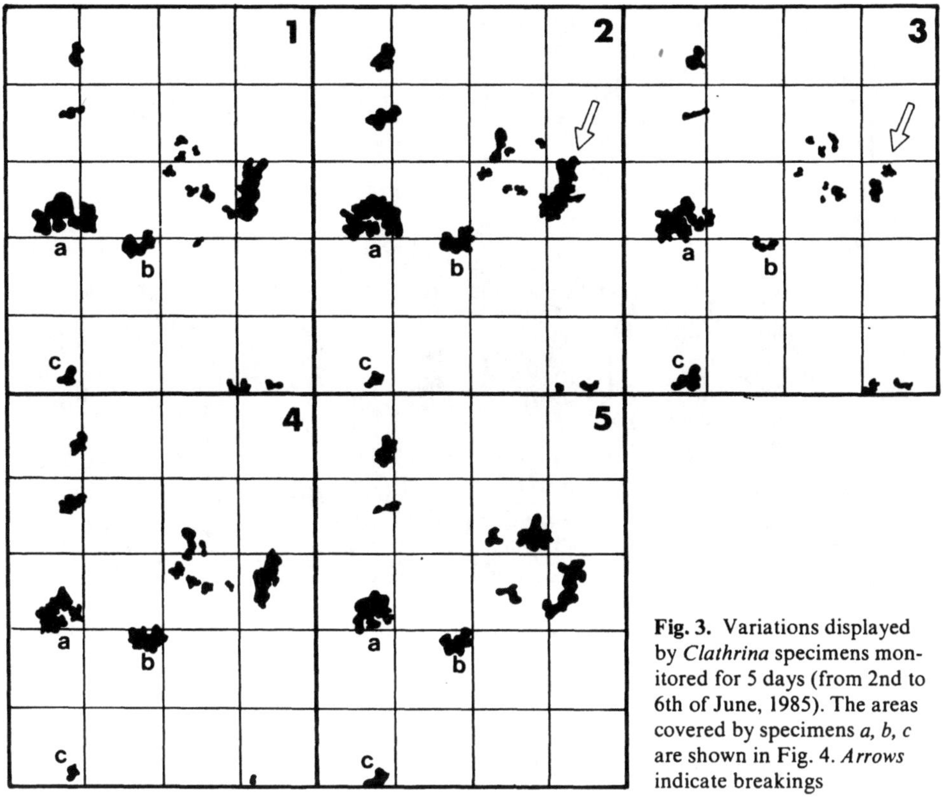


**Fig. 3.** Variations displayed by *Clathrina* specimens monitored for 5 days (from 2nd to 6th of June, 1985). The areas covered by specimens *a, b, c* are shown in Fig. 4. *Arrows* indicate breakings

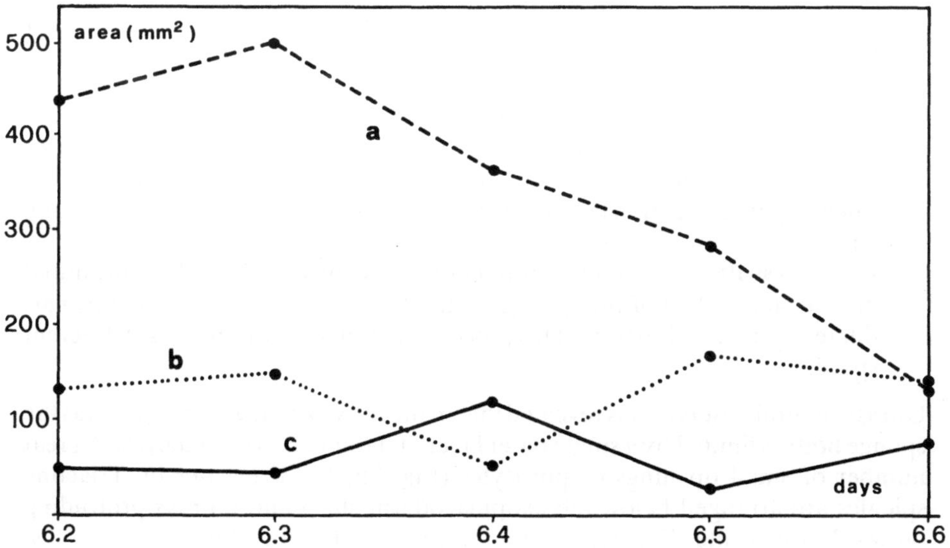


**Fig. 4.** Variations in the areas covered by three *Clathrina* specimens (*a,b,c,* see also Fig. 3) monitored for 5 days

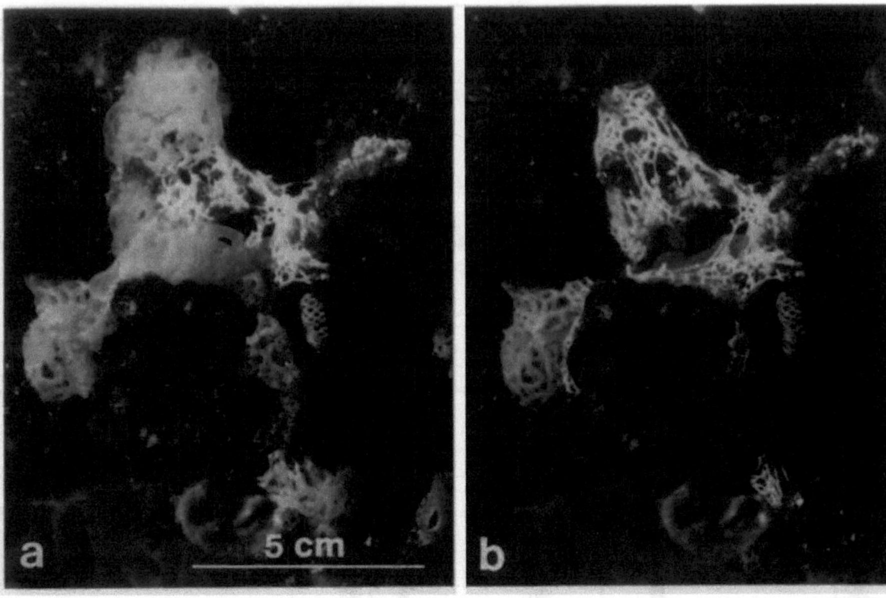


**Fig. 5** The same *Clathrina* specimen: **a** expanded form (6/26/1985, 17.30 h); **b** contracted form (6/27/1985, 12.30 h)

Four areas may be distinguished in the monitored specimen (Fig. 6):

- Area 1: a swift contraction (1 hour) follows the initial expanded state (lasting 13 h and 30 min). This part of the sponge remains contracted until the end of the experiment.
- Area 2: it stays contracted throughout the whole time of the experiment.
- Area 3: the initial expanded condition (2 h) is followed by a slow and incomplete contraction taking 2 h; 6 h later a slow expansion process begins (13 h 30 min). The expanded condition persists for 5 h 30 min and precedes a new slight contraction lasting 4 h, followed by a stronger contraction lasting 3 h.
- Area 4: a swift and complete contraction taking place in 2 h follows the initial dilated state (11 h) of the sponge. After 9 h 30 min of contraction this part dilates again in 10 h 30 min. This process takes place in two phases at different speeds.

Ultrastructural observations revealed that in the fully expanded form (Fig. 7a) the sponge body is limited by a single outer layer of flattened exopinacocytes. A great number of small openings or porocytes (Fig. 7b) lets the water in. Triactine spicules are arranged below the exopinacoderm sheet and no rays protruding outwards are visible (Fig. 7c). A very thin, poorly developed mesohyl with few cells separates the outermost sponge surface from the internal cavity, which is lined by a single layer of choanocytes characterized by the typical collar fringe

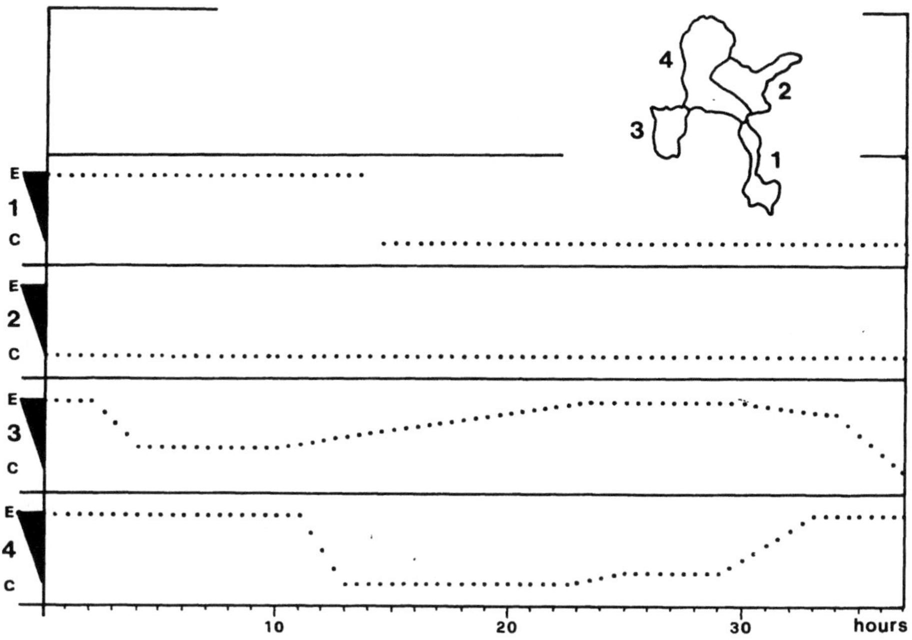


**Fig. 6.** Expansion (*E*) and contraction (*C*) displayed by a *Clathrina* specimen (subdivided into four areas: *1–4*) monitored for 37 h

surrounding the lower portion of the flagellum (Fig. 7d). These cells form little clumps where they are more tightly packed, thus realizing a discontinuous layer (Fig. 7e).

The contraction of the tubes leads to important morphological rearrangements. The external surface of the less expanded tubes is covered with roundish pinacocytes characterized by a significantly reduced spreading (Fig. 8a). Inhalant pores are inconspicuous and spicules bend so that their rays protrude outwards (Fig. 8b). The atrial cavity is lined by other cells, besides the choanocytes, arranged to border the body wall protrusions (Fig. 8c). These cells are scattered among the choanocytes and sometimes push up above them. Their diameter changes according to their degree of spreading. The mesohyl fibrillar matrix presents a compact architecture and supports the layer of spicules and cells which borders the atrial cavity (Fig. 8d).

Strongly contracted tubes are characterized by the folding of choanoderm which, by trigerring the above cells, greatly reduces the sponge cavity (Fig. 8e), while the exopinacoderm remains unfolded. Each fold, which consists of choanocytes and other cell types differing in shape and size, extends towards the atrial cavity (Fig. 8f).

The passage from a less contracted stage to a more contracted one is closely related to the folding of the innermost epithelium.

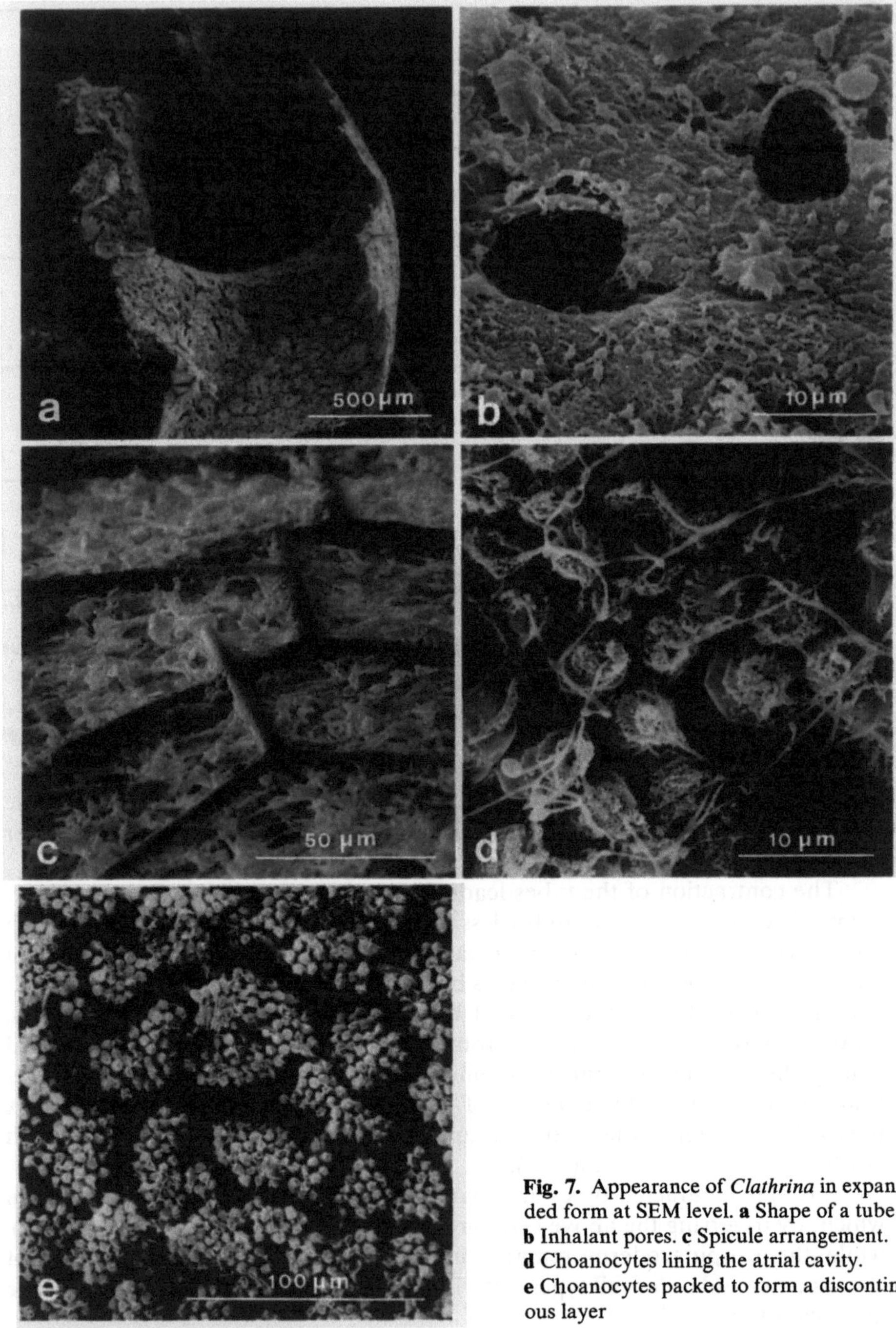


**Fig. 7.** Appearance of *Clathrina* in expanded form at SEM level. **a** Shape of a tube. **b** Inhalant pores. **c** Spicule arrangement. **d** Choanocytes lining the atrial cavity. **e** Choanocytes packed to form a discontinous layer

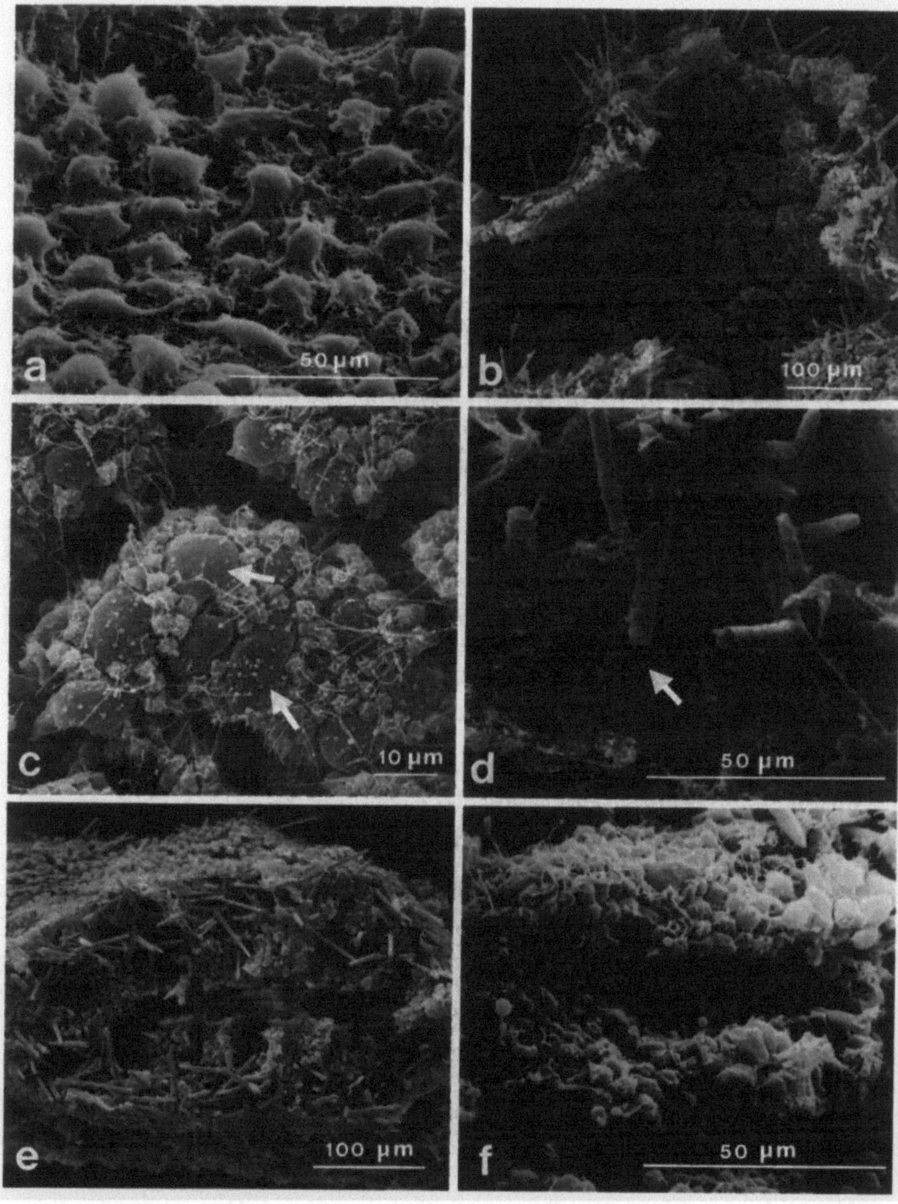


**Fig. 8.** Structural rearrangement involved in the contraction process of *Clathrina* observed at SEM level. **a** Pinacocytes. **b** Cut of a tube showing atrial cavity protrusions and spicular rays protruding from sponge body. **c** Wide cells (*arrows*) scattered among choanocytes lining atrial cavity protrusions. **d** Fibrillar matrix (*arrow*) supports skeleton and atrial cavity protrusions. **e** Cut of a tube showing choanoderm folding. **f** Atrial folds covered by different cell types

## Discussion

The ability of the tissue to contract is a common feature in sponges, resulting in a decrease of the size of the animal (see review in Simpson 1984). Recently Pansini and Pronzato (1990) showed that quite evident, short-lasting contractions take place in some Demospongiae such as *Oscarella lobularis* and *Acanthella acuta*.

Our observations on *Clathrina clathrus* pointed out the ability of this calcareous sponge to regulate pumping rate by means of ostial temporal closure, exopinacoderm contraction, and choanoderm folding. Such rearrangements imply variations in size and in the external morphology, which together bear witness to the high plasticity of the species. In addition, single definite individuals exist for a matter of hours or days in *Clathrina*, owing to the concurrence of growth, reduction, fragmentation, and merging processes.

The existing studies on the dynamics of calcareous sponges refer to relatively short cycles (1 or 2 years only) and not to long-term monitoring. Notwithstanding the length of our investigations, no seasonal variations were recorded for *Clathrina clathrus*, at least in the station monitored at depth of 28 m. *Clathrina coriacea*, on the contrary, presents a specific seasonal variability (Sarà 1958; Labate 1966, 1968; Johnson 1979). In the same locality *C. cerebrum* also shows a clear seasonality with a remarkable drop in late spring and summer months (unpubl. data). *C. clathrus* actually shows a certain decrease at the upper levels – about 2 to 15 m depth – during winter, but such regression was never recorded in the monitored station at 28 m.

Annual variations in the development of calcareous sponges were reported by Burton (1949), Sarà (1964), Johnson (1979) and others and were related to differences in water temperature (Johnson 1979) or to very harsh winters (Juniper and Steele 1979). *Clathrina clathrus* annual variations, on the other hand, recorded at a depth where rather steady conditions are expected, appear to be bound to a pluriennial development cycle which is probably independent of climatic conditions. The clear decrease in the total area covered by *Clathrina*, beginning in 1984 and lasting 3 years at least, may be due to the age of the population.

The longevity of *C. clathrus* is difficult to ascertain since it was impossible to monitor single specimens for more than 3 years. In that period the individual may change its shape, dilate or contract, fragment or coalesce, continuing, however, to occupy the same restricted area. Obviously, as is often the case with marine invertebrates, the evaluation of a single specimen is quite subjective.

No evidence of interspecific competition was found. One reason for this is that *Clathrina* variations (in covered areas and in numbers of specimens) are swifter than those displayed by other organisms (e.g., Bryozoans) that may affect encrusting Porifera populations (Pansini and Pronzato 1990), making it almost impossible to discriminate these interactions.

Intraspecific relations, on the contrary, are continuous and, owing to the very high plasticity of the species, they do not assume the character of competition.

The absence of predation appears to be normal, since Nudibranchs feeding on calcareous sponges are few and extremely rare in the Mediterranean (Cattaneo Vietti and Balduzzi 1988).

The high plasticity of *C. clathrus* also involves some structural rearrangements of the sponge body. The process of contraction and dilation is completely asynchronous, since different areas of the same specimen behave independently. The contraction process appears to be much swifter (1–3 h) than the dilation process (13 h 30 min). The state of complete contraction prevails over the expanded state in terms of time. Indeed, for 25% of the time the sponge body is expanded, for 46.6% it is contracted and for the remaining 28.4% it is shifting from one form to the other.

The shift from the expanded form to the contracted one and back emphasizes the high tissue plasticity occuring in *C. clathrus*. Unlike *Clathrina coriacea* and *C. blanca*, where tissue contraction takes place before the death of the sponge (Johnson 1979), in *C. clathrus* such a phenomenon represents a dynamic process which occurs alternatively in the same individual. Factors inducing the shift are still undetectable, though environmental causes seem to be highly improbable since different areas of the same specimen investigated behave independently.

Tube contraction was reported by Minchin (1900) in *Clathrina coriacea*, involving a peculiar structural modification of choanocytes, and a repositioning of porocytes.

Spontaneous wave contractions were reported for several sponge species (Rasmont and de Vos 1974; Bond and Harris 1988). Time-lapse microcinematography in young *Ephydatia fluviatilis* revealed the main role played by choanocytes and pinacocytes in this process (De Vos and Van de Vyver 1981). This species contracts and dilates rhythmically (Kilian and Wintermann-Kilian 1979) and such behaviour involves choanocyte chamber modifications (De Vos and Van de Vyver 1981); no rhythmical pulsations were observed in *C. clathrus* in our studies.

According to Pavans de Ceccatty (1974), pinacoderm plays an important role in cell conduction and communication. Indeed the same functional role may be envisaged in *Clathrina clathrus* where pinacocyte contraction or dilation could affect choanoderm folding, causing the size reduction of the atrial cavity. Spontaneous propagating contractions have been observed for the pinacoderm of *Ephydatia fluviatilis* (Pottu-Boumendil and Pavans de Ceccatty 1976). These contractions involve expansion and compression of both mesohyl and chambers, thus affecting water-pumping activity. This process, together with the state of contraction of inhalant pores and oscules, suggests that water transport inside the sponge body might also be less efficient. In *Clathrina clathrus*, in addition, the folding of the choanoderm brings the choanocytes located on opposite folds closer to one another, thus probably affecting their flagellar activity. Contraction and flagellar beating could both be involved in controlling water transport. Propagative contractile responses are known in sponges (Pavans de Ceccatty 1960, 1969; Reiswig 1971; Mackie 1979) and in invertebrate epithelia as well (Mackie 1965). In *C. clathrus* pinacoderm contraction or dilation might prompt cell movement and repositioning, as suggested by the varying appearance of the sponge organization observed under SEM. Significant difficulties emerge in identifying the specific cell types occurring among choanocytes when tubes greatly contract.

According to Minchin (1900) in the contracted tubes of *C. coriacea* porocytes push over to the choanocytes arranged in three or four layers. In asconoid calcareous sponges, porocytes are an integral part of the exopinacoderm and are morphologically similar to exopinacocytes (Jones 1966). As a consequence, the epithelial features of the cells lining the atrial cavity when the tubes contract could support the notion that they migrate from the outermost sheet. It is not surprising that mesohyl cells may advance and associate together in order to border choanoderm folds when tubes are strongly contracted. This repositioning may explain the different appearance and the high concentration of the cells delimiting the folds.

In conclusion, our monitoring of *C. clathrus* at 28 m shows an absence of seasonality, thus revealing that this species behaves differently from other congeneric species. In the process of contraction and dilation a control of the sponge body as a whole is absent. Both the contraction of the inhalant pores and the reduction in size of the atrial cavity certainly affect water transport. This new condition may thus represent a metabolically less active state of healthy specimens especially since the sponge remains in the contracted state for most of the time.

## References

Bond C, Harris AK (1988) Locomotion of sponges and its physical mechanism. J Exp Zool 246:271–284

Burton M (1949) Observations on littoral sponges, including supposed swarming of larvae, movement and coalescence in mature individuals, longevity and death. Proc Zool Soc Lond 118:893–915

Cattaneo Vietti R, Balduzzi A (1988) Relationship between radular morphology and food in the Doridina (Mollusca: Nudibranchia). In: Edmunds M (ed) Evolutionary Biology of Opistobranchs 9th Int Malacological Congr, Edinburgh 1986: 15–23

Cicogna F, Pronzato R (1985) A time-lapse photography equipment for investigation of macrozoobenthos activity. Rapp Comm Int Mer Medit 29 (6):183–184

De Vos L, Van de Vyver G (1981) 'Etude de la contraction spontanée chez l' éponge de eau douce *Ephydatia fluviatilis* cultivée in vitro. Ann Soc R Zool Belg 3:21–31

Johnson MF (1978) Studies on the reproductive cycles of the calcareous sponges *Clathrina coriacea* and *C. blanca*. Mar Biol 50:73–79

Johnson MF (1979) Recruitment, growth, mortality and seasonal variations in the calcareous sponges *Clathrina coriacea* (Montagu) and *C. blanca* (Miklucho-Maclay) from Santa Catalina Island, California. In: Lévi C, Boury-Esnault N (eds) Biologie des Spongiaires. Colloq Int CNRS Paris pp 325–334

Jones WC (1966) The structure of the porocytes in the calcareous sponge *Leucosolenia complicata* (Montague). J R Microsc Soc 85:53–62

Juniper AJ, Steele RD (1969) The intertidal sponges of the Portsmouth area. J Nat Hist 3:153–163

Kilian EF, Wintermann-Kilian G (1979) Mouvement cellullaire et contraction chez *Spongilla lacustris* et *Ephydatia fluviatilis*. In: Lévi C, Boury-Esnault N (eds) Biologie des Spongiaires. Colloq Int CNRS Paris 291:137–143

Labate M (1966) Variazioni nel tempo del popolamento di Poriferi di una grotta superficiale. Boll Zool 33:173–174

Labate M (1968) Variazioni temporali in un popolamento di Poriferi. Archo Oceanogr Limnol 16:63–80

Mackie GO (1965) Conduction in the nerve-free epithelia of siphonophores. Am Zool 5:439–453

Mackie GO (1979) Is there a conduction system in sponges? In: Lévi C, Boury-Esnault N (eds) Biologie des Spongiaires. Colloq Int CNRS Paris 291:145–151

Minchin EA (1900) In: Lankester ER (ed) The Porifera and Coelenterata. A treatise on zoology, Part II. Adam and Charles Black, London, pp 1–178

Pansini M, Pronzato R (1990) Observations on the dynamics of a Mediterranean sponge community. In: Rützler K, Macintyre VV, Smith K (eds) New perspectives in sponge biology. Proc III Int Conf on the Biology of Sponges. Smithsonian Inst Press, Washington DC (in press)

Pavans de Ceccatty M (1960) Les structures cellulaires de type nerveux et de type musculaire de l' éponge siliceuse *Tethya lyncurium* Lmk. C R Acad Sci Paris 251:1818–1819

Pavans de Ceccatty M (1969) Les systemes activité motrice, spontanées et provoquées, des éponges. C R Acad Sci Paris 269:596–599

Pavans de Ceccatty M (1974) Coordination in sponge. The foundations of integration. Am Zool 14:895–903

Pottu-Boumendil J, Pavans de Ceccatty M (1976) Mouvements cellulaires et morphogenèse de l'éponge *Ephydatia mulleri* Lieb. Bull Soc Zool Fr 101:31–38

Rasmont R, De Vos L (1974) 'Etude cinématographique de la gemmulation d'une éponge d'eau douce: *Ephydatia fluviatilis*. Arch Biol (Bruxelles) 85:329–341

Reiswig H (1971) In situ pumping activities of tropical demospongiae. Mar Biol 9:38–50

Sarà M (1955) La nutrizione dell' ovocita in Calcispongie Omoceli. Ann Ist Mus Zool Univ Napoli 7:1–30

Sarà M (1958) Studio sui Poriferi di una grotta di marea del Golfo di Napoli. Arch Zool Ital 43:203–280

Sarà M (1960) Poriferi del litorale dell' isola d' Ischia e loro ripartizione per ambienti. Pubbl Staz Zool Napoli 31:421–472

Sarà M (1964) Distribuzione ed ecologia dei Poriferi in acque superficiali della Riviera Ligure di Levante. Arch Zool Ital 49:181–248

Sarà M (1974) Sexuality in the Porifera. Boll Zool 41:327–348

Simpson TL (1984) The cell biology of Sponges. Springer, Berlin Heidelberg New York Tokyo, pp 662

Tuzet O (1964) L'origine de la lignée germinale chez les vertébrés et chez quelques groupes d'invertébrés. In: Wolff E (ed) Origine de la lignée germinale Herman, Paris, pp 79–111

# (Paleo-)Ecology and Diagenesis of Sponges

# Introduction

(Paleo-) Ecology of sponges is a vast area and therefore only few aspects are investigated. During the conference in Berlin more than half of the talks dealt with these questions. Aut- and synecological parameters in Recent sponges in many cases have been less frequently investigated, and the ecological parameters are known from few taxa only. Some of these questions were discussed during the conference in Berlin. Remarkable contributions in this volume were made to population dynamics (Pronzato and Manconi), to the role in their aquatic ecosystems, to symbiosis (Saller), commensalism and parasitism (Corriero, Pronzato and Sarà), and to biogeographic and bathymetric distribution patterns (de Weerdt, Tabachnik). Further papers pertain to changes in the morphology of skeletal hard parts and vitality based on seasonal cycles (Barthel, Jones).

Besides further additional ecological information presented in several chapters of this volume (presented in Parts I and II; e.g. Reitner and Keupp,), within this main part, only studies are considered which predominantly deal with ecological aspects. Thus, six chapters in this section deal with the above mentioned questions and allow an overview of current ecological research.

Paleoecological studies of fossil sponges concentrate on two topics:

1. Reconstruction of paleoenvironmental conditions of different sponge groups (e.g. "Lithistids", Hexactinellids, "Sclerosponges") with paleogeographical, sedimentological, and geochemical methods. These methods help to reconstruct former hydrodynamic conditions, sediment transport, and the role of organisms fixing sediments. Growth forms, size, and morphology of the hard parts of sponges are probably controlled by these factors.
2. Paleontologists are able to study changes in ecosystems and taxa over time. A main question is the paleobathymetric distribution of hexactinellids and lithistids during Mesozoic. During Late Triassic and Jurassic many taxa of these groups are widely distributed in shallow marine environments (tidally influenced to shallow subtidal environments, reefs). Within Cretaceous a rapid change occurred in their habitat. Since this time hexactinellids are mainly found only in deeper and cooler water, the lithistids are restricted to deeper shelf areas (100–400 m), and the sclerosponges are restricted to tropical and Mediterranean cryptic and dim-light environments. No recognizable reasons are known why these groups have moved from shallow to deeper water. A causal connection between long-term environmental changes might be possible, e.g. $CO_2$ cycle and the Ca:Mg ratio in seawater.

The main goal of the investigation of sponge synecology is to study their role in modern and fossil buildups (reefs, mud mounds). Since the Early Paleozoic, sponges with rigid primary and secondary skeletons have been prominent framebuilders of mounds and reefs. This evolution starts with archaeocyathids, sponge-like calcareous organisms in the Lowermost Cambrian but affinities to discrete sponge taxa are still doubtful. Stromatoporoids (mainly demosponges with calcareous skeletons) are dominant framebuilders in Silurian and Devonian carbonate platforms. Chaetetid mounds (most of the organisms with chaetetid skeletons are also demosponges) are common during the Late Paleozoic and Mesozoic (West and Kershaw, Reitner, this Vol.). Rigid calcareous skeletons of thalamid sponges play an important role in Permian and Triassic buildups (Senowbari-Daryan, this Vol.). Mud mounds dominated by rigid siliceous sponges of both hexactinellids and "lithistid" demosponges are located during the Jurassic on shallow shelfs related to the Tethyan Ocean. With the increasing importance of scleractinian corals in reef environments, rigid sponges retired into deeper habitats during the Cenozoic. "Soft" sponges without a rigid skeleton play an important role in modern reefs and shallow-water sediments as reef dwellers and sediment binders (Wiedenmayer 1977).

The evolution of reef ecosystems through time has been marked by global environmental changes (Late Devonian, Permian/Triassic, Triassic/Jurassic, Aptian/Albian, Cretaceous/Tertiary). All of these interruptions evolve new sponge groups.

Of particular interest at the conference in Berlin were siliceous sponge buildups from the Jurassic. These investigations have a long tradition and facies models of these buildups do exist (Roll 1933; Wagenplast 1972, a.o.). Today, however, environmental conditions and bathymetric positions of these sponge mounds are controversial (Gaillard 1983; Schorr and Koch 1985; Wirsing and Koch 1986; Kott 1989; Gygi and Persoz 1986). The taxonomy of siliceous sponges in these mounds is in most cases unclear and also under discussion. Scientific teams in Warszawa, Lyon, Erlangen, Berlin, Bochum, München, Stuttgart, etc. are working on these Jurassic mounds in Germany, France, Spain, Poland, and Morocco. These current activities are reflected in the dominance of papers dealing with these topics (A. Pisera, Deusch et al., J. Trammer, B. Lang, T. Brachert).

*Diagenetic Problems*

Work with fossil sponges includes the problems of its primary preservation potential. It also depends on sedimentological factors and diagenetic processes. Taxonomy of fossil sponges and paleoecological interpretations of sponge communities in buildups make the understanding of these factors necessary. To reconstruct diagenetic histories of skeletons, sedimentological, geochemical, and mineralogical methods must be employed. The close relationship of diagenetic aspects with environmental interpretations is seen in the papers of Brachert and Zimmerle, who demonstrate the importance of these factors.

The "Deutsche Forschunsgemeinschaft" (FRG) is currently financing research projects on the "Evolution of reefs through time", closely linked with Ginsburg's "Cretaceous global bioevents". Sponges are, within these projects and based on their importance as framebuilders through time, important research topics. This program advances the cooperation between paleontologists and neontologists.

## References

Gaillard C (1983) Les Biohermes a Spongiaires et leur Environment dans l'Oxfordien du Jura Méridional. Docum Lab Géol Lyon 90:515 pp

Gygi RA, Persoz F (1986) Mineralostratigraphy, litho- and biostratigraphy combined in correlation of the Oxfordian (Late Jurassic) formations of the Swiss Jura range. Eclogae Geol Helv 79:385–454

Kott R (1989) Facies und Geochemie des Treuchtlinger Marmors (Unter- und Mittel-Kimmeridge, Südliche Frankenalb). Berliner Geowiss Abh A 111:115 pp

Roll A (1933) Form, Bau und Entstehung der Schwammstotzen im süddeutschen Malm. Paläont Z 16:197–246

Schorr M, Koch R (1985) Fazieszonierung eines oberjurassischen Algen-Schwamm-Bioherms (Herrlingen, Schwäbische Alb). Facies 13:227–270

Wagenplast P (1972) Ökologische Untersuchung der Fauna aus Bank- und Schwammfazies des Weißen Jura der Schwäbischen Alb. Arb Inst Geol Paläont Univ Stuttgart NF 67:1–99

Wiedenmayer F (1977) Shallow-water sponges of the western Bahamas.Experientia Suppl 28:287 pp

Wirsing G, Koch R (1986) Algen-Schwamm-Bioherme des Flachwasser-Bereichs (Schwäbische Alb, Weißjura Delta 3). Facies 14:285–308

# Adaptation of the Hexactinellid Sponges to Deep-Sea Life

K.R. Tabachnick

## Introduction

Hexactinellida differ strongly from other sponges. In recent investigations, their taxonomic status has been raised from a class to the subphylum level (Symplasma; Mackie et al. 1983), or even to that of a distinct phylum (with the same name; Bergquist 1985). They possess remarkable external morphological features and have a syncytial tissue organization. Recent representatives mainly inhabit deep-sea waters.

Burton (1928a) distinguished some features that differentiate deep-sea sponges from shallow-water ones: symmetry, regularity, and stability of the body, absence of pigmentation, and development of special anchoring structures. Recent data on hexactinellid sponges supplement the observations of Burton.

## Body Shape and Symmetry of Hexactinellid Sponges

Most Hexactinellida have radial symmetry. They are glass-, sack-, barrel-, tube- or fungus-like (Fig. 1a; Koltun 1967). *Hyalonema* has four-rayed symmetry corresponding to its four atrial cavities. According to the functional anatomy, the upper and lower parts of a sedentary organism exist under different external conditions, on the contrary, the conditions are similar in all radial directions perpendicular to the main body axis of the organism (Beklemishev 1964). In addition to active feeding, an advantage of the tube-like body is the ability of passive feeding, which takes place even when water flow is unpredictable (Vogel 1974). Earlier, Bidder (1923) presumed that passive feeding would be the only manner of feeding for Hexactinellida, but now intense active filtration has been proven in experiments (Mackie et al. 1983). Moreover, passive feeding and its improvement are especially important for adaptation to deep-sea conditions with limited food resources. An interesting feature of the hexactinellid sponges is the presence of bilateral (sagittal) symmetry, which is almost always absent in Demospongiae. *Hyalonema* possesses a thin stalk, which can curve under the influence of water flow (Fig. 1h; Heezen and Schneider 1966). This ability is useful under conditions of unpredictable water flow. The curving of tube-like sponge growth is expected to occur under the influence of constantly directed water flow. Some curved specimens were found in the radial species of *Euplectella* and

J. Reitner and H. Keupp (Eds.)
Fossil and Recent Sponges

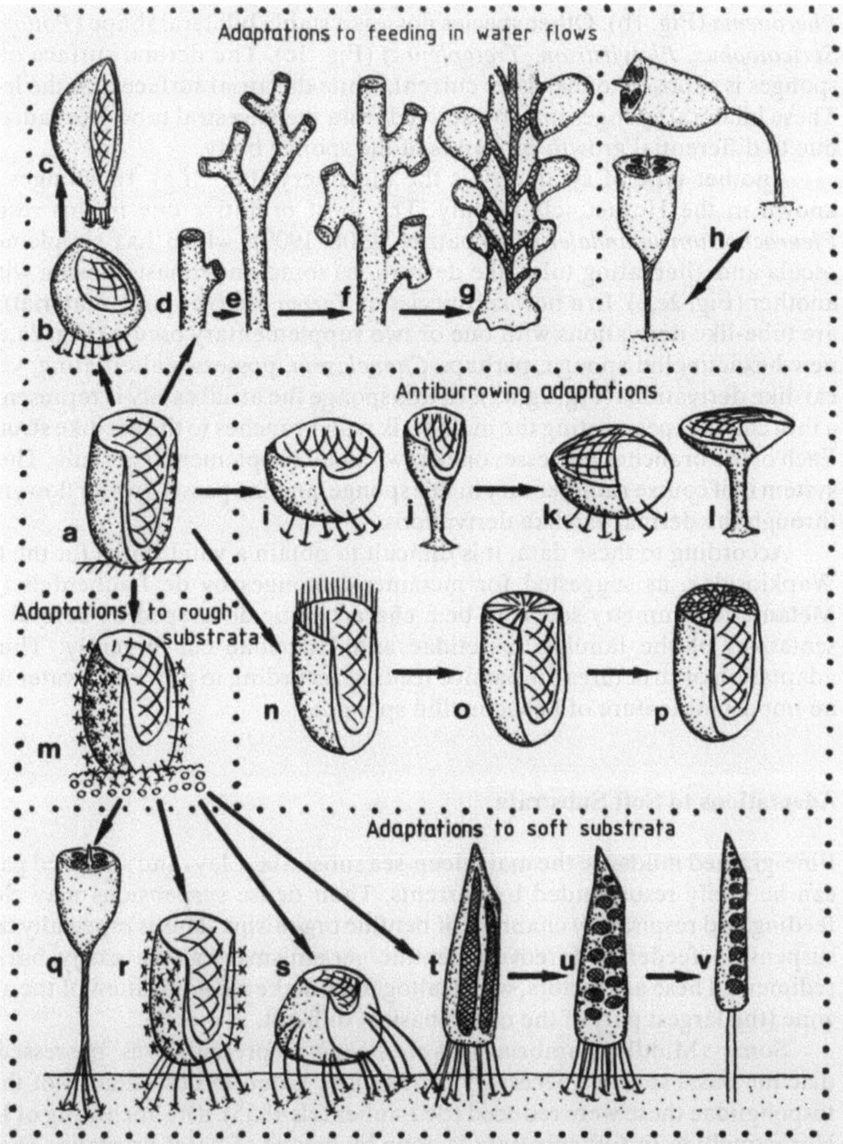


**Fig. 1.** The most important adaptations of the hexactinellid sponges. For explanation see text. The curving of *Hyalonema* according to water flow. (Heezen and Schneider 1966). *Dotted* dermal surface; *hatched* atrial surface

*Pheronema* (Fig. 1b). Other species possess a stable bilateral shape (*Polioplogon, Sericolophus, Platylistrum, Tretopleura*) (Fig. 1c). The dermal surface of these sponges is exposed to the water current, while the atrial surface is at the lee side. These bilateral sponges may be derived from the ancestral tube-like radial form due to differential growth processes of the sponge body.

Another type of symmetry is the metamery (Fig. 1f,g). In sponges this is known in the Hexactinellida only. The most primitive one in this respect is *Pleurochorium anandalei* (Kirkpatrick 1908–1909), which has supplementary oscula and alternating tube-like derivations sometimes anastomosing with one another (Fig. 2c,b). In a new subspecies of *Farrea occa* (unpubl. material), there are tube-like derivations with one or two supplementary oscula (Fig. 2a,d,e). A new hexactinellid sponge, perhaps *Chonelasma*, possesses alternating, vertical, ear-like derivations (Fig. 2g,h,i). In this sponge the atrial cavity is represented by a thin channel penetrating the main stalk with branches to the ear-like structures. Each of the branches possesses one or two small supplementary oscula. The atrial system is of course rudimentary in this sponge, and the passing water flows mainly through the dermal ear-like derivations.

According to these data, it is difficult to obtain a valid status for the family Wapkiosidae, as suggested for metameric sponges by de Laubenfels (1955). Metameric symmetry seems to be a characteristic developed by several representatives of the families Euretidae and Farreidae convergently. Thus, the adaptation of structures for passive feeding according to prevalent water flows is an important feature of hexactinellid sponges.

## Adaptations to Soft Substrata

Fine-grained muds are the main deep-sea substrata. Clay- and silt-sized particles can be easily resuspended by currents. Their dense suspensions may clog the feeding and respiratory channels of benthic organisms. This is especially true for suspension feeders. Moreover, benthic organisms can be easily buried by sediment. These are factors, which altogether make a colonization of the abyssal zone (the largest part of the ocean basins) difficult.

Some (Middle Cambrian) *Protospongia* representatives possessed long diactine basal spicules necessary for fixation. Later, in descendants of the Protospongiidae these were reduced (de Laubenfels 1955). The anchoring of Recent Hexactinellida in soft substrata is done by means of basal prostalia. These are developed from hexactine or pentactine spicules by elongation of the proximal ray and reduction of the other ones. Finally, the spicules protude from the dermal

**Fig. 2.** Metameric sponges: **a** *Farrea occa* (n.ssp.) (unpubl.). **b,c** *Pleurochorium annandalei* (n.ssp.). **d,e** *Farrea* sp. **f** Tube-like *Chonelasma*. **g,h,i** *Chonelasma*, perhaps with ear-like derivations. **h** Supplementary oscula can be seen near the main axis, **d,e,h,i** Underwater photos from the mid-Pacific, depth about 1500 m

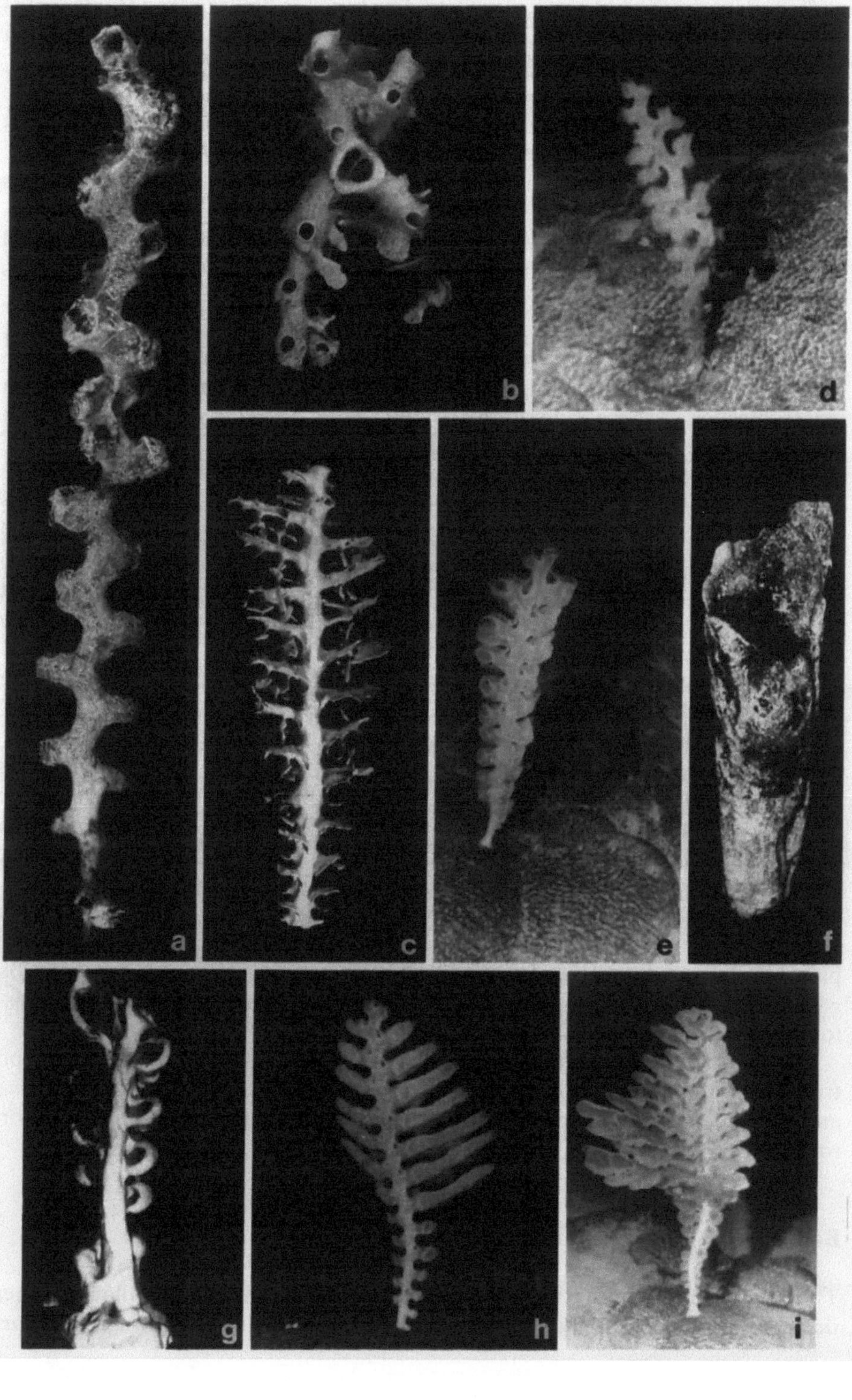
a
b
c
d
e
f
g
h
i

surface. Such anchor-like spicules were developed in the families Hyalonematidae, Pheronematidae, and some Euplectellidae (Fig. 1q-v). Similar adaptation occurred with hypodermal spicules in some Rosselliidae (*Rossella, Aulorossella, Lophocalyx*). In these sponges long tufts of prostale spicules developed in the basal part of the body. These tufts of prostalia may interconnect distally (without fusion) into one single tuft (e.g., some *Lophocalyx* species).

It was very interesting to find the sponges mentioned above on dense porous or wrinkled substrata. *Semperella* was found on a basalt ground with a ferromanganese covering; *Poliopogon* and some of the new Hyalonematidae and Pheronematidae on pumice stones. In *Lophocalyx* all hypodermal pentactines are equal to one another without any remarkable elongation of the basal ones. One specimen was attachhed to a dead dictyonine hexactinellid by means of these spicules (Fig. 1m). All of these forms are characterized by a complex of morphological preadaptations for a future colonization of soft substrata, on which representatives of these genera are actually found. Thus, adaptation to life on a rough surface encourages the development of morphological peculiarities for life on soft substrata.

The "iceberg" adaptation, flattening, and reduction in weight are also important for life on soft substrata (Thayer 1975). A great enlargement of basal tufts in the dead lower part of the organism is known in *Pheronema, Euplectella* and others. Dictyonine heavy sponges with rigid skeletons are unknown on soft grounds. The special adaptations of dictyonine hexactinellids are discussed by Finks (1970) and Reif and Robinson (1976).

The adaptations against burial by sediment is also very important. One possibility is raising the body high above the substrata on a long stalk, e.g., *Hyalonema, Monorhaphis, Caulophacus*, etc. Another adaptation is the protection of the atrial cavity from sweepings. The most primitive way of protection is by reducting the atrial cavity; the atrial surface is turned outside, so that both the osculum and atrial cavity disappear (e.g. some *Pheronema*- and *Caulophacus* species, Fig. 1i-l). Thus, the inverted atrial surface provides protection. Other Hexactinellida possess sieve plates of different origin. *Euplectella* and *Regadrella* have sieve plates covering the oscula, or they possess rows of long spicular rays protuding over the atrial cavities (Fig. 1n-p; Ijima 1901). The origin of the sieve plates in *Aphrocallistes* is unknown. A sponge previously known as *Poliopogon gigas* Schulze (1887) has an interesting sieve plate consisting of tufts of anastomosing, unfused spicules.

In spite of all these adaptations, only *Hyalonema* and *Monorhaphis* live on true fine-grained muds, while others prefer sediments consisting of *Globigerina* and *Pteropoda* shells or mixed rough substrata. Most Hexactinellida settle on rocks only.

## Body Size

The size of adult hexactinellid sponges normally differs from that of other deep-sea animals. Usually their size is 10–50 cm, sometimes up to 1 m; *Monor-*

*haphis chuni* possesses an axial spicule 3 m in length (Koltun 1967). According to deep-sea photos, *Chonelasma* is about 2 m in length and 1.5 × 0.5 m in width (Fig. 2h,i).

An ancestral form of all Hexactinellida, *Protospongia*, possessed thin walls of a single layer of unfused four-rayed spicules. The whole body measured less than 30 mm in diameter (Finks 1960). In the course of evolution, a thickening of the walls and an enlargement of the body took place. The transition from four- to six-rayed spicules proved to be the basis for future transformations.

The problem of individuality is much easier for Hexactinellida than for other sponges. Hexactinellida are poorly differentiated in colonial organization, and it can be assumed that individual and single specimens are identical. Thus, one may conclude that the giant size of hexactinellids is a feature of a long phylogenetic history, as in many other groups. The enlargement of deep-sea animals results in a prolongation of embryogenesis and many larvae, which is necessary due to the difficult conditions for the larvae. Such organisms are unsuitable for future progress in evolution because of the very low speed of generation change; hence, they have a low capacity of adaptation to changing environments (Schmalhausen 1969). This might be a prerequisite for deep-sea life with its more stable conditions. The increase in body size in phylogenesis is usually related to a prolongation of growth by means of recapitulation.

## Recapitulation

The only work on the embryology of the Hexactinellida is a publication of Okada (1928). According to Okada, the youngest "embryo" of *Farrea sollassii* possesses six stauractines (four-rayed spicules). In the advanced larva, the oxystauractines enlarge in size, apparently without increasing in number. Later, microscleres and hexactinic spicules appear, and in the basal part of the body the fusion of hexactines begins (Fig. 3f-h).

Other known larva are from *Leucopsacus orthodocus* Ijima (1903) (Fig. 3c,d), *Leucopsacus ingolfi* Burton (1928b) (Fig. 3e), *Vitrolulla fertilis* Ijima (1904) (Fig. 3i-k). All of them have stauractines. The stauractines may be present in all hexactinellid "embryos".

Protospongia also had stauractinic spicules only (Finks 1960, 1970). The presence of these spicules in larvae of Recent sponges supports Haeckel's concept of recapitulation of ancestral characters in embryogenesis. Another feature of recapitulation in *Farrea sollassii* is the beginning fusion of spicules in the basal part of a young sponge. Later, the fusion spreads upwards. The fusion of spicules in the basal part of the body is present even in Hexactinellida with mainly unfused skeletons (*Regadrella*, Caulophacus, and others). The basal fusion improves the adaptation of a sponge fixed on rocks to an environment with high hydrodynamic energy (Reif and Robinson 1976).

There are two types of recapitulation. One is acceleration, i.e., precipitation of the somatic growth with a stable rate of development of the reproductive system. Acceleration, however, is hardly possible due to the limited food re-

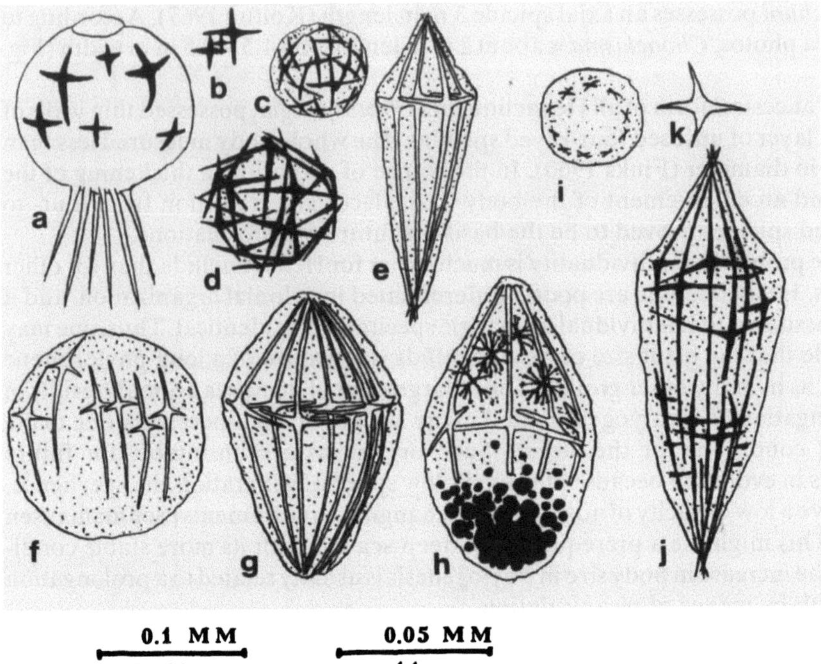


**Fig. 3. a,b** *Protospongia* and its stauractines (Finks 1970). Larvae of the hexactinellid sponges: **c,d** *Leucopsacus ortodocus* (Ijima 1903). **e** *Leocopsacus ingolfi* (Burton 1928b). **f,g,h** Stages of *Farrea sollasii* (Okada 1928). **i,j,k** Stages and a spicule of *Vitrollula fertilis* (Ijima 1904)

sources in the habitats of Hexactinellida. The second is hypermorphosis, i.e., stable somatic growth whereby the development of the reproductive system is delayed (Raff and Kaufman 1983). The latter is in contrast to neotheny proposed by Koltun (1971). Neotheny, according to Koltun, is a source of polymorphism, which is usual in deep-sea Demospongiae and Calcarea. In contrast, hypermorphosis facilitates speciation; the differences between species become noticeable in the late stages of development.

The most visible feature of hypermorphosis is the unproportional external shape of an organism, because the increase in the absolute size of an animal will bring about a change in the relative size of the part involved (Schmalhausen 1969). In the Hexactinellida the growth potential between the lower (basal) part of a sponge, when it is dead, and the living and growing upper part (e.g. Hyalonema, Chonelasma) is distinguished. Hence, after some time the lower part can no longer carry the living and growing upper part, which leads to the destruction of whole organism.

## Bathymetric Distribution of the Hexactinellida

In accordance with all these facts, the Hexactinellida favor deep water. Hexactinellid sponges in Recent habitats and paleontological history are most similar to the Brachiopoda. A catastrophic faunal change at the Paleozoic-Mesozoic boundary is clearly observed in the Hexactinellida (Finks 1960). At that time, as shown for the Brachiopoda, the dominance of thick-walled algae cells in the suspension caused great changes in the feeding matter of suspension feeders (Zezina 1985). It was proven for Recent brachiopods that some of them feed on algae, while others prefer colloidal organic matter and hence avoid regions with a high abundance of phytoplankton (McCommon 1969).

For the Hexactinellida, it could be expected that the absence of a well-developed special excretion system may lead to an obstruction of the body aggravated by a poor immune system. During in vitro aggregation, membrane recognition processes permit different syncytia and cells to fuse with fagocytic active syncytia of aggregation bodies at a rate considerably above that observed in other sponges (Pavans de Ceccatty 1982). Contrary to the demosponges, the aggregation factor in Hexactinellida functions unspecifically and agglutinates cells from different sponge classes, even mammalian cells (Müller et al. 1984). It leads to genetic homogenization. To avoid this, the Hexactinellida prefer deep-sea environments with their lack of living organic matter.

Another reason for the deep-sea life of the Hexactinellida is the competition among sedentary organisms for substrata (more important than the competition for food resources). Other sponges have strong passive defence systems. They use strong poisons and a well-developed immune system to protect themselves against fusion, overgrowth, and predation. The Hexactinellida have a poor system of passive defence, so they are forced to avoid the more convenient shallow-water regions. Hence, the Hexactinellida prefer the deep-sea regions, where the density of macrobenthic organisms is low. These facts, however, do not explain the existence of significant shallow-water settlements in the Antarctic, the presence of hexactinellid sponges in the vicinity of the West Canadian seashore, or some occurrences near the Azores and, perhaps, Indonesia.

All of the advantageous adaptations in recent adult representatives of the Hexactinellida are achieved due to their hexactinic spicules, which are considered ancestral. The stereometric shape itself is a result of the square symmetry of the axial filaments (Reiswig 1971), thus showing the capacity for numerous future transformations and modifications, such as fusions, ray reductions, ray branchings, etc.

The syncytial tissue organization provides these transformations, because intertissue spicule secretion can be prolonged for a considerable time. For example, the stalk of *Monorhaphis*, the largest siliceous structure in living beings, consists of a single spicule up to 3 m in length and about 8.5 mm in diameter.

*Acknowledgments.* The editors and the author thank Dorte Mehl (pro: Salomon) (FU Berlin) for the special review of the manuscript.

## References


Beklemishev VN (1964) The principles of comparative anatomy of invertebrate animals, vol I. Nauka, Moscow, pp 1-437 (in Russian)

Bergquist PR (1985) Poriferan relationships. In: Conway S, Morris SC, George JD, Gibson R, Platt HM (eds) The origin and relationships of lower invertebrates, vol 28. Clarenden, Oxford, pp 14-27

Bidder GP (1923) The relation of the form of a sponge to its currents. Q J Microsc Sci 67, NS:293-323

Burton M (1928a) A comparative study of the characteristics of shallow-water and deep-sea sponges, with notes on their external form and reproduction. Q J Microsc Club Lond 16:49-70

Burton M (1928b) Hexactinellida. Danish Ingolf Exp, vol 6, H Hagerup. Cøpenhagen, pp 1-18

de Laubenfels MW (1955) Archaeocyata and Porifera. In: Treatise on invertebrate paleontology. Geol Soc Am Univ Kansas Press, part E, pp 22-122

Finks RM (1960) Late Paleozoic sponge faunas of the Texas region. Bull Am Mus Nat Hist 120: 1-160

Finks RM (1970) The evolution and ecologic history of sponges during Paleozoic times. Symp Zool Soc, vol 25. Academic Press, London, pp 3-22

Heezen BC, Schneider ED (1966) Sediment transport by the Antarctic bottom currents on the Bermuda rise. Nature 211, N 5049:611-612

Ijima I (1901) Studies on the Hexactinellida. Cont. I. (Euplectellidae). J College Sci Tokyo 15:1-299

Ijima I (1903) Studies on the Hexactinellida. Cont 3. (Placosoma, a new euplectellid; Leucopsacacidae and Caulophacidae). J College Sci Tokyo 18:1-124

Ijima I (1904) Studies on the Hexactinellida. Cont 4. (Rossellidae). J College Sci Tokyo 18:1-307

Kirkpatrick R (1908) Description of a new dictionine sponge from the Indian Ocean. Rec Ind Mus Calcutta, 1908-1909 2:21-24

Koltun VM (1967) Hyalospongiae or Hexactinellida from north and far-east seas of the USSR. Nauka, Leningrad, pp 1-124 (in Russian)

Koltun VM (1971) The neotheny in sponges. In: The problems of zoology Leningrad. Proc Zool Inst Acad Sci, pp 43:45 (in Russian)

Mackie GO, Lawn ID, Pavans de Ceccatty M (1983) Studies on hexactinellid sponges. 2. Excitability, conduction and coordination of responses in *Rhabdocalyptus dawsoni* (Lambe 1873). Philos Trans R Soc Lond, Ser B 301, N 1107:401-418

McCommon HM (1969) The food of articulate branchiopods. J Paleontol 43, N 4:976-985

Müller WEG, Conrad J, Zahn RK, Steffen R, Uhlenbruck G, Müller J (1964) Cell adhesion molecule in the hexactinellid *Aphrocallistes vastus*. Species-unspecific aggregation factor. Differentiation 26, N I:30-35

Okada Y (1928) On the development of a hexactinellid sponge, *Farrea sollasii*. J Fac Sci Tokyo 4:1-27

Pavans de Ceccatty M (1982) In vitro aggregation of syncitia and cells of a hexactinellid sponge. Dev Comp Immunol 6, N 1:15-22

Raff RA, Kaufman TC (1983) Embryos genes and evolution. The developmental-genetic basis of evolutionary change. Macmillan, New York

Reif WE, Robinson JA (1976) On functional morphology of the skeleton in lichnise sponges. Hexactinellida. Paläontol Z 50, N 1/2:57-69

Reiswig HM (1971) The axial symmetry of sponge spicules and its philogenetic significance. Cah Biol Mar 12, N 4:505-514

Schulze FE (1887) Report on the Hexactinellida. Challenger Rep 21:1-513

Schmalhausen JJ (1969) The problems of Darwinism. Nauka, Moscow, pp 1-493 (in Russian)

Thayer CW (1975) Morphologic adaptation of benthic invertebrates to soft substrata. J Mar Res 33, N 2:177-189

Vogel S (1974) Current-induced flow through the sponge, *Halichondria*. Biol Bull 147:443-456

Zezina ON (1985) The recent brachiopods and the problems of bathyal zone in the ocean. Nauka, Moscow, pp 1-247 (in Russian)

# Influence of Different Current Regimes on the Growth Form of *Halichondria panicea* Pallas

D. Barthel

## Introduction

In assessing the influence of currents on the life of an adult sponge, two different aspects have to be taken into consideration. On the one hand, supply of oxygen and particulate and dissolved food and removal of metabolic wastes from the vicinity of the sponge are improved with increasing current velocities. Besides this, the obstruction of the ostia by sedimentation of suspended particles like silt or detritus is impeded by strong water movement. On the other hand, higher current velocities impose elevated mechanical stress on the sponge and may require increased structural resistance. Consequently, intertidal communities in surf-pounded areas consist mainly of encrusting forms (Ayling 1981), while high erect forms dominate in calmer waters. The richest sponge faunas are usually found in areas with medium hydrodynamic stress, as Laubenfels (1950a,b) demonstrated in the Bermudas, where maximum species abundance occurred at current speeds of about 3 km/h.

In the evolution of sponges, the necessity to both profit from the current and to resist its mechanical stress has led to a number of very specialized growth forms, like in the cup-shaped *Calyx lieberkühni* or the stalked *Clathrina blanca*. These growth forms minimize the fraction of exhaled water re-entering the sponge (Bidder 1923; Warburton 1960) so that even habitats with low current velocities can be utilized effectively. In contrast to those species confined to areas with certain current patterns by their genetically fixed body form, the demosponge *Halichondria panicea* Pallas (1766) is a cosmopolitan, surviving in a wide range of habitats with varying current regimes. Its growth form ranges from thinly encrusting in the intertidal surf zone (e.g., Burton 1948) to globose, lumpy forms in areas of medium currents (e.g., Barthel 1986, Peattie and Hoare 1981) and erect ramose forms in calm waters (Peattie and Hoare 1981; Barthel this chap). According to Vethaak et al. (1982), *H. panicea* has a lower tolerance to siltation than the closely related and often sympatric species *H. bowerbanki* Burton (1930). For *H. panicea* in Torch Bay, Alaska, which offers habitats along a wide wave force gradient, Palumbi (1986) reports a number of structural adaptations to wave force such as variations of length and width of spicules, tear resistance etc. The question ensuing is whether the obvious polymorphy of *H. panicea* in the Baltic Sea is the only or main adaptation to different current regimes, or whether other structural differences, namely differences in the amount of both organic and

J. Reitner and H. Keupp (Eds.)
Fossil and Recent Sponges

inorganic skeletal material and shape, size, and number of oscula can be found as well.

## Material and Methods

Sponges were sampled at two locations in the Baltic Sea and two locations in the North Sea.

In the Western Baltic Sea, sponges were sampled by divers from two different locations. Location 1, Surendorf, is on the southern side of the entrance of Eckerförder Bucht. Here *H. panicea* grows in abundance on the sunken remainders of a World War II torpedo station. This artificial hard substrate supplies conditions similar to those of a rocky coast with small channels, crevices, and miniature caves. Within close proximity, microhabitats with average high current velocities (ledges, channels) and with almost stillwater conditions ("caves") are found. The second location, Kleverberg Ost, is situated at the northeastern entrance of Kiel Fjord, where dense populations of red macroalgae form the preferred substrate for *Halichondria panicea* between 2 and 10 m water depth (Barthel 1986). The current stress at this location can be regarded as medium to strong, but part of the mechanical force exerted by sudden wave action is compensated for by the swaying movement of the substrate algae.

For a more detailed assessment of growth form variation, *H. panicea* were also collected from two North Sea locations. They were collected by hand in the Königshafen area on the island of Sylt, where they grow intertidally on *Mytilus edulis* banks and on a local variety of *Fucus vesiculosus*. This habitat is characterized by strong surf and wave action due to the tidal cycle. The second North Sea location, Jordsand, is situated between the islands of Sylt and Röm, and is strictly subtidal. The area is governed by strong tidal currents; small bolders and banks of *Mytilus edulis* are used as a substrate by *H. panicea*. This area was sampled by dredging. The North Sea *H. panicea* were inspected for overall growth form only.

In the laboratory, color, size, growth form and number, shape, and size of the oscula were recorded. Wet weight (WW), dry weight (DW) and organic content as ash-free dry weight were determined according to methods previously described (Barthel 1986). For parallel determination of organic skeletal material, sponges were divided into two portions, one of which was lyophylized and ashed as usual. The other was macerated with a 2% solution of chinolinole sulfate (Arndt 1938) for 48 h at room temperature to remove the organic cellular material. The amount of organic skeletal material, i.e., collagen and spongin, was calculated as the difference in AFDW without and with maceration.

## Results

### *Growth Form*

As was expected, the growth form of *H. panicea* varies from station to station according to ambient current conditions: Baltic Sea, Kleverberg Ost, subtidal, swept by medium to strong currents. The sponges grow epiphytic on red algae of the genera *Phyllophora* and *Phycodrys*, and form extensive populations. Part of the current shear stress is probably compensated for by the movement of the substrate algae, but still the sponges are exposed to some degree of hydrodynamism. The growth form is thickly encrusting to globose, following more or less the growth of their algal substrate (Fig. 1a). Baltic Sea, Surendorf, subtidal. At this station a wide range of growth forms is found within close distance, depending on the position. At position A, the rim of a ledge in a small canal swept by strong currents, *Halichondria panicea* takes on an encrusting form as often observed in the North Sea intertidal (see below). At the more sheltered position B, the sponges exhibit a lumpy-globose form comparable to that found in the red algae zone of Bokniseck. The site is exposed to daylight and the sponges partly utilize the concrete itself, but partly also small red algae as substrate. At the completely sheltered, "cave"-like positions D,C and E, where the sponges are directly attached to the concrete surface, long slender, partly anastomosing branches extend upwards from a relatively small base (Fig. 1b). Oscula are conspicuous at the tips of the branches (Fig. 1b, arrow), but can be found distributed over the whole surface as well.

North Sea, Königshafen, intertidal. This station is exposed to strong wave force coupled to the tidal rhythm. Here, *H. panicea* grows as mostly this encrustations on the valves of *Mytilus edulis* and the thalli of the brown seaweed *Fucus vesiculosus* (Fig. 1c). The color is predominantly light yellow, despite the fact that they are living in the euphotic zone, where they usually harbor algae in their tissue (e.g., Christensen 1985), indicating that the sponges are living mainly on the shaded parts of the *Fucus*. During ebb tide the *Fucus* flattens out, covering the substrate, thus helping to prevent exposure to air and subsequent dessication of the sponges.

North Sea, Jordsand, subtidal. This station is governed by a regime of alternating tidal currents. At this station *H. panicea* uses subtidal mussel banks and small stones as substrates. Colonies are massive, with numerous ridgelike outgrowths and small oscular chimneys distributed over the whole upper surface (Fig. 1d). Very often, living *Mytilus edulis* and *Balanus crenatus* are embedded into the colonies and fragments of their shells are distributed in the sponge tissue.

## *Oscula*

In order to see whether the number of oscula necessary to drain a given amount of sponge tissue would vary between specimens of different growth form, oscula were counted in animals of the lumpy/globose type (Surendorf, position B) and the ramose type (Surendorf, position C). The encrusting variety was not analyzed, as it was not possible for the divers to remove the sponges from the substrate undamaged.

The results compiled in Table 1 show that those sponges originating from a low current environment apparently need more oscula to drain their water-conducting system than animals living in an area with higher hydrodynamism. Statistical analysis with the two-tailed Student's t-test showed the difference to be

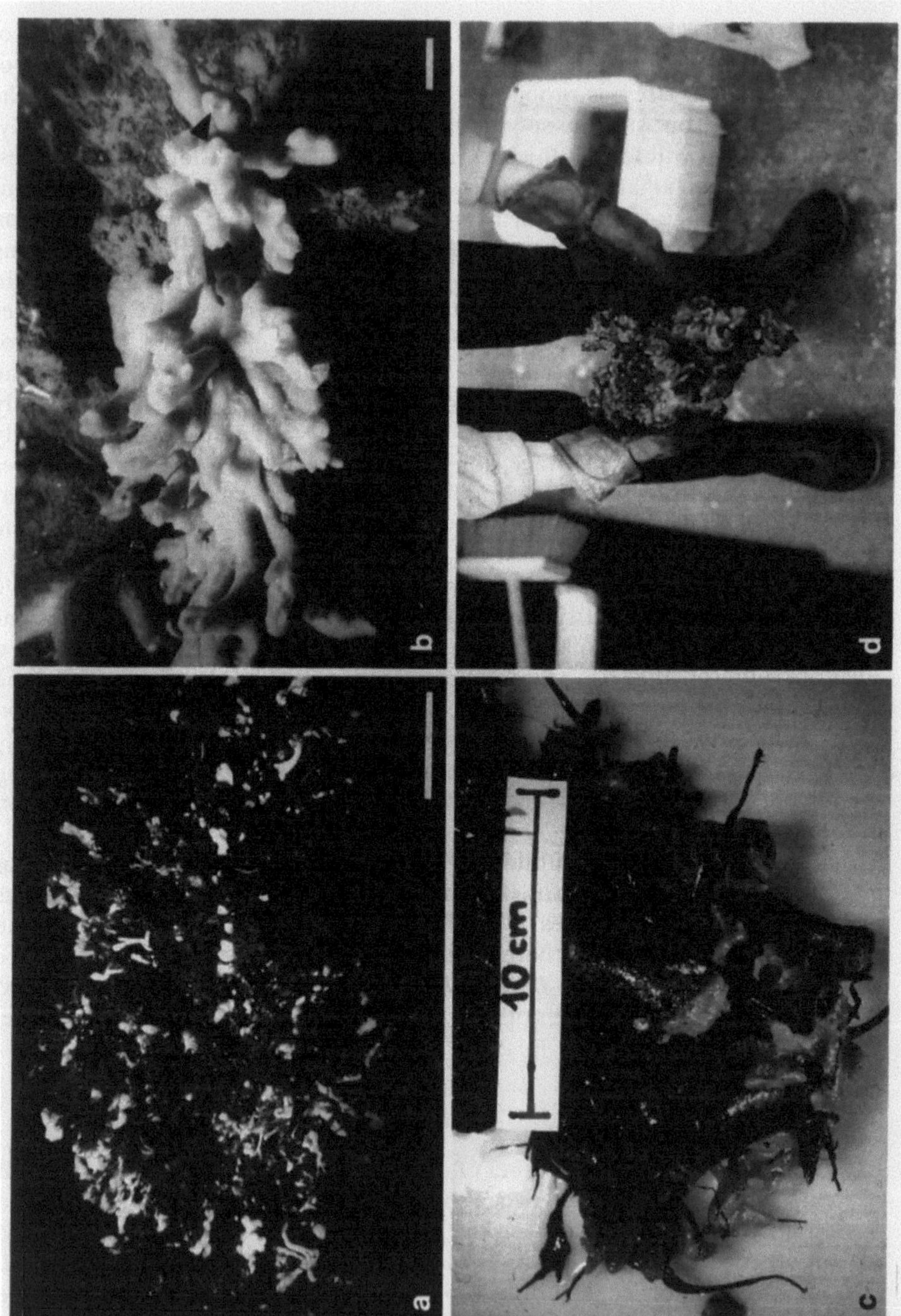
b
d
a
c
10 cm

**Table 1.** Oscular characteristics in sponges from the Surendorf station. B: medium to high current velocities, C: low current velocities

| Position | No. of oscula per g DW | $S_x$ | n | Range of oscular diameter (mm) |
|---|---|---|---|---|
| B | 15 | 7 | 19 | 1–4 |
| C | 26 | 7 | 13 | 0.5–2 |

significant with p = 0.0001%. Average oscular diameter was estimated in both groups. It ranged from 0.5 to 2 mm in the ramose growth form and from 1 to 4 mm in the globose growth form. In both groups oscula were partly level with the sponge surface, partly raised as chimneylike protrusions from the main body.

### *Tissue Composition*

The composition of the sponge tissue at the different locations was determined in order to see whether higher average current velocities and thus higher mechanical stress would lead to a more pronounced reinforcement of the skeleton with spicules and/or whether a greater part of the organic substance would be devoted to collagen/spongin. The results compiled in Table 2 show that sponges exposed to high mechanical stress (Surendorf, location A) contain more spicules (58.5% DW) than sponges from locations with medium hydrodynamism (Surendorf location B: 47.2% DW, Kleverberg Ost: 47.0% DW) and those from very sheltered microhabitats (Surendorf location C: 49.1% DW, D: 42.5% DW and E: 51.5% DW). The amount of organic skeletal material, i.e., collagen and spongin, varied as well, but with the exception of the most exposed location A in Surendorf, where the total sceletal material amounted to 88.5% DW, spicules plus collagen/spongin made up about 80% of the total dry weight at the different stations. The overall differences between locations were not significant. Metabolically active cellular material accounts for only about 20% of the dry mass of *Halichondria panicea* at the time of the investigation, and the fraction can be much smaller during the winter season (Barthel unpubl.).

◄

**Fig. 1. a** *Halichondria panicea* on red algae, Kleverberg Ost, Kiel Bight, 8 m water depth. Growth form is thickly encrusting/globose, colony shape resulting mainly from the shape of the substrate algae. Bar = 10 cm. **b** *Halichondria panicea* on concrete in a sheltered cave-like microhabitat at the Surendorf station (locations C, D and E). Slender branches extend fom a small base about 10–15 cm up into the water column. *Arrow head* Oscula at the tip of the branches. Bar = 2 cm. **c** *Halichondria panicea* from the intertidal North Sea. Königshafen, island of Sylt. Thinly encrusting colonies on *Fucus vesiculosus* and *Mytilus edulis*. **d** *Halichondria panicea* from the subtidal North Sea, Jordsand area between the islands Sylt and Röm, 2 m water depth. The colony is massive with ridge-like structures and sturdy chimney-like oscula on the upper side. Height of main colony 10–15 cm

**Table 2.** Composition of tissue of *Halichondria panicea* from different locations in Kiel Bight. Characterization of stations see text

| Series | Degree of exposure to waves | Growth form | Ash (%DW) | Spongin/ collagen (%DW) | Total skeletal material (%DW) | Cells (%DW) | n | $S_x$ |
|---|---|---|---|---|---|---|---|---|
| Sur A | + + + | Thinly encrusting | 58.5 | 30.0 | 88.5 | 11.5 | 4 | 2.5 |
| Sur B | + + | Lumpy/ globose | 47.2 | 35.2 | 82.4 | 17.6 | 12 | 7.9 |
| Sur C | + | Ramose | 49.1 | 31.1 | 80.2 | 19.8 | 17 | 8.5 |
| Sur D | + | Ramose | 42.9 | 37.1 | 80.0 | 20.0 | 9 | 11.4 |
| Sur E | + | Ramose | 51.5 | 27.0 | 78.5 | 21.5 | 16 | 7.5 |
| Kleverb | + + | Thickly encrusting on algae | 47.0 | 31.6 | 78.6 | 21.4 | 14 | 8.3 |

+ + +: high degree of exposure to wave action, + +: exposure to medium wave action, +: sheltered in miniature caves and crevices. August 6/7, 1986.

## Discussion

The observations and data presented here show that the sponge *Halichondria panicea* can react with a variety of morphological and structural changes to the different current regimes encountered in the Kiel Bight area, inspite of the low and unstable salinity (average about 15 ppt) constituting an additional stress factor. These results generally confirm previous results (Peattie and Hoare 1981; Vethaak et al. 1982; Palumbi 1986) from other areas. In assessing the degree of colony morphology variation (colony shape, number and size of oscula) and variation in tissue composition, the morphological changes seem to be more drastic. However, Palumbi (1984, 1986) documented more pronounced structural differences along the wave force gradient in the Alaskan Torch Bay. It is interesting to note that on average the oscular diameter found by Palumbi for sponges from a high wave force location is 0.5–0.3 mm, that from a low wave force environment is 0.7–0.4 mm, and number of oscula per cm² is 13–16 for the first and 5.4–3.8 for the latter group, i.e., sponges living in a high wave force environment have more, but smaller oscula. The opposite is the case in this study: *Halichondria panicea* from sheltered locations have significantly more oscula with average smaller diameter than those exposed to stronger wave force. Comparing values for spicule content as % total dry weight, a higher amount of spicules is found for more exposed sponges in both studies, but the overall amount of spicules is much lower in the Baltic Sea specimens: Palumbi (1986) gives 77.1% DW for high wave force sponges as opposed to 65.2% DW in low wave force sponges, while in the Baltic Sea the values are 58.5% DW and a minimum of 42.9% DW, respectively.

These differences and the greater average oscular diameter in Baltic Sea *Halichondria* clearly indicate that there are wave force gradients in both areas, but the general hydrodynamic level at the Atlantic site is probably much higher, necessitating higher amounts of reinforcing spicules in the sponge tissue. The results on numbers of oscula under different current regimes still remain contradictory, unless we assume that *H. panicea* changes its growth strategy at relatively low overall current velocities, forming small oscula frequently during growth instead of fewer large ones, and only is forced to construct very small oscula in an environment calling for high tear resistance, i.e., small internal cavities. To ensure sufficient drainage of the sponge tissue, these small oscula of course have to be more numerous again, which agrees with Palumbi's hypotheses.

In this context, the question of the efficiency of certain oscular construction types has to be considered. Several authors (e.g., Bidder 1923; Vogel 1974 and Vogel and Bretz 1972) reasoned that the chimneylike shape of oscula induces enhanced "passive flow" (Vogel 1974) through the sponge by tapping the laminar flow above the immediate sponge surface, thereby acting as "drag motors" of flow. This would be most effective if *Halichondria panicea* had an essentially open canal system, which it probably does not. The exhaling part of the canal system is sealed off from the inhaling part by the choanocyte chambers which are at the same time the internal pumping units maintaining water flow through the sponge. This implies that the pumping activity of the choanocyte chambers is the limiting factor for the internal flow rate. Any current suction or drag at the oscula exceeding this flow would either be resisted by this seal and thus left unused, or would damage the choanocyte chambers; these cannot technically be compared to valves able to open or close according to external pressure or suction. Thus, the main effect of raised oscula and/or a ramifying sponge shape seems to be the prevention of exhaled water re-entering the ostia again.

Finally, in interpreting the connection of current regime and sponge shape and tissue composition it is important to note that both factors may also vary with season in *H. panicea* (Barthel 1988; unpubl. data).

## Conclusions

The degree of polymorphism found in *Halichondria panicea* is much higher than observed in most other sponges, but morphological variations have been found in other species as well. Parker (1910) mentions that *Stylotella heliophila* Wilson (1911) forms massive colonies at sites exposed to air from time to time and attains a "long-fingered" shape at strictly subtidal locations, and Wilkinson and Vacelet (1979) found variations in *Verongia aerophoba* under different conditions of light and current. Together with the capacity to harbor symbiotic algae (Christensen 1985; Eimhjellen 1967), a good structural adaptation to high loads of suspended particles (Reiswig 1973) and tissue sloughing as an effective mechanism to counteract settlement of micro-fouling organisms (Barthel and Wolfrath 1988), the high degree of polymorphism observed in *H. panicea* accounts for the extreme success of this organism in most coastal areas of the Northern Atlantic area.

*Acknowledgments.* Harald and Ragnhild Asmus from the List station of the Biologische Anstalt Helgoland and the crew of the research vessel MYA helped in collecting the North Sea specimens. S.A. Gerlach, N. Weissenfels, and P. Langenbruch helped with critical comments on the subject. The study was in part financed by the Deutsche Forschungsgemeinschaft grant Ba 913/1.

## References

Arndt G (1938) Chinosol als Mittel zur Herstellung von Spongienmazerations-Skeletten. Sitzungsber Ges Naturf Freunde Berl 1938:4–7 and 1939:94–96

Ayling AL (1981) The role of biological disturbance in temperate subtidal encrusting communities. Ecology 62:830–847

Barthel D (1986) On the ecophysiology of the sponge *Halichondria panicea* in Kiel Bight. I. Substrate specificity, growth and reproduction. Mar Biol 32:291–298

Barthel D (1988) On the ecophysiology of the sponge *Halichondria panicea* in Kiel Bight. II. Biomass, production, energy budget and integration in environmental processes. Mar Ecol 43:87–93

Barthel D, Wolfrath B (1988) Tissue sloughing in the sponge *Halichondria panicea*: a fouling organism prevents being fouled. Oecologia (Berl) 78:357–360

Bidder GP (1923) The relation of the form of a sponge to its currents. Q J Microsc Sci 266:293–323

Burton M (1948) Observations on littoral sponges, including the supposed swarming of larvae, movement and coalescence in mature individuals, longevity and death. Proc Zool Soc Lond 118:893–915

Christensen T (1985) *Microspora ficulinae*, a green alga living in marine sponges. Br Phycol J 20:5–7

de Laubenfels MW (1950a) The porifera of the Bermuda Archipelago. Trans Zool Soc Lond 27:1–154

de Laubenfels MW (1950b) An ecological discussion of the sponges of Bermuda. Trans Zool Soc Lond 27:155–201

Eimhjellen KE (1967) Photosynthetic bacteria and carotenoids from a sea sponge *Halichondrium panicea*. Acta Chem Scand 21:2280–2281

Palumbi SR (1984) Tactics of acclimation: Morphological changes of sponges in an unpredictable environment. Science 225:1478–1480

Palumbi SR (1986) How body plans limit acclimation: Responses of a demosponge to wave force. Ecology 67(1):208–214

Parker GH (1910) The reactions of sponges, with a consideration of the origin of the nervous system. J Exp Zool 8(1):1–41

Peattie ME, Hoare R (1981) The sublitoral ecology of Menai Strait. II. The sponge *Halichondria panicea* (Pallas) and its associated fauna. Estuarine Coastal Shelf Sci 13:621–635

Reiswig H (1973) Population dynamics of three Jamaican demospongiae. Bull Mar Sci 23:191–226

Vethaak AD, Cronie RJA, Van Soest RWM (1982) Ecology and distribution of two sympatric, closely related sponge species, *Halichondria panicea* (Pallas 1766) and *H. bowerbanki* Burton, 1930 (Porifera, Demospongiae), with remarks on their speciation Bijdr Dierk 52(2):82–102

Vogel S (1974) Current-induced flow through the sponge, *Halichondria*. Biol Bull 147:443–456

Vogel S, Bretz W (1972) Interfacial organisms: passive ventilation in the velocity gradients near surfaces. Science (USA) 175:210–211

Warburton FE (1960) Influence of currents on the form of sponges. Science 132:89

Wilkinson CR, Vacelet J (1979) Transplantation of marine sponges to different conditions of light and current. J Exp Mar Biol Ecol 37:91–104

# The Sponge Fauna Associated with *Arca noae* L. (Mollusca, Bivalvia)

G. Corriero, R. Pronzato, and M. Sarà

## Introduction

Studies on the epibiotic fauna associated with bivalve molluscs have shown that sponges are often the most common organisms living on mollusc shells. Yet the structure of sponge fauna inhabiting this kind of substratum appears very different from those observed on natural rock.

Sponges can occupy large areas of shell surface (Bakus 1966; Forester 1979). Only few shelled molluscs can be covered by many species of sponges, as observed in *Pinna nobilis* (Corriero and Pronzato 1987), but, more frequently, they are colonized by one or few species. Bergquist (1978), for instance, reports that *Jophon minor* is the only sponge settling on the shells of *Chlamis zealandiae*, and Bloom (1975) reports that *Chlamis varia* and *Chlamis hastata hericia* are inhabited by only *Myxilla incrustans* and *Mycale adherens*, respectively. Examples of specific mutualism have also been given (Forester 1979; Forbes 1964; Vance 1978).

Sponge communities on rocky substrata generally show high values of species richness (Pansini and Pronzato 1973; Sarà et al. 1978), and only in few cases does a marked dominance of some species occur (Sarà 1970). In mature communities, the persistence of species is high (Pansini and Pronzato 1990), while in the first phases of substratum colonization the recruitment is very slow, with a high degree of stochasticity in the evolution of species settlement (Pansini and Pronzato 1981; Vacelet 1981).

Literature sources provide little information on direct comparisons between the Porifera associated with mollusc shells and those living on the neighboring rocky substrata. Indeed they often report data on the association between sponges and motile bivalve molluscs (Forester 1979; Pitcher and Butler 1987), or data on mollusc populations living far from the coast (Corriero and Pronzato 1987).

*Arca noae* L. is a byssate bivalve very common on Mediterranean infralittoral rocky substrata. The association of this mollusc with the demosponge *Crambe crambe* is well known also to divers, and has been studied by Melone (1969). The purposes of the present work are: to verify the existence of a selectivity mechanism regulating the colonization of this substratum by the neighboring sponge fauna; to compare the settlement on ark populations in geographically different areas; and, finally, to detect the possible occurrence of mutualistic interferences between this mollusc and its epibiotic sponges.

J. Reitner and H. Keupp (Eds.)
Fossil and Recent Sponges

## Material and Methods

Some populations of *Arca noae* from three different localities of the Mediterranean sea were studied.

### *Addaura* (North-Western Sicilian Coast)

Forty-eight individuals of *Arca noae* were collected by SCUBA on a rocky wall at a depth of 1–2 m. Thirty-six control samples were randomly collected from the rocky substratum of the same area for a comparison with the sponge fauna living on the mollusc shells. The size of each control rocky sample (20 $cm^2$) was equal to the mean size of the ark shells.

Because of the small dimensions of the control samples, an extra quantitative collection was carried out on a large area in order to list the sponge fauna living in this locality.

The sponges living on the arks and on the rocky control samples were mapped and their area was calculated with a graphic digitizer interfaced with a personal computer. With the same method the surface area of each shell was calculated (Fig. 1, right).

Data concerning the species distribution and percent covering of sponges on all the shell and control rocky samples were analyzed by the similarity index (Kulczynski 1927). Values of percent covering of single species on the whole examined area, for both ark and control rocky sample sets, were utilized for a Motomura (see Daget 1979) abundance distribution analysis (rank-log of % covering).

In order to evaluate differences in the distribution of Porifera on the different parts of the shells, each mollusc has been subdivided into a large front area and

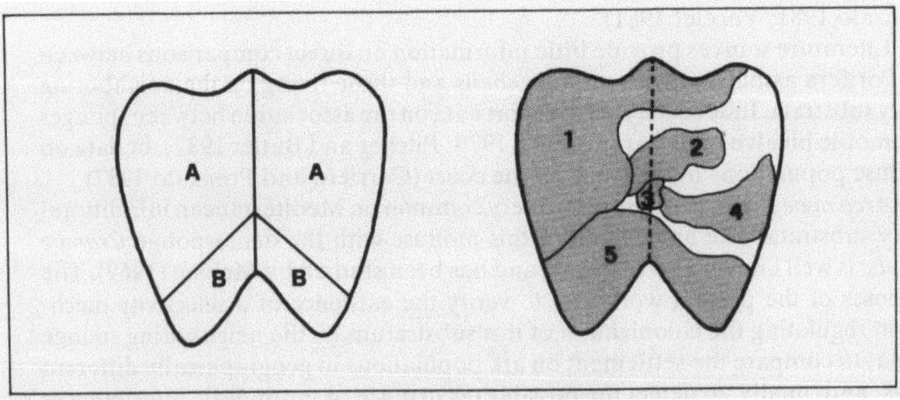


**Fig. 1.** *Right.* Diagram showing the locations of epibiontic sponges on a specimen of *Arca noae* from Addaura. *1 Stylopus dujardinii; 2 Anchinoe paupertas; 3 Anchinoe tenacior; 4 Antho involvens; 5 Crambe crambe. Left.* Front area (*A*) and back area (*B*) in *Arca noae* specimens

a back one. The latter, which may be influenced by the feeding currents generated by *Arca noae*, was estimated to be one-fifth of the total shell surface (Fig. 1, left). The percent area covered by each sponge species was calculated on both the front and back areas.

### *Portofino* (Eastern Ligurian Coast)

Twenty-six individuals of *Arca noae* were collected from the Promontory of Portofino at a depth of 5 to 20 m. Maps of arks and of their epibiotic Porifera were made by the same technique utilized for the Addaura samples. A quantitative list of the Porifera living in this area, reported by Sarà et al. (1978), was employed for a comparison with the sponge fauna from Addaura.

### *Ustica Island* (Southern Tyrrhenian Sea)

Fifty-two specimens of *Arca noae* were examined in situ in the marine reserve of the Island of Ustica at a depth of 3 to 20 m. Fragments of the most common epibiotic species were collected for identification, while the covered area was estimated in situ.

## Results

### *Addaura*

Nineteen species of Porifera were found on the shells of *Arca noae*. The number of species per specimen varied from 0 to 5. One species, *Crambe crambe*, was present on more than 50% of the examined shells; six other species, *Myxilla rosacea, Antho involvens, Anchinoe fictitius, Anchinoe tenacior, Hymedesmia* sp., and *Spongia virgultosa*, were present on about 10% of the total examined molluscs, the others being very sporadic. The total covering value on the whole sampled surface was 29.8% (Table 1). The covering value for each shell varied from 0 to 85.7%.

The control rocky samples yielded 27 species of Porifera. The number of species per sample varied from 0 to 4. Only one species (*Ircinia variabilis*) was present on about 10% of the examined standard samples. The total covering value on the whole sampled surface was 40.9% (Table 1). The covering values for each sample varied from 0 to 144.4%.

Twelve species were common to both sample sets, seven were present on the arks only and 15 on the control rocky samples only.

The samples without sponges were 6 out of the 48 examined arks, and 11 out of the 36 control samples. This different ratio between colonized and empty samples was significant at the chi-square test ($p < 0.05$).

**Table 1.** Number of specimens and percent covering of Porifera living on *Arca noae* and on the control rocky samples

| | *Arca Noae* | | Control rocky samples | |
|---|---|---|---|---|
| Number of samples | 48 | | 36 | |
| Number of nonempty samples | 42 | | 25 | |
| Total sampled area ($cm^2$) | 1151.69 | | 720 | |
| Total sponge species number | 19 | | 27 | |
| Sum of percent coverings | 29.78 | | 40.94 | |
| | Specimens No. | Covering | Specimens No. | Covering |
| 1 *Sycon* sp. | 3 | 0.01 | 1 | 0.01 |
| 2 *Jaspis johnstonii* (Schmidt) | – | – | 2 | 1.49 |
| 3 *Prosuberites epiphytum* (Lamarck) | – | – | 1 | 0.18 |
| 4 *Terpios fugax* Duchassaing & Michelotti | 3 | 0.14 | 1 | 0.05 |
| 5 *Cliona schmidti* (Ridley) | – | – | 2 | 2.98 |
| 6 *Cliona vastifica* Hancock | – | – | 3 | 3.35 |
| 7 *Cliona vermifera* Hancock | – | – | 2 | 1.08 |
| 8 *Cliona viridis* (Schmidt) | – | – | 3 | 7.37 |
| 9 *Cliona* sp. | – | – | 1 | 1.39 |
| 10 *Thoosa mollis* Topsent | – | – | 1 | 1.65 |
| 11 *Chondrilla nucula* Schmidt | – | – | 1 | 1.23 |
| 12 *Chondrosia reniformis* Nardo | – | – | 1 | 1.16 |
| 13 *Timea crassa* (Topsent) | – | – | 2 | 1.37 |
| 14 *Batzella inops* Topsent | 1 | 0.55 | – | – |
| 15 *Mycale rotalis* (Bowerbank) | 3 | 2.10 | 1 | 0.42 |
| 16 *Crambe crambe* (Schmidt) Thiele | 26 | 16.55 | 4 | 3.66 |
| 17 *Phytheas rosea* (Topsent) | – | – | 1 | 0.30 |
| 18 *Lissodendoryx isodictyalis* (Carter) | 2 | 0.22 | 3 | 1.39 |
| 19 *Myxilla rosacea* (Lieberkuhn) | 4 | 1.15 | 1 | 0.37 |
| 20 *Tedania anhelans* (Lieberkuhn) | 2 | 1.88 | – | – |
| 21 *Hymedesmia* sp. 1 | 5 | 0.87 | – | – |
| 22 *Hymedesmia* sp. 2 | 2 | 0.05 | – | – |
| 23 *Stylopus dujardini* (Boverbank) | 2 | 0.76 | – | – |
| 24 *Anchinoe fictitius* (Bowerbank) | 5 | 0.85 | 1 | 1.28 |
| 25 *Anchinoe paupertas* Bowerbank) | 5 | 0.38 | 3 | 0.58 |
| 26 *Anchinoe tenacior* Topsent | 7 | 1.55 | 2 | 2.04 |
| 27 *Antho involvens* (Schmidt) | 4 | 1.14 | 3 | 0.92 |
| 28 *Haliclona* sp. | 2 | 0.22 | – | – |
| 29 *Hippospongia communis* (Lamarck) | – | – | 1 | 0.58 |
| 30 *Spongia officinalis* L. | 1 | 0.54 | – | – |
| 31 *Spongia virgultosa* (Schmidt) | 5 | 0.65 | 1 | 0.63 |
| 32 *Cacospongia scalaris* Schmidt | – | – | 1 | 0.66 |
| 33 *Ircinia variabilis* (Schmidt) | 3 | 0.18 | 4 | 3.67 |
| 34 *Dysidea* sp. | – | – | 1 | 1.15 |

All the Porifera species found on ark and control rocky samples were also present in the extra collection carried out in the studied area, where a total of 43 species of Porifera were collected.

Boring sponges (Clionidae) were very abundant on the rocky substrata, (six species covered 17.8% of the total control samples), but they were completely absent from the ark shells. *Crambe crambe*, on the contrary, was much more frequent and abundant on the mollusc (25 specimens covering 16.55% of the total surface of the examined shells) than on the control samples (four specimens covering 3.66% of the total examined surface). The differential distribution of this species was highly significant at the chi-square test ($p < 0.01$). Sometimes this sponge had a cushion-like shape on the shells, while it was always encrusting on the rocky substrata. The other species of Porifera present on shells were mainly encrusting. No branching forms were present, and only few species had a cushion-like shape. Other organisms of the sessile zoobenthos (e.g., bivalves, ascidians, hydrozoans, bryozoans, and barnacles) were observed sporadically on the shells of *Arca noae*.

The similarity indices calculated for all non-empty samples showed a more homogeneous sponge fauna for the ark samples than for the rocky ones: over 50% of the ark samples were indeed associated together (Fig. 2, gray cluster); the control rocky samples did not show any significant similarities and were randomly associated.

The Motomura abundance distribution, calculated for the two sets of samples (including the empty ones), allowed better synthesis of the differences in species richness and abundance, confirming the higher abundance values and the larger number of species for the control rocky set (Fig. 3A) than for the ark set (Fig. 3B). It also pointed out the lack of strongly dominant species on the rocky samples, while on the arks *Crambe crambe* was very abundant and few species were rare.

The number of sponge species does not appear significantly related to the dimensions of ark specimens (Fig. 4).

The Porifera distribution on the two different parts of the shells showed that *Crambe crambe* clearly prefers the back area of the host, since it occupied 42% of

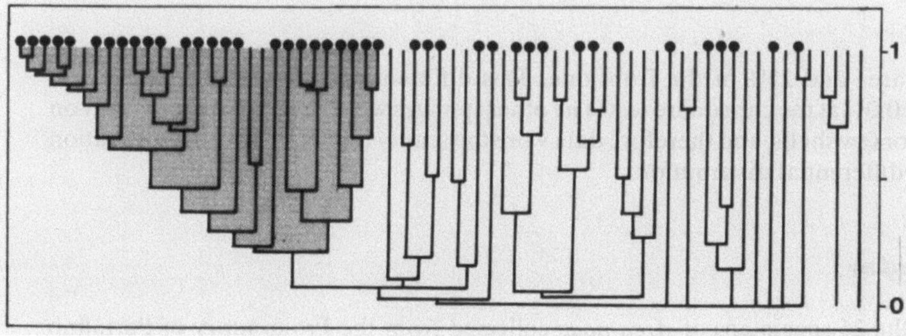


**Fig. 2.** Dendrogram based on Kulczynski similarity indices among all the nonempty shells (*dots*) and the control rocky samples at Addaura. The *gray* cluster covers over 50% of ark samples

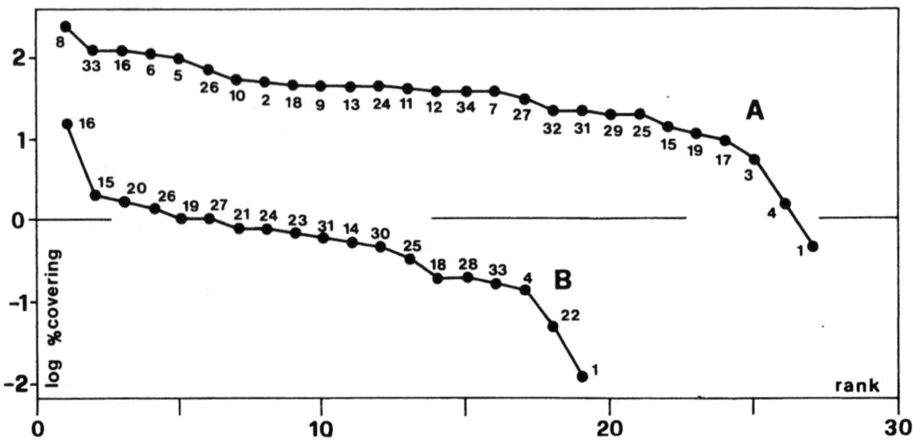


**Fig. 3.** Motomura abundance distribution analysis utilizing the percent covering values on the whole examined area at Addaura for both control rocky (**A**) and ark (**B**) samples. Porifera species are denoted by the same *numbers* as in Table 1

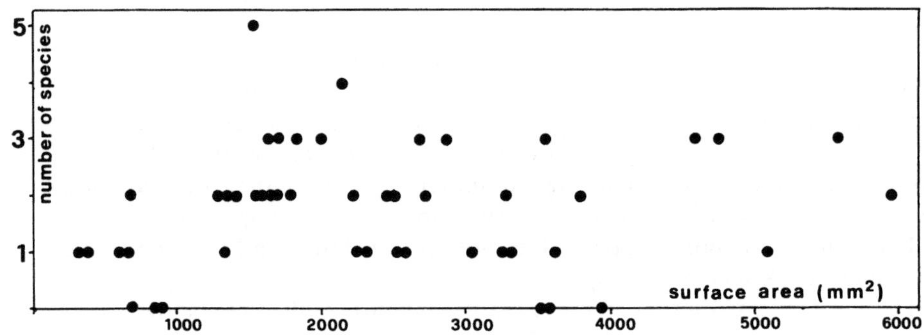


**Fig. 4.** Relationship between the size of *Arca noae* specimens and the number of species of epibiotic Porifera (Addaura). No significant correlation exists between the two parameters

this area and 23% of the front one. This difference resulted highly significant ($p < 0.01$) at the chi-square test. The other sponges were generally present only on one or few shells, and therefore data were too scanty for a statistical corroboration of a differential distribution.

## *Portofino*

On the 26 specimens of *Arca noae* collected from the Promontory of Portofino only one species of Porifera (*Crambe crambe*) was found. Only one mollusc showed no sponge settlement, while on the others *Crambe crambe* covered from 64 to 100% (medium value 90%) of the shell surface. *Crambe crambe* specimens living on the shells showed a different development than on the rocky substrata. Indeed, the epibiotic specimens showed a cushion-like shape, and sometimes the tendency to a branched one, whereas the specimens living on the rocky substrata showed a marked tendency to the encrusting shape.

Seventy-three species of sponges were present in the collecting area (Sarà et al. 1978). A comparison with the Addaura sponge fauna showed 18 common species from a total number of 98. Fifty-five were present only at Portofino and 25 only at Addaura. Among the species associated with the arks living at Addaura, some (like *Anchinoe tenacior* and *Tedania anhelans*) showed very low abundance values at Portofino, while others (e.g., *Myxilla rosacea*) were completely absent. On the rocky substrata, *Crambe crambe* showed higher covering values in Portofino (3.8%) than in Addaura (2.3%).

### *Ustica Island*

In this locality six species of Porifera were commonly found on the shells of *Arca noae*. Among these, *Crambe crambe* was the most frequent and settled on about 50% of the examined molluscs with high covering values. The other five species (*Anchinoe tenacior*, *Lissodendoryx isodyctialis*, *Terpios fugax*, *Myxilla* sp., and *Spongia virgultosa*) settled on about 10% of the molluscs. Other sporadic species living on the shells were observed; boring sponges were not present.

## Discussion

The comparison between ark and rock samples at Addaura points out the selectivity of *Arca noae* for sponge settlement. In fact, the ark samples show a smaller number of species and lower covering values than the control rocky samples. Moreover, the sponge fauna associated with *Arca noae* is more homogeneous. A selective mechanism is also pointed out by the absence of correlation between the number of sponge species and the size of the ark specimens. This is in contrast with the literature data on rocky substrata (Pansini et al. 1977), where a sharp increase in the number of sponge species was observed, as a function of the increase of the examined surface, when the size of sampled area is similar to that of ark specimens.

The examined populations of *Arca noae* show marked differences in their epibiotic fauna of Porifera. The low species richness characterizing *Arca* shells at Addaura is even more evident at Portofino, where only one species was found on the collected shells. This particular feature is probably due to the absence of efficient competitors of *Crambe crambe* at Portofino. Indeed, some of the sponge species living on *Arca noae* at Addaura are absent or very rare in the rocky substrata of the Portofino area. Therefore the occurrence of a situation of selectivity in the southern localities (Addaura and Ustica) and of exclusivity in the northern one (Portofino) could be related with the structure of the sponge communities probably depending on the different latitude of the considered areas.

*Crambe crambe* is the most common species living associated with *Arca noae*, and it shows higher covering values on the shells than on rocky substrata. This is in agreement with other studies about associations with molluscs, where the sponges are more common on the shells than on the neighboring substratum (Pitcher and Butler 1987; Melone 1969).

*Crambe crambe* seems to benefit by the association with *Arca noae*. It seems to prefer the back area of the ark, where the water flow generated by the host is greater. Forester (1979) reports that *Halichondria panicea*, associated with the scallop *Chlamis varia*, shows an orientation of its growth with respect to the prevailing currents induced by the mollusc host, gaining an additional supply of suspended nutrient. *Crambe crambe* does not show any particular orientation of its growth, but a remarkable tendency to a massive shape in the specimens living on *Arca noae*. The massive shape might be due to the favorable trophic conditions induced by the mollusc host. Moreover, *Crambe crambe* could benefit its host against possible predators (e.g., starfish) sealing the shell of the mollusc, as suggested by Forester (1979). From this point of view, the mollusc might also profit by the preferential location of the sponge in the back area of the shell, as observed for the specimens collected at Addaura. Indeed, under disturbance, *Arca noae* lets the ventral side of its shell adhere to the substratum, thus exposing to possible predators only the back part of the shell.

## References

Bakus GJ (1966) Some relationships of fishes to benthic organisms on coral reefs. Nature (Lond) 210:280–284

Bergquist PR (1978) Sponges. Hutchinson, London

Bloom SA (1975) The motile escape response of a sessile prey: a sponge-scallop mutualism. J Exp Mar Biol Ecol 17:311–321

Corriero G, Pronzato R (1987) Epibiontic sponges on the bivalve *Pinna nobilis*. Mar Ecol Prog Ser 35:75–82

Daget J (1979) Les modèles mathématiques en écologie. Masson, Paris

Forbes M (1964) Distribution of the commensal oyster, *Ostrea permollis*, and its host sponge. Bull Mar Sci 14:719–750

Forester AJ (1979) The association between the sponge *Halichondria panicea* (Pallas) and scallop *Chlamis varia* (L): a commensal-protective mutualism. J Exp Mar Biol Ecol 36:1–10

Kulczynski S (1927) Die Pflanzenassoziationen des Pieninen. Bull Int Acad Pol Sci Lett Classe Sci Math Nat 2:57–203

Melone N (1969) Associazione della Demospongia *Crambe crambe* (Schmidt) Thiele con i lamellibranchi *Spondylus gaederopus* ed *Arca noae*. Natura (Milan) 60:195–198

Pansini M, Pronzato R (1973) Il coralligeno di Bogliasco ed il suo popolamento di poriferi. Boll Mus Ist Biol Univ Genova 41:5–34

Pansini M, Pronzato R (1981) Etude des Spongiaires de substrats artificiels immergés durant quatre ans. Vie Milieu 31:77–82

Pansini M, Pronzato R (1990) Observations on the dynamics of a Mediterranean sponge community. In: Rützler K, Macintyre VV, Smith KP (eds) New perspectives in sponge biology. Smithsonian Int Press, Washington DC (in press)

Pansini M, Pronzato R, Leone E (1977) Area minima in un popolamento di Poriferi infralitorali. In: Cinelli F, Fresi E, Mazzella L (eds) Atti del IX Congresso della Società Italiana di Biologia Marina. Lito "La Seppia", Firenze, pp 331–335

Pitcher CR, Butler AJ (1987) Predation by asteroids, escape response and morphometrics of scallops with epizoic sponges. J Exp Mar Biol Ecol 112:233–249

Sarà M (1970) Competition and cooperation in sponge populations. Symp Zool Soc Lond 25:273–284
Sarà M, Balduzzi A, Boero F, Pansini M, Pessani D, Pronzato R (1978) Analisi di un popolamento bentonico di falesia del Promontorio di Portofino: dati preliminari. Boll Mus Ist Biol Univ Genova 46:119–137
Vacelet J (1981) 'Etude qualitative et quantitative des salissures biologiques de plaques expérimentales immergeés en pleine eau. 6 – Les éponges. Tethys 10:165–172
Vance RR (1978) A mutualistic interaction between a sessile marine clam and its epibionts. Ecology 59:679–685

# Monthly Variations in the Size of Spicules of the Haplosclerid Sponge, *Haliclona rosea* (Bowerbank)

W.C. Jones

## Introduction

Seasonal variation in spicule size in a marine siliceous sponge from temperate waters was first demonstrated by Hartman (1958). He noted that the spicules in *Haliclona canaliculata* were thicker in specimens collected in the winter and suggested that the difference was due to the seasonal variation in silicate concentration in the sea water. Later, Stone (1970) investigated the variation in the averages of 40 lengths and 40 widths of styles from the tips of papillae excised each month from ten specimens of *Hymeniacidon perleve* and found that they decreased from mid-winter peaks to mid-summer minima and back again. The silicate concentration of the sea water, measured at high water in earlier years, had undergone a similar fall and rise, though the rise occurred about a month after, and the fall about a month before, the corresponding changes in spicule dimensions. Nevertheless, the correlation was significant. The coefficients of variation for length (CV-L) and width (CV-W) also varied, the range of values respectively being 28.4–17.6 and 55.6–31.7, with the maxima occurring in June. A significant inverse correlation between the spicule means and the sea temperature was also found.

Somewhat different results were obtained in studies of haplosclerid sponges from North Wales (Jones 1984). Here the lengths and widths of 100 spicules taken at random were measured per specimen and it was found that when the data for all samples of haplosclerid sponges collected at approximately the same times of the year were aggregated, the average CV-W was markedly less in May than at other times. Moreover, taking the data for *Reneira rosea* and *Gellius angulatus* separately, it was shown that from the low values in May, the CV-W rose to a maximum in July. In both species, sections revealed that thin oxea became incorporated in the spicule framework in July and August.

Investigations on *Haliclona elegans* (Jones 1987a) indicated that both the CV-W and the CV-L decreased in June due to a cessation in spicule production, which was probably associated with reproductive activity (oogenesis in particular). In May both CV's had risen from the winter level and in July the values were again higher, leading to maxima in August. The appearance of the skeleton in sections of this species (readily identified by the presence of characteristic "slime strands") was markedly different in specimens collected in the autumn from that in spring specimens.

J. Reitner and H. Keupp (Eds.)
Fossil and Recent Sponges

Further studies with *Haliclona* (= *Reniera*) *rosea* (Jones 1987b) confirmed that the mean CV-W and mean CV-L varied from May to late August. Only the months from May to September were included in these studies, because the data were derived only from embryo-containing specimens. The presence of aggregated embryos of appropriate size increased confidence in the identification of this species, particularly vis-a-vis *H. fistulosa*. However, from these data the local maximum lengths and widths of the oxea of *H. rosea* were established and these, together with the other diagnostic features for this species (absence of "slime strands", primary bundles with no more than three spicules abreast, porous surface texture, oscula simple or on low mounds) can now be used to include samples derived from specimens that do not contain embryos. The variations in spicule parameters throughout a complete year are the subject of this chapter.

Histological studies of both *elegans* and *rosea* have revealed that gametogenesis occurs in most specimens in the early summer and have confirmed that oogenesis involves the incorporation of enormous numbers of trophocytes. The spherical reproductive bodies, referred to as "embryos" in the earlier papers (Jones 1987a,b), in some cases at least, would have been developing egg cells. Only when the characteristic "laryal" spicules are present can they be called "embryos" with certainty. However, it is clear that the expenditure of so much energy on oogenesis is bound to affect sponge growth and spicule production.

The effect of silicate concentration and temperature on spicule growth and production have been experimentally investigated (Jones 1979). More recently it has been shown that increased silicic acid concentration resulted in an increase in spicule width, but not length, in *Suberites domuncula* (Simpson, Gil, Connes, Diaz and Paris 1985). Temperature was inversely correlated with spicule width, but showed little correlation with spicule length, excepting the toxa, in *Microciona prolifera* (Simpson 1978).

Recently, Palumbi (1986) investigated the effect of wave force on *Halichondria panicea* and found an increase in spicule number and size in specimens from areas of high wave action. *Haliclona rosea*, however, is a soft sponge, occurring sublittorally or under boulders in pools at low levels on the shore. Skeletal variation in response to wave action is less likely than in *Halichondria panicea*, which can better withstand exposure on the shore.

## Materials and Methods

The sponges were collected at Church Island and Rhosneigr, Anglesey, North Wales, at low tides. Usually they were fixed in 90% industrial spirit on site. Spicule samples were prepared by boiling small pieces of the sponges in fuming nitric acid, washing several times with tap water and allowing a few drops of the spicule suspension to spread over a microscope slide and evaporate to dryness. D.P.X and then a coverslip were added. The lengths and widths of 100 spicules taken in order along a randomly selected transect across the slide were measured using a digitizer and PDP-11 computer. Statistical analysis was carried out with the aid of MINITAB. Further details have already been published (Jones 1984; 1987a,b).

The data derived from specimens collected at about the same date regardless of year were combined to form 14 groups of samples. The date ascribed to each group is the weighted mean, calculated from the actual dates of collection and number of specimens taken on each date. The number of specimens in each group is given in brackets after the dates on Fig. 6.

Local silicate concentrations were derived from the graph published by Ewins and Spencer (1967). Mean temperatures were provided by staff at the Marine Sciences Laboratories, Menai Bridge, Anglesey. Included are the temperature variations published by Walker (1985). The temperatures were taken off the pier at Menai Bridge, Anglesey, during the period 1983 to 1987, which corresponds more or less to the period over which many of the sponges were collected.

## Observations

### *Monthly Changes in the Mean Coefficients of Variation* (*CV*)

The graphs in Fig. 1 show the mean CV-L and mean CV-W for each of the 14 groups of samples. One way analysis of variance (AOVONEWAY) confirmed that there were statistical differences in means between some of the groups (pooled STDEV = 2.155; df = 12; F = 2.05; P = 0.023 for CV-L: pooled STDEV = 7.462; F = 8.64; P = 0.000 for CV-W). Using the group standard deviations, the 95% confidence limits for the means of the individual groups were also determined, and these are indicated on the graphs, excepting the November groups, which included only one specimen. Graphs of the silicate concentration and mean temperature of the local sea water are also plotted.

As reported elsewhere (Jones 1984, 1987b), the values for CV-W are much greater than those for CV-L. The spicule widths are much smaller than the lengths. Also there are greater errors possibly in measuring width from edge to edge across the middle of the spicule than in measuring length from end-point to end-point. Spicule widths furthermore tend to be more influenced by external conditions, as mentioned in the Introduction.

The values of CV-W from day 160 (June 9th) to day 241 (August 29th) are significantly higher than the rest. The short plateau from June 9th to July 8th (day 189) is suggestive (there was a big fall in May and early June in *H. elegans* – Jones 1987a), but more samples would be required to confirm its reality. Changes in mean CV-L to some extent reflect those for CV-W, but the changes are mostly small in relation to the 95% confidence limits. However, the high values on 31st July (day 212) and 9th June (day 160) are significantly different from the low values on 21st February (day 52) and 22nd October (day 295).

The degrees of correlation between the variation in mean CV-W, mean CV-L, temperature and silicate concentration were calculated using the Pearson product moments. The correlation coefficients are shown in Table 1.

As expected, temperature and silicate concentration are inversely correlated, but not strongly so. Temperature is quite strongly correlated with mean CV-W,

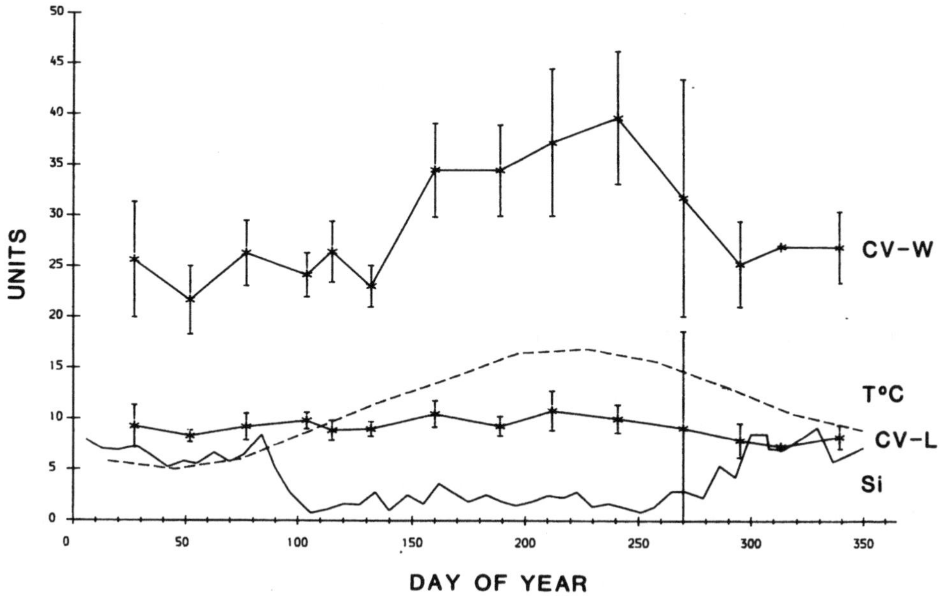


**Fig. 1.** Graphs showing the changes in the mean coefficient of variation for width (*CV-W*) and the mean coefficient of variation for length (*CV-L*) during the year, together with the 95% confidence limits (excepting day 313, for which there was only one sample). Variation in the silicate concentration (Si; units = μg at/l) and mean temperature (T°C) are also shown

**Table 1.** Correlation coefficients

| | Si | T°C | MCV-L | MCV-W | ML | MW | MSD-L | MSD-W |
|---|---|---|---|---|---|---|---|---|
| T°C | -0.603 | | | | | | | |
| MCV-L | -0.574 | 0.384 | | | | | | |
| MCV-W | -0.436 | 0.823 | 0.634 | | | | | |
| ML | 0.284 | -0.588 | -0.181 | -0.489 | | | | |
| MW | -0.025 | -0.586 | 0.149 | -0.487 | 0.730 | | | |
| MSD-L | -0.439 | 0.112 | 0.896 | 0.390 | 0.272 | 0.476 | | |
| MSD-W | -0.430 | 0.371 | 0.764 | 0.619 | 0.133 | 0.360 | 0.798 | |
| MCC | -0.181 | 0.225 | 0.743 | 0.510 | -0.051 | 0.040 | 0.694 | 0.560 |

Key:
Si = silicate concentration in μg at/l (Menai Straits);
T = mean temperature (Celsius) (Menai Straits);
MCV-L = mean coefficient of variation for spicule length per group;
MCV-W = mean coefficient of variation for spicule width per group;
ML = mean spicule length per group;
MW = mean spicule width per group;
MSD-L = mean standard deviation for length per group;
MSD-W = mean standard deviation for width per group;
MCC = mean correlation coefficient (L:W) per group.

but only poorly with mean CV-L. Mean CV-W and mean CV-L are not particularly well correlated. Both mean CV-L and mean CV-W are inversely correlated with silicate, but again not strongly so.

The CV is calculated using the expression:

$$CV = 100 \times \text{standard deviation/mean}$$

A study of the corresponding variations in the mean size and standard deviation would thus facilitate an explanation of the variations in CV.

### *Changes in Mean Width and Length*

For each sample, the mean values of length and width (ML, MW) of the 100 spicules measured were calculated. Average values of these means for each of the 14 groups of samples are given in the graphs in Figs. 2 and 3, together with the confidence limits of the averages based on the group standard deviations. AOVONEWAY (pooled STDEV = 8.64; F = 4.61; P = 0.000 for ML: pooled STDEV = 0.6873; F = 13.72; P = 0.000 for MW) confirmed that the groups are not all from the same population. The spicule number was the same in every sample, so that the average of the mean values is the same as the average calculated from the raw aggregated data in each group.

AOVONEWAY on the raw combined data gives much closer 95% confidence limits for the means than those calculated from the groups of mean values,

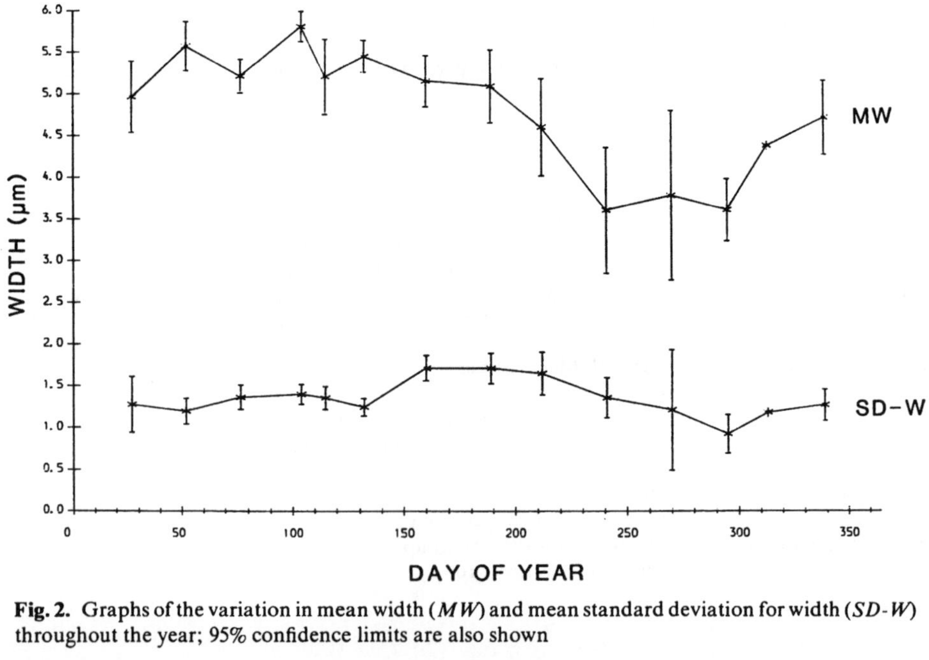


**Fig. 2.** Graphs of the variation in mean width (*MW*) and mean standard deviation for width (*SD-W*) throughout the year; 95% confidence limits are also shown

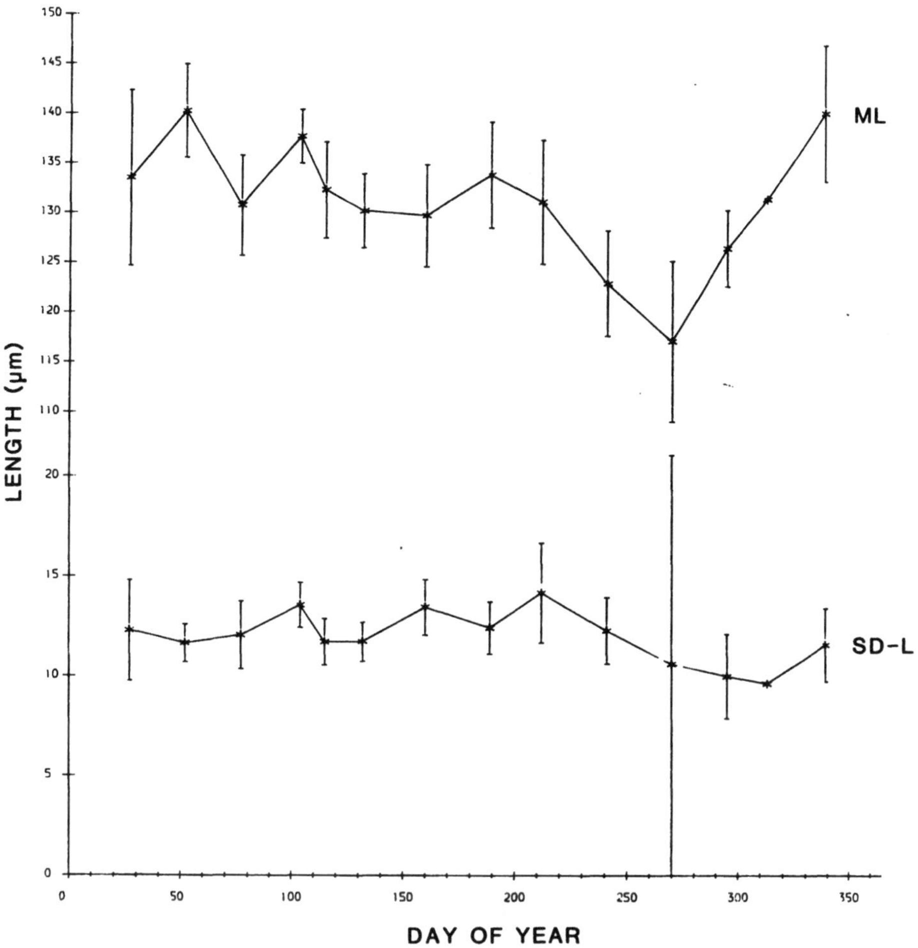


**Fig. 3.** Graphs of the variation in mean length (*ML*) and mean standard deviation for length (*SD-L*) throughout the year; 95% confidence limits are also shown

because pooled standard deviations are used. However, one cannot assume that the raw data conform to the normal distribution in each and every group. It was the sample means that were used in the calculation of CV; consequently the wider confidence limits shown on the graphs are more relevant.

Comparing the graphs for MW and ML, both show peaks on day 52 (21st February) and day 104 (14th April) and minima on day 77 (18th March) and from days 241–270 (29th August–27th September). The two graphs are not identical, however, ML peaking again on day 189 (8th July), and day 339 (5th December), and MW peaking on day 132 (12th May). Also the ML is low on day 160 (9th June), whereas the MW is low on day 115 (25th April).

Table 1 gives the correlation coefficients between MW and ML and between these parameters and other variables. It can be seen that MW and ML are quite well correlated one with the other, as would be expected (Jones 1984). Possibly the coefficient is not closer to 1 because length and width may not be influenced to the same degree by temperature or silicate concentration. However, in *rosea* the correlation coefficients with temperature for both dimensions are the same, indicating that length is influenced inversely by temperature. They are also similar in relation to silicate concentration, but unexpectedly small, bearing in mind the marked effect of silicate on spicule width. However, the samples would usually have included spicules that had formed at an earlier stage in the life history of the sponge. Only exceptionally would they have contained only spicules formed just before the time of fixation. Thus changes in environmental factors affecting spicule dimensions would normally have to persist for some time before the samples would contain sufficient newly forming spicules for the means to be appreciably affected. The graphs reveal that the silicate concentration fell suddenly between day 77 (18th March) and day 104 (14th April) and rose again between day 270 (September 27th) and day 295 (October 22nd), whereas the MW and ML fell steadily from day 132 (12th May) to day 241 (29th August) and started rising after October 22nd. There was a delay of about 30 days between expected cause and effect therefore. Similarly one can detect an apparent phase lag of 30–50 days between the mean size and temperature.

MW and ML are both poorly correlated with mean CV-L, but better correlated with mean CV-W. Standard deviations for length (SD-L) and mean lengths are much larger than the corresponding parameters for width, so that changes in MW would have a relatively greater effect on CV-W than would corresponding changes in ML have on CV-L. And since MW is quite well correlated with ML, changes in ML would show some correlation with CV-W.

### *Changes in the Mean Standard Deviations for Width and Length*

Graphs showing the variation in the mean values of the standard deviations (MSD-L and MSD-W) of the samples in each group are given in Figs. 2 and 3, together with the 95% confidence limits. Again AOVONEWAY (pooled STDEV = 2.668; F = 2.19; P = 0.015 for L: pooled STDEV = 0.3029; F = 7.66; P = 0.000 for W) indicated that the means were not all the same.

The October 22nd MSD-L (day 295) was significantly different from the means for April 14th (day 104), June 9th (day 160) and July 31st (day 212): for W, the MSD-W for June 9th and July 8th (day 189) were greater than the rest, while the MSD-W for October 22nd was lower than the means for March 18th (day 77), April 14th (day 104) and 25th (day 115), May 12th (day 132) and August 29th (day 241). Comparing the graphs for MSD in Figs. 2 and 3, one can see that the two are in general agreement, and this is confirmed by the high correlation coefficient (Table 1). In the period April 14th to 31 July there was greater variation in the spicule dimensions, whereas in October the variation was less. The cause of these differences should be intelligible after a study of the frequency distributions.

The graph for MSD-L is very like that for mean CV-L and the agreement is confirmed by the high correlation coefficient (0.896). There is less agreement between the MSD-W and MCV-W (0.619), the former falling sooner after the July peaks. As both SD and CV are measures of variation, the good correlation between them is only to be expected.

The MSD-L and MSD-W values are poorly correlated with ML and MW. The values are low in October and high on April 14th in all four cases, but while one would expect greater variation with increasing mean size, it is clear that other factors are at work. Again the frequency distributions should throw light on the problem.

The mean SD's are inversely, but poorly, correlated with silicate concentration. Correlation with temperature is positive but very poor. Again a lag between cause and effect is probably responsible.

### *Correlation of Spicule Length with Width*

While one would expect that length and width would be well correlated with each other (Jones 1984), the correlation coefficients between L and W of the samples were found to vary considerably. The values obtained for the 168 samples were normally distributed, the mode being in the range 0.7 to 0.8. One sample had a coefficient above 0.9 (0.905; August 1st) and another a value below 0.1 (0.056; June 11th). Plots of width against length in the two cases are given in Fig. 4. It can be seen that poor correlation results from an absence of small and large spicules in the sample. When spicule production has been minimal or the sample does not include fully grown spicules formed at a time of high silicate concentration and low temperature, one would expect the correlation coefficient to be close to zero. Spicule size might also be influenced by genotype. Some samples in the groups for October, March, May and June had length maxima below 140 μm, like some samples in the groups from July to September, whereas other samples in almost all of the groups had maxima exceeding 170 μm.

The regression equation for width on length for the sample with the highest correlation coefficient (0.905) is:

$$W = -9.63 + 0.105 \times L$$

and for length on width:

$$L = 100 + 7.83 \times W.$$

The monthly variations in the mean correlation coefficients (MCC) between length and width are shown by the graph in Fig. 5. AOVONEWAY (pooled STDEV = 0.1459; F = 2.77; P = 0.002) confirms that some means are significantly different from the rest. The MCC for October 22nd (day 295) is lower than the means for December 5th (day 339), January 27th (day 27), April 14th (day 104), June 9th (day 160) July 31st (day 212) and August 29th (day 241). The mean for April 25th (day 115) is lower than for June 9th and July 31st. The low values in October and on April 25th do not arise from a paucity of juvenile

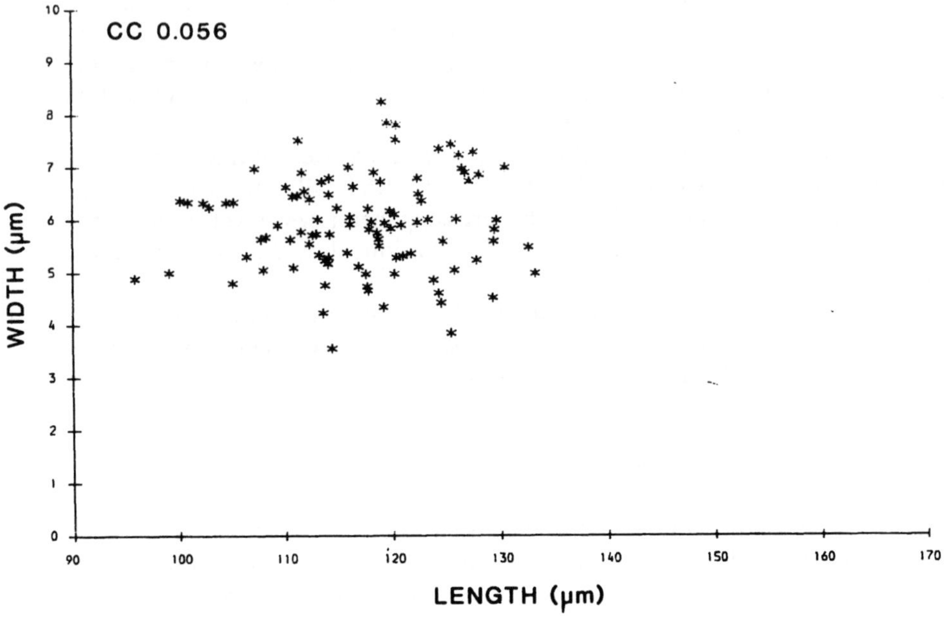


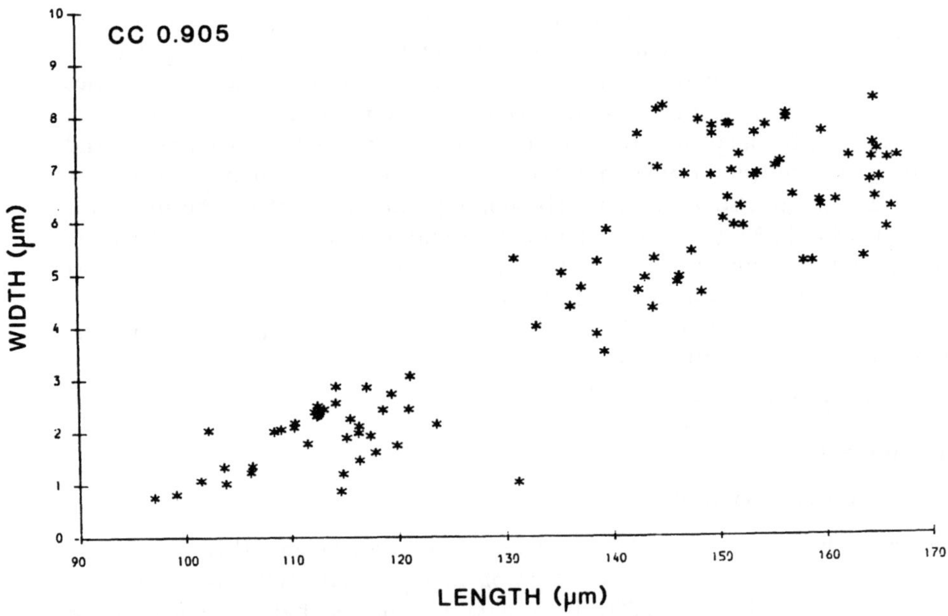


**Fig. 4.** Plots of spicule width against spicule length for a sample in which the correlation coefficient was 0.056 (above) and 0.905 (below)

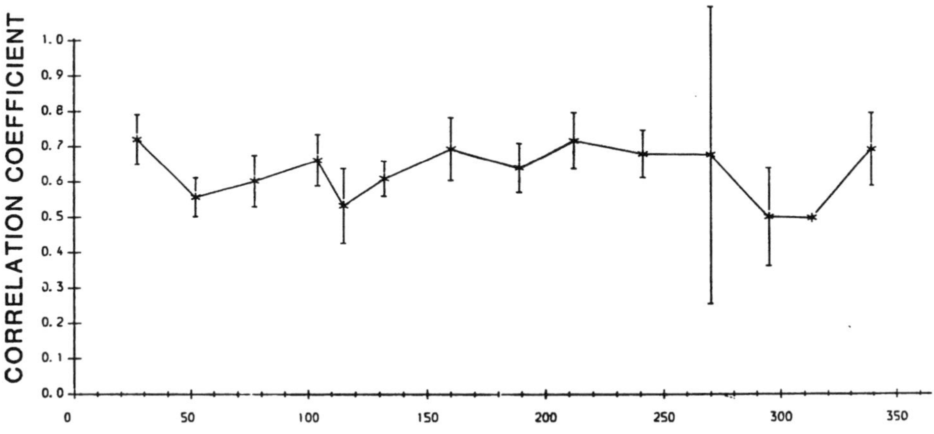


**Fig. 5.** Graph showing the variation in the mean correlation coefficient between width and length throughout the year; 95% confidence limits (excepting for day 313) are also depicted

spicules in the samples (see histograms below), but rather from low maximum values of length (163 μm for October and 169 μm for 25th April).

Table 1 indicates that MCC is moderately well correlated with MSD-L and MSD-W, but not with ML or MW. The greater the SD, the greater the size range for L and W, and hence the better the CC. Correlation of MCC with both silicate concentration and temperature is poor, whereas with MCV-L and MCV-W it is better. SD-L and SD-W are quite well correlated with respectively CV-L and CV-W, both being measures of variation; hence the moderate correlation between CC and CV.

## *Histograms for Lengths and Widths*

Figure 6 indicates the relative frequencies of different size categories in the combined raw data for each group. The size intervals are 1 μm for width and 10 μm for length. Because spicule growth involves the deposition of the spicopal on an organic axial filament of uniform width and characteristic length, the spicule lengths and widths are never less than about 60 μm and 0.5 μm respectively.

The frequency distributions would not be expected to conform to the normal distribution, because the presence of developing spicules in the samples would introduce negative skewness. During periods of zero spicule production, however, the distributions should approximate to the normal.

The Mann-Whitney two sample rank test for the difference between population medians carried out on the 14 groups of aggregated data taken in order indicate that highly significant differences occur, with the exception of June 9th-July 8th and September 27th-October 22nd for width, and May 12th-June 9th

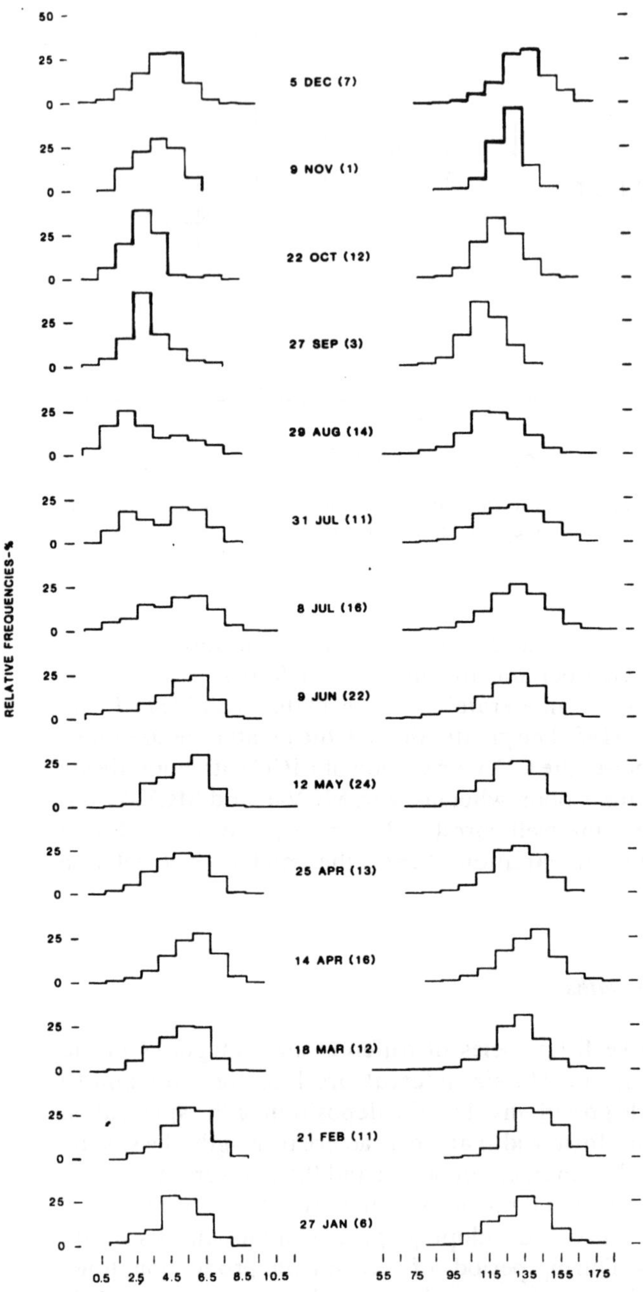


**Fig. 6.** Relative frequency histograms for width and length for the 14 groups of samples. The mean date of collection for each group is given, together with the number of specimens in each group (*in parentheses*). Intervals of 1 μm (width) and 10 μm (length) were used and the mid-points are indicated along the *abscissa*

for length. AOVONEWAY likewise suggests that the means are different, excepting June-July and August-October for width and May-June for length, although, as stated earlier this test is strictly not applicable since the distributions do not all conform to the normal.

The chi-squared test, comparing the categories of raw data in the histograms taken in pairs one after another, indicated that the differences between the frequency distributions were significant at $P < 0.001$, with the exception of the width histograms for November and December. This is hardly surprising considering that the November group comprised only one sample. In the chi-squared test the categories at either end of the distributions with 0 or very few spicules were neglected to avoid having "expected" frequencies of $< 5$ in more than 20% of the categories, or $< 1$ in any category.

The length and width histograms for January 27th may be taken as a starting point. Both are unimodal, the mode for length being in the 130–140 μm interval and that for width in the 4–5 μm interval. The distributions are not symmetrical, there being an apparent shortage of spicules of 3.5 μm width and 125 μm length.

Comparing these distributions with those for February 21st (25 days later), one sees that in the latter the mode for width has shifted to the 5.5 μm interval and that the intervals to the right of the mode have all increased in height, whereas those to the left have all decreased. As regards length, the mode is still at 135 μm, but again the intervals to the right have increased, while those to the left have decreased. Clearly spicule production has declined between January 27th and February 21st, the samples of the latter having proportionally less smaller and more larger spicules.

Between February 21st and March 18th (25 days later) the converse has occurred, for the categories up to 4.5 μm width and 135 μm length have all increased, while those above 4.5 μm and 135 μm have decreased. Spicule production had thus increased.

Between March 18th and April 14th (interval 27 days) spicule production appears to have slackened; the 0.5 and 1.5 categories show little change, but now there are fewer spicules in each of the 2.5, 3.5, 4.5 and 5.5 categories and more in the 6.5 (now the mode) and higher categories. Spicules shorter than 80 μm are not present and there are fewer in the length intervals up to 145 μm, which has become the mode, while to the right of the mode there are increases in all categories.

The same alternation in pattern continued from April 14th to April 25th (interval 11 days), the modes having moved back respectively to 5.5 and 135 μm and the levels having all risen to the left and fallen to the right of the modes. Clearly, spicule production had increased during this period.

Between April 25th and May 12th (interval 17 days) it seems that spicule production declined somewhat; for width, the levels were slightly lower in the 0.5–2.5 and 4.5, and higher in the 5.5 and 6.5 (the mode), categories, suggesting fewer juvenile spicules. However, for length there were increases in the 105 and 115, and decreases in the 145 and 155 μm, categories, which is inconsistent. The 3.5, 7.5 and higher intervals for width maintained the same levels, so that possibly there is here an indication of spicules tending to achieve shorter lengths and widths, as a result of the diminished silicate concentration in the sea water.

In June 9th (28 days later) the smaller categories have relatively more spicules than on May 12th. The levels are much higher from 0.5 to 2.5 μm as regards width and from 65 to 105 μm as regards length. Spicule production was in progress. However, a small secondary mode has appeared at 1.5 μm and the levels are lower, from 3.5–5.5 and 105–125 μm respectively, suggesting that spicule production had increased shortly before June 9th and after a period of diminished production. The principal modes are in the same intervals and the categories to the right of the modes are much the same or slightly less in height.

The width histogram for July 8th (29 days later) is distinctly bimodal, with a new mode at 3.5. The levels at 5.5 and 6.5 μm have diminished in height, as have those of the higher categories. However, there are fewer spicules in the 0.5–2.5 categories, suggesting a diminished spicule production. This is confirmed by the length histogram, in which the categories up to 125 μm show relatively fewer spicules while those to the right of the mode show relatively more. The additional mode can be explained by the formation of spicules of smaller thickness than those produced in the winter.

The bimodality is more evident in the width histogram for July 31st (23 days later). Both width modes have shifted to the left into the 2.5 and 5.5 μm intervals respectively. The levels have decreased for the categories above 5.5 μm. For length, while the levels are the same for 75–95 μm, those for 105 and 115 μm have increased. Spicule production seems to have continued, but thinner, shorter spicules are being formed.

By August 29th (29 days later) the histograms have become quite different. Now the 2.5 μm mode predominates and the mode for length has moved to the 115 μm interval. Spicule production appears to have increased, the fall in levels in categories above 4.5 μm and 135 μm indicating that the samples contain relatively few large, thick spicules.

From August to September 27th (interval 29 days) there appears to have been a recovery to a limited extent in the fully grown spicule size. The mode for width has moved from 2.5 to 3.5 μm and while the length mode is still at 115 μm, its height is lower and there has been an increase in the 125 μm category. Spicule production had increased because there are fewer spicules in the higher categories for both length and width. However, recently formed spicules had grown to a larger size, as is evident from the decrease in levels of the first three categories for both length and width.

From September 27th to October 22nd (interval 25 days) there was little change, but the trend towards recovery appears to have continued. The mode for length has moved to the 125 μm interval and the levels have increased in the categories from 135–165 μm. As regards width there was an increase in the 4.5 μm category.

Only one specimen was included in the November 9th group, so that one cannot draw reliable conclusions. However, the November histograms are in between those for October and December, indicating that the recovery trend was continuing. The modes for November have shifted to 4.5 and 135 μm respectively.

December 5th has modes at 5.5 μm (very slightly above the level for 4.5 μm) and 145 μm, a little above the 135 μm level. The distributions are negatively skewed with uniformly increasing steps to the left and uniformly decreasing steps to the right of the modes. Compared with the histograms for January there were more spicules in the 3.5 and less in the 6.5 and 7.5 μm categories, but less in the 105–125 μm and more in the 145–165 μm categories. Such differences are difficult to explain, unless variation due to small sampling is the cause.

Three main conclusions may be drawn from the changes in frequency distributions. First, the rate of spicule production is not constant throughout the year. The evidence suggests possibly a rhythmic rise and fall in production, at least for much of the year. Secondly, the histograms show a tendency for thin, fully grown spicules to be produced during the summer. This has been confirmed by examination of sections of the sponges. These exhibit thin spicules incorporated in the skeletal framework in specimens collected from June to September (Jones 1987b). Thirdly, it is clear that external conditions are influencing spicule length as well as width, despite the experimental evidence to the contrary, obtained using other species.

### *The Relation Between Width and Length in March and August*

Figure 7 shows plots of width against length for all of the raw data in the March and August groups respectively. Whereas the points are more or less concentrated in one dense area in the graph for March, centred about the 5.5 μm: 135 μm point, an additional dense area is present on the August graph, centred at about 2.5 μm: 115 μm. The two dense areas here correspond to the modes of the histograms referred to above. They do not appear to be in line with the general trend of the points, suggesting that the width:length relationship for the spicules formed in the late summer months is not precisely the same as that for the spicules formed in late winter and early spring.

Both graphs reveal a great deal of scatter. Thus, on the August graph widths of approximately 0.5 μm were associated with lengths of from 65 to 105 μm, and spicules of about 120 μm length have widths ranging from 1 to 8 μm. Less variation was shown by the plots for individual samples in Fig. 4. Here lengths of 120 μm were associated with widths of about 3.5 to 8 μm in one case and about 1.5 to 3 μm in the other. Presumably much of the additional variation in the graphs for the aggregated data is genotypic in origin, because the specimens were derived from only two collecting sites, indistinguishable on average by spicule size. With so much variation, it is clear why the correlation coefficient (W:L) should never have a value of 1.0. Spicule dimensions depend upon the length of the axial organic filament when silicification commences, and this can vary from 65 to 105 μm judging from the August example in Fig. 7. Presumably they also depend upon whether the organic axial filament continues growing while silicification proceeds (giving rise to the "open" type of spicule), or whether the ends of the axial canal become "closed" by spicopal during growth (Jones 1979).

The band of points in Fig. 7 (August) tapers to the left near the origin, because of the variable length of the organic axial filament when silicification begins. It also tends to taper to the right at the top. Linear regression analysis is not

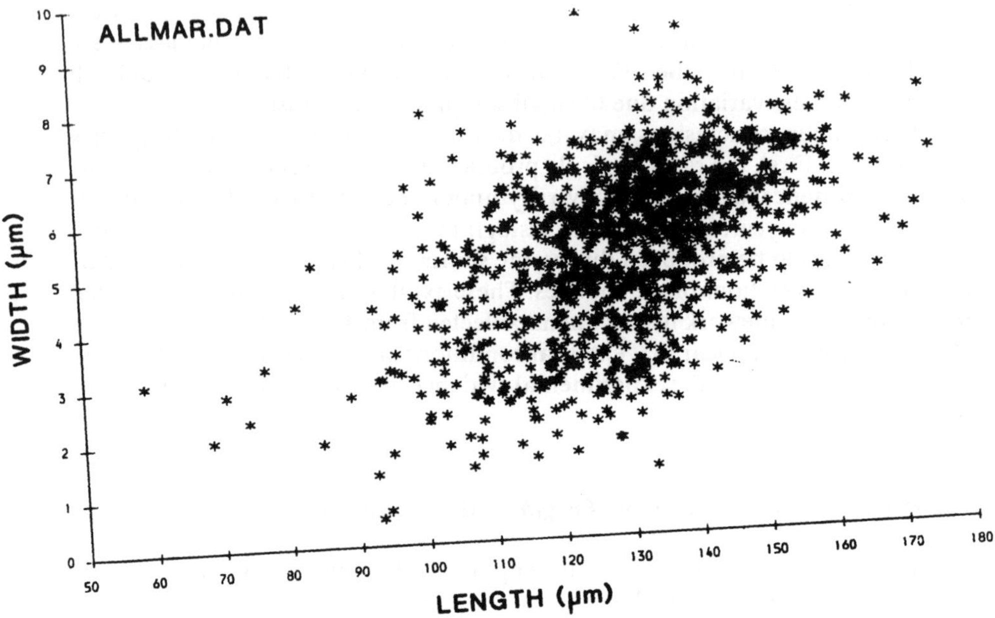


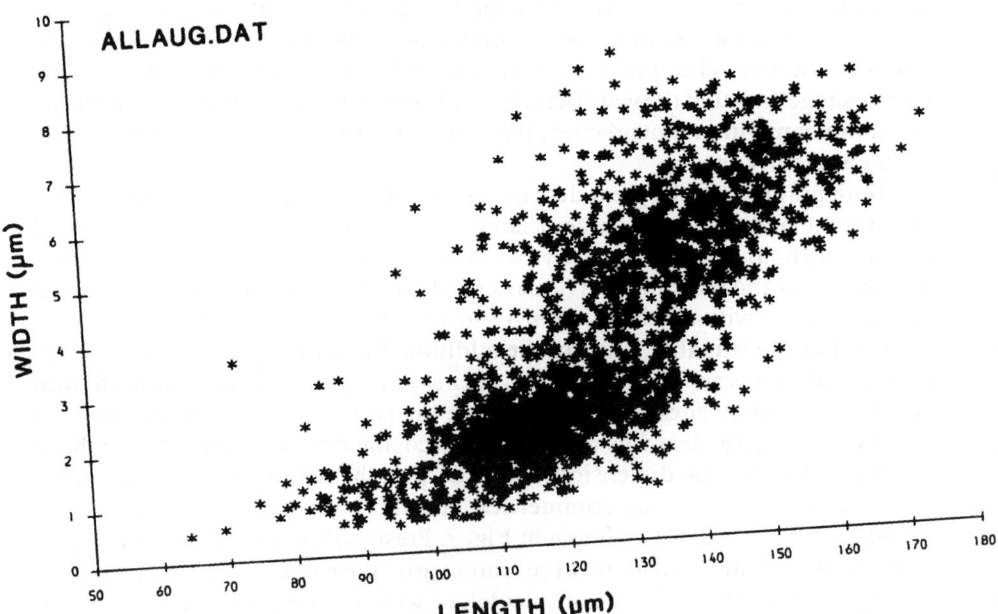


**Fig. 7.** Plots of width against length for all spicules in the March group (*above*) and August 29th group (*below*)

facilitated by this lack of uniformity. However, the regression equations for the two graphs in Fig. 7 have been calculated. Respectively the equations for width on length are:

$$W = -1.10 + 0.0482 \times L$$
$$W = -7.18 + 0.0876 \times L$$

and for length on width are:

$$L = 104 + 5.06 \times W$$
$$L = 102 + 5.80 \times W.$$

The correlation coefficients (W:L) are 0.494 and 0.713, respectively.

## Discussion

It has been shown that significant differences in the CV-W and CV-L occur, depending on the date of collection of the sponge specimens. It is unlikely that such CV's would be an aid to species identification of temperate sponges in consequence. The variation is caused by changes in both the mean size and mean standard deviation, the latter having the greater effect.

Correlation between MW or ML and the silicate concentration was very poor, contrary to the results of Stone (1970). However, he used the tips of the papillae of *Hymeniacidon perleve* as his source of spicule samples and as these would presumably have been newly grown, the spicules would have better reflected the degree of silicate concentration at the date of collection. It is still difficult to understand how the recovery (increase) in spicule dimensions could have preceded the rise in silicate concentration, but possibly the data on the latter concerned a region insufficiently close to that from which the sponges were collected. A lag between change in spicule parameters and temperature or silicate concentration is more intelligible, bearing in mind that *H. rosea* can live for more than a year, so that some specimens at least would yield samples containing spicules formed during different seasons.

Clear evidence that spicule length, as well as width, is influenced by external conditions is shown by the inverse correlation of ML with temperature and the close correlation between length and width (Table 1), by the histograms (Fig. 6) and by the plots in Fig. 7 (August). When thin, fully grown spicules are formed during the summer, they tend to be also shorter. This is somewhat unexpected in view of the experimental evidence, albeit on different siliceous sponge species, published by Simpson et al. (1985) and Simpson (1978). The relation between width and length is not precise, however, in that the organic axial filament can have a variable length when silicification begins.

The evidence for an apparent rhythmical production of spicules over much of the year and even when the silicate concentration is low, is of interest and needs further investigation. Preferably the same specimens should be monitored at intervals to confirm the rhythmicity of spicule production. A decline of spicule production in June was detected in specimens of *Haliclona elegans* (Jones 1987a), and was attributed to the onset of sexual reproduction. Gametogenesis and embryo formation would induce a heavy drain on the energy resources of the sponge, so that a concomitant decline or cessation of growth and spicule production would be expected. Whether the changes in the latter would continue to be rhythmic for more than a limited period is questionable, however.

Knowledge of the seasonal variations in spicule size and in the appearance of the skeletal framework in sections is of considerable importance to sponge taxonomy in view of the use made of skeletal characteristics in sponge identification and classification. The appearance of the skeleton as seen in vertical sections and surface slices can vary greatly, as shown by the plates already published for *H. elegans* (Jones 1987a) and *H. rosea* (Jones 1987b).

## References

Ewins PA, Spencer CP (1967) The annual cycle of nutrients in the Menai Straits. J Mar Biol Assoc UK 47:533–542

Hartman WD (1958) Natural history of the marine sponges of Southern New England. Peabody Mus Nat Hist Yale Univ Bull 12:1–155

Jones WC (1979) The microstructure and genesis of sponge biominerals. In: Levi C, Boury-Esnault N (eds) Biologie des Spongiaires. Colloq Int CNRS No 291:425–447

Jones WC (1984) Spicule dimensions as taxonomic criteria in the identification of haplosclerid sponges from the shores of Anglesey. Zool J Linn Soc 80:239–259

Jones WC (1987a) Seasonal variations in the skeleton and spicule dimensions of *Haliclona elegans* (Bowerbank) sensu Topsent (1887) from two sites in North Wales. In: Jones WC (ed) European contributions to the taxonomy of sponges. Sherkin Island Marine Station Publ 1:109–129

Jones WC (1987b) Skeletal variation in embryo-containing specimens of *Haliclona rosea* (Bowerbank) from Anglesey, North Wales. In: Vacelet J, Boury-Esnault N (eds) Taxonomy of Porifera from the NE Atlantic and Mediterranean Sea. NATO ASI Ser G: Ecol Sci 13:101–124

Palumbi SR (1986) How body plans limit acclimation: responses of a demosponge to wave force. Ecology 67:208–214

Simpson TL (1978) The biology of the marine sponge *Microciona prolifera* (Ellis and Solander). III. Spicule secretion and the effect of temperature on spicule size. J Exp Mar Biol Ecol 35:31–42

Simpson TL, Gil M, Connes R, Diax J-P, Paris J (1985) J Morphol 183:117–128

Stone AR (1970) Seasonal variations of spicule size in *Hymeniacidon perleve*. J Mar Biol Assoc UK 50:343–348

Walker G (1985) The cypris larva of *Sacculina carcini* Thompson (Crustacea: Cirripedia: Rhizocephala). J Exp Mar Biol Ecol 93:131–145

# Vicariance Biogeography Using North Atlantic Chalinidae (Demospongiae)

W.H. de Weerdt

## Introduction

The present chapter shows that it is possible to apply the methods of vicariance biogeography to North Atlantic shallow-water Chalinidae (Haplosclerida, Demospongiae). Sponges are known to be difficult tools for biogeographic studies. This is primarily due to their complicated taxonomy, which in turn may be ascribed to their very old geological age, and the apparent evolutionary trend towards simplicity (Vacelet 1985). This applies especially to chalinid species, which are of a discouraging simplicity for the taxonomist. The reason why this group has been selected for biogeographical analyses following the strict "rules" of vicariance biogeography is simply the fact that I am most familiar with these sponges, albeit predominantly with the North Atlantic species. Through taxonomic studies based on type material and freshly collected specimens, and by scrutinizing all the references in the literature, it appeared possible to establish the North Atlantic distribution of the chalinid species with a sufficiently high degree of reliability (de Weerdt 1986, 1987, 1989). The next steps towards analyses of the historical relationships of the North Atlantic endemic areas based on analyses of the phylogenetic relationships of the chalinid species and their distribution patterns appeared to be well justified and a novelty (Zandee pers. commun.). The fully elaborated analyses will be presented in a forthcoming paper (de Weerdt 1989). Here, an example of the procedure is given, based on reduced but still real data. Vicariance biogeography (also called historical biogeography, cladistic biogeography or area cladistics) combines the phylogenetic relationships of organisms with their distributional patterns, in order to infer the historical relationships of areas of endemism. It is the direct result of the principles of phylogenetic systematics as formulated by Hennig (1966), and the recognition of congruent, but disjunct distribution patterns in several unrelated organisms by authors ("generalized tracks"; cf. Rosen 1975; Wiley 1981, and references in Humphries and Parenti 1986).

Vicariance biogeography is based on the assumption that organisms and the biota inhabited by them share the same history, namely vicariance events (fragmentation of an ancestral biota into two or more isolated biotas). Evolutionary novelties, normally acquired by members of a population but mixed up with the other members through gene flow, may lead to new, genetically isolated species when a vicariance event has fragmented the ancestral biota (Wiley 1981).

J. Reitner and H. Keupp (Eds.)
Fossil and Recent Sponges

The two daughter species will be more closely related to each other than each of them is to a third species, and the same holds for the areas inhabited by them. Analysis of the historical relationships of areas is, therefore, possible through the phylogenetic analysis of the relationships of species or higher taxa.

The principles of vicariance biogeography were for the first time formulated by Platnick and Nelson (1978). A methodological application on poecillid fishes of Central America has been carried out by Rosen (1978) (for discussions concerning Rosen's method see Nelson 1984; Humphries and Parenti 1986; Zandee and Roos 1987).

The procedure leading to a cladistic biogeographic analysis follows the next general steps:

1. Taxonomic revision (if necessary) of the groups under study.
2. Establishment of monophyly and analysis of the phylogenetic relationships of the groups, resulting in cladograms of species or higher taxa.
3. Establishment of the distributions of the taxa, and of areas of endemism ("the smallest relevant biogeographical entity", cf. Zandee and Roos 1987)
4. At least two phylogenetic analyses of unrelated taxa, inhabiting the same endemic areas, are necessary in order to reach to the recognition of a general pattern in the distributions of the taxa.
5. Replacement of the taxon names in the cladograms by their areas. This procedure leads to area cladograms.
6. Generation of a general area cladogram. The general area cladogram is a hypothesis on the historical relationships of the areas, and is composed of the recent areas at the tips of the branches, and of composite ancestral areas at the internal nodes.
7. Comparison of the general area cladogram with paleogeographical, paleoclimatological and paleoecological data.
8. Evaluation of the individual taxon cladograms in order to establish congruent patterns with the general area cladogram ("vicariance fit"), and incongruent patterns which need ad hoc explanations (dispersal, extinction, etc.).

The generation of a general area cladogram is the most crucial step in the procedure, and is strongly influenced by assumptions concerning widespread species (species occupying more than one endemic area, cf. Nelson and Platnick 1981; Wiley 1987a, and Zandee and Roos 1987). "Assumption zero" is followed here (cf. Zandee and Roos 1987), which is the assumption that when a species or higher taxon occupying more than one endemic area turns out to be polytypical, the resulting two species or higher taxa will always be more closely related to each other than each of them is to a third taxon. The same applies to the areas occupied by them. In other words, the phylogenies as analyzed by the investigator are primarily considered unambiguous. An important conseguence of this assumption is the fact that composite distributional types (i.e., distributions including more than one endemic area) of both recent and ancestral taxa can be used as "fixed" building blocks for the construction of a general area cladogram.

Under other assumptions (cf. Nelson and Platnick 1981), the phylogenies which include widespread species are considered doubtful, and composite distributional types are not treated as fixed units. This may lead to partly or completely unresolved general area cladograms (Zandee and Roos 1987).

## Phylogeny and Vicariance Biogeography of North Atlantic Chalinidae

Recent taxonomic studies on North Atlantic shallow-water Chalinidae (currently known as Haliclonidae) by the present author have included phylogenetic analyses of the species relationships within the family (de Weerdt 1987, 1989). On the basis of presumed phylogenetic characters eight monophyletic groups have been established, three of which belong to the nominal genera *Haliclona* Grant, *Acervochalina* Ridley and *Dendroxea* Griessinger. For the analyses of the phylogenetic relationships of the species groups, the two-step procedure of outgroup comparison and parsimony (finding a cladogram with a minimal number of evolutionary steps) has been followed. For literature concerning this subject one is referred to Watrous and Wheeler (1981), Maddison et al. (1984), and Wiley, (1987b).

The characters used for the analyses were confined to morphological characters of the mature sponges, e.g., the habit, ectosomal and choanosomal skeletal architecture, size, and shape of the megascleres, the amount of spongin, consistency, color, the presence of fistular outgrowths, etc. Many of these characters or character states are not confined to chalinid sponges, but occur outside the group as well. Parallelism may quite well be a common feature in sponges, at least when such simple characters as skeletal architecture and color are concerned. Reversals and homoplasious developments of all character states were, therefore, accepted, except for the microscleres (sigmata, toxa, raphides, microxea). The first assumption allows for an analysis using Wagner parsimony (losses and gains are equally probable, cf. Kluge and Farris 1969; Felsenstein 1985). The second assumption is based on the idea that complex features are lost more easily than gained (Ross 1974; Farris 1977; Felsenstein 1985; Pimentel and Riggins 1987). Microscleres are considered to fall into this category of evolutionary characters, although they are certainly not all equally complex.

For the establishment of character polarity within the Chalinidae, the Niphatidae, Callyspongiidae, Petrosiidae and Oceanapiidae were used as outgroups. The phylogenetic analyses were carried out using the computer program Tree Tools (Ellis 1986). Tree Tools utilizes Wagner parsimony only, but it allows for all possible transformations series in multistate characters (cf. Mickevich 1982), including closed loops, and this situation was quite regularly met with.

For the analyses all well-known North Atlantic chalinids were used (57 species), based on the recent taxonomic studies of Griessinger (1971, Mediterranean), van Soest (1980, West Indies), Boury-Esnault and Lopes (1985, Azores), de Weerdt (1986, North East Atlantic), de Weerdt and van Soest (1986, southeastern part of the North Atlantic), and van Lent and de Weerdt (1987, Banyuls,

Mediterranean). For a detailed treatment of all characters and the phylogenetic analyses one is referred to de Weerdt (1989).

According to the distribution of the well-known chalinids in the North Atlantic, the following areas of endemism could be established: Arctic (one endemic species), North West Atlantic (one endemic species), Caribbean (nine endemic species), the southeastern part of the North Atlantic (including the Azores, Madeira, Canary Islands, Cape Verde Islands, and the northwest coast of Africa (five endemic species), Mediterranean (15 endemic species), and eastern Boreal (one endemic species).

Of these areas, the Caribbean, eastern Boreal, Mediterranean and the southeastern part of the North Atlantic are used here in the presented example of vicariance biogeography. Two of the eight presumed monophyletic chalinid species groups are used, viz. the *acervochalina* group (species currently known as *Acervochalina* or under the invalid genus name *Chalinula*, cf. de Weerdt 1986) and the *fistulosa* group (species currently known under the invalid genus name *Pellina*, cf. de Weerdt 1986). Only part of the original cladograms are used here in order to show that the construction of a general area cladogram, on the basis of small data sets, can be carried out manually, thus without using a computer. The procedure is based on the component-compatibility method developed by Zandee and Roos (1987), for which the program CAFCA (Zandee 1987) is available.

In Fig. 1a and b the reduced cladograms of the *acervochalina* and *fistulosa* groups are shown. The synapomorphous characters defining the *fistulosa* group as a monophyletic group are: (1) Massive habit (unique within the Chalinidae); (2) firm/friable consistency (unique within the Chalinidae); (3) choanosomal skeleton a rigid, subisotropic reticulation (unique within the Chalinidae); (4) no microscleres (not unique).

Characters 5b and 5c (orange-red and light-orange color, respectively) are part of a transformation series with more states (by outgroup comparison purple

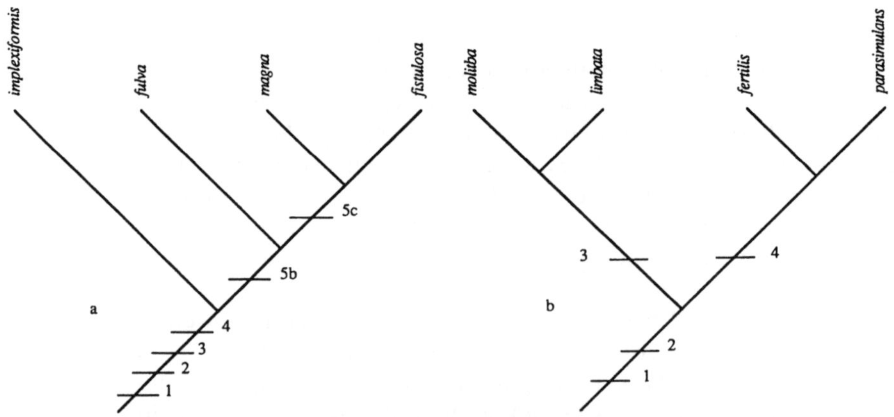


**Fig. 1.** Cladograms of the *fistulosa* **(a)** and *acervochalina* **(b)** groups

is presumed to be the plesiomorphous state 5a). Here, only the relevant states are shown.

The characters defining the *acervochalina* group as a monophyletic group are: (1) Secondary lines more than one spiculum long. Unique within the Chalinidae, but the state occurs in the other haplosclerid families as well. By parsimony it has been generated that the state is a reversal in the *acervochalina* group. It may, however also be considered as an "underlying synapomorphy" (Saether 1983). (2) No ectosomal skeleton (not unique). Character 3 is an synapomorphy of *A. molitba* and *A. limbata* and is a high variability in spicule size and amount of spongin. Character 4 is a synapomorphy of *A. fertilis* and *A. parasimulans* and is a spicule length over 120 μm (not unique).

In Fig. 2a and b the distributions of the species are added to the cladograms, by which they are converted into species area cladograms. The internal nodes, which represent presumed ancestral species, are given the letters a-f; the ancestral distributional types follow directly from the recent distributional types by adding the ranges of the daughter species (cf. Wiley 1981; Zandee and Roos 1987).

From the recent and ancestral distributional types the following eight components (building blocks for construction of the general area cladogram) can be recognized (see also Table 1): 1–4: the single areas 1 (Caribbean), 2 (southeastern part of the North Atlantic), 3 (Mediterranean), and 4 (Boreal); 5: 2,3; 6: 3,4; 7: 2,3,4 and 8: 1,2,3,4 (the entire ancestral area, consisting of all recent areas). Under Assumption zero, only the components which are compatible (mutually ex- or inclusive) may be combined to form a general area cladogram. Two alternative general area cladograms can be formed, viz:

1. ((2,3)4)1)
2. ((3,4)2)1)

The next step in the procedure is a feedback of the two general area cladograms to the individual species area cladograms, in order to establish the numbers of "fit" and "nonfit" of each species area cladogram with the general area clado-

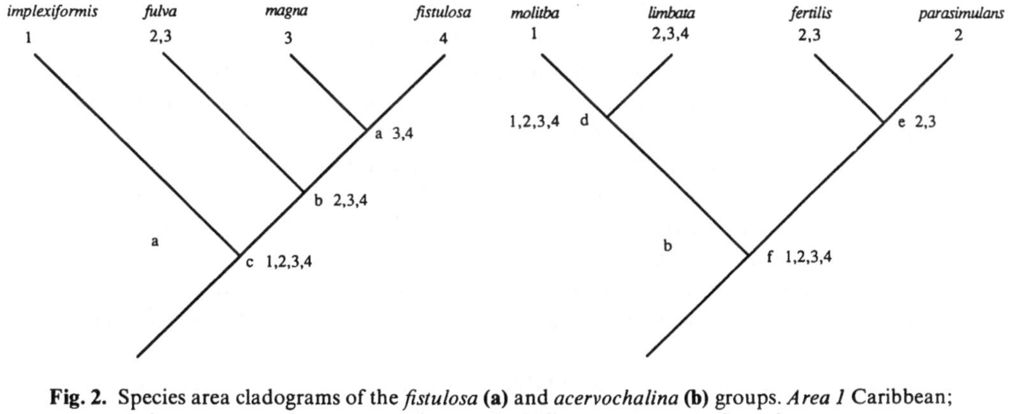


**Fig. 2.** Species area cladograms of the *fistulosa* (**a**) and *acervochalina* (**b**) groups. *Area 1* Caribbean; *area 2* southeastern part of North Atlantic; *area 3* Mediterranean; *area 4* Boreal

**Table 1.** Components (distributional types), which can be recognized from the species area cladograms of the *fistulosa* and *acervochalina* groups

| # | Component | Species |
|---|---|---|
| 1 | 1 | *H. implexiformis; A. molitba* |
| 2 | 2 | *A. parasimulans* |
| 3 | 3 | *H. magna* |
| 4 | 4 | *H. fistulosa* |
| 5 | 2,3 | *H. fulva; A. fertilis;* e |
| 6 | 3,4 | a |
| 7 | 2,3,4 | *A. limbata;* b |
| 8 | 1,2,3,4 | c; d; f |

Area 1 = Caribbean; area 2 = southeastern part of the North Atlantic; area 3 = Mediterranean; area 4 = Boreal.

grams. The cladogram with the largest number of "fit" will be selected as the best general area cladogram. This can most easily be done by making a binary data matrix, consisting of all taxa (recent and ancestral) of the species cladograms and the distributional types of all these taxa (Table 2). Each distributional type fitting with a component with one of the general area cladograms is given + 1, nonfitting distributional types are given –1. It is clear that the distributional types consisting of single areas, as well as those consisting of the entire ancestral area, always fit with each general area cladogram. Therefore, they do not contribute to the final result, and could have been left from the data matrix. The result is evident: only one species has a distributional type not fitting with the general area cladogram # 1 (ancestral species a of the *fistulosa* group), whereas three species do not fit with general area cladogram #2. General area cladogram # 1 is therefore selected as the best hypothesis on the historical relationship between areas 1 (Caribbean), 2 (southeastern part of the North Atlantic), 3 (Mediterranean) and 4 (Boreal) (Fig. 3). This hypothesis can be tested by other phylogenetic and biogeographical studies on unrelated organisms inhabiting the same areas.

Through comparison of general area cladogram # 1 with paleogeographical and paleoclimatological data, it is possible to recognize the entire ancestral area as the Western Tethys Sea, before the separation of South America and Africa. The vicariance event leading to the separation of component 8 into component 1 (Caribbean) and component 7 (an ancestral area consisting of the southeastern part of the North Atlantic, the Mediterranean and the Boreal areas) is presumed to be the Cretaceous-early Tertiary widening of the South Atlantic. Through the increasing distance between South America and Africa the southern shallow-water dispersal route along the shelf seas disappeared, and the former continuous biota became fragmented into an eastern (Atlantic-Mediterranean-Indo-West Pacific) fauna and a western Atlantic (Caribbean-eastern Pacific) fauna. A generalized track consisting of closely related eastern and western Atlantic (Caribbean) species in several unrelated groups has already been established by Rosen (1975), and is confirmed by the present analysis.

**Table 2.** Evaluation of the two general area cladograms

| Species | Areas | General area cladogram | | | |
|---|---|---|---|---|---|
| | | #1 | | #2 | |
| | 1 2 3 4 | Fit | Nonfit | Fit | Nonfit |
| *H. implexiformis* | 1 0 0 0 | +1 | | +1 | |
| *H. fulva* | 0 1 1 0 | +1 | | | -1 |
| *H. magna* | 0 0 1 0 | +1 | | +1 | |
| *H. fistulosa* | 0 0 0 1 | +1 | | +1 | |
| a | 0 0 1 1 | | -1 | +1 | |
| b | 0 1 1 1 | +1 | | +1 | |
| c | 1 1 1 1 | +1 | | +1 | |
| *A. molitba* | 1 0 0 0 | +1 | | +1 | |
| *A. limbata* | 0 1 1 1 | +1 | | +1 | |
| *A. fertilis* | 0 1 1 0 | +1 | | | -1 |
| *A. parasimulans* | 0 1 0 0 | +1 | | +1 | |
| d | 1 1 1 1 | +1 | | +1 | |
| e | 0 1 1 0 | +1 | | | -1 |
| f | 1 1 1 1 | +1 | | +1 | |

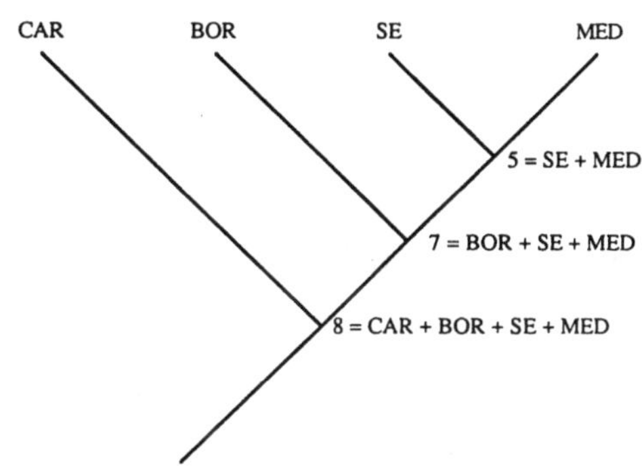


**Fig. 3.** General area cladogram representing the presumed historical relationships between the Caribbean (*CAR* area 1), Boreal (*BOR* area 4), the southeastern part of the North Atlantic (*SE* area 2) and the Mediterranean (*MED* area 3). The numbers *5, 7* and *8* correspond to the component numbers derived from the individual species area cladograms (see Table 1 and text)

More or less simultaneously with the continental drift there was a worldwide climatic cooling during the Cretaceous period, which has led to faunal shifts through replacement by better-adapted species and extinctions of lesser-adapted species (Ekman 1953; Kauffman 1973). It is still ambiguous whether the eastern Boreal area obtained its endemic character during this period, or later as a result of the Pleistocene glaciations (Ekman 1953; Pielou 1979). The general area cladogram suggests the first possibility, because the vicariance at the branching point at component 5 is presumed to be an event earlier than the Pleistocene glaciations. However, more studies on unrelated organisms are needed to test this hypothesis.

Component 5 is presumed to be the Atlantic-Mediterranean connection prior to the Miocene (Messinian) salinity crisis in the Mediterranean basin. During the salinity crisis the Strait of Gibraltar was a landbridge through the isostatic lowering of the Mediterranean sea level as a result of the extremely high evaporation in the Mediterranean basin (Maldonado 1985). This event is presumed to be the vicariance event leading to separation of component 5 into components 2 (south-eastern part of the North Atlantic) and 3 (Mediterranean). There is no fossil evidence to corroborate the hypothesis that chalinid sponges were able to survive the extreme conditions in the Mediterranean during the salinity crisis. However, euryhaline corals (especially *Porites* with aberrant growth forms), bryozoans, and several other marine organisms among which also boring sponges, flourished well in the western part of the Mediterranean during the salinity crisis (Esteban 1979 [1980]; Esteban et al. 1978; Chevalier 1961). It is, therefore, quite plausible that euryhaline chalinds survived and speciated in the (western) Mediterranean during this period.

The Azores, Canary Islands, and Cape Verde Islands originated more or less simultaneously with the Miocene salinity crisis (Dillon and Sougy 1974; Ridley et al. 1974), and the present-day Mediterranean-Mauritanian affinities with respect to sponge faunas can therefore not be placed earlier than the Miocene. It has already been remarked by Boury-Esnault and Lopes (1985) that the eastward surface current at Gibraltar is a strong inhibiting factor for colonization of the Azores, and that there must be a historical explanation for the faunal resemblance. It is here suggested that colonization of the Macaronesian islands has taken place shortly after the salinity crisis, when the surface current has been temporarily directed westward caused by the decreased salinity in the Mediterranean (Maldonado 1985).

## Evaluation of the Individual Species Area Cladograms

The last step in the procedure is evaluation of the individual species area cladograms in order to establish which species conform to the presumed vicariance events, and which species need ad hoc explanations because of their deviating distribution patterns. In the here presented example there is only one distribution pattern which needs an ad hoc explanation, viz. the distribution of the hypothesized ancestor of *H. magna* and *H. fistulosa* in the *fistulosa* group (ancestral species a of Fig. 2a). Its distributional pattern is compatible with component 7 of the general area cladogram (areas 2, 3 and 4), and it is therefore the most parsimonious solution to assume that extinction from area 2 (southeastern part of the North Atlantic) has taken place here.

## Discussion

Sponge biogeography is still in the state of infancy, which appears among other things from the fact that authors of standard biogeographical works (Ekman 1953, and more recently, Briggs 1974; Pielou 1979) have avoided basing their geographical discussions on sponge distributions. The problems in sponge systematics, caused by the morphological plasticity of sponges, inconsistent treatment of type material by authors, poor descriptions, misinterpretation of literature, etc., combined with poor sampling record, are known to be responsible for this.

Recent taxonomic studies including biogeographical sections, like those of Boury-Esnault and Lopes (1985) and Desqueyroux and Moyano (1987), are certainly more trustworthy than those of, for instance, Burton (e.g., 1932, 1956) and Hechtel (1976), which are almost entirely based on a compilation of names. However, to my knowledge, none of these recent studies has gone beyond the descriptive phase, which is only the first step in biogeographical analyses (Ball 1975). I claim to go further than that, by making use of methods which are, in my opinion, the only reliable tools to reach to the analytical part of biogeography (see also Ball 1975). Being hypotheses by themselves, cladograms of species and/or areas of endemism can be falsified by new characters, new sampling records, etc., and there is no risk of being inductive. The here applied component-compatibility method of Zandee and Roos (1987) is only one of the available methods to construct a general area cladogram, and different opinions concerning this method can be found in a recent Issue of Systematic Zoology (e.g., Page 1988).

Vicariance biogeography is a new scientific development with few applications throughout (e.g., Nelson 1984; Wiley 1988), and none of these are based on marine organisms. My attempts, using a difficult group of sponges and "difficult" (largely contiguous) areas, will therefore "au fond" be subject to criticism and/or (hopefully) improvement, by myself or others. Nevertheless, I hope that, even through presentation of these preliminary results, the practice of vicariance biogeography will meet with some enthusiasm among spongologists.

*Acknowledgments.* I would like to thank Rob van Soest for the many discussions, his enthusiasm and encouragement, and for his critical reading of the manuscript. M. Zandee (Institute for Theoretical Biology, University of Leiden, the Nertherlands) made me conscious of the fact that my phylogenetic and biogeographic studies of the chalinids were useful tools for the program CAFCA. Klaus Ruetzler and an anonymous referee provided valuable comments on the manuscript. I am grateful to Joachim Reitner and Helmut Keupp for inviting me to participate in the conference.

## References


Ball IR (1975) Nature and formulation of biogeographical hypotheses. Syst Zool 24 (4):407–430

Boury-Esnault N, Lopes T (1985) Les Démosponges littorales de l'Archipel des Açores. Ann Inst Océanogr 61 (2):149–225

Briggs JC (1974) Marine Zoogeography, 1. Marine faunal geographical distributions. McGraw-Hill, New York, pp 461

Burton M (1932) Sponges. DISCOVERY Rep 6:327–392

Burton M (1956) The sponges of West Africa. Atlantide Rep 4:111–147

Chevalier JP (1961) Recherches sur les Madréporaires et les formations récifals miocènes de la Méditerranée occidentale. Mem Soc Geol Fr NS 93:1–562

Desqueyroux R, Moyano H (1987) Zoogeografia de Demospongias Chilenas. Bol Soc Biol Concepcion Chile 58:39–66

de Weerdt WH (1986) A systematic revision of the north-eastern Atlantic shallow-water Haplosclerida (Porifera, Demospongiae), part II: Chalinidae. Beaufortia 36 (6):81–165

de Weerdt WH (1987) The shallow-water Haplosclerida (Porifera, Demospongiae) of the north east Atlantic Ocean: taxonomy, phylogeny and historical biogeography PhD thesis, pp 243

de Weerdt WH (1989) Phylogeny and vicariance biogeography of North Atlantic Chalinidae (Haplosclerida, Demospongiae). Beaufortia 39 (3):55–95
de Weerdt WH, van Soest RWM (1986) Marine shallow-water Haplosclerida (Porifera) from the south-eastern part of the North Atlantic Ocean
Dillon WP, Sougy JMA (1974) Geology of West Africa and Canary and Cape Verde Islands. In: Nairn AEM, Stehli FG (eds) The ocean basins and margins, vol 2. Plenum, New York, pp 598
Ekman S (1953) Zoogeography of the sea. Sidgwick & Jackson, London, pp 417
Ellis WN (1986) TreeTools, a set of programs for the construction of Wagner Networks and Trees for the Macintosh 512 computer (distributed by the author)
Esteban M (1979) [1980] Significance of the Upper Miocene coral reefs of the western Mediterranean. Palaeogeog Palaeoclimatol Palaeoecol 29:169–188
Esteban M, Calvet F, Dabrio CJ, Barón A, Giner J, Pomar L, Salas R, Permanyer A (1978) Aberrant features of the Messianian coral reefs, Spain. Acta Geol Hisp 1:20–22
Farris JS (1977) Phylogenetic analysis under Dollo's Law. Syst Zool 26:77–88
Felsenstein J (1985) Confidence limits on phylogenies: an approach using the bootstrap. Evolution 39:783–791
Griessinger J-M (1971) Étude des Réniérides de Méditerranée (Démosponges, Haplosclérides). Bull Mus Natl Hist Nat (3) 3 (Zool 3): 97–182
Hechtel GJ (1976) Zoogeography of Brazilian marine Demospongiae. In: Harrison FW, Cowden RR (eds) Aspects of sponge biology. Academic Press, New York, pp 354
Hennig W (1966) Phylogenetic systematics. Univ Illinois Press, Urbana, pp 163
Humphries CJ, Parenti LR (1986) Cladistic biogeography. Clarendon Oxford, pp 98
Kauffman EG (1973) Cretaceous Bivalvia. In: Hallam A (ed) Atlas of palaeobiogeography. Elsevier Scientific, Amsterdam, pp 531
Kluge AG, Farris JS (1969) Quantitative phyletics and the evolution of anurans. Syst Zool 18:1–32
Maddison WP, Donoghue MJ, Maddison DR (1984) Outgroup analysis and parsimony. Syst Zool 33:83–103
Maldonado A (1985) Evolution of the Mediterranean basins and a detailed reconstruction of the Cenozoic paleoceanography. In: Margalef R (ed) Key environments; western Mediterranean. Pergamon, Oxford, pp 17–59
Mickevich MF (1982) Transformation series analysis. Syst Zool 31 (4):461–478
Nelson (1984) Cladistics and biogeography. In: Duncan T, Stuessy TF (eds) Cladistics: perspectives on the reconstruction of evolutionary history. Columbia University Press, New York, pp 312
Nelson G, Platnick NI (1981) Systematics and biogeography; cladistics and vicariance. Columbia University Press, New York, pp 567
Page RDM (1988) Quantitative cladistic biogeography: constructing and comparing area cladograms. Syst Zool 37 (3):254–270
Pielou EC (1979) Biogeography. John Wiley, New York, pp 351
Pimentel RA, Riggins R (1978) The nature of cladistic data. Cladistics 3 (3):201–209
Platnick NI, Nelson G (1978) A method of analysis for historical biogeography. Syst Zool 27:1–16
Ridley WI, Watkins ND, MacFarlane DJ (1974) The Oceanic Islands: Azores. In: Nairn AEM, Stehli FG (eds) The ocean basins and margins, vol 2. Plenum, New York, pp 598
Rosen DE (1975) A vicariance model of Caribbean biogeography. Syst Zool 24 (4):431–464
Rosen DE (1978) Vicariant patterns and historical explanation in biogeography. Syst Zool 27:159–188
Ross HH (1974) Biological systematics. Addison-Wesley Reading, Massachusetts, pp 345
Saether OA (1983) The canalized evolutionary potential: inconsistencies in phylogenetic reasoning. Syst Zool 32 (4):343–359
van Lent F, de Weerdt WH (1987) The haplosclerid sponge fauna (Porifera, Demospongiae) from Banyuls-sur-mer (Mediterranean) with the description of a new species. In: Vacelet J, Boury-Esnault N (eds) Taxonomy of Porifera from the north east Atlantic and Mediterranean Sea. NATO-ASI-Serie G: Ecol Sci, vol 13. Springer, Berlin Heidelberg New York Tokyo, pp 332
van Soest RWM (1980) Marine sponges from Curaçao and other Caribbean localities. Part III. Haplosclerida. Stud Fauna Curaçao Other Caribb Isl 62 (104):1–174
Vacelet J (1985) Coralline sponges and the evolution of Porifera. In: Morris SC, George JD, Gibson R, Platt HM (eds) The origins and relationships of lower invertebrates. The Systematic Assoc, Special vol 28. Clarendon, Oxford, pp 1–13

Watrous LE, Wheeler QD (1981) The out-group comparison method of character analysis. Syst Zool 30 (1):1–11
Wiley EO (1981) Phylogenetics: the theory and practice of phylogenetic systematics. John Wiley, New York, pp 439
Wiley EO (1987a) Methods in vicariance biogeography. In: Hovenkamp P, Gittenberger E, Hennipman E, de Jong R, Roos MC, Sluys R, Zandee M (eds) Systematics and evolution: a matter of diversity. Utrecht University, pp 341
Wiley EO (1987b) Approaches to outgroup comparison. In: Hovenkamp P, Gittenberger E, Hennipman E, de Jong R, Roos MC, Sluys R, Sandee M (eds) Systematics and evolution: a matter of diversity. Utrecht University, pp 341
Wiley EO (1988) Vicariance biogeography. Annu Rev Ecol Syst 19:513–542
Zandee M (1987) A computing environment for cladistic analysis. Preliminaries of a user manual for CAFCA/PC. In: Hovenkamp P, Gittenberger E, Hennipman E, de Jong R, Roos MC, Sluys R, Zandee M (eds) Systematics and evolution: a matter of diversity. Utrecht University, pp 341
Zandee M, Roos MC (1987) Component-compatibility in historical biogeography. Cladistics 3 (4):305–332

# Colonization, Life Cycles and Competition in a Freshwater Sponge Association

R. PRONZATO and R. MANCONI

## Introduction

Freshwater sponges have been deeply studied under natural and experimental conditions, with particular attention to life histories, growth sexual, and asexual reproduction, population dynamics, standing crop, clearance rates, feeding, and symbiosis (Simpson and Gilbert 1973, 1974; Van de Vyver and Willenz 1975; Frost and Williamson 1980; Wilkinson 1980; Frost et al. 1982; Harsha et al. 1983; Rader and Winget 1985; Francis and Poirrier 1986; Saller 1988). In spite of this, little is known about colonization strategies and spatial competition (Frost et al. 1982; De Nagy 1978).

Studies on temporal evolution of sponge associations and communities generally deal with marine sponges (Hartman 1957; Rützler 1965; Sarà 1970), and few of them are based on long-term observations (Pansini and Pronzato 1990).

Marine benthic biocoenoses of hard substrata in shallow water, where sponges are abundant, are characterized by a high spatial density of organisms linked by a complex network of relationships. Space is an important limiting resource in these environments (Jackson 1977); moreover, a small surface area could exhibit a large number of microhabitats depending on changes in light, water movement, etc. The relationships among associated species under such complex conditions are difficult to define, and chance is often regarded as the element that can give rise to a sponge association (Corriero and Pronzato 1987). Nevertheless, studies on the temporal variations of the microdistribution of some littoral sponges (Labate 1966; Sarà 1970) permit the interpretation of interspecific relationships as competition.

Artificial substrata were utilized to study the successive stages of colonization, but sponges are not efficient colonizers. Their settlement is backward and may be influenced by disturbances caused by other organisms. As a consequence, after some years of submersion, sponge associations seem to be not yet entirely structured (Pronzato 1972; Pansini and Pronzato 1981; Sutherland 1981).

Since sessile zoobenthos is represented by few taxa in freshwater habitats, sponges usually have plenty of space for their settlement. Long-term observations on sponge populations living in this environment provide information on their life cycles and colonization strategies and data for better understanding their morphofunctional adaptations.

J. Reitner and H. Keupp (Eds.)
Fossil and Recent Sponges

Our goal is to study the temporal evolution of an association – free from interference with other sessile animals – between *Spongilla lacustris* and *Ephydatia fluviatilis*, which contemporaneously colonized a homogenous hard substratum.

## Study Area and Methods

The study area is located in the town of Ronco, along the Scrivia, a eutrophic stream in the Po river basin (Liguria, Italy). Sources of organic waste determine in this area high turbidity values in summer. The Scrivia basin is 605 km$^2$ wide and its water flow ranges 28.90 m$^3$/s in November to 2.99 m$^3$/s in August (mean values over 30 years). During our investigations, water temperature ranged from 28.5°C in July to – 1°C in January (Fig. 1).

The available substrata for sponge settlement in the area investigated are pebbles, rocks at the bottom of the stream, and the piers of a bridge (depth ranges from 10 to 150 cm). In the present study in order to have standard experimental conditions, we consider only the piers, which have homogeneous substrate made of granitic blocks and concrete. In particular, we refer to six identical vertical surfaces parallel to the water flow. The submerged area of each is about 150,000 cm$^2$, the total length of the pillar being 20 m, and the distance between two opposite surfaces 10 m (Fig. 5).

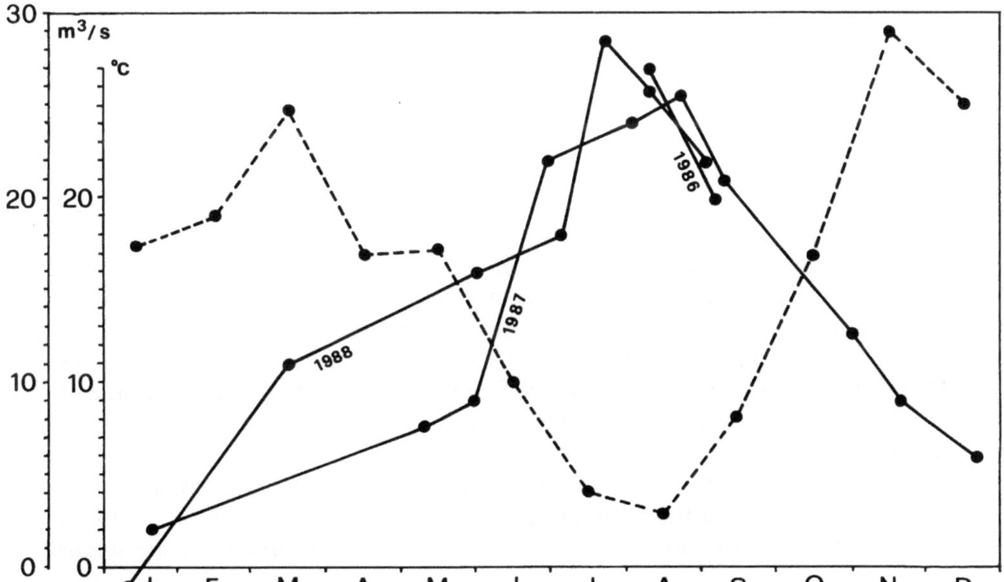


**Fig. 1.** *Dashed line* water flow in the Scrivia stream (average over 30 years, 1931–1960). *Continuous line* water temperature values from 1986 to 1988

The beginning of our observations was in September 1985, and this study was performed for 4 years up to 1988. The specimens of *S. lacustris* and *E. fluviatilis* were mapped frequently, photos taking to follow sponges growth and their changes in shape and color. For each pier surface, the area covered by the sponges was measured and the number of specimens calculated.

## Results

### *Life Cycles*

In winter and early spring, the two populations consist of carpets of gemmules covering the recorded surfaces. The onset of *S. lacustris* varies according to years: it begins in May (1986), April (1987), July (1988), whereas *E. fluviatilis* sets in constantly in July. Both species reach their maximum values of size, number of specimens, and area covered on substrata in August. The end of their degenerative phase occurs in November. Even if the cycle of *S. lacustris* begins some months earlier, the two species overlap for a long time (Figs. 2,3).

When fully grown, *S. lacustris* forms finger-like projections (1 cm diameter; 20 cm length) from an encrusting base that can reach 1 m in diameter; color, due to symbiosis with zoochlorellae, is emerald green. *E. fluviatilis* is cushion-shaped with an irregular surface, and its color is constantly light brown; the diameter of the specimens usually ranges from 1 to 20 cm.

Sexual reproduction has never been observed in the two studied populations.

### *Population Dynamics*

The trend of *S. lacustris* was very similar in the first two years: the maximum values were 51,600 $cm^2$ of area cover (18 specimens) in July 1986 and 55,200 $cm^2$ (24 specimens) in September 1987. In 1988, a considerable increase occurred with a maximum of 141,800 $cm^2$ (86 specimens) in August, due to colonization of new surfaces (Figs. 2,3,4).

*E. fluviatilis* showed a different trend, with a value of 2450 $cm^2$ (20 specimens) in August 1986 and a remarkable increase in August 1987, with 42,250 $cm^2$ (523 specimens). This quantitative increase was produced by the colonization of all the available surfaces (Figs. 2,3,4). A reduction in the number of specimens (154) and in the covered area (10,600 $cm^2$) occurred in 1988, probably because of the settlement of *S. lacustris* on surfaces C and F, previously occupied only by *E. fluviatilis*.

At full growth *S. lacustris* can cover wide areas on the six examined surfaces (max 141,800 $cm^2$) but the number of specimens is restricted to a maximum of 86 (Table 1). On the contrary, *E. fluviatilis* shows a larger number of specimens (max 569) with a smaller area cover (max 42,250 $cm^2$) (Table 2). The mean area covered by each specimen is so 1648.8 $cm^2$ for *S. lacustris* and 74.2 $cm^2$ for *E. fluviatilis*.

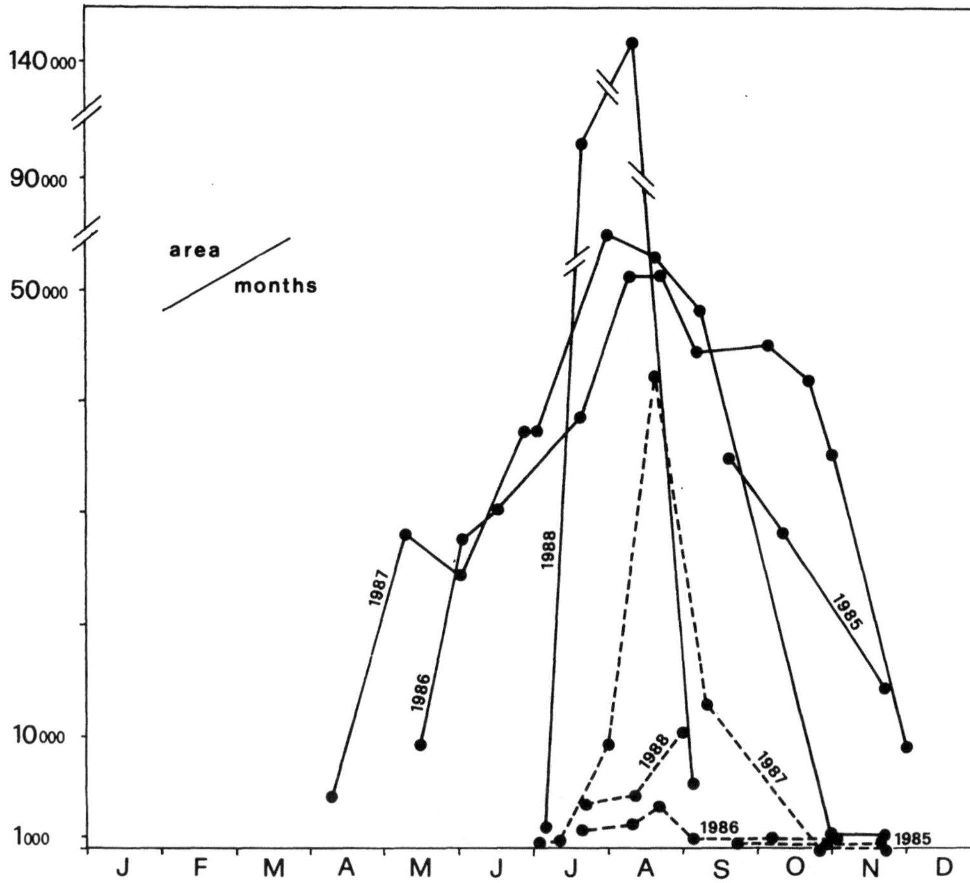


**Fig. 2.** Area cover in cm² during four vegetative cycles (1985–1988). *E. fluviatilis* is indicated by *dashed lines* and *S. lacustris* by *continuous lines* (For details see Table 1,2)

## *Colonization*

Both species settled in the considered area during 1985. *S. lacustris* was present on surface B, *E. fluviatilis* on surface A (Fig. 4). However, in the following years, the colonization process concerned all the other surfaces, with differences between the two species.

In 1986, *E. fluviatilis* expanded on surface A and colonized also surface B, while in 1987 it extended on the remaining surfaces, increasing also in area cover and number of specimens. In 1988, it seemed to be quantitatively reduced, even if it was still present on the six monitored surfaces (Fig. 4).

*S. lacustris* showed a different colonization pattern: in 1986 and 1987, it expanded on surface B, occupying then surface D in 1987. It colonized two other substrata, C and F in the following year (1988) reaching its maximum values of

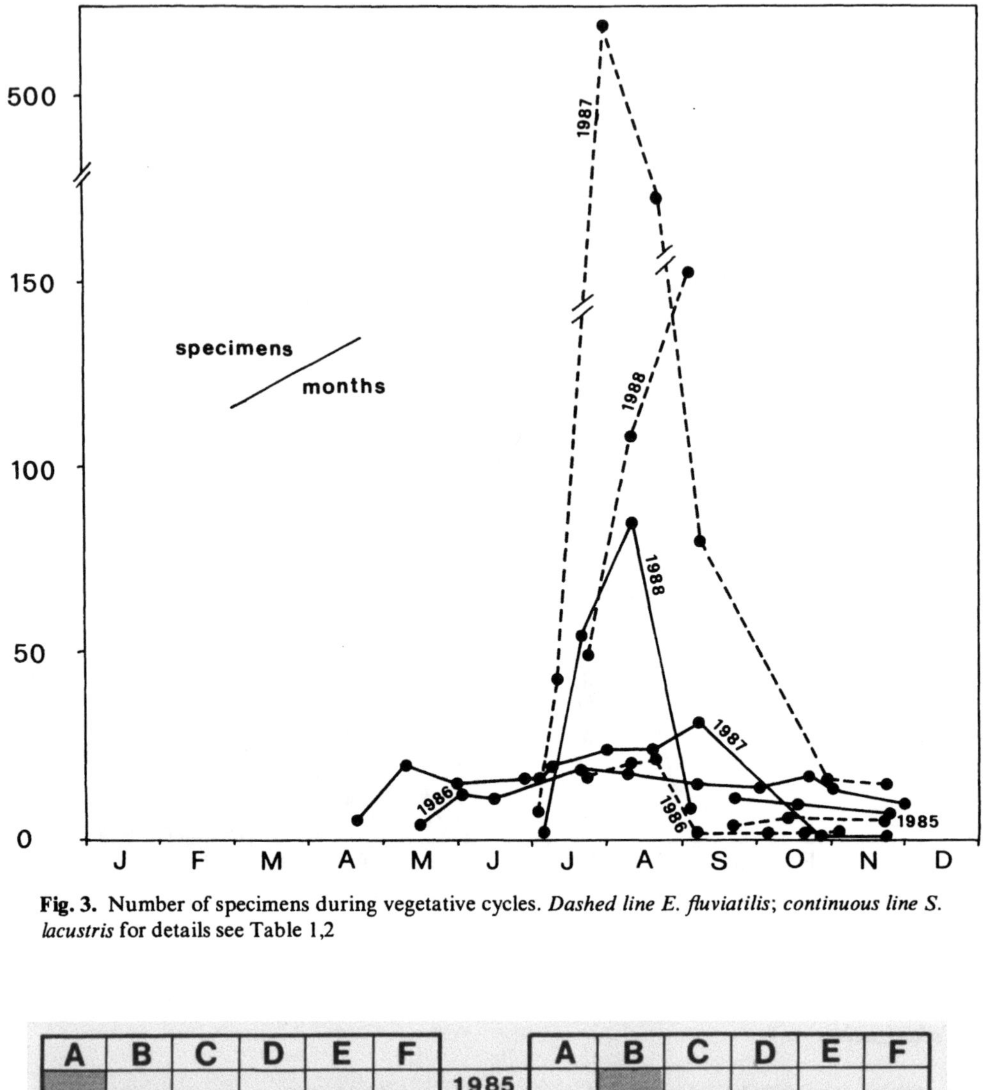


**Fig. 3.** Number of specimens during vegetative cycles. *Dashed line E. fluviatilis*; *continuous line S. lacustris* for details see Table 1,2

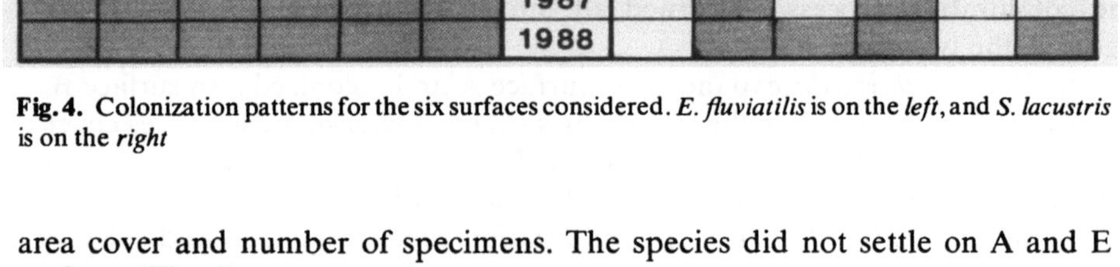


**Fig. 4.** Colonization patterns for the six surfaces considered. *E. fluviatilis* is on the *left*, and *S. lacustris* is on the *right*

area cover and number of specimens. The species did not settle on A and E surfaces (Fig. 4).

**Table 1.** *S. lacustris*: area cover values ($cm^2$) and number of specimens (*in parentheses*). A to F: surfaces of the bridge piers

| No. | Date | Total | | A | | B | | C | | D | | E | | F | |
|---|---|---|---|---|---|---|---|---|---|---|---|---|---|---|---|
| 1 | 21.9.85 | 34,800 | (11) | – | – | 34,800 | (11) | – | – | – | – | – | – | – | – |
| 2 | 13.10.85 | 28,200 | (10) | – | – | 28,200 | (10) | – | – | – | – | – | – | – | – |
| 3 | 24.11.85 | 14,400 | (7) | – | – | 14,400 | (7) | – | – | – | – | – | – | – | – |
| 4 | 8.4.86 | – | – | – | – | – | – | – | – | – | – | – | – | – | – |
| 5 | 16.5.86 | 9,600 | (4) | – | – | 9,600 | (4) | – | – | – | – | – | – | – | – |
| 6 | 2.6.86 | 27,600 | (13) | – | – | 27,600 | (13) | – | – | – | – | – | – | – | – |
| 7 | 14.6.86 | 30,000 | (11) | – | – | 30,000 | (11) | – | – | – | – | – | – | – | – |
| 8 | 19.7.86 | 38,400 | (19) | – | – | 38,400 | (19) | – | – | – | – | – | – | – | – |
| 9 | 8.8.86 | 51,600 | (18) | – | – | 51,600 | (18) | – | – | – | – | – | – | – | – |
| 10 | 21.8.86 | 51,600 | (18) | – | – | 51,600 | (18) | – | – | – | – | – | – | – | – |
| 11 | 6.9.86 | 44,400 | (15) | – | – | 44,400 | (15) | – | – | – | – | – | – | – | – |
| 12 | 5.10.86 | 45,000 | (14) | – | – | 45,000 | (14) | – | – | – | – | – | – | – | – |
| 13 | 19.10.86 | 42,000 | (18) | – | – | 42,000 | (18) | – | – | – | – | – | – | – | – |
| 14 | 31.10.86 | 35,400 | (14) | – | – | 35,400 | (14) | – | – | – | – | – | – | – | – |
| 15 | 1.12.86 | 9,000 | (10) | – | – | 9,000 | (10) | – | – | – | – | – | – | – | – |
| 16 | 18.2.87 | – | – | – | – | – | – | – | – | – | – | – | – | – | – |
| 17 | 18.4.87 | 4,800 | (5) | – | – | 4,800 | (5) | – | – | – | – | – | – | – | – |
| 18 | 9.5.87 | 28,200 | (20) | – | – | 28,200 | (20) | – | – | – | – | – | – | – | – |
| 19 | 31.5.87 | 24,600 | (15) | – | – | 24,600 | (15) | – | – | – | – | – | – | – | – |
| 20 | 28.6.87 | 37,200 | (16) | – | – | 37,200 | (16) | – | – | – | – | – | – | – | – |
| 21 | 2.7.87 | 37,200 | (16) | – | – | 37,200 | (16) | – | – | – | – | – | – | – | – |
| 22 | 10.7.87 | 42,000 | (20) | – | – | 42,000 | (20) | – | – | – | – | – | – | – | – |
| 23 | 31.7.87 | 55,200 | (24) | – | – | 51,600 | (21) | – | – | 3,600 | (3) | – | – | – | – |
| 24 | 18.8.87 | 53,400 | (24) | – | – | 51,600 | (23) | – | – | 1,800 | (1) | – | – | – | – |
| 25 | 7.9.87 | 48,600 | (31) | – | – | 27,600 | (17) | – | – | 21,000 | (14) | – | – | – | – |
| 26 | 27.10.87 | 1,200 | (1) | – | – | 1,200 | (1) | – | – | – | – | – | – | – | – |
| 27 | 20.11.87 | 1,200 | (1) | – | – | 1,200 | (1) | – | – | – | – | – | – | – | – |
| 28 | 18.12.87 | – | – | – | – | – | – | – | – | – | – | – | – | – | – |

**Table 1.** (*Continued*)

| No. | Date | Total | | A | | B | | C | | D | | E | | F | |
|---|---|---|---|---|---|---|---|---|---|---|---|---|---|---|---|
| 29 | 12.1.88 | – | – | – | – | – | – | – | – | – | – | – | – | – | – |
| 30 | 14.3.88 | – | – | – | – | – | – | – | – | – | – | – | – | – | – |
| 31 | 18.4.88 | – | – | – | – | – | – | – | – | – | – | – | – | – | – |
| 32 | 1.6.88 | – | – | – | – | – | – | – | – | – | – | – | – | – | – |
| 33 | 4.7.88 | 1,800 | (1) | – | – | 1,800 | (1) | – | – | – | – | – | – | – | – |
| 34 | 20.7.88 | 92,400 | (77) | – | – | 42,600 | (36) | 25,200 | (22) | 23,400 | (17) | – | – | 1,200 | (2) |
| 35 | 10.8.88 | 141,800 | (86) | – | – | 64,200 | (27) | 39,000 | (31) | 32,600 | (23) | – | – | 6,000 | (5) |
| 36 | 3.9.88 | 6,000 | (8) | – | – | 2,400 | (3) | – | – | 2,400 | (3) | – | – | 1,200 | (2) |

**Table 2.** *E. fluviatilis*: area cover values ($cm^2$) and number of specimens (*in parentheses*). A to F: surfaces of the bridge piers

| No. | Date | Total | | A | | B | | C | | D | | E | | F | |
|---|---|---|---|---|---|---|---|---|---|---|---|---|---|---|---|
| 1 | 21.9.85 | 100 | (4) | 100 | (4) | – | – | – | – | – | – | – | – | – | – |
| 2 | 13.10.85 | 120 | (6) | 120 | (6) | – | – | – | – | – | – | – | – | – | – |
| 3 | 24.11.85 | 150 | (6) | 150 | (6) | – | – | – | – | – | – | – | – | – | – |

| | | | | | | | | | | | | | | | | | |
|---|---|---|---|---|---|---|---|---|---|---|---|---|---|---|---|---|---|
| 4 | 8.4.86 | – | – | – | – | – | – | – | – | – | – | – | – | – | – |
| 5 | 16.5.86 | – | – | – | – | – | – | – | – | – | – | – | – | – | – |
| 6 | 2.6.86 | – | – | – | – | – | – | – | – | – | – | – | – | – | – |
| 7 | 14.6.86 | – | – | – | – | – | – | – | – | – | – | – | – | – | – |
| 8 | 19.7.86 | 1,500 | (18) | 1,200 | (15) | 300 | (3) | – | – | – | – | – | – | – | – |
| 9 | 8.8.86 | 1,900 | (19) | 1,500 | (16) | 400 | (3) | – | – | – | – | – | – | – | – |
| 10 | 21.8.86 | 2,450 | (20) | 1,950 | (17) | 500 | (3) | – | – | – | – | – | – | – | – |
| 11 | 6.9.86 | 250 | (2) | 250 | (2) | – | – | – | – | – | – | – | – | – | – |
| 12 | 5.10.86 | 250 | (2) | 250 | (2) | – | – | – | – | – | – | – | – | – | – |
| 13 | 19.10.86 | 250 | (2) | 250 | (2) | – | – | – | – | – | – | – | – | – | – |
| 14 | 31.10.86 | 250 | (2) | 250 | (2) | – | – | – | – | – | – | – | – | – | – |
| 15 | 1.12.86 | – | – | – | – | – | – | – | – | – | – | – | – | – | – |
| 16 | 18.2.87 | – | – | – | – | – | – | – | – | – | – | – | – | – | – |
| 17 | 18.4.87 | – | – | – | – | – | – | – | – | – | – | – | – | – | – |
| 18 | 9.5.87 | – | – | – | – | – | – | – | – | – | – | – | – | – | – |
| 19 | 31.5.87 | – | – | – | – | – | – | – | – | – | – | – | – | – | – |
| 20 | 28.6.87 | – | – | – | – | – | – | – | – | – | – | – | – | – | – |
| 21 | 2.7.87 | 100 | (8) | 100 | (8) | – | – | – | – | – | – | – | – | – | – |
| 22 | 10.7.87 | 500 | (41) | 400 | (39) | 100 | (2) | – | – | – | – | – | – | – | – |
| 23 | 31.7.87 | 11,500 | (569) | 4,100 | (191) | 2,750 | (126) | 1,200 | (110) | 2,600 | (70) | 200 | (13) | 650 | (59) |
| 24 | 18.8.87 | 42,250 | (523) | 23,400 | (170) | 300 | (5) | – | – | 15,800 | (95) | – | – | 2,750 | (253) |
| 25 | 7.9.87 | 11,250 | (81) | 10,250 | (42) | – | – | – | – | 300 | (7) | – | – | 700 | (32) |
| 26 | 27.10.87 | 10 | (15) | 10 | (15) | – | – | – | – | – | – | – | – | – | – |
| 27 | 20.11.87 | 10 | (15) | 10 | (15) | – | – | – | – | – | – | – | – | – | – |
| 28 | 18.12.87 | – | – | – | – | – | – | – | – | – | – | – | – | – | – |
| 29 | 12.1.88 | – | – | – | – | – | – | – | – | – | – | – | – | – | – |
| 30 | 14.3.88 | – | – | – | – | – | – | – | – | – | – | – | – | – | – |
| 31 | 18.4.88 | – | – | – | – | – | – | – | – | – | – | – | – | – | – |
| 32 | 1.6.88 | – | – | – | – | – | – | – | – | – | – | – | – | – | – |
| 33 | 4.7.88 | – | – | – | – | – | – | – | – | – | – | – | – | – | – |
| 34 | 20.7.88 | 3,900 | (49) | 1,350 | (21) | – | – | 300 | (4) | 1,400 | (13) | – | – | 850 | (11) |
| 35 | 10.8.88 | 4,800 | (109) | 900 | (13) | 700 | (17) | 850 | (5) | 1,500 | (12) | 50 | (8) | 800 | (54) |
| 36 | 3.9.88 | 10,600 | (154) | 20 | (1) | 260 | (6) | 800 | (14) | 5,260 | (23) | 900 | (21) | 3,360 | (89) |

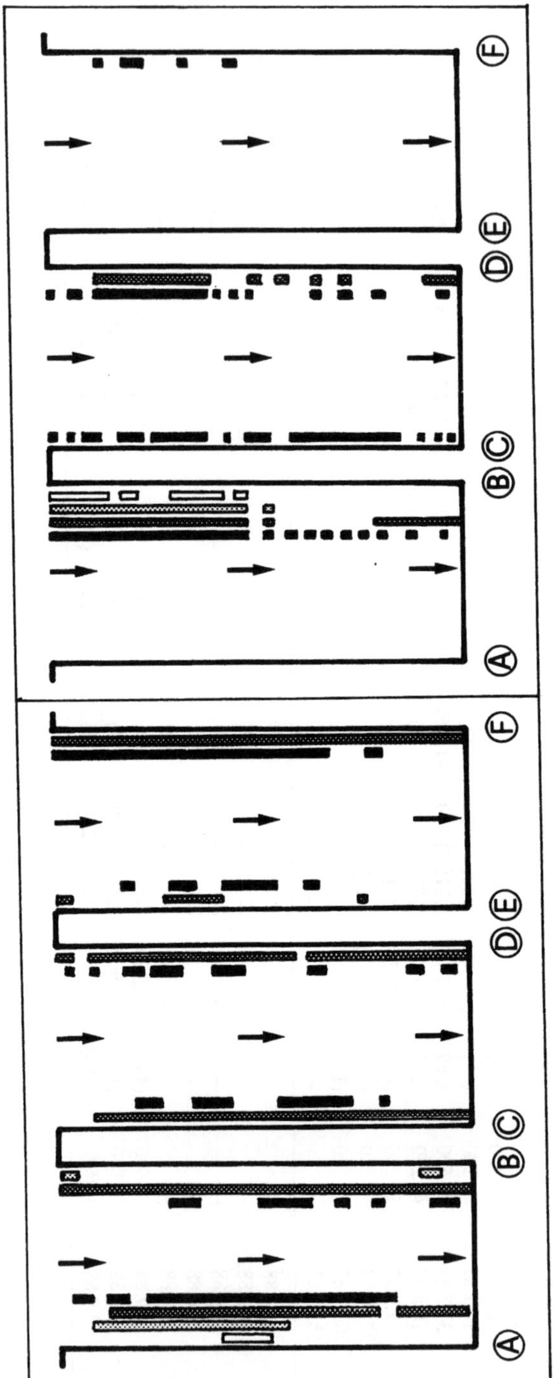


**Fig. 5.** Distribution of *E. fluviatilis* (*left*) and *S. lacustris* (*right*) on the surfaces of the bridge piers over the 4-year period (*white* 1985; *light gray* 1986; *gray* 1987; *black* 1988). *Arrows* indicate the water flow direction

### *Spatial Competition*

Three years after the first observation, the two species coexisted on four of the six surfaces considered, while on the remaining two surfaces only *E. fluviatilis* was settled.

Surface A: was occupied by *E. fluviatilis* in the first year. This species gradually spread on the whole surface, remaining the sole present species up to 1988. The maximum values reached on this surface are: 23,400 $cm^2$ of area cover and 170 specimens (Fig. 5).

Surface B: in the first year, only *S. lacustris* settled on it, but, in the following years (1986–88), the two species coexisted on the whole surface. *S. lacustris* covered a large surface with few specimens (64,200 $cm^2$ per 27 specimens). *E. fluviatilis*, on the contrary covered a smaller surface with a larger number of specimens (2750 $cm^2$ per 126 specimens) (Fig. 5).

Surface C and F: were devoid of sponges at the beginning, since settlements of *E. fluviatilis* on the whole surfaces occurred in the third year (1987), only with limited area cover (C = 1200 $cm^2$: F = 2750 $cm^2$), and a large number of specimens (110 on C and 253 on F). Only in the fourth year (1988) did *S. lacustris* colonize these surfaces, with high values of area cover (C = 39,000 $cm^2$; F = 6000 $cm^2$), small numbers of specimens (31 on C and 5 on F), and restricted distribution (Fig. 5).

Surface D: was colonized by both species in the third year (1987), with an area cover of 15,800 $cm^2$ and 95 specimens for *E. fluviatilis*, whereas *S. lacustris* was represented by 14 specimens covering an area of 21,000 $cm^2$. Whilst the former spread on the surface, the distribution of the latter appeared more restricted. In the fourth year, *S. lacustris* was clearly dominant, being present on the whole surface, even if in patches, with an area cover of 32,600 $cm^2$ and 23 specimens, while *E. fluviatilis* showed a reduction both in area cover (5260 $cm^2$) and in the number of specimens (23) (Fig. 5).

Surface E: *E. fluviatilis* in the third year occupied this substratum with a small number of specimens (13), covering a small surface (200 $cm^2$); in 1988, the situation was similar: 900 $cm^2$ and 21 specimens. The distribution of the species was limited to a small portion of the substratum (Fig. 5).

In most cases, the two neighboring species were not in contact, but in some sites different specimens were very close to each other; in general, *E. fluviatilis* was regularly distributed on the entire surface considered, while *S. lacustris* formed discrete patches (Fig. 5).

## Discussion

The contemporaneous settlement of *S. lacustris* and *E. fluviatilis* during 1985 and the fact that the two species chose different surfaces (A and B) may be regarded as casual, since it is clear that the habitat features (e.g., substratum, light, water movement, and food availability) were suitable for both species. The subsequent colonizations of all the six surfaces supports this consideration.

The analysis of four successive vegetative cycles, which allowed *E. fluviatilis* to colonize all the available surfaces in the third year, suggests that the ability of this species as a colonizer is considerable. *S. lacustris* seems to be ineffective in the colonization process, as shown by its succession pattern; in fact, in the fourth year it was still absent from surfaces A and E.

Nevertheless, *S. lacustris* when reaching a substratum on which *E. fluviatilis* has already settled (e.g., surface D), usually occupied after some time a considerable area, thus causing a quantitative reduction of former colonizer.

*S. lacustris* shows a higher effectiveness in spatial competition, while *E. fluviatilis* exhibits a remarkable capability of colonizing new substrate, also confirmed by its cosmopolitan distribution. In addition, *S. lacustris* appears to be favored by its longer growth period.

In general, no sponge shows a clear-cut dominance in interspecific competition, especially in communities composed of a large number of species, whose associations are characterized by a marked stability (Sarà 1970; Pansini and Pronzato 1990). In less complex associations (i.e., with a smaller number of species), competition and cooperation phenomena could play an important role for equilibrium conditions (Sarà 1970). Nevertheless, sponges could compete for space through a nonlinear hierarchy acting as a network of interspecific interferences (Buss 1979); this mechanism allows the coexistence of different species on the same substratum.

On the contrary, in the case of an association between two direct competitor species, one with higher ability for colonization and the other for competition, temporal evolution could cause a gradual reduction in one species and the final success of the second one only (Fig. 6) (Whittaker 1970).

Our considerations on *S. lacustris* and *E. fluviatilis* suggest that the two species can undoubtedly live together, and that each of them can be the first to colonize one of the six surfaces (Fig. 6).

The higher ability of *S. lacustris* in space competition seems to be balanced and controlled by the presence of *E. fluviatilis*. *S. lacustris* usually settles on a new surface, by numerous and constantly small specimens, whose number can strongly increase. Therefore, the development strategy of *S. lacustris* is based on size, while that of *E. fluviatilis* is based on number of specimens.

The equilibrium between the two species is also maintained by the water flow. The high, finger-like processes of *S. lacustris* are particularly exposed to the water movement action, which in a summer flood may eradicate a large number of specimens (as occurred in late August 1988). By contrast, *E. fluviatilis* seems to be able to stand even strong currents.

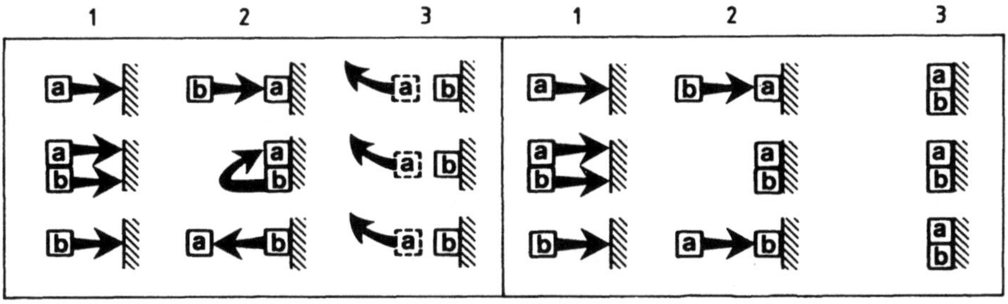


**Fig. 6.** Competition patterns between two species: *a* (colonizer) and *b* (competitor) for three different possible colonizations: (1) *a* arrives first, (2) *a* and *b* arrive contemporaneously, and (3) *b* arrives first. Expected evolution (*left*): the competitor dominates the colonizer. Observed evolution (*right*): the colonizer, *E. fluviatilis*, and the competitor, *S. lacustris*, settle on the same substratum and coexist

Short-lasting observations may lead to regarding this type of disturbance as the principal factor influencing and modifying the population dynamics and the quantitative ratio between the two species. On the contrary, long-term observations allow to consider the biological features of each species as the most important factors affecting the onset of synecological relationships.

The morphofunctional adaptations and the biological and ecological features of *S. lacustris* and *E. fluviatilis* are very different. The long finger-like projections permit the former to increase the filtering surface and to take the water far from the pier surface, while its cushion shape allows the latter to exploit the water layer close to the substratum.

The two species show similar clearance rates and their coexistence suggests a nondirect competition for food, presumably due to species-specific particle selection (Francis and Poirrier 1986). Zoochlorellae play an important role in the growth process of *S. lacustris*, and their presence increases its growth rate (Frost and Williamson 1980). The vegetative growth rate can start in early spring, most likely due to the autotrophic condition of this species; conversely, the considered population of *E. fluviatilis* does not show any presence of zoochlorellae, and relies for food on the environmental particle concentration.

Since an ecological niche is besides defined by food, habitat space, and time of activity (niche axes: Connell 1978), while morphological features, specimen abundance and frequency, the ability to withstand changes in environmental conditions also contribute to determinating the niche of a species: the dynamic equilibrium between *S. lacustris* and *E. fluviatilis* sharing a common environment suggests the existence of a niche diversification preventing a direct competition.

## References

Buss LW (1979) Habitat selection, directional growth and spatial refuges: why colonial animals have more hiding places. In: Larwood G, Rosen BR (eds) Biology and systematics of colonial organisms. Academic Press, London, pp 459–497
Connell JH (1978) Diversity in tropical rain forest and coral reefs. Science 199:1302–1310
Corriero G, Pronzato R (1987) Epibiontic sponges on the bivalve *Pinna nobilis*. Mar Ecol Prog Ser 35:75–82
De Nagy GS (1978) Life cycle, growth and substrata competition of freshwater sponges in a sphagnum bog-pond. Thesis. Dartmouth College, Hannover, New Hampshire, USA
Francis JC, Poirrier MA (1986) Particle uptake in two freshwater sponge species, *Ephydatia fluviatilis* and *Spongilla alba* (Porifera: Spongillidae). Trans Am Microsc Soc 105:11–20
Frost TM, Williamson CE (1980) In situ determination of the effect of symbiotic algae on the growth of the freshwater sponge *Spongilla lacustris*. Ecology 61:1361–1370
Frost TM, De Nagy GS, Gilbert JS (1982) Population dynamics and standing biomass of the freshwater sponge *Spongilla lacustris*. Ecology 63:1203–1210
Harsha RE, Francis JC, Poirrier MA (1983) Water temperature: A factor in the seasonality of two freshwater sponge species. *Ephydatia fluviatilis* and *Spongilla alba*. Hydrobiologia 102:145–150
Hartman WD (1957) Ecological niche differentiation in the boring sponges (Clionidae). Evolution 11:291–297
Jackson JBC (1977) Competition on marine hard substrata: the adaptive significance of solitary and colonial strategies. Am Nat 111:743–767
Labate M (1966) Variazioni nel tempo del popolamento di Poriferi di una grotta superficiale. Boll Zool 33:181
Pansini M, Pronzato R (1981) Etude des Spongiaires des substrats artificiels immergés durant quatre ans. Vie Milieu 31:77–82
Pansini M, Pronzato R (1990) Observations on the dynamics of a Mediterranean sponge community. In: Rützler K, Macintyre VV, Smith KP (eds) New perspectives in sponge biology. Proc 3rd Int Conf on Biology of Sponges, Woods Hole. Smithsonian Institution Press, Washington (in press)
Pronzato R (1972) I Poriferi del fouling del porto di Genova. Boll Mus Ist Biol Univ Genova 40:89–98
Rader RB, Winget RN (1985) Seasonal growth rate and population dynamics of a freshwater sponge. Hydrobiologia 123:171–176
Rützler K (1965) Systematik und Ökologie der Poriferen aus litoral. Schattengebieten der Nordadria. Z Morphol Ökol Tiere 55:1–82
Saller U (1988) Oogenesis and larval development of *Ephydatia fluviatilis* (Porifera, Spongillidae). Zoomorphology 108:23–28
Sarà M (1970) Competition and cooperation in sponge populations. Symp Zool Soc Lond 25:273–284
Simpson TL, Gilbert JJ (1973) Gemmulation, gemmule hatching, and sexual reproduction in freshwater sponges, I. The life cycle of *Spongilla lacustris* and *Tubella pennsylvanica*. Trans Am Micros Soc 92:422–433
Simpson TL, Gilbert JJ (1974) Gemmulation, gemmule hatching, and sexual reproduction in freshwater sponges, II. Life cycle events in young, larva-produced sponges of *Spongilla lacustris* and an unidentified species. Trans Am Microsc Soc 93:39–45
Sutherland JP (1981) The fouling community at Beaufort, North Carolina: a study in stability. Am Nat 118:499–519
Van de Viver G, Willenz P (1975) An experimental study of the life-cycle of the freshwater sponge *Ephydatia fluviatilis* in its natural surroundings. Wilhelm Roux'Arch 177:41–52
Whittaker RH (1970) Communities and ecosystems. Macmillan, London, pp 162
Wilkinson CR (1980) Nutrient translocation from green algal symbionts to the freshwater sponge *Ephydatia fluviatilis*, Hydrobiologia 75:241–250

# Chaetetid Habitats

R.R. WEST and S. KERSHAW

## Introduction

While genera of the order Chaetetida are known from the Ordovician to the Eocene, they are not a conspicuous component of the benthic invertebrate assemblages of these rocks with the exception of rocks of the Middle Carboniferous. The *Treatise* (F Supplement 1, Hill 1981) lists 20 Paleozoic genera (in six families) and seven post-Paleozoic genera (in two families). Of the six Paleozoic families, four are queried, indicating doubt as to their status and phylogenetic placement. Post-Paleozoic chaetetids are best known from the Jurassic and Cretaceous rocks of France and Italy. As noted above, chaetetids were only conspicuous during the Middle Carboniferous (Bashkirian through Moscovian in Russia and Morrowan into the Missourian in North America).

During the Middle Carboniferous chaetetids were one of the dominant reef mound builders (Heckel 1974 and West 1988). Even so, chaetetids did not produce any reefs like the stromatoporoid reefs of the Devonian or the sponge-algal reefs of the Permian. These chaetetid, and other Carboniferous organic structures are more accurately referred to as reef mounds, as defined by James (1984), with the best-developed structures in the Middle Carboniferous being chaetetid biostromes (West 1988). Because of the relatively minor role of chaetetids in the geologic record and the more or less insignificance of the Middle Carboniferous reef mounds, chaetetids have not been well studied, except for the work of Sokolov in Russia. Our studies, to date, have concentrated on Paleozoic chaetetids, especially those of the Middle Carboniferous in North America, and our comments here are based on these studies. As such, we will be speaking *only* of these Carboniferous chaetetids with the hope that our observations and ideas will be applicable to other chaetetid occurrences.

## Growth Forms of Chaetetid Colonies

As with other colonial marine invertebrates (bryozoans and corals), the variation in the colony form of chaetetids is great (Kershaw and West 1988). Tabular, domical, columnar, "basket-shaped", digitate, and colony complexes occur. However, it appears that there are only three basic types, that is geometric categories, into which all chaetetid colonies can be placed as long as we recognize

J. Reitner and H. Keupp (Eds.)
Fossil and Recent Sponges

that "basket-shaped" and digitate are special cases of columnar and colony complexes respectively. These three basic colony geometries are illustrated in Fig. 1A to C with sketches made from outcrop photographs. Most colonies are tabular to domical and/or small columnar forms; however, in the Atokan of central Texas, Winston (1963) has reported columnar chaetetid colonies over 3 m high and up to 1 m in diameter. Sketches, also made from outcrop photographs, of "basket-shaped" and digitate colonies and colony complexes are illustrated in Fig. 1D to F.

While there are three basic colony shapes, studies of polished surfaces indicate that the basic building block of all chaetetid colonies is a thin, laminar sheet. Colony shape, then, is initially determined by characteristics of the substrate and/or growth of this initial layer. Growth may be vertical and uniform, not very common in the specimens studied to date, or vertical and nonuniform, a slightly more common condition, but still rare. Vertical, nonuniform growth occurs when some part of the initial layer, usually the center, grows upward more rapidly that the other parts of the layer. Thus, nonuniform growth could produce a domical or columnar colony while uniform vertical growth would produce a tabular colony. Growth can also be oblique, again an uncommon occurrence in the specimens studied to date, such that the calicles grow upward at any number of angles oblique to the substrate, but ultimately growth was normal to the substrate. Reasons for oblique growth are not clear at this time. A fourth type of growth is one involving multiple growth centers, a more or less common occurrence in specimens studied to date. This type of growth could occur when the initiating entities of chaetetid colonies settled very close to one another. Such a situation produces a very complex and involved initial layer because of competition for space. When there are conspicuous irregularities in the surface on which the initiating thin, laminar layer coats, that is small topographic "highs", the initial layer drapes over them, resulting in a domical, and often, a columnar colony, depending on other environmental conditions. This style of colony initiation is the most common of the five types recognized in the specimens studied to date. The initial thin, laminar growth layer and the affects of the substrate and other environmental parameters is considered by Kershaw and West (1987, 1988).

## Chaetetid Substrates

Substrates on which chaetetids grew were (1) hard, (2) inferred to be firm, and (3) loose and coarse grains, in other words, only loose fine-grained substrates (fluid muds) were not colonized. Within these three categories of substrates, the greatest diversity of substrate types is found within the hard category. Hard substrates include corals, both solitary rugose and tabulates such as *Syringopora* and *Aulopora*; brachiopods, particularly *Composita* and *Neospirifer* and some productids; bellerophontid gastropods; nautiloid cephalopods; crinoid columnals; oncolites (algal-coated brachiopods and rock fragments); and shingles (pebbles and cobbles) of rock fragments. Some examples of these substrates

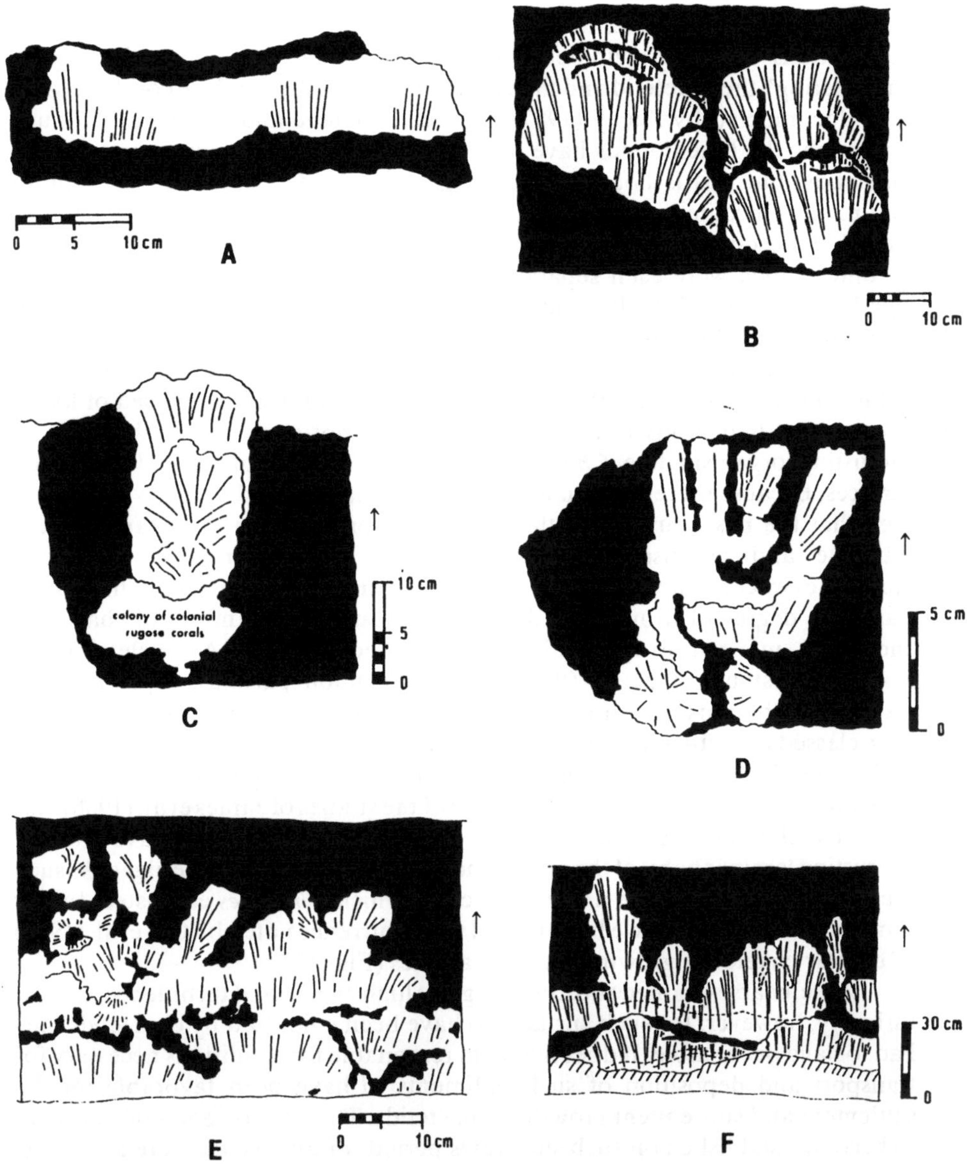


**Fig. 1.** Chaetetid growth forms. **A** Tabular chaetetid colony that grew on, and was covered by, skeletal wackestone, from the Blackjack Creek limestone (Kansas). **B** Domical chaetetid colonies with interruption partings, from the Amoret limestone (Kansas). **C** Columnar chaetetid colony that grew on a colony of colonial rugose corals, from the Marble Falls limestone (Texas). **D** Basket-shaped (flared) chaetetid colony, from the Middle Magdalena Group (Texas). **E** Digitate chaetetid colony, from the Middle Magdalena Group (Texas). **F** Complex of chaetetid colonies (tabular to high domical and columnar) from the Amoret limestone (Kansas). *Arrows* indicate stratigraphic up direction

are shown in Fig. 2A to E. In the case of tabulate corals, there can be several "generations" of overgrowth between the chaetetid and the coral, with the coral "smothering" the chaetetid and then a chaetetid "smothering" the coral. This has been documented by Suchy (1987) between *Aulopora* and *Chaetetes*. Not uncommonly, the brachiopods have been algal-coated before being encrusted by chaetetids, but this is not the case for gastropods, cephalopods, and crinoid columnals. Oncolites are a hard substrate frequently encrusted by chaetetids, indeed some domical chaetetid colonies, when sectioned, show two or more oncolites which were each separately encrusted by chaetetids and then fused together into one colony by continued chaetetid growth (Fig. 2E). Mathewson (1977) reported chaetetids encrusting shingles (pebbles/cobbles) of silty shale.

Substrates classed as firm are inferred to have been firm because as a sediment (before some lithification) they would have been composed of loose, fine grains (fluid muds in some cases). It does not seem reasonable that an encrusting organism like *Chaetetes* would have been able to colonize such surfaces. Because there are numerous examples of chaetetids in growth position such that the basal surface of the colony conforms to the topography of the substrate and because the rock composing the substrate is a skeletal mudstone/wackestone, shale, or organic parting, it is concluded that there was some degree of firmness to the colonized surface. Examples of chaetetids encrusting inferred firmgrounds are illustrated in Fig. 1A (skeletal mudstone/wackestone), Fig. 1E (thin clay, interruption parting), Fig. 2F (algal wackestone/boundstone), and Fig. 3A (black shale and skeletal wackestone). We have classed the substrate in Fig. 2F as an algal wackestone/boundstone because of the abundance of pylloid algal blades in the rock. It is quite possible that this substrate was an algal boundstone in light of the studies of James et al. (1988) that suggested that some of the Late Paleozoic phylloid algae may have had the encrusting laminar habit of the extant encrusting Peyssonneliaceae algae. As such they would have provided a firm surface on which *Chaetetes* encrusted. In this same substrate category we have placed the occurrence noted by Ota et al. (1969) of tabular chaetetids growing on stromatolites (Fig. 3B).

Loose, coarse-grained substrates suggest high energy and turbulence, but low turbidity, however, for such surfaces to have been colonized by chaetetids they had to have been stable. It does not seem reasonable for conditions necessary for transport and deposition of such sediments to have been favorable for the settlement and subsequent growth of chaetetids. Thus, for colonies of *Chaetetes* to become established on such substrates periods of quiescence were necessary. At the same time, once the chaetetid was established, it had to have been able to tolerate the relatively high energy and turbulence of such environments. In our studies, to date, three rock types are classes as loose, coarse-grained substrates: (1) fusulinid packstones/grainstones, (2) oolitic and algal-coated grainstones, and (3) arkoses. Figure 3C-E presents sketches, made from outcrop photographs of examples of chaetetids that grew on such substrates. In most cases, the colonies are relatively small with tabular and domical forms, or complex colonies (of these two forms) being dominant. Colonies that grew on the arkose or fusulinid packstones/grainstones have very ragged margins and numerous growth inter-

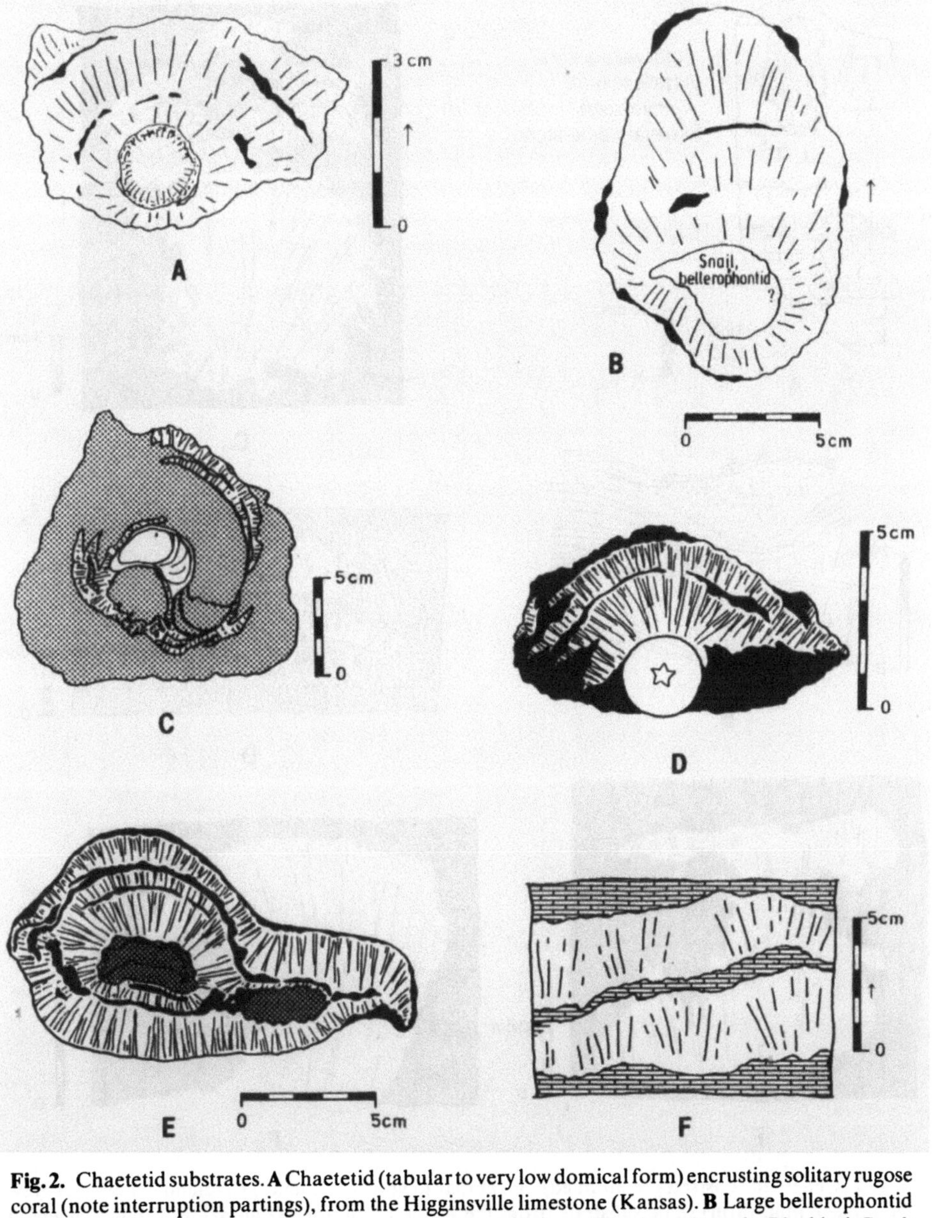


**Fig. 2.** Chaetetid substrates. **A** Chaetetid (tabular to very low domical form) encrusting solitary rugose coral (note interruption partings), from the Higginsville limestone (Kansas). **B** Large bellerophontid snail served as a substrate for what became a high domical chaetetid colony, from the Blackjack Creek limestone (Kansas). **C** Thin tabular chaetetids encrusted on the shell of a coiled nautiloid cephalopod, from the Amoret limestone (Kansas). **D** Tabular to low domical chaetetid colony encrusted on part of the stem of a crinoid, from the Pennsylvanian (Kansas). **E** complex of chaetetid encrustations on oncolites (note interruption partings), the oncolite on the *left* formed around the shell of a productid brachiopod *outlined by the dark line* within the oncolite, from the Amoret limestone (Kansas). **F** Tabular chaetetid colonies interlayered with algal wackestone/boundstone, from the Amoret limestone (Kansas). *Arrows* indicate stratigraphic up direction

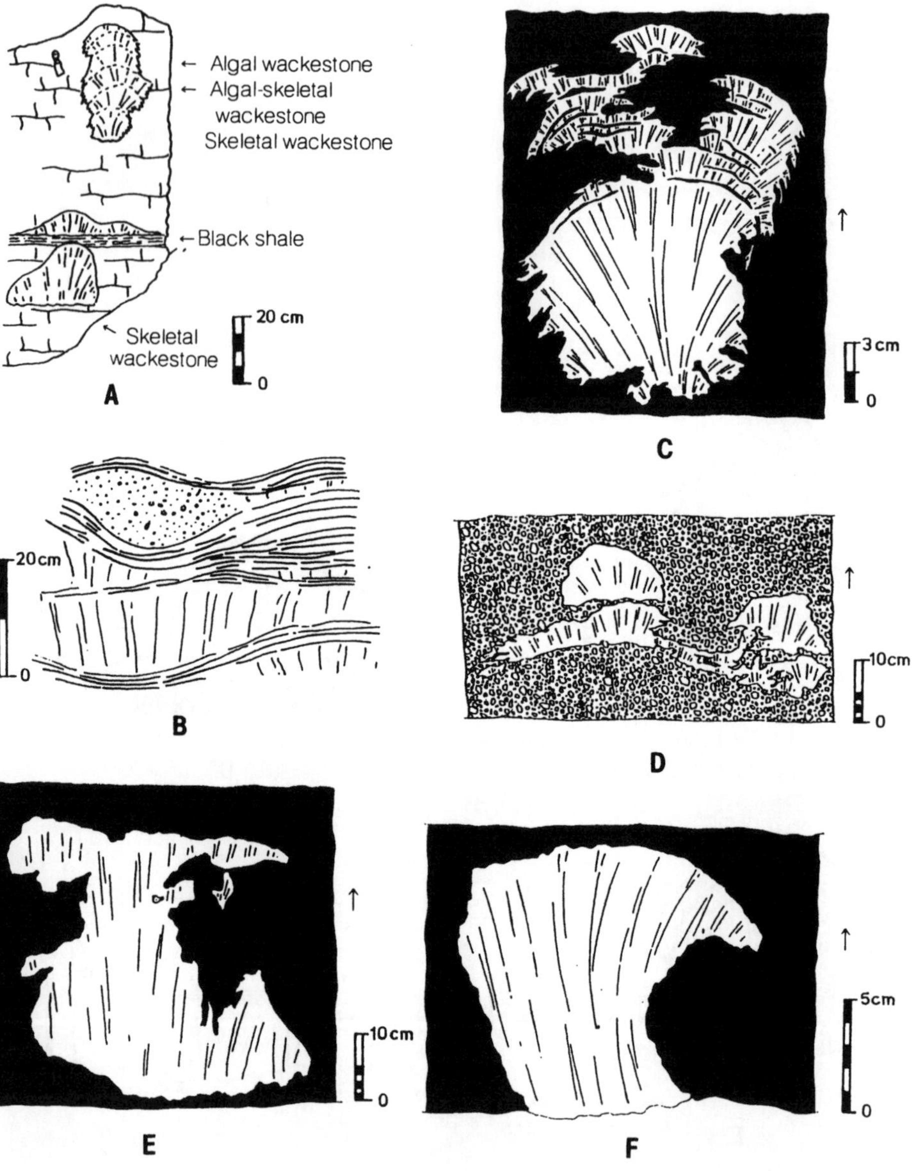


**Fig. 3.** Chaetetid substrates (continued). **A** Tabular to very low domical chaetetid colony that grew on a black mud (firm) shale substrate, from the Amoret limestone (Kansas). **B** Tabular chaetetid colonies interlayered with algal stromatolites. (Ota et al. 1969, Fig. 5, p. 8). **C** Ragged, domical chaetetid colony that grew on, and was covered by, fusulinid grainstone/packstone, from the Higginsville limestone (Kansas). **D** Tabular to low domical chaetetid colonies that grew on, and were covered by, arkose, from the Minturn Formation (Colorado). **E** Partially silicified chaetetid colony that grew on, and was covered by an oolitic grainstone, from the Lower Magdalena Group (La Tuna Formation, Texas). **F** Same substrate and enclosing lithology, as well as the same locality, as for specimen illustrated in **E**, note here the irregular base, smooth margins, and upward flare of the colony. *Arrows* indicate stratigraphic up direction

ruptions. Draping of the sediment over the colonies, as shown by tracing laminae within the fusulinid packstones/grainstones, suggest that the colonies stood about 15 cm above the adjacent substrate surface (Suchy 1987). The ragged margins, growth interruptions, and low colony relief indicate that the sediment accumulated as the colonies grew, and as long as the colony grew as fast, or faster than, the rate of sediment accumulation, it would escape smothering by the sediment. Not infrequently, the colony was smothered and another colony would start during the next period of quiescence. On the other hand, colonies that grew on an oolitic or algal-coated grain substrate are low domes (hemispherical) with relative smooth margins, and rather irregular bases (Fig. 3F). This suggests that they developed completely during periods of quiescence. Return of less stable conditions, more turbulence, is evidenced by overturned domical colonies at some localities (Fig. 4B).

Thus, *Chaetetes* was capable of inhabiting a wide variety of substrates under a rather wide range of environmental conditions, indeed one might think of them as rather eurytopic within the marine environment. If they were eurytopic, one then wonders why they were not more successful and why they did not build larger and more conspicuous reef structures. Perhaps the increase in phylloid algae during the Carboniferous (West 1988) prevented them from dominating the shallow epicontinental seas which record part of the history of the Carboniferous Period.

## Other Important Environmental Parameters

After initial colonization, environmental factors, other than substrate composition, became important influences in the habitat of *Chaetetes*. One factor that was important, or appears to have been important in determining the growth form, is irregularities in the substrate surface. Some data suggest that nonuniform vertical growth may produce domical and/or columnar colonies, but to our knowledge, there is no intrinsic (genetic) reason for (1) upward growth to exceed the rate of lateral growth or (2) one part of the initial encrusting layer to grow upward faster than any other part of that layer. We have the general impression that domical and columnar colonies exist because of an irregularity (topographic "high") in the substrate. In Fig. 4A the oncolite directly below the high domical colony existed as such an irregularity, and when encrusted by the initial chaetetid layer, which is continuous along the entire length/width of the colony, that part of the colony on the topographic "high" was more advantageously situated. It was higher above the seafloor/water interface, where it was more in the "stream" of water moving over the sea bed. As such it grew faster producing a colony as shown. The relatively smooth colony margins suggest that the environment was rather quiet following the development of the oncolites and "firming up" of the substrate, however it was probably still within wave base.

On the other hand, there are occurrences that suggest much more agitated, and somewhat cleaner, water. Figure 4B (from Connolly et al. 1989) documents domical chaetetid colonies with smooth margins that grew on skeletal wacke-

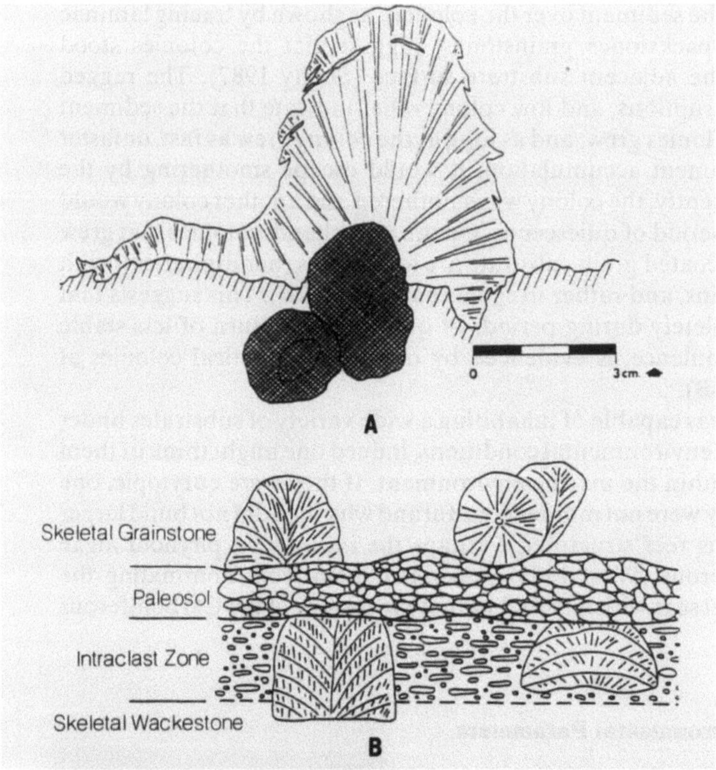


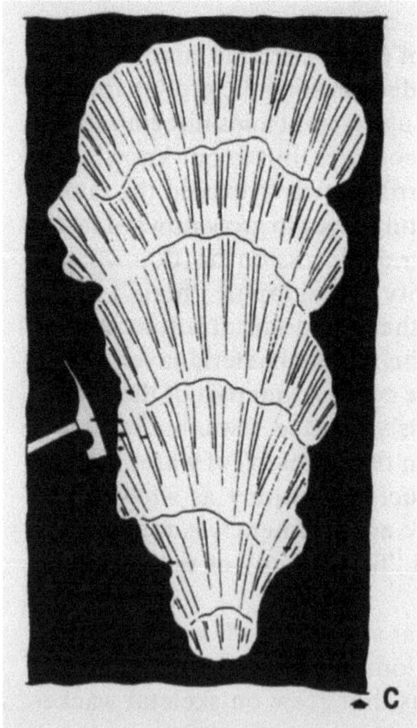


**Fig. 4.** Other aspects of chaetetid habitats. **A** Influence of substrate irregularities on the growth form of chaetetid colonies, from the Amoret limestone (Kansas), *dark lines* are the nuclei of the oncolites. **B** Domical chaetetid colonies with smooth margins that grew in a turbulent environment represented by the intraclast bed, from the La Tuna Formation (Texas), note also colonization on the paleosol surface and the subsequent rolling and continued growth of domical colonies as illustrated by the specimen in the *upper right*. Intraclast bed is about 20 cm thick, (Connolly et al. 1989). **C** Columnar chaetetid colony that flares upward suggested that it was growing into shallower water, from the Amoret limestone (Kansas), *hammer head* is 15 cm long. *Arrows* indicate stratigraphic upward direction

stones and intraclasts and were then truncated by erosion. Following erosion, a paleosol developed on the eroded surface which was colonized by domical chaetetid colonies (with smooth margins) during the next sea level rise (transgression). These latter colonies grew upward into an environment recorded by a skeletal grainstone. All of these features suggest very high energy in very shallow water; in fact, conditions close to intertidal. A more detailed discussion of this occurrence is given by Connolly et al. (1989).

Lambert and Stanton (1986) suggested that the lateral flaring of chaetetid colonies in the Atokan rocks of the Hueco Mountains in West Texas was a response by the colony to sea level, that is, as the colonies grew upward toward sea level they flared outward such that the tops of the colonies were at sea level (note the upper parts of the colonies in Fig. 1D).

At the other end of the spectrum, that is deep, quiet water, are the columnar colonies. Winston (1963), in his discussion of the over 3-m-tall chaetetid colonies in the Atokan rocks of Central Texas, interpreted their environment as an area of sediment by-pass in deep water. Thus they were below wave base, but as some of these large columnar colonies have been knocked over and in part broken, the environment was probably above storm wave base. Figure 4C is a sketch of a columnar colony with a smooth margin that appears to flare upward, suggesting, according to Lambert and Stanton (1986), that it may have grown, in its later stages, within wave base near sea level.

Tabular chaetetid colonies, the third basic colony geometry, appear to be confined to very shallow water, well within wave base. Additionally, the environment of these colonies was rather turbid and quiet as supported by their occurrence in skeletal wackestones (Fig. 1A); different types of mudrocks, including black shales (Fig. 3A); and arkoses (Fig. 3D). Tabular colonies also occur associated with algal wackestones/boundstones (Fig. 2F) and laminar stromatolites (Fig 3B). Occurrences with algae and stromatolites also indicates very shallow or clear water or both. While chaetetids occur in rocks containing algae and stromatolites, they appear to be, more or less, mutually exclusive, in that tabular chaetetid colonies occur interlayered with algal wackestones/boundstones or laminar stromatolites. These observations tend to suggest that when the water was shallow and clear photosynthesizing organisms dominated the environment and when sunlight was restricted by suspended particles (high turbidity) or high sedimentation tabular chaetetid colonies developed. In deeper waters, below the photic zone, high domical to columnar chaetetid colonies developed depending on, in part, substrate irregularities.

In addition to substrate type and character, other parameters that had an affect on chaetetid growth were substrate irregularity, water depth, water turbidity, water turbulence, and normal and storm wave base.

## Conclusions

The habitat of chaetetids was such that they occasionally developed into reef mounds (bioherms and biostromes). These reef mounds are dominated by tabular colonies which, in some cases, develop into high domical (Fig. 1F), and occasionally columnar colonies (Fig. 4C), depending on the variation in the environmental parameters noted above. The general habitat and range of environmental parameters for chaetetids is similar to that reported for demosponges in general by Wiedenmayer (1980). Chaetetids appear to have developed the largest colonies in light-free areas where there was little or no sedimentation. Such a habitat is very similar to one reported by Hartman (1983) for what were then called sclerosponges, but are now included with the coralline demosponges (Reitner 1986). Coralline demosponges (sclerosponges) are the extant forms most similar, morphologically, to chaetetids. Tabular chaetetid colonies, the dominant growth form in the Middle Carboniferous rocks of North America, have, not only a general morphological, but also a habitat similarity with the extant Pacific tabulate sponges *Acanthochaetetes wellsi*, in that the latter occurs in very shallow subtidal to intertidal environments (Saunders and Thayer 1987).

## References


Connolly WM, Lambert LL, Stanton RJ Jr (1989) Paleoecology of Lower and Middle Pennsylvanian (Upper Carboniferous) Chaetetes in North America. Facies 20:139–168

Hartman WD (1983) Modern and ancient Sclerospongiae. In: Rigby JK, Stearn CW (eds) Sponges and spongiomorphs, notes for a short course, University of Tennessee Department of Geological Sciences. Stud Geol 7, pp 116–129

Heckel PH (1974) Carbonate buildups in the geologic record: a review. In: Laporte LF (ed) Reefs in time and space. Society of Economic Paleontologists and Mineralogists Special Publication 18, pp 90–154

Hill D (1981) Rugosa and Tabulata. In: Teichert C (ed) Treatise on invertebrate paleontology, Part F, Coelenterata, Supplement 1, Univ Kansas Press, Lawrence, Kansas and Geol Soc Am, Boulder, Colorado, pp F1–F762

James NP (1984) Reefs. In: Walker RG (ed) Facies models. Geoscience Canada Reprint Series 1, 2nd edn, pp 229–244

James NP, Wray JL, Ginsburg RN (1988) Calcification on encrusting aragonite algae (Peyssonneliacea): implications for the origin of Late Paleozoic reefs and cements. J Sediment Petrol 58:291–303

Kershaw S, West RR (1987) Functional morphology of Pennsylvanian chaetetids in North America. Annual Meeting Palaeontological Association, abstract (first reserve paper)

Kershaw S, West RR (1988) Preliminary studies of chaetetid growth form. Fifth Int Symp on Fossil Cnidaria, including Archaeocyatha and Spongiomorphs. Int Assoc for the study of fossil Cnidaria and Porifera, Queensland Museum Brisbane, abstract, pp 45

Lambert LL, Stanton RJ Jr (1986) Environmental controls on *Chaetetes* growth morphology. Geol Soc Am, Abstracts with Program 18:665

Mathewson JE (1977) Chaetetids and their palaeoenvironments in the Amoret Limestone Member (Desmoinesian) of Labette County, Kansas. Kansas State University unpublished MS thesis, pp 148

Ota N, Sugimura A, Ota M (1969) Reef deposits in the *Millerella* zone of the Akiyoshi Limestone Group. Palaeontol Soc Jpn, Spec Pap 14:12

Reitner J (1986) Fossil and modern calcareous demosponges ("stromatoporoids," "chaetetids," "sphinctozoans" pars): systematics and polyphyletic origin. Fourth North American Paleontological Convention, Abstracts with Program, Boulder, Colorado, pp A38

Saunders WB, Thayer CW (1987) A cryptic intertidal brachiopod/sclerosponge community in Palau, West Caroline Islands. Geological Society of America, Abstracts with Program 19:829

Suchy DR (1987) Regional stratigraphic setting and paleoecology of a chaetetid "reef" in the Higginsville-Houx limestone (Pennsylvanian) of southeast Kansas. MS Thesis, Kansas State University, pp 286 (unpubl)

West RR (1988) Temporal changes in Carboniferous reef mound communities. Palaios 3:152–169

Wiedenmayer F (1980) Shallow-water sponges of the Bahams. In: Hartman WD, Wendt JW, Wiedenmayer F (eds) Living and fossil sponges; notes for a short course, Sedimenta VIII. University of Miami, Miami, Florida, pp 146–168

Winston D II (1963) Stratigraphy and carbonate petrology of the Marble Falls Formation, Mason and Kimble Counties, Texas. PhD Thesis, University of Texas (Austin), pp 344 (unpubl)

*Acknowledgements.* We are grateful to J. Reitner and H. Keupp for the invitation and financial assistance to participate in this Symposium. We acknowledge the Donors of the Petroleum Research Fund, administered by the American Chemical Society (ACS-PRF Grant 17375-AC2), the Earth Sciences Division (Stratigraphy and Paleontology Program) of the National Science Foundation (EAR-8816678), The Kansas Geological Society, Kansas State University, and the West London Institute (UK) for support of this research. The helpful comments we received from M. Connelly, A. Fagerstrom, C. Maples, and R. Stanton will be incorporated in the revision of this paper.

*Note added in proof by authors.* Chaetetids are coralline demosponges and while clonal they are not colonial *senso stricto*. The terms colonial and colony in this paper are NOT MEANT TO IMPLY that a sponge is an aggregate of individuals as is the case of some corals and bryozoans.

# Eastern Tethyan Sponge Buildups at the Close of the Paleozoic (Uppermost Permian, Sichuan/China)

J.W. Reinhardt

## Introduction

The Permian-Triassic boundary is generally considered to mark one of the major evolutionary crises or global mass extinction events during the history of life on earth. The erathem boundary between the Paleozoic and the Mesozoic (approx. 245 Ma) has been in the focus of numerous paleontological, sedimentological, and geochemical studies for decades (see, e.g., Logan and Hills 1973).

Due to the overall regressive character of the latest Permian, however, outcrops exhibiting a continuous deposition of marine and preferentially fossiliferous sediments are rare. Depositional breaks and erosion commonly separate Middle/Upper Permian from Lower Triassic rocks.

A number of locations in southern China are thought to exhibit continuous sedimentary records of the Permian-Triassic transition within the Eastern Tethyan Ocean (see, e.g., Rui et al. 1983; Sheng et al. 1984). Carbonates of the Uppermost Permian Changhsingian stage (Changxing formation) bear the youngest and final fusulinid faunas known to date (Rui 1979; Rui and Sheng 1981; Sheng and Rui 1984). They are described as being conformably overlain by marine sedimentary rocks of lowermost Triassic age (Fan et al. 1980; Rui et al. 1983).

This paper is intended to draw attention to sponge buildups which have recently been discovered and described from the subsurface and several outcrops of the above-mentioned Uppermost Permian Changxing formation limestones (Chen et al. 1985; Qiang et al. 1985; Reinhardt 1988a).

## Late Permian Reef Communities – New Clues for the "P/T Boundary Problem"?

As reef or buildup environments usually consist of rather complex faunal and floral communities which act as sensitive indicators of environmental changes, the Late Permian sponge buildups of southwestern China are considered to be of particular interest for the ongoing discussion on the "P/T boundary event". These reefs, together with a few counterparts on the Greek island of Skyros, are of prime importance for the study of the evolution of fossil reefs (Flügel and Reinhardt in press). They represent the last known reef ecosystems of the Paleozoic era and

J. Reitner and H. Keupp (Eds.)
Fossil and Recent Sponges

were followed by a first-order extinction event and a period of about 8 million years in the Lower Triassic from which no reefs are known worldwide (Flügel and Stanley 1984).

Thus, the reef communities of the Uppermost Permian could bear essential evidence for one (or more!) of the numerous postulated "event scenarios" for this period (see, e.g., Holser and Magaritz 1987).

## Geological Setting and Paleogeography

The sponge buildups of Sichuan and Hubei provinces are located at the margins and on the interior of an extensive Permo-Triassic intracratonic carbonate platform known as the East-Sichuan/West-Hubei Platform (SHP) (Fig. 1). During its time of development, the SHP was situated on the western part of the Yangzi Plate (South China Craton) which formed part of the continental shelf of Paleo-Eurasia and the Eastern Tethyan Ocean.

While Fan et al. (1982) described a reef location in eastern Hubei as a platform margin buildup with distinctive forereef, reef core, and backreef facies development, Reinhardt (1988a) described stratigraphic reef mounds with very similar faunas, however, lacking this clear zonation from southeastern Sichuan. The latter, a group of at least six buildups, known as the Beibei buildup group, were positioned along the southwestern margin and upper slope of the SHP (Fig. 1).

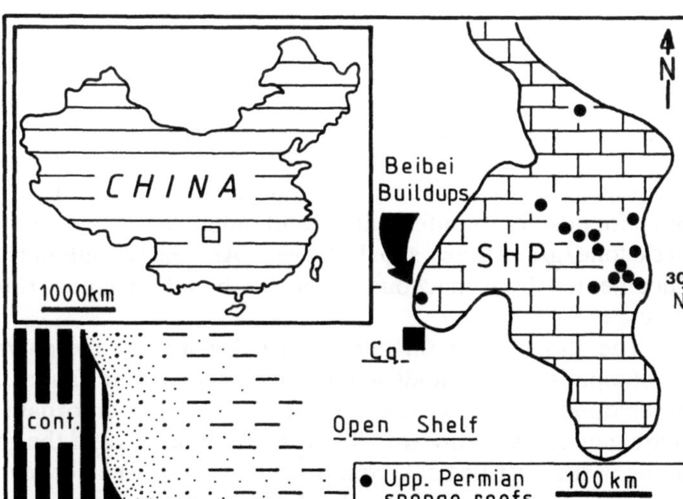


**Fig. 1.** Location map of the studied Late Permian sponge reefs of the East Sichuan/West Hubei carbonate platform (*SHP*) on the Eastern Tethyan Yangzi plate in southern China (*Cq* Chongqing city)

## Buildup Geometry and Sedimentology

Basing on the results of personal fieldwork, only the stratigraphic reef mounds of the Beibei area in southeastern Sichuan (Fig. 1) will be discussed in the following. They are situated about 35 km north-northeast of the city of Chongqing, along the western flank of the Guanyinxia anticline.

The sponge buildups form light to medium gray thick-bedded to massive limestone complexes with heights of 10–80 m and basal widths of 40–140 m. The limestones lack the abundant chert nodules characteristic of the bioclastic wackestones of the adjacent medium-bedded interbuildup facies. Buildup and interbuildup facies gradually interfinger and are not separated by sharp facies boundaries, primary reef flanks, or onlap structures. Furthermore, the lack of typical reef talus or reef debris deposits indicates that the buildups most probably showed a relatively gentle primary relief with low-angle flanks in an environment of medium to low water agitation.

Today's buildup morphology (flanks dipping 25–35° towards thinner interbuildup facies) can largely be considered as the result of a facies-selective diagenesis and compaction.

The buildups exhibit a vertical succession of sponge boundstone lenses or stacked reef patches recalling the "saccoliths" described by Smith (1981). These sponge saccoliths of 3–15 m width and 0.5–4 m height developed in several laterally correlateable horizons forming an intrabiohermal cyclicity of facies. The cyclicity is interpreted as reflecting minor relative sea-level fluctuations during the growth of the Late Permian buildups (Fig. 2), whereby the flourishing phases of the sponge faunas presumably reflect deeper environments at/below normal wave base (Reinhardt 1988b).

### *Basal Stabilization Stages*

Higher-energy accumulations of bioclastic sands and coral rubble, intercalated with occasional shell tempestites, form the foundation facies of the buildups (Fig. 3). Carbonate lithoclasts and reworked fragments of crinoids, brachiopods, foraminifera, algae and molluscs are the main constituents of this packstone/grainstone shoal facies. An early submarine lithification is documented by isopachous fibrous carbonate cements rimming interskeletal pores.

The shoals were further stabilized against wave energy and erosion by brachiopods and crinoids acting as pioneer sessile faunas. Subsequently, the brachiopods were also encrusted by fistuliporid bryozoans which increase in frequency and size upwards. These faunal changes and the gradual increase and fining of the micritic background sediment suggest a deepening of the colonized shoal environment in time. Bioclastic wackestones and floatstones with brachiopods, crinoids, bryozoans and a rich assemblage of nodosariid and fusulinid foraminifera are characteristic for this late stabilization stage.

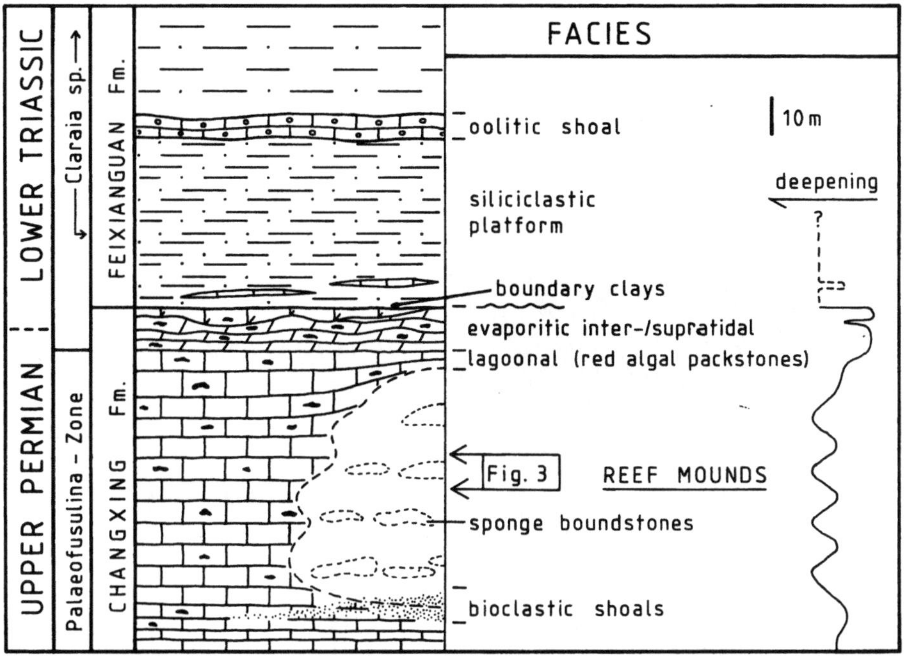


**Fig. 2.** Generalized section indicating lithology, stratigraphy, and depositional facies of the Late Permian buildups and the P/T transition in the area of study (Guanyinxia anticline, SE Sichuan)

## *Colonization Stages and the Organic Framework*

The boundstones representing the colonization stages of the buildup biota usually evolve gradually from the underlying facies of stabilizing faunas (and not-preserved floras! (Fig. 3).

Coralline sponges are the most important and dominant organisms within the boundstones of the described buildups. Although sponges can occasionally be found as bioclasts and isolated specimens throughout the buildup facies, they reach their highest diversity and density only in a few laterally correlateable horizons of sponge boundstone saccoliths. Here, they occur as large solitary individuals or in groups and colonies, forming the actual organic framework of the reefs (not comparable to rigid wave-resistant frameworks of modern reefs!). The dominating genus in terms of distribution and importance as framebuilder is *Amblysiphonella*. Sponges also carry an important role as encrusters, binders, and stabilizers within the framework.

The rich and diverse assemblage of sponges includes the following genera (minimum number of identified species in parentheses):

a) Sphinctozoans
*Amblysiphonella* (2), *Colospongia* (1), *Cystauletes* (1), *Imbricoelia* (1), *Parauvanella* (1), *Sollasia* (1)

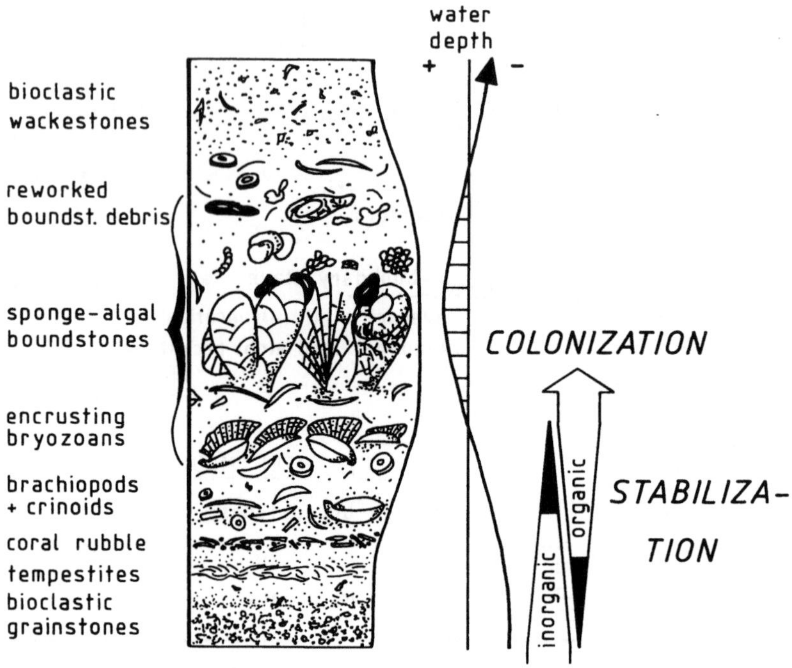


**Fig. 3.** Idealized section with typical facies units of a sponge boundstone saccolith, characteristic of the massive buildup facies of the studied Upper Permian buildups (SE Sichuan)

b) Inozoans
*Peronidella* (2), *Coelocladia* (1)
c) Sclerosponges/spongiomorphs
"Stromatoporoids" (Ceratoporellida): *Disjectoporida* (1), *Praeceratoporella* (1)
"Chaetetids" (Tabulospongida): *Permosoma* (1), gen. indet. (2)

Very occasionally, thin-stemmed rugose corals (*Plerophyllum*, *Waagenophyllum*) can be found as possible secondary framebuilders within the boundstones. The coral rubble found within the buildup foundation facies (discussed above) suggests the existence of coral "reef" patches, possibly in higher-energy environments of the carbonate platform.

Within the reef framework, additional important epibiotic encrusters and stabilizers are *Tubiphytes obscurus* Maslov, *Archaeolithoporella hidensis* Endo (both systematically problematical organisms), bryozoans (*Fistulipora*), sessile foraminifera, and sessile polychaete worms.

The most important preserved reef dwellers are fusulinids and smaller foraminifera (e.g., a highly diverse nodosariid assemblage), brachiopods (*Richthofenia*, *Oldhamina*, spiriferids), echinoderms (crinoids, echinids and holothurians), bryozoans (*Acanthocladia*, *Fistulipora*, *Rhombopora*), gastropods, bivalves, nautilids, ammonoids and ostracods. Calcreous algae are represented predominantly by red algae (*Gymnocodium*, *Permocalculus*) and occasional dasycladacean green algae (*Mizzia*).

## Biostratigraphy

The exact biostratigraphical age of the buildups in the Beibei area was determined with the help of fusulinids (det. F. Kahler, Klagenfurt 1987) (Fig. 2). The assemblage includes *Palaeofusulina wangi* Sheng, *P. mutabilis* Sheng, *P. nana* Likharev, *Reichelina changhsingensis* Sheng et Chang, *R. media* Miklukho-Maklay and *Parareichelina* cf. *mia* Kosninc. This clearly indicates deposition during the upper part of the *Palaeofusulina* zone [i.e. *Palaeofusulina sinensis* subzone (Rui 1979; Sheng and Rui 1984)], which corresponds to the youngest known Upper Permian.

## End-Permian Termination of Reef Growth and the "P/T Event"

The final stage of sponge boundstone saccoliths is conformably overlain by 5–15 m of shallow lagoonal fusulinid-bearing red algal (gymnocodiacean) pack-/wackestones. The high frequency/low diversity red algal assemblage is interpreted as reflecting a slightly restricted lagoonal setting (increasing salinity?). Caliche nodule extraclasts are observed as detritus originating from an adjacent supratidal environment. This evaporitic inter-/supratidal facies is documented in the overlying 1–5-m-thick package of dolomitic limestones and dolomites. They bear a rich inventory of characteristic features such as embryonic tepees, domal stromatolitic carbonates, and leached evaporites (calcite pseudomorphs after anhydrite, halite and gypsum crystals, and aggregates). An exact age of this transitionary facies has not been determined to date, as it only yields a very low diverse gastropod/ostracod fauna, lacking biostratigraphical indicators.

Subaerial exposure led to at least one (presumably relatively short) depositional break. Signs of erosion and karstification of the Uppermost Permian can be observed in a number of sections in the area of study (Reinhardt 1988a,b).

Laminated mudstones, shales and siltstones of the Feixianguan formation drape the carbonates of the Changxing formation (Fig. 2). Their lowermost Triassic age (Griesbachian) is indicated by the characteristic index fossils of the bivalve genus *Claraia*, appearing only several meters above the lithological P/T boundary.

Attention is drawn to one to two thin layers of yellowish-gray polymict illite-montmorillionite clays overlying the evaporitic and/or eroded Uppermost Permian carbonates in various locations of the studied area (Reinhardt 1988a, Fig. 28). The 1–6-cm-thick "boundary clays" at the base of the Feixianguan formation have been reported from numerous P/T boundaries in southern China (Rui and Jiang 1984; Sheng et al. 1984). Recent geochemical studies indicate that they are altered ash originating from extensive highly silicic volcanic eruptions

(Clark et al. 1986; Zhou and Kyte 1988; one of the analyzed boundaries of the latter study, Liangfenya section, lies only a few km away from the sponge reefs described above).

The massive input of silicic volcanic material into the environment at this stage of the P/T transition possibly explains the origin and distribution of the abundant early diagenetic chert nodules and bands within the Changxing formation. The cherts are found only within the bedded inter-reef facies and increase in number and size upwards towards the P/T boundary. Early lithification obviously prevented the silica-enriched pore fluids from percolating in the chert-free massive buildup facies.

Furthermore, large amounts of volcanic ash ejected into the atmosphere would have strongly filtered sunlight for a period long enough to considerably affect many photosynthesizing organisms (phytoplankton, zooxanthellae, marine and terrestrial floras), trophic chains, and symbiotic faunal/floral associations.

This might well have led to the end-Permian extinctions (perhaps in combination with weakened populations due to the reduced area and quality of habitat (Schopf 1974) during the documented synchronous regression).

Organisms which were possibly less affected by the sudden demise of marine and terrestrial floras (in marine environments, e.g., deposit feeders or less-specialized suspension filterers) were obviously more likely to survive the mass extinction event. Among these were conodonts (Clark et al. 1986), bivalves, and apparently a considerable number of sponges and reef-builders common to both Upper Permian and Middle/Upper Triassic reefs (Flügel and Stanley 1984).

## Conclusions

1. Shallow marine carbonates of Uppermost Permian age bear reef mounds with a rich and diverse faunal and floral assemblage along the upper slope and margin of the E-Sichuan/W-Hubei platform of southwestern China.
2. Fusulinids clearly indicate that the reef-bearing strata belongs to the youngest known Permian (upper *Palaeofusulina* zone).
3. Coralline sponges are abundant and dominating framebuilders within the boundstones of the buildups. They also act as important encrusters and stabilizers within the organic framework.
4. An increase of faunal diversity and complexity within the reef communities can be observed towards the P/T boundary.
5. Sea-level fluctuations during the Late Permian are documented as an intrabiohermal cyclicity of facies. An overall regressive trend culminates in form of restricted lagoonal and evaporitic inter-/supratidal carbonates just below/at the P/T boundary. At least one hiatus within the evaporitic supratidal facies and local karstification is documented.
6. No obvious "warnings" of a catastrophic mass extinction event could be found within the complex reef assemblages. Termination of reef growth can

mainly be attributed to paleogeographical facies changes (marine regression).

7. The end-Permian regression obviously reduced the area and quality of shallow-marine habitats (Schopf 1974). This possibly resulted in weakened faunal and floral populations.
8. Acidic volcanic eruptions, identified from the geochemistry of the widely distributed "illite-montmorillionite boundary clays" (Clark et al. 1986; Zhou and Kyte 1988), are considered to have had the strongest impact on all photosynthesizing organisms, trophic chains, and symbiotic faunal/floral associations.
9. The combined effect of massive ejecta-rich volcanic eruptions during a time of considerable marine regression is interpreted as having been the most likely cause for mass extinctions at the P/T boundary of southern China.

*Acknowledgments.* I thank the authorities of the Chengdu College of Geology (Sichuan, China) and the Ministry of Geology and Mineral Resources in China (Beijing) for permission to carry out field work in Sichuan during spring and summer of 1986. I am particularly indebted to Prof. Dr. Peng Dajun (Chengdu) and Prof. Dr. Wu Xichun (Chengdu) for their supervision and kind support in China, to Prof. Dr. Erik Flügel (Erlangen) for support and supervision during all stages of my research, to Prof. Dr. F. Kahler (Klagenfurt) for fusulinid identification, Dr. Baba Senowbari-Daryan (Erlangen) for assistance with sponge identification and to Prof. Dr. Jobst Wendt (Tübingen) for critical discussions in the field and as a reviewer of an initial version of the manuscript.

The study is part of the research project Evolution of Reefs (Fl 42/49), supported by the Deutsche Forschungsgemeinschaft.

## References

Chen JG, Zhao XW, Zhang YB (1985) Distribution of the reefs in Changxing Formation of Upper Permian in Sichuan Basin and its relation to oil and gas. Nat Gas Ind 5:10–18

Clark DL, Wang CY, Orth CJ, Gilmore JS (1986) Conodont survival and low iridium abundances across the Permian-Triassic boundary in south China. Science 233:984–986

Fan JS, Sun YY, Chang SX (1980) On the Permian-Triassic boundary in southern China. Sci Geol Sin 2:112–124

Fan JS, Ma X, Zhang W (1982) The Upper Permian reefs in West Hubei, China. Facies 6:1–14

Flügel E, Reinhardt JW Uppermost Permian reefs in Skyros (Greece) and Sichuan (China): implications for the Late Permian extinction event. Palaios (in press)

Flügel E, Stanley GD (1984) Re-organization, development and evolution of post-Permian reefs and reef-organisms. Palaeontol Am 54:177–186

Holser WT, Magaritz M (1987) Events near the Permian-Triassic boundary. Mod Geol 11:155–180

Logan A, Hills LV (eds) (1973) The Permian and Triassic systems and their mutual boundary. Can Soc Petrol Geol Mem 2, Calgary

Qiang ZT, Guo YH, Zhang F, Yan CT, Zheng JF (1985) The upper Permian reef and its diagenesis in Sichuan Basin. Oil Gas Geol 6:82–90

Reinhardt JW (1988a) Uppermost Permian reefs and Permo-Triassic sedimentary facies from the southeastern margin of Sichuan Basin, China. Facies 18:231–288

Reinhardt JW (1988b) End-Paleozoic regression and carbonate to siliciclastic transition: a Permian-Triassic boundary on the Eastern Tethyan Yangzi Platform (Sichuan, China). Zentralbl Geol Palaeontol Teil I:7–8

Rui L (1979) Upper Permian fusulinids from western Guizhou. Acta Palaeontol Sin 18:271–297

Rui L, Jiang NY (1984) Lithofacies and biofacies of latest Permian and earliest Triassic at the bordering region of Jiangsu, Zhejiang and Anhui provinces. Acta Palaeontol Sin 23:286–298

Rui L, Sheng JZ (1981) On the genus *Palaeofusulina*. Geol Soc Am Spec Pap 187:33–37
Rui L, Jiang NY, Cheng CZ (1983) Latest Permian and earliest Triassic sedimentary provinces and their facies types in South China. Sci Sin Ser B 26:1099–1108
Schopf TJM (1974) Permo-Triassic extinctions: relation to sea-floor spreading. J Geol 82:129–143
Sheng JZ, Rui L (1984) Fusulinaceans from Upper Permian Changhsingian in Mingshan coal field of Leping. Jiangxi. Acta Micropal Sin 1:30–48
Sheng JZ, Chen CZ, Wang YG, Rui L, Liao ZT, Bando Y, Ishii KI, Nakazawa K, Nakamura K (1984) Permian-Triassic boundary in middle and eastern Tethys. J Fac Sci Hokkaido Univ Ser IV 21:133–181
Smith DB (1981) Bryozoan-algal patch-reefs in the Upper Permian Lower Magnesian limestone of Yorkshire, northeast England. Soc Econ Paleontol Mineral Spec Publ 30:187–202
Zhou L, Kyte FT (1988) The Permian-Triassic boundary event: a geochemical study of three Chinese sections. Earth Planet Sci Lett 90:411–421

# Pharetronids in Triassic Reefs

P. RIEDEL and B. SENOWBARI-DARYAN

## Introduction

During the Phanerozoic, different groups of sponges played more or less important roles in reef structures, acting as reef builders, reef dwellers, or as encrusting and stabilizing organisms of the primary reef frame (Fig. 3b). Sponges were also important pioneers, for example acting as sediment fixers, and provided substrate for other reef builders (e.g., corals).

Particularly in Permian and Triassic reefs most fabrics are dominated by sponges. They have been attributed to the Pharetronida, an order introduced by Zittel (1878) and characterized as calcareous sponges with a rigid skeleton composed of parallel calcareous fibers embedded in a calcareous matrix. Since Steinmann's "Pharetronen-Studien" (1882) this order of calcareous sponges has been subdivided into two suborders: the segmented "Sphinctozoans" (Figs. 2b; 3a,b) and the massive, unsegmented "Inozoans" (Fig. 2a). Although the pharetronids seem to be a uniform systematic group, later investigations by Ott (1967) gave strong evidence for the polyphyletic origin of the "sphinctozoans" (see also Senowbari-Daryan this Vol.).

## Problems

Compilations of stratigraphic and regional occurrences of the Triassic pharetronids reported here (Tables 1–4) are based on reports in the literature and on our studies of Triassic reefs of the alpine-mediterranean realm (Fig. 1). The "sphinctozoan" faunas are well documented, at least from the western Tethys, and more than 100 Triassic species are known. However, studies at the species level from other regions are rare.

The "inozoan" faunas are poorly investigated. Only the one compilation by Dieci et al. (1968) has documented the unsegmented pharetronids of the Cassian Beds (southern Alps).

We have based our study at the generic level, because of the available data. Some genera are shown in inverted commas (Tables 1, 3) because, in our opinion, the affiliations of some species to these genera are doubtful.

Of course, it would be more satisfactory to show development of Triassic pharetronids during short time spans. Exact biostratigraphic dates for the Triassic

J. Reitner and H. Keupp (Eds.)
Fossil and Recent Sponges

Table 1. Stratigraphic ranges of the "sphinctozoan" genera during the Triassic

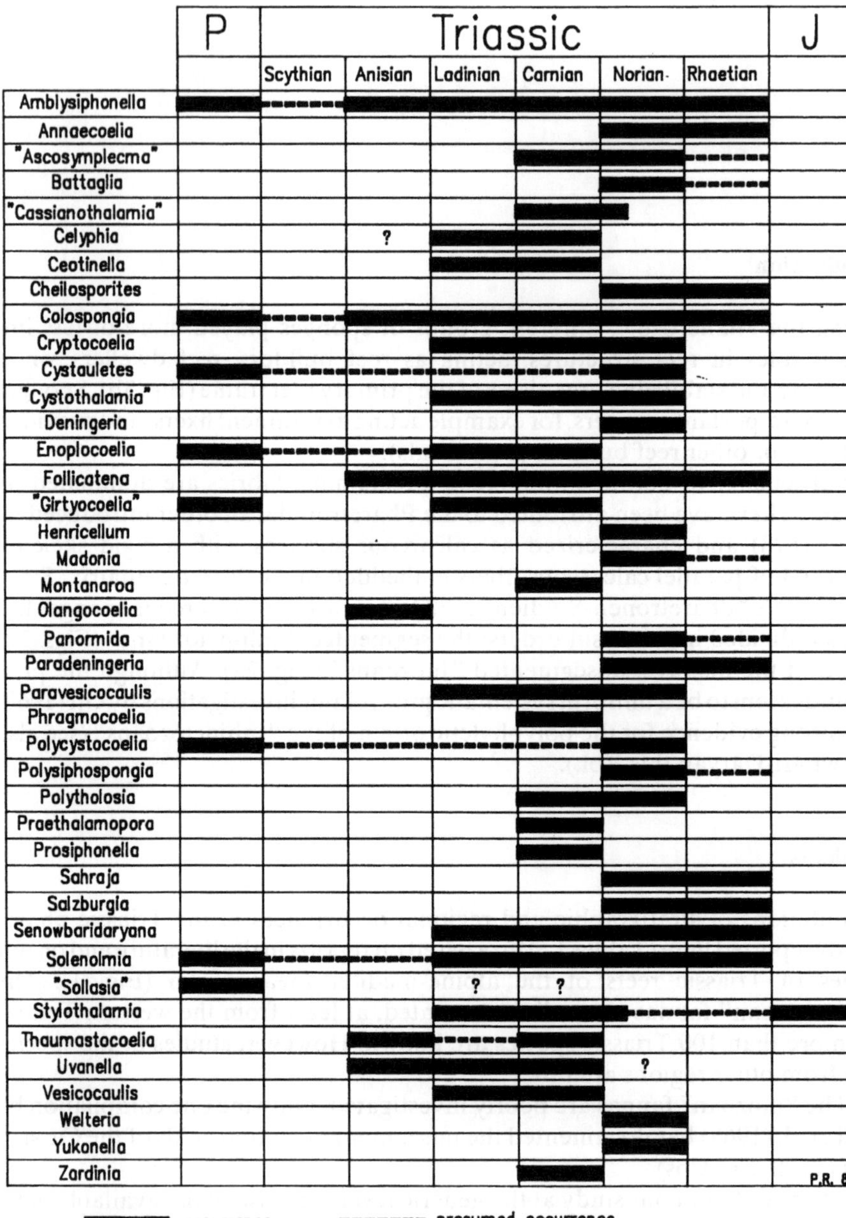

Table 2. Stratigraphic ranges of the "inozoan" genera during the Triassic

| | P | Triassic | | | | | | J |
|---|---|---|---|---|---|---|---|---|
| | | Scythian | Anisian | Ladinian | Carnian | Norian | Rhaetian | |
| Brevisiphonella | | | | | ■ | | | |
| Calicia | | | | | ■ | | | |
| Corynella | ■ | | | | ■ | ■ | ■ | ■ |
| Eudea | | | | | ■ | | | |
| Euepirrhysia | | | | | ■ | | | |
| Himatella | ■ | | | | ■ | | | |
| Hodsia | | | | | | ■ | ■ | |
| Holocoelia | | | | ■ | ■ | | | |
| Leiospongia | | | | | ■ | | | |
| Molengraaffia | | | | | ■ | ■ | ■ | |
| Peronidella | | | ■ | – – – | ■ | ■ | ■ | ■ |
| Polysiphonella | | | | | ■ | | | |
| Precorynella | | | | ■ | ■ | | | |
| Reticulocoelia | | | | | ■ | ■ | | |
| Sestrostomella | | | | ■ | ■ | | | ■ |
| Solenocoelia | | | | | | ■ | | |
| Stellispongia | ■ | | | | ■ | | | ■ |

■ occurrence  – – – presumed occurrence

Table 4. Regional occurrences of the Triassic "inozoan" genera

| | Northern Alps | Southern Alps | Carpathians | Julian Alps | Bakony/Hungary | Sicily/Italy | Hydra/Greece | Taurids/TR | Pamir/USSR | Himalaya/India | Sichuan/China | Indonesia | Nevada/USA | Oregon/USA | Yukon/CDN | Andes/Peru |
|---|---|---|---|---|---|---|---|---|---|---|---|---|---|---|---|---|
| Brevisiphonella | | ■ | | | | | | | | | | | | | | |
| Calicia | ■ | | | | | | | | | | | | | | | |
| Corynella | ■ | ■ | ■ | ■ | ■ | | | | ■ | | | ■ | | | | ⧗ |
| Eudea | | ■ | | | | | | | | | | | | | | |
| Euepirrhysia | | ■ | | | | | | | | | | | | | | |
| Himatella | | ■ | | | ■ | | | | | | | ■ | | | | |
| Hodsia | | | | | | | | | | | | | | | | |
| Holocoelia | ■ | | | | | | | ■ | | | | | | | | |
| Leiospongia | | ■ | ■ | | ■ | | | | | | | ■ | | | | |
| Molengraaffia | ■ | | ■ | | ■ | | | | ■ | | | ■ | | | | |
| Peronidella | ■ | ■ | ■ | | ■ | ■ | ■ | | | ■ | ■ | ■ | | ■ | ⧗ | |
| Polysiphonella | | ■ | | | | | | | | | | | | | | |
| Precorynella | ⧗ | ■ | | ■ | | | | | ■ | | | | | | ⧗ | |
| Reticulocoelia | | ■ | | | | | | ■ | | | | | | | | |
| Sestrostomella | | ■ | ■ | ■ | | | | | | | | | | | | |
| Solenocoelia | | | | | | | | ■ | | | | | | | | |
| Stellispongia | ■ | ■ | | | ■ | | | | | | | ■ | | | | ■ |

■ occurrence  ⧗ presumed occurrence

Table 3. Regional occurrences of the Triassic "sphinctozoan" genera

| | Northern Alps | Southern Alps | Carpathians | Julian Alps | Bakony/Hungary | Sicily/Italy | Hydra/Greece | Taurids/TR | Pamir/USSR | Himalaya/India | Sichuan/China | Indonesia | Nevada/USA | Oregon/USA | Yukon/CDN | Andes/Peru |
|---|---|---|---|---|---|---|---|---|---|---|---|---|---|---|---|---|
| Amblysiphonella | ■ | ■ | ■ | | ■ | ■ | ■ | ■ | ■ | | ■ | ■ | | ■ | | ■ |
| Annaecoelia | ■ | | | ■ | | ■ | | | | | | | | | | |
| "Ascosymplecma" | | ■ | | | | | | ■ | | ■ | | | ■ | | ■ | ■ |
| Battaglia | | | | ■ | | ■ | | | | | | | | | | |
| "Cassianothalamia" | | ■ | | | | | | ■ | | | | | | | | |
| Celyphia | ■ | ■ | ■ | | ■ | ■ | | ■ | | | | | | | | |
| Ceotinella | ■ | | ■ | ■ | | ■ | ■ | | | | | | | | | |
| Cheilosporites | ■ | | ■ | ■ | | ■ | | ■ | | | | | | | | |
| Colospongia | ■ | ■ | ■ | ■ | ■ | ■ | ■ | ■ | ■ | | ■ | ■ | | ■ | ■ | |
| Cryptocoelia | ■ | ■ | ■ | ■ | | ■ | ■ | ■ | | | | | | | ■ | |
| Cystauletes | ■ | | | | | ■ | | | | | | | | | | |
| "Cystothalamia" | ■ | ■ | ■ | ■ | | ■ | ■ | ■ | ■ | | | | | | | |
| Deningeria | ■ | | | | | | | | ■ | | ■ | ■ | | | | |
| Enoplocoelia | | ■ | | | | | | | | | | | | | | |
| Follicatena | ■ | | ■ | ■ | | ■ | ■ | | ■ | | | | | ■ | ■ | |
| "Girtyocoelia" | ■ | | ■ | ■ | | ■ | | | | | | | | | | |
| Henricellum | | | | | | | | | | | ■ | ■ | | | ■ | |
| Madonia | | | | | | ■ | | | | | | | | | | |
| Montanaroa | | ■ | | | | | | | | | | | | | | |
| Olangocoelia | | ■ | | | | | | | | | | | | | | |
| Panormida | ⧗ | | | | | ■ | | | | | | | | | | |
| Paradeningeria | ■ | | | ■ | | ■ | | ⧗ | | ■ | | | | ■ | ■ | |
| Paravesicocaulis | ⧗ | | ■ | ■ | | ■ | ■ | | | | | | | | | |
| Phragmocoelia | ■ | | | | | | | | | | | | | | | |
| Polycystocoelia | | | | | | | | | | | | | | | ■ | |
| Polysiphospongia | | | | | | ■ | | | | | | | | | | |
| Polytholosia | ■ | | | | | ■ | | | ■ | | | | ■ | | ■ | ■ |
| Praethalamopora | | ■ | | | | | | | | | | | | | | |
| Prosiphonella | | ■ | | | | | | | | | | | | | | |
| Sahraja | | | | | | | | | ■ | | ■ | | | | | |
| Salzburgia | ■ | | | | | | | | | | | | | ⧗ | ■ | |
| Senowbaridaryana | ■ | | ■ | | | ■ | ■ | ⧗ | ■ | | ■ | | | | | |
| Solenolmia | ■ | ■ | ■ | ■ | | ■ | ■ | ■ | ■ | ■ | | | | | ■ | |
| "Sollasia" | | | ⧗ | | | ⧗ | | ⧗ | | | | | | | | |
| Stylothalamia | ■ | | ■ | ■ | | ■ | ■ | ■ | | | | | | | | |
| Thaumastocoelia | | ■ | | | ■ | | | | | | | | | | | |
| Uvanella | ■ | ■ | ■ | ■ | | ■ | ■ | ■ | | | | | | | ■ | |
| Vesicocaulis | ■ | ■ | ■ | ■ | | ■ | ■ | ⧗ | | | | | | | | |
| Welteria | ■ | | | ■ | | ■ | | | | | ■ | ■ | | | | |
| Yukonella | | | | | | | | | | | | | | | ■ | |
| Zardinia | ■ | ■ | | ■ | | | ■ | | | | | | | | | P.R. 88 |

■ occurrence ⧗ presumed occurrence

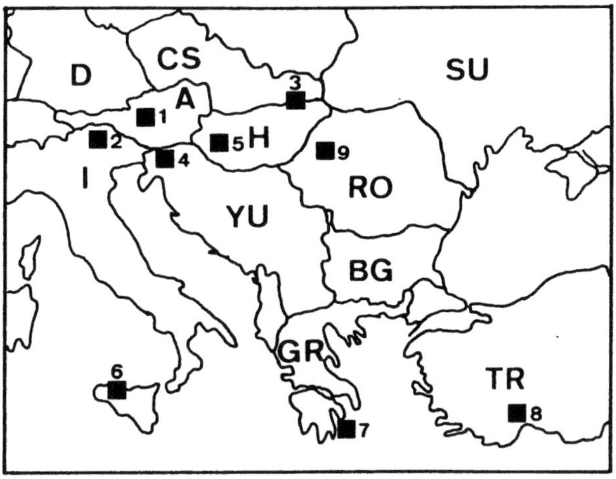


**Fig. 1.** Regions of the alpine-Mediterranean realm with sites containing Triassic pharetronids. *1* northern Alps; *2* southern Alps; *3* western Carpathians; *4* Julian Alps; *5* Bakony Mountains; *6* Sicily; *7* Hydra; *8* western Taurids; *9* Apuseni Mountains/Carpathians

reefs are not available because of the lack of index fossils and/or because of the lack of investigations. So we will refer this study to stage level as these age determinations are comparatively certain.

## Reef Development During the Triassic

The Triassic, particularly in the alpine-mediterranean realm of the Tethys, is commonly represented by shallow water limestones that accumulated on large platforms, which are rimmed by reefs. We will focus on the alpine region, for many reef studies have been carried out in this region (Fig. 1; see also regional occurrences in Tables 3 and 4).

In the Lower Triassic (Scythian) no reefs or reef organisms have been reported yet. Oldest reef structures from the Triassic are known from the Anisian of the southern Alps (Bechstädt and Brandner 1970) and from the Carpathians (Scholz 1972).

The more widespread reef and platform development of the Wetterstein Limestone lasted from the Ladinian to the Lower Carnian (Cordevolian). According to Lein et al. (in press), it can be subdivided into two independent phases, at least in the Northern Calcareous Alps. Each represents a regressive cycle that ended with subaerial exposure of the shallow water carbonates. The Middle Carnian Reingrabener Wende (Schlager and Schöllnberger 1973), a transgressional phase, enabled the formation of the Tisovec Limestone platform which extended up to the Upper Carnian (Tuvalian).

In some places in the Alps, parts of the Carnian are represented by Leckkogel Beds, which are sponge-rich carbonate banks within clastic, basinal beds at the slope of the Tisovec platform (Lein 1980); by sponge biostromes within clastic beds (Turnsek et al. 1982); or by "Cipit" boulders, which are allochthonous boulders of reef limestone within clastic strata.

# Pharetronids in Triassic Reefs

P. Riedel and B. Senowbari-Daryan

## Introduction

During the Phanerozoic, different groups of sponges played more or less important roles in reef structures, acting as reef builders, reef dwellers, or as encrusting and stabilizing organisms of the primary reef frame (Fig. 3b). Sponges were also important pioneers, for example acting as sediment fixers, and provided substrate for other reef builders (e.g., corals).

Particularly in Permian and Triassic reefs most fabrics are dominated by sponges. They have been attributed to the Pharetronida, an order introduced by Zittel (1878) and characterized as calcareous sponges with a rigid skeleton composed of parallel calcareous fibers embedded in a calcareous matrix. Since Steinmann's "Pharetronen-Studien" (1882) this order of calcareous sponges has been subdivided into two suborders: the segmented "Sphinctozoans" (Figs. 2b; 3a,b) and the massive, unsegmented "Inozoans" (Fig. 2a). Although the pharetronids seem to be a uniform systematic group, later investigations by Ott (1967) gave strong evidence for the polyphyletic origin of the "sphinctozoans" (see also Senowbari-Daryan this Vol.).

## Problems

Compilations of stratigraphic and regional occurrences of the Triassic pharetronids reported here (Tables 1–4) are based on reports in the literature and on our studies of Triassic reefs of the alpine-mediterranean realm (Fig. 1). The "sphinctozoan" faunas are well documented, at least from the western Tethys, and more than 100 Triassic species are known. However, studies at the species level from other regions are rare.

The "inozoan" faunas are poorly investigated. Only the one compilation by Dieci et al. (1968) has documented the unsegmented pharetronids of the Cassian Beds (southern Alps).

We have based our study at the generic level, because of the available data. Some genera are shown in inverted commas (Tables 1, 3) because, in our opinion, the affiliations of some species to these genera are doubtful.

Of course, it would be more satisfactory to show development of Triassic pharetronids during short time spans. Exact biostratigraphic dates for the Triassic

J. Reitner and H. Keupp (Eds.)
Fossil and Recent Sponges

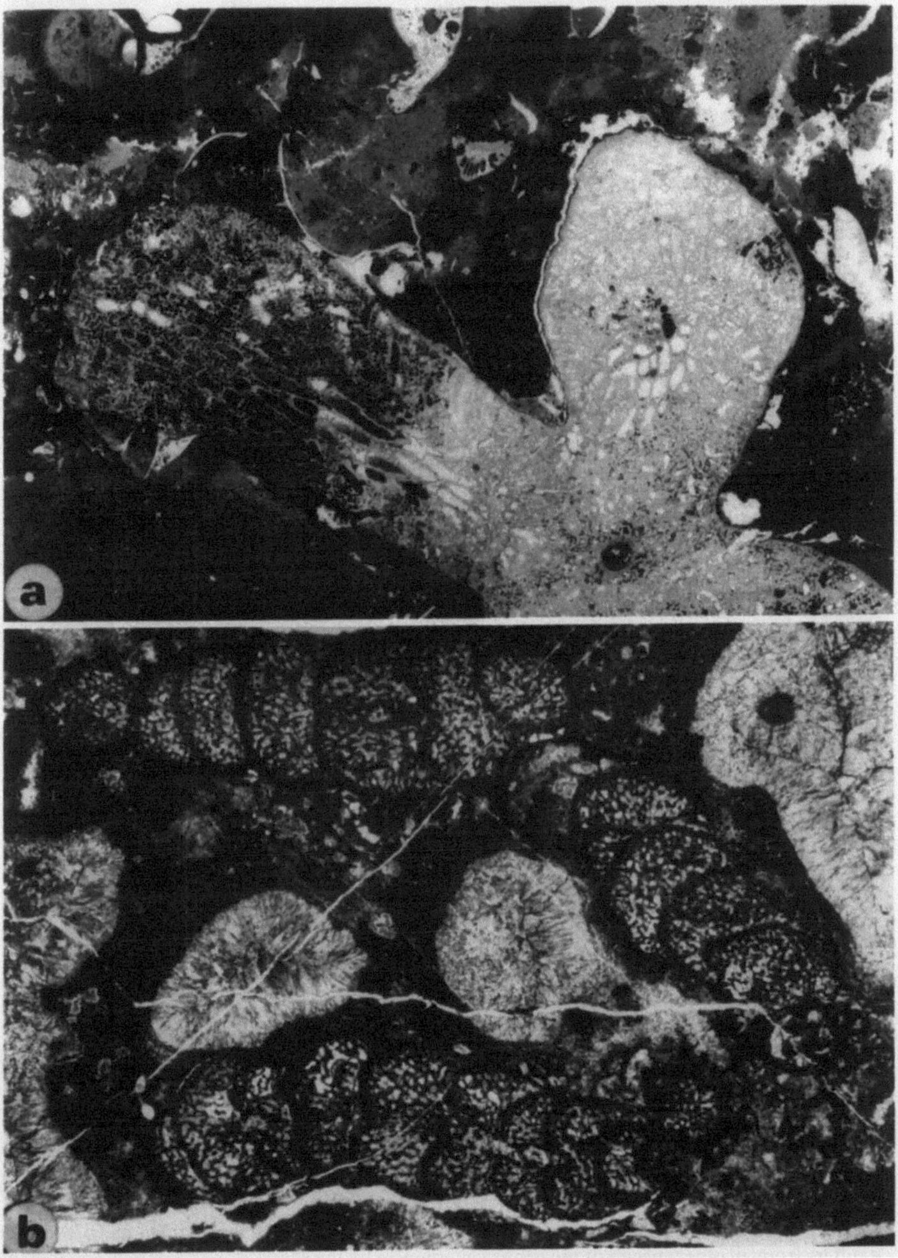

**Fig. 2. a** *Sestrostomella robusta* Zittel occurs here as a bioclast in a "Cipit" boulder from the Cassian Formation. This branching specimen is partly recrystallized. At the *left* the reticular filling tissue and the tubes of the spongocoel are still recognizable. Alpe di Specie, southern Alps; Carnian; scale 2×. **b** *Solenolmia manon* (Münster) is a very common baffler in Ladinian and Carnian reefs. The longitudinal sections of the two individuals show the reticular filling tissue. In some of the chambers of these "sphinctozoans" the central channel is visible. Pokljuka, Julian Alps; Carnian; scale 3.5×

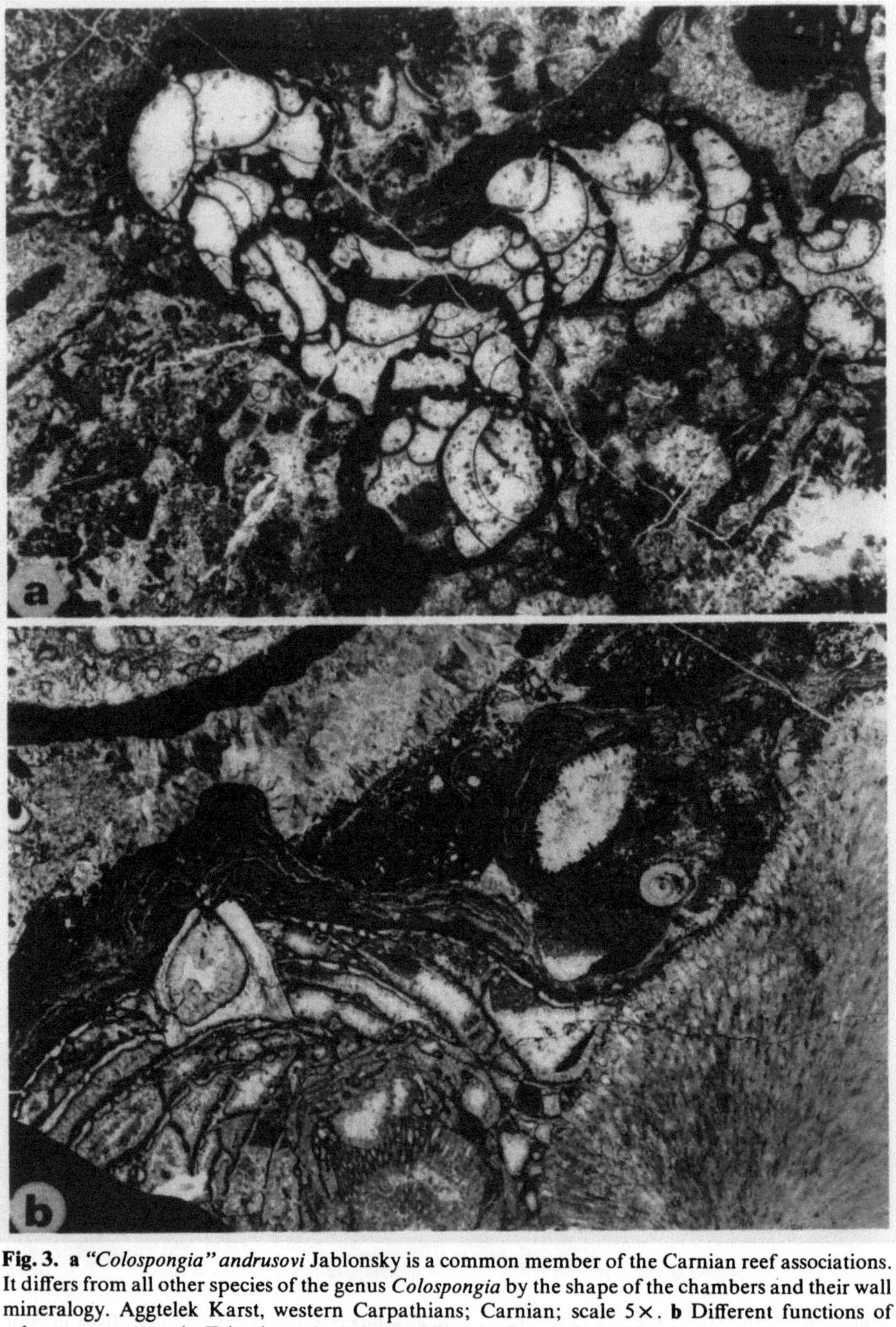

**Fig. 3. a** *"Colospongia" andrusovi* Jablonsky is a common member of the Carnian reef associations. It differs from all other species of the genus *Colospongia* by the shape of the chambers and their wall mineralogy. Aggtelek Karst, western Carpathians; Carnian; scale 5×. **b** Different functions of calcareous sponges in Triassic reefs. Substrate for initial growth of this porate, irregularly shaped sponge is a small chaetetid. The sponge which has overgrown it is also acting as a stabilizing organism of the primary frame, as it encrusts the neighboring chaetetid, too. The sponge itself serves as substrate for a barnacle-like organism, but this does not harm it for the sponge partly integrates the shell into its body. Finally, the sponge, the shell, and the big chaetetid are encrusted by thin layers of *Spongiostromata*. Cozzo di Lupo, Sicily; Norian; scale 2×

reef biota. This already indicates the decline of the pharetronids (Fig. 4) and the success of the corals, which since then have become very important reef-building organisms. Among the pharetronids the "sphinctozoans" dominate.

## Development of the Pharetronids During the Triassic

From the Upper Carboniferous until the end of the Permian a first radiation of pharetronid sponges, especially of the "sphinctozoans", occurs but most of these genera became extinct at the end of the Permian. During the Triassic another radiation of pharetronids can be observed, for 33 new "sphinctozoan" and 14 "inozoan" genera appear (Fig. 4).

As of now, no sponges are known from the Lower Triassic (Scythian). First occurrences of Triassic pharetronids are in the Anisian (Tables 1, 3). Probably due to the state of research (and preservation?) only seven genera (six "sphinctozoans" and one "inozoan") have been described. Two of them are conservative types like *Amblysiphonella* and *Colospongia*, which occur in Permian and Triassic faunas all over the world (Table 2). It is our opinion that *Follicatena*, which extends from the Anisian to the Rhaetian in Table 1, may include two quite similar though distinct "sphinctozoan" genera. *Follicatena* sensu Ott (1967) is restricted to the Anisian, Ladinian, and Carnian, whereas Norian and Rhaetian forms should be attributed to a different genus because of differences in skeletal mineralogy. In addition to these genera, Bechstädt and Brandner (1970) described a probable representative of *Celyphia*.

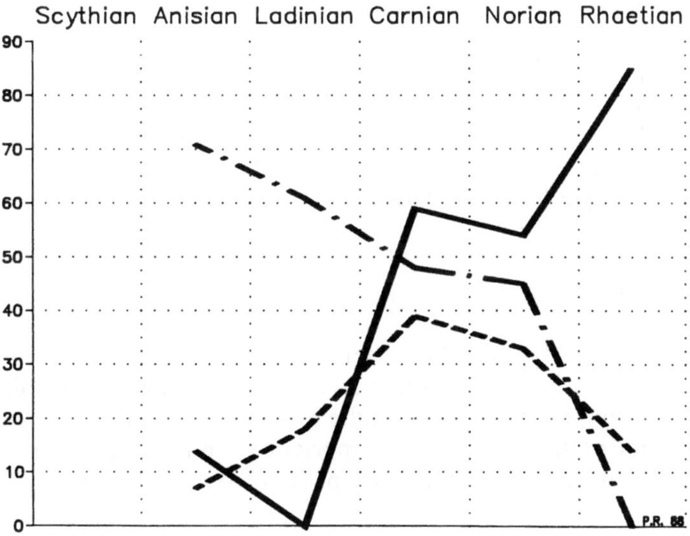


**Fig. 4.** The plot shows, at stage level, for the Triassic the total number of genera (*dashed line*), the percentage of extinct genera (*solid line*) and the percentage of new genera (*dash-dot line*)

In the Ladinian, 13 new genera appear, two of which have identifiable ancestors in the Permian (*Solenolmia*, *Enoplocoelia*; Table 1). Two new "inozoan" genera are known from the Upper Ladinian (Table 2). Ranges of all these taxa extend up to the end of the Carnian. "*Cystothalamia*" and "*Girtyocoelia*" are shown in quotation marks in Table 1 because the Triassic species attributed to these genera have different skeletal mineralogy than the Permian taxa. Occurrences of the Permian "*Sollasia*" in the Triassic, marked by question marks in Table 1, are uncertain, because the Triassic specimens do not show close affinities to Permian ones. *Cryptocoelia*, "*Cystothalamia*", *Paravesicocaulis* extend up to the end of the Norian and *Senowbaridaryana* and *Solenolmia* extend up to the end of the Rhaetian. *Stylothalamia* also occurs in the Lower Norian (Lacian 1) "Cipit"-limestones of the western Taurids (Turkey) and has been described from the Lower Jurassic of the Southern Alps (Beccarelli Bauck 1986) and of Morocco (Schroeder 1984). According to Schroeder (1984) Triassic and Jurassic species are identical, although there is a gap in the record of *Stylothalamia* from the Middle Norian to the end of the Rhaetian.

The climax of the diversity of Triassic pharetronids is in the Carnian (39 genera), where eight "sphinctozoan" and twelve "inozoan" genera originated (Fig. 1). At the Carnian-Norian boundary, six "sphinctozoan" and eight "inozoan" genera became extinct. Only three Carnian "sphinctozoan" genera extend up to the Norian, and only four Carnian "inozoan" genera have been found in the Norian, Rhaetian, and Jurassic, respectively (see Table 2). Three Carnian "inozoans" seem to have identifiable ancestors in the Permian fauna described by Termier et al. (1977) from Djebel Tebaga (Tunisia).

A major change in pharetronid faunas occurs at the Carnian-Norian boundary, for nearly all Carnian species (at least 90%) became extinct. An obvious characteristic of Norian "sphinctozoans" is their probable aragonitic skeleton, whereas Carnian sponges also had Mg-calcitic walls. Our recent investigations on lowermost Norian (Lacian 1) "sphinctozoans" show that there was not an abrupt change in the wall mineralogy then (see Senowbari-Daryan, this Vol.), and that some Carnian species reach up to the Lacian 1 (Table 1).

Fifteen genera originated in the Norian. All of them becomes extinct near the end of the Triassic. Nine occur in the Norian and Rhaetian as well, but no new genus appears in the Rhaetian.

## Conclusions

Pharetronid sponges sensu Steinmann are the most important faunal elements in Triassic reefs. They are reef builders, reef dwellers, and stabilizing organisms. In the struggle for space within the reefs, pharetronids were very successful, but at the end of the Triassic their importance diminished, because from then on corals were more successful, perhaps related to their probable symbiosis with zooxanthellae.

The greatest diversity of the Triassic pharetronids appears to have developed in the Carnian and Norian (but this may be due to more intensive studies of these reefs!).

Extinction of most pharetronids at the Permian-Triassic boundary and their absence in the Lower Triassic (Scythian) is followed by two more crises in the development of the pharetronids. Nearly all pharetronid species became extinct near the Carnian-Norian boundary, although 16 of 58 genera survived. Our recent investigations on Lower Norian faunas give evidence for slow replacement of Carnian taxa by Norian forms.

At the Triassic-Jurassic boundary all Triassic "sphinctozoans" except one species became extinct. Only 4 of 17 genera of "inozoans" occur in both Triassic and Jurassic assemblages.

The extinctions of the pharetronid genera at the beginning and the end of the Triassic correspond well with extinction patterns of other groups and of reef faunas as plotted by Raup and Boyajian (1988).

*Acknowledgments.* The Deutsche Forschungsgemeinschaft supported this study within the projects Evolution of Reefs (Fl 42/49-2) and Karn/Nor-Türkei (Fl 42/53-1). We thank the director of these projects Prof. Dr. E. Flügel for encouragement and discussion.

## References

Becarelli Bauck L (1986) Stylothalamien aus dem unterjurassischen Misone-Kalk der Südalpen, Italien. Palaeontogr Abt A 192:1-13

Bechstädt T, Brandner R (1970) Das Anis zwischen St. Vigil und dem Höhlensteintal (Pragser- und Olanger Dolomiten, Südtirol). Festbd Geol Inst 300 Jahr-Feier Univ Innsbruck: 9-103

Dronov VI, Gazdzicki A, Melnikova GK (1982) Die triadischen Riffe im südöstlichen Pamir. Facies 6:107-128

Dieci G, Antonacci A, Zardini R (1968) Le spugne cassiane (Trias mediosuperiore) della regione dolomitica attorno a Cortina d'Ampezzo. Bull Soc Paleontol Ital 7/2:94-155

Dullo WC, Flügel E, Lein R, Riedel P, Senowbari-Daryan B (1987) Algen, Kalkschwämme und Mikroproblematika aus unterkarnischen Riffkalken des Bosruck-Gipfels (Nördliche Kalkalpen, Österreich). Jahrb Geol Bundesanst 129:525-543

Flügel E (1981) Paleoecology and facies of Upper Triassic reefs in northern calcareous Alps. In: Toomy DF (ed) European fossil reef models. SEPM Spec Publ 30:291-359

Flügel E, Lein R, Senowbari-Daryan B (1978) Kalkschwämme, Hydrozoen, Algen und Mikroproblematika aus den Cidaris-Schichten (Karn, Ober-Trias) der Mürztaler Alpen (Steiermark) und des Gosaukammes. Mitt Ges Geol Bergbaustud Österr 25:153-195

Lein R (1980) Leckkogelschichten – ein neuer Terminus für eine schwammreiche Plattformrandentwicklung im ostalpinen Karn (Obertrias). Sitzungsber Österr Akad Wiss Math-Naturwiss Kl 1980:107-119

Lein R (1985) Das Mesozoikum der Nördlichen Kalkalpen als Beispiel eines gerichteten Sedimentationsverlaufes infolge fortschreitender Krustenausdünnung. Arch Lagerstforsch Geol Bundesanst 6:117-128

Lein R, Riedel P, Dullo WC (in press) Fossile tidal flat pools aus der obertriadischen Karbonatplattform des Hochschwab (Mürzalpendecke, Steiermark). Jahrb Geol Bundesanst

Ott E (1967) Segmentierte Kalkschwämme (Sphinctozoa) aus der alpinen Mitteltrias und ihre Bedeutung als Riffbildner im Wettersteinkalk. Bayer Akad Wiss Math-Nat Kl Abh N F 131:1-96

Raup DM, Boyajian GE (1988) Patterns of generic extinction in the fossil record. Paleobiology 14/2:109-125

Schauer M (1983) Zur Altersstellung obertriadischer Dachsteinriffkalke. Anz Österr Akad Wiss Math-Naturwiss Kl 1983:127-137

Schlager W, Schöllberger W (1973) Das Prinzip stratigraphischer Wenden in der Schichtfolge der Nördlichen Kalkalpen. Mitt Geol Ges Wien 66-67:165-193

Scholz G (1972) An Anisian Wetterstein limestone reef in North Hungary. Acta Mineral-Petrogr Szeged 20/2:337-362

Schroeder R (1984) Revision von *Stylothalamia columnaris* (Le Maitre) 1935 (Sphinctozoa, Porifera) aus dem Lias von Marokko. Paläontol Z 58/1-2:33-39

Stanton RJ, Flügel E (1987) Palecology of Upper Triassic reefs in the Northern Calcareous Alps: reef communities. Facies 16:157-186

Steinmann G (1882) Pharetronen-Studien. Neues Jahrb Miner Geol Paläontol 1882/II:139-191

Termier H, Termier G, Vachard D (1977) Monographie Paleontologique des Affleurements Permiens du Djebel Tebaga (Sud Tunisien). Palaeontographica Abt A 156:1-109

Turnsek D, Ramovs A (1987) Upper Triassic (Norian-Rhaetian) reef buildups in the Northern Julian Alps (NW Yugoslavia). Razprave IV Razreda SAZU 28:27-67

Turnsek D, Buser S, Ogorelec B (1982) Carnian coral-sponge reefs in the Amphiclina beds between Hudajuzna and Zakriz (Western Slovenia). Razprave IV Razreda SAZU 24:1-48

Zankl H (1969) Der Hohe Göll, Aufbau und Lebensbild eines Dachsteinkalk-Riffes in der Obertrias der nördlichen Kalkalpen. Abh Senckenberg Naturforsch Ges 519:1-123

Zittel KA (1878) Studien über fossile Spongien. III. Abteilung: Monactinellidae, Tetractinellidae und Calcispongiae. K Bayer Akad Wiss II. Cl 13: II. Abt 11-48

# Sponge Bioherms from Epicontinental Triassic Formations of Upper Silesia (Southern Poland)

A. Bodzioch

## Introduction

Occurrences of sponge bioherms in epicontinental Triassic formations of Europe have been recorded until now only in a few localities in the western part of Upper Silesia (Fig. 1). They occur in the middle and upper part of the Karchowice Beds, which are in this region the uppermost lithostratigraphical unit of the Lower Muschelkalk (Middle Triassic). These beds are correlated with the Illyrian (Zawidzka 1975) on the basis of conodont stratigraphy (Fig. 2).

Sponge bioherms from the Triassic of Upper Silesia have already been studied in the biostratinomical aspect (Bodzioch, in press); consequently, this chapter will focus mainly on the taxonomy of the sponges and on ecological and paleogeographical aspects.

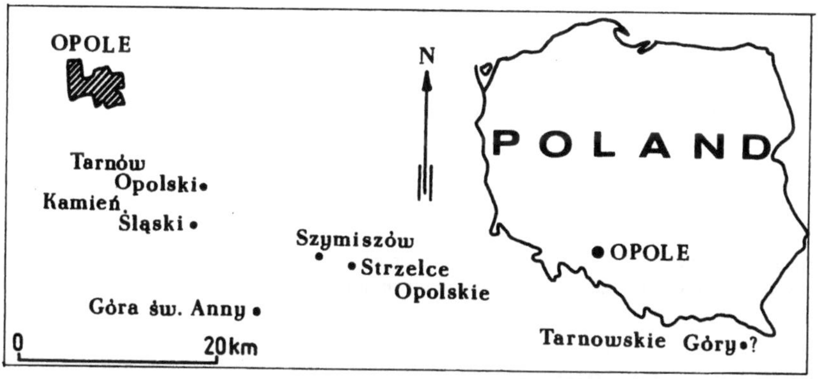


**Fig. 1.** Localities of the sponge bioherms in the western part of Upper Silesia. ? = Nonexisting outcrops from which sponges were described in the old German literature. German synonyms of Polish geographical names. (After Assmann 1944): *Opole* Oppeln; *Tarnów Opolski* Tarnau; *Kamień Śląski* Gross Stein; *Strzelce Opolskie* Gross Strehlitz; *Szymiszów* Schimischow; *Góra św. Anny* Annaberg; *Tarnowskie Góry* Tarnowitz

J. Reitner and H. Keupp (Eds.)
Fossil and Recent Sponges

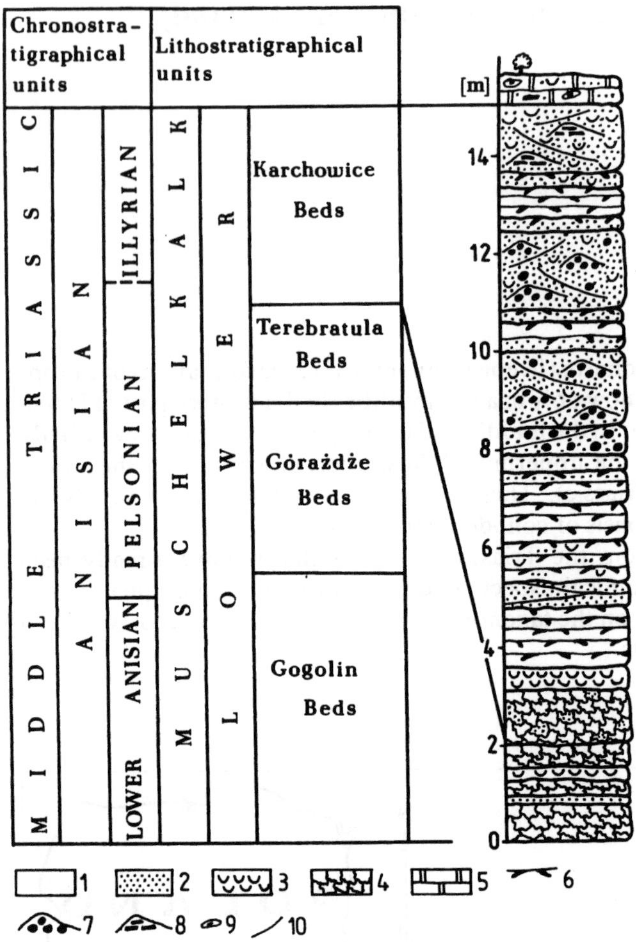


**Fig. 2.** Stratigraphic position of the Triassic sponge bioherms in the western part of Upper Silesia. *1* Pelitic limestones; *2* detrital limestones; *3* bioclastic limestones; *4* crumpled limestones; *5* dolomites; *6* burrows of Enteropneusta *7* sponge bioherms; *8* coral bioherms; *9* oncoids; *10* erosional channels. The section has been generalized from all outcrops (Fig. 1)

## Description of the Bioherms

The bioherms are of small dimensions, up to 3 m in diameter at the base and up to 1 m high (Fig. 3). They occur irregularly in the central parts of echinoderm-sponge biostromes. Sponges occur individually or in well-knit colonies. Crinoids, echinoids, brachiopods, pelecypods, gastropods, polychaetes, and corals accompany them. The sponges do not dominate, either in the number of individuals or the number of species. Their dominance is marked only in biomass, which has been measured as space occupied by particular taxa on surfaces of bioherm sections.

**Fig. 3.** Sponge bioherms (*left and right of the hammer*) from the Karchowice Beds, Szymiszów

The bioherms are built of siliceous sponges, which make them different from analogous bioherms from the Alpine Triassic, that are composed mainly of calcareous sponges (cf. Dieci et al. 1970; Bosselini and Rossi 1974; Turnšek et al. 1982; Stanton and Flügel 1987), and they appear somewhat later.

## Description of Sponges

*"Tremadictyon" roemeri* (Eck 1865), known earlier both from the German Triassic (Eck 1865; Assmann 1926; Schmidt 1928; Rauff 1937) and Alpine Triassic (Tichy 1975), has been identified in the collected material, together with two problematic species, mentioned in the proceedings abstract as Dictyida Species A, but described here as Hexactinellida species A and B.

Phylum: Porifera Grant 1872
Class: Hexactinellida Schmidt 1870

*Tremadictyon roemeri* (Eck 1865) (Fig. 4)

1865 – *Scyphia Römeri* sp. nov., H. Eck, p. 85 and 123, Pl. I, Fig. 1.
1926 – *Scyphia Römeri*, P. Assmann, p. 506.
1928 – *Scyphia Roemeri* Eck, M. Schmidt, p. 114, Fig. 208.

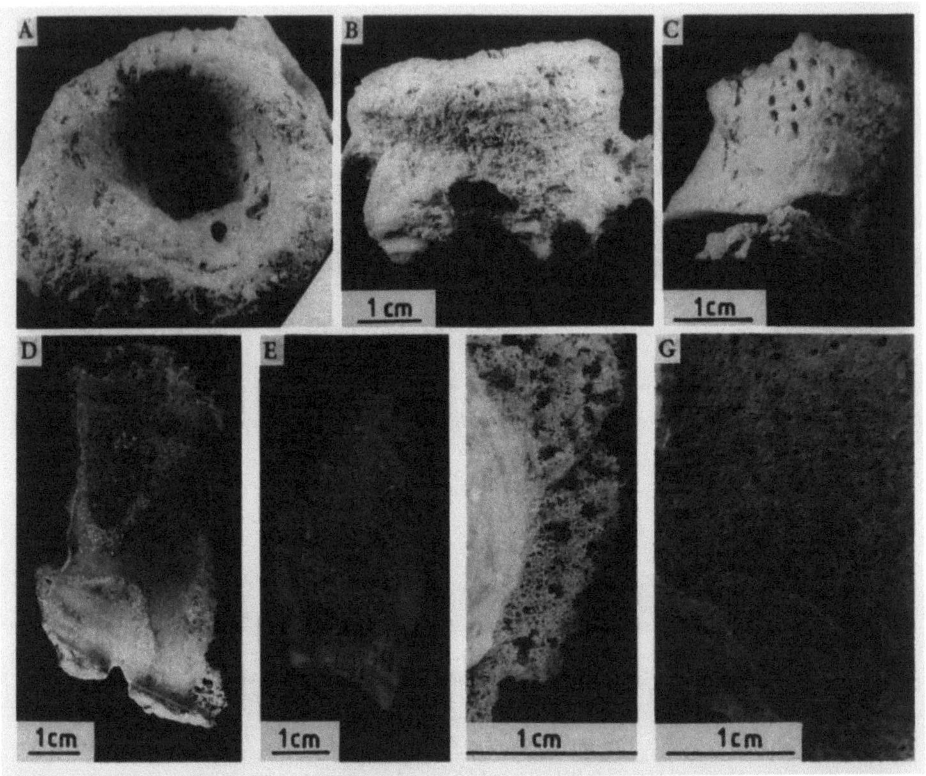


**Fig. 4.** *"Tremadictyon" roemeri* (Eck) from the Karchowice Beds. **A** Cloacal view, Strzelce Opolskie. **B** The same specimen, side view. **C** Ostia of inhalant canals, Szymiszów. **D** Axial section through silificied sponge, Strzelce Opolskie. **E** As **D**. **F** a part of rigid internal skeleton, Góra św. Anny. **G** Exhalant canals, Szymiszów

1937 — *Tremadictyon roemeri* n. sp., H. Rauff in Assmann 1937, p. 7, Pl. I, Figs. 1–12.
1937 — *Scyphia* (*Tremadictyon*?) *roemeri* (Eck), H. Rauff in Assmann 1937, p. 10, Pl. II, Figs. 1–7.
1975 — *Tremadictyon* cf. *roemeri* (Eck), G. Tichy, p. 68, Figs. 1–5.

Holotypus: H. Eck 1865, p. 85 and 123, Pl. I, Fig. 1.
Locus typicus: Tarnowskie Góry, Górny Śląsk (= Upper Silesia, = Oberschlesien).
Stratum typicum: Karchowice Beds (= Karchowitzer Schichten).
Occurrences: Szymiszów, Strzelce Opolskie, Góra św. Anny.
Material: Four almost completely preserved specimens and two fragments.
Depository: Author's collection.

## Description

Euryproct forms, shaped like vases (Fig. 4A-E). Cloaca deep, with oval cross section, entering the root part. Wall thick (for biometric data see Table 1), and approximately proportional to diameter of the cloaca. On the outer wall, circular or oval-shaped pores of inhalant canals (Fig. 4B,C) are arranged without order or in alternate rows. Pores of exhalant canals not so evident and of smaller dimensions (Fig. 4G). Canal system simple, consists of radial, blind inhalant and exhalant canals. Big, loose hexacts occur in the external part of walls. Rigid internal skeleton is built by well-fused hexacts, creating a very irregular skeletal network of cubic outline (Fig. 4F). Nodes of the skeletal network are cross-shaped, filled. Regularity of the skeletal network is disturbed by numerous deflections of hexact rays, supplementary beams, envelopes round nodes, and small appendices, which diminish the space of the network meshes and round off their original tetragonal outline.

**Table 1.** Biometric data of *"Tremadictyon" roemeri* (Eck)

| Specimen Fig. 4 | H (mm) | D (mm) | T (mm) | IMD (mm) | EMD (mm) | DIM (mm) | DEM (mm) | DN (mm) |
|---|---|---|---|---|---|---|---|---|
| A-B | 38 | 45 | 10 | 0.969 | – | 2.085 | – | – |
| C | 28 | 27 | 7 | 1.386 | – | 2.018 | – | – |
| D | 74 | 33 | – | – | – | – | – | – |
| F | – | – | 10 | 1.303 | – | 2.000 | – | 0.400 |
| G | – | – | 15 | – | 0.600 | – | 1.350 | – |

H, height; D, maximum diameter of the cloaca; T, thickness of the wall; IMD, mean diameter of ostia of inhalant canals; EMD, mean diameter of ostia of exhalant canals; DIM, mean distance between inhalant canals; DEM, mean distance between exhalant canals; DN, mean distance between nodes of the skeletal network.

Specimens shown in Fig. 4A-B, C, D, and E are most similar to the forms described by Rauff (1937) as *Tremadictyon roemeri*. The specimen in Fig. 4F has been included in the same species on the grounds of its shape, structure of internal skeleton and canal system. The specimen shown in Fig. 4G is deformed and has been included on the grounds of shape and size of exhalant canals and thickness of the wall.

All specimens in my collection are silicified, as were those described in the old German literature. Therefore, their certain affiliation with the genus *Tremadictyon* or even the order Dictyida is problematical, because the structure of the rigid internal skeleton is not quite clear.

*Hexactinellida* sp. A (Fig. 5)
Euryproct forms, shaped like vases, often with reflexed rims. Cloaca deep, wall thin (for biometric data see Table 2). Pores of inhalant and exhalant canals round

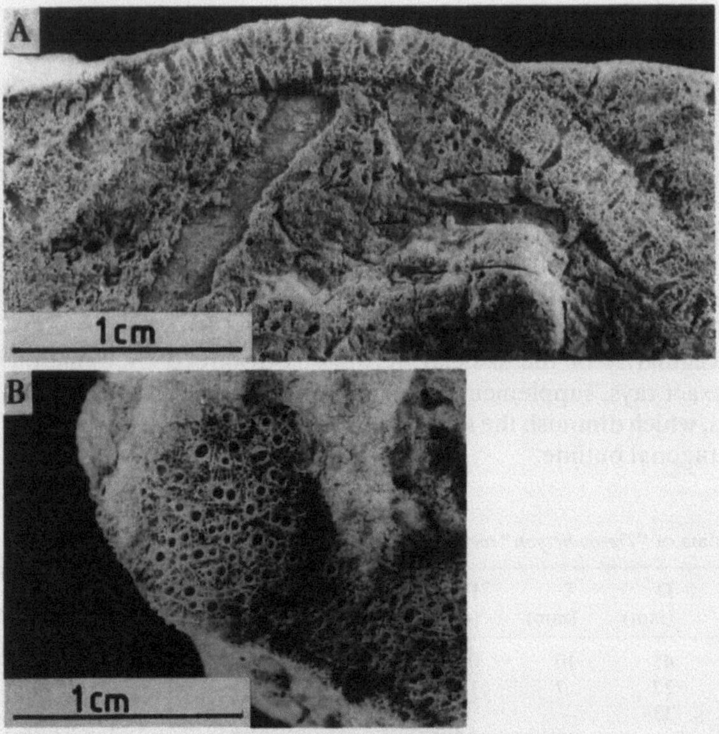


**Fig. 5.** *Hexactinellida* sp. A. **A** Cross-section through the wall, Szymiszów. **B** Outer surface of the wall, Szymiszów

**Table 2.** Biometric data of *Hexactinellida* sp. A (compiled from all specimens). Symbols as in Table 1

| T (mm) | IMD (mm) | EMD (mm) | DIM (mm) | DEM (mm) | DN (mm) |
|---|---|---|---|---|---|
| 2–5 | 0.528 | 0.555 | 1.142 | 1.022 | 0.282 |

or somewhat oval and small in a delicate skeletal network (Fig. 5B). Canals are arranged without order or in short rows, and their sizes show low variability. The canal system and the structure of the rigid internal skeleton (Fig. 5A) are similar to "*Tremadictyon*" *roemeri* (Eck).

Material: Six fragments.
Occurrences: Szymiszów, Tarnów Opolski, Strzelce Opolskie.
Depository: Author's collection.

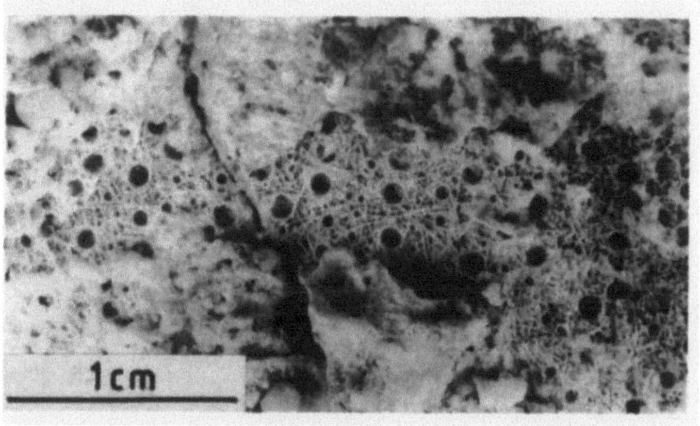


**Fig. 6.** *Hexactinellida* sp. B, Szymiszów

**Table 3.** Biometric data of *Hexactinellida* sp. B. Symbols as in Table 1

| T (mm) | IMD (mm) | EMD (mm) | DIM (mm) | DEM (mm) | DN (mm) |
|---|---|---|---|---|---|
| 2–5 | 0.565 | 0.758 | 1.250 | 1.673 | 0.556 |

*Hexactinellida* sp. B (Fig. 6)
Dimensions, shapes, canal system and structure of the rigid internal skeleton similar to Hexactinellida sp. A, but with greater variability of diameters of canals and pores Fig. 6. Moreover, original spiculation is preserved on the surface of the wall. Biometric data are similar to those of Hexactinellida sp. A (Table 3).

Material: One fragment.
Occurrence: Szymiszów.
Depository: Author's collection.

### *Discussion*

Among features differentiating the described sponges, each has a certain limitation as to taxonomic value. In the case of thin-walled specimens, shape cannot be a valuable criterion because completely preserved specimens have not been found, whereas some of the described fragments show mechanical damage such as disruption or crushing of the wall. Therefore, shapes of those sponges may be accidental, connected with their post-mortem history.

Loose hexacts have been recorded only in the case of thick-walled specimens. This leads to the conclusion that thin-walled specimens are devoid of the external skeleton. Therefore, thickness of the wall may be also dependent upon the processes of fossilization.

The mean size of canals and pores are larger in thick-walled specimens than in thin-walled specimens. Size may be connected with thickness of the wall.

Thick-walled specimens have preserved their external skeletons, so they have loose hexacts and their canal ostia are relatively big. Thin-walled specimens do not have a preserved external skeleton; they lack loose hexacts, and their canal ostia are relatively small.

Thus, the only distinct feature that is independent of the state of preservation is variability of size and placement of ostia. Consequently, it seems that this feature and the still little understood structure of the rigid internal skeleton have the greatest importance in the taxonomy of Triassic hexactinellids from Upper Silesia.

## Ecology

Contemporary distribution of Hexactinellida, from tropical to arctic waters (cf. Laubenfels 1955 and 1957; Reid 1968) and at depths from about 4000 m (cf. Hartman 1980) to the tidal zone (Burton 1948), does not allow the application of actualistic methods of reconstruction of life habitat to the described sponges. Biostratinomical investigations of the sponge-bearing formations of the Karchowice Beds (Bodzioch, in press) lead to the following conclusions:

1. Sponge bioherms developed over a depth range from several meters to tens of meters (maximum), within the realms of storm wave action.
2. Very shallow depths were probably the cause of bottom-water turbulence and of proper oxygenation of the sponges.
3. Fossil assemblages accompanying the sponges indicate normal salinity and high temperature of sea water.
4. The sponges grew on the coarse-grained, echinoderm-shell debris, in conditions of variable energy and slow sedimentation, which, in combination with fossil assemblages, indicate an open shelf environment.

## Paleogeography

The bioherms occur in the narrow zone, oriented NWW-SEE (Fig. 1), and most probably in Illyrian times formed some kind of shoal barrier that separated a near-shore lagoon (SSW from this zone) from the open epicontinental sea (NNE from this zone). They developed on the echinoderm-shell sands and are followed by coral bioherms, described by Morycowa (1974). Such a situation, common in European Jurassic formations (cf. Wendt 1980), indicates a gradual increase in environmental energy. In this case such an increase may be linked with shallowing of the sedimentary basin. In consequence, such facies development is a proof of postulated earlier regression of the Triassic epicontinental sea from the Upper Silesian region (Senkowiczowa 1962), caused by tectonic rebuilding of German Basin.

## References

Assmann P (1926) Die Fauna der Wirbellosen und die Diploporen der oberschlesischen Trias mit Ausnahme der Brachiopoden, Lamellibranchiaten, Gastropoden und Korallen. Jahrb Preuss Geol Landesanst 46:504–527

Assmann P (1944) Die Stratigraphie der oberschlesischen Trias. Der Muschelkalk. Abh Reichsamt Bodenforsch NF 208:1–125

Bodzioch A Biostratinomy and sedimentary environment of echinoderm-sponge biostromes from the Karchowice Beds Lower Muschelkalk of Upper Silesia, southern Poland. Ann Soc Geol Polon (in press)

Bosselini AA, Rossi D (1974) Triassic carbonate buildups of the Dolomites, Northern Italy. Soc Econ Paleontol Mineral Spec Publ 18:209–233

Burton M (1948) Ecology of sponges. Nature 162:73–74

de Laubenfels MW (1955) Porifera. In: Moore RC (ed) Treatise on invertebrate paleontology, Part E. The University of Kansas Press, Lawrence, p 122

de Laubenfels MW (1957) Sponges of the Post-Paleozoic. In: Ladd HS (ed) Treatise on marine ecology and paleoecology, vol 2, Paleoecology. Mem Geol Soc Am 67:771–772

Dieci G, Antonacci A, Zardini R (1970) Le spugne cassiane Trias medio-superiore della regione dolomitica attorno a Cortina d'Ampezzo. Boll Soc Paleontol Ital 7:94–155

Eck H (1865) Ueber die Formationen des bunten Sandsteins und des Muschelkalks in Oberschlesien und ihre Versteinerungen. Berlin

Hartman WD (1980) Systematics of Porifera. Sedimenta 8:24–51

Morycowa E (1974) Koralowce z warstw karchowickich z Kamienia Śląskiego koło Opola. Przewodnik 46 Zjazdu PTG, Opole, pp 163–165

Rauff H (1937) Spongien. In: Assmann P (ed) Revision der Fauna der Wirbellosen der oberschlesischen Trias. Abh Geol Landesanst NF 170:7–14

Reid REH (1968) Bathymetric distribution of Calcarea and Hexactinellida in the present and past. Geol Mag 105:546–559

Schmidt M (1928) Die Lebewelt unserer Trias. Hohenlohe'sche Buchhandlung F. Rau, Ohringen, pp 461

Senkowiczowa H (1962) Wpływ fauny alpejskiej w osadach retu i wapienia muszlowego na obszarach Polski. Księga pamiątkowa ku czci prof J Samsonowicza. Wyd Geol, Warszawa, pp 239–257

Stanton RJ, Flügel E (1987) Paleoecology of Upper Triassic reefs in the Northern Calcareous Alps: Reef Communities. Facies 16:157–186

Tichy G (1975) Der erste körperlich erhaltene Kieselschwamm (*Tremadictyon* cf. *roemeri*) aus der Trias (Anis) der Alpen. Verh Geol Bundesanst 2-3:67–73

Turnšek D, Buser S, Ogorelec B (1982) Carnian coral-sponge reefs in the Amphiclina Beds between Hudajužna and Zakriž (western Slowenia). Dissert Acad Sci Artium Slovenica, Classis 4, 24:1–48

Wendt J (1980) Sponge-reefs of the German Upper Jurassic. Sedimenta 8:122–134

Zawidzka K (1975) Conodont stratigraphy and sedimentary environment of the Muschelkalk in Upper Silesia. Acta Geol Pol 25:217–256

# Upper Jurassic Sponge Megafacies in Spain: Preliminary Report

A. PISERA

## Introduction

The Upper Jurassic sponge megafacies (with siliceous sponges) extends in Europe from Romania through Poland, Germany, Switzerland, and France to Spain. This facies, unique in the fossil record, has in recent times gained considerable interest. So far the best-known megafacies development is in Southern Germany, where it has also the largest stratigraphical range from Oxfordian to Tithonian (see Wagenplast 1972; Nitzopoulos 1974; Gwinner 1976; Flügel and Steiger 1981; Schorr and Koch 1985; Wirsing and Koch 1986, and others, also literature therein). A recent and thorough monograph of sponge megafacies in France was presented by Gaillard (1983). Trammer (1982) investigated the taxonomy of the Lower Oxfordian sponges from Poland, giving short characteristics of all aspects of the megafacies. Data on sponge megafacies in Spain, however, are scarce and dispersed in the general geological literature. Because of Spain's paleogeographical position as the southernmost area of occurrence, the study of all aspects (paleontology, sedimentology and stratigraphy) seems important. This chapter presents, therefore, a general introductory report on the Upper Jurassic sponge megafacies in Spain, based both on earlier published data and on my own observations. Detailed study of sponge taxonomy, paleoecology, and functional morphology, including also some sedimentological and paleogeographical problems, is now being carried out.

## Previous Studies

The Upper Jurassic limestones and marls rich in sponges were described by Riba (1959) from Montes Universales (Celtiberic area, see Figs. 1,2) in his general geological monograph. Later, Behmel (1970) for Prebetic; and El Khoudary (1974) for Celtiberic regions gave the stratigraphical and lithological characteristics of the Upper Jurassic deposits. They mentioned also the most important fossils among the sponges. Stratigraphy and sedimentology of the Upper Jurassic deposits were the subjects of papers by Geyer et al. (1975) and Benke (1981). The only papers directly concerning sponges are those by Moret (1926) and Hérenger (1941).

J. Reitner and H. Keupp (Eds.)
Fossil and Recent Sponges

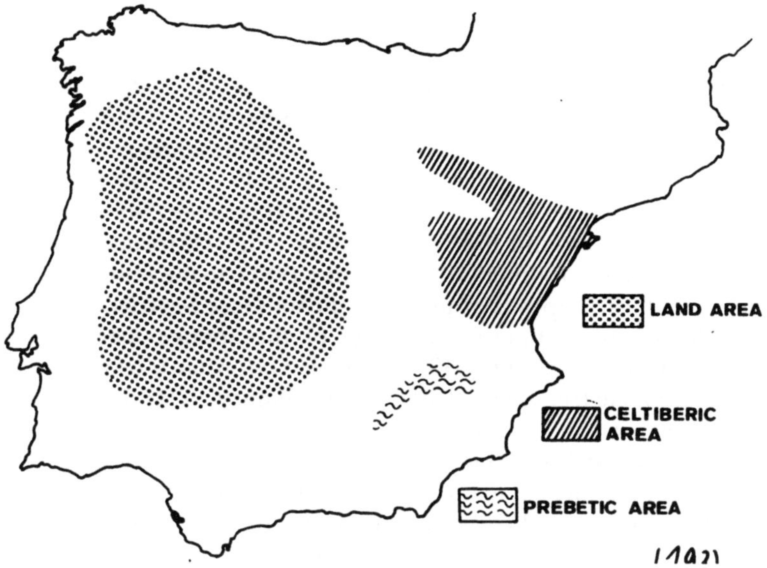


**Fig. 1.** Paleogeographic situation of the Upper Jurassic (Oxfordian) sponge facies in Spain (based on various sources). *Land area* corresponds to the Meseta Block

**Fig. 2.** Location of the investigated sections with the sponge facies

## Paleogeographic Situation and Stratigraphy

Sponge megafacies occurs in both Celtiberic and Prebetic units, the best development being evident in Montes Universales (Celtiberic area) near Albarracin, Teruel, and Cañete (see Fig. 2). From these areas comes also the best and richest sponge fauna. In the Prebetic area, rocks of the facies are poorly exposed and contain sponges less frequently. The sponge megafacies had developed on a wide shelf sea around the Meseta Block (Fig. 1), forming during the Upper Jurassic time the continental area (Benke 1981), to the south and southeast occur slope and basin deposits of the Subbetic unit (see Seyfried 1978).

Maximum stratigraphical range of the sponge megafacies is from the *plicatilis* to *planula* zone (Middle-Upper Oxfordian), usually only to the *bimmamatum* zone (see Behmel 1970; El Khoudary 1974). In the Calanda section, limestones rich in sponges (lithistids) appear once more in the Kimmeridge portion of the column (Geyer and Pelleduhn 1979). The occurrence of mixed sponge-coral bioherms in the Upper Kimmeridge of Jabaloyas was also reported (Giner and Barñolas 1979); however, I have found that the sponges in those reefs are represented only by calcareous forms.

## Lithology and Petrography

Sponge megafacies in Spain is developed mostly as bedded limestones with subordinate marls intercalations (Figs. 3, 4A-D). Contrary to the other areas of their occurrence in Europe, bioherms are rare and small, and limited to some sections only. In the Celtiberic unit they were noted only in the Arroyo Cerezo section, where they are (Fig. 4A) up to 1 m high and 2 m in diameter. In the Prebetic unit bioherms of similar or smaller size occur in Pozo Cañada and Fuente Alamo. Common in biohermal rocks, apart from siliceous sponges, are serpulids, bryozoans, and solitary corals (most probably ahermatypic *Trochocyathus* sp.: Prof. E. Roniewicz pers. commun.) (Fig. 5A,B). An important role is played by stromatolitic (of cyanobacterial, not algal genesis) encrustations (Figs. 5A,6F). During growth, all these biogenic components formed some kind of framework with many constructional cavities which contain internal sediments (Fig. 6B); the matrix is developed as tuberolitic biomicrite or biomicrites. Biohermal rocks may be classified as biolithites. No reef detritus has been found around these structures.

Bedded limestones and marls, however, form the overwhelming majority of the sponge facies rocks, and are developed as irregularly bedded, often "crumpled" to nodular limestones with common thin marl intercalations (Figs. 3,4A-E). Both lithologies contain sponges. Limestones are usually thin-bedded, 10 to 40 cm in thickness, rarely represented by 2-m-thick beds (Figs. 3,4A,B). Except for mummified sponges (Fig. 5F-G) in various amounts, they contain also very common ammonites and crinoids, and echinoid ossicles. Brachiopods, bivalves, and bryozoans are much less common. Microfauna is represented by both benthic and pelagic (protoglobigerinid) foraminifers. Common in some

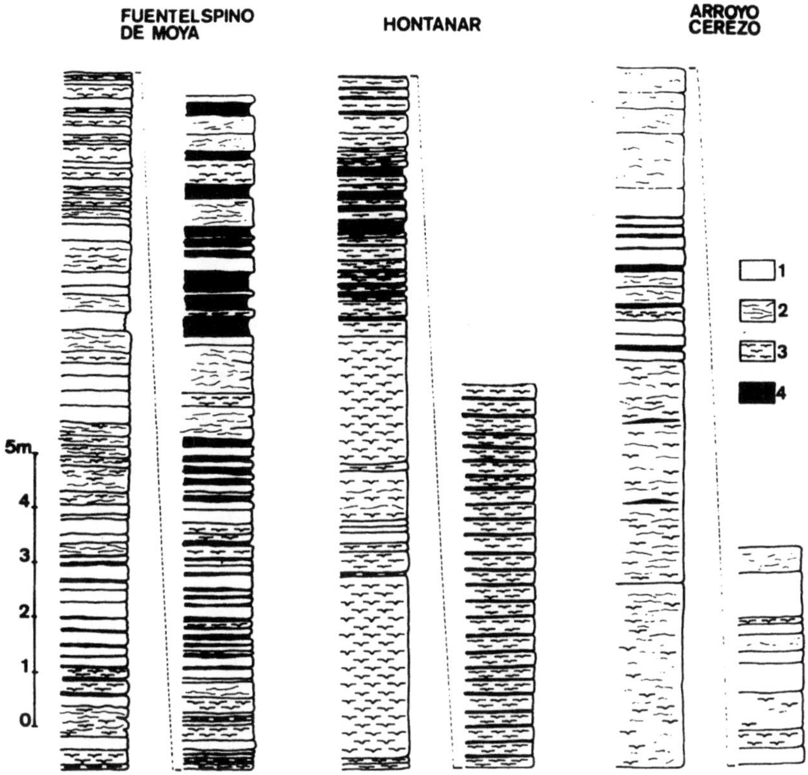


**Fig. 3.** Lithological sections of some typical outcrops of the sponge megafacies in Celtiberic unit. *1* Limestones without sponges. *2* "Crumpled" limestones. *3* Limestones with sponges. *4* Marls with sponges. (After Matyja and Pisera unpubl.)

sections (Griegos and Frias de Albarracin) is fine biogenic detritus and quartz (Fig. 6D). One of the most important components volumetrically is tuberoids (see Fritz 1958) (Fig. 6C). Generally, bedded limestones represent a spectrum of tuberolitic biomicrites and micrites. Such lithologies suggest quiet and/or deep-water environment.

Despite general similarities of sponge-bearing rocks, the details of the sequence developed vary in various areas (Figs. 3,4A-E). In the northern part of the Celtiberic unit the sponge sequence starts with a layer of marls (Fig. 4E) with numerous dish-shaped sponges (Ricla, La Almunia, Aguilon) and in the whole sequence marls have an important share. The central and southern part of the Celtiberic unit is characterized by the presence of one or two thick beds of limestones (up to 2 m) in the sole of the sequence (Figs. 3,4A-B). The sequence contains less marls, and condensed horizons are common. Sponge facies here reaches the greatest thickness. The Prebetic unit is characterized by a lesser thickness of sponge-bearing rocks; red coloration is common and nodularity well expressed (Fig. 4D). Bioherms are also common. All these differences may be

**Fig. 4.** Lithology of sponge megafacies. **A** Arroyo Cerezo section, note the presence of small bioherms. **B** Frias de Albarracin, well-bedded limestones with marl intercalations, note the presence of thick limestone beds at the sole of the section. **C** Fuentelspino de Moya section, well-bedded limestones with thin intercalations of marls visible. **D** Pozo Cañada section, visible are red nodular limestones which contain numerous dish-shaped sponges. **E** Aguilon section, visible marl layer (containing numerous dish-shaped sponges), which forms the first bed of the sponge facies; it lay over the top of condensed horizon (hammer) **F** Large overturned dish-shaped sponge on the top of the bed. Aguilon

**Fig. 5.** Lithology and petrography of the sponge-bearing limestones. **A** Biolithite. Fuente Alamo section, note the presence of serpulids and stromatolitic encrustations; in the middle solitary coral visible. **B** Biolithite, Fuente Alamo, visible numerous serpulids encrusting sponges. **C** Tuberolitic limestone, Pozo Cañaola. **D** Tuberolitic limestone, Fuentelspino de Moya. **E** Tuberolitic limestone with numerous bioclasts (crinoid ossicle in the center) Gea de Albarracin. **F** Biomicrite with numerous sponge mummies and bioclasts, Gea de Albarracin. **G** Biomicrite with numerous sponge mummies, Ricla. Bar = 1 cm

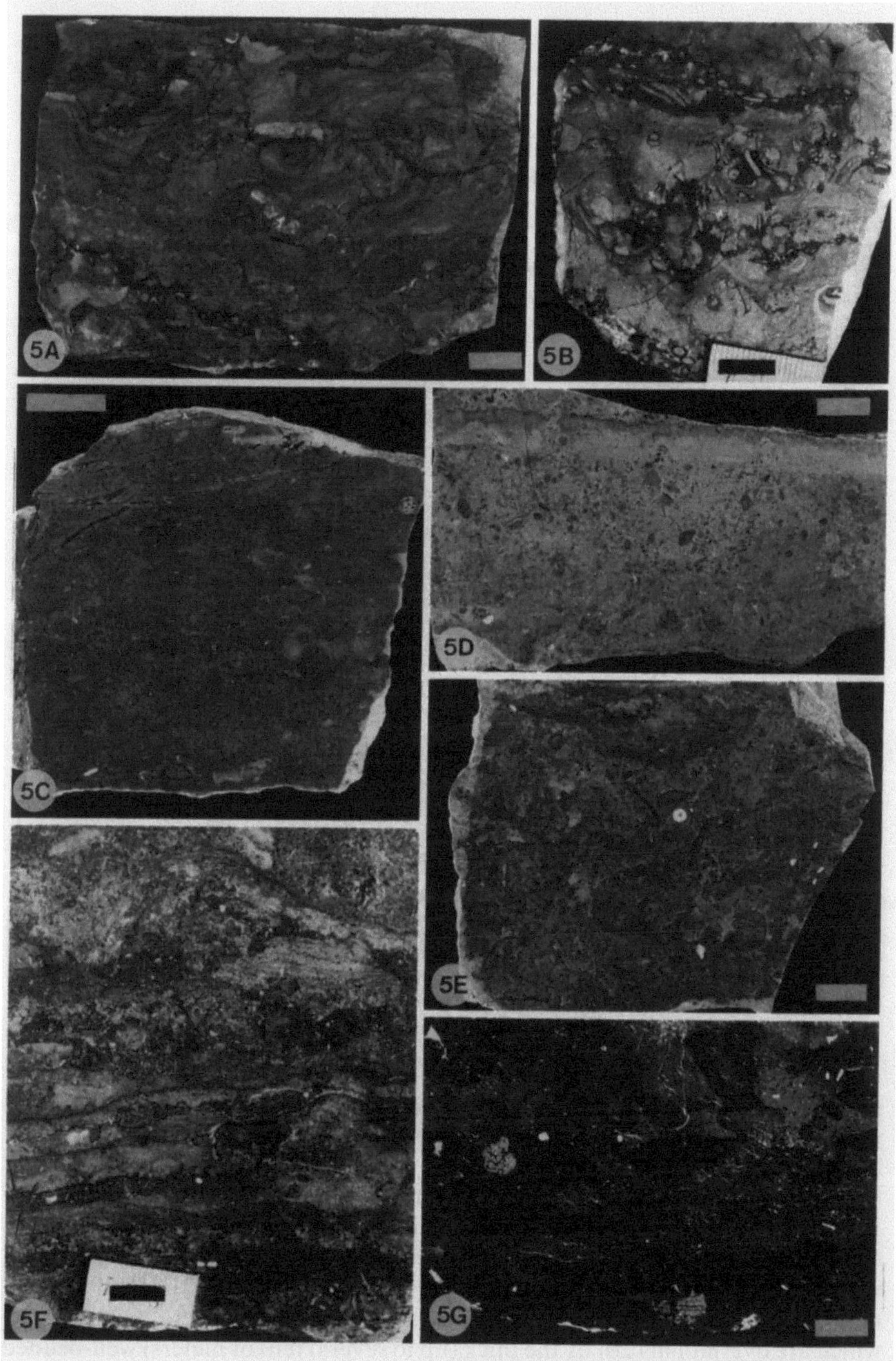
5A
5B
5C
5D
5E
5F
5G

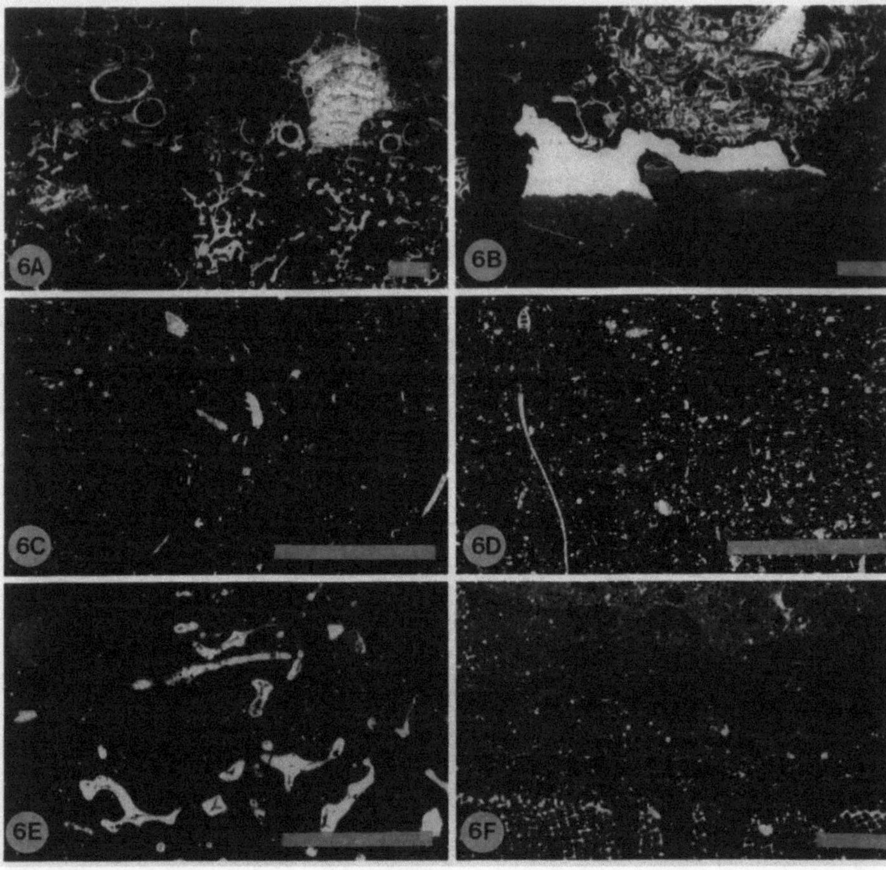


**Fig. 6.** Petrography of the sponge-bearing limestones. **A** Biolithite, visible calcified sponge skeleton, serpulid tubes and bryozoans, Fuente Alamo. **B** Biolithite, geopetal infilling of the constructional cavity, Fuente Alamo. **C** Tuberolitic biomicrite, Fuentelspino de Moya. **D** Biomicrite with numerous tiny bioclast and quartz grains, Griegos. **E** Typical preservation of calcified sponge spicules, Fuente Alamo. **F** Stromatolitic crust on the dictyid sponge, Fuente Alamo. Bar = 2 cm

related to distance from the shore and or/basin and amount of terrigenous material delivered. The influence of some bottom relief cannot be excluded, however.

Generally, sponge facies represents the lowest part of the shallowing Upper Jurassic sequence starting with the condensed and/or discontinuity horizon comprising part of the Callovian and the Lower Oxfordian, developed as ferruginous crust and/or ferruginous oolites (see Geyer et al. 1975; Benke 1981, and others). Sponge-bearing rocks are overlain by gray marls (several meters thick) which are overlain in turn by the thick (up to several tens of meters) well-bedded intercalating marls and limestones. The sequence is completed with shallow water rocks such as oolites, oncolitic limestones, and coral reefs (Riba 1959; Geister and

Geyer 1968; Behmel 1970; El Khoudary 1974; Benke 1981). The thickness of the sponge-bearing rocks varies from 8 to 23 m in the Celtibeic area, being smaller (not exceeding 10 m) in the Prebetic unit. In sections of the central Celtiberic area (Teruel and Cañete regions, see Fig. 2) several condensed and/or discontinuity horizons were noted within the sponge facies. They are developed at the top of some beds and show accumulation of sponges, ammonites, and other macrofauna with traces of corrosion. Usually these horizons are red or brown in color. Such condensed and/or discontinuity surface occurs at the top of sponge facies.

## Sponges

Practically all sponges from the Middle to Upper Oxfordian sponge-bearing limestones and marls are hexactinellid and lithistid sponges. Only few specimens of calcareous sponges were found in the uppermost part of some sections. The skeleton of sponges is always calcified and usually relatively poorly preserved (Fig. 6F), thus making identification even of some beautifully externally preserved specimens difficult.

The most common (in percent number of specimens) are dish-shaped sponges (Fig. 7A), which have usually over 50% share in the assemblage. The second place is occupied by tube-shaped sponges, much less common are vase-shaped forms. All other shapes represented, i.e., spherical, branched, mushroom, occur in negligible number of specimens (Fig. 7A). No encrusting

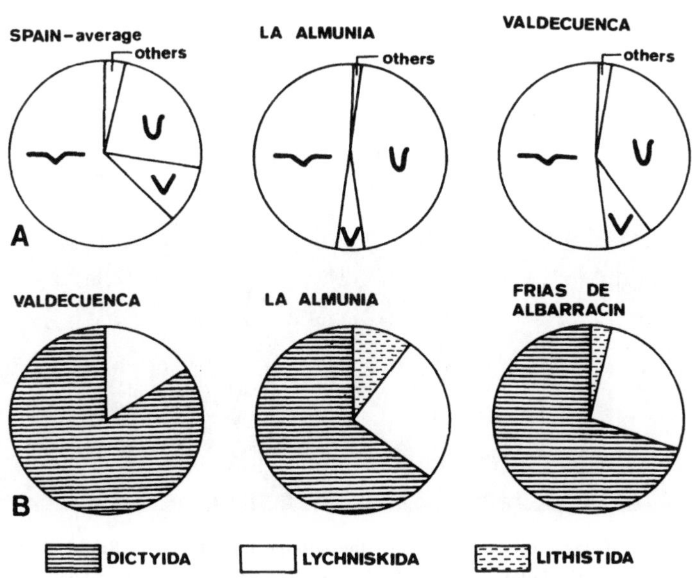


**Fig. 7.** Composition of the sponge assemblage in relation to sponge shapes (**A**) and taxonomy (**B**) (in perçent of number of specimens)

forms were noted. According to Trammer's (1983) model such frequency distribution of sponge shapes indicates very quiet (and perhaps deep) environment.

Taxonomically the most common are (percent share in number of specimens) dictyid sponges, which dominate over lychniskids (Fig. 7B). Lithistid sponges are totally absent or very rare and occur only in some sections (La Almunia) in more important quantities. The whole assemblage is dominated by different species of *Tremadictyon* (dictyid) *Placotelia* (lychniskid), *Thyroidium* and *Stauroderma* (dictyids) (see Fig. 8A,C-E). Not common but very characteristic and reaching a

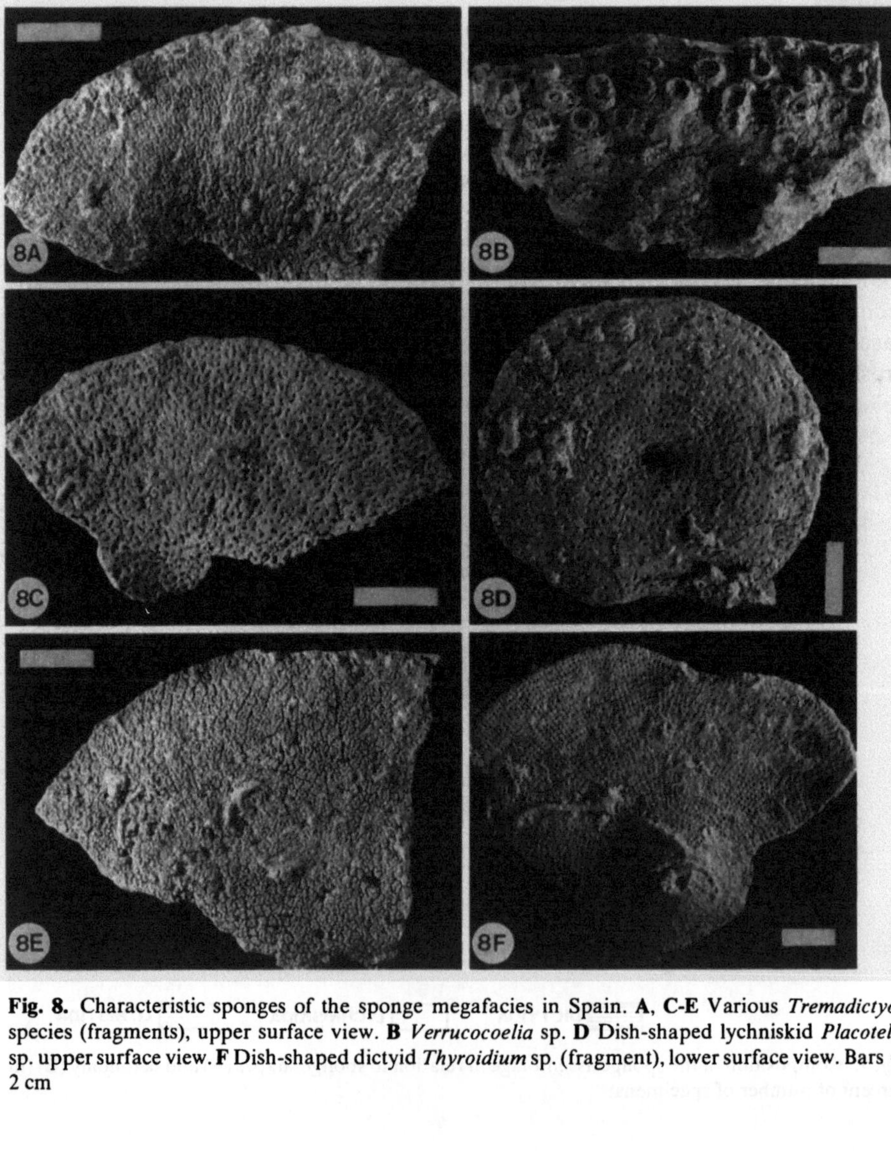


**Fig. 8.** Characteristic sponges of the sponge megafacies in Spain. **A**, **C-E** Various *Tremadictyon* species (fragments), upper surface view. **B** *Verrucocoelia* sp. **D** Dish-shaped lychniskid *Placotelia* sp. upper surface view. **F** Dish-shaped dictyid *Thyroidium* sp. (fragment), lower surface view. Bars = 2 cm

large size is dictyid *Verrucocoelia* (Fig. 8B). Such a taxonomic composition with the dominance of hexactinellids and absence of calcareous sponges also suggests deep-water environments (Reid 1968).

Most sponges are overturned, and dish-shaped sponges (even very large, see Fig. 4F) rarely occur in a living position. Only in a small part of the sponges have encrustations by various macrofauna such as serpulids and brachiopods been found.

In fact, no important differences have been noted in taxonomic composition of the sponge assemblages in various sections. The differences are rather in percent share of the particular species in various samples. An exception is the La Almunia section, where marls are dominant and the faunal assemblage is generally similar to the German Lochen facies.

When comparing the taxonomic composition of the sponge assemblage from the same stratigraphical levels from Spain, Poland, and Germany, it seems that only a few new forms occur in Spain. Others, which are common in Spain, occur also in other areas but with smaller frequency. Such a pattern of distribution may be explained, for example, by climatic differences between considered areas rather than by entirely different depositional environments.

## Discussion

Sponge megafacies in Spain shows some peculiarities in comparison with other occurrences in Europe. It represents condensed deposits, if one compares their thickness, not exceeding 23 m for the Middle and Upper Oxfordian, with the several-hundred-meters-thick (see Kutek et al. 1977) deposits of the same stratigraphical levels in Poland. The rarity and small size of the bioherms also make an important difference, as well as the existence of condensed horizons and/or nondeposition or even erosion events within the formation. These features are more similar to the Middle Jurassic sponge facies in Spain (see Scheer 1988; Wilde 1988). Composition of the sponge assemblage is, however, similar to the other areas of the Upper Jurassic sponge megafacies occurrence if one considers only the presence or absence of taxa. On the other hand, the share of sponges in the number of specimens differs strongly. Morphological classes of sponges which probably indicate ecological conditions are, however, similar to the Schwäbische Alb. This may indicate similar depositional environments, and differences may be caused entirely by climatic conditions (i.e., temperature).

Characteristic are also lithological characters such as red coloration of rocks in some southern sections, as well as the nodular character of many beds in the Prebetic unit. Taking all this into consideration, and regarding the paleogeographic position (close to the slope and basin deposits), as well as the absence of traces of wave action (except of condensed horizon which may represent storm effects), ubiquitous ammonites, common globigerinids, and the absence of calcareous sponges – relatively deep-water depositional environments (with slow deposition rate) may be assumed for these rocks. This indicates also the external part of the wide shelf close to the slope, far from the land, and without important terrigenous influences but under pelagic control. This conclusion agrees well with the model of sponge megafacies development proposed for France by Gaillard (1983), but is at the same time in total disagreement with the recently proposed model of Brachert (1986).[1]

## Final Remarks

As the present chapter is the only preliminary report, no final conclusions may be presented for the different aspects of sponge megafacies. This study forms only a part of a larger project concerning the whole area of sponge megafacies occurrence in Europe. We hope to reconstruct the details of the megafacies environment and controlling factors by comparing the results of investigations concerning stratigraphy and sedimentology, taxonomy, and functional morphology of sponges and associated fauna from different areas of occurrence.

*Acknowledgments.* Field work in Spain was possible thanks to financial support from the National Geographic Society to Dr. J. Ghiold (LSU). Dr. B.A. Matyja (Warsaw University) helped in section measurements and discussed the stratigraphy. I am indebted to all these persons and institutions. Special thanks go to Prof. S. Calzada and the staff of the Museo y Laboratorio de Geologia del Seminario de Barcelona for their invaluable help during my work in Spain.

## References

Behmel H (1970) Beiträge zur Stratigraphie und Paläontologie des Juras von Ostspanien. V. Stratigraphie und fazies im präbetischen Jura von Albacete und Nord-Murcia. Neues Jahrb Geol Paläontol 137:1–102

Benke K (1981) Die Dogger/Malm-Wende in den NW Keltiberischen Ketten (Spanien) und angrenzenden Gebeiten – Sedimentologie, Stratigraphie und Paläogeographie. Facies 4:95–164

Brachert TC (1986) Kontinuierliche und diskontinuierliche Sedimentation im süddeutschen Oberjura (unteres Kimmeridge; Ludwag/Oberfranken, Nördliche Frankenalb). Facies 15:39–45

El Khoudary RH (1974) Beiträge zur Stratigraphie und Paläontologie des Juras von Ostspanien. VI. Untersuchungen im Oberjura der südwestlischen Iberischen Kordillere unter besonderer Berücksichtigung der Mikrofauna (Provinz Teruel und Ricon de Ademuz). Neues Jahrb Geol Paläontol Abh 144:296–341

Flügel E, Steiger T (1981) An Upper Jurassic sponge-algal buildup from the northern Frankenalb, W-Germany. In: Toomey DF (ed) European fossil reef models. Soc Econ Paleontol Mineral Spec Pap 30:371–397

Fritz G (1958) Schwammstozen, Tuberolithe und Schuttbreccien im Weissen Jura der Schwäbischen Alb. Eine vergleichende petrogenetische Untersuchung. Arb Geol-Paläontol Inst Tech Hochsch Stuttgart NF 13:1–119

Gaillard C (1983) Les bioherms à spongiaires et leur environment dans l'Oxfordien du Jura Meridional. Doc Lab Géol Lyon 90:515

Geister J, Geyer OF (1968) Beiträge zur Stratigraphie und Paläontologie des Juras von Ostspanien. IV. Der Jura der Sierra de Corbera (Prov. Valencia) Neues Jahrb Geol Paläontol Abh 131:310–336

Geyer OF, Pelleduhn R (1979) Sobre la estratigrafia y la facies espongiolitica del Kimmeridgense de Calanda (Provincia de Teruel). Cuad Geol 10:67–71

[1] According to my investigations in the Schwäbische Alb Brachert's model does not hold true, also for the Oxfordian sponge-bearing rocks in this area.

Geyer OF, Behmel H, Hinkelbein K (1975) Beiträge zur Stratigraphie und Paläontologie des Juras von Ostspanien. VII. Die Grenzoolithe im Jura von Ostspanien. Neues Jahrb Geol Paläontol Abh 145:17–57

Ginyer J, Barñolas A (1979) Las constructiones arrecifales del Jurasico superior de la Sierra de Albarracin (Teruel). Cuad Geol 10:73–81

Gwinner MP (1976) Origin of the Upper Jurassic Limestones of the Swabian Alb (Southwest Germany). Contrib Sedimentol 5:75 (Schweizerbart)

Hérenger L (1941) Contribution à l'étude des spongiaires du Jurassique et du Crétacé de catalogne. Trav Lab Géol Fac Sci Univ Grenoble 23:1–56

Kutek J, Wierzbowski A, Bednarek J, Matyja BA, Zapaśnik T (1977) Z problematyki stratygraficznej osadów górnojurajskich Jury Polskiej. Przegląd Geol 8–9:438–445

Moret L (1926) Sur quelques Spongiaires de Catalogne (Argovien, Sénonien, Eocène). Bull Soc Cienc Nat Barcelona "Club Montagnec" 9

Nitzopoulos G (1974) Faunistisch-ökologische, stratigraphische und sedimentologische Untersuchungen am Schwammstozen-Komplex bei Spielberg am Hahnenkamm (Ob Oxfordien, Südlische Frankenalb). Stuttg Beitr Naturk B16:143

Reid REH (1968) Bathymetric distribution of Calcarea and Hexactinellida in the present and in the past. Geol Mag 105:546–559

Riba AO (1959) Estudio geológico de la Sierra de Albarracin. Monografias del Cientif Instituto "Lucas Mallada", Madrid 16, pp 283

Scheer U (1988) Influence of the paleogeographic position and sea-level changes on spongiolitic limestones in the Lower Bajocian (*discites*- to *humphresianum*-zone; Middle Jurassic) in Northern Spain, Berl Geowiss Abh A100:36

Schorr M, Koch R (1985) Fazieszonierung eines oberjurassischen Algen-Schwamm-Bioherms (Herrlingen, Schwäbische Alb). Facies 13:227–270

Seyfried H (1978) Der subbetische Jura von Murcia (Südost-Spanien). Geol Jahrb B29 3:201

Trammer J (1982) Lower to Middle Oxfordian sponges of the Polish Jura. Acta Geol Pol 32:1–39

Trammer J (1983) The relation of the morphological type of a sponge to water turbulence. Speculation Sci Technol 6:143–146

Wagenplast P (1972) Ökologische Untersuchungen der Fauna aus Bank-und Schwammfazies des Weissen Jura. Arb Inst Geol Paläontol Univ Stuttg NF 67:1–99

Wilde S (1988) The Upper Bajocian-Lower Bathonian (Middle Jurassic) of the northwest Iberian Range (Spain) – a spongiotuberolitic platform environment. Berl Geowiss Abh Reihe A 100:52

Wirsing G, Koch R (1986) Algen-Schwamm-Bioherme des Flachwasser-Bereiches (Schwäbische Alb. Weissjura Delta 3). Facies 14:285–308

# The Spongiolithic Facies in the Upper Jurassic of Spain

M. Deusch, A. Friebe, and M. Krautter

## Introduction

Sponge-bearing limestones occur repeatedly in the Middle and Upper Jurassic of Spain (Fig. 1). They are comparable with sponge formations of Central Europe (eastern France, Switzerland, southern Germany, southern Poland).

Mallada (1885) described some Oxfordian sponges. Much later, Herenger (1942) described Oxfordian sponges from the southwestern Celtiberian Range. Until today, however, no detailed study of the spongiolithic facies or of the paleoecology of sponges has been carried out. The Demosponges (Lithistida) and Hyalosponges (Dictyida and Lychniskida) are preserved in a calcareous form. A great part of these "sponge mummies" are covered with calcareous crusts, which are the result of precipitation, maybe as a consequence of decay processes. Calcareous crusts have been the subject of a number of studies (Fritz 1958; Wagenplast 1972; Nitzopoulos 1974; Gwinner 1976; Krumbein 1978; Flügel and Steiger 1981). The Jurassic fossil sponges are colonized by other organisms, which are mostly on the lower side of the sponges: foraminifera, other sponges, corals, brachiopods, bryozoans and serpulids (very often). The sponges are often toppled over as a result of lateral currents.

## Bajocian

The sponge facies of Bajocian age (Hinkelbein 1969, 1975) is now being studied by A. Friebe in the southern and central Celtiberian Range. The sponge facies ranges stratigraphically from the biozone of *Stephanoceras humphriesianum* to the biozone of *Parkinsonia parkinsoni* (Fig. 2). Exceptionally, the first sponges can appear in the upper part of the biozone of *Otoites sauzei* and the last ones can extend into the Lower Bathonian (Hinkelbein 1975).

In the Montes Universales de Albarracin, spongiolithic bioherms predominate in the Bajocian. Only in the biozone of *subfurcatum* are there biostromes with facies transitions to biodetritic bedded limestones. The sponge-bearing limestones reach a thickness of about 40–50 m. In the adjacent northern and eastern areas of the Montes Universales de Albarracin there are interdigitations of the biohermal facies with spongiolithic biostromes or tube-

J. Reitner and H. Keupp (Eds.)
Fossil and Recent Sponges

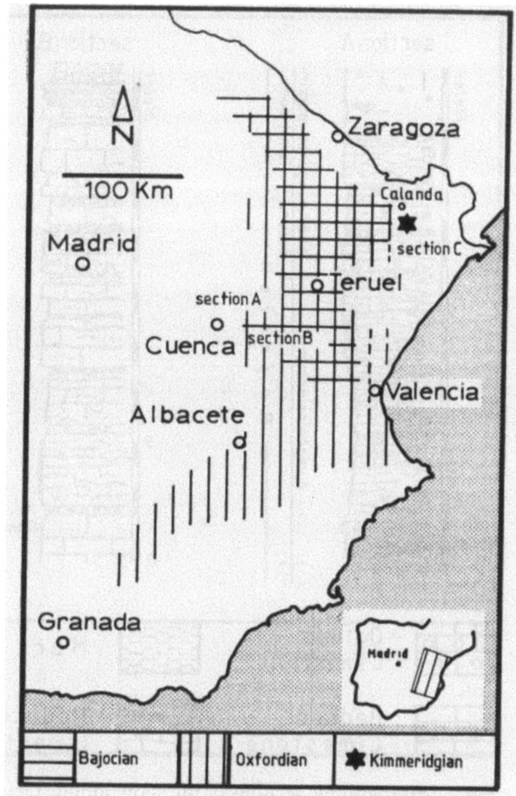


**Fig. 1.** The spongiolithic facies in the Iberian Mountains

rolithic carbonates, which are 10–50 m thick. Biomicritic non-sponge-bearing bedded limestones with thicknesses of up to 120 m adjoin in the western areas.

The mud mounds consist of small bioherms whose thicknesses of up to 7 m have been described from the Rio Ebron Vally between Valdecuenca and El Cuervo. The sponge fauna consists of dictyid and lychniskid sponges, which predominate in the lower parts, as well as lithistids, which are frequent in the upper part of the sections. All sponges are preserved in calcareous form. The sponges colonized crypto-hardgrounds (formed by blue green algae, foraminifers and serpulids), and biodetritic ground. The sponge mummies are often covered by thin calcareous crusts. The accompanying fauna in the bioherms consists of corals (infrequent), brachiopods (very abundant), pelecypods (abundant), gastropods and belemnites (both less abundant), and echinids (less abundant). Within the spongiolithic buildups of the bioherms, boundstones and grain/packstones are abundant. Outside these buildups, mudstones as well as packestones-wackestones are frequent. In the Upper Bajocian, where the sponge facies ends, redeposited beds can be found. The microfacies type "floatstone" is abundant here.

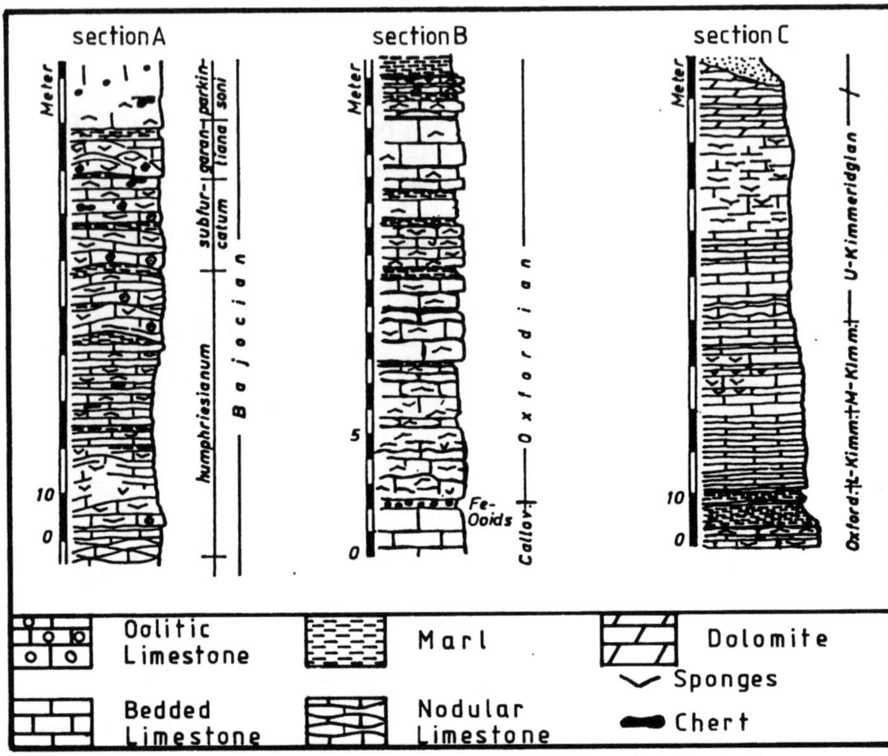


**Fig. 2.** Stratigraphic sections of the spongiolithic facies

The biostromes of the Montes Universales de Albarracin (the biozone of *subfurcatum*) and of their adjacent northern and eastern areas are characterized by calcareous and marly layers with more or less isolated embedded sponges. Here, too, there are tuberoliths. The sponge fauna consists mainly of cup-shaped specimens, accompanied by brachiopods, pelecypods, ammonites, etc. Chert and glauconite are more abundant in the biostromal facies than in the biohermal facies.

The question about the source aspects of the spongiolithic facies of Bajocian age can possibly be answered by comparing the biological and lithological features of bioherms and biostromes. So the part of carbonate is bigger in the bioherms than in the biostromes where also the part of marl predominates. The accompanying fauna (as described above) shows the change from a lower energetic environment of the bioherms where porifera find more optimal conditions to a higher energetic environment of the biostroms. The Bajocian sponge layers were deposited in an open shelf environment of moderate depth.

## Oxfordian

Since the last century the Oxfordian sponge facies of the Celtiberian Range has been known (Dereims 1898; Riba Arderiu 1959; Geister and Geyer 1968; Hinkelbein 1969; Geyer 1971; El Khoudary 1974; Gomez 1979; Deusch 1986; Ott 1987, and others). Likewise, the sponge facies of the Prebetic Provinces have been described by Behmel (1970); Lopez Garrido (1971); Geyer (1971); Geyer et al. (1974), and others. A detailed study of both regions is currently being carried out by M. Deusch.

The sponge-bearing facies of Oxfordian age ranges stratigraphically from the biozone of *Perisphinctes plicatilis* to the biozone of *Idoceras planula* – at least in certain sections. The sponge-bearing limestones reach a thickness of up to 30 m. Biohermal structures are not known; nevertheless, sponges are so abundant as to sometimes be lithogenous. Although the outer shape of the sponges is very well preserved, it is difficult to determine their skeletal structures, due to the complete calcification which they suffered. An approximate determination of the sponges is possible with the help of thin sections.

Figure 3 shows the share of the various morphotypes (of 300 specimens), and Fig. 4 shows the percentage of the taxonomic groups Dictyida, Lychniskida and Lithistida in the Celtiberian Region. The accompanying macrofauna is characterized by ammonites, and to a lesser degree, by brachiopods and crinoids (Fig. 5). The sponges are covered by thin calcareous crusts which show no discernible internal structure. The crusts were formed on the upper side of the sponges and

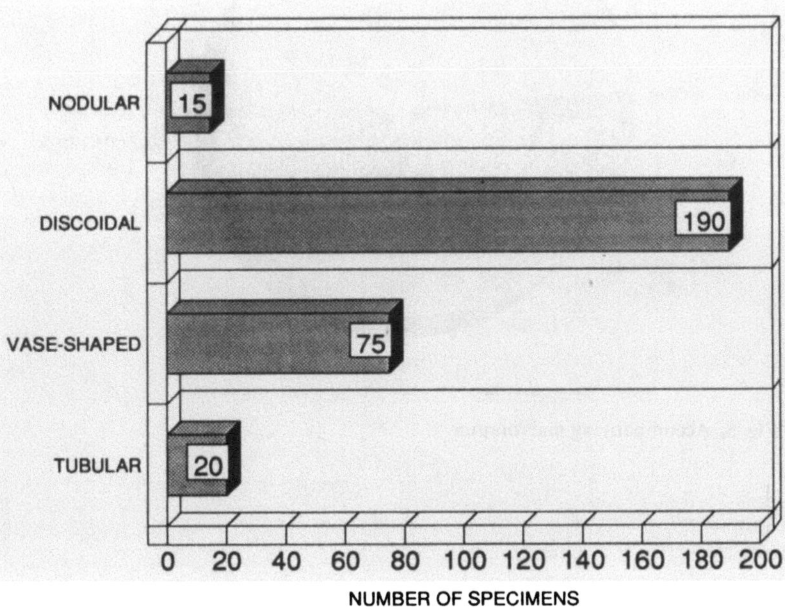


**Fig. 3.** Share of morphotypes (300 specimens)

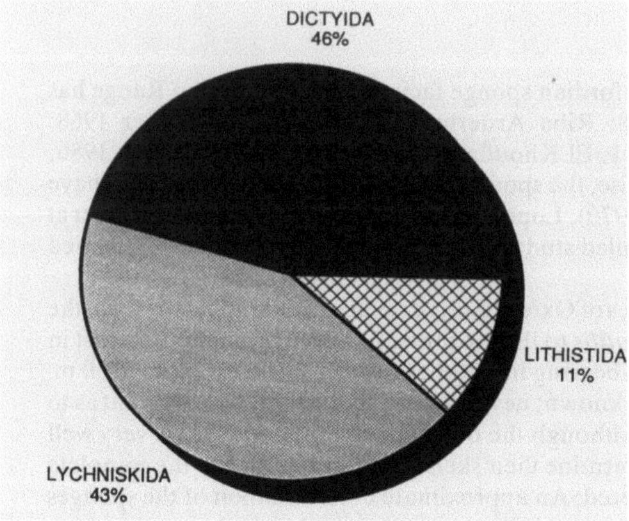


**Fig. 4.** Percentage of sponge orders

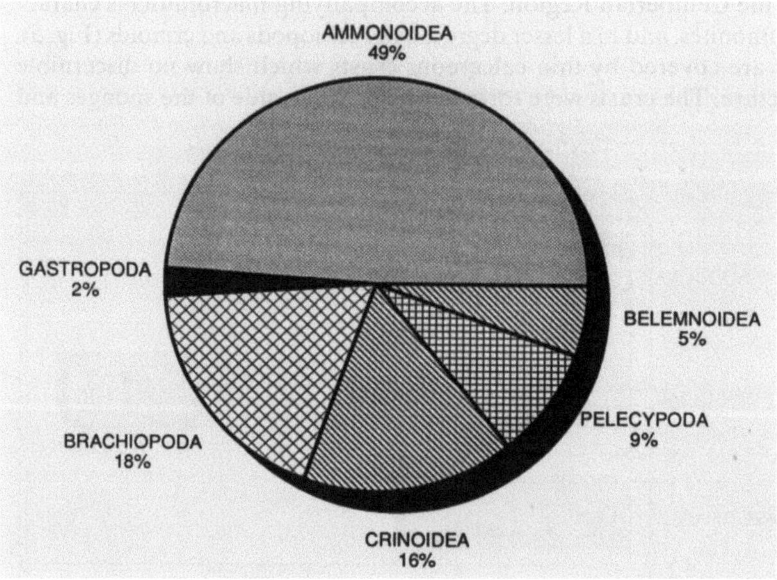


**Fig. 5.** Accompanying macrofauna

consist of serpulids, bryozoans, brachiopods and (rarely) sessile foraminifera (*Nubeculinella*). As to the embedding of the sponges, one must distinguish between the group of the discoidal sponges on the one hand and the vase-shaped, tubular, or nodular sponges on the other hand. Eighty percent of the first group are buried in an inverted position, whereas only 45–55% of the second group lie upside-down. Only 5% of all sponges are buried in their life position.

Figure 6 shows the distribution of six microfacial types as seen in practically all sections of the Celtiberian Region. In the middle and upper parts of the formation biomicrite is predominant. It seems that wackestones are abundant in the middle part, and packstones in both the lower and upper parts. The accompanying fauna and the biostratonomical and microfacial aspects of the Oxfordian sponge facies indicate that the environment of origin was an open shelf with large sponge platforms at moderate depth.

## Kimmeridgian

Up to the present, sponge-bearing limestones in Spain had only been described from the region between Calanda and Puertos de Beceite (Province of Teruel) in the Iberian Range (Fezer and Geyer 1988; Geyer 1963; Geyer and Pelleduhn 1981; Pelleduhn 1981). Except for a short remark by Bulard (1972), the first mention of this occurrence was made by Rückert (1977). In the well-studied section of Calanda, the sponge-bearing facies is divided in two levels: an older one

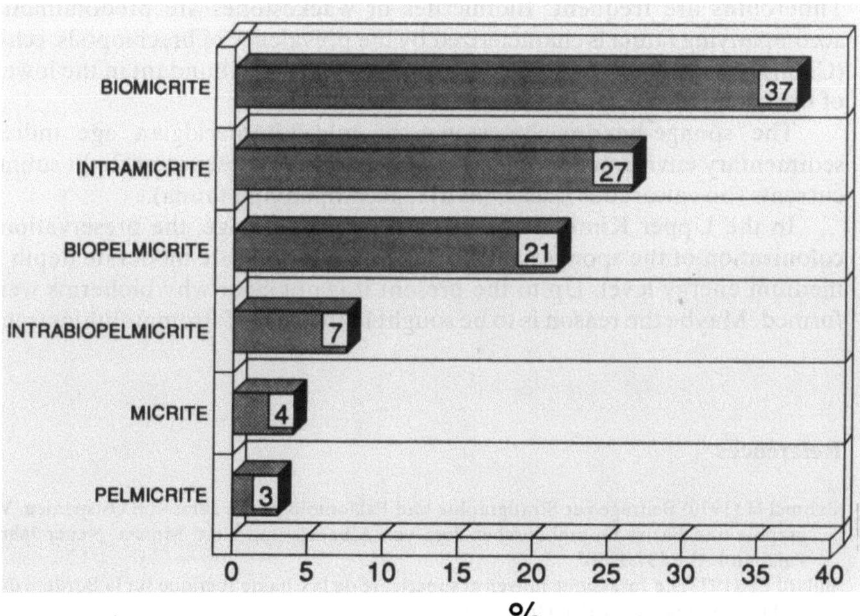


**Fig. 6.** Microfacial types of the Oxfordian sponge facies

(Kimmeridgian IV/V), biozone of *Taramelliceras strombecki* and a younger one (Kimmeridgian VI), biozone of *Hybonoticeras beckeri*.

In the spongiolithic facies of Middle Kimmeridgian age there are pseudo-biohermal structures (10 to 12 m diameter, 5 to 6 m high). These bioherm-like structures pass laterally into stratified limestones and marly limestones. The spongiolithic buildups are stratified because they consist of biostromal beds which grew one upon another. Their bioherm-like appearance is due to weathering processes. Because of their incompetent behavior, these pseudo-bioherms have been pressed into the underlying beds. The taxonomic determination of the dictyid and lithistid sponges is difficult. The cup- and plate-shaped sponges have diameters of up to 30 cm. Colonization by other organisms is not very clear (serpulids, bryozoans). The Middle Kimmeridgian sponge-bearing limestones are micrites or mudstones. Biogenetic material is therefore not very abundant. It would seem that the scarcity of bioclasts is not primary, but rather the result of diagenetic processes.

The spongiolithic character of the Upper Kimmeridgian limestones is more marked, as opposed to the Middle Kimmeridgian facies. The sponge-bearing limestones reach thicknesses of up to 25 m, biohermal structures are not found here. Although the strata are well bedded, they have a bulky appearance. Single layers have a thickness of 25 to 50 cm. Discoidal sponges, reaching diameters of 100 cm, predominate. These "giant" sponges have walls of 5 cm thickness and more. In the upper part of the section, cup-shaped sponges are more frequent. The big discoidal sponges are buried in their life position, their size having prevented their being toppled over. They have calcareous crusts on their upper sides. Tuberoliths are frequent. Biomicrites or wackestones are predominant. The accompanying fauna is characterized by the prevalence of brachiopods, echinoids (Cidariidae) and ammonites. Limide pelecypods are abundant in the lower part of the spongiolithic level.

The sponge-bearing limestones of mid-Kimmeridgian age indicate a sedimentary environment of deeper water and the presence of slight submarine currents (no calcareous crusts, matrix, accompanying fauna).

In the Upper Kimmeridgian the fossil assemblage, the preservation, and colonization of the sponges and the microfacies indicate moderate depth and a medium energy level. Up to the present it is not clear why bioherms were not formed. Maybe the reason is to be sought in the lack of stromatolithic structures.

## References

Behmel H (1970) Beiträge zur Stratigraphie und Paläontologie des Juras von Ostspanien. V Stratigraphie und Fazies im präbetischen Jura von Albacete und Nord-Murcia. Neuer Jahrb Geol Paläontol Abh 137:1–102

Bulard PF (1972) Le Jurassique moyen et supérieure de la Chaine Ibérique sur la Bordure du bassin de l'Ebre (Espagne). Thèse Univ Nice

Dereims A (1898) Recherches géologiques dans le sud de l'Aragon. Thèse Univ Paris Lille

Deusch M (1986) Der "Obere Grenzoolith" im SW-Keltiberikum. Dipl Arbeit Univ Stuttgart

El Khoudary RH (1974) Beiträge zur Stratigraphie und Fazies des Juras von Ostspanien. VI. Untersuchungen im Oberjura der südwestlichen Iberischen Kordillere unter besonderer Berücksichtigung der Mikrofauna (Provinz Teruel und Rincon de Ademuz). Neuer Jahrb Geol Paläontol Abh 144:296-341

Fezer R, Geyer OF (1988) Der Oberjura von Calanda im nordöstlichen Keltiberikum (Provinz Teruel, Spanien). I. Stratigraphie. Arb Inst Geol Paläontol Univ Stuttg

Flügel E, Steiger T (1981) An Upper Jurassic Sponge-Algal Buildup from the Northern Frankenalb, West Germany. In: Toomey DF (ed) European Fossil Reef Models. Soc Econ Paleontol Mineral Spec Publ 30:371-397

Fritz GK (1958) Schwammstotzen, Tuberolithe und Schuttbreccien im Weissen Jura der Schwäbischen Alb. Arb Geol Paläontol Inst TH Stuttg 13:1-119

Geister J, Geyer OF (1968) Beiträge zur Stratigraphie und Paläontologie des Juras von Ostspanien. IV. Der Jura der Sierra de Corbera (Provinz Valencia). Neuer Jahrb Geol Paläontol Abh 131:310-336

Geyer OF (1963) Beiträge zur Stratigraphie und Paläontologie des Juras von Ostspanien. I. Eine Ammonitenfauna aus dem Unterkimmeridgium der Sierra de Montenegrelo (WSW Tortosa, Provinz Tarragona). Neuer Jahrb Geol Paläontol Abh 118:182-196

Geyer OF (1971) Zur paläobathymetrischen Zuverlässigkeit von Ammonoideen-Faunen-Spektren. Palaeogeogr Palaeoclimatol Palaeoecol 10:265-272

Geyer OF, Pelleduhn R (1981) Sobre la estratigraphia y la facies espongiolithica del Kimmeridgiense de Calanda (Provincia de Teruel). Cuad Geol 10:Granada

Geyer OF, Behmel H, Hinkelbein K (1974) Beiträge zur Stratigraphie und Paläontologie des Juras von Ostspanien. VII. Die Grenzoolithe im Jura von Ostspanien. Neuer Jahrb Geol Paläontol Abh 145:17-57

Gomez JJ (1979) El Jurásico en facies carbonatadas del sector levantino de la Cordillera Ibérica. Semin Estratigr 4 Madrid

Gwinner MP (1976) Origin of the Upper Jurassic Limestones of the Swabian Alb (southwest Germany). In: Füchtbauer H (ed) Contributions to sedimentology 5 Schweizerbart'sche, Stuttgart, pp 1-75

Hérenger L (1942) Contribution a l'étude des Spongiaires au Jurassique et du Crétacé de Catalagne. Trav Lab Geol Fac Sci Univ Grenoble XXIII

Hinkelbein K (1969) El Triasico y el Jurasico de los Alrededores de Albarracin. "Teruel" 41:35-75

Hinkelbein K (1975) Beiträge zur Stratigraphie und Paläontologie des Juras von Ostspanien. VIII. Stratigraphie und Fazies im Mitteljura der zentralen Iberischen Ketten. Neuer Jahrb Geol Paläontol Abh 148:139-184

Krumbein WE (1978) Algal mats and their lithification. In: Krumbein WE (ed) Environmental biochemistry and geomicrobiology, the aquatic environment: vol 1. Science Publ Inc, Ann Arbor, Michigan, pp 209-225

Lopez Garrido AC (1971) Geologia de la Zona Prebética, al NE de la Provincia de Jaén. Tesis doctoral Univ Granada

Mallada L (1885) Sinopsis de las especies fosiles que se han encontrada en Espagna. Terreno Mesozóico. (Sistemas Triásico y Jurásico). Madrid

Nitzopoulos G (1974) Faunistisch-ökologische, stratigraphische und sedimentologische Untersuchungen am Schwammstotzenkomplex bei Spielberg am Hahnenkamm (Ob Oxfordien, Südliche Frankenalb). Stuttg Beitr Naturkd B 16:1-143

Ott P (1987) Schwammkalk-Profile im Oxfordium des Spanischen Keltiberikums. Dipl Arbeit Univ Stuttgart

Pelleduhn R (1981) Stratigraphie und Fazies des Oberjuras bei Calanda (Provinz Teruel, Spanien). Dipl Arbeit Univ Stuttgart

Riba Arderiu O (1959) Estudio geológico de la Sierra de Albarracin. Inst "Lucas Mallada" Invest Geol 16: p 283

Rückert A (1977) Die Ammonitenfauna des Unterkimmeridgiums bei Alcaniz (Provinz Teruel, Spanien). Dipl Arbeit, Univ Stuttgart, FRG

Wagenplast P (1972) Ökologische Untersuchung der Fauna aus Bank- und Schwammfazies des Weissen Jura der Schwäbischen Alb. Arb Geol Paläontol Inst Univ Stuttgart NF 67:1-99

# Ecologic History of the Oxfordian Sponge Assemblage in the Polish Jura Chain

J. Trammer

## Introduction

The Polish Jura Chain is among those areas in Europe (Fig. 1) where the Upper Jurassic strata contain sedimentary rocks rich in siliceous sponges (= sponge megafacies of Matyja 1976; = Malm reef belt of Wilson 1975; see also Gaillard 1983, Fig. 130). The sponge megafacies varies in its stratigraphic range. In Germany, for example, it ranges from the Oxfordian through the Tithonian (see, e.g., Ziegler 1977; Meyer 1977, Fig. 1). In the Polish Jura Chain, the sponge megafacies is confined to the Oxfordian.

## Oxfordian Lithology

The Oxfordian includes several lithological complexes in the Polish Jura Chain; not all of these complexes contain sponges, though sponges occur in bedded rocks as well as in bioherms of various sizes (Fig. 2).

Except for some clearly shallow-water rocks of the Oxfordian/Kimmeridgian transition, the sponge-bearing rocks of the Polish Jura Chain are considered to have originated in open-shelf environment (Kutek et al. 1984, Fig. 1), generally below the storm wave base. This is suggested by the absence of sedimentary structures indicative of wave action on the bottom sediment, and by the fauna which is typical of deeper shelf environments (Trammer 1982, p. 15).

## The Investigated Sections

I have collected sponges in 15 sections in the Polish Jura Chain. Their geographic location is shown in Fig. 3, whereas their stratigraphic position and lithology are shown in Figs. 2 and 4.

## General Characteristics of the Sponge Assemblage

Totally, 68 sponge species have been found in the investigated sections (Fig. 5). They will be discussed in a separate publication (Trammer, in prep.). These are

J. Reitner and H. Keupp (Eds.)
Fossil and Recent Sponges

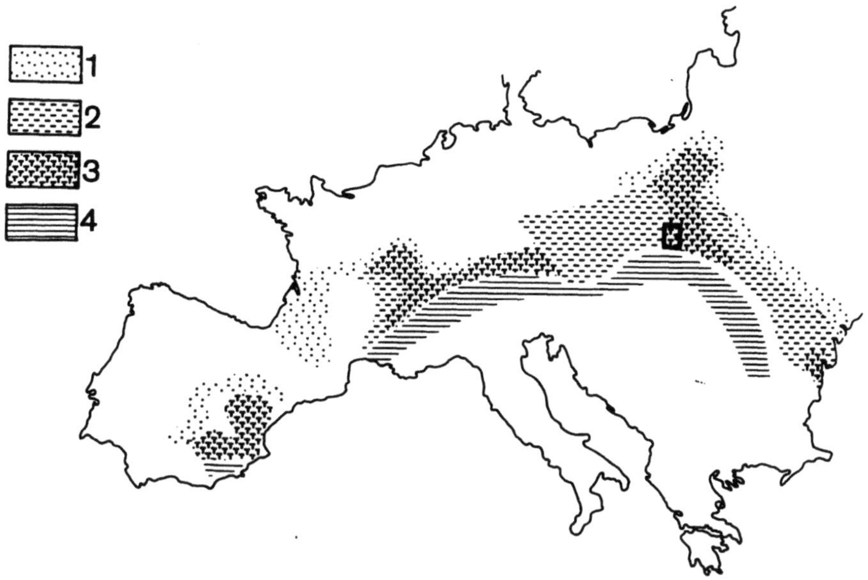


**Fig. 1.** Paleogeographic position of the Upper Jurassic facies in Europe. *1* Shallow-water deposits; *2* probable extent of the sponge facies; *3* recognized range of the sponge facies; *4* Tethyan deposits. Rectangled is the investigated area (see Fig. 3). (After Matyja 1976)

mostly well-known taxa, reported previously from the sponge megafacies of various European countries (see Goldfuss 1833; Zittel 1877, 1878; Quenstedt 1878; Oppliger 1897, 1907, 1915, 1926; Kolb 1910; Siemiradzki 1913; Antonescu 1929; Schrammen 1937; Trammer 1982; Gaillard 1983; Müller 1984; Meyer et al. 1987, and others).

The majority of the species belong to the Hyalospongea (47 spp.); 16 species belong to the Demospongea (primarily "Lithistida"), four to the Calcispongea, and one to the "Sclerospongea". Some poorly preserved sponges could not be precisely identified and are here described as Dictyida gen. et sp. indet. (Fig. 5).

Only 11 species are common in the analyzed sponge assemblage. They account for 74% of the total number of specimens, and they represent the genera *Platychonia*, *Cnemidiastrum*, *Hyalotragos*, *Cylindrophymma*, *Craticularia*, *Tremadictyon*, *Xenoschrammenum*, *Ordinatus*, and *Cypellia* (Fig. 5).

## Ecologic History

A peculiar trend is discernible in the Oxfordian sponge assemblage of the Polish Jura Chain (Fig. 6). At the beginning of the Oxfordian, the assemblage was dominated by lithistids which accounted for 96% of the total number of specimens. Later, the share of the Hyalospongea gradually increased at the expense of the Lithistida, and hyalosponges accounted at the end of the Oxfordian for 90% of the total number of specimens.

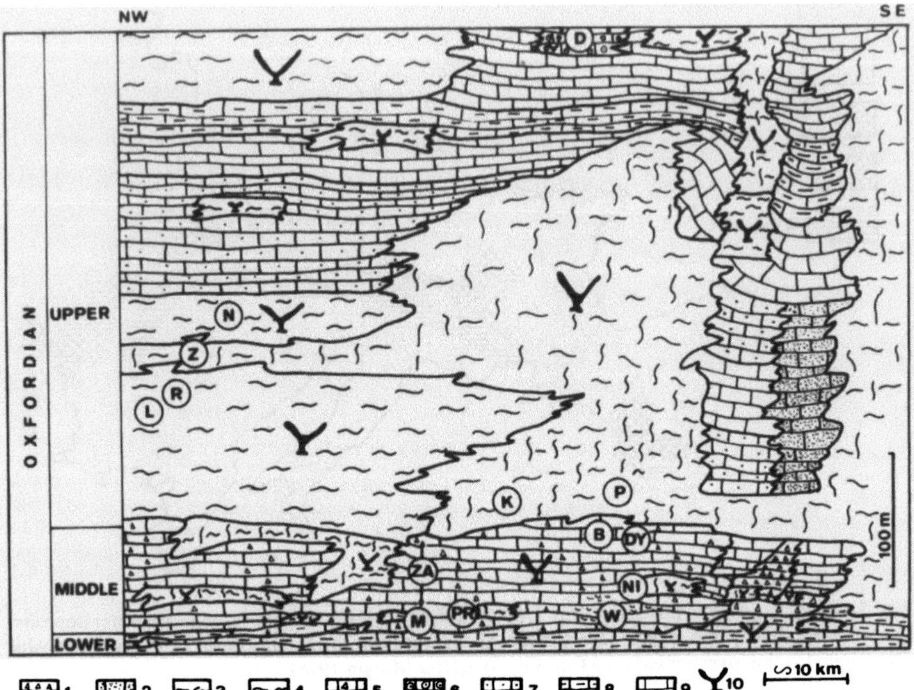


**Fig. 2.** Lithologic units of the Oxfordian in the Polish Jura Chain, showing the range of sponge-bearing units (see 10 in explanations) *1* Detrital limestones; *2* nondetrital grained limestones; *3* sponge or algal-sponge massive limestones; *4* chalky limestones; *5* platy limestones with sponges and tuberoliths, locally also sponge crumpled limestones; *6* crinoid limestones with corals; *7* friable micritic limestones; *8* marls and marly limestones; *9* micritic limestones; *10* idealized sponge to denote the sponge-bearing lithologic units. Localities studied: *M* Mirów; *PR* Prędziszów; *W* Wysoka; *NI* Niegowonice; *ZA* Zawodzie; *DY* Dymnik; *B* Birów; *K* Kielniki; *P* Podzamcze; *L* Lisowice; *R* Raciszyn; *Z* Zalesiaki; *N* Niwiska; *D* Dobrogoszczyce. Drawn on the basis of data presented by Kutek et al. (1977) and Brunsz (1987)

**Fig. 4.** Stratigraphical position of the investigated exposures. *1* Algal-sponge massive limestones; *2* chalky limestones; *3* crinoid limestones with corals; *4* platy limestones with sponges and tuberoliths; *5* sponge crumpled limestones; *6* sponge marly limestones and marls. Set after Łacwik (1970), Bednarek (1974), Brochwicz-Lewiński (1976), Wierzbowski (1978), Wierzbowski et al. (1983), Brunsz (1987), and Irmiński (1987)

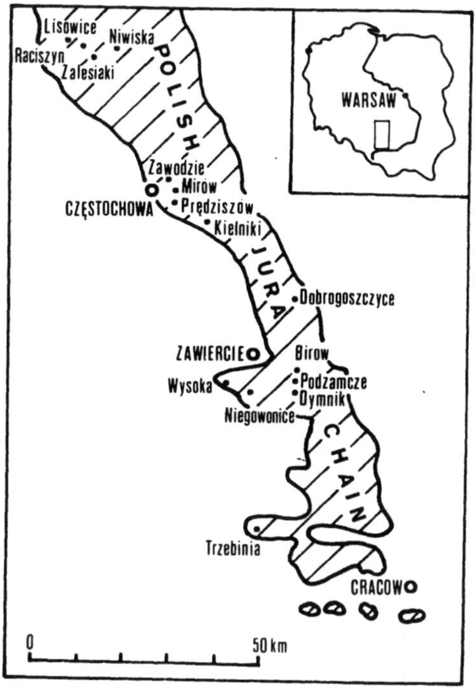


**Fig. 3.** Investigated exposures of the Oxfordian within the occurrence zone (*hatched*) of Upper Jurassic deposits in the Polish Jura Chain; *inset* shows the position of the area in Poland

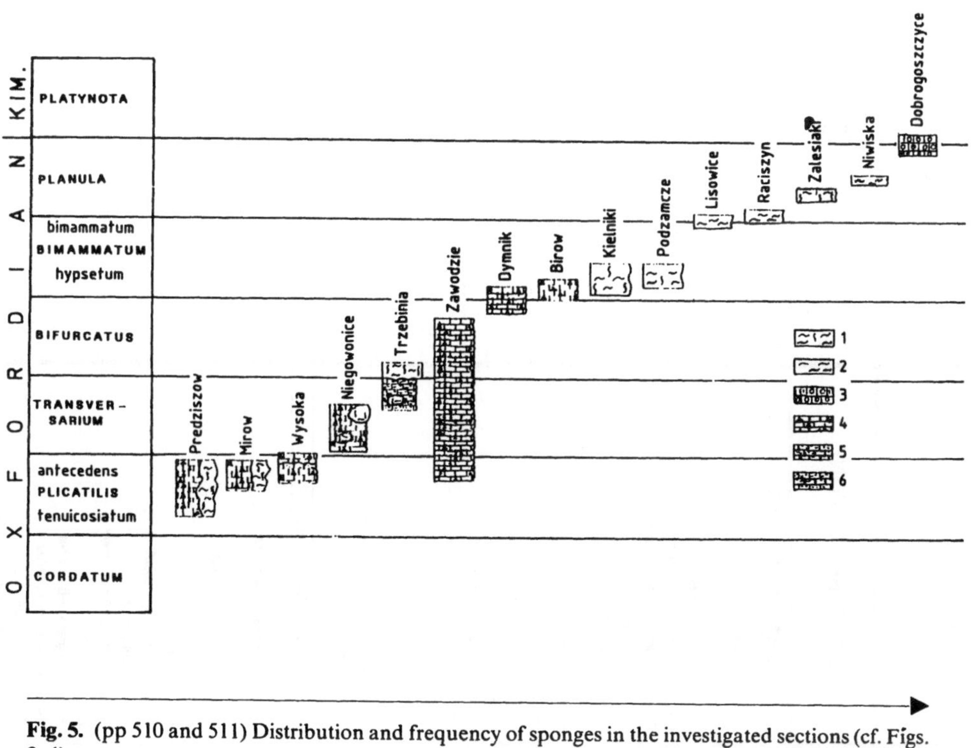


**Fig. 5.** (pp 510 and 511) Distribution and frequency of sponges in the investigated sections (cf. Figs. 2–4)

| LOCALITY | PREDZISZOW | | MIROW | WYSOKA | NIEGOWONICE | TRZEBINIA | | ZAWODZIE | | | DYMNIK | BIROW | KIELNIKI | PODZAMCZE | LISOWICE | RACISZYN | ZALESIAKI | NIWISKA | DOBROGOSZCZYCE |
|---|---|---|---|---|---|---|---|---|---|---|---|---|---|---|---|---|---|---|---|
| ZONE or SUBZONE | TENUI. | ANT. | ANT. | ANT. | TRANS. | TRANS. | BIF. | ANT. | TRANS. | BIF. | BIF.or HYP. | HYP. | HYP. | HYP. | BIM.or PLAN. | PLAN. | PLAN. | PLAN. | PLAN.or PLAT. |
| **DEMOSPONGEA / total /** | 93 | 18 | 11 | 94 | 56 | 148 | 8 | 12 | 6 | 12 | 23 | 9 | 14 | 31 | 20 | 23 | 33 | 10 | |
| CHORISTIDA | | | | | | | | | | | | | | | | | | | |
| *Ophiodesia* sp. | | | | | | | | | | | | | | | | | | 1 | |
| LITHISTIDA | | | | | | | | | | | | | | | | | | | |
| RHIZOMORINA | | | | | | | | | | | | | | | | | | | |
| *Platychonia schlotheimi* | 58 | 12 | 7 | 41 | 34 | 35 | 3 | 9 | 6 | 10 | 7 | 7 | 2 | 14 | 7 | 11 | 18 | 4 | |
| *Cnemidiastrum stellatum* | 6 | 3 | | 1 | | 9 | 2 | | | | | | | 1 | | 1 | 2 | 1 | |
| *C. rimulosum* | 11 | | | 37 | 6 | 25 | | | | | 3 | | 1 | 1 | | | 1 | 1 | |
| *Hyalotragos patella* | 3 | 1 | 1 | 5 | 4 | 11 | | 1 | | | 6 | 6 | 4 | 4 | 6 | 2 | 3 | 1 | |
| *H. pezizoides* | 3 | 1 | 2 | 7 | 6 | 49 | 2 | 2 | | 1 | 6 | 5 | 4 | 4 | 2 | 7 | 3 | | |
| *H. radiatum* | 2 | | | | | 2 | | | | | 1 | 1 | 1 | 1 | | | | | |
| *Jereica* sp. | | | | | 2 | | | | | | | | | | | | | | |
| *Leiodorella tubata* | | | | | | | | | | | | | 1 | | | | | | |
| *L.* sp. | | 1 | | 1 | | | | | | | | | | | | | | 1 | |
| *Hyalospongia rugosa* | | | | | | | | | | | | | | | | 1 | | | |
| *Bothrolemma osculifera* | | | | | 1 | | | | | 1 | | | | | | | | | |
| *Reiswigia ramosa* | 10 | | | | | | | | | | | | | | | | | | |
| Rhizomorina gen. et sp. indet. | | | | | 2 | | | | | | | | | | | | 6 | | |
| ANOMOCLADINA | | | | | | | | | | | | | | | | | | | |
| *Cylindrophyma milleporata* | | | 1 | 2 | 1 | 16 | 1 | | | | | | 1 | 6 | 4 | 1 | | 1 | |
| *Melonella radiata* | | | | | | | | | | | | | | | 1 | | | | |
| *Lecanella* sp. | | | | | | 1 | | | | | | | | | | | | | |
| **HYALOSPONGEA / total /** | 10 | 11 | 18 | 20 | 41 | 73 | 4 | 3 | 11 | 7 | 21 | 13 | 99 | 34 | 44 | 50 | 58 | 41 | 2 |
| LYSSAKIDA | | | | | | | | | | | | | | | | | | | |
| *Stauractinella jurassica* | | | | | | 1 | | | | 2 | | | | | | | | 2 | |
| DICTYIDA | | | | | | | | | | | | | | | | | | | |
| *Craticularia parallela* | 4 | 1 | | | 9 | 16 | | | | | 3 | 1 | 9 | | | 6 | 2 | 5 | |
| *C. paradoxa* | | 1 | | | 5 | 4 | | | | | | | | | | | 1 | 1 | |
| *C. clathrata* | | | | | | | | | | | 1 | 1 | 3 | 1 | | | | | |
| *Paracraticularia tubifera* | | | | | | | | | | | | | | | 1 | | | | |
| *P. procumbens* | | | | | | | | | | | | | 3 | | | | | | |
| *Thyroidium schweiggeri* | | | | | | | | | | | | | 3 | 1 | | 1 | | 1 | |
| *T. lineatum* | | | | | | 1 | | | | | | | 7 | 2 | | 2 | | | |
| *Pycnocalyptra* aff. *calyx* | | | | 1 | | | | | | | | | | | | | | | |
| *Pachyascus punctatus* | | | | | | 1 | | | | | | | | | | | | | |
| *Psephosyllogus* sp. | | | | | | | | | | | | | | | 1 | | | | |
| *Tremadictyon reticulatum* | 1 | 2 | 2 | | 2 | 12 | 1 | | | | 9 | | 45 | 19 | 7 | 6 | 13 | 4 | 1 |
| *T. radicatum* | | | 3 | 4 | 4 | | | | | | | | | | | | | | |

| | | | | | | | | | | | | | | | | | | | |
|---|---|---|---|---|---|---|---|---|---|---|---|---|---|---|---|---|---|---|---|
| *T. rugatum* | | | | | | | | | | | | | | | 4 | 1 | 2 | 2 | |
| *T.* sp. | | | 4 | 3 | | | | 2 | 1 | 1 | | 2 | | | | | | | |
| *Stauroderma lochense* | | | | | 1 | | | | | | 1 | | 2 | 1 | 1 | 3 | | 1 | |
| *Xenoschrammenum alternans* | | | | | | 1 | | | | | | 3 | 6 | 2 | | 1 | 16 | 4 | |
| *X. punctatum* | | | 1 | | | 1 | | | 1 | | | | 2 | | | | 3 | | |
| *Walcotella pertusa* | 1 | | | 1 | | 5 | | | | | 1 | 1 | 3 | | | | | | |
| *Ordinatus texturatus* | 1 | 3 | 2 | | 7 | 7 | | 1 | | | | | 2 | 7 | 3 | 2 | 6 | | |
| *Erineum minutum* | | | | | | 1 | | | | | | | | | | | | | |
| *Feifelia favositides* | | | | | | 1 | | | | | | | | | | | | | |
| *Casearia* sp. n. | | | | | | 1 | | | | | | | | | | | | | |
| *Porospongia marginata* | | | | | | | | | | | 1 | | | | 4 | | | | |
| *Multiloqua fungiformis* | | | | | | | 1 | | | | | | | | | | | | |
| *Linonema calyx* | | | | | | | | | | | | | | | | 1 | | | |
| *Ramispongia ramosa* | | | | 1 | | | | | | | | 1 | 1 | | 1 | 1 | | 1 | |
| *Verrucocoelia gregaria* | | | | | | | | | 1 | | | | | | | 3 | | | |
| *Dactylocalyx* sp. | 1 | | | | | | | | | | | | | | | | | | |
| *Polygonatium sphaeroides* | | | | | | 2 | | | | | | | | | | | | | |
| *Polygonatium* sp. n. | | | | | | 1 | | | | | | | | | | | | | |
| Dictyida gen. et sp. indet. | | | 6 | 2 | 6 | 3 | 2 | | 7 | 3 | | 1 | | | 10 | 6 | 11 | 10 | 1 |
| LYCHNISKIDA | | | | | | | | | | | | | | | | | | | |
| *Cypellia rugosa* | 2 | 1 | | 5 | 3 | 14 | | | 2 | 1 | 6 | 3 | 5 | | 8 | 11 | 2 | 6 | |
| *C. inberbis* | | | | | | | | | | | | | | | 1 | | | | |
| *C. trabeculata* | | | | | | | | | | | | | | | 1 | 1 | 1 | | |
| *Paracypellia prolifera* | | 1 | | | 2 | | | | 1 | | | | 1 | | | | | 1 | |
| *Placotelia marcoui* | | 1 | | | | | | | | | | | | | | | | | |
| *P. rugatum* | | | | | 1 | | | | | | | | | | | | | | |
| *P.* sp. | | | | | | | | | | | | | | | | | | 2 | |
| *Sporadopyge speciosa* | | | | | | | | | | | | | | | 1 | | | | |
| *Coscinaulus* aff. *micropora* | | | | 1 | | | | | | | | | | | | | | | |
| *Sporadopyle obliqua* | | 1 | | | | | | | | | | | 1 | | | 1 | | | |
| *Pachyteichisma lamellosa* | | | | | | | | | | | | | 4 | | | | 1 | | |
| *P. gresslyi* | | | | | | | | | | | | | 1 | | | | | | |
| *Trochobolus dentatus* | | | | | | | | | | | | | 1 | | | | | | |
| *T. texatus* | | | | | 1 | 1 | | | | | | | | | | | | | |
| *Pachyrhachis* sp. | | | | | | | | | | | | | | 1 | | | | | |
| Lychniskida gen. et sp. indet. | | | | 2 | | | | | | | | | | | | 4 | 1 | | |
| **DEMOSPONGEA or HYALOSPONGEA** | | | | | | | | | | | | | | | | | | | |
| Gen. n., sp. n. | | | | | | | | | | | | | 1 | | | | | | |
| **CALCISPONGEA / total /** | | | | | | | | | | | | | | | 3 | 3 | | | 22 |
| PHARETRONIDA | | | | | | | | | | | | | | | | | | | |
| *Blastinia costata* | | | | | | | | | | | | | | | 1 | | | | |
| *Corynella quenstedti* | | | | | | | | | | | | | | | 1 | | | | |
| *Peronidella cylindrica* | | | | | | | | | | | | | | | 1 | 1 | | 1 | 4 |
| *Radicispongia radiciformis* | | | | | | | | | | | | | | | | 1 | | | |
| Pharetronida gen. et sp. indet. | | | | | | | | | | | | | | | | 1 | | | 18 |
| **SCLEROSPONGEA / total /** | | | | | | | | | | | | | | | 1 | 2 | 9 | | |
| *Neuropora* sp. | | | | | | | | | | | | | | | 1 | 2 | 9 | | |
| **Total sponges** | 103 | 29 | 29 | 114 | 97 | 221 | 12 | 16 | 19 | 19 | 44 | 32 | 114 | 65 | 68 | 78 | 100 | 52 | 24 |

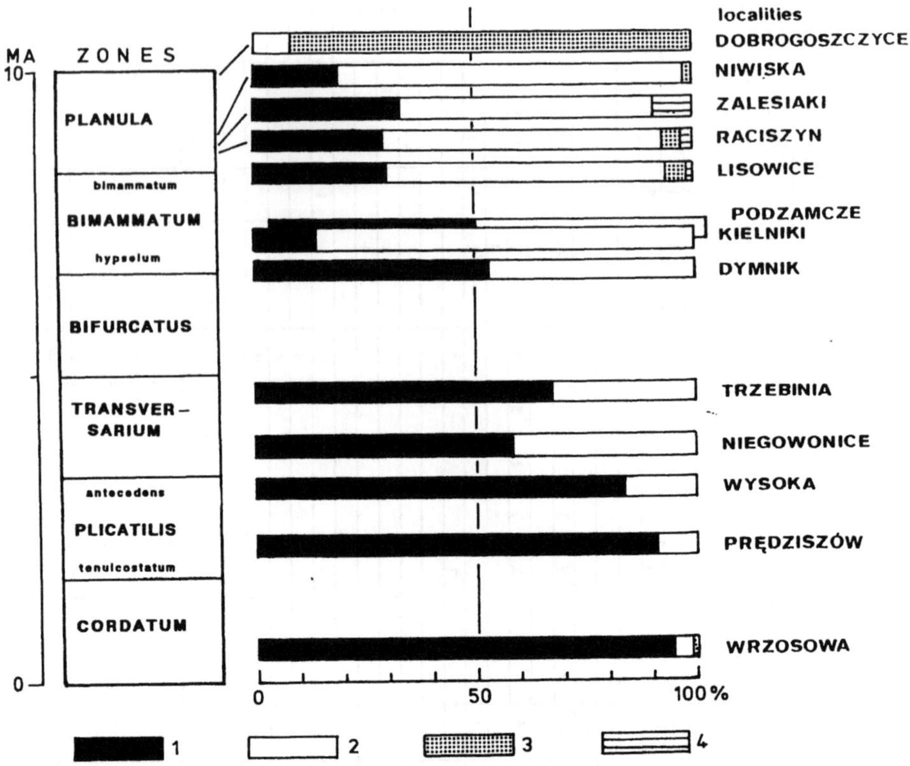


**Fig. 6.** Changes in frequency distribution of sponge taxa from the Oxfordian of the Polish Jura Chain, based on relative abundance of individuals (in percent). *1* Lithistida; *2* Hyalospongea; *3* Pharetronida; *4* Sclerospongiae. Drawn on the basis of data presented in Fig. 5 and, for the locality Wrzosowa, after Trammer (1982)

The species composition of the assemblage, however, remained almost constant throughout the entire Oxfordian (Fig. 7). What changed at the species level is the dominance hierarchy in the assemblage. Species representative of the Hyalospongea, e.g., *Tremadictyon reticulatum*, were extremely rare at first but becoming more and more common with time. Conversely, such lithistid species as *Cnemidiastrum stellatum* and *Hyalotragos patella* were dominant at first but rare toward the end of the Oxfordian. Thus, even though the Oxfordian lasted for about 10 Ma (Odin 1984), the remarkable change in the sponge assemblage was primarily ecological, rather than evolutionary, in nature.

This trend in composition of sponge assemblage ended at the Oxfordian Kimmeridgian transition, when shallow-water, coral-bearing crinoid limestones became dominant (Fig. 4) and brought also predominance of the Pharetronida among sponges (Figs. 5,6).

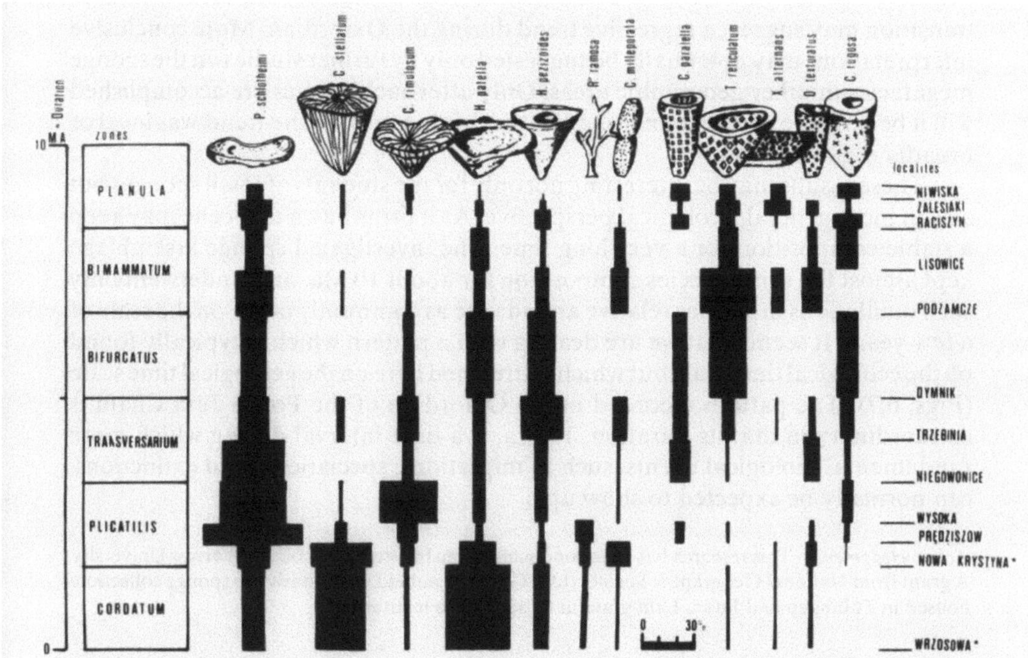


**Fig. 7.** "Spindle diagrams" illustrating the variation in relative abundance of selected sponge species during the Polish Oxfordian, to show that the trend depicted in Fig. 6 concerns a constant group of species. The specimens determined as "Dictyida gen. et sp. indet." and those belonging to very rare species are left out account. Drawn on the basis of data presented in Fig. 5 and, for the localities Wrzosowa and Nowa Krystyna, after Trammer (1982)

## Discussion

When considered jointly, the data presented in Figs. 6 and 7 strike me as rather surprising. On the one hand, the clearcut trend evident in Fig. 6 forces me to envisage a substantial environmental change during the Oxfordian, which turned a habitat that was initially favorable to lithistids into one that was more suitable for hyalosponges. On the other hand, however, Fig. 7 clearly shows that all the species underwent only a change in abundance, but persisted throughout the Oxfordian. Species abundance changes comparable to those observed over 10 Ma of the Oxfordian occur today, in Recent marine environments, on the scale of a few years and without any major environmental change (Petersen 1915; Ziegler 1972, Fig. 210). The persistence of sponge species composition throughout the Oxfordian forces me therefore to envisage a generally stable environmental setting.

I can find two explanations to account for these seemingly contradictory inferences. Either the long-term trend was induced by a very slow and gradual climatic change; or by a very slow and gradual shallowing, as the appearance of shallow-water rocks higher in the section at the Oxfordian/Kimmeridgian transition may suggest a regressive trend during the Oxfordian. More conclusive interpretations may potentially be suggested only by further studies on the sponge megafacies in other geographic areas. Only after such studies are accomplished will it be possible to determine whether the driving force of the trend was local or broadly regional.

These results may be interesting not only for the students of fossil sponges but also in more general, ecological perspective. As it turns out, a taxocene may keep a stable composition for a very long time. The investigated sponge assemblage kept almost the same species composition for about 10 Ma, and underwent only such oscillations in species relative abundance as commonly occur on the scale of a few years. It seems that we are dealing with a pattern which is typically found on the ecological time scale but which is stretched here on the geological time scale (Figs. 6,7). The pattern recorded in the Oxfordian of the Polish Jura Chain is extraordinary in that its duration, 10 Ma, is a time interval during which more fundamental biological events, such as migrations, speciations, and extinctions, can normally be expected to show up.

*Acknowledgments.* This research has been conducted in the Institute of Geology, Warsaw University. A grant from National Geographic Society (to J. Ghiold) enabled me to study the sponge collections housed in Tübingen and Basel. I am grateful to both these institutions.

## References

Antonescu CS (1929) Spongieri Jurasici din Dobrogea. Acad Rom Mem Sect Stii 5:481–490

Bednarek J (1974) Budowa geologiczna strefy wychodni górnojurajskich między Zawierciem, Łazami i Pilicą. Thesis, University of Warsaw

Brochwicz-Lewiński W (1976) Oxfordian of the Częstochowa area. II. Lithostratigraphy versus biostratigraphy. Bull Acad Pol Sci Ser Sci Terre 24:47–53

Brunsz M (1987) Analiza facjalna i stratygrafia jury górnej pomiędzy Niegową a Dzibicami na Jurze Polskiej. Thesis, University of Warsaw

Gaillard C (1983) Les biohermes a spongiaires et leur environment dans l'Oxfordien du Jura Meridional. Doc Lab Geol Lyon 90:1–515

Goldfuss A (1833) Petrefacta. Arnz, Düsseldorf

Irmiński W (1987) Jura Górna okolic Niegowonic. Thesis, University of Warsaw

Kolb R (1910) Die Kieselspongien des schwäbischen weissen Jura. Palaeontographica 57:141–256

Kutek J, Wierzbowski A, Bednarek J, Matyja A, Zapaśnik T (1977) Notes on the Upper Jurassic in the Polish Jura Chain. Przegl Geol 1977:438–445

Kutek J, Matyja A, Wierzbowski A (1984) Late Jurassic biogeography in Poland and its stratigraphical implications. In: Michelsen O, Zeiss A (eds) International Symposium on Jurassic Stratigraphy. Geological Survey of Denmark, Copenhagen, pp 743–754

Lacwik H (1970) Biostratygrafia i sedymentologia oksfordu ostańca Wysokiej. Thesis, University of Warsaw

Matyja A (1976) Oksford południowo-zachodniego obrzezenia Gór Swiętokrzyskich. Thesis, University of Warsaw

Meyer RK (1977) Mikrofazies im Ubergangsbereich von der Schwammfazies zur Korallen-Spongiomorphiden-Fazies im Malm/ Kimmeridge-Tithon/ von Regensburg bis Kelheim. Geol Jahrb A37:33–69

Meyer RK, Müller W, Übelacker L (1987) Die Schwamm- und Begleitfauna eines Schwammriffes des Unteren Malm in der Oberpfalz. Geol Bl NO-Bayern 37:63–88

Müller W (1984) Die Kalkschwämme der Unterordnung Inozoa Steinmann aus dem Oberen Jura von Württemberg /SW-Deutschland/. Stuttg Beitr Naturkd B100:1–85

Odin GS (1984) Geochronology of the Jurassic time: status in 1984. In: Michelsen O, Zeiss A (eds) International Symposium on Jurassic Stratigraphy. Geological Survey of Denmark, Copenhagen, pp 767–776

Oppliger F (1897) Die Juraspongien von Baden. Abh Schweiz Paläontol Ges 24:1–58

Oppliger F (1907) Spongien aus dem Argovien I /Birmensdorferschichten/ des Department du Jura, Frankreich. Abh Schweiz Paläontol Ges 34:1–19

Oppliger F (1915) Kieselspongien des schweizerischen weissen Jura. Abh Schweiz Paläontol Ges 40:1–84

Oppliger F (1926) Die Spongien der Birmensdorferschichten des schweizerischen Jura. Abh Schweiz Paläontol Ges 46:1–76

Petersen CG (1915) On the animal communities of the sea bottom in the Skagerrak, the Christiania Fiord and the Danish Waters. Rep Dan Biol Stat 23:3–28

Quenstedt FA (1878) Petrefactenkunde Deutschlands, Bd 5: Die Schwämme. Fues's Verlag, Leipzig, 612 pp

Schrammen A (1937) Die Kieselspongien des Oberen Jura von Süddeutschland. B. Besonderer Teil. Palaeontographica A85:1–114

Siemiradzki J (1913) Les spongiaires Jurassiques de la Pologne. Paleontol Pol 1:1–50

Trammer J (1982) Lower to Middle Oxfordian sponges of the Polish Jura. Acta Geol Pol 32:1–39

Trammer J Oxfordian sponges of the Polish Jura Chain (in press)

Wierzbowski A (1978) Ammonites and stratigraphy of the Upper Oxfordian of the Wieluń Upland, Central Poland. Acta Geol Pol 28:299–333

Wierzbowski A, Matyja A, Ślusarczyk-Radwan D (1983) New Data on Upper Jurassic strata in the Wielun Upland and vicinities of Burzenin and their economic value. Kwartalnik Geol 27:517–533

Wilson JL (1975) Carbonate facies in geologic history. Springer, Berlin Heidelberg New York

Ziegler B (1972) Einführung in die Paläobiologie. Teil 1. Allgemeine Paläontologie. Schweizerbart'sche, Stuttgart

Ziegler B (1977) The "White" /Upper/ Jurassic in Southern Germany. Stuttg Beitr Naturkd B26:1–79

Zittel KA (1877) Studien über fossile Spongien. I. Hexactinellidae. Abh K Bayer Akad Wiss Math Phys Kl 13:1–63

Zittel KA (1878) Studien über fossile Spongien. Zweite Abteilung: Lithistidae. Abh K Bayer Wiss Math Phys Kl 13:66–153

# Baffling, Binding, or Debris Accumulation? Ecology of Upper Jurassic Sponge-Bacterial Buildups (Oxfordian, Franconian Alb, Southern Germany)

B. LANG

The life conditions within Upper Jurassic siliceous sponge bioherms in southern Germany and in particular the question of whether the fossil preservation of massive bioherm buildups really depends on the sponges themselves, have been a matter of controversial discussions for many years (for the occurrence of European Oxfordian sponge buildups approximately parallel to paleolatitude see Fig. 1; Fig. 2 shows the development of bioherms in southern Germany through time).

The central part of a typical Oxfordian bioherm of the Franconian Alb consists of boundstones with sponges and laminated peloidal crusts growing on their upper surfaces. The whole bioherm is completely micritic and no traces of winnowing can be seen. The siliceous sponges are generally preserved as complete sponge bodies (Fig. 4.3,4). As can be seen in thin sections, nearly all the sponge skeletons and skeletal debris are filled with micritic peloids (20–100 $\mu$m diameter) which are cemented by a very fine grained sparite (Fig. 4.2). This interspicular peloidal sediment was cemented immediately after the death and decay of the sponges: this is verified by sponge borings (also in this case caused by *Aka* De Laubenfels, see Keupp and Reitner, this Vol.) which transect the peloidal sediment along with the spicular skeleton of the host sponge; thus the spicular skeleton must have been partly calcified before being bored. Other evidence is that the small debris of sponge skeletons filled with peloidal sediment was frequently transported to the bioherm margins; third evidence is that even sponges with a skeleton consisting of unlinked spicules (e.g., *Discispongia unica* KOLB) are preserved as complete fossils because their skeleton is held together by a micritic to pelsparitic matrix.

The peloidal sediment stabilized the skeletal framework as well as the external shape of the sponges, therefore it can be called a true peloidal cement (the cemented sponges may be called "sponge mummies" because of the perfect preservation of their shape). The siliceous spicules were dissolved and the resulting mold was cemented with calcite soon after peloidal cementation, otherwise no spicular structure would have been preserved (Brachert et al. 1986). However, some sponges have been preserved with a still siliceous skeleton, which allows etching and precise identification of the fossils; this preservation type is very frequent at some localities (for comparison – the early silica dissolution of sponges in recent coral reefs, but rather good preservation in muddy back reef sediments have been described by Land 1976; see T. Brachert, this Vol.).

J. Reitner and H. Keupp (Eds.)
Fossil and Recent Sponges

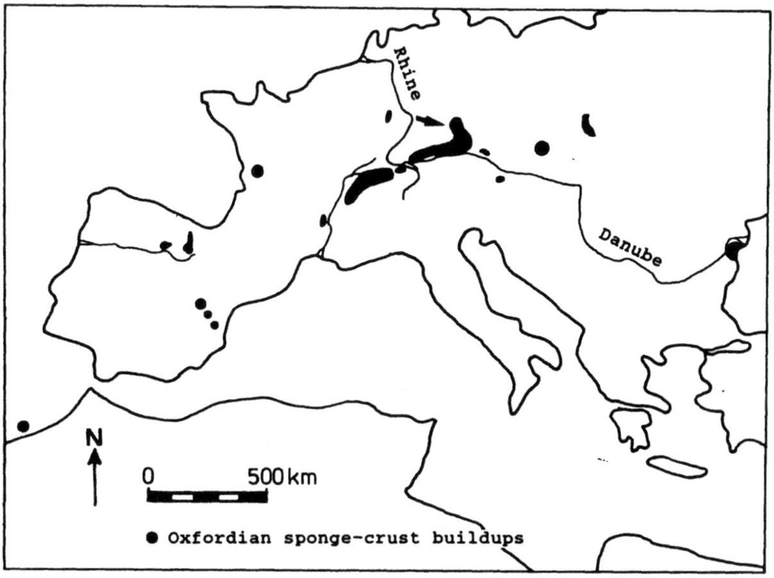


**Fig. 1.** The occurrences of Oxfordian sponge-crust bioherms transect Europe from east to west and follow approximately the 30° paleolatitude within the Tethys. The sponge facies extends into the Atlantic basin off the shore of Morocco (westernmost occurrence). *Arrow* study area in the northern Franconian Alb (southern Germany)

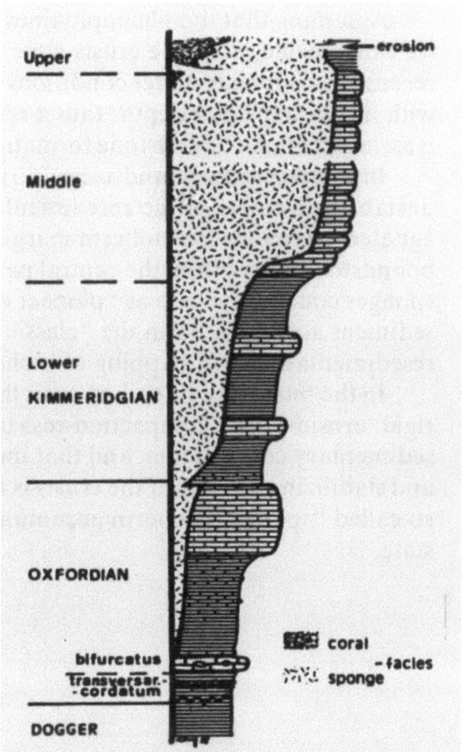


**Fig. 2.** Development of sponge bioherm facies versus stratified facies in the northern Franconian Alb through time. The Lower Oxfordian, which is condensed, and the lowermost Kimmeridgian consist of marly limestones with glauconite, ammonites, and sponge debris. Not to scale

A bacterially induced origin of the peloidal cement is probable since Chafetz (1986) found calcified bacterial sheaths in Holocene to Pleistocene peloids and emphasized that extracellular bacterial HMC-precipitation was common for microenvironments rich in organic matter (see Krumbein 1979); in the fossil example, the decaying sponge soft tissues provided enough organic matter. However, almost every sponge bioherm seems to provide a slightly different kind of fossil diagenesis triggered by an individual depositional and diagenetic history (Lang and Steiger 1985, see T. Steiger, this Vol.).

In the center of a bioherm, the laminated peloidal crusts growing on the sponge surfaces constitute up to 50% of the rock volume and bind the sponge mummies one to another. The crusts were also cemented synsedimentarily, because they are settled by incrusting foraminifera, bryozoans, and polychaete worms and are bored by sponges exactly in the same way as the calcified sponge mummies. Recent laminated peloidal crusts with an identical texture were described by Marshall (1983) from micritic low energy sediments of the Great Barrier Reef; he assumed a microbiologic origin for such crusts. Recently, Prof. W.E. Krumbein (University of Oldenburg) made photoacoustic, spectroscopic studies of the fossil crusts of my samples (with this method one can see even small amounts of photosynthetic metabolic products). He found bacterial phaeophytines, which are specific metabolic products of anoxygenic phototrophic sulfur bacteria, which also live in the anoxic subsurface layers of recent shallow water microbial mats (Gerdes and Krumbein 1987).

Assuming that the phaeophytines could be preserved in the rock until now, we can conclude that the crusts were generated by sulfur bacteria; they live in recent environments under conditions of infrared light which is rapidly absorbed with increasing water depth: thus a relatively shallow, quiet water environment is assumed for the boundstone formation in the fossil example (Lang 1987, 1988).

In contrast to the boundstones, no peloidal crusts could grow on the obviously unstable substrates of the resedimentation environment (possibly sporadically agitated water) on the bioherm margins and in the bioclastic marl layers of the boundstone area within the central part of the bioherm (Fig. 4.5). Only siliceous sponges could grow here as "pioneer organisms" (Fig. 3). So, only in this case is sediment accumulated in the "classic manner" by sponge colonization and by resedimentation and trapping of bioherm debris (see Lang 1989).

In the bioherm center, however, the crusts and sponge mummies construct a rigid, erosion- and compaction-resistant bindstone texture: however, the synsedimentary cementation, and that means the binding, precipitation of cement, and stabilizing activity of the crusts is the reason why the sediment masses of the so-called "sponge" bioherm accumulate (Fig. 3) and are preseved in the fossil state.

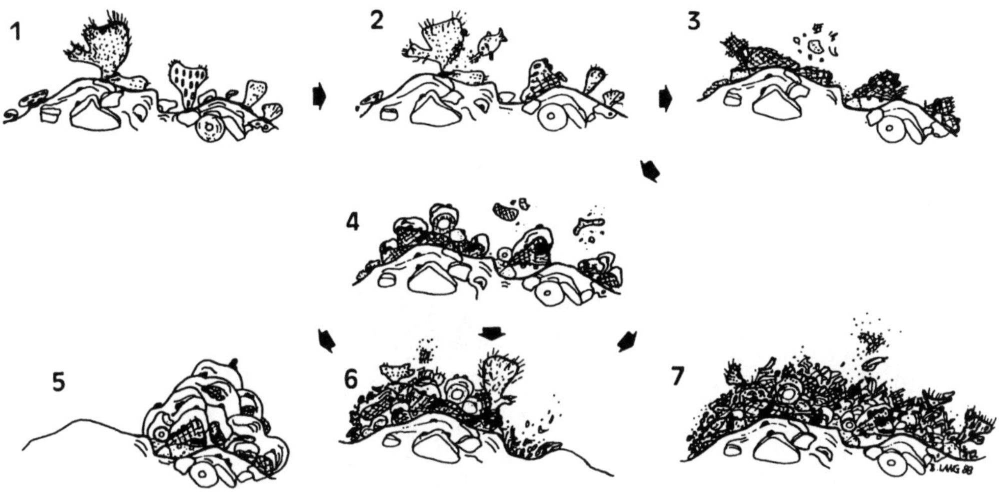


**Fig. 3.** Model for the biostratinomy of the dwellers of the boundstone facies of the central part of the sponge bioherm showing stages from living sponges and crusts to their syndiagenetic cementation and fossil preservation. **1** Sponges grow on a substrate of prelithified crusts and sponges and even in the marginal resedimentation environment. **2** Maceration: after the death and decay of their basal fixing tissue, the sponges fall to the ground and their siliceous skeleton is uncovered by maceration of soft tissues by bacterial decay. **3** Mummification: rapid precipitation of a bacterially induced interspicular peloidal cement results in rigid "sponge mummies" with a high fossilization potential. (Siliceous spicules may already be dissolved when still exposed to sea water). $H_2S$ is produced which is needed by the bacteria in the next stage: **4** Crustal growth: laminated peloidal bacterial crusts grow on the surfaces of the sponge mummies and their fragments; crusts and sponges are bored by sponges and settled by epizoans. **5** Bindstone formation: the sponges are connected by laterally extended crusts to a rigid bindstone texture in an environment with low sedimentation rates within the photic zone. If sufficient time is available, this bacterial activity finally leads to the formation of the massive bioherm structure. **6** Sponge settlement: increasing input of parautochthonous bioclastic debris inhibits crustal growth but even on an unstable substrate the living conditions for siliceous sponges are good. **7** Resedimentation: repeated reworking of bioherm debris on the bioherm margins allow no crustal growth and provide only poor life conditions for sponges. The packstone texture resulting from resedimentation is easily compacted as compared to the boundstones of the central bioherm

**Fig. 4.** Facies types and fossil preservation of the Oxfordian sponge bioherms of northern Franconia (southern Germany). **1** Sponge-crust facies from the bioherm center, a siliceous sponge skeleton (*lower right*) is overgrown by a peloidal crust, which is settled by the agglutinating polychaete *Terebella lapilloides* (*arrow*). Us126. × 20. **2** The skeleton of a lychniskid sponge in siliceous preservation with axial canals. To the *left*, the spicules are replaced by a late fine-grained pelsparite (peloidal cement, *arrow*). Us121. × 50. **3** A hexactinellid sponge mummy is particularly well-preserved and has retained its fragile stauractinal external skeleton (*light horizontal lines, arrow*). Us359. × 15. **4** Well-preserved siliceous sponges (*A*) indicate good life conditions for sponges even in a resedimentation environment on the bioherm margin (floatstone facies), where sponge debris (*B*) were sedimented and only small bacterial crusts (*C*) could develop. Peel, negative print. Us20. × 2. **5** The resedimentation of parautochthonous bioherm debris prevailed at the bioherm margins (packstone facies, e.g. sponge debris (*A*), oncoids (*B*), and crust debris. Only tiny crusts could develop on small shell debris (*C*), if they developed at all. Peel, negative print. Us77. × 6

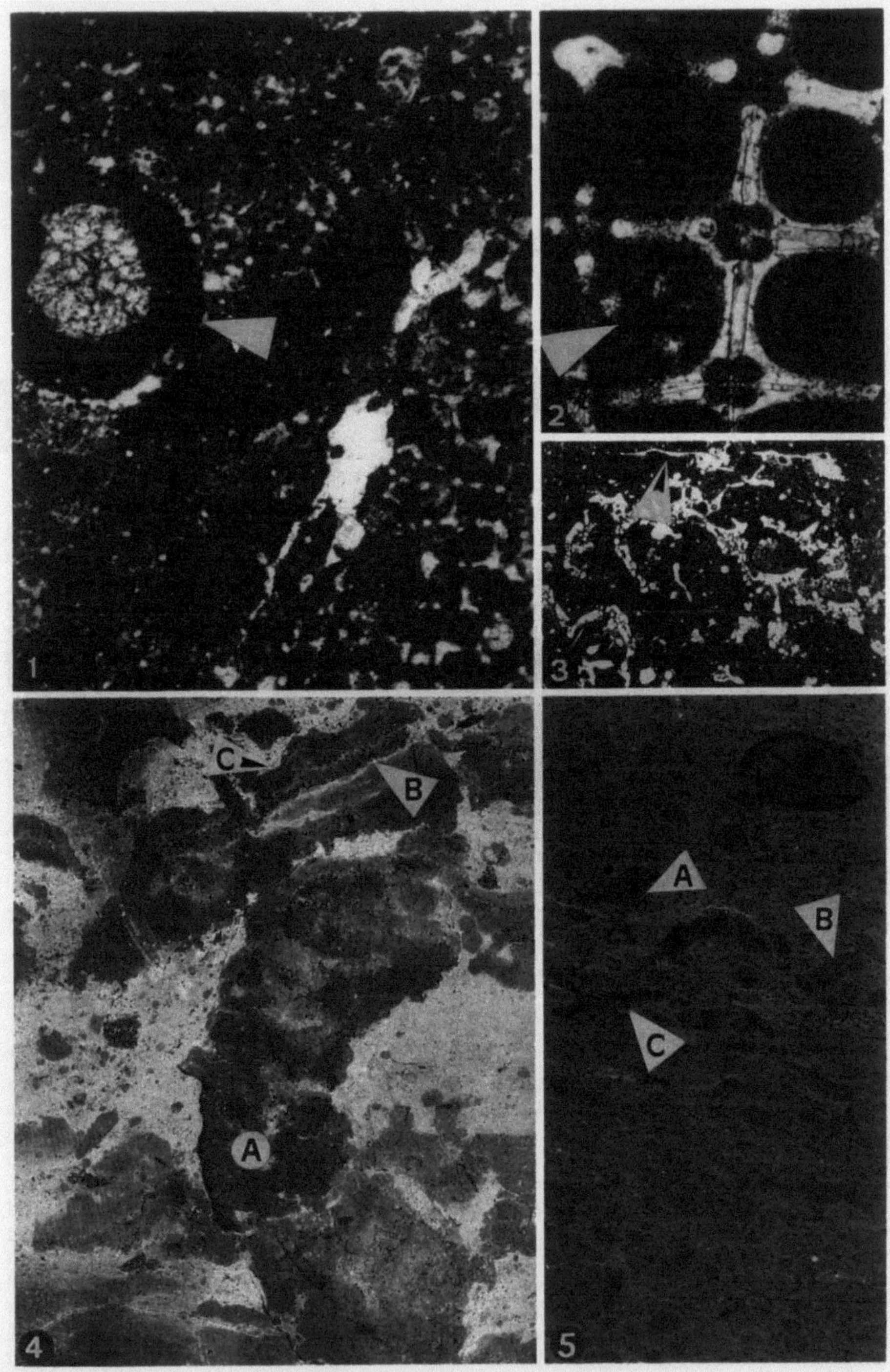
1
2
3
4
C
B
A
5
A
B
C

## References

Brachert TC, Dullo WC, Lang B, Stoffers P (1986) Kieselschwämme vom Ventotene- und Vavilov-Seamound. In: Laschek D (ed) Forschungsfahrt Sonne 41, Hymas 1. Karlsruhe, pp 95–105

Chafetz HS (1986) Manne peloids: a product of bacterial induced precipitation of calcite. J Sediment Petrol 56:812–817

Gerdes G, Krumbein WE (1987) Biolaminated deposits. Lect Notes Earth Sci 9:193

Krumbein WE (1979) Photolithotrophic and chemoorganotrophic activity of bacteria and algae as related to beachrock formation and degradation (Gulf of Aqaba, Sinai). Geomicrobiol 1:139–203

Land LS (1976) Early dissolution of sponge spicules from reef sediments, North Jamaica. J Sediment Petrol 46:457–462

Lang B (1987) Die Schwamm-Biohermfazies der Nördlichen Frankenalb (Steinbruch Urspring, Oxford, Malm): Mikrofazies, Palökologie, Paläontologie. Diss Univ Erlangen, pp 198

Lang B (1988) Bakterienkrusten und Sturmsedimente: Ablagerungszyklen oberjurassischer Schwammsedimente (Oxford, Nördliche Frankenalb). Bochumer Geol Geotechn Arb 29:107–108

Lang B (1989) Upper Jurassic sponge bioherm facies of the northern Franconian Alb (Urspring; Oxfordian): Microfacies, Palecology, Paleontology. Facies 20:199–274

Lang B, Steiger T (1985) Paleontology and diagenesis of Upper Jurassic siliceous sponges from the Mazagan Escarpment. Oceanol Acta, 1984, Submersible Cyana Studies of the Mazagan Escarpment (Moroccan continental.margin) CYAMAZ cruise, pp 93–100

Marshall JF (1983) Submarine lithification in a high-energy platform reef: One Tree Reef, Southern Great Barrier Reef. J Sediment Petrol 53:1133–1149

# Fauna Associated with the Sponges *Aplysina archeri* and *Aplysina lacunosa* in a Coral Reef of the Archipiélago de Los Roques, National Park, Venezuela

E. VILLAMIZAR and R.A. LAUGHLIN

## Introduction

The tropical marine communities are considered among the most diverse on earth. The variety of colors, shapes, and sizes shown by the fauna of these communities are immense. The coral and sponge communities are among the most conspicuous. In spite of this, little is known about the ecology of marine sponges, as research has been hindered by taxonomic problems and difficulties that arise in their quantitative evaluation.

Because of their sessile condition and porous construction, sponges are used by other animals as shelter (Randall and Hartman 1968; Tyler and Böhlke 1972; Rützler 1978). Other sessile animals such as cnidarians, briozoans, and cirripeds can grow on the surface of sponges. The water channels are inhabited by a great variety of animals, especially annelids and crustaceans.

The sponges may provide support, protection against desiccation, wave action, or predation, food supply by water currents around the sponge, other sponge residents and surface organic detritus (Frith 1976). The existence of a water current passing through the sponge body satisfies the nutritional and respiratory needs of the commensals and also conveys chemical or hydrodynamic information which guide the larval stages to the host (Bergquist 1978).

Research has concentrated on determining the species composition of the associated fauna of various sponge species (Pearse 1950; Fishelson 1966; Long 1968; Randall and Hartman 1968; Tyler and Böhlke 1972; Jimenez 1974; Frith 1976; Rützler 1976) and on evaluating the possible factors that determine their community structure (Schoener 1974; Uebelacker 1977; Peattie and Hoare 1981; Westinga and Hoetjes 1981). Other studies have tried to focus on the types of relationships (symbiosis, mutualism, commensalism, parasitism, etc.) that exist between the sponges and their dwellers (Daüer 1973; Bloom 1975; Uebelacker 1978; Forester 1979; Frith 1979).

Knowledge on the animal communities associated to marine sponges of Venezuela is totally lacking (Villamizar 1985). The purpose of this paper is to study the animal communities associated with two dominant species of sponges characterized by different degrees of structural complexity, in a fringing coral reef at the Archipiélago de Los Roques, Venezuela and to evaluate the possible relationships between the community parameters of the fauna and the structural complexity of the sponges.

J. Reitner and H. Keupp (Eds.)
Fossil and Recent Sponges

## Study Area

The sponges were collected from a well-developed fringing coral reef adjacent to Dos Mosquises Sur Cay at the southwest corner of the Archipiélago de Los Roques (Fig. 1). The Archipelago is a reef complex situated 150 km north of the central coast of Venezuela between 11°44′45″ and 11°58′36″ north and 66°32′42″ and 66°57′27″ west. The reef extends from the coastline to a depth of 30 m, followed by a smooth sandy slope. In this reef the sponges represent a very conspicuous and abundant group.

## Materials and Methods

The species *A. archeri* and *A. lacunosa* were selected for study for being the most frequent tubular sponges along the depth gradient. Both species were most abundant at depths between 14–30 m with maximum values found at 21 m and 18 m for *A. archeri* and *A. lacunosa*, respectively (Alvarez and Diaz 1985). All sponge specimens were collected at depths of 3 to 45 m in a random fashion although an effort was made to collect samples of different volumes. Four strata were determined according to depth and reef topography in order to facilitate data analysis; the strata were: stratum 1, from 3 to 8 m, stratum 2, from 9 to 20 m, stratum 3, from 21 to 30 m and stratum 4, at depths below 30 m.

Sampling was conducted from September 1982 to July 1983. Each sponge collected was covered with a bag of a fine mesh size, cut at its base with a knife and then taken to the surface. For each sponge the number of branches and the depth at which it was collected were registered in situ on a writing plexiglass board.

For each sponge collected the following variables were measured in the laboratory (Fig. 2a,b and c): Sponge length (SL), maximum perimeter ($P_{max}$), minimum perimeter ($P_{min}$), upper diameter of the sponge (UD), osculum diameter (OD), length of the atrium (AL), maximum thickness of the wall (MaxT), minimum thickness of the wall (MinT), wet weight (weight of the sponge completely saturated with water), dry weight (weight of the sponge after it was dried for 8 h at 50°C), and volume. The volume was determined for being a parameter easily measured and it represents the amount of useful space available for the associated fauna. Sponge volume was measured in a volumometer especially constructed to divert, into a graduate cilinder, the volume of sea water displaced by the sponge. Three types of water volume displacements were measured: (1) the volume of the sponge completely saturated ($V_s$), (2) the volume of the sponge tightly wrapped with a thin aluminum foil to prevent the water from entering the atrial cavity ($V_{sc}$), and (3) the volume displaced after the sponge was oven dried ($V_d$). To calculate the total water volume that the sponge can contain, two types of estimations were made: (1) Volume of atrium ($V_a$), and (2) volume of the channel system and meanders of the external wall ($V_{cm}$). Both determinations were made by calculating the differences of water volumes as follows:

$$V_a = V_{sc} - V_s \quad (1)$$
$$V_{cm} = V_s - V_d. \quad (2)$$

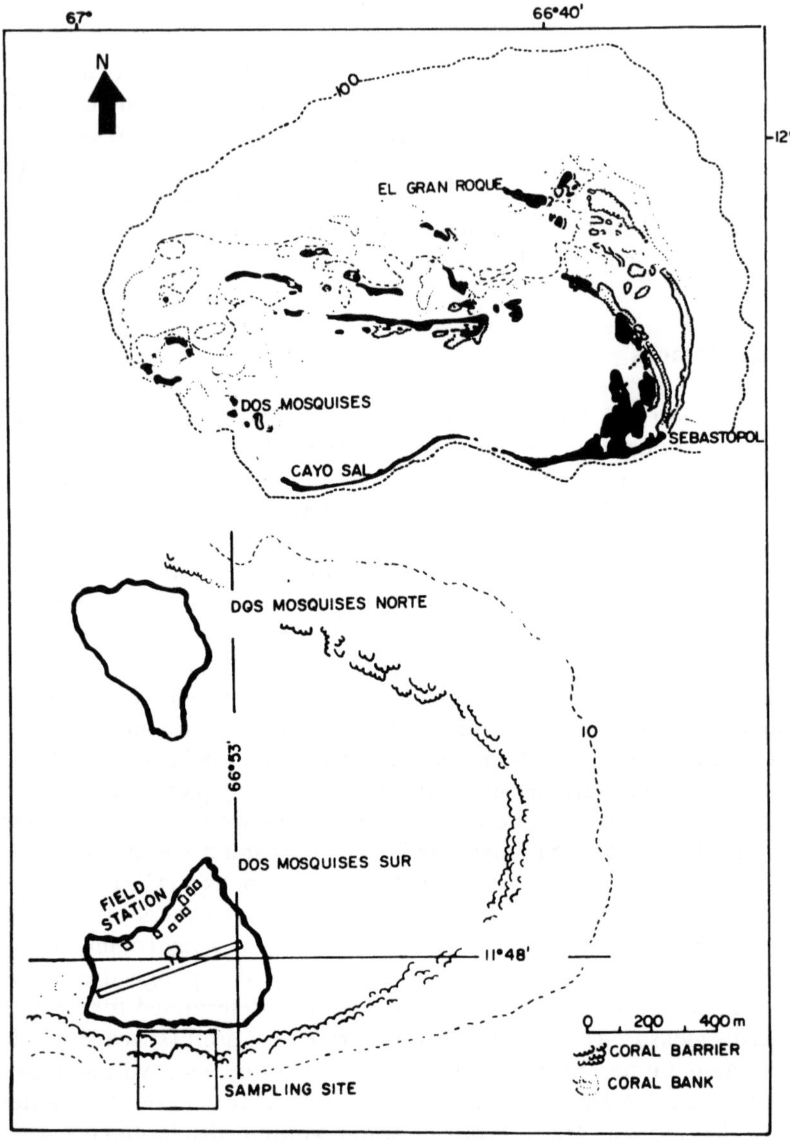


**Fig. 1.** The Archipiélago de Los Roques, National Park (*top*), Dos Mosquises Sur Cay, Study area (*bottom*) with indication of the sampling site

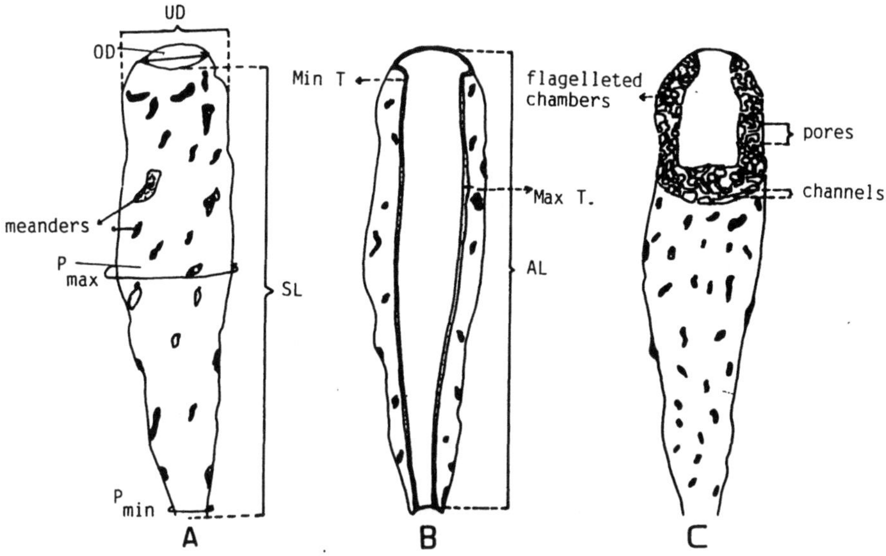


**Fig. 2.** External surface (**A**), longitudinal section (**B**), and internal structure of a tubular sponge (**C**), with description of the morphological measurement taken

Finally, by adding the volume of the atrium and the volume of the channels and meanders of the external surface, the total volume $V_t$ that the sponge contains was obtained; so

$$V_t = V_a + V_{cm}. \quad (3)$$

The removal of the associated fauna from each sponge was carefully made with tweezers and scissors. The bags used to collect the sponges were washed thoroughly and the water filtered through a sieve of 500 $\mu$m. The animals were placed in 5% formalin, with the exception of the fishes which were preserved in a 10% formalin sea water solution. The animals were separated carefully by groups and counted on a cross-sectioned glass dish under a stereoscopic microscope. Finally, the dry weight of each associated species for each sponge sample was determined.

## Data Analysis

An attempt was made to determine the rank of the associated animal species. A single criterium of ranking was established based on the Importance Value Index (IVI) for each species, which considers three percentages simultaneously. It is expressed as:

$$IVI = \%FO + \%RA + \%DW, \quad (4)$$

where:

%FO is the percentage value of the relative frequency of each species,
%RA is the percentage value of the relative abundance of each species,
%DW is the percentage value of dry weight of each species.

To compare the taxonomic composition of the animal communities of *A. archeri* and *A. lacunosa* a similarity analysis was conducted considering: (a) all the species found and (b) three categories or species groups or as referred to by Sneath and Sokal (1973), Operational Taxonomic Units. The latter analysis was performed to facilitate statistical analysis due to the large number of species found and their sharp differences in frequency of occurrence. The three categories are:

- "Very rare" species, < 5% of frequency of occurrence,
- "Rare" species, 5–20% of frequency of occurrence, and
- "Common" species, > 20% of frequency of occurrence.

Frequency of occurrence of a species is defined here as the percentage of sponges with a particular animal species.

For the similarity analysis mentioned above, two different indexes were calculated:

1. The Sorensen Similarity Index (Sorensen 1948):
   This index estimates the similarity between both sponge communities based only in the types of species presents. It is expressed as:

$$S_{i,j} = \frac{a}{a+b+c}, \tag{5}$$

   where
   a is the number of common species present in both sponges,
   b is the number of species present in *A. archeri*, and
   c is the number of species present in *A. lacunosa*.

2. The Pearson Correlation Coefficient (Pearson 1900):
   It estimates the degree of proportionality between the abundance of the animal species of both communities, and is expressed as:

$$r_{jk} = \frac{\sum_{i=1}^{n} (X_{i,j}-X_j)(X_{i,k}-X_k)}{\sum_{i=1}^{n} (X_{i,j}-X_j)^2 \sum_{i=1}^{n} (X_{i,k}-X_k)^2}, \tag{6}$$

   where
   $X_{i,j}$ is the abundance of the i animal species in *A. archeri*,
   $X_{i,k}$ is the abundance of the i animal species in *A. lacunosa*,
   $X_j$ is the average value of abundance of all animal species found in *A. archeri*,
   $X_k$ is the average value of abundance of all animal species found in *A. lacunosa*, and
   n is the overall number of species found.

The relationship between the length of the sponge and volume was established by a simple linear regression analysis.

To determine which type of sponge measurement and/or environmental variable (temporal, spatial) showed the closest relationship to number of species, number of individuals and dry weight of the associated fauna, simple correlation and regression analyses and stepwise multiple regression analysis were performed. The independent variables were: sponge length, total volume, volume of atrium, volume of channels and meanders, depth and month. The analyses were made with three types of mathematical models: (1) Untransformed model (S/V) (Connor and McCoy 1979), (2) Exponential or Semilogarithmic Model (S/logV) (Gleason 1925), and (3) Power Function Model (log S/log A) (Preston 1962; Kilburn 1966) which is often approximated by a double logarithmic transformation (Connor and McCoy 1979).

The overall diversity (Shannon and Weaver 1963) and equitability indexes (Pielou 1966) of the animal community associated with *A. archeri* and *A. lacunosa* were estimated. Values of diversity and equitability for each sample were also estimated.

For all statistical tests a significance level of 95% was considered ($p < 0.05$).

## Results

### *Sample Size*

Totals of 98 and 100 specimens of *A. archeri* and *A. lacunosa* were collected, respectively. The sample size was determined by the relationship of the accumulative animal species number vs. the accumulative total volume of the sponges collected (Fig. 3). The samples of *A. archeri* taken were considered sufficient as the curve stabilized at 47.8 l of total volume. The situation for *A. lacunosa* was different; the curve never stabilized, although the slope changed at some levels specially at the point of 13 l and 86 animal species. For a certain value of sponge volume the accumulative species number of *A. lacunosa*, was considerably higher than that obtained for *A. archeri*.

### *Morphometric Characteristics of the Sponges*

The external morphological characteristics of *A. archeri* and *A. lacunosa* are very similar. The most significant difference between the two lies in their texture. The external surface of *A. archeri* is conulose and very uniform without folds or prominences of any type, while that of *A. lacunosa* present a great number of deep meanders or lagoons, of different shapes and sizes that make it very irregular.

The morphometric analysis conducted on both species shows that *A. archeri* has a larger and more variable total length (8.20–175.00 cm) than *A. lacunosa* (7.50–65.00 cm). Hence, the average values of volume of atrium, volume of

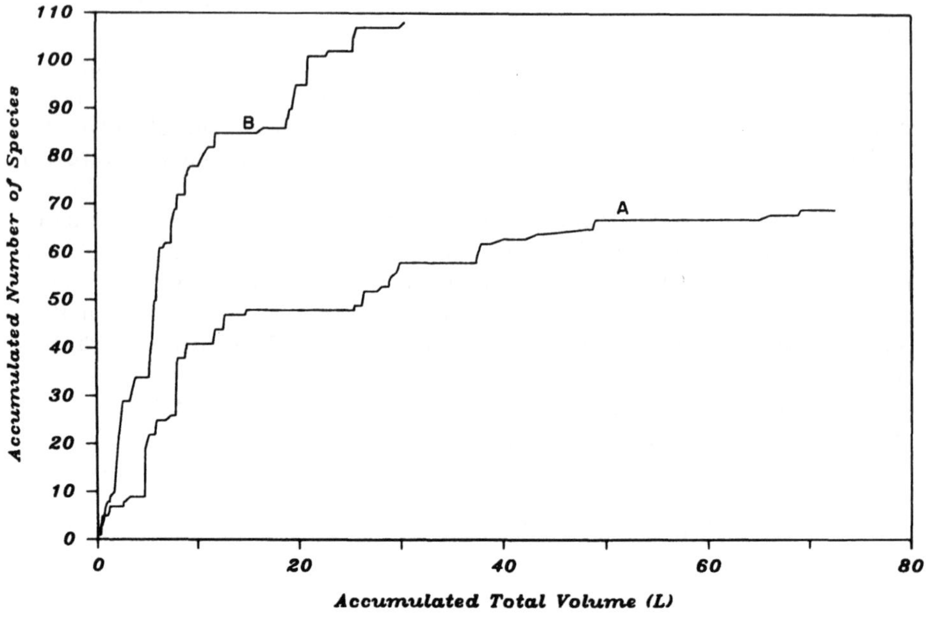


**Fig. 3.** Accumulated number of species vs. accumulated total volume of the spoonges *A. archeri* (*A*) and *A. lacunosa* (*B*) collected in a fringing coral reef of the Dos Mosquises Sur Cay, Los Roques, Venezuela

channels and meanders and total volume are also significantively higher (Table 1).

The linear regression analysis between length and total volume for each species, however, shows a higher slope (29.99 ml/cm) for *A. lacunosa* (Table 2), indicating a larger increase in volume per unit of length in this species. Similar results were obtained when length was related to the other two types of volume, obtaining, in all cases, correlation, and regression coefficients statistically significant ($p < 0.05$).

Generally, the individuals or branches of these sponges tend to group in colonies. The individual number per colony can affect the composition and abundance of the associated fauna. The colonies of *A. lacunosa* sampled had, in average, a higher number of individuals ($\bar{x}$ = 7.57 individuals/colony) than the colonies of *A. archeri* ($\bar{x}$ = 5.56 individuals/colony).

## *Composition of the Animal Community Associated with the Sponges*

Totals of 53 species and 1001 individuals were found in *A. archeri* (N = 98), and 139 species and 5694 individuals in *A. lacunosa* (N = 100). The species found were mainly crustaceans, ophiuroideans, and molluscs although some species of fishes were also encountered.

**Table 1.** Morphometric characteristics of the sponges *A. archeri* (N = 98) and *A. lacunosa* (N = 100) collected at a reef in Dos Mosquises Sur Cay (Los Roques, Venezuela) from 3 to 45 m. (see Fig. 2 for graphic descriptions of measurements).

| | Sponge species | |
|---|---|---|
| | *A. archeri* | *A. lacunosa* |
| Length (cm) | 45.71 ± 27.19 (8.20–175)[a] | 24.39 ± 11.42 (7.50–65) |
| Upper diameter (cm) | 4.32 ± 1.97 (1.20–13) | 4.48 ± 1.12 (2.00–6.84) |
| Osculum diameter (cm) | 2.58 ± 1.41 (0.62–7.73) | 2.19 ± 0.76 (0.53–4.30) |
| Maximum perimeter (cm) | 17.07 ± 5.72 (7.30–40) | 21.97 ± 5.72 (10.10–36.00) |
| Minimum perimeter (cm) | 11.09 ± 2.38 (4.70–19) | 13.51 ± 3.93 (4.50–27) |
| Maximum thickness of the choanosome wall (cm) | 1.62 ± 0.50 (0.76–3.30) | 2.77 ± 0.93 (1.20–6.10) |
| Minimum thickness of the choanosome wall (cm) | 0.68 ± 0.23 (0.26–1.99) | 0.89 ± 0.41 (0.33–2.00) |
| Volume of atrium (ml) | 388.21 ± 452.07 (5.0–3120) | 315.55 ± 324.50 (25–1965) |
| Volume of channels and meanders (ml) | 234.38 ± 302.52 (4.0–2159.5) | 202.98 ± 233.37) (3.0–1806) |
| Total volume (ml) | 614.11 ± 693.7 (29.5–4452) | 518.53 ± 471.02 (37.0–2551) |

[a] Av. ± SD (Min-Max).

**Table 2.** Correlation (r) and simple linear regression coefficients (b) between the different types of volume, and the length of *A. archeri* (N = 98) and *A. lacunosa* (N = 100).

| | Sponge species | |
|---|---|---|
| | *A. archeri* | *A. lacunosa* |
| Volume of atrium vs. | r = 0.869 | r = 0.724 |
| length | b = 14.45 | b = 20.56 |
| Volume of channels and meanders vs. | r = 0.87 | r = 0.64 |
| length | b = 8.67 | b = 9.42 |
| Total volume vs. | r = 0.89 | r = 0.78 |
| length | b = 22.86 | b = 29.99 |

The 147 different species are distributed in 8 Phyla, 17 Classes, 30 Orders, and 63 Families. The most common families found were: for the fishes, Gobiidae, Clinidae, and Apogonidae, for the crustaceans, Palaemonidae, Alpheidae and Xanthidae; for the molluscs, Cerithiidae, Magilidae, Semelidae, Carditidae, Potamididae, Buccinidae, Columbellidae, Rissoidae, Triphoridae, Caecidae, and for the ophiuroids, Ophiotrichidae and Amphiuridae.

A total of 48 species was common to *A. archeri* and *A. lacunosa*, 13 were found exclusively in *A. archeri* and 86 in *A. lacunosa*. Of the common species, the fishes was a very constant group in both sponges. The crustaceans common to both sponges were the species of greatest abundance and total dry weight, particularly the Palaemonidae and Alpheidae. Only 21% of the molluscs (91 species) were common to both sponges and belonged mainly to the Gasteropoda. All the species of Ophiotrichidae were common to both sponges but the diversity and relative biomass of their populations varied considerably. Of the remaining groups, only some species of Foraminifera, one species of Platyhelminthes, and one species of anemone were present in both sponge species.

The shrimp of the species *Periclimenaeus bredini*, the crab Melybia thalamita the molluscs *Alvania portoricana*, *Daphnella morra*, *Cymatium americanum*, *Triphora* c.f. *decorata*, *Barleeia* c.f. *acuta*, *Olivella nivea*, *Malleus candeanus*, *Musculus lateralis*, *Chlamys ornata*, the ophiuroids *Ophiotrix brachyactis*, and the mollusc of the genus *Cardita* are new records for Los Roques (Zoppi 1967; Chace 1972; Tello 1975; Rodriguez 1980; Grajal and Laughlin 1984).The mollusc *Caecum nitidum* is a new record for Venezuela. The fish *Kaupichthys nuchalis* represents a new family, genus and species for Venezuela (F Cervigón, pers. commun. 1985).

The relative percentages of number of species, number of individuals and dry weight, of the most important taxa found in *A. archeri* and *A. lăcunosa* are shown in Fig. 4. The molluscs were the group with the largest number of species in both sponges (76.51%), while the crustaceans were the group with the highest percentage values of number of individuals (67.27%). For the dry weight parameter, the fishes (50.45%) and ophiuroids (42.68%) obtained the highest values in *A. archeri* and *A. lacunosa*, respectively, while the crustaceans were ranked second for this parameter in both sponges.

Rarely were the associated species found during all sampling periods. Most of them occurred in only 1-monthly sample and a large number was sampled on alternate months. Only five species occurred in all months sampled in both sponges.

The analysis of the species types that occurred along the depth gradient in both sponges indicated a characteristic zonation for the different taxa (fishes, Ophiuroidea, Mollusca, etc.). The great majority of the species found along the entire depth gradient belonged to the fishes, Crustacea and Ophiuroidea, whereas those found in sponges at the deeper stratum belonged mainly to the Mollusca, indicating a change in the species composition along the depth gradient.

*Species Rank*. Only the species with IVI higher than 10% were listed (Tables 3 and 4). Twelve species in *A. archeri* and 28 in *A. lacunosa* had IVI values higher than 10%. The fish *Gobiosoma horsti*, and the shrimps *Periclimenes* sp., and *Periclimenaeus bermudensis* occupied the number 1, 2, and 3 ranks in *A. archeri* (Table 3), while the crab *Pilumnus* sp., the shrimp *Synalpheus townsendi* and the Amphiuridae 1, occupied the 1, 2 and 3 ranks in *A. lacunosa* (Table 4).

### *Richness, Abundance, and Dry Weight of the Species*

The estimated values (absolute and relative) of the parameters: species number, number of individuals, and dry weight of the animal communities are shown in Table 5. Six important results merit attention:

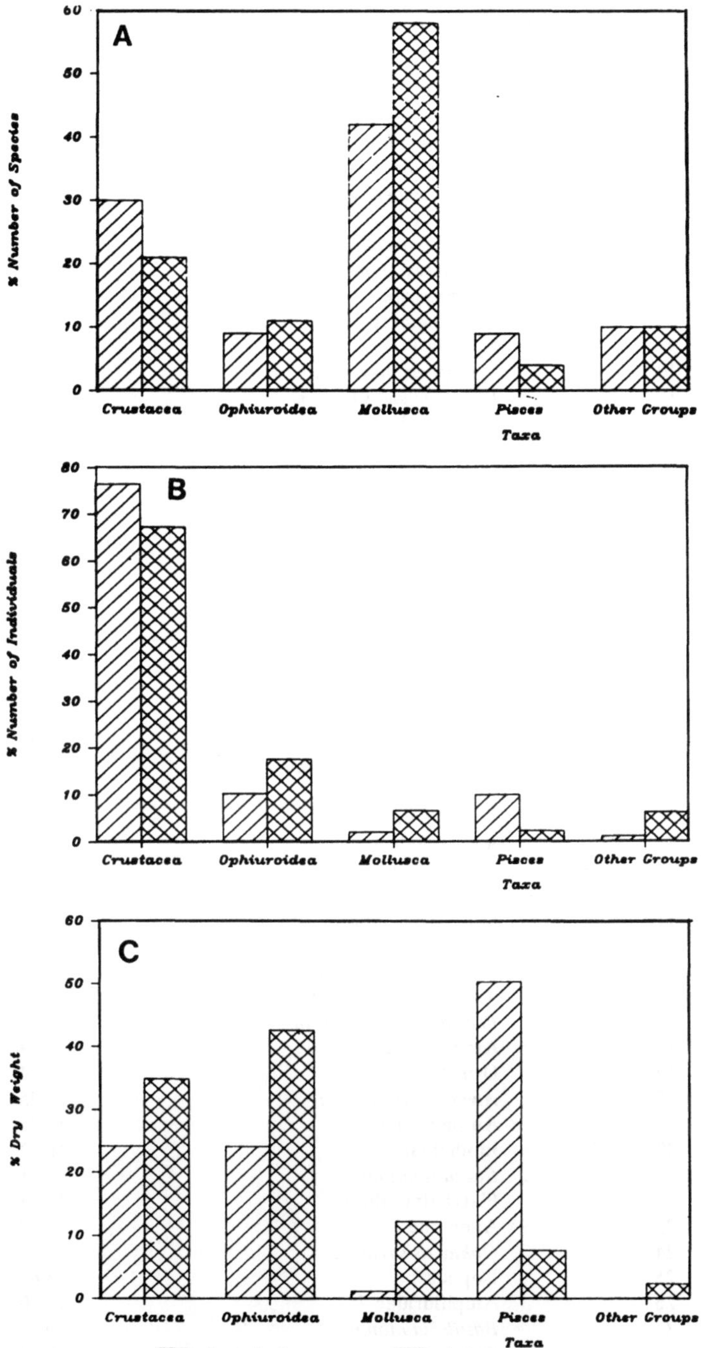


**Fig. 4.** Percentage of number of species (**A**), number of individuals (**B**), and dry weight (**C**) of the animal groups associated to the sponges *A. archeri* and *A. lacunosa*

**Table 3.** Rank of species found in *A. archeri* according to the index of importance value (IVI)

| Rank | Species | IVI (%) |
|---|---|---|
| 1 | *Gobiosoma horsti* | 125.56 |
| 2 | *Periclimenes* sp. | 93.76 |
| 3 | *Periclimenaeus bermudensis* | 68.26 |
| 4 | *Ophiotrix suensoni* | 55.04 |
| 5 | *Periclimenaeus perlatus* | 36.18 |
| 6 | *Phaeoptyx pigmentaria* | 32.10 |
| 7 | *Pilumnus* sp. | 26.54 |
| 8 | *Leptochelia* sp. | 15.77 |
| 9 | Isopoda 1 | 13.75 |
| 10 | *Synalpheus townsendi* | 13.73 |
| 11 | *Periclimenaeus bredini* | 13.21 |
| 12 | Amphipoda 1 | 12.34 |

**Table 4.** Rank of species found in *A. lacunosa* according to the index of importance value (IVI)

| Rank | Species | IVI (%) |
|---|---|---|
| 1 | *Pilumnus* sp | 120.64 |
| 2 | *Synalpheus townsendi* | 90.77 |
| 3 | Amphiuridae 1 | 63.69 |
| 4 | *Leptochelia* sp | 58.55 |
| 5 | *Periclimenaeus bermudensis* | 52.84 |
| 6 | Amphiuridae 6 | 48.63 |
| 7 | *Synalpheus minus* | 45.67 |
| 8 | Amphipoda 1 | 42.70 |
| 9 | *Periclimenaeus perlatus* | 41.38 |
| 10 | *Starksia hassi* | 40.06 |
| 11 | *Gobiosoma horsti* | 36.48 |
| 12 | Foraminifera 1 | 35.30 |
| 13 | *Periclimenaeus bredini* | 30.32 |
| 14 | *Alpheus cristulifrons* | 26.64 |
| 15 | *Periclimenes pedersoni* | 20.00 |
| 16 | Amphiuridae 8 | 19.59 |
| 17 | Ostracoda 1 | 18.94 |
| 18 | *Alvania portoricana* | 18.75 |
| 19 | Amphiuridae 2 | 16.96 |
| 20 | Idotheidae 1 | 16.12 |
| 21 | *Caecum nitidum* | 15.71 |
| 22 | Platyhelminthe 1 | 15.63 |
| 23 | Anemone 1 | 14.80 |
| 24 | *Ophiactis savingyi* | 13.98 |
| 25 | Isopoda 1 | 13.63 |
| 26 | Amphiuridae 7 | 12.10 |
| 27 | *Batillaria minima* | 10.37 |
| 28 | *Pisania pusio* | 10.24 |

**Table 5.** Totals of species number, number of individuals, dry weight of associated animals, and average, standard deviation, minimum and maximum values of the parameters of the animal communities found in the sponges *A. archeri* (N = 98) and *A. lacunosa* (N = 100)

| | Sponge species | |
|---|---|---|
| | *A. archeri* | *A. lacunosa* |
| Number of species | 53 | 139 |
| Number of individuals | 1001 | 5694 |
| Total dry weight (g) | 33.9 | 73.2 |
| Number of species/ sample | 3.63 ± 2.34[a] (0–12) | 11.65 ± 7.38 (1–35) |
| Number of species/ ml | 0.0158 ± 0.0224 (0.0020–0.1560) | 0.0335 ± 0.0268 (0.0050–0.1100) |
| Number of individuals/ sample | 11.51 ± 18.10 (1–91) | 56.96 ± 62.21 (2–318) |
| Number of individuals/ ml | 0.0272 ± 0.0335 (0.0010–0.2000) | 0.1296 ± 0.1199 (0.0004–0.6000) |
| Dry weight/sample | 0.3410 ± 0.4560 (0.0001–2.4560) | 0.7320 ± 1.0100 (0.0030–6.3600) |
| Dry weight/ml | 0.0009 ± 0.0019 (0.0010–0.0180) | 0.0016 ± 0.0014 (0.0001–0.0088) |

[a] Av ± SD (Min-Max).

1. The species total occurring in *A. lacunosa* is almost three times greater than in *A. archeri.*
2. The number of individuals found in *A. lacunosa* is five times greater than in *A. archeri.*
3. The total dry weight value of the animal community in *A. lacunosa* is twice that of *A. archeri.*
4. The average values of species number, number of individuals and dry weight per sample and per milliliter are larger in the animal community associated to *A. lacunosa.*
5. The standard deviation of these variables are also greater in *A. lacunosa.*
6. The number of individuals per sample (sponge) or volume in *A. lacunosa* is five times greater than in *A. archeri,* while the dry weight per sample or volume in *A. lacunosa* is only twice that of *A. archeri.*

The diversity values of the animal communities of *A. archeri* and *A. lacunosa* were 1.1446 and 1.4758, respectively, and the equitability values were 0.6607 for *A. archeri* and 0.6791 for *A. lacunosa*. Hence, the diversity was significantly higher in *A. lacunosa,* while equitability values were similar for both sponges.

Positive and significant correlation and regression coefficients ($p < 0.05$) were found between the richness and diversity variables for both sponges, and were particularly high for *A. lacunosa* ($r = 0.879$, $b = 23.88$). The relationship between the equitability and diversity variables is less evident. A positive and significant correlation, however, was found only for *A. archeri* ($r = 0.717$, $p < 0.05$).

***Similarity Analysis***

The species composition of both sponge communities is very similar according to the common species category (> 20 %FO) ($S_{i,j}$ = 0.96971), that is, 97% of the common species are in both sponges. On the contrary, both species do not share the "very rare" species ($S_{i,j}$ = 0.2858). As expected, intermediate values of similarity were obtained when both species are compared according to the "rare" species category ($S_{i,j}$ = 0.6786). When all the species are taken into account (overall similarity) the index of similarity was also of intermediate magnitude ($S_i$ = 0.5024).

On the other hand, considering the entire depth gradient there was no relation in the abundances of the species categories between the two species of sponges ($p < 0.05$). When the depth gradient was divided into strata, however, a strong and negative proportionality was only found between the abundances of the "common" species occurring below a depth of 30 m (stratum 4) ($r = -1$, $p < 0.05$). When all species are considered, a slight negative correlation between the abundances of the species that occurred between 21 and 30 m in depth was found (stratum 3).

***Relationships Between the Sponges Volume and the Parameters of the Animal Communities***

Three mathematical models were used to determine the model that defines best the relationship between the biological parameters of the animal communities and volume, with special emphasis given to the species-volume relationship.

The parameters number of species and number of individuals were strongly related to the various estimates of sponge volumes, as supported by the positive and significant correlations found ($P < 0.05$) (Table 6).

Overall, the first model, the untransformed, produced the best fit of the data, particularly for *A. archeri*. For *A. lacunosa* there was a slight increases in the correlation coefficients with the logarithmic model but the differences were not significant ($p < 0.05$).

Although the absolute values of the correlation coefficients for the number of species according all the models are relatively small (0.335–0.536), they are considered significantly due to the large sample size. For the number of individuals the absolute values of the coefficients were slightly higher (0.428–0.664).

For all cases the regression coefficients were higher in *A. lacunosa* (Table 6), indicating that each unitary increase of total volume produces a larger increase in the number of species and in the number of individuals in *A. lacunosa* than in *A. archeri*. In biological terms this means, that *A. lacunosa* can hold more species and individuals per unit of volume than *A. archeri*.

A relationship between the total dry weight of the individuals and the sponge volume was found, however no clear tendency was observed in the model that produces the best fit of these data.

**Table 6.** Correlation (r) and regression (b) coefficients between number of species, number of individuals and the different estimates of volume for *A. archeri* and *A. lacunosa* for the untransformed model ($p < 0.05$)

| | | Untransformed $S = K + Z.V$ | |
|---|---|---|---|
| | | *A. archeri* | *A. lacunosa* |
| Number of species vs. $V_a$ | r | 0.442 | 0.456 |
| | b | 0.002 | 0.010 |
| Number of species vs. $V_{cm}$ | r | 0.536 | 0.466 |
| | b | 0.004 | 0.021 |
| Number of species vs. $V_t$ | r | 0.510 | 0.516 |
| | b | 0.002 | 0.008 |
| Number of individuals vs. $V_a$ | r | 0.628 | 0.482 |
| | b | 0.021 | 0.083 |
| Number of individuals vs. $V_{cm}$ | .r | 0.558 | 0.664 |
| | b | 0.028 | 0.222 |
| Number of individuals vs. $V_t$ | r | 0.609 | 0.612 |
| | b | 0.014 | 0.079 |

$V_a$ = Volume of atrium.
$V_{cm}$ = Volume of channels and meanders.
$V_t$ = Total volume.

### *Multiple Analysis*

For "species number" ($N_{sp}$) a significant multiple regression was found. For *A. archeri*, the volume of channels and meanders ($V_{cm}$) was the first variable introduced in the analysis, and it accounts for 28% of the data dispersion on species number; the second variable introduced was "month" (M) but its incorporation accounted for only an additional 2% ($N_{sp} = 1.80 + 0.004\ V_{cm} + 0.17M$). For *A. lacunosa* "total volume" ($V_t$) was the first variable incorporated into the analysis and accounted for 26% of the total variation in species number. The second variable introduced was depth (D) improving in 5% the fit of the equation ($N_{sp} = 13.52 + 0.01V_t - 0.09D$).

As far as number of individuals ($N_{in}$) is concerned, in *A. archeri* only "length of the sponge" (SL) was incorporated into the equation, thus resulting in a simple linear equation ($N_{in} = 6.47 + 0.37SL$). The regression coefficient obtained points out that length accounts for 41% of the total variance of number of individuals. For *A. lacunosa*, two variables were introduced, volume of channels and meanders and the total volume, the first one accounts for 43% of the total variability in the number of individuals and the second one accounts for an additional 1.7% of the variance ($N_{in} = 9.02 + 16V_{cm} + 0.03V_t$).

***Discussion***

The results of this study corroborate the importance of the sponges as biotopes.

It is possible to consider the organisms living in the sponges as integral components of an ecological community. In ecological research, five characteristics of communities have been evaluated and studied: (1) The diversity of species, (2) the forms and structures of growth, (3) the dominance of species, (4) the relative abundance of the species, and (5) the trophic structure. In the present study emphasis was made on the diversity and patterns of relative abundance of the species associated to sponges.

As demonstrated in this work, it is impossible to establish a single value of similarity between the animal communities associated to *A. archeri* and *A. lacunosa*. The species had to be grouped into categories according to their patterns of occurrence in the sponges and to the stratum of depth at which they were found. This arrangement produces more logical and interpretable results. In conclusion, when only the presence or absence of the animal species in the sponges are considered, without taking into account their abundances, a very high similarity is found, especially at the level of the "common" species category for both sponge communities. On the contrary, when the degree of proportionality between the abundance of the species is considered no type of relation was found, even between the "common" species. This discordance disappears when the abundance of the species of each category in both communities at each depth stratum are compared. Thus, the "common" species generate some degree of relationship, even though this relationship is inverse and only for one depth stratum. It is possible to interpret this as the existence of a certain degree of selectivity of the animals for a particular sponge species occurring at these depths.

The difference between the animal composition and the faunistic density between two (or more) sponges is a consequence of several factors: (1) Types of sponges. The form of growth, the external morphological characteristics (texture, consistency, cuticle, etc.), the internal structure (channels, internal surface area, etc.), the exudation of biologically active metabolites of the sponge. (2) The type of association between the host and their guests. Symbiosis, mutualism, commensalism, parasitism, or simple inquilinism.

The sponges studied belong to the same genus and have the same growth form (tubulars) and similar consistency. The major difference is the texture. *A. lacunosa* presents a system of channels and meanders that provide the species with a complex kind of habitat for the animal dwellers. It appear that none of the two sponge species exudate chemical substances that have repulsive or attractive action on the possible guests, but as this possibility was not researched it cannot be totally discarded.

For *A. lacunosa* no particular volume was found at which the number of animal species stabilizes. Two reasons can be proposed to explain this result: (1) The existence of a high degree of randomness in the composition of the associated animal community of this sponge, hence for the continued and unexpected occurrence of rare species in the samples, and (2) an insufficient sample size.

Some of the species found in this study were associated to only one of the sponges. Several of these associates, however, seem not to be important in terms of the frequency of occurrence, abundance, and dry weight. The shrimp *Stenopus scutellatus* was the only species exclusively found in *A. archeri* that is considered important according to these criteria (rank = 14). This shrimp is lodged in the wall of the atrial cavity and this association could be interpreted as mutualism, as the species is a cleaner shrimp that cleans the atrium walls of the sponge from organic residuals.

A larger quantity of species considered as important were found associated to *A. lacunosa*. Some Crustacea as *Munida irrasa*, *Paraliomera longimana* and *Mithrax ruber* were found in the internal lagoons and meanders of the wall of the sponges. Some isopods and ostracods were found on the external surface of the sponges. Among the molluscs, species such as *Pisania pusio*, *Batillaria minima*, gasteropod 6, *Musculus lateralis*, *Malleus candeanus*, bivalve 18, and the trumpet (Caecidae) were obtained from different areas of the sponges and, in occasions, were associated to algae. The brittle star *Ophiactis savingyi* and many other species of this group were found inside the meanders and lagoons of this sponge, which in many cases are internally interconnected.

In both the animal communities of *A. archeri* and *A. lacunosa* there was no particular dominance of an animal species. On the other hand, a large number of species were considered as "rare". The maximum Index of Importance Value (IVI) determined was 125.26% in *A. archeri* and 120.64% in *A. lacunosa*, confirming that none of the species of these communities has an absolute dominance. The results of this analysis, however, does not exclude the existence of intermediate values of dominance for some of the species.

The animal communities associated to the sponges *A. archeri* and *A. lacunosa* have a structure similar to other natural communities, that is "a high number of species with few individuals and low dry weight, and very few species with a high number of individuals and dry weight". The species that contribute with high values in number of individuals and dry weight form the group of the codominant species. It is possible that their success depends on their greater competitive abilities that result in an efficient exploitation of the environmental resources.

The animal community associated to *A. lacunosa* is, on the average, more diverse than the animal community associated to *A. archeri*, and also the most variable. The increase in the diversity variable for *A. lacunosa* is mainly due to the increase in the species number of such communities, while in *A. archeri* both variables, richness and equitability, define the increase in the diversity values. There are few available data on the diversity of other animal communities associated to benthic organisms that could permit somewhat valuable comparisons of the results of this study. The sharp differences in diversity between the associated fauna of both sponge communities can be attributed to the particular morphological and structural characteristics of each sponge. The higher increments in volume per unit of length in *A. lacunosa* reflects the higher structural complexity of the sponge due to the large number of meanders and lagoons of the choanosome wall. These lagoons generate a space system of great structural complexity that give the sponge a larger superficial area and volume. The shape and distribution of the various spaces within the sponge, which in turn contain the water, are of essential importance in the structure of these communities as they provide a very propitious habitat for the associated organisms.

On the other hand, many of the animal species associated to *A. lacunosa* (e.g., micromolluscs) are smaller and of lesser dry weight than the animal species of *A. archeri* and this is partially the reason why the dry weight per sample or volume in *A. lacunosa* is only twice that of *A. archeri.*

In this study the sponges were considered as excellent material to study the species-area (species-volume) relationship due to three reasons: (1) for their diversity of sizes, (2) for their isolated condition, and (3) for lodging an enormous variety of organisms. Thus, the sponges can be seen as isolated habitats of "living islands" lodgers of a great variety of organisms.

An interesting result of this study was that for any accumulative total volume, *A. lacunosa* always had a higher accumulative number of species than *A. archeri*, suggesting that the species number and volume relationship do not take place equally for both sponges.

The three types of volume (atrium, channels and meanders, total volume) and the species number were related. Generally, a certain tendency between both variables was found. In this study, the correlation values of both variables varied between 0.395–0.536. A comparative analysis between the results of this and other studies conducted in sponge habitats is presented.

Uebelacker (1977) found highly significant correlations between the size (mass, volume, surface, area, length and circumference) of the sponge *Gelliodes digitalis* ($N = 29$) from the Bahamas, and the number of cryptofaunal species associated with each sponge, obtaining relatively high correlation coefficient values (0.61–0.71). Between the untransformed model and the logarithmic model, the latter produced the best fit in his study for all the relationships established.

Westinga and Hoetjes (1981) studied the relationship between the intra-sponge fauna and the volume of the sponge *Spheciospongia vesparia* ($N = 34$) from Curacao and Bonaire. They describe this relationship as an exponential function ($a = 9.55*b^{0.22}$). The relationship was valid only in the region of 0 to 14 l., and the correlation coefficient obtained was very high ($r = 0.86$). It is evident that the correlation coefficients found in the present study ($0.39 < r < 0.53$) for all the models used were significantly smaller. A series of possible explantions can be suggested. The existence of a great dispersion of the points at small values of volume should be taken into account. It is possible that the relationship between the number of species and the sponge volume is not valid for specimens of small sizes (probably less age) for which a minimum value exists of the volume from which this relationship starts. Then, the inclusion of sponges of all sizes into the analysis should decrease the degree of correlation between both variables. On the other hand, perhaps the relationship is not linear, hence the linear models used in this study were not optimal to describe the relationship.

Two additional factors should be taken into account when comparing results of all these studies. The first relates to the morphological differences among the sponge species, and the second to the total number of samples taken.

Specimens of *Speciospongia vesparia* show a semispherical and compact form, *Gelliodes digitalis* a vasiform shape, and *A. archeri* and *A. lacunosa* a tubular form. The relationship between the species number and volume is affected, to a certain degree, by the growth form of the sponge.

The increase of volume in masive sponges is proportional to the increase of volume of the channels, pores and cavities within the sponge. In tubular sponges the situation varies, the increase of volume depends on the increase of the volume of the atrium. The atrium is a large cavity which is not used by most of the associated fauna; in the present study only two shrimps, two fishes and two ophiuroids occupied the atrium cavity. Consequently, when the volume of the atrium is the major contributor to the values of total volume of the sponge, the relationship between the number of species and volume is not so clear. In this study the increase in the volume of the atrium per unit of length was larger than the increase in volume of channels and meanders.

The sample sizes used in studies of the species-area relationship are generally low (10–35 samples). Perhaps the high values of correlation between number of species and area found in several studies are related to the low sample size. A large sample size can increase the dispersion of the data on "number of species" (noise) reducing the correlation values. For *S. vesparia* (Westinga and Hoetjes 1981) and *G. digitalis* (Uebelacker 1977) a much lower number of samples was used than in the present study.

The real form of the species-volume relationship could not be established with certainty in this study. The untransformed and the power mathematical models produced the best fits for *A. archeri* and *A. lacunosa*, respectively. Hence, it is impossible to propose a single best model for both sponges. Each species could be visualized as a particular "living habitat island" (with a determined degree of structural complexity) where the relationship number of species-volume could take different forms and thus be modeled by particular mathematical functions. The existence of such different forms result, at least in part, from the theoretical basis of how the relationship is viewed and interpreted. Two principal hypotheses have been suggested: The first, proposed by Williams (1964) is termed "Habitat diversity hypothesis", and it proposes that when the amount of area sampled is increased, new habitats with their associated species are incorporated, and thus the number of species increases with area. The second hypothesis is termed the "area per se", developed by Preston (1962) and McArthur and Wilson (1967) and is derived as a prediction of the equilibrium theory of island biogeography. This hypothesis explains the species number as a function of the immigration and extinction rates. An alternative to both hypotheses is that the number of species is controled by passive sampling from the species pool, larger areas receiving larger samples than the smaller ones, which results in more species (Osman 1977).

Concerning the biological basis of the relationship some observations are important. The increase in volume per unit of length is larger in specimens of *A. lacunosa*. Furthermore, an increase in length in *A. lacunosa* generates new microhabitats. According to these results it is logical to assume that the theoretical basis of the species-area relationship in *A. lacunosa* lies in the higher number and diversity of habitats. Other information on extinction and colonization rates and the effect of the sampling methodology, however, should also be evaluated.

In order to substantiate what has been mentioned above, a comparison of the published work on the relationship between number of species and sponge volume is given in Table 7. The comparison, however, must be viewed with caution as there is marked variation in sample sizes, growth forms (massive, incrustant, fistulose, ramose, vasiform, and tubular), and methodology.

In absolute terms, *A. archeri* and *A. lacunosa* shelter the largest number of species, particularly the latter, which holds about three times the number of species than any other sponge. In relative terms ($N_s/l$) and based on total volume sampled, the tubular sponges, *A. archeri* and *A. lacunosa*, suffer a sharp drop in importance, with 0.9 and 2.5 species per liter sampled. When the volume of the channels and meanders is considered, however, the tubular sponges regain their importance particularly *A. lacunosa* with 6.36 species per liter. This last estimation is justified by the fact that in tubular sponges the volume of the atrium

**Table 7.** Comparison of number of sponge samples (N), total volume for all samples (T. Vol.), total number of species associated ($N_s$), and number of species per unit volume (l) of the host sponge ($N_s$), obtained in different studies

| Sponge species | N | T. Vol. (l) | $N_s$ | $N_s/l$ | Reference |
|---|---|---|---|---|---|
| *Ircinia strobilina* (M) | 3 | 203.250 | 40 | 0.1970 | Pearse (1950) |
| *Ircinia variabilis* (M) | 6 | 3.598 | 16 | 4.4470 | Rützler (1976) |
| *Ircinia muscarum* (M) | 4 | 10.840 | 14 | 1.2920 | Rützler (1976) |
| *Ircinia fasciculata* (M) | 2 | 1.400 | 13 | 9.2800 | Rützler (1976) |
| *Ircinia oros* (M) | 1 | 1.186 | 6 | 5.0590 | Rützler (1976) |
| *Halichondria panicea (I)* | 32 | 0.104[a] | 68 | – | Long (1968) |
| *Halichondria panicea* (F,R) | 116 | 10.958 | 36 | 3.28000 | Frith (1976) |
| *Halichondria panicea* (M) | 41 | – | 50 | – | Peattie and Hoare (1981) |
| *Sphecíospongia vesparia* (M) | 3 | 285.000 | 39 | 0.13700 | Pearse (1932) |
| *Sphecíospongia vesparia* (M) | 2 | 25.740 | 11 | 0.42700 | Pearse (1950) |
| *Sphecíospongia vesparia* (M) | 34 | 179.645 | 24 | 7.48520 | Westinga and Hoetjes (1981) |
| *Suberites lata* (M) | 14 | 0.503[a] | 25 | 3.5500 | Long (1968) |
| *Microciona prolifera* (R) | 15 | 0.594[a] | 52 | 5.8360 | Long (1968) |
| *Haliclona rubens* (R) | 4 | 29.000 | 16 | 0.55200 | Pearse (1950) |
| *Hippospongia comunis* (M) | 40 | 51.222 | 25 | 0.48800 | Rützler (1976) |
| *Spongia zimocca* (M) | 2 | 0.415 | 13 | 31.32500 | Rützler (1976) |
| *Hymeniacidon perleve* (I) | 60 | 5.115 | 31 | 6.06100 | Frith (1976) |
| *Mycale macilenta* (I) | 29 | 3.978 | 23 | 5.78200 | Frith (1976) |
| *Gelliodes digitalis* (V) | 29 | – | – | – | Uebelacker (1977) |
| *Aplysina archeri* (T) | 98 | 60.182 | 53 | 0.90000 | Present chapter |
| *Aplysina lacunosa* (T) | 99 | 51.853 | 134 | 2.50000 | Present chapter |
| *Aplysina archeri* (T) | 98 | 22.948[b] | 43 | 1.8000 | Present chapter |
| *Aplysina lacunosa* (T) | 99 | 20.420[b] | 130 | 6.3600 | Present chapter |

[a] Average volume of the samples.
[b] Volume of channels and meanders for all samples. (M) massive sponge; (R) ramose sponge; (T) tubular sponge; (F) fistulose sponge; (I) incrustant sponge; (V) vasiform sponge.

contributes significantly to the total sponge volume but not to the total number of species. This is specifically true for *A. archeri*, a sponge that presents a very large atrial cavity.

The effect of growth form on the number of species that a sponge can shelter can further be illustrated with the comparison of the various growth forms of *Halichondria panicea* shown in Table 7. The incrustant and massive forms hold a significantly larger number of species than the fistulose and ramose forms. It is recommended that further studies on the relationship of species vs. volume take into account tubular as well as incrustant growth forms, as literature on these are scarce.

As the most important conclusion it can be said that the volume a sponge can contain and the way this volume is distributed within the sponge are key factors determining the number of species that can be found within sponges.

Finally, the inclusion of some types of volume into the multiple linear regression equations corroborate the great importance of this parameter in the structure of the animal communities associated to the sponges studied.

*Acknowledgments.* We thank the staff of the Foundación Cientifica Los Roques at Dos Mosquises, and at Caracas, for their collaboration and enthusiasm throuhout the project. This project was made possible in part by a grant of CONICIT. We are specially grateful to Drs. H. Keupp and J. Reitner of the Intitut für Paläontologie Freie Universität of Berlin, who helped us participate in the Fossil and Recent Sponges International Conference, Berlin 1988; and to the Embassy of the German Federal Republic in Venezuela. Our most sincere thanks to Lic. Bladimir Rodriguez for his constant and inconditional help during the realization of this study and in the preparation of this manuscript.

## References

Alvarez MB, Diaz MC (1985) Las esponjas de un arrecife coralino en el Parque Nacional Archipiélago de los Roques: I, Taxonomia. II, Estructura ecologica: 1-216 (Tesis de Licenciatura, Universidad Central de Venezuela, Facultad de Ciencias, Escuela de Biologia, Caracas)

Bergquist PR (1978) Sponges. University of California Press, Berkeley

Bloom SA (1975) The motile scape response of a sessile prey: A sponge-scallop mutualism. J Exp Mar Biol Ecol 17(3):311-321

Chace FA (1972) The shrimps of the Smithsonian-Bredin Caribbean Expedition with a summary of the West Indian Shallow-water species (Crustacea: Decapoda: Natantia). Smithson Contrib Zool 98(i-x):1-179

Connor EF, McCoy ED (1979) The statistics and Biology of the species-area relationship. Am Nat 113(6):791-817

Daüer DM (1973) Polychaete fauna associated with Gulf of México sponges. Fl Sci 36(2-4):192-196

Fishelson L (1966) *Spirastrella inconstans* Dendy (Porifera) as an Ecological niche in the littoral zone of the Dahlak Archipelago (Eritrea). Bull Sea Fish Res Stn Israel 41:17-25

Forester AJ (1979) The association between the sponge *Halichondria panicea* (Pallas) and scallop *Chlamys varia* (L.): A commensal-protective mutualism. J Exp Mar Biol Ecol 36:1-10

Frith DW (1976) Animals associated with sponges at North Hayling, Hampshire. J Linn Soc (Zool) 58:353-362

Frith DW (1979) A preliminary analysis of the associates of amphipods *Microdeutopus damnoniensis* (Bate), *M. anomalus* (Rathke) and *Corophium sextoni* (Crawford) with sponges *Halichondria panicea* (Pallas) and *Hymeniacydon perleve* (Montagu). Crustaceana 32(2):113-118

Gleason HA (1925) Species and area. Ecology 6:66-74

Grajal A, Laughlin RA (1984) Decapod crustaceans inhabiting live and dead colonies of three species of *Acropora* in the Roques Archipelago, Venezuela. Bijdr Dierkd 54(2):220–230
Jimenez FA (1974) Biotic community inside the sponge *Laxosuberites* sp. (Porifera: Demospongiae). Philipp Sci 10:123–124
Kilburn PD (1966) Analysis of the species-area relation. Ecology 47:831–843
Long ER (1968) The associates of four species of marine sponges of Oregon and Washington. Pac Sci 22(3):347–351
MacArthur RH, Wilson EO (1967) The theory of island biogeography. Princeton Universtiy Press, Princeton NJ
Osman RW (1977) The establishment and development of a marine epifaunal community. Ecol Monogr 41:37–63
Pearse AS (1932) Inhabitants of certain sponges at Dry Tortugas. Pap Tortugas Lab 28(7):117–124
Pearse AS (1950) Notes on the inhabitants of certain sponges at Bimini. Ecology 31:149–151
Pearson K (1900) On the criterion that a given system of deviations from the probable in the case of a correlated system of variables is such that it can be reasonably supposed to have arisen from random sampling. Phil Mag Ser 5, 50:157–175
Peattie ME, Hoare R (1981) The sublittoral ecology of the Menai Strait II. The sponge *Halichondria panicea* and its associated fauna. Est Coast Shelf Sci 13(6):621–635
Pielou EC (1966) Shannon's formula as a measure of specific diversity: its use an disuse. Am Nat 100:463–465
Preston FW (1962) The canonical distribution of commonness and rarity. Ecology 43:185–215, 410–432
Randall IE, Hartman WD (1968) Sponge-feeding fishes of the West Indies. Mar Biol 1(3):216–225
Rodriguez G (1980) Crustaceos decápodos de Venezuela. Instituto Venezolano de Investigaciones Cientificas (IVIC)
Rützler K (1976) Ecology of Tunisian commercial sponges. Tethys 7(2–3):249–264
Rützler K (1978) Sponges in coral reefs. In: Stoddart DR, Johannes RE (eds) Coral Reefs: Research methods. Monographs on oceanographic methodology 5. UNESCO, pp 299–313
Schoener A (1974) Experimental zoogeography: Colonization of marine mini-islands. Am Nat 108(964): 715–738
Shannon CE, Weaver W (1963) The mathematical theory of communication. University of Illinois Press, Urbana
Sneath PH, Sokal RR (1973) Numerical taxonomy. WH Freeman, San Francisco
Sorensen T (1948) A method of establishing groups of equal amplitude in plant society based on similarity of species content. K Dan Vidensk Selsk 5:1–34
Tello J (1975) Catálogo de la fauna venezolana, VIII Mollusca. Publicaciones de la comision organizadora de la III Conferencia de las Naciones Unidas sobre el derecho del mar. Caracas
Tyler JC, Böhlke JE (1972) Records of sponge-dwelling fishes, primarily of the Caribbean. Bull Mar Sci 22(3):601–643
Uebelacker JM (1977) Cryptofaunal species/area relationship in the coral reef sponge, *Gelliodes digitalis.* In: Taylor D (ed) Proc: 3rd Int Coral Reef Symp 1. Biology, May 1977 Miami, Florida
Uebelacker JM (1978) A new parasitic polychaetous annelid (Arabellidae) from the Bahamas. J Parasitol 64(1): 151–154
Villamizar E (1985) Fauna asociada a las esponjas *Aplysina archeri* y *Aplysina lacunosa* a lo largo de un gradiente de profundidad en el Parque Nacional Archipiélago de Los Roques, Venezuela. Tesis de Grado Universidad Central de Venezuela
Westinga E, Hoetjes PC (1981) The intrasponge fauna of *Spheciospongia vesparia* (Porifera, Demospongiae) at Curacao and Bonaire. Mar Biol 62:139–150
Williams CB (1964) Patterns in the balance of nature. Academic Press, London
Zoppi E (1967) Contribución al estudio de los Equinodermos de Venezuela. Acta Biol Venez 5(17):267–324

# Environmental Control on Fossilization of Siliceous Sponge Assemblages: A Proposal

T.C. BRACHERT

## Introduction

The depositional environment of Jurassic siliceous sponge limestones in Germany is still a subject of controversial debate. Environmental reconstructions are based on two different approaches. The first is faunal assemblage analyses of ammonites and sponges. Among the latter, hexactinellids are thought to prefer deeper waters whereas lithistids should be able to inhabit higher energetic environments. Based on this philosophy a maximum water depth of about 100–200 m is assumed (Ziegler 1967, 1987). The occurrence of photoautotrophic bacteria (Lang 1988) proves the photic zone.

The second approach is petrography of carbonate cements and probable evaporite pseudomorphs. These provide some evidence for local subaerial exposure of sponge buildups (Koch and Schorr 1986; Koch and Schweizer 1986) and general shallow water conditions. The major part of the rocks is micritic, however, making analysis of early diagenetic carbonate cements very difficult.

Hiller (1964) observed a general decrease of sponge spicules showing silica relicts through time. Nevertheless the diagenesis of the siliceous sponge skeletons and spicula has never been used in detail for environmental considerations. Opal-A is a silica modification which is metastabile under shallow marine and meteoric conditions (Knauth 1979). Therefore it is used here to show its reactions in different depositional environments.

In most cases sponges are preserved as calcite pseudomorphs today. This was first described by Fritz (1958) concerning sponges with "pseudopelletoid preserved internal structure", so called Kalkmumien (calcareous mumies). Gaillard (1983) gave a flow scheme of diagenetic processes with different preservational stages.

## Material and Methods

The study is based on 810 samples from surface outcrops in the Fränkische Alb area. From these 340 acetate peels or large scale thin sections (10 × 15 cm) were prepared and examined on microfacies and diagenetic characteristics. Investigations of micrite textures were carried out by SEM methods on broken and etched surfaces. They were cleaned with destilled water and pressured air before coating.

J. Reitner and H. Keupp (Eds.)
Fossil and Recent Sponges

## Case Study: Upper Jurassic of S. Germany

The Upper Jurassic (Oxfordian and Kimmeridgian) of S. Germany (Fig. 1) was chosen for this study because it reveals different constructional sponge facies through time (Fig. 2). These are related to different paleogeographic settings which show the evolution from an epicontinental basin stage to a carbonate platform. The Oxfordian is characterized by carbonate-marl alternations, interrupted by "sponge reef tracts" (von Freyberg 1966). The dimensions of these buildups are very different; they reach from 1 m to hundreds of meters. This situation continues in Lower Kimmeridgian times, but now locally sponge biostromes occur. During the Middle Kimmeridgian biostromes are characteristic, whereas micrite-marl alternations are subordinate. Another important

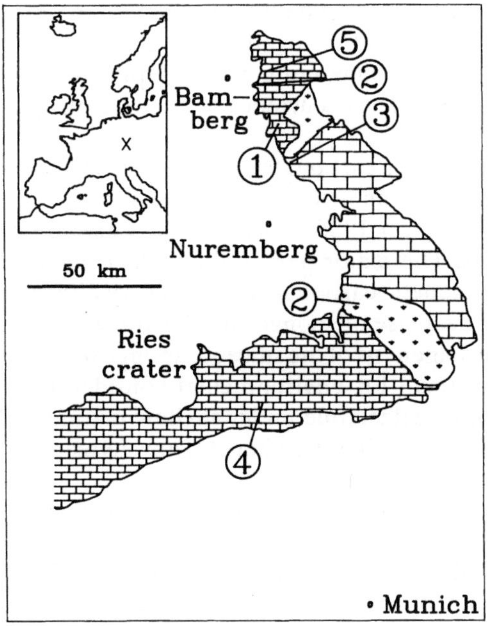


**Fig. 1.** Upper Jurassic outcrops in the Frånkische Alb area. *Numbers* refer to sections mentioned in the text (*1* Ebermannstadt; *2* Ludwag Quarry E of Bamberg and Mittersthal Quarry SE of Nuremberg; *3* Gräfenberg Quarry; *4* Konstein section; *5* Würgau section) and Fig. 2

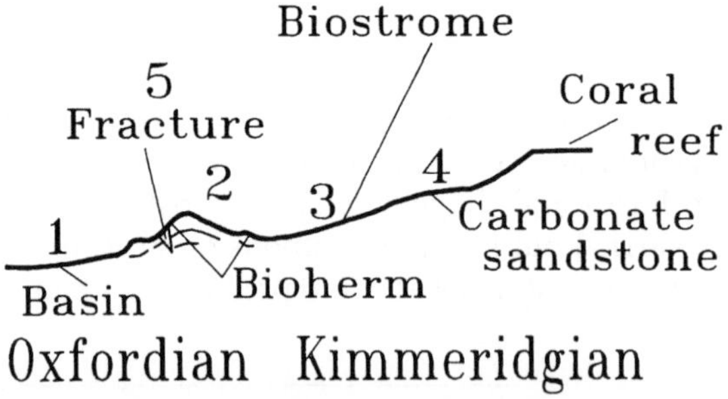


**Fig. 2.** Paleogeographic setting of sponge-bearing facies in the Upper Jurassic of S. Germany. *Number 1–5* refer to text and Fig. 1. *1* basin; *2* multi- and single-generation sponge bioherm; *3* sponge biostrome; *4* carbonate sandstone; *5* condensed section (fracture)

facies group are carbonate sands with siliceous sponges and a few stony corals. Together with coral bioherms they are thought to represent the most shallow marine platform environment (Fig. 2).

For detailed descriptions of the Upper Jurassic sponge algal facies and stratigraphy, see Gwinner (1976) and Zeiss (1977).

### *Shelf-Basin Micrites*

Shelf-basin carbonates comprise uniform micrite-marl alternations. Each bed can be traced for more than a hundred kilometers (Schmidt-Kaler 1962). These deposits are free of sponges. Typical biota are ammonites, belemnites, and benthonic foraminifera. Planctonic organisms occur subordinately. Most sedimentary fabrics are destroyed by bioturbation (Ricken 1985).

In the neighborhood of sponge buildups single sponges and sponge debris ("tuberoids" of Fritz 1958) may occur ("intermound facies").

Each local basin has its own but very constant bedding rhythm (von Freyberg 1966). Marls become more important in deeper waters and the typical bedding rhythm is lost.

### *Sponge Bioherms*

The sponge buildup facies is an association of cyanobacterial crusts (thrombolites and stromatolites of Wirsing 1988) and siliceous sponges. The ratio of hexactinellids and lithistids changes with time, but regional patterns exist also. An ideal bioherm succession has been described by Flügel and Steiger (1981) from an Oxfordian buildup. A cycle begins with micritic resediments, brachiopods, and sponge debris. The second stage is characterized by an overall growth of cup- and dish-shaped sponges, which are sometimes still in life position. The sponges are successively replaced by peloidal cyanobacterial crusts which dominate the last stage. The matrix always remains micritic.

The buildups exhibit different geometries depending on wether they comprise single or multigeneration mounds. Single generation bioherms are lenticular and 1–2 m in diameter whereas the latter are often elongated and composed of several garland-like layers of 1–2 m thickness. They may be more than 100 m long. For detailed descriptions see also Brachert (1986), Lang (1988), Wirsing (1988), and Nitzopoulos (1973). In the case of increased terrigenous input cyclicity becomes lost and hexactinellid sponges dominate, while calcareous crusts are thin ("Schwamm-Mergel"). Mound-dwelling organisms are bryozoans, inozoans, brachiopods, and benthonic foraminifera as well as various echinoderms.

### *Sponge Biostromes: Treuchtlingen Facies*

Sponge biostromes exhibit the same cyclic pattern as bioherms (Kott 1989) but every bed is flat lying and can be traced over tens of kilometers. It is an equivalent of the shelf-basin micrite facies during times of lower sea level. This facies is restricted to the Middle Kimmeridgian.

### *Carbonate Sandstones*

The Treuchtlingen facies grades laterally and vertically into bioclastic and sometimes oolithic grainstones. Oolite grains are in many cases consistent with pelagic oolites (Jenkyns 1972): Micritic cortex and globigerinid or planctonic echinoderm nuclei. Sponges and "stromatolitic lamination" are common, as reported from Italian examples (Massari and Dieni 1983), while single corals occur additionally. Foraminifera are dominated by trocholinids. Sponges are poorly preserved but may be essentially lithistids. The absence of green algae suggests a depth of deposition some tens of meters below sea level (Jenkyns 1972). Hardgrounds (Fig. 3G) and endoclasts probably indicate temporary condensation events, as vadose cementation is absent. Dike filling crystal silt is common, also dogtooth cements and syntaxial cements on echinoderm grains.

## Diagenesis of the Siliceous Skeleton

The skeleton of a living siliceous sponge is opal-A (Simpson 1984). It is an amorphous and hydrated silica phase. The solubility is controlled by its water

**Fig. 3.** Preservational stages of siliceous sponge fossils in thin section. Bar always is 1 mm. **A** Hexactinellid sponge from intermound setting. The spicules are siliceous and axial canals are filled by ferroan hydroxides. Between spicules pelletoid micrite. Lu.63. **B** Siliceous sponge from bioherm. The spicule molds are typically filled with two-generation calcite cement. Some of them are filled with internal sediments after the formation of A-cement (fibrous precursor?). Lu.124. **C-D** Lithistid sponges from somewhat shallower bioherm (or biostrome) settings. Increased silica dissolution rates from **B** to **C** and **D** with respect to micrite formation and lithification lead to poorly preserved spicules resulting in clotted structure. Lu.114. **E** Sandstone-type preservation. Sponge completely replaced by fine granular spar with common concentric cherts (*black arrowhead*). Micritic infilled dyke running *from upper right to left* emplaces chert nodules (*white arrowhead*). Thus it proves early diagenetic silification. Note shells of endolithic bivalves being excluded from chertification. Ko.35. **F** Hexactinellid sponge clast within synsedimentary fracture. The outline of the spicules is enlarged by carbonate dissolution (arrowhead). The molds were later filled by first a calcite rim cement and secondly by the dyke filling micrite. The small dyke partially filled with micrite points to repeated fracturation. Wü.3. **G** Hardground in skeletal grainstone facies overgrown by sessile foraminifera and thrombolite crusts. It is cutting through carbonate grains as *Tubiphytes*. Ko.39. **H** Siliceous sponge void from biostrome (sandstone-type preservation). Void formation is due to dissolution of the siliceous skeleton prior to the lithification of lime mud in between the spicules. A rigid shelter against collapsing is provided by encrusting foraminifera on the former sponge surface. Ko.15

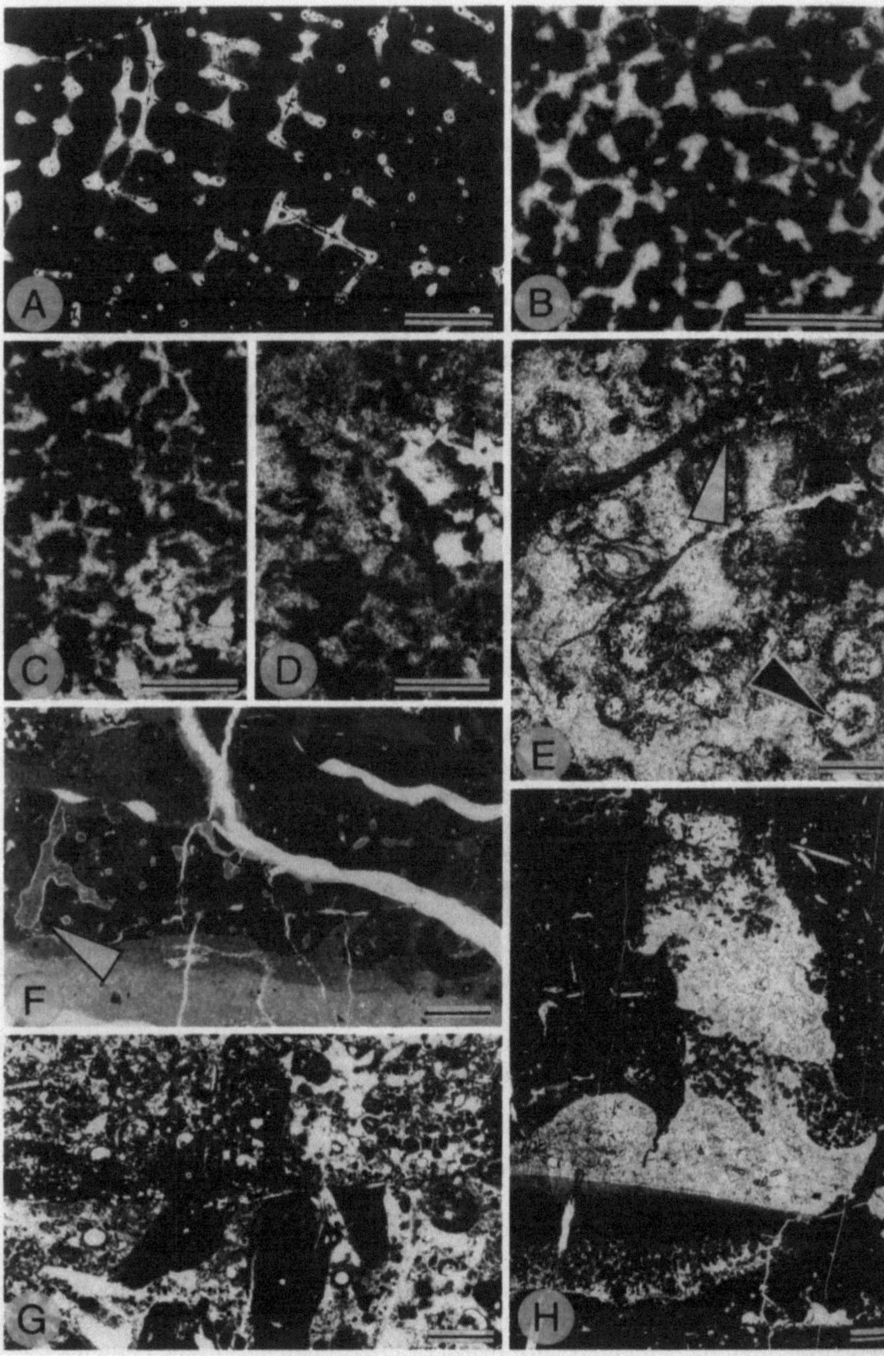
A
B
C
D
E
F
G
H

content, chemical inhibitors (Berner 1980), and environmental conditions. High water percentages, temperature, and increasing turbulence increase dissolution rates, whereas Mg-ions cause precipitation (Laschet 1984). Stability is further reduced under full marine or freshwater conditions (Knauth 1979).

The distribution of siliceous spicules in recent carbonate sediments exhibits analog patterns: in high energy carbonates as reefs spicules are quickly dissolved while they are still siliceous, although corroded, in shallow marine lagoonal carbonates (Land 1976; Rützler and Macintyre 1978). Rapid silica dissolution in deep marine sponges from the Mediterranean has been reported by Froget (1976) and Allouc (1987). It is caused by the homothermal water column of this marginal sea (Brachert et al. 1987) producing a deep silica corrosion zone (Berger 1975) which is normally typical for surface waters.

These features contrast with the reaction of carbonate phases to environmental changes. Aragonite and Mg-calcite are most stable in warm shallow marine waters whereas they get dissolved in meteoric as well as cold $CO_2$-rich marine waters.

The preservation style of sponge spicules varies strongly in the investigated Upper Jurassic material. For detailed descriptions of the single types see Brachert (1986).

### *Sponges in Intermound Settings and Basins*

Preservation is best within this low energy environment. Skeletons may be partly or totally replaced by calite (Fig. 3A). Silica relicts are altered to microquartz and show all taxonomic features. The axial canal, however, may be enlarged and cemented with micrite or ferroan hydroxides. Alternatively, it may be siliceous and the spicule itself enlarged by preferred silification around it. Sometimes massive cherts develop. Carbonate pseudomorphs of the skeleton exhibit drusy or fine two-generation granular cements ("A" with fibrous precursor?). In these cases the axial canal became lost during diagenesis or is at least enlarged, because the opaline skeleton is most susceptible to dissolution around the axial canal (Simpson 1984). The outline of the spicules, however, is well preserved, except for the formation of scattered pyrite cubes. The matrix between the spicules is micritic or micropelletoid. Micritic worm tubes are common.

### *Sponges in Bioherms and Biostromes*

According to the paleogeographic setting of the buildups, various preservation patterns may be developed. They range from well-preserved sponges of the intermound type in deep-water mounds to individuals completely replaced by sparry calcite (Fig. 3B-D) but intact surfaces in somewhat shallower settings. The skeletons may be cemented with drusy or coarse calcite mosaic cements ($>30$ $\mu$m). Close to pressure-solution reactates spicules may be dolomitic. The matrix between the spicules is microspar (clotted structure) or fine-grained mosaic cement ($< 30\ \mu m$) and the outline of the spicules is not well defined. Crystals are coarsest where spicules were thickened. Chertification may occur. Single individuals may show several preservational stages.

### *Sponges in Carbonate Sandstones*

Depending on strong wave action, sponges are fragmented frequently. The sponge skeletons themselves constitute big molds with basal internal sediments, showing sometimes calcitic spicula relicts. Voïd-filling cements are, first, a fibrous equant rim, second, a radiaxial calcite spar followed by coarse blocky mosaic cements (Fig. 3H). A rigid shelter against collapsing was promoted by incrustations of sessile foraminifera and peloidal stromatolithic crusts. An other type consists of fine granular spar. It can also be seen in bioherms of the Lower Kimmeridgian (ki 1.3) (Fig. 3E).

### *Sponges in Condensed Sections*

Condensed sections are restricted to times of rapid sea level rises as the Kimmeridgian 1.2/1.3 boundary but may also occur in fractures of drowned bioherms. Skeletons are embedded in homogenous micrite but spicules are irregularly enlarged due to carbonate dissolution (Fig. 3F). Nevertheless the outline of the spicule molds is well defined, as is typical of shelf-basin micrites. Simultaneously, aragonitic shells are also dissolved. Both spicule molds and aragonite molds may be filled with gray micrite, often forming geopetal fabrics and remaining pore space is cemented by fine-grained blocky calcite. Early meteoric cement is absent.

## Discussion

The preservational types of siliceous sponge fossils are not randomly distributed. Well-preserved skeletons, sometimes with silica relicts, are restricted to muddy environments as low energy buildups (Oxfordian) and intermound basins in the Lower Kimmeridgian. In this case spicules are embedded in micrite, providing an exact and rigid form for spicule pseudomorphs (Fig. 3B). Micrite formation may be induced by microbes (Narbonne and Dixon 1985; Brachert et al. 1987). Extensive microbial activity additionally modifies the type of preservation, producing a micropelletoid matrix. Spatial differences in microenvironment can be deduced from single individuals showing several preservational stages, as well as from several sponges per thin section having different styles of preservation.

Silification is confined to bryozoans, brachiopods, belemnites, and echinoderm ossicles as well as some foraminifers, whereas primary aragonitic and siliceous biota are commonly excluded. In massive chert nodules surrounding sponges, spicules are often still calcite (Schorr and Koch 1985: Pl. 31/7), whereas micrite is completely replaced by microquartz. If silification is not controlled by different porosities, this implies probably a grain size controlled silification, as suggested by Beurer (1971) and Maliva and Siever (1988).

In some cases (fractures, unconformities) spicule molds are irregularly enlarged (Fig. 3F) and shells of aragonitic molluscs are dissolved. They are both filled with the same gray silt and fine granular spar which also occur in vertical dikes and moldic porosity of the micrite sediment (Fig. 3F). As carbonates are not corroded in the shallow marine environment, I assume early diagenetic influence of meteoric waters or more probably subsolution in condensed sections, because meteoric cements are absent.

In Kimmeridgian rocks, sponge preservation is worst, particularly in grainstone shoals. This is due to silica dissolution, which takes place faster than carbonate cementation. Sponges are only preserved if they are coated by sessile foraminifera or micritic stromatolites forming a rigid shelter prohibiting them from total collapse (Fig. 3H). The remaining voids are filled by different generations of calcite cement: first fibrous equant rim cement and radiaxial fibrous rims and secondly blocky mosaic cements.

A second type of silification is restricted to sponges completely replaced by fine granular spar. Frequently small radially structurated cherts with concentric stainings occur ($< 950\,\mu m$), excluding shells of boring bivalves (aragonite!) which implies early diagenetic Mg-calcite cements. This is evident from the fact that bivalves never bore into siliceous material (Palmer and Fürsich 1981) and selective silification of calcite is ascribed to its Mg content by various authors (Folk 1975). Associated with aragonite dissolution sedimentary dikes cross calcified sponges and cherts (Fig. 3E) giving evidence of silification before aragonite dissolution, as is also seen in low energy intermound carbonates. However, within the latter aragonite dissolution is related to physicochemical compaction producing mollusc steinkerns. Simultaneously, chalcedony may be precipitated in compaction fractures and molds of former aragonitic (?) (replaced siliceous) spicules.

## Conclusions

Siliceous sponges of the Upper Jurassic of S. Germany are in most cases more or less completely depleted in silica. As silica dissolution is increased in warm agitated waters, fossilization of sponges is mainly controlled by hydrodynamic forces (Fig. 4). Rapid micrite cementation and slow silica dissolution results in well-preserved siliceous sponge fossils. This is true in protected areas such as intermound basins or deep water mounds. Carbonate cementation is accelerated due to microbial activity, whereas dissolution of opaline silica is reduced by lack of water movement and probably reduced temperatures.

As water turbulence/temperature is increased, silica dissolution is accelerated with respect to micrite formation, producing poorly preserved sponges or even prohibiting fossilization. Fossil sponge assemblages therefore do not necessarily reflect the living community and their completeness is a function of

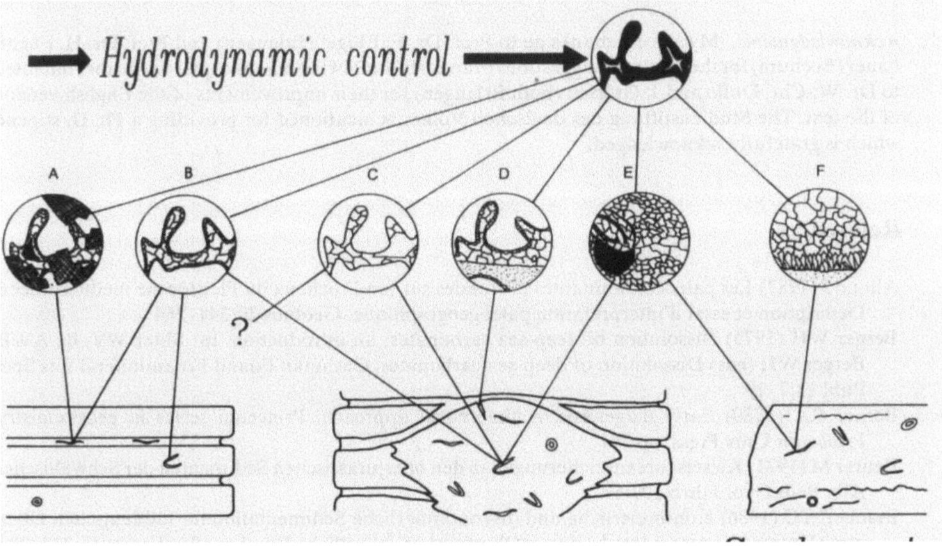


**Fig. 4.** Hydrodynamic control of siliceous sponge fossilization. Opaline silica dissolution of sponge spicules is fastest in warm agitated waters. Preservation is best in low-energy intermound and basin carbonates whereas it is worst in carbonate sandstones. Diameter of circles is 1 mm. **A,B** Sponges from intermound settings and deep water bioherms. The outline of the spicule molds is well defined and shows an infilling by two-generation granular cements. Scattered pyrite cubes truncate spicules in A. **C** Sponge from bioherm. The siliceous spicules are replaced by coarse calcite spar or dolomite rhombs. **D** Sponge from condensed section. The outline of the spicules is well defined but their shape is modified by carbonate dissolution. Partly infillings with gray micrite are typical. **E** Sponge from shallow water bioherm/biostrome or sandstone. The siliceous skeleton is replaced by granular calcite cement. Small radially structurated cherts with concentrically arranged dark stainings are typical. Silifications exclude shells of boring bivalves. **F** Sponge from carbonate sandstone. Voids resulting from dissolution collapse of the sponge skeleton show micritic internal sediments and equant pore-lining carbonate cements.

the environment. In other words, the mesh of the sieve of diagenesis (Dullo 1988) varies from place to place and produces various spatial fractions for different areas.

As the opaline silica is mobilized during different stages of diagenesis in different environments, silification occurs heterochronously. Early diagenetic cherts are found in skeletal grainstones, whereas silica precipitation is found in spicule molds and fractures resulting from physicochemical compaction in low energy environments as sponge-bearing marls ("Schwamm-Mergel") or deep-water bioherms.

Thus it is proposed in this chapter that the types of siliceous sponge fossils reflect the environmental transition from a deeper water ramp stage in an epeiric basin to a shallow marine carbonate platform.

*Acknowledgments.* My sincere thanks go to Prof. Dr. E. Flügel (Erlangen) and Prof. Dr. H. Füchtbauer (Bochum) for their helpful suggestions to an early draft of the manuscript. I am further indebted to Dr. W.-Chr. Dullo and J. Grötsch (both Erlangen) for their improvements of the English version of the text. The Studienstiftung des deutschen Volkes is mentioned for providing a Ph. D. stipend which is gratefully acknowledged.

## References

Allouc J (1987) Les paleocommunautes profondes sur fond rocheux du Pleistocene mediterraneen. Description et essai d'interpretation paleogeographique. Geobios 20:241-264

Berger WH (1975) Dissolution of deep-sea carbonates: an introduction. In: Sliter WV, Be AWB, Berger WH (eds) Dissolution of deep-sea carbonates. Cushman Found Foraminiferal Res Spec Publ 13:7-10

Berner RA (1980) Early diagenesis. A theoretical approach. Princeton series in geochemistry, Princeton Univ Press, pp 241

Beurer M (1971) Kieselsäureanreicherungen in den oberjurassischen Sedimenten der Schwäbischen Alb. Beih Geol Jahrb 109:69

Brachert TC (1986) Kontinuierliche und diskontinuierliche Sedimentation im süddeutschen Oberjura (Unteres Kimmeridge; Ludwag/Oberfranken, Nördliche Frankenalb). Facies 15:233-284

Brachert TC, Dullo W-C, Stoffers P (1987) Diagenesis of siliceous sponge limestones from the Pleistocene of the Tyrrhenian Sea (Mediterranean Sea). Facies 17:41-50

Dullo W-C (1988) Paläontologie, Fazies und Diagenese der Pleistozänen Riffterrassen an der nördlichen Ostküste des Roten Meeres (Saudi Arabien). Thesis, Erlangen University, FRG, pp 199 (unpubl)

Flügel E, Steiger T (1981) An Upper Jurassic sponge-algal buildup from the north-eastern Frankenalb (W. Germany) Soc. Econ Mineral Paleontol Spec Publ 30:371-397

Folk RL (1975) Fossils by length-slow chalcedony and associated dolomitization. J Sed Petrol, Discussion 45:952

Fritz GK (1958) Schwammstotzen, Tuberolithe und Schuttbrekzien im Weißen Jura der Schwäbischen Alb. Eine vergleichende petrogenetische Untersuchung. Arb Geol Paläontol Inst TH Stuttg NF 13:118

Froget C (1976) Observations sur l'alteration de la silice et des silicates au cours de la lithification carbonatee (Region Siculo-Tunisienne). Geol Mediterr 1976:219-226

Gaillard C (1983) Les biohermes a spongiaires et leur environment dans l'Oxfordien du Jura meridional. Doc Lab Geol Lyon 90:515

Gwinner MP (1976) Origin of the Upper Jurassic limestones of the Swabian Alb (SW Germany). Contrib Sediment 5:75

Hiller K (1964) Über die Bank- und Schwammfazies des Weißen Jura der Schwäbischen Alb (Württemberg). Arb Geol Paläontol Inst TH Stuttg NF 40:190

Jenkyns HC (1972) Pelagic "oolites" from the Tethyan Jurassic. J Geol 80:21-33

Knauth LP (1979) A model for the origin of chert in limestone. Geology 7:274-277

Koch R, Schorr M (1986) Diagenesis of Upper Jurassic sponge-algal reefs in SW Germany. In: Schroeder JH, Purser BH (eds) Reef diagenesis. Springer, Berlin Heidelberg New York Tokyo, pp 224-244

Koch R, Schweizer V (1986) Mikrofazies, Diagenese und Bildungsraum oberjurassischer Schwamm-Algen-"Riffe" der Schwäbischen Alb. Heidelb Geowiss Abh 6:247-263

Kott R (1989) Fazies und Geochemie des Treuchtlinger Marmors (Unter- und Mittelkimmeridge, Südliche Frankenalb). Berl Geowiss Abh Reihe A 111:115

Land LS (1976) Early dissolution of sponge spicules from reef sediments, North Jamaica. J Sediment Petrol 46:967-969

Lang B (1988) Die Schwamm-Biohermfazies der Nördlichen Frankenalb (Steinbruch Urspring, Oxford, Malm): Mikrofazies, Palökologie, Paläontologie. Thesis, Erlangen University, FRG (unpubl)

Laschet C (1984) On the origin of cherts. Facies 10:257-289

Maliva RG, Siever R (1988) Mechanism and controls of silification of fossils in limestones. J Geol 96:387-398
Massari F, Dieni I (1983) Pelagic oncoids and ooids in the Middle-Upper Jurassic of eastern Sardinia. In: Peryt TM (ed) Coated grains. Springer, Berlin Heidelberg New York, pp 367-376
Narbonne GM, Dixon OA (1985) Upper Silurian lithistid sponge reefs on Somerset island, arctic Canada. Sedimentology 31:25-50
Nitzopoulos G (1973) Faunistisch-ökologische, stratigraphische und sedimentologische Untersuchungen vom Spielberg am Hahnenkamm (Ob. Oxforiden, südliche Frankenalb). Thesis, TU Berlin, FRG, pp 155
Palmer J, Fürsich FT (1981) Ecology of sponge reefs from the Upper Bathonian of Normandy. Palaeontology 24:1-23
Ricken W (1985) Epicontinental marl-limestone alternations: event deposition and diagenetic bedding (Upper Jurassic, SW Germany). In: Bayer U, Seilacher A (eds) Evolutionary and sedimentary cycles. Springer, Berlin Heidelberg New York Tokyo, pp 127-162
Rützler K, Macintyre IG (1978) Siliceous sponge spicules in coral reef sediments. Mar Biol 49:147-159
Schmidt-Kaler H (1962) Zu Ammonitenfauna und Stratigraphie des Malm Alpha und Beta in der Südlichen und Mittleren Frankenalb. Erlanger Geol Abh 43:12
Schorr M, Koch R (1985) Fazieszonierung eines oberjurassischen Algen-Schwamm-Bioherms (Herrlingen, Schwäbische Alb). Facies 13:227-270
Simpson TL (1984) The cell biology of sponges. Springer, Berlin Heidelberg New York Tokyo, pp 662
von Freyberg B (1966) Der Faziesverband im unteren Malm Frankens: Ergebnisse der Stromatometrie. Erlanger Geol Abh 62:112
Wirsing G (1988) Mikrofazielle und diagenetische Entwicklung einer oberjurassischen Karbonatabfolge (Weißjura beta bis delta 4, Mittlere Schwäbische Alb) mit Berücksichtigung der chemischen Diagenese ausgewählter karbonatischer Bestandteile. Heidelb Geowiss Abh 15:337
Zeiss A (1977) Jurassic stratigraphy of Franconia. Stuttg Beitr Naturkd Ser B 31:32
Ziegler B (1967) Ammoniten-Ökologie am Beispiel des Oberjuras. Geol Rundsch 56:439-464
Ziegler B (1987) Der Weiße Jura der Schwäbischen Alb. Stuttg Beitr Naturkd Ser C 23:71

# Stratigraphic Distribution, Lithological Paragenesis, Depositional Environments and Diagenesis of Fossil Siliceous Sponges in Europe

W. Zimmerle

## Introduction

Sedimentary rocks bearing remnants of siliceous sponges are a peculiar rock type in the Phanerozoic. A brief review on ecological aspects of modern sponges can guide to a better interpretation of fossil sponges.

De Laubenfels (1955) summarized the ecological aspects of Recent sponge deposits: Sponges live in fresh, brackish, or marine waters. Rich sponge faunas occur in the zone of low-water spring tide, frequently offshore from rivers. Sponges normally require a solid substratum; they are destroyed by silt, but not by deposition of coarse sand. Currents of 2–3 km/h are optimum; both stagnant water and violent currents are adverse. Sponges spread into new areas rather slowly, even in terms of geological time. Sponge skeletons consist of discrete mineral elements called spicules either of $SiO_2$ or $CaCO_3$.

Siliceous sponge deposits form within the limits of the continental shelf and slope (down to a depth of 400 m). Mass development of siliceous sponges is observed in relatively cold, turbulent waters of normal salinity. Sponges are common but accessory elements of Recent reef faunas (Rigby 1971). They occur in regions with and without volcanic activity. Data on fossil sponges in sediments is summarized by Reid (1978).

The aim of this chapter is to compile petrographic data on depositional environment, lithological paragenesis, and diagenesis of siliceous sponge-bearing sedimentary rocks in Europe.

## Samples and Methods of Study

Rock samples examined petrographically and depicted (Fig. 1) were collected from the Lower Carboniferous of the Rheinische Schiefergebirge and Harz Mountains, from the Upper Jurassic (Oxfordian Heersum Beds), and from the Lower Cretaceous (Aptian/Albian) of NW Germany as well as from the Middle Eocene (Lutetian) of the Rheingraben. Diagenetic features were studied by means of the petrographic microscope and scanning electron microscope.

J. Reitner and H. Keupp (Eds.)
Fossil and Recent Sponges

## Stratigraphic Occurrences of Siliceous Sponges in Europe

In Table 1 occurrences of siliceous sponges in Europe are compiled distinguishing between occasional and massive occurrences with spongolites. The columns "Central Europe" include: the Federal Republic of Germany, the former German Democratic Republic, Switzerland, and Austria; and "East Europe": the ČSFR, Poland, and the USSR. The stratigraphy is subdivided into (1) Precambrian and Paleozoic, (2) Mesozoic, and (3) Cenozoic.

### *Precambrian and Paleozoic*

Metamorphosed traces of presumably monaxial siliceous sponges were found in quartz phyllites of the Precambrian Villé Series from the Vosges Mountains (Doubinger and von Eller 1963). Sponge spicules from metamorphosed "Precambrian rocks", described from Brittany in W France (Cayeux 1929, pp. 52–65), are now considered metamorphosed Late Paleozoic.

Minor occurrences of siliceous sponges are known from the Cambrian to the Upper Devonian of Europe (Hinde 1887–1912; Katzer 1902; Schwarzbach 1948; Wurm 1961; Geyer 1962; Reif 1967; Wuttke and Kott 1987).

- Cambrian (Great Britain, Norway, France, Spain, Sardinia)
- Ordovician (Great Britain, Sweden, USSR-Estonia)
- Silurian (Great Britain, FRG)
- Lower Devonian (ČSFR, FRG).

In the Upper Devonian/Lower Carboniferous of the Federal Republic of Germany (Rheinische Schiefergebirge, Harz Mts., Frankenwald), former German Democratic Republic (Thüringer Wald), Great Britain, Poland (Holy Cross Mts.), France (Montagne Noir), and Belgium, isolated and loosely disseminated, rather minute, siliceous sponge spicules occur in cherts and siliceous shales; whole sponges are rare (Hinde 1887–1912; Cayeux 1929; Wurm 1961; Geyer 1962; Alexandrowicz 1978; Hurcewicz and Czarniecki 1985; Hurcewicz 1985).

These siliceous sediments belong to the starved-basin association that is mainly composed of laminated dark shales, siliceous shales, and cherts with thin intercalations of tuffs, sandstones, and limestones. Radiolarian cherts are normally associated with the sponge-bearing sediments. The siliceous sponge spicules occur either solitary or in lenticular accumulations (Fig. 1a). Spongolites are known from Great Britain.

The originally opaline sponge spicules have been transformed into quartz. Along stylolites spicules are partly pressolved (Fig. 1b). Silica is supplied from the decay of pyroclastic debris and; presumably, from silica exhalations (Kubanek and Zimmerle 1986; Zimmerle 1986).

Sponge beds, up to 105 m thick, are known from Namurian calcareous cherts of Great Britain and Ireland (Hinde 1887–1912; p. 98ff.; Hey 1956) and lacustrine spiculites from the coal-bearing Stephanian of SE France (Cayeux 1929, Pl. 17, Figs. 64,65).

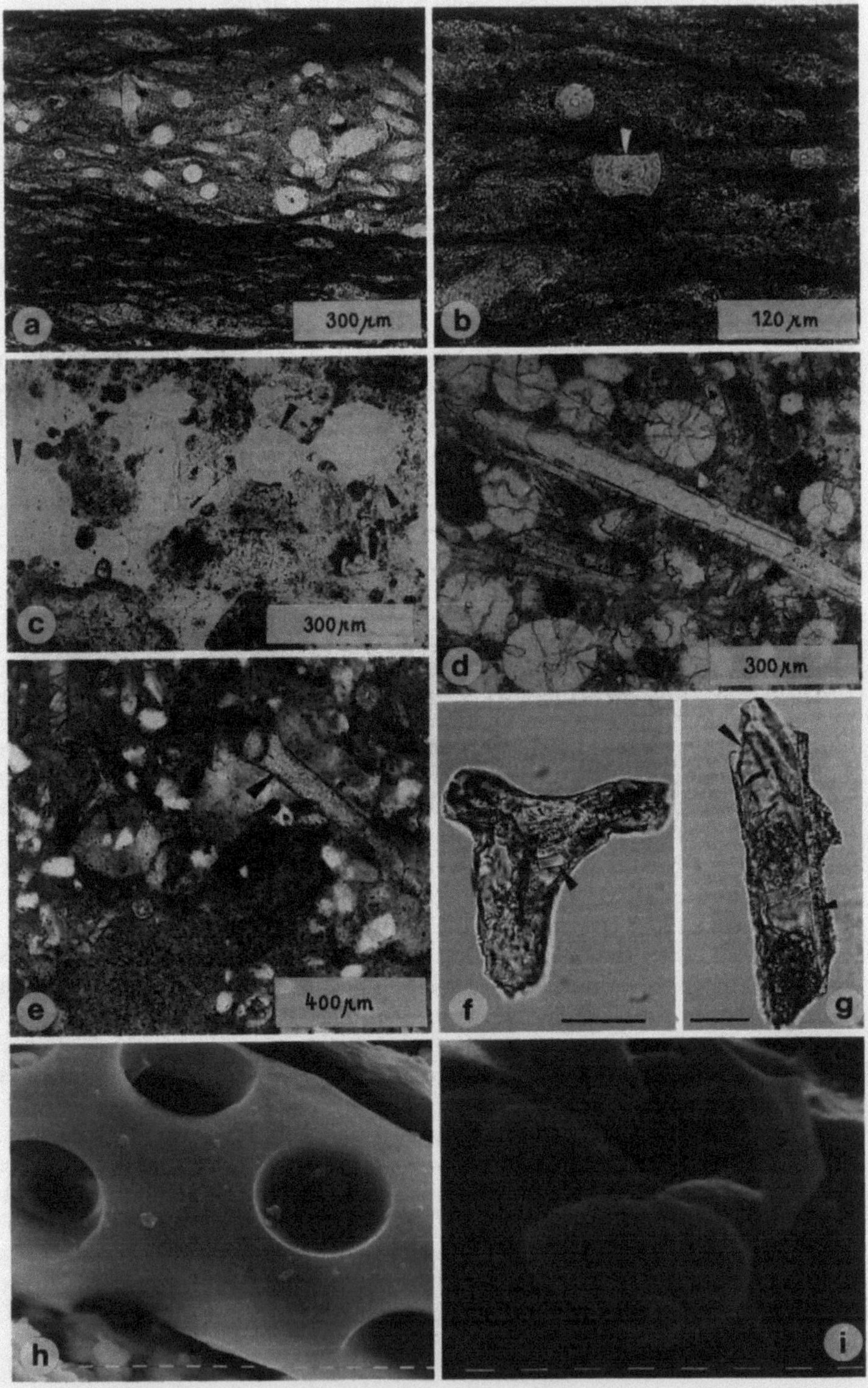
a
300 µm
b
120 µm
c
300 µm
d
300 µm
e
400 µm
f
g
h
i

Spectacular spiculites are found in the cherty Upper Permian Kapp Starostin Formation (Lowell 1968) of Spitsbergen (Hurcewicz 1982). From the Rotliegendes lacustrine deposits of the Saar-Nahe Basin no freshwater sponges have been recorded.

## *Mesozoic*

*Triassic.* In the Muschelkalk (middle Muschelkalk), siliceous sponge spicules are restricted to a single horizon (Schmidt 1928; 115; Rauff 1937, pp. 7–14), also characterized by hornstones. In the Alpine Triassic siliceous sponge spicules are of stratigraphic value: in the middle Anisian, lower Carnian, and higher Norian (Mostler 1971). In the Dolomites, NE Italy, sponge-algal-bryozoan mounds of Anisian age contain radiolarians and siliceous sponges; sponges also occur in the Carnian of Hungary (Boda 1964). In the Muschelkalk of Spain siliceous sponge spicules occur in small reefs (Esteban Cerda et al. 1977).

*Jurassic.* In Great Britain (Hinde 1887–1912) siliceous sponge spicules are found in the Lower Liassic Limestone (Hettangian-Pliensbachian) and in southern France (Jullian 1953, Figs. 9,10,12–14) in Pliensbachian limestones.

At the northern continental margin of the western Tethys, siliceous sponges and isolated spicules occur in pelagic marine marlstones and limestones of the lower Jurassic. In the cherty Allgäu Beds and in the cherty Lombardic Limestone, both of Hettangian-Sinemurian age, whole sponges and isolated spicules are found (Wiedenmayer 1980a). Sponge-bearing gaizes are interstratified in the Lower Jurassic cherty sponge limestone (Spongien-Kieselkalk) (Geyer 1962, p. 57) and in the Fleckenmergel (Hagn 1955).

In the Domerian of the High Tatra, Poland, and in the Hettangian-Lotharingian of the Sierra Nevada, SE Spain, siliceous sponge spicules are common, together with radiolarians.

**Fig. 1.** Characteristic photomicrographs of siliceous sponge spicules from NW German occurrences and their diagenetic alterations. **a** Lenticular accumulation of sponge spicules in siliceous shale. Lower Carboniferous, W Heimbergskopf, Harz Mts. **b** Siliceous sponge spicules pressolved at argillaceous partings (*white arrowhead*). Lower Carboniferous, W Heimbergskopf, Harz Mts. **c** Round to oval cross-sections (*black arrowheads*) of siliceous sponge microscleres in a well-sorted fine-grained spiculite. Upper Jurassic (Oxfordian), Heersum Beds, Hildesheimer Wald. **d** Spiculite replaced by sparry calcite. Note radiate structure of microscleres. Upper Jurassic (Oxfordian), Heersum Beds, Hildesheimer Wald. **e** Opaline sponge spicules (*black arrowheads*) in phosphorite nodules partly replaced by phosphate and partly transformed into clinoptilolite. Lower Cretaceous, Aptian, Sarstedt claypit. **f,g** Turbid fragments of siliceous sponge spicules partially replaced by transparent clinoptilolite (*black arrowheads*). Lower Cretaceous, Aptian, Sarstedt claypit. **f** Bar = 50 μm. **g** Bar = 50 μm. **h** Siliceous freshwater sponge spicule with large conical corrosion pits (compare with Martini and Rietschel 1978, Figs. 1–7). Middle Eocene, Lutetian, Messel Oil Shale openpit mine, Scale in 1-μm bars. **i** Progressive corrosion of an opaline freshwater sponge spicule (*S*) and neoformation of opal CT spherules (but no lepispheres). Middle Eocene, Lutetian, Messel Oil Shale open-pit mine, Scale in 1-μm bars

**Table 1.** Stratigraphic distribution of sponge-bearing sedimentary rocks in the Phanerozoic of Europe

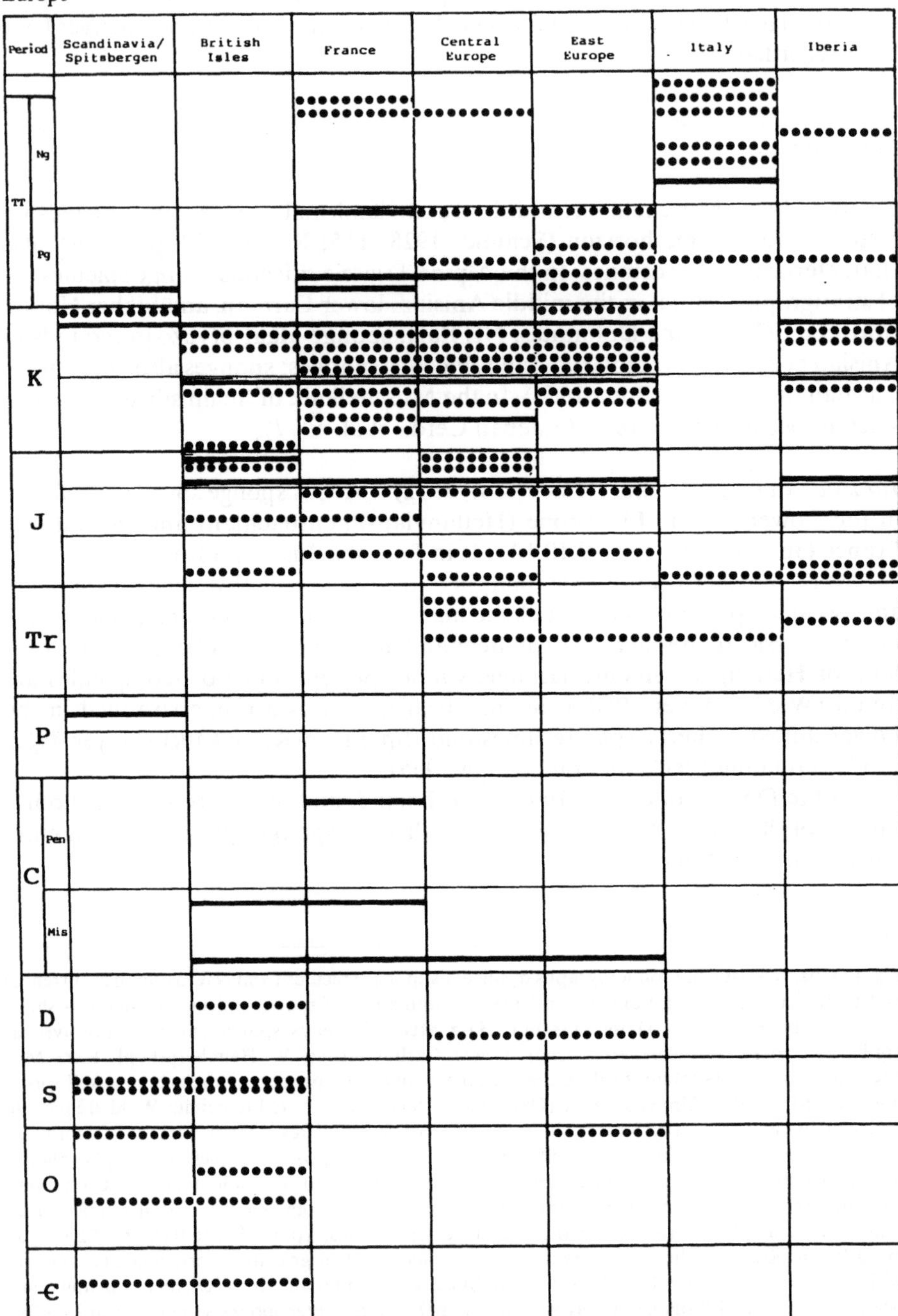


Solid bar = abundant siliceous sponge spicules and/or spongolites.
Interrupted bars = mainly siliceous sponge spicules observed.

In the Inferior Oolite (Aalenian-Bajocian) of W England siliceous sponges were observed by Hinde (1887–1912) and Richardson and Thacker (1920); they are more frequent in marine sediments. Spongolites occur in the Sponge Beds at the top of the Inferior Oolite. In the Bajocian of the Languedoc, France, spiculitic limestones are present (Jullian 1953, Fig. 16). In the Bathonian of the Normandy, France, sponge-bryozoan reefs occur, in the Callovian of the Languedoc spiculitic marly limestones.

Siliceous sponges and spicules, partly replaced by phosphate and associated with radiolarians, are contained in phosphorite nodules in argillaceous shallow-marine sediments of Callovian age in S Germany. The sponges, however, are not preserved in the argillaceous host rock, the so-called Ornatenton (Riegraf 1986). Algal-spongiolitic biostromes and thin sponge layers occur throughout the Bajocian and Bathonian in the Iberian Range of Spain (Scheer 1988; Wilde 1988).

The Upper Jurassic, especially the Oxfordian/Kimmeridgian, was a time of intense siliceous sponge development in argillaceous sediments as well as in reef limestones all over Europe. Precursors of massive sponge development are thin lenticular bodies, the so-called Sponge Marls (Schwammergel), in Middle Oxfordian marls and limestones of the Franconian Alb (Klieber 1983).

More peripheral are remarkable sponge occurrences in a 2-m-thick limestone of the Corallian Beds (Upper Oxfordian) of England (Milodowski and Wilmot 1984) and in coeval mudstones (Chowdhury 1982), which are characterized by a volcanogenic mineral paragenesis and diagenesis with smectite, zeolite, biotite, and apatite. "*Rhaxella*" Cherts of Oxfordian age are found (Arkell 1933, pp. 414, 431) in spiculitic calcarenites and cherty limestones (Arngrove stone and Lower Calcareous Grit). In Great Britain and in the North Sea area the Kimmeridge Shale, the famous source rock of North Sea oil, is characterized by the massive occurrence of *Rhaxella perforata*, dispersed in the shale matrix.

The Oxfordian of the Wiehengebirge, NW Germany, consisting of a nearshore sedimentary rock sequence (Heersumer Schichten and Unterer Korallenoolith) is characterized by "*Rhaxella*" spiculites, up to 5 m thick (Klassen 1984, pp. 387–395, 412, Anl. jw 2 profile), likewise the Oxfordian of the Hildesheim, Hannover and Ith area as well as of the Elsfleth 1 well near Bremen. The well-sorted, fine-grained "*Rhaxella*" spiculites are characterized by round to reniform microscleres, called Sterraster. The microscleres are preserved as chalcedony (Fig. 1c) or replaced by sparry calcite (Fig. 1d). Through diagenetic dissolution of the microscleres pinpoint porosity, typical of this spiculitic rock type, developed (Gramann 1962).

Cayeux (1929, pp. 289–296) referred to "spongiolithes à spicules globuleux de l'Oxfordien des Ardennes".

A striking development of siliceous sponges is observed in the calcareous sponge-algal mounds, sponge-algal crusts and sponge-coral bioherms sensu Wilson (1975) of Oxfordian-Kimmeridgian age in the Swabian and Franconian Alb of S Germany and in the Jura Mountains of Switzerland and France. This sedimentary rock suite has been described in great detail (Fritz 1958; Gwinner 1958; Beurer 1971; Müller 1978; Gaillard 1983; Meyer and Schmidt-Kaler 1983; Schorr and Koch 1985; Koch and Schorr 1986; Ricken 1986; Lang 1987, 1988a,b;

Wirsing 1988). The single bioherm/reef bodies are calcareous lenticular masses, 0.1–20 m high and 1–60 m in lateral extent; they contain siliceous sponges and sponge spicules and have been deposited in a quiet shelf environment, mostly in the photic zone. Spongiolithic chalcedony concretions also contain siliceous sponge spicules and fragments of sponge skeletons (Beurer 1971, p. 4).

The Oxfordian Jasna Góra Beds of Poland are "part of the European Upper Jurassic sponge megafacies" (Trammer 1982, 1988). The siliceous sponges are found in bedded limestones; bioherms occur only locally.

In the northern Tethys Upper Jurassic dark marly shales and radiolarites are characterized by biosiliceous components, among others also siliceous sponge spicules (Diersche 1980; Faupl and Beran 1983). Ulrich (1960, pp. 108–109, 127–132) referred to gaize-like cherty limestones of Oxfordian/Kimmeridgian and Tithonian (in the Aptychenkalk) age, both in the Karwendel Trough, Tyrol, Austria. Sponge-bearing limestones occur in the Oxfordian/Kimmeridgian of the Iberian Range (Geyer 1962; Deusch et al. this Vol.) and of the Pyrenees, siliceous sponge-bearing cherts in allodapic limestones of the Betic Cordillera, Spain (Bustillo and Ruiz-Ortiz 1987; Deusch et al. this Vol.).

The Upper Jurassic reefs/bioherms of Spain, France, Switzerland, S Germany, Poland, and Romania form a continuous belt at the northern periphery of the Tethys (Trammer 1982, Fig. 6).

Freshwater sponge spicules are found in cherts of the Purbeck Beds, S England (Greensmith 1978, p.180).

*Cretaceous.* The widespread occurrence of siliceous sponges continues in Cretaceous times.

Valanginian pyritic marlstones contain sponge spicules in northern Tethys ("fosse vocontienne") (Lagneau-Hérenger 1962, p. 215).

In the Hauterivian of NW Germany the Noricum Sandstone and the Gildehaus Sandstone are marly, fine-grained spicular sandstones to very fine-sandy marlstones (Kemper 1976, pp. 61–65). In the Hauterivian of the "fosse vocontienne" siliceous sponges are also found (Lagneau-Hérenger 1962, p. 215; Charollais et al. 1967).

Rock-forming siliceous sponges are developed in the Hauterivian/ Barremian between Weser River and Goldenstedt, NW Germany, in a subsurface belt at the northern rim of the Lower Saxony Basin.

Spiculitic cherty limestones and marly limestones of Barremian age were observed by Hagn (1955, Pl. 30, Fig. 1, Pl. 42, Fig. 1). Some siliceous sponges are also found in the Barremian of SE France (Langneau-Hérenger 1962, p. 216).

Aptian cherts (0.1–1 m in thickness) in the Lower Greensand in S England and clays and marls in NW Germany, contain admixture of siliceous sponge spicules (Hinde 1886, p. 447; Krüger 1982; Zimmerle 1982a, p. 82). The Upper Aptian dark gray clays of the Sarstedt claypit are characterized by original sediment particles of sand grade, by a considerable proportion of fecal pellets, a subordinate amount of volcanogenic particles, and by marked amounts of siliceous fossils such as radiolarians, diatoms, and sponge spicules. In phosphorite nodules siliceous sponge spicules are well preserved in shape, but partly replaced by phosphate (Fig. 1e). Spicules from the embedding clay are partially replaced by clinoptilolite (Fig. 1f,g).

In the Lower Saxony Basin siliceous sponge spicules are confined to the southern flank as a product of upwelling cold waters in a shallow-marine environment. Also in the "fosse vocontienne", SE France, and in Catalonia, Spain, siliceous sponges and spicules are observed (Langneau-Hérenger 1962, p. 216).

Albian sediments, in S England the Upper Greensand (Hinde 1886, 447; Reid 1968; Greensmith 1978, p. 181), in the Ardennes of France the Gaize de l'Argonne, in Belgium the Meule de Bracquegnies (Cayeux 1929, p. 254), and in NW Germany the Hils Sandstone and the Flammenmergel, are characterized by the sporadic occurrence of siliceous sponges and an abundance of loose siliceous sponge spicules (Lehmann 1954; Jordan 1968; Georgi 1976; Neuss 1979; Ernst and Schmid 1979, p. 17; Krüger 1982; Zimmerle 1982a,b). Most sponge horizons are composed of detached, free spicules; entire sponges are rare, with the exception of spongolites in the Meule de Bracquegnies. Recrystallization of biogenic silica is advanced, replacement of siliceous sponge spicules by phosphate not uncommon. Most sediments are glauconitic.

Lithostratigraphic equivalents of the Flammenmergel with phosphorites, glauconitic spongolites, and gaize beds are widespread in the Upper Albian of the Polish Lowlands (Sokołowski 1976, pp. 549,562,564, Table 43).

The Flammenmergel, a lithostratigraphic unit of Aptian to Turonian age, forms a narrow E-W belt N of the Variscian chains extending from Great Britain to the USSR. Highly bioturbated spongolitic argillaceous or marly sediments (gaize or opoka) predominate. This peculiar sediment is characterized by varying silt and sand admixtures, considerable glauconite content, bioturbation, high silica content (opal, opal CT, and quartz), zeolites, and phosphorites (Grigjalis et al. 1978; Gluschko et al. 1980; Kemper and Zimmerle 1982). Direct vestiges of volcanic activity during Flammenmergel deposition, namely tuffs, are sparse (Keller et al. 1989). Indirect indications of remote volcanogenic influence such as the presence of zeolites (clinoptilolite and analcite) are observed locally (Klassen 1984, p. 440). Siliceous, mainly autochtonous concretions, rich in siliceous sponge remains, are common. The inner surface area of the Flammenmergel is large. Opaline silica is not only the main constituent of the spicules, but also a cryptocrystalline component of the rock matrix. In a peculiar way silica encrusts clay minerals of the matrix, forming minute globules. The siliceous sponge fragments are preferentially dissolved, leading to secondary porosification. Thus, in some oil fields of E Europe the Flammenmergel becomes a reservoir rock.

In the Tethyan realm, Aptian/Albian sediments all over Europe are characterized by admixtures of siliceous sponges and spicules. Rat (1959) and Pascal (1985) pointed to the geology of the Vasco-Cantabrian Basin of N Spain where siliceous sponges abound in Aptian/Albian carbonate reefs (Urgonian Facies), they are less frequent in the Cenomanian platforms (Reitner 1986, p. 192, Fig. 3; 1987a, p. 215; 1987b; Reitner and Engeser 1987, p. 17; Koch and Reitner 1988). In the "fosse vocontienne" siliceous sponges occur (Lagneau-Hérenger

1962, p. 218). In the Penninic "Rechnitzer Schiefergebirge", Austria, pyritized siliceous sponge spicules, presumably of Lower Cretaceous / Upper Cretaceous age, were observed (Schönlaub 1973). The Albian of the Carpathian Flysch in S Poland contains bluish spongolitic cherts and gaize beds (Lgota Beds) within the Veřovice black shales (Sokołowski 1976, p. 667, Table 53). The spiculitic beds are associated with shales rich in organic matter, high in Mn (up to 1.06%), and rich in Mo and Cu (Gucwa and Wieser 1980, pp. 40/41). Some of these shales are potential hydrocarbon source rocks. Intense submarine volcanism, occurring in the E Alps and E Carpathians during Albian/Cenomanian, is supposed to have furnished silica, directly or indirectly, through the decay of volcanic debris. The thick-bedded, mainly fine-grained graded gaize sandstones of the Polish Carpathians, were formed through erosion of sponge biostromes.

Upper Cretaceous sediments of Europe are generally more marly to calcareous than those of the Lower Cretaceous. The classical chalk (Schreibkreide) in Denmark, Sweden, and northernmost Germany (Ruegen Island) contains siliceous sponge spicules, besides radiolarians and diatoms (Kayser 1924, p. 183).

In the Chalk interval (Cenomanian-Campanian) of Great Britain siliceous sponges and their detached spicules are abundantly represented, but not so well preserved (Hinde 1880; Reid 1968).

In France, siliceous sponges and sponge spicules have been reported from the Cenomanian-Maastrichtian (Cayeux 1929, p. 288; Lagneau Hérenger 1959; 1967; Defretin-Lefranc 1960), especially the "spongolithes du Campanien de Langeais". In Belgium spongolitic gaizes are known as "Têtes de chat" of Turonian age and "Smectique" of Campanian age (Cayeux 1929, p. 270, Plate 16, Figs. 58,61).

In NW Germany siliceous sponges and sponge spicules are widespread, minor or common biogenic constituents in sandstones as well as in shallow marine light-colored limestones, marlstones and claystones of Cenomanian, late Turonian, Coniacian, Santonian, and Campanian age and in solid flint (Feuerstein) nodules (Schrammen 1910, 1912a,b, 1924; Prescher 1959, p. 80; Nestler 1961; H. Jordan 1974, p. 27).

In the transgressive Campanian, sponge development climaxed again. Spongolitic marlstones are known from the Hanover area (Ernst and Schmid 1979, p. 59), from the Münsterland (Klassen 1984, p. 450; Gasse et al. 1988a,b), and from neritic marls, partly spongolitic, and litoral detrital limestones of the Subhercynian Cretaceous Trough (Ulbrich 1974).

In S Germany the Turonian and Cenomanian around Regensburg is known to contain spiculitic sandstones and marlstones (Oschmann 1958, pp. 88–89, Pl. 3, Fig. 1; Wagner 1963a,b). Also the sandy to calcareous Upper Cretaceous of the Polish Lowlands is characterized by spongolites and spongolitic gaizes at the base of the sandy Cenomanian (Sokołowski 1976). Hurcewicz (1966; 1968) reviewed the Upper Cretaceous siliceous sponge faunas of Poland with maxima in Campanian and Maastrichtian marls. In a phosphatized greensand of Maastrichtian age siliceous sponges are abundant (Machalski and Walaszczyk 1987).

The Upper Cretaceous of the Tethyan realm contains siliceous sponges and sponge spicules. However, occurrences like that of a spiculitic cherty limestone

(Zementmergel-Gruppe) of Coniacian-Maastrichtian age (Hagn 1955, Pl. 36, Fig. 1) are rare. In the Carpathian Flysch of Poland, siliceous shales and marlstones are abundant in the lower Upper Cretaceous; spongolites and siliceous sponge spicules occur mainly in the Cenomanian and Turonian (Sokołowski 1976, p. 660, Table 53). As in the Albian, the biosiliceous rocks are associated with manganese nodules (Burtan and Turnau-Morawska 1978).

Upper Cretaceous rudist reefs and reefoid structures (Cenomanian-Campanian/Maastrichtian), 1000–2000 m thick, and characterized by Hippurites and Radiolites as main reef-building organisms, form a continuous belt at the northern flank of the Tethys from Portugal, Spain, France, Switzerland to Austria. Siliceous sponges occur still as minor reef-building organisms (Bouroullec and Deloffre 1974, Tables 2,3,4, and 5). They do not reach the importance of siliceous sponges in late Jurassic time. In the Vasco-Cantabrian Basin and in the Estella Basin of N Spain, siliceous sponges were observed in carbonate rock series from Coniacian to Upper Campanian age (Engeser 1985, p. 127; Schwentke and Wiedmann 1985, p. 46).

### *Cenozoic*

Tertiary Siliceous sponge occurrences seem to be concentrated in Paleocene-Eocene and Oligocene-Miocene and younger deposits.

In the Paleocene of Denmark, characterized by pyroclastic sedimentary components, "siliceous sponges are common in all parts of the series, rarest in the Kerteminde Marl. Usually the sponge spicules are agglomerated in separate layers which afterwards have been silicified. They are preserved in the form of opal or chalcedony" (Gry 1935, pp. 145/146).

The volcanogenic Moler Formation of Denmark contains sediments interstratified with siliceous sponge remains, likewise the London Clay in Great Britain.

Characteristic occurrences of siliceous sponge remains, still preserved as opal, have been described from the Paleocene (Landenién) and Lower Eocene (Ypresien) of France (Cayeux 1929, p. 274).

The volcanogenic Lower Eocene Tarras Clay, N Germany, a stratigraphic equivalent of the London Clay and Ypres Clay, contains in phosphorite nodules phosphatized sponge spicules (Paproth and Zimmerle 1980, pp. 89/90).

Layers of siliceous freshwater sponges are intercalated in the highly bituminous Messel Oil Shale of Middle Eocene age (Lutetian). The sponge spicules are 160–250 $\mu$m long and perforated (Heil 1964; Müller et al. 1982). The enclosing oil shale is composed of smectitic clay with minor admixtures of quartz and feldspar, siderite, and phosphate. The opaline spicules are easily dissolved (Fig. 1h), opal CT spherules of 5 $\mu$m in diameter (Fig. 1i) formed authigenically. The environment of deposition of the Messel Oil Shale is lacustrine. The sediment is derived from weathered and decayed pyroclastics of basanite-nephelinite composition having released plenty of silica together with postulated silica exhalations. The spicular layers can be interpreted as incipient spongolites.

Of a similar paragenetic association, comparable and coeval with the Messel deposits, are siliceous sponges from the famous fossil deposit of Bolca, Italy (Malaroda 1954; Menin 1972). Siliceous freshwater sponges also abound in the volcanogenic Paleogene of the ČSFR (Obrhelová and Obrhel 1987, p. 710).

Sponge spicules are abundant in the Paleogene of the S Ural (Koltun 1960). In the Caucasus, Eocene flysch deposits contain individual suites of siliceous rocks including spongolites (Nalivkin 1973, p. 620).

In the late Tertiary, siliceous sponge occurrences are rather sparse. Siliceous sponges occur in Middle Oligocene marly claystones of NW Germany (Engel 1961). Upper Miocene freshwater sponge spicules were observed in the Fichtelgebirge and in the diatomites from Beuern near Giessen, Rhön region (Reul 1954). Recent freshwater sponges live in shallow lakes (in part Schwarzwässer) and in creeks of low current velocity. So-called freshwater "spongo-diatomites", thoroughly opaline, rich in organic matter, and resembling paper shales, occur in the Oligocene (Aquitanian) of the Puy-de Dôme area, France (Cayeux 1929, pp. 391/392).

Ivanik and Maslun (1977) and Gluschko et al. (1980, p. 97) referred to siliceous sponge remains which abound in the Oligocene siliceous menilite shale of shallow-marine origin of the Carpathians, USSR. Spongolites are also widespread in the Oligocene of the Caucasus (Ruchin 1958, p. 100).

In Spain, excavating sponges occur in deeper water reef mounds of middle Eocene age (Keupp and Reitner 1988). Siliceous sponges in the Miocene (Tortonian) of the Sorbas basin, Andalucia, are closely associated with tectonic activities (Ott d'Estevou and Termier 1978; Ott d'Estevou et al. 1981). Miocene (Tortonian) and Pliocene siliceous sponge faunas have been found in Italy (Malfatti 1900; Rovereto 1901).

Early diagenesis of siliceous sponge limestones from the Pleistocene of the Tyrrhenian Sea was described by Brachert et al. (1987).

Freshwater sponges show a wide geographic and stratigraphic distribution (e.g., Messel, Bolca, ČSFR and SE France) during Tertiary time. And siliceous sponge spicules climax markedly in the Paleocene/Eocene; they occur also in the Upper Oligocene (Hiltermann 1984, Table 1). These two culminations are approximately coeval with the peaks of volcanic activity in central and eastern Europe.

## Diagenesis of Opaline Sponge Spicules

Diagenetic processes affecting opaline sponge spicules comprise (1) dissolution of opal, (2) transformation of unstable opal into more stable silica modifications such as opal-CT, chalcedony, and quartz, and (3) partial or complete replacement of the unstable opal by the adjacent sedimentary rock matrix or by other minerals.

Dissolution and dissolution processes were discussed by numerous authors (Hinde 1887–1912, part III, p. 192; Cayeux 1929, p. 302ff.; H. Jordan 1968, p. 418; Kamatani 1971; R. Jordan 1974, pp. 26/27; Land 1976; Eisenack 1978, p. 651; Martini and Rietschel 1978; Mann 1979, p. 110; Wiedenmayer 1980b,

p. 108ff.; Flügel 1982, p. 296; Narbonne and Dixon 1984, p. 96; Lang and Steiger 1984; Taguchi et al. 1986; Reitner 1987a,b; Reitner and Engeser 1987; Brachert 1988).

In Recent porous fore-reefs of Jamaica Land (1976) observed early dissolution of opaline sponge spicules which is more intense than in the pelitic back-reef sediments. Likewise Reitner (1986, Fig. 3) observed early diagenetic dissolution of siliceous sponge scleres by marine phreatic waters in Late Albian diapir-influenced reef atolls from the Vasco-Cantabrian Basin, N Spain.

Dissolution of a freshwater sponge is shown in Fig. 1h,i. Bacterial lithification may stabilize the partially dissolved sponge.

In its final stage dissolution leads to casts and molds (Hansen et al. 1960, p. 7; Rigby 1986, p. 5). The dissolution of microscleres leaves pinhole-like cavities such as those observed in the Kimmeridgian Heersum Beds. Complete diagenetic dissolution of siliceous sponges may be the cause for the apparent underrepresentation of fossil sponges in ancient sediments, especially in reefs. This trend is also evidenced by the fact that biosiliceous fossils are well preserved in phosphate nodules, but not in the surrounding shales, e.g. in the Lower Carboniferous (Zimmerle et al. 1980, p. 42), in the Middle Jurassic (Riegraf 1986) or in the Lower Eocene (Paproth and Zimmerle 1980, 90).

Dissolution of opaline sponge spicules might generate secondary porosity and thus produce potential hydrocarbon reservoirs, comparable with the reservoirs caused by dissolution of framework chert beneath an unconformity at the prolific North Slope hydrocarbon province, Alaska (Shanmugam and Higgins 1988).

Silica derived from sponge remains through complete dissolution or replacement does often silicify the enclosing host rock anew (R. Jordan 1974, p. 26; Füchtbauer 1974, p. 233; Wiedenmayer 1980b, p. 110). Flint nodules are claimed to have been formed through the dissolution of biogenic silica.

Transformation from unstable silica into stable silica modifications was described in detail by Dapples (1967, pp. 115–116), v. Rad and Rösch (1974), Flörke (1976), and Mann (1979, pp. 137–143, Pl. 3,6,7). Silica diagenesis of sponges normally proceeds in Cenozoic sediments as a discontinuous age-, burial-, and facies-dependent maturation process from the instable opal-A via metastable opal-CT to stable quartz. An abrupt reduction in porosity is associated with the different silica phase transformations (Isaacs 1981).

Opaline silica of sponge spicules is commonly replaced by the pelitic host sediment and by minerals such as calcite, pyrite, glauconite, phosphate, zeolite, barite, and celestite. Pelitic host sediments, claystones, shales, or calcilutites can occasionally replace siliceous sponges immediately after deposition (Wiedenmayer 1980b, p. 109).

Replacement by calcite is widespread and frequent (Langneau-Hérenger 1962, p. 22; Menin 1972, p. 65; Wiedenmayer 1980b, p. 109; Zimmerle 1982b, p. 186, Pl. 4.7–8; Lang and Steiger 1984, p. 96). Often pyrite replaces siliceous sponge spicules early in diagenesis. (Langneau-Hérenger 1962, p. 233; Wiedenmayer 1980b, p. 112; Flügel 1982, p. 295; Engeser 1985, p. 169; Schwentke and Wiedmann 1985, p. 47; Riegraf 1986; Rigby 1986, p. 5; Reitner

1987b, p. 219; Reitner and Engeser 1987, p. 17; Wuttke and Kott 1987, p. 54; Brachert 1988; Steiger 1988). Glauconite starts normally replacing opaline silica from the central axial channel (Carozzi 1960, p. 51; Eisenack 1978, p. 650ff., Figs. 29–33).

Early phosphatization of siliceous sponges and sponge spicules is widespread. It leads to good preservation of the internal fossil structures (Fig. 1e). Phosphatization of siliceous sponge was described by numerous authors (Cayeux 1929; Pożaryski 1947, p. 48; Carozzi 1960, pp. 389–390; Lagneau-Hérenger 1962, p. 23; Wiedenmayer 1980b, p. 112; Zimmerle 1982b, p. 181; Riegraf 1986).

Partial replacement by clinoptilolite, similar to the replacement in radiolarians, was observed in Aptian/Albian siliceous sponges of NW Germany (Zimmerle 1982b, p. 182). Late diagenetic replacement by barite was reported by Riegraf (1986) and by celestite by v. Bülow and Schulz (1978).

Diagenetic alteration features of sponge spicules such as translucency, surface textures, and crystallinity with increasing depth of burial were described from volcanogenic Tertiary sediments of the Shinjo Oil Field, NE Japan (Taguchi et al. 1986). Minerals like clinoptilolite, mordenite, and cristobalite are associated only with opal-A sponge spicules.

Observations on the trace element content of siliceous sponges are sparse. Ruzhentsev (1985) found that Jurassic-Cretaceous siliceous deposits of fold belts and oceans have locally a higher content of metals such as iron and manganese, i.e. a high Fe-Mn-Ti modulus, and relatively high concentrations of Ba, Cu, Pb, Zn, Ge, Mo, and Ni. Also uranium and/or phosphorus are associated. Gucwa and Wieser (1980) showed that spongolites may not only contain a high Mn content (up to 1.06%), but also high V and Cu. Germanium affects silica deposition in sponges (Simpson 1981).

Silica of opaline biogenic components like sponge spicules is primarily sourced from land-derived weathering products brought into a basin by rivers either as particulate matter or in solution, from halmyrolytic weathering on the ocean floor or lake ground, and/or from direct volcanogenic exhalations on the seafloor. Furthermore, biogenic silica can be recycled.

According to R. Jordan (1974, pp. 26, 27) natural brines facilitate the decomposition of clay minerals on the seafloor, and thus lead first to the additional release of silica and second to the preservation of biosiliceous components as shown in the DSDP wells in the Gulf of Mexico (Sarnthein, pers. commun.).

Deep-seated faults and fault zones with ascending hot and aggressive brines may enhance the diagenetic dissolution and alteration of siliceous sponge-bearing sediments. On the basis of Eh and pH Teodorovich (1954) assigned for sediments containing siliceous sponge spicules a pH range of 5.0 to 8.0 for weakly oxidizing and 6.6 to 7.2 for oxidizing zones.

Perfect preservation of sponges and their remains in ancient sedimentary rocks is the exception; it seems to be enhanced by early silicification (Whittington 1980, p. 132; Wuttke and Kott 1987).

## Discussion

Accumulations of siliceous sponges and sponge spicules are found in different lithologies and environments of deposition: in bioherms and reefs, in inter-reef muds, in shelf sandstones, and in pelagic pelites comprising claystones, marlstones, siliceous shales, and cherts, often in condensed sections.

Lower Carboniferous cherts with sponge spicules are a good example of a tuffo-siliceous rock association occurring in ancient geosynclinal belts (Burlin et al. 1976). Wiedenmayer (1980b, p. 109) claimed that the most extensive spiculitic deposits occur in stagnant basins of all ages. His claim has to be taken with caution because the Upper Jurassic sponge bioherms did not form in stagnant basins; the development of siliceous sponges culminated during Middle/Upper Jurassic and Cretaceous times.

Spongolitic sediments are commonly associated with following mineral and rock parageneses: smectitic clays, poorly crystallized silica, zeolites such as clinoptilolite or analcite, glauconite, phosphates, and volcanogenic sediments.

Spatial association and genetic interrelation between sponge-bearing sediments or spongolites and volcanic activity can be observed in many sedimentary sequences (Wiedenmayer 1980b, p. 109). In some, however, this interrelation is not so obvious.

An instructive case history in which there is no direct hint at the derivation of an abundant silica supply for sponge growth is the algal-sponge bioherm/reefs of late Jurassic age in the Swabian-Franconian Alb. Numerous studies have dealt with paleontology and sedimentology, a few only with the regional geology and other geological aspects. Beurer (1971, p. 42) having studied flint concretions ("Kieselknollenkalke") referred to the origin of the silica: "Kieselsäurereiche, vulkanische Exhalationen als Kieselsäurequelle stehen für den oberjurassischen Sedimentationsraum Schwabens nicht zur Diskussion. Für die Herkunft der Kieselsäure aus halmyrolytischen Vorgängen oder diagenetischen Tonumwandlungen ergaben die Untersuchungen keine Anhaltspunkte". Since then, our knowledge of the regional geology has advanced. Hints at Europe-wide volcanic activity, partly rather concealed, during Oxfordian/Kimmeridgian reef growth have become known recently:

– a glauconite horizon with a volcanic fingerprint, the so-called Glaukonitbank at the $\delta_3/\delta_4$ Malm boundary (Knoblauch 1963, p. 63ff.; Knoblauch and Weiskirchner 1979).
– an unusual smectite content in the Solnhofen Lithographic Limestone (Bausch 1980, p. 21).
– peculiar lithology of highly siliceous, bituminous dark marlstones between the reefs.

The unstable tectonic setting during Oxfordian/Kimmeridgian was associated with volcanic activity in the Tethys (Diersche 1980, pp. 142–147; Nalivkin 1973, pp. 664–667) and with rift volcanism on the NW European platform. Thus, Beurer's statement has to be revised.

On the other hand, siliceous freshwater sponge spicules are found in very close association with volcanic debris and diatoms in thin silt-size ash layers deposited in Quaternary lakes of the High Plain of Bogotá (Riezebos and Zimmerle 1989).

In the Upper Oxfordian (Corallian) of England small coral bioherms were smothered by volcanic ash falls which later were transformed into clay layers (Ali 1977). Corallian Beds hosting *Rhaxella perforata* also contain admixtures of volcanic detritus (Chowdhury 1982; Milodowski and Wilmot 1984). Smothering by ashes could easily be responsible also for extinguishing the major reefs in Permian (see Reinhardt 1988), Triassic, and Jurassic times elsewhere, heralding the so-called reef crises. However, bentonized clay layers, "geologically aged" into other clay minerals, are extremely difficult to identify (Chowdhury 1982, p. 492). Moreover, vestiges of hydrothermal activity or volcanic centers are difficult to locate. Mostly, volcanic centers have not been preserved and their existence may be inferred only from the sediments containing volcanic material (Khvorova 1978). Additionally, volcanogenic silica is undoubtedly "mixed with silica of other origins and "deprived of individuality" in the dynamic reservoir of the world ocean" (Burlin et al. 1976, p. 12).

The Upper Cretaceous silicisponge fauna is one of the most prolific and diverse in earth history. These optimal conditions for sponge formation are again accompanied by vestiges of cryptovolcanic activities throughout the Upper Cretaceous of Europe (Zimmerle, 1989).

Spiculitic rocks are either rich in organic matter or they are associated with sediments rich in organic matter up to 30% $C_{org}$ (Gluschko et al. 1980; Wiedenmayer 1980a, p.. 109; Bazhenova and Burlin 1983). In Turonian spongolitic siliceous marls of the Polish Carpathian Flysch siliceous flagellates, blue algae, and sponges are "the basic substratum of kerogen and bitumens" (Gucwa and Wieser 1980).

The main phase of petroleum generation in siliceous rock sequences occurs with incipient catagenesis. Diagenetic transformation comprising the formation of microstructure and recrystallization leads also to the early formation of hydrocarbon reservoirs such as in Jurassic argillaceous silicites of W Siberia and Japan, in Neogene silicites of the Carpathians (Menilite Shale), E Sakhalin and NE Japan, and in the diatomites of the famous Monterey Formation in California (Burlin 1982; Bazhenova and Burlin 1983; Taguchi et al. 1986). Tertiary siliceous shales, especially the so-called menilite shales of E Europe, are the major hydrocarbon source rock in the Carpathian oil province. They are characterized by volcanic debris and fragments of siliceous sponges (Gluschko et al. 1980 and Burlin 1982). Comparable oil- and gas-producing siliceous rocks occur in Sakhalin, Taiwan, and Japan. For instance, siliceous shales from the tuffaceous-diatomaceous Kirasi sequence of SW Sakhalin are composed of opaline diatom frustules and sponge spicules (30–70%) as well as of volcanic glass (up to 20%). Such rocks, e.g. the Monterey Formation, California, may constitute both source and reservoir beds, especially in fractured late Mesozoic and Cenozoic sequences.

With increasing recrystallization of the silica, specific inner surface, adsorption capacity, and catalytic effects decrease.

The culminations of siliceous sponge developments, namely in the Lower Carboniferous and in the Jurassic/Cretaceous, fall into the range of so-called episodes (North 1980, p. 323ff) which are characterized by (1) strong deformation and extensive tectonic phases, (2) major transgressions, (3) increased metamorphism and metallogenesis, and (4) augmented magmatic and volcanic activity. The famed oil source sediments of the North Sea, central Europe, and W Siberia are black shales of Toarcian and Kimmeridgian age.

From the viewpoint of plate tectonics (Siever 1979) sponge-bearing sediments are mainly found in the Intra-Plate (Craton) and Rift Valley environments; the presence of siliceous sponges strongly affects the diagenetic evolution of these sediments.

The association of siliceous sponge deposits and volcanogenic sediments is well established in numerous examples. Thus, it might be justified in cases of older Paleozoic rocks in which only sponges have survived to a marked extent, such as in the Middle Cambrian Burgess Shale or in the Lower Devonian Hunsrück Slate to postulate some concealed volcanic interference for these deposits, even if the purely mineralogical evidence like minerals and rock fragments diagnostic of volcanic derivation is completely extinguished by thorough diagenetic alteration. The same argumentation may apply for the Chattian of NW Germany where only sponge spicules are found (Hiltermann 1984, Table 1), but barely sedimentary traces of volcanic activity (Zimmerle et al. 1990).

In summary, there is a whole bundle of known as well as unknown factors controlling the growth and fossilization of siliceous sponges; but siliceous sponges definitely preserve an environmental signal as to the abundant supply of silica and other peculiarities of water chemistry, water temperature, current regime, climate, and metabolism. The high concentration of silica in sea water is certainly not the only controlling factor, but its presence seems to have a favorable influence. Moreover, Pleistocene sediments from the SW Pacific show an interrelation between atmospherically transported volcanic dust and the "diversity" of siliceous microfauna, supporting the hypothesis of volcanic enhancement of organic activity probably the increase of corresponding micronutrients and trace metals (Huang et al. 1974).

## Conclusions

Ancient deposits of siliceous sponges and/or siliceous sponge spicules are commonly stratiform, of limited thickness (mm to m dimensions), but of wide lateral extent (e.g., Oxfordian, Albian, Campanian, and Eocene of Europe). In bioherms and reefs, their distribution is patchy.

The growth of solitary siliceous sponges does not require an increased supply of silica; silica dissolved in sea water appears to be sufficient. The prolific production of siliceous sponges in the past, however, would seem to have required more specific conditions, especially for the formation of spongolites.

In the Phanerozoic siliceous sponge spicules and spongolites are not evenly distributed throughout the stratigraphic column. There are stratigraphic intervals in which they predominate, presumably indicating a higher supply of silica. The widespread lateral extent of sponge occurrences, e.g. in the Upper Jurassic of the Tethys, in the Lower and Upper Cretaceous of Europe, reflects a uniform water chemistry and free water exchange.

The high supply of silica for building sponge skeletons is thought to be derived from decaying volcanic glass particles and pyroclasts as well as from concealed silica exhalations. Times of high silica supply in the past are mostly times of increased volcanic activity, in part of the cryptovolcanic type. Jurassic and Cretaceous are times of cryptovolcanic phenomena, i.e. the volcanic influence is less evident than in other periods of earth history (Zimmerle 1985, 1989).

Siliceous sponge-bearing sediments are commonly characterized by the presence of minor amounts of amorphous and/or cryptocrystalline silica, zeolites, and glauconite. Present and ancient siliceous sponges are very sensitive to diagenetic dissolution and alterations. During diagenesis the opaline silica of sponge skeletons is transformed, generally with age and depth of burial, into opal CT, chalcedony, and quartz. Moreover, pyritization and phosphatization of sponge spicules – and thus a higher preservation potential – is common. Partial replacement of opaline silica by clinoptilolite is less frequent.

Brines seem to enhance preservation of Recent sponges. R. Jordan (1974) stressed the importance of fossil brines in diagenesis. Complete replacement of siliceous sponge skeletons by the enclosing sediment or their complete dissolution is widespread (Rigby 1986; Lang and Steiger 1984, p. 96ff.). It is likely that siliceous sponge spicules and fragments, unstable under certain physico-chemical conditions, have been entirely dissolved in older rocks such as in the Paleozoic.

Sedimentary rocks with siliceous sponge spicules, especially spongolites, show extraneous petrophysical properties, namely primary as well as secondary porosities, low permeabilities, and large inner surface areas. As such some of them are suited as reservoirs for water or hydrocarbons. The source rock potential, however, is usually low because the organic matter content is below 1% $C_{org}$.

For the manufacture of light-weight building material Turonian spongolitic silty rocks are exploited (Svoboda et al. 1966, p. 511).

*Acknowledgments.* Dr. E. Kemper, Bundesanstalt für Geowissenschaften und Rohstoffe, and F. Elstner, BEB Erdgas und Erdöl GmbH, both Hannover, furnished valuable information on early Cretaceous siliceous sponge occurrences in the Federal Republic of Germany. BEB Erdgas und Erdöl GmbH, Hannover, permitted the publication of the oral communication by F. Elstner. Prof. Dr. H. Keupp and Dr. J. Reitner critically read the manuscript and gave helpful advice. To all I extend my sincere thanks.

## References

Alexandrowicz SW (1978) Sponge spicules from the Lower Carboniferous of the Olkusz area. Bull Acad Pol Sci, Sér Sci Terre 26:87–94

Ali OE (1977) Jurassic hazards to coral growth (Correspondence). Geol Mag 114:63–64

Arkell WJ (1933) The Jurassic System in Great Britain. Clarendon, Oxford, p 681

Bausch W (1980) Tonmineralprovinzen in Malmkalken. Erlanger Forschungen, Reihe B, Naturwiss Med 8: p 78

Bazhenova OK, Burlin YK (1983) Geochemistry of sedicahytes and petroleum generating potential of siliceous sediments (Poster). 11th Int Meeting on Organic Geochemistry, The Hague, Sept. 12-16, 1983, p 105

Beurer M (1971) Kieselsäureanreicherungen in den oberjurassischen Sedimenten der Schwäbischen Alb. Beih Geol Jahrb 109: p 69

Boda J (1964) Catalogus Originalium Fossilium Hungariae. Pars Zoologica M All Földtani Intézet, Budapest, p 229

Bouroullec J, Deloffre R (1974) Essai d'utilisation statistique des microfacies dans le Crétacé Moyen d'Aquitaine occidentale (France). Bull Cent Rech Pau 8:469-509

Brachert T (1988) Environment controls siliceous sponge fossilization (Abstract). In: Keupp H, Reitner J (eds) Fossil and Recent Sponges Int Conf Berlin 1988, Sept. 26-28. Berl Geowiss Abh (A) 10:7

Brachert TC, Dullo W-Chr, Stoffers P (1987) Diagenesis of siliceous sponge limestones from the Pleistocene of the Tyrrhenian Sea (Mediterranean Sea). Facies 17:41-50

Burlin JK (1982) Über die Erdölführung der Kieselgesteine. Izv Akad Nauk Cerija Geol 8:134-142 (in Russian)

Burlin JK, Bazhenova OK, Karnyushina EE, Konyukhov AI (1976) The problem of oil formation in siliceous rocks of geosynclinal areas. Vestnik Moskovskogo Universiteta. Geologiya 31:12-24 (English translation)

Burtan J, Turnau-Morawska M (1978) Biochemiczne skały krzemionkowe zachodnich karpat Fliszowych. Pol Akad Nauk, Oddziat w Krakowie Komisja Nauk Geologicznych, Pr Geol 111: p 31

Bustillo MA, Ruiz-Ortiz PA (1987) Chert occurrences in carbonate turbidites: examples from the Upper Jurassic of the Betic Mountains (southern Spain). Sedimentology 34:611-621

Carozzi AV (1960) Microscopic sedimentary petrography. Wiley, New York, p 485

Cayeux L (1929) Les roches sédimentaires de France. Roches siliceuses. Mém Carte Géol Fr, Imprimerie Nationale, Paris, p 774

Charollais J, Rietschel S, Rosset J (1967) Sur un gisement d'eponges d'âge hauterivien-barrémien dans la chaine des Aravis (Haute-Savoie). Trav Lab Géol Grenoble 43:63-69

Chowdhury AN (1982) Smectite, zeolite, biotite and apatite in the Corallian (Oxfordian) sediments of the Baulking area in Berkshire, England. Geol Mag 119:487-496

Dapples EC (1967) Diagenesis of sandstones. In: Larsen G, Chilingar GV (eds) Diagenesis in sediments, Developments in sedimentology, vol 8. Elsevier, Amsterdam, 91-125

Defretin-Lefranc S (1960) Contribution à l'étude des spongiaires siliceux du Crétacé supérieur du Nord de la France. Thèse Fac Sci Lille, p 178

De Laubenfels MW (1955) Porifera. In: Moore RC (ed) Treatise on Invertebrate Paleontology. Part E, Archaeocyatha and Porifera. Geol Soc Am Lawrence, Kansas, E21-E112

Diersche V (1980) Die Radiolarite des Oberjura im Mittelabschnitt der Nördlichen Kalkalpen. Geotekt Forsch 58: p 213

Doubinger J, von Eller J-P (1963) Présence de spongiaires dans les schistes précambriens métamorphiques des Vosges. Bull Serv Carte Géol Als Lorr 16:111-123

Eisenack A (1978) Beitrag zur Glaukonit-Forschung. Neues Jahrb Geol Paläont Monatsh, 641-656

Engel G (1961) Fund eines hexactinelliden Schwammes im Oligozän des Emslandes. Ber Naturhist Ges Hannover 105:7-10

Engeser T (1985) Sedimentologische, fazielle und tektogenetische Untersuchungen in der Oberkreide des Basko-kantabrischen Beckens (Nordspanien). Naturwiss Diss, Tübingen 15.02. 1984, p 235

Ernst G, Schmid F (1979) Die Oberkreide des Raumes Braunschweig-Hannover unter besonderer Berücksichtigung des Cenoman und Turon. Exkursionsführer Geländetagung "Kreide Niedersachsen" der Subkommission für Kreide-Stratigraphie 1.-5. Mair 1979, Berlin Braunschweig Hannover, p 75

Esteban Cerdá M, Calzada S, Via Boada L (1977) Ambiente deposicional de los yacimientos fosiliferos del Muschelkalk Superior de Alcover – Mont Ral. Cuard Geol Ibérica 4:189-200

Faupl P, Beran A (1983) Diagenetische Veränderungen an Radiolarien- und Schwammspicula-führenden Gesteinen der Strubbergschichten (Jura, Nördliche Kalkalpen, Österreich). Neues Jahrb Geol Paläont Monatsh 129-140

Flörke OW (1976) Intergrowth and twinning in opal-CT lepispheres. Contrib Mineral Petrol 58:235-242

Flügel E (1982) Microfacies analysis of limestones. Transl of a rev and expanded version of: Mikrofazielle Untersuchungsmethoden von Kalken. Springer, Berlin Heidelberg New York, p 633
Fritz GK (1958) Schwammstotzen, Tuberolithe und Schuttbreccien im Weißen Jura der Schwäbischen Alb. Arb Geol Paläont Inst TH Stuttg NF 13: p 118
Füchtbauer H (1974) Sediments and sedimentary rocks. 2nd revised and enlarged edition. Schweizerbart, Stuttgart, p 464
Gaillard C (1983) Les biohermes à spongiaires et leur environment dans l'Oxfordien du Jura Méridional. Doc Lab Géol Lyon 90, p 515
Gasse W, Göcke R, Hilpert K-H (1988a) Oberkretazische Spongien des NW-Münsterlandes – ein Überblick. Beitr Elektronenmikroskop Direktabb Oberfl 21:385–396
Gasse W, Göcke R, Hilpert K-H (1988b) Hexactinellida and Lithistida from the Campanian of the NW Münsterland with a special note to the genus *Becksia* Schlüter, 1868 (Upper Cretaceous, NW Germany) (Abstract). In: Keupp H, Reitner J (eds) Fossil and Recent Sponges Int Conf Berlin 1988, Sept 26–28. Berl Geowiss Abh (A) 10:14–15
Georgi K-H (1976) Mikrofaunistisch-lithologische Untersuchungen der Hilssandstein-Region (Apt/Alb) im Raum Salzgitter-Goslar. Mitt Geol Inst Tech Univ Hannover 13:5–112
Geyer OF (1962) Über Schwammgesteine (Spongiolith, Tuberolith, Spiculit und Gaizit). In: Geyer OF (ed) Hermann-Aldinger-Festschrift. Schweizerbart, Stuttgart, 51–59
Gluschko VV, Maksimov AV, Reifman LM (1980) Besonderheiten der Karpaten-Entwicklung im Oligozän. Sow Geol 12:89–100 (in Russian)
Gramann F (1962) Schwamm-Rhaxen und Schwamm-Gesteine (Spongiolithe, Spiculite) aus dem Oxford NW-Deuschlands. Geol Jahrb 80:213–220
Greensmith JT (1978) Textbook of petrology: 2 Petrology of the sedimentary rocks. 6th edn completely rev, Allen and Unwin, London, p 241
Grigjalis AA, Mertinene RA, Akimez WS, Gluschko WW, Wischnjakow IB, Poluchtowitsch BM, Senjkowski JN, Gawrilischin WI, Pasternak SI (1978) Paläogeographie des Albs auf dem westlichen Territorium des europäischen Teils der UdSSr. Z Angew Geol 24:451–456
Gry H (1935) Petrology of the Paleocene sedimentary rocks of Denmark. Dan Geol Unders Raekke 2, 61: p 171
Gucwa I, Wieser T (1980) Geochemia i mineralogia skat osadowych Fliszu Karpackiego zasobnych w materie organiczną. Geochemistry and mineralogy of sedimentary rocks of the Carpathian Flysch rich in organic matter. Polska Akad Nauk, Oddzial W Krakowie, Pr Mineral 69: p 42
Gwinner MP (1958) Schwammbänke, Riffe und submarines Relief im oberen Weißen Jura der Schwäbischen Alb (Württemberg). Geol Rundsch 47:402–418
Hagn H (1955) Fazies und Mikrofauna der Gesteine der bayerischen Alpen. Int Sediment Petrogr 1: p 174
Hansen K, Heller E, Larsen G (1960) Sedimentary and economic geology of Denmark. Int Geol Congr, XXI Session, Norden 1960, Guide to excursions Nos A 47–48 and C 42–43, p 54
Heil R (1964) Kieselschwamm-Nadeln im Ölschiefer der Grube Messel bei Darmstadt. Notizbl Hess Landesamt Bodenforsch 92:60–67
Hey RW (1956) Cherts and limestones from the Crow Series near Richmond, Yorkshire. Proc Yorksh Geol Soc 30:289–299
Hiltermann H (1984) Tertiär. In: Klassen H (ed) Geologie des Osnabrücker Berglandes. Naturw Museum, Osnabrück, 463–496
Hinde GJ (1880) Fossil sponge spicules from the upper chalk. Found in the interior of a single flint-stone from Horstead in Norfolk. Phil Diss Munich 1880: Wolf, p 83
Hinde GJ (1886) On beds of sponge-remains in the Lower and Upper Greensand of the South of England. Philos Trans R Soc Lond 176:447–448
Hinde GJ (1887–1912) A monograph of the British fossil sponges vol 1, Sponges of Palaeozoic and Jurassic strata 1887–1912, Palaeontol Soc Lond: p 264
Huang TC, Fillon RH, Watkins ND, Shaw DM (1974) Volcanism and siliceous microfaunal "diversity" in the southwest Pacific during the Pleistocene period. Deep Sea Res 21:377–383
Hurcewicz H (1966) Siliceous sponges from the Upper Cretaceous of Poland. 1 Tetraxonia. Acta Paleontol Pol 11:15–129
Hurcewicz H (1968) Siliceous sponges from the Upper Cretaceous of Poland. 2 Monaxonia and Triaxonia. Acta Paleontol Pol 13:3–96

Hurcewicz H (1982) Permian sponges from brachiopod cherts at Hornsund, Spitsbergen. Acta Paleontol Pol 27:85–114

Hurcewicz H (1985) Gabki z osadów franu w Wietrzni i kowali w Górach Świetokrzyskch. Kwartalnik Geol 29:271–300

Hurcewicz H, Czarniecki S (1985) Lyssakidae sponges from the Carboniferous Limestone and the Culm of southern Poland and their environmental differentiation. Ann Soc Geol Pol 55:333–354

Isaacs CM (1981) Porosity reduction during diagenesis of the Monterey Formation, Santa Barbara coastal area, California. In: Garrison RE, Douglas RG (eds) The Monterey Formation and related siliceous rocks of California. Society of Economic Paleontologists and Mineralogists, Los Angeles, 257–271

Ivanik MM, Maslun NV (1977) Kremnevye mikroorganizmy i ich ispol'zovanie dlja rasčlenenija paleogenovych otloženij Predkarpat'ja (Kieselige Mikroorganismen und ihre Benutzung für die Gliederung des Paläogen des Karpatenvorlandes). Naukova Kiev dumka

Jordan H (1968) Gliederung und Genese des Flammenmergels (Alb) in Hils- und Sackmulde (Süd-Hanover). Z Dtsch Geol Ges 1965, 117:391–424

Jordan R (1974) Salz- und Erdöl/Erdgas-Austritt als Fazies bestimmende Faktoren im Mesozoikum Nordwest-Deutschlands. Geol Jahrb A 13: p 64

Jullian Y (1953) Presentation de microfacies jurassiques du Languedoc. C R XIX Sess Congr Géol Int Alger 1952, Sect XIV, fasc XVI, Alger

Kamatani A,(1971) Physical and chemical characteristics of biogenous silica. Mar Biol 8:89–95

Katzer F (1902) Geologie von Böhmen. 2. Aufl I Taussig, Prag, p 96

Kayser E (1924) Lehrbuch der geologischen Formationskunde. II. Band, Jura- Kreide- Tertiär- und Quartärformation. 6 u 7 Auflage. Enke, Stuttgart, p 657

Keller S, Kemper E, Koch R, Mutterlose J, Nöltner T, Owen HG, Spaeth C, Stoffers P, Walter P, Zimmerle W (1989) Das Unter- und Mittelalb von Alfeld (Leine). Geol Jahrb A 113:223–316

Kemper E (1976) Geologischer Führer durch die Grafschaft Bentheim und die angrenzenden Gebiete mit einem Abriss der emsländischen Unterkreide. Das Bentheimer Land Nr 64, 5 ergänzte Aufl, Verlag Heimatverein Grafschaft Bentheim eV Nordhorn-Bentheim, p 206

Kemper E, Zimmerle W (1982) Facies patterns of Cretaceous/Tertiary subtropical and Aptian/Albian boreal upwelling systems (Great Syrian Desert and NW Germany). In: Suess E, Thiede J (eds) Coastal upwelling: Its sediment record. Part B Sedimentary Records of Ancient Coastal Upwelling Proc NATO Advanced Res Inst Vilamoura, Sept 1981. Plenum, New York, p 610

Keupp H, Reitner J (1988) The excavating sponge Aka (Haplosclerida, Demospongiae) within the fossil record (Poster). In: Keupp H, Reitner J (eds) Fossil and Recent Sponges. Int Conf Berlin 1988, Sept 26–28. Berl Geowiss Abh (A) 100, 20

Khvorova IV (1978) Advances and objectives in the study of volcanic-sedimentary rocks. Lithology and mineral resources, 546–552. Engl transl from: Litologiya i Polesne Iskopaemye 5:50–57

Klassen H (1984) Geologie des Osnabrücker Berglandes. Text, Naturwiss Mus, Osnabrück, p 672

Klieber D (1983) Zur Ökologie der Schwammergel im unteren Malm des Dillbergs/Opf. (mittlerer Frankenalb). Geol Bl NO-Bayern 32:130–145

Knoblauch G (1963) Sedimentpetrographische und geochemische Untersuchungen an Weißjurakalken der geschichteten Fazies im Gebiet von Urach und Neuffen. Diss Math Naturw Fakultät Univ Tübingen, p 106

Knoblauch G, Weiskirchner W (1979) Ein Beitrag zur Frage der Sedimentation im Malm der Schwäbischen Alb. Jahresber Mitt Oberrhein Geol Ver NF 61:321–329

Koch R, Schorr M (1986) Diagenesis of Upper Jurassic sponge-algal reefs in SW Germany. In: Schroeder JH, Purser BH (eds) Reef diagenesis. Springer, Berlin Heidelberg New York Tokyo, 224–244

Koch T, Reitner J (1988) Sponge mud mounds from the middle Albian of La Gandarra (northern Spain) (Poster). In: Keupp H, Reitner J (eds) Fossil and Recent Sponges Int Conf Berlin 1988, Sept 26–28. Berl Geowiss Abh (A) 10:21

Koltun VM (1960) Spicule analysis and its use in geology. Izv Akad Nauk SSSR Ser Geol 1960:73–77

Krüger S (1982) Die Makrofossilien des späten Apt und frühen Alb in Nordwestdeutschland. 8.1. Die Porifera des späten Apt in Nordwestdeutschland. Geol Jahrb A 65:499–511

Kubanek F, Zimmerle W (1986) Tuffe und kieselige Tonschiefer aus dem tieferen Unterkarbon der Bohrung Adlersberg (West-Harz). Geol Jahrb D78:207–268

Lagneau-Hérenger L (1959) Les spongiaires siliceux du Crétacé supérieur francais. C R 84[e] Congrès Sociétés Savantes de Paris et des Départments tenu à Dijon, en 1959, Section des Sciences, Sous-Section de Géologie, Colloque sur le Crétacé supérieur Francais, Paris, 399–413
Lagneau-Hérenger L (1962) Contribution à l'étude des spongiaires siliceux du Crétacé inférieur. Mém Soc Géol Fr NS 95: p 252
Lagneau-Hérenger L (1967) Les spongiaires Turoniens du Synclinal d'Eygaliers. Trav Lab Géol Grenoble 43:79–93
Land LS (1976) Early dissolution of sponge spicules from reef sediments, North Jamaica. J Sediment Petrol 46:967–969
Lang B (1987) Die Schwamm-Biohermfazies der Nördlichen Frankenalb (Steinbruch Urspring, Oxford, Malm): Mikrofazies. Palökologie, Paläontologie, Diss Univ Erlangen, p 198
Lang B (1988a) Bakterienkrusten und Sturmsedimente: Ablagerungszyklen oberjurassischer Schwammbioherme (Oxford, Nördliche Frankenalb) (Zusammenfassung). 3. Treffen deutschsprachiger Sedimentologen 23–26 Mai 1988 Bochum. Bochumer Geol Geotech Arb 29:107–108
Lang B (1988b) Baffling, binding or debris accumulation? Ecology of Jurassic sponge-bacterial buildups (Oxfordian, Franconian Alb) (Abstract). In: Keupp H, Reitner J (eds) Fossil and Recent sponges. Int Conf Berlin 1988, Sept 26–28. Berl Geowiss Abh (A) 10:22
Lang B, Steiger T (1984) Paleontology and diagenesis of Upper Jurassic siliceous sponges from the Mazagan Escarpment. Oceanol Acta 1984, Submersible Cyana studies of the Mazagan Escarpment (Moroccan continental margin). Cyamaz cruise 1982, 93–100
Lehmann WS (1954) Die Oberstufe des Oberalbs (Flammenmergel) im subherzynen Becken und im westlichen Harzvorland. Diss TH Braunschweig, p 133
Lowell JD (1968) Upper Palaeozoic and Lower Mesozoic stratigraphy of south-western Nordaustlandet, Spitsbergen. Geol Mag 105:348–355
Machalski M, Walaszczyk I (1987) Faunal condensation and mixing in the uppermost Maastrichtian/Danian Greensand (Middle Vistula Valley, Central Poland). Acta Geol Pol 37:75–91
Malaroda R (1954) Il Luteziano di Monte Postale (Lessini medi). Mem Ist Geol Min Univ Padova 19: p 108
Malfatti P (1900) Contributo alla spongiofauna del Cenozoico italiano. Palaeontogr Ital 6:267–302
Mann U (1979) Mineralogisch-sedimentologische Untersuchungen von Tiefseeablagerungen am ostasiatischen Kontinentalrand unter besonderer Berücksichtigung frühdiagenetischer Veränderungen an biogenem Opal. Deep Sea Drilling Project Leg 56 and Leg 57. Inaug Diss Naturw Gesamtfakultät Ruprecht-Karl-Universität Heidelberg, p 146
Martini E, Rietschel S (1978) Lösungserscheinungen an Schwammnadeln im Messeler Ölschiefer (Mittel-Eozän). Erdoel Erdgas Z 94:94–97
Menin A (1972) Silicosponge nell'Eocene medio della Valle del Chiampo (M Lessini, Vicenza). Univ Ferrara Annali, NS, Sez 9, 5 2:63–69
Meyer RKF, Schmidt-Kaler H (1983) Erdgeschichte sichtbar gemacht – Ein geologischer Führer durch die Altmühlalb. Bayer Geol Landesamt, München, p 260
Milodowski AE, Wilmot RD (1984) Diagenesis, porosity and permeability in the Corallian Beds (Upper Oxfordian) from the Harwell Research Site, South Oxfordshire, UK. Clay Minerals 19:323–341
Mostler H (1971) Häufig Keit und Bedeutung von Schwamm spiculae in triassischen Mikrofaunen. Geol Paläontol Mitt Innsbruck 1, 11, 1–19
Müller W (1978) Beobachtungen zur Ökologie von Kieselspongien aus dem Weiβen Jura der Schwäbischen Alb. Stuttg Beitr Naturk Ser B, 37:1–15
Müller WEG, Zahn RK, Maidhof A (1982) Spongilla gutenbergiana n.sp, ein Süβwasserschwamm aus dem Mittel-Eozän von Messel. Senckenb Lethaea 63:465–472
Nalivkin DV (1973) Geology of the USSR translated from the Russian by Rast N. Oliver & Boyd, Edinburgh, p 855
Narbonne GM, Dixon OA (1984) Upper Silurian lithistid sponge reefs on Somerset Island, Arctic Canada. Sedimentology 31:25–50
Nestler H (1961) Spongien aus der weiβen Schreibkreide (Unt. Maastricht) der Insel Rügen (Ostsee). Paläontol Abh I, Teil 1d: 1–70

Neuss P (1979) Zur Biostratigraphie und Fazies der Unterkreide-Serien (Hauterivium-Aptium) im Eisenerz-Tagebau "Morgenstern" N Goslar (SE-Niedersachsen). Mitt Geol Inst Tech Univ Hannover 17:155–222
North FK (1980) Episodes of source-sediment deposition. Z The episodes in individual close-up. J Petrol Geol 2, 3:323–338
Obrhelová N, Obrhel J (1987) Paläoichthyologie und Paläoökologie des kontinentalen Tertiärs und Quartärs der CSSR. Z Geol Wiss Berl 15:709–731
Oschmann F (1958) Erläuterungen zur Geologischen Karte Bayern 1:25 000, Blatt Nr 7038 Bad Abbach. Bayer Geol Landesamt, München
Ott d'Estevou P, Termier G (1978) Etude préliminaire des spongiaires tortoniens du bassin de Sorbas (Espagne méridionale). Indications bathymétriques. Bull Soc Géol Fr Sér 7, 20:315–318
Ott d'Estevou P, Termier G, Termier H (1981) La spongiofaune Néogène de Sorbas (Andalousie orientale, Espagne). Géol Méditerr 8:61–78
Paproth E, Zimmerle W (1980) Stratigraphic position, petrography, and depositional environment of phosphorites from the Federal Republic of Germany. Meded Rijks Geol Dienst 32–1, 14:81–95
Pascal A (1985) Les systèmes biosédimentaires urgoniens (Aptien-Albien) sur la marge nord ibérique. Mém Géol Univ Dijon 10: p 569
Pozaryski W (1947) A phosphate deposit of the north-eastern periphery of the Swiety Krzyz Mountains. Serv Géol Pol, Inst Géol Pol Bull 27:45–56
Prescher H (1959) Geologie des Elbsandsteingebirges. Eine Einführung. T. Steinkopff, Dresden, p 187
Rat P (1959) Les pays crétacés basco-cantabriques (Espagne). Publ Univ Dijon 18: p 525
Rauff H (1937) Beitrag über die Spongien. In: Assmann P (ed) Revision der Fauna der Wirbellosen der oberschlesischen Trias. Abh Preuß Geol L-Anst, NF 170:7–14
Reid REH (1968) Hexactinellid faunas in the Chalk of England and Ireland. Geol Mag 105:15–22
Reid R (1978) Sponges in sediments. In: Fairbridge RW, Bourgeois J (eds) The encyclopedia of sedimentology. Encyclopedia of Earth Sciences, VI:759–760, Dowden, Hutchinson & Ross, Stroudsburg, Penn
Reif WE (1967) Schwammspicula aus dem Weißen Jura Zeta von Nattheim (Schwäbische Alb). Palaeontographica Abt A, Stuttgart 127:85–102
Reinhardt JW (1988) Sponge buildups close to the Permian-Triassic boundary crisis: SE Sichuan, China (Poster). In: Keupp H, Reitner J (eds) Fossil and Recent Sponges Int Conf Berlin 1988, Sept 26–28. Berl Geowiss Abh (A) 10:26
Reitner J (1986) A comparative study on the diagenesis in diapir-influenced reef atolls and a fault block reef platform in the Late Albian of the Vasco-Cantabrian Basin (North Spain). In: Schroeder JH, Purser BH (eds) Reef diagenesis. Springer, Berlin Heidelberg New York, 186–209
Reitner J (1987a) Mikrofazielle, palökologische und paläogeographische Analyse ausgewählter Vorkommen flachmariner Karbonate im basko-kantabrischen Strike Slip Fault-Becken-System (Nordspanien) an der Wende von der Unterkreide zur Oberkreide. Doc Nat 40: p 239
Reitner J (1987b) Euzkadiella erenoensis n gen n sp eine Stromatopore mit spikulärem Skelett aus dem Oberapt von Ereño (Prov Guipuzcoa, Nordspanien) und die systematische Stellung der Stromatoporen. Paläontol Z 61:203–222
Reitner J, Engeser TS (1987) Skeletal structures and habitats of Recent and fossil Acanthochaetetes (subclass Tetractinomorpha, Demospongiae, Porifera). Coral Reefs 6:13–18
Reul K (1954) Spongilliden-Erde im Miocän des Fichtelgebirges. Ein Vergleich mit der miocänen Kieselgur von Beuern. Geol Bl NO-Bayern 4:14–20
Richardson L, Thacker AG (1920) On the stratigraphical and geographical distribution of the sponges of the Inferior Oolite of the west of England. Proc Geol Assoc Lond 31:161–186
Ricken W (1986) Diagenetic bedding. A model for marl-limestone alternations. Lecture Notes in Earth Sciences 6. Springer, Berlin Heidelberg New York Tokyo, p 210
Riegraf W (1986) Callovian (Middle Jurassic) radiolaria and sponge spicules from Southwest Germany. Stuttg Beitr Naturkd Ser B 123: p 31
Riezebos PA, Zimmerle W (1989) Biosiliceous components and their frequency variations in some Quaternary ash layers from the Eastern Cordillera, Colombia (S.A.). Geol Norandina 11:41–50
Rigby JK (1971) Sponges and reef and related facies through time. Proc North American Paleontol Convention, Field Museum of Natural History, Chicago Sept 5–7, 1969, 1374–1388. Allen, Lawrence Kansas

Rigby JK (1986) Sponges of the Burgess shale (Middle Cambrian) British Columbia. Paleontogr Can 2: p 105
Rovereto G (1901) Briozoi, anellidi e spugne perforanti. Palaeontogr Ita, Mem Palaeontol 7:219–234
Ruchin LB (1958) Grundzüge der Lithologie. Lehre von den Sedimentgesteinen. Akademie Verlag, Berlin, p 806
Ruzhentsev SV (1985) Jurassic-Cretaceous siliceous deposits of fold belts and oceans. Dokl Earth Sci Sect 274:105–106 (English translation, Washington DC)
Scheer U (1988) Influence of the paleogeographic position and sea-level changes on spongiolitic limestones in the Lower-Bajocian (discites-to-humphriesianum-zone; Middle Jurassic) in Northern Spain (Abstract). In: Keupp H, Reitner J (eds) Fossil and Recent sponges. Int Conf Berlin 1988, Sept 26–28. Berl Geowiss Abh (A): 10:36–37
Schmidt M (1928) Die Lebewelt unserer Trias. F Rau, Öhringen, p 461
Schönlaub HP (1973) Schwamm-Spiculae aus dem Rechnitzer Schiefergebirge und ihr stratigraphischer Wert. Jahrb Geol BA Wien 116:35–49
Schorr M, Koch R (1985) Fazieszonierung eines oberjurassischen Algen-Schwamm-Bioherms (Herrlingen, Schwäbische Alb). Facies zonation of an Upper Jurassic-algal-sponge bioherm (Swabian Alb, Germany). Facies 13:227–270
Schrammen A (1910) Die Kieselspongien der oberen Kreide von Nordwestdeutschland. I. Teil. Tetraxonia, Monaxonia, und Silicea incert sedis. Palaeontographica Suppl-Bd V:1–96
Schrammen A (1912a) Die Kieselspongien der oberen Kreide von Nordwestdeutschland. Palaeontographica Suppl-Bd V:281–385
Schrammen A (1912b) Ergebnisse meiner Bearbeitung der kretazischen Kieselspongien von Nordwestdeutschland. Inaug Diss naturw Fak Univ Tübingen. Palaeontographica Suppl-Bd V:337–385
Schrammen A (1924) Die Kieselspongien der oberen Kreide von Nordwestdeutschland. III. und letzter Teil. Mit Beiträgen zur Stammesgeschichte. Monogr Geol Palaeontol Ser 1. Borntraeger, Berlin, 1, 2: p 159
Schwarzbach M (1948) Schwämme. In: Rüger L (ed) Naturforschung und Medizin in Deutschland 1939–1946. Geol und Paläont 48: p 213
Schwentke W, Wiedmann J (1985) Oberkreide-Entwicklung und Diapirismus im Estella-Becken (Nordspanien). Facies 12:1–74
Shanmugam G, Higgins JB (1988) Porosity enhancement from chert dissolution beneath Neocomian unconformity: Ivishak Formation, North Slope, Alaska. Am Assoc Petrol Geol Bull 72:523–535
Siever R (1979) Plate-tectonic controls on diagenesis. J Geol 87:127–155
Simpson TL (1981) Effects of germanium on silica deposition in sponges. In: Simpson TL, Volcani BE (eds) Silicon and siliceous structures in biological systems. Springer, Berlin Heidelberg New York, 527–550
Sokołowski S (1976) Geology of Poland vol 1, Stratigraphy, Part 2, Mesozoic. Wydawnictwa Geologiczne, Warsaw, p 859
Steiger T (1988) Calcification and calzitization of Jurassic siliceous sponges depend on the influence of carbonate producing cyanobacteria (Abstract). In: Keupp H, Reitner J (eds) Fossil and Recent Sponges Int Conf Berlin 1988, Sept 26–28. Berl Geowiss Abh (A) 10:43
Svoboda J, Beneš K, Dudek A, Dvořák J, Havlena V, Havlicěk V, Holubec J, Horný R, Chaloupský J, Chlupáč I, Klein V, Kodym O jun, Kopecký L, Malecha A, Malkovský M, Odehnal L, Polák A, Pouba Z, Sattran V, Soukup J, Škvor V, Tásler R, Václ J, Weiss J, Žebera K (1966) Regional geology of Czechoslovakia. Part I. The Bohemian Massif. In: Zoubek V (ed) Geol Surv Czech, Prague, p 668
Taguchi K, Shimoda S, Itihara Y, Imoto N, Ishiwatari R, Shimoyama A, Akiyama M, Suzuki N (1986) Relationship of organic and inorganic diagenesis of Neogene Tertiary rocks, northeastern Japan. Roles of Organic Matter in Sediment Diagenesis. SEPM Spec Publ 38:47–64
Teodorovich GI (1954) Sedimentary geochemical facies along the profile of the oxidation-reduction potential. Dokl Akad Nauk SSSR 96:3
Trammer J (1982) Lower to Middle Oxfordian sponges of the Polish Jura. Acta Geol Pol 32:1–39
Trammer J (1988) Ecologic history of the Oxfordian sponge assemblage in the Polish Jura Chain (Abstract). In: Keupp H, Reitner J (eds) Fossil and Recent Sponges Int Conf Berlin 1988, Sept 26–28. Berl Geowiss Abh (A) 10:44–45

Ulbrich H (1974) Die Spongien der Ilsenburg-Entwicklung (Oberes Unter-Campan) der Subherzynen Kreidemulde. Freiberger Forsch Hefte C291 Paläontol: p 121
Ulrich R (1960) Die Entwicklung der ostalpinen Juraformation im Vorkarwendel zwischen Mittenwald und Achensee. Geol Bavar 41:99–151
von Bülow W, Schulz W (1978) Oberkretazische Schwämme in Coelestin-Erhaltung als Geschiebe. Z Geol Wiss 6:1219–1229
von Rad U, Rösch H (1974) Petrography and diagenesis of deep-sea cherts from the central Atlantic. In: Hsü KJ, Jenkyns HC (eds) Pelagic sediments: on land and under the sea. Int Assoc Sedimentol Spec Publ 1:327–347
Wagner W (1963a) Die Schwammfauna der Oberkreide von Neuburg (Donau). Palaeontographica A 122:166–250
Wagner W (1963b) Kieselschwämme und Schwamm-Ökologie im Korallenkalk des oberen Malm von Laisacker bei Neuburg a.d. Donau. Mitt Bayer Staatssamml Paläont Hist Geol 3:1–20
Whittington HB (1980) The significance of the fauna of the Burgess Shale, Middle Cambrian, British Columbia. Proc Geol Assoc Lond 91:127–148
Wiedenmayer F (1980a) Spiculites and sponges in the lower Jurassic of the western Tethys. In: Hartmann WD, Wendt J, Wiedenmayer F (eds) Living and fossil sponges. Notes for a short course. Sedimenta 8:135–145
Wiedenmayer F (1980b) Diagenesis of spicules. In: Hartmann WD, Wendt J, Wiedenmayer F (eds) Living and fossil sponges. Notes for a short course. Sedimenta 8:108–112
Wilde S (1988) Das Bathonium und Callovium der nordwest-iberischen Ketten (Jura; Spanien): Sequenzen einer sich differenzierenden Epikontinental-Plattform. Bochumer Geol Geotech Arb 31: p 271
Wilson JL (1975) Carbonate facies in geologic history. Springer, Berlin Heidelberg New York, p 471
Wirsing G (1988) Mikrofazielle und diagenetische Entwicklung einer oberjurassichen Karbonatfolge (Weißjura beta bis delta 4, Mittlere Schwäbische Alb) mit Berücksichtigung der chemischen Diagenese ausgewählter karbonatischer Bestandteile. Heidelb Geowiss Abh 15: p 337
Wurm A (1961) Geologie von Bayern I. Frankenwald, Münchberger Gneismasse, Fichtelgebirge, Nördlicher Oberpfälzer Wald 2. Aufl Bornträger, Berlin, p 555
Wuttke M, Kott R (1987) Zur Genese der Silifizierung und Pyritisierung von Spongienweichteilen aus dem Hunsrückschiefer. Paläontologische Gesellschaft 75 Jahre 1912–1987, 57. Jahrestagung in Clausthal-Zellerfeld, Institut für Geologie und Paläontologie 2–4 10 1987:54
Zimmerle W (1982a) Sedimentologische Dünnschliff-Analyse der dunklen Tonsteine von Ober-Apt und Unter-Alb (Niedersächsisches Becken). Geol Jahrb A 65:63–109
Zimmerle W (1982b) Die Phosphorite des nordwestdeutschen Apt und Alb. Geol Jahrb A 65:159–244
Zimmerle W (1985) New aspects on the formation of hydrocarbon source rocks. Geol Rundsch 74:385–416
Zimmerle W (1986) Gesteinstypen und kleindimensionale Sedimentstrukturen im tieferen Unterkarbon der Bohrungen Adlersberg, Bullars, Eselsberg und Spiegeltal im West-Harz. Geol Jahrb D 78:95–205
Zimmerle W (1989) Vestiges of volcanic activity in Cretaceous sediments of Europe. Proc 3rd Int Cretaceous Symposium, Tübingen 26. 8–8 9 1987:951–987
Zimmerle W, Kuster H, Nöltner T (1990) Sedimente des höheren Oligozäns bis tiefstes Miozäns mit vulkanoklastischen Einschaltungen aus dem Raum zwisehen Weser-Aller und Elbe. Veröff Übersee Mus A 10:119–177
Zimmerle W, Gaida K-H, Gedenk R, Koch R, Paproth E (1980) Sedimentological, mineralogical and organic-geochemical analyses of Upper Devonian and Lower Carboniferous strata of Riescheid, Federal Republic of Germany. Meded Rijks Geol Dienst 32–5:34–43

# Subject Index

MIX
Papier aus verantwortungsvollen Quellen
Paper from responsible sources
FSC® C105338

If you have any concerns about our products,
you can contact us on
**ProductSafety@springernature.com**

In case Publisher is established outside the EU,
the EU authorized representative is:
**Springer Nature Customer Service Center GmbH**
**Europaplatz 3, 69115 Heidelberg, Germany**

Printed by Libri Plureos GmbH
in Hamburg, Germany